MW00764536

To Joe,

Welcome to the Origin family.

Ray

June 2021.

Wireless AI

With this groundbreaking text, discover how wireless AI can be used to determine position at centimeter level, sense motion and vital signs, and identify events and people. Using a highly innovative approach that employs existing wireless equipment and signal processing techniques to turn multipaths into virtual antennas, combined with the physical principle of time reversal and machine learning, it covers fundamental theory, extensive experimental results, and real practical use cases developed for products and applications. Topics explored include indoor positioning and tracking, wireless sensing and analytics, wireless power transfer and energy efficiency, 5G and next-generation communications, and the connection of large numbers of heterogeneous IoT devices of various bandwidths and capabilities. Demo videos accompanying the book online at www.cambridge.org/WirelessAI enhance understanding of these topics.

Providing a unified framework for wireless AI, this is an excellent text for graduate students, researchers, and professionals working in wireless sensing, positioning, IoT, machine learning, signal processing, and wireless communications.

K. J. Ray Liu is Christine Kim Eminent Professor of Information Technology in the Department of Electrical and Computer Engineering at the University of Maryland, College Park. A highly cited researcher, he is a Fellow of the IEEE and the AAAS, IEEE Vice President, Technical Activities, and a former President of the IEEE Signal Processing Society. He is a recipient of the 2016 IEEE Leon K. Kirchmayer Award, the IEEE Signal Processing Society 2014 Society Award, and the IEEE Signal Processing Society 2009 Technical Achievement Award. He has also coauthored several books, including *Cooperative Communications and Networking* (Cambridge University Press, 2008).

Beibei Wang is Chief Scientist in Wireless at Origin Wireless, Inc., and is also affiliated with the University of Maryland. She has been a recipient of the Outstanding Graduate School Fellowship, the Future Faculty Fellowship, and the Dean's Doctoral Research Award from the University of Maryland, and also received the Overview Paper Award from the IEEE Signal Processing Society in 2015. She coauthored *Cognitive Radio Networking and Security: A Game-Theoretic View* with K. J. Ray Liu (Cambridge University Press, 2010).

'Wireless AI is changing the world by enabling wireless sensing and tracking to an unprecedented level. This is the first book on major breakthroughs in this emerging field. A must read!'

Sadaoki Furui, *Toyota Technological Institute at Chicago*

'*Wireless AI* is an exciting and timely book that provides the reader with the background and material needed to not only ride the wave of technological advancement but also contribute to it. Paradigm-shifting advancements, like time-reversal, cloud-RAN, motion detection and localization, and waveform designs are described in detail. *Wireless AI* is an innovative text that is sure to help engineers and students contribute to the rapidly evolving fields of wireless sensing and communications.'

Wade Trappe, *Rutgers University*

'...an excellent book on wireless AI, with unique and comprehensive coverage, for both researchers and practitioners.'

Geoffrey Li, *Georgia Institute of Technology*

Wireless AI

Wireless Sensing, Positioning, IoT, and Communications

K. J. RAY LIU
University of Maryland and Origin Wireless, Inc.

BEIBEI WANG
Origin Wireless, Inc.

CAMBRIDGE
UNIVERSITY PRESS

University Printing House, Cambridge CB2 8BS, United Kingdom

One Liberty Plaza, 20th Floor, New York, NY 10006, USA

477 Williamstown Road, Port Melbourne, VIC 3207, Australia

314–321, 3rd Floor, Plot 3, Splendor Forum, Jasola District Centre, New Delhi – 110025, India

79 Anson Road, #06–04/06, Singapore 079906

Cambridge University Press is part of the University of Cambridge.

It furthers the University's mission by disseminating knowledge in the pursuit of education, learning, and research at the highest international levels of excellence.

www.cambridge.org
Information on this title: www.cambridge.org/9781108497862
DOI: 10.1017/9781108597234

First published 2019

Printed in the United Kingdom by TJ International Ltd, Padstow Cornwall

A catalogue record for this publication is available from the British Library.

Library of Congress Cataloging-in-Publication Data
Names: Liu, K. J. Ray, 1961– author. | Wang, Beibei, author.
Title: Wireless AI : wireless sensing, positioning, IoT, and communications / K. J. Ray Liu, University of Maryland, College Park, and Origin Wireless, Inc., Maryland, Beibei Wang, Origin Wireless, Inc., Maryland.
Description: New York : Cambridge University Press, 2019. | Includes index.
Identifiers: LCCN 2019009295 | ISBN 9781108497862 (hardback)
Subjects: LCSH: Wireless communication systems. | Artificial intelligence. | Internet of things. | Indoor positioning systems (Wireless localization)
Classification: LCC TK5103.2 .L59 2019 | DDC 006.3–dc23
LC record available at https://lccn.loc.gov/2019009295

ISBN 978-1-108-49786-2 Hardback

To the Origin Wireless team for the exciting journey of scientific discovery and the dream of making the world a better place.

Contents

Preface

We have many senses. We can see with our eyes, but only to the limit of line-of-sight. We can hear with our ears, we can smell, we can taste, and we can touch. These are the five senses that we rely upon in our daily life and living.

As the smart phone and smart Internet of Things (IoT) devices are connected with radio frequency signals, Wi-Fi is ubiquitous everywhere indoors, long-term evolution (LTE) is available in almost every corner of the world, and future 5G systems will be more powerful. We rely on wireless radios for communications, chatting, surfing Internet, seeing each other via Facetime, sending text messages, etc., while we are at home, driving, eating, or on the move. Wireless radios enable us many new modern conveniences to the point that we simply cannot live without them.

But can wireless radios offer us a new sense – a "sense" that can help us track people and devices, monitor our environments, detect our activities, even beyond the limit of our vision, hearing, and touch? Indeed, if there is such a "sense," it can qualify as a new intelligence that has been a fantasy for many for so long. No wonder the whole world has been trying to uncover such a new breakthrough to make this dream possible.

In fact, when one refers to "wireless," it is no longer in the narrow sense of communications. It has been for so long that we have been only concerned with the messages sent to us. We try to remove interference, equalize the channel, decrypt the code, and decode the message. Yet we have ignored (or are simply unaware) that the radio signals come with them containing information about our environment and activities. If we can make sense of the radio signals, as if we are evolving to a new sixth sense that allows us to sense/detect/track/recognize our environments and activities and communicate, it is in essence a new intelligence. The information analytics, signal processing, and machine learning that enable such a new intelligence constitute an emerging field of wireless artificial intelligence (AI).

So what is wireless AI? It is to use wireless/radio waves/signals to make sense of our environments, detect/monitor our activities, track and locate users/objects, connect "things" together and empower them, and offer a platform for future communications.

But how can we accomplish that vision? There have been many approaches in recent years in the research community attempting to unlock the secret. In this book, we would like to offer our view by combining the physical principle of time reversal and signal processing/information science to answer some difficult questions that seemingly no better solutions were obtained over the attempts of the last three decades. We have developed a unifying framework to enable the wireless AI dream as we envision.

When one uses radio signals, multipaths always come with it, especially in indoor environments. It has been a long time that we consider multipaths as interferences, noise, or simply nuisance. Previous attempts have always been trying to take them out or at least to neutralize or compensate for their "bad" effects, but there has been too little or no effect because the profiles of multipaths change from location to location, and how can one tell which multipath is a good one and which is not? Under such a thinking paradigm, the struggle continues, and the problem remains.

With the ever-increasing large bandwidth, more and more multipaths can be seen. Each multipath can be viewed as a virtual antenna/sensor located at the direction where it comes from with a distance equal to the speed of light times the time of arrival. Therefore it is as if there are a tremendous number of virtual antennas surrounding us, appearing at our disposal on demand.

The question is, how do we harvest multipaths? Two approaches are by increasing power and bandwidth. The larger the transmitting power, the more radio waves can bounce back and forth around the environment, therefore the more observable number of multipaths. However, oftentimes such a transmitting power is limited by regulations or standards. The other means is to increase the bandwidth in that the larger the bandwidth, the better the time resolution to reveal more multipaths.

Each of these multipaths is in essence a degree of freedom, ready for any use. But how can we control them to serve our purpose? As they are surrounding us virtually, we have to resort to the physics in that we have to generate a waveform to reach out to "them" and control them to achieve the desired effect. One such physics is the principle of time reversal, where we use the time-reversal waveform to control the multipaths to generate the well-known focusing effect.

We have found that by using enough bandwidth in a typical indoor environment, such a focusing effect can be reliably produced. By using the 5 GHz ISM band, for example, we can produce a focusing ball of about 1–2 cm in diameter. If using the 60 GHz band, then it will goes down to the millimeter level. Such an effect serves as the fundamental basis for us to be able to perform indoor positioning with the unprecedented accuracy of centimeter/millimeter, under both line-of-sight and non-line-of-sight conditions.

In fact, with the use of machine learning and signal processing, a revolutionary AI platform can be built to enable many cutting-edge Internet of Things applications that have been envisioned for a long time, but have never been achieved.

This book, *Wireless AI: Wireless Sensing, Positioning, IoT and Communications*, aims at providing comprehensive coverage of fundamental issues that form an artificial intelligence platform that consists of many radio analytic engines for a wide range of applications, including the world's first-ever centimeter-accuracy indoor positioning/tracking, wellness monitoring, home/office security, radio human biometrics, health care, wireless charging, and 5G communications. A goal of the book is to provide a bridge between advanced scientific research and practical industry design and implementation to offer readers a glimpse of what the future wireless AI can achieve.

We first start from the principle of time reversal and effective bandwidth in Chapter 1 to lay out the fundamental concepts for the rest of the book. In Part I, we address the issues of indoor positioning and tracking. Chapter 2 demonstrates that the use of time reversal at 5 GHz with the entire 125 MHz ISM band can produce a focusing ball of 1–2 cm, which translates to an indoor positioning scheme of the same accuracy. Note that the pinpoint locationing of time-reversal focusing effect regardless of any indoor conditions such as line-of-sight and non-line-of-sight inherently implies that the notion of walls and obstacle no longer exists. It is as if there are no walls nor obstacles in the space. In Chapter 3, we show how to use standard Wi-Fi devices to achieve the same centimeter accuracy by leveraging multiple antennas to achieve a large effective bandwidth, and in Chapter 4, we leverage frequency hopping for a large effective bandwidth to again achieve centimeter accuracy of positioning. Note that if 60 GHz Wi-Fi devices are used, the focusing ball will be at the millimeter range in diameter, and therefore one can expect a millimeter level of accuracy. Then, in Chapter 5, we present our discovery that when the number of multipaths is large enough, the focusing ball's energy distribution follows a Bessel function. Therefore, it is location independent, and one can use such a principle to track users with decimeter accuracy without any training or mapping. One just needs to know the starting point and a map to be able to track an unlimited number of users.

In Part II, the focus is on wireless sensing and analytics. One can imagine there is a time-reversal space, and every channel impulse response has a definite focusing ball location. Let one open a door to obtain an impulse response and then close the door to have the other one. If one can tell both locations at the time-reversal space, then in essence one can tell if the door is open or closed. Chapter 6 illustrates such a basic principle for wireless event detection in indoor environments. In Chapter 7, we extend such a concept by developing a statistical model to improve robustness. Next, in Chapter 8, we further extend to recognize humans by developing radio human biometrics. The human body contains over 70% water, and therefore we all uniquely deflect/distort/absorb radio waves impinging on us in a unique way. Such a subtle difference allows us to distinguish different people. Then, in Chapter 9, we discuss how to pick up one's breathing rate from Wi-Fi signals. Even though breathing is a tiny motion, it embeds to the radio waves the periodic motion of chest movement, which can be used to estimate breathing rate. Motion is not periodic but yet can be detected as well. We show that motion detection can be done with very high accuracy and low false alarm in Chapter 10. The last chapter of this part estimates speed without any wearable devices. A statistical theory for EM waves is developed in Chapter 11 to serve as the foundation for speed estimation so that no active device is needed.

In Part III, the wireless power transfer and energy efficiency using time reversal are presented. First the energy efficiency of the time-reversal technique is shown in Chapter 12 to argue that it is ideal for green technology. In Chapter 13, we propose a new waveform called power waveforming, other than time reversal waveform, to achieve optimal wireless power transfer. Further, in Chapter 14 we extend power waveforming for multiple antenna scenarios to jointly work with beamforming to significantly improve performance.

Following the preceding discussion on indoor positioning/tracking, wireless sensing, and power transfer, one may ask if the principle of time reversal can also be leveraged for communications, especially for 5G and future generations of wireless communications. In fact, one can easily link the concepts between massive MIMO and time-reversal massive multipaths. We have seen that time reversal can produce a focusing ball by leveraging a large number of multipaths. And that is exactly what massive MIMO is doing. When there is no multipath outdoors, one has to rely on real antennas, many of them, to create "multipaths" so that by proper precoding, one can control massive MIMO to essentially produce a focusing ball at a desired location. The difference is that the time-reversal technique controls massive multipaths as virtual antennas/sensors to achieve the massive MIMO effect.

In Chapter 15, we first introduce the concept of time-reversal division multiple access, which takes advantage of the focusing effect for multiple access. Because the focusing effect has a unique strong–weak resonance effect, in Chapter 16 an adaptive algorithm is presented to combat such an effect. Then we show in Chapter 17 that the time-reversal massive multipaths effect is indeed an equivalence to the massive MIMO effect with the difference in leveraging virtual antennas. When it comes to communications, the optimal waveform is no longer the time-reversal waveform. It is because what is concerned is the signal-to-noise-ratio. Therefore, in Chapter 18, we consider waveform designs for various scenarios with a comparison to beamforming. Then in Chapter 19, we consider how the spatial focusing effect can be leveraged for networking design, and finally we introduce the tunneling effect of the time-reversal principle for the cloud radio access network.

Finally in Part V, our focus turns to how time reversal can be used to connect a large number of heterogeneous IoT devices of various bandwidths and capabilities. Chapter 21 gives an overview of how the time-reversal technique can make an impact on IoT. In Chapter 22, we illustrate that time reversal is ideal for the connection of IoT devices of various heterogeneous bandwidths and standards without any need of transcoding or complex transform.

This book is intended to be a textbook or a reference book for researchers, practitioners, or graduate students working in wireless sensing, positioning/tracking, and communications. We hope that the comprehensive coverage of the wireless AI that enables us to infer/decipher our environments and activities will make this book a useful resource for readers who want to understand this emerging technology, as well as for those who conduct research and development in this field.

This book could not have been made possible without the research contributions by the following people: Chen Chen, Yan Chen, Feng Han, Yi Han, Chunxiao Jiang, Meng-Lin Ku, Hung-Quoc Lai, Hang Ma, Zhung-Han Wu, Qinyi Xu, Yu-Hang Yang, and Feng Zhang. Also special thanks to the Origin Wireless team for their enlightening of the future of wireless AI. We also would like to thank all the colleagues whose works enlighten our thoughts and research that made this book possible. We can only stand on the shoulders of giants.

1 Principles of Time Reversal and Effective Bandwidth

With the proliferation of Internet of Things (IoT) applications, billions of household appliances, phones, smart devices, security systems, environment sensors, vehicles and buildings, and other radio-connected devices will transmit data and communicate with each other or people, and everything will be able to be measured and tracked all the time. Among the various approaches to measure what is happening in the surrounding environment, wireless sensing has received increasing attention in recent years because of the ubiquitous deployment of wireless radio devices. In addition, human activities affect wireless signal propagations, therefore understanding and analyzing how wireless signals react to human activities can reveal rich information about the activities around us. As more bandwidth becomes available in the new generation of wireless systems, wireless sensing will make many smart IoT applications only imagined today possible in the near future. That is because when bandwidth increases, one can see many more multipaths, in a rich-scattering environment such as in indoors or metropolitan area, which can be treated as hundreds of virtual antennas/sensors. In order to control the virtual antennas and make good use of the multipaths, one can resort to the physical principles of radio propagation. Inspired by the high-resolution spatial-temporal resonance of time-reversal (TR) phenomenon, one can develop various types of wireless artificial intelligence (AI) based on the multipath channel profiles. In addition, by the spatial-temporal resonance, TR is also a good candidate platform for future 5G communications. In this chapter, we will lay out the concept of treating multipaths as virtual antennas/sensors, use the basic physical principle of TR to control them, and see how to achieve a large effective bandwidth built from smaller ones.

1.1 Introduction

With the development of wireless technologies in the era of the IoT, people are paying more and more attention to understanding the who, what, when, where, and how of everything surrounding them with wireless technologies. Human activities can affect wireless signal propagations surrounding them, and information about their activities is in turn embedded in the signals. This makes one wonder whether one can extract meaningful information through wireless sensing by analyzing various features embedded in wireless signals. By deploying wireless transceivers indoors, macro changes due to human activities and moving objects can be extracted from the wireless signals, which can help infer the real-time location of a moving object [1–8], detect an event [9–18], and facilitate applications in manufacturing asset tracking, intelligent transportation, and home/office security systems. In addition, microchanges generated by gestures [10] and vital signals [19–21] can also be captured without requiring people to wear any device, which is especially useful for providing assistance to the disabled and elderly people in smart home applications.

The performance of wireless sensing depends greatly on the richness of information that can be extracted from the radio signals, while the information richness is often dictated by the channel bandwidth through which the radio signals are transmitted. Due to the limited bandwidth in the past, only a limited number of multipaths can be seen, and not much information can be revealed in the past. With more and more bandwidth available for the next generation of wireless systems, many more smart IoT applications and services only imagined today may be possible in the near future because richer information becomes available with a wider bandwidth. For example, one can see many more multipaths indoors with a much larger bandwidth, which can serve as hundreds of virtual antennas/sensors ready to aid us for many applications.

How to control the virtual antennas to meet our needs for smart IoT applications? We have to resort to physics to do so, and TR phenomenon is a good starting point [22]. TR technique treats each path of the multipath channel as a distributed virtual antenna and provides a high-resolution spatial-temporal resonance, commonly known as the focusing effect [23–25]. In physics, the TR spatial-temporal resonance can be viewed as the result of the resonance of electromagnetic (EM) field in response to the environment [26]. When the propagation environment changes, the involved multipath signal varies correspondingly, and consequently the spatial-temporal resonance also changes. Inspired by the fundamental physical principle of TR, various types of analytics, referred to as wireless AI, can decipher radio waves to reveal the activities surrounding us. The wireless channel state information can be developed to enable many cutting-edge IoT applications that have been envisioned for a long time, but have never been achieved.

In this chapter, we will present the fundamental concept of TR. We will first discuss the impact of bandwidth on the multipath channel state information (CSI) and the principles of TR that can fully harvest the multipath CSI. Then, we discuss how to achieve a large effective bandwidth by exploiting various types of diversity.

1.2 Multipaths as Virtual Antennas

In wireless communications, when a signal emitted from a transmitter (TX) is reflected or scattered by a scatterer, an attenuated copy of the original signal is generated and reaches the receiver (RX) through a different path. The phenomenon that a signal is received by two or more paths is well known as the multipath propagation. As depicted in Figure 1.1, where each scatterer is marked by a star, the arrow from the TX directly to the RX represents the line-of-sight (LOS) path, while the other arrows represent paths reflected and scattered by scatterers. All paths together form a multipath channel between the TX and the RX [27]. As two or more copies of the original signal arrive at

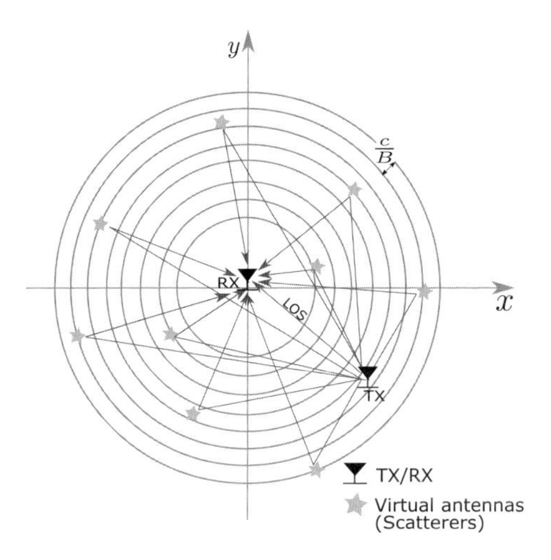

Figure 1.1 Illustration of multipath as virtual antenna.

the receiver that may be added in a noncoherent way, multipaths can cause destructive interference and degrade the performance of communication.

However, viewed from another perspective, the scatterers in the environment in fact act as virtual antennas/sensors that can be leveraged to offer some desirable outcomes. Just imagine that everyday human activities with motion and body movements affect wireless signal propagation surrounding us and thus change the channel profiles, and information about these activities is embedded in the signals. When signals are bounced back and forth by the scatterers, multiple "replicas" are generated, which contains enriched meaningful information about our activities. Each of such multipaths is in essence a degree of freedom naturally existing in our surrounding environment. They can be considered as tens or hundreds of virtual antennas ready to serve us on demand. In other words, our environment provides a high degree of freedom by means of radio multipath propagations, ready for our uses.

Now how do we harvest multipaths? The transmission power and bandwidth are two key components to consider [28, 29]. On the one hand, increasing the transmission power leads to a higher signal-to-noise ratio (SNR) and thus more observable multipath components. On the other hand, the spatial resolution in resolving independent multipath components is determined by the transmission bandwidth. Due to the limited bandwidth, which is equal to the channel sampling rate, the multipaths with propagation delay difference less than a channel sampling period T_{sample} will merge into a single tap. Thus, as noted in Figure 1.1, the resolution to separate radio paths with different lengths in a multipath propagation is limited to $cT_{sample} = c/B$, with c being the speed of light and B being the bandwidth. Therefore, the larger the bandwidth, the better the spatial resolution and thus the more multipaths can be resolved. An example of multipath channel profiles captured under different bandwidths from LTE, Wi-Fi, and the entire ISM 5 GHz band, at the same location in a rich-scattering environment is demonstrated in Figure 1.2. When the bandwidth is 20 MHz (as in LTE), only 5

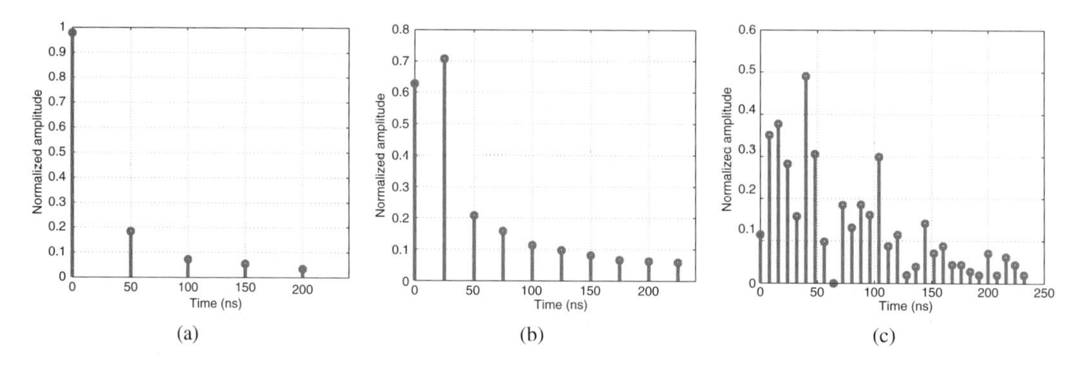

Figure 1.2 Multipath channel vs. bandwidth. (a) Measured channel under 20 MHz bandwidth (LTE standard). (b) Measured channel under 40 MHz bandwidth (the IEEE 802.11n standard). (c) Measured channel under 125 MHz bandwidth (entire ISM 5 GHz band) [27]. We first measure a sample channel impulse response in a typical indoor environment us ing a time-reversal prototype [30] with a bandwidth of 125 MHz in the ISM 5 GHz band. Then different filters with a bandwidth of 20 MHz, 40 MHz, and 125 MHz, respectively, are applied to the measure channel profile. For a linear time-invariant system, the filtering operation on the receiver side is equivalent to that on the transmitter side. Therefore, the filtered channel impulse response is equivalent to that measure with the same bandwidth as the filter.

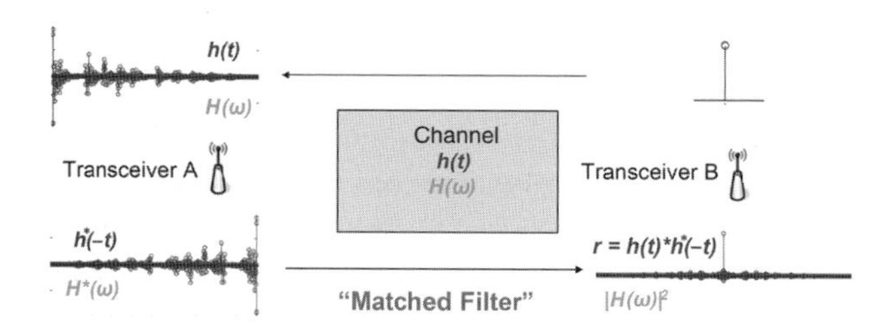

Figure 1.3 The time-reversal signal processing principle.

multipaths can be resolved; when the bandwidth increases to 40 MHz (as in Wi-Fi) about 10 multipaths can be resolved. When the bandwidth further increases to 125 MHz, around 30 multipaths with clear details of difference can be resolved, which clearly shows an increasing number of multipaths as bandwidth increases.

How to utilize the multipaths as virtual antennas/sensors? We find that a good starting point is to resort to the physics of TR and its focusing effect. As shown in Figure 1.3, in TR with two transceivers A and B, transceiver B first sends a channel probing signal (e.g., an impulse) to transceiver A, at which the multipath channel state information (CSI) [31] can be estimated. Then transceiver A time-reverses (and conjugates if the signal is complex) the received waveform and transmits the time-reversed version of waveform back to transceiver B. It has been well known [25–28] that the convolution of the time-reversed waveform and the channel can generate a unique peak at the specific receiver's location, called the spatial focusing effect (more details will be discussed in the next section). This indicates that the multipath channel profile works as a unique and location-specific signature, and the spatial focusing effect only happens when the channel can "match" the time-reversed waveform. By comparing the multipath CSI with a set of time-reversed CSI precollected at multiple known locations, one can infer the current location of a device, and this idea can be applied to assist positioning. Because this works for both line-of-sight and non-line-of-sight conditions, it is as if the notion of walls and obstacles of indoors has disappeared. That has been well captured in the CSI.

Because each multipath profile is in essence a focusing point on the "time-reversal logical space," if an event happens such as a door opens or closes that affects the multipath, as a result the multipath profile is now mapped to another focusing point. If one can perform analytics or machine learning to distinguish both events, then one shall be able to infer what has happened. With this notion, one can further design various types of analytics based on the multipath CSI, which we refer to as *wireless AI*. By fully exploiting the rich multipath CSI, wireless AI can decipher the propagation environment, reveal subtle information on various human activities, as if a new extended sixth human sense. Wireless AI can enable many cutting-edge IoT applications, such as accurate indoor positioning, tracking, wireless event detection, human recognition, vital signs monitoring, wireless power transfer, and 5G communications, as we will illustrate in this book.

1.3 Time-Reversal Principle

TR is a fundamental physical phenomenon that takes advantage of the unavoidable but rich multipath radio propagation environment to create a spatial-temporal resonance effect, the so-called focusing effect. Let us imagine that there are two points, A and B, within the space of a metal box. When A emits a radio signal, its radio waves bounce back and forth within the box, some passing through B. After a certain time, the energy level decreases and is no longer observable. Meanwhile, B can record the multipath profile of the arriving waves as a distribution in time. Then, such a multipath profile is time reversed (and conjugated) by B and emitted accordingly, the last first and the first last. With channel reciprocity, all the waves, following the original paths, will arrive at A at the same particular time instant, adding up in a perfectly constructive way. This is called the focusing effect. In essence, it is a resonance effect taking place at A stimulated by B using the time-reversed multipath profile through the interaction with the box as demonstrated in Figure 1.4. Mathematically, the TR effect is simply for the environment to serve as the computer to perform a perfect deconvolution; in essence, the environment behaves like a matched filter.

The spatial focusing effect of TR is fundamentally due to the decreased correlation between the channel states of two different locations. As discussed in the previous section, the multipaths can be viewed as virtual antennas, and as the transmission bandwidth increases, more multipaths can be resolved in a rich scattering environment. By utilizing the TR technique, the signal energy can concentrate on an intended location. On the other hand, the massive number of (physical) antennas in a massive MIMO (multiple-input, multiple-out) system can create the CSI with a large dimension for each

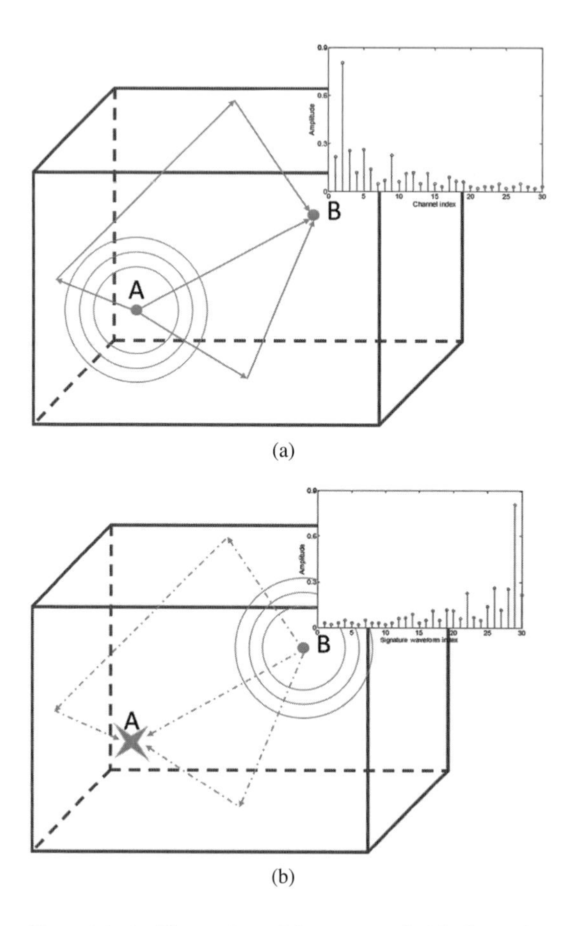

Figure 1.4 An illustration of time reversal: (a) channel probing phase and (b) data transmission and focusing phase.

location. By utilizing a matched filter-based precoder or equalizer, the signal energy can also concentrate on a corresponding location [32].

To illustrate the spatial focusing effect of both systems with different bandwidth/number of antennas, we conduct a simulation based on ray-tracing techniques in a discrete scattering environment. As shown in Figure 1.5, 400 scatterers are distributed randomly in a $200\lambda \times 200\lambda$ area, where λ is the wavelength corresponding to the carrier frequency of the system. The wireless channel is simulated by calculating the sum of the multipaths using the ray-tracing method given the locations of the scatterers. Without loss of generality, we use a single-bounce ray-tracing method to calculate the channels for both the TR system and the massive MIMO system on the 5 GHz ISM band. We select the reflection coefficients of the scatterers to be i.i.d. complex random variables with uniform distribution in amplitude $[0, 1]$ and phase $[0, 2\pi]$. For the massive MIMO system, the linear array is configured with the line facing the scattering area, and the interval between two adjacent antennas is $\lambda/2$. The distance from the transmitter and the intended location is chosen to be 500λ for both systems.

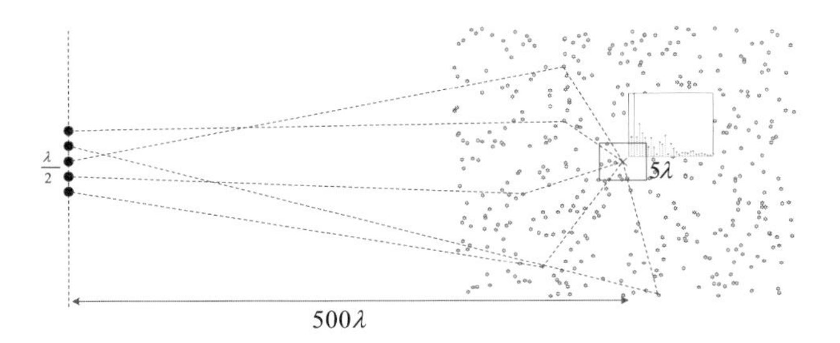

Figure 1.5 Simulation setup for validating spatial focusing effect.

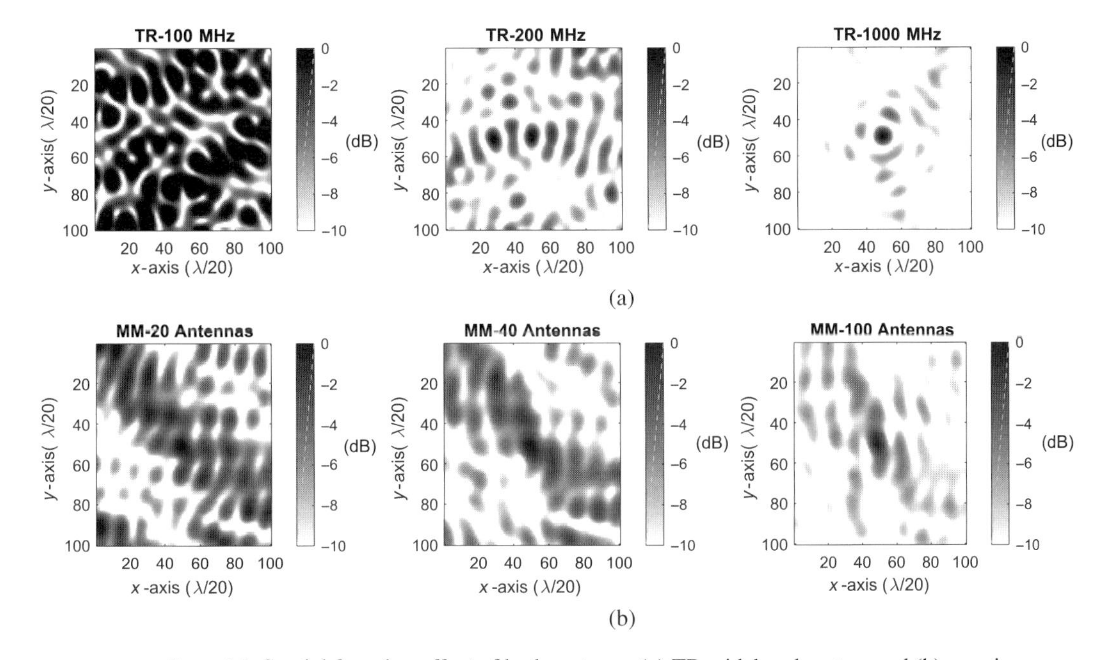

Figure 1.6 Spatial focusing effect of both systems: (a) TR wideband system and (b) massive MIMO system.

In the simulations, we adjust the transmitting bandwidth of the TR system and the number of antennas in the massive MIMO system to show their impact on the spatial focusing effect. The transmitter of the TR system transmits with bandwidths ranging from 100 MHz to 1 GHz with one antenna, where a wider bandwidth can resolve more CSI taps and increase the degree of freedom of the system. The number of antennas in the massive MIMO system is selected from 20 to 100 with bandwidth fixed at 1 MHz in the simulation. We select the matched filter waveform and beamforming weights in the TR system and the massive MIMO system, respectively.

We consider the received energy strength in a $5\lambda \times 5\lambda$ area around the location of the intended user. Figure 1.6 shows the simulation results for both systems with a single realization of the channel and scatterer distribution, and we normalize the maximum received energy to 0 dB. We can see that the energy focusing effect becomes more obvious at the intended location with the increase in the bandwidth and the number of

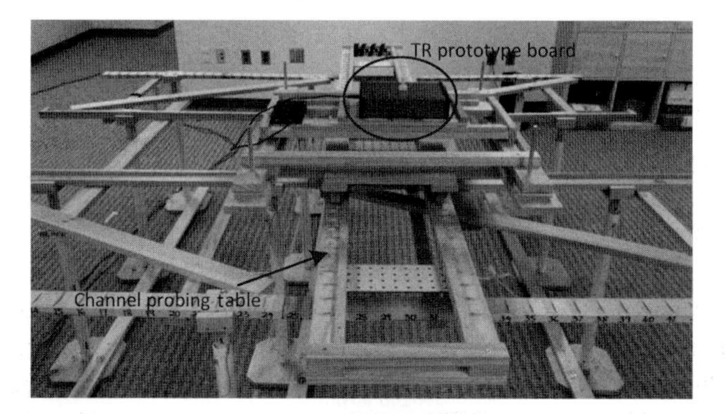

(a)

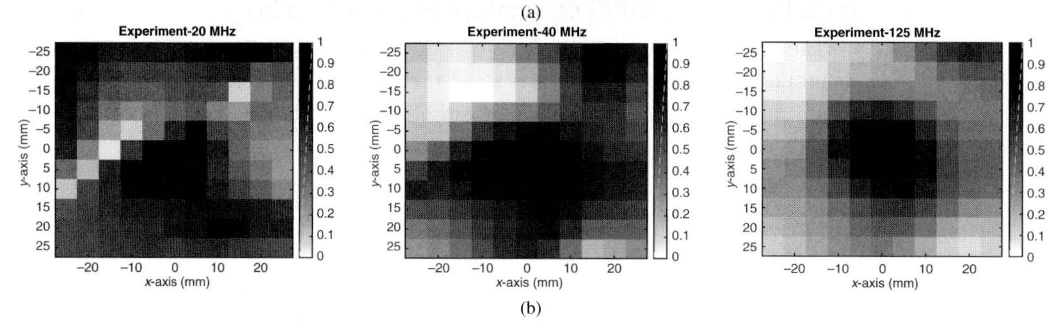

(b)

Figure 1.7 Spatial focusing effect of TR wideband system prototype: (a) TR wideband system prototype and (b) experiment results.

antenna, which is the result of a larger degree of freedom to concentrate the energy only at the intended location.

To further verify the spatial focusing effect, we built a prototype of TR wideband system on a customized software-defined radio (SDR) platform, as shown in Figure 1.7(a). The hardware architecture combines a specific designed radio-frequency board covering the ISM band with 125 MHz bandwidth, a high-speed Ethernet port, and an off-the-shelf user-programmable module board. In this experiment, we measure the CSIs of a square region with a dimension of 5 cm × 5 cm on a channel probing table, which is located in a typical office environment as shown in Figure 1.7. The intended location is chosen to be the center of the measured region, and the corresponding normalized field strength is shown in Figure 1.7(b). We can see that the TR transmission can generate a clear energy focusing around the intended location, even under a not-so-wide bandwidth of 125 MHz.

The research of TR dates back to the 1950s, where TR was utilized to compensate the phase-delay distortion that appears during long-distance transmissions of slow-speed pictures over telephone lines [33]. It has also been used to design noncausal recursive filters to equalize the ghosting artifacts of analog television signals caused by multipath propagation [34].

It was observed in a practical underwater propagation environment [35] that the energy of the TR acoustic waves from transmitters could be refocused only at the intended location with very high spatial resolution. The spatial and temporal focusing feature can also be used for radar imaging and acoustic communications. Note that the resolution of spatial and temporal focusing highly depends on the number of multipaths. To be able to harvest a large number of multipaths, large bandwidth and thus high sampling rate is required, which was difficult or even impossible to achieve in the past. Fortunately, with the advance of semiconductor technologies, broadband wireless technology has become available in recent years, and exploiting TR effect has also become possible in wireless radio systems. Experimental validations of the TR technique with electromagnetic (EM) waves have been conducted [36], including the demonstration of channel reciprocity and spatial and temporal focusing properties. Combining the TR technique with ultra-wideband (UWB) communications has been studied with the focus on the bit error rate (BER) performance through simulations [37]. A system-level theoretical investigation and comprehensive performance analysis of a TR-based multiuser communication system was conducted [38], where the concept of time-reversal division multiple access (TRDMA) was proposed. Also, a TR radio prototype was built to conduct TR research and development [39].

When applying the TR technique in wireless communications, if the transmitted symbol duration is larger than (or equal to) the channel delay spread, the time-reversed waveform can guarantee the optimal BER performance by virtue of its maximum signal-to-noise ratio (SNR) property. However, if smaller, which is generally the case in high-speed wireless communication systems, the delayed versions of the transmitted waveforms will overlap and thus interfere with each other. Such intersymbol interference (ISI) can be notably severe and cause crucial performance degradation, especially when the symbol rate is very high. The problem becomes even more challenging in a multiuser transmission scenario, where the interuser interference (IUI) is introduced due to the nonorthogonality of the channel impulse responses among different users. To address this problem, one can utilize the degrees of freedom provided by the environment, i.e., the abundant multipaths, to combat the interference using signature waveform design techniques. The basic idea of signature waveform design is to carefully adjust the amplitude and phase of each tap of the signature waveform based on the channel information such that the signal at the receiver can retain most of the useful signal while suppressing the interference as much as possible. Moreover, with random scatterers, TR can achieve focusing that is far beyond the diffraction limit [40], which is half wavelength.

1.4 Principle of Effective Bandwidth

As discussed earlier in this chapter, the multipath channel profile works as a unique and location-specific signature/or fingerprint, and the spatial focusing effect only happens when the channel can "match" the time-reversed waveform. By comparing the multipath CSI with a set of time-reversed CSI precollected at multiple known locations, one can infer the current location of a device, and this idea can be applied to assist positioning.

However, as the maximum bandwidth in mainstream Wi-Fi devices is only either 20 or 40 MHz, bandwidth limitation becomes the main reason that prevents the forming of a clear and precise spatial focusing effect. As shown in Figure 1.7(b) with 20 or 40 MHz bandwidth, there is no obvious spatial focusing effect, and a large region of nearby locations is ambiguous with the target location. Enlarging the bandwidth shrinks the area of ambiguous regions. When the bandwidth increases to 125 MHz, the ambiguous region is reduced to a ball of 1 cm radius, which implies centimeter accuracy in localization. The experiment results motivate us to formulate a large effective bandwidth by exploiting diversities to facilitate high-accuracy indoor localization and other wireless AI based on CSI fingerprints. In this following, we will discuss the concept of effective bandwidth and how to achieve a large effective bandwidth using Wi-Fi positioning as an example.

To characterize the similarity between CSIs collected at the same or different locations, the time-reversal resonating strength (TRRS) of the focusing effect can be defined as [41]

$$\gamma[\mathbf{H},\mathbf{H}'] = \left(\frac{\eta}{\sqrt{\Lambda}\sqrt{\Lambda'}}\right)^2, \tag{1.1}$$

with

$$\eta = \max_{\phi}\left|\sum_{k=1}^{K} H_k H_k'^{*} e^{-jk\phi}\right|, \quad \Lambda = \sum_{k=1}^{K} |H_k|^2, \quad \Lambda' = \sum_{k=1}^{K} |H_k'|^2, \tag{1.2}$$

where $\mathbf{H}$ and $\mathbf{H}'$ represent two fingerprints, K is the total number of usable subcarriers, H_k and H_k' are the CSIs on subcarrier k, η is the modified cross-correlation between $\mathbf{H}$ and $\mathbf{H}'$ with synchronization error compensated, and Λ, Λ' are the channel energies of $\mathbf{H}$ and $\mathbf{H}'$, respectively. Realizing that the Wi-Fi receiver may not be fully synchronous with the Wi-Fi transmitter due to mismatches in their radio-frequency front-end components [42], an additional phase rotation of $e^{-jk\phi}$ is employed to counteract the phase distortions incurred by the synchronization errors in the calculation of η, where ϕ can be estimated and compensated using Algorithm 1, shown later. Equation (1.1) implies that TRRS ranges from 0 to 1. More specifically, a larger TRRS indicates a higher similarity between two CSIs and thus the two associated locations.

1.4.1 Increasing Effective Bandwidth via Diversity Exploitation

Two different diversities exist in the current Wi-Fi system, i.e., frequency diversity and spatial diversity. According to IEEE 802.11n, 35 Wi-Fi channels are dedicated to Wi-Fi transmission in 2.4 GHz and 5 GHz frequency bands with a maximum bandwidth of 40 MHz. The multitude of Wi-Fi channels leads to frequency diversity in that they provide opportunities for Wi-Fi devices to perform frequency hopping when experiencing deep fading or severe interference. On the other hand, spatial diversity can be exploited on multiple-input-multiple-out (MIMO) Wi-Fi devices, which is a mature technique that greatly boosts the spectral efficiency. MIMO has not only become an essential component of IEEE 802.11n/ac but has also been ubiquitously deployed on numerous commercial Wi-Fi devices. For Wi-Fi systems, both types of diversity can be harvested to provide a fingerprint with much finer granularity and thus lead to less ambiguity in comparison with the fingerprint measured with a bandwidth of only 40 MHz.

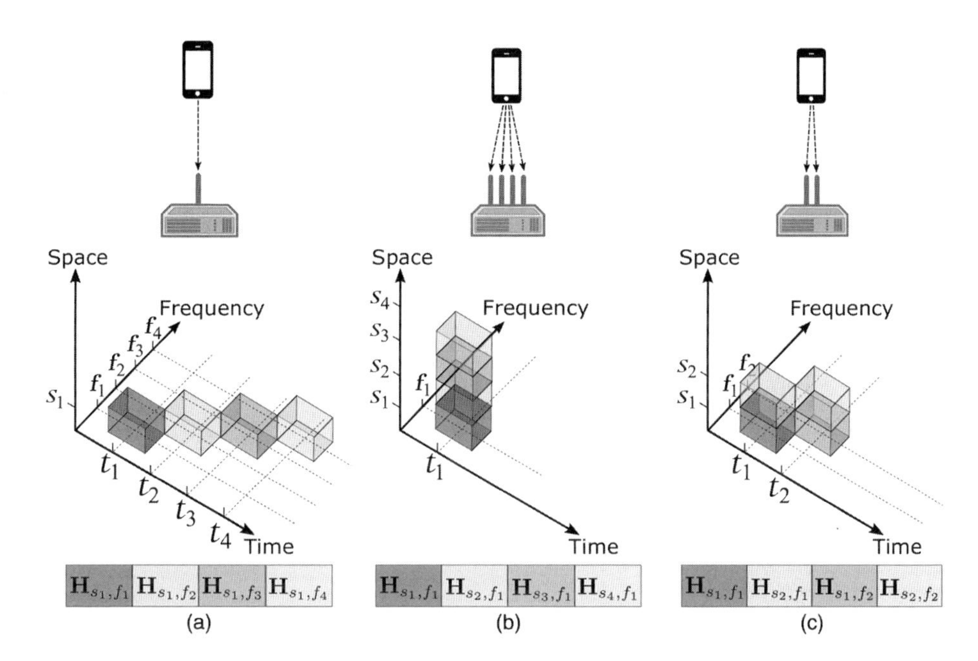

Figure 1.8 Leveraging frequency and spatial diversities in Wi-Fi to achieve large effective bandwidth: (a) frequency diversity, (b) spatial diversity, and (c) frequency–spatial diversity.

Figure 1.8 shows the general principle of creating a large effective bandwidth by exploiting the frequency and spatial diversities either independently or jointly. Because Wi-Fi devices can work on multiple Wi-Fi channels, one can exploit the frequency diversity by performing frequency hopping to obtain CSIs on different Wi-Fi channels. As demonstrated in Figure 1.8(a), CSIs on four different Wi-Fi channels are concatenated together to formulate a fingerprint of a large effective bandwidth. Despite the fact that the frequency diversity can be exploited on a single-antenna Wi-Fi device, it is time consuming to perform frequency hopping. For time efficiency, spatial diversity can be exploited on multiantenna Wi-Fi devices. For a Wi-Fi receiver with four antennas, e.g. in Figure 1.8(b), CSIs on the four receiving antennas can be combined together to formulate the fingerprint with a large effective bandwidth. Figure 1.8(c) shows an example of utilizing both the frequency and spatial diversities, where CSIs on two Wi-Fi channels and from two receiving antennas are combined into the fingerprint.

For a Wi-Fi system, the spatial diversity is determined by the number of antenna links, while the frequency diversity is dependent on the number of available Wi-Fi channels. Denoting the maximum spatial diversity by S, the maximum frequency diversity by F, and the bandwidth for each Wi-Fi channel by W, the effective bandwidth can be calculated as $S \times F \times W$.

1.4.2 Calculation of TRRS with Diversity Exploitation

In the following, we discuss the calculation of TRRS when both of the frequency and spatial diversities are available.

For Wi-Fi devices with a spatial diversity of S and a frequency diversity of F, the CSI measurements can be written as $\overline{\mathbf{H}} = \{\mathbf{H}_{s,f}\}_{s=1,2,\cdots,S}^{f=1,2,\cdots,F}$, where $\mathbf{H}_{s,f}$ stands for the CSI measured with the sth antenna link on the fth Wi-Fi channel, denoted as the virtual link (s, f). $\overline{\mathbf{H}} = \{\mathbf{H}_{s,f}\}_{s=1,2,\cdots,S}^{f=1,2,\cdots,F}$ can provide a fine-grained fingerprint with an effective bandwidth of $S \times F \times W$. Consequently, TRRS in (1.1) can be extended to the fine-grained fingerprint $\overline{\mathbf{H}}$ and $\overline{\mathbf{H}'}$, with η and Λ, Λ' modified as

$$\eta = \sum_{s=1}^{S}\sum_{f=1}^{F} \eta_{s,f}, \ \Lambda = \sum_{s=1}^{S}\sum_{f=1}^{F} \Lambda_{s,f}, \ \Lambda' = \sum_{s=1}^{S}\sum_{f=1}^{F} \Lambda'_{s,f}, \tag{1.3}$$

where

$$\eta_{s,f} = \max_{\phi} \left| \sum_{k=1}^{K} H_{s,f,k} H'^{*}_{s,f,k} e^{-jk\phi} \right| \tag{1.4}$$

represents the modified cross-correlation on the virtual link (s, f), and $\Lambda_{s,f} = \sum_{k=1}^{K} |H_{s,f,k}|^2$, $\Lambda'_{s,f} = \sum_{k=1}^{K} |H'_{s,f,k}|^2$ are the channel energies of $\mathbf{H}_{s,f}$ and $\mathbf{H}'_{s,f}$ on the virtual link (s, f), respectively.

Algorithm 1 elaborates on the calculation of $\gamma[\overline{\mathbf{H}}, \overline{\mathbf{H}'}]$. As shown in Algorithm 1, steps 4–9 are used to calculate the channel energies on the virtual link (s, f), while steps 10–14 are targeted to compute the modified cross-correlation of two CSIs

---

Algorithm 1 Calculating TRRS by exploiting diversities

Input: $\overline{\mathbf{H}} = \{\mathbf{H}_{s,f}\}_{s=1,2,\cdots,S}^{f=1,2,\cdots,F}, \overline{\mathbf{H}'} = \{\mathbf{H}'_{s,f}\}_{s=1,2,\cdots,S}^{f=1,2,\cdots,F}$

Output: $\gamma[\overline{\mathbf{H}}, \overline{\mathbf{H}'}]$

1: $\Lambda = 0, \Lambda' = 0, \eta = 0$
2: **for** $s = 1, 2, \cdots, S$ **do**
3:   **for** $f = 1, 2, \cdots, F$ **do**
4:     $\Lambda_{s,f} = 0, \Lambda'_{s,f} = 0$
5:     **for** $k = 1, 2, \cdots, K$ **do**
6:       $\Lambda_{s,f} \leftarrow \Lambda_{s,f} + |H_{s,f,k}|^2$
7:       $\Lambda'_{s,f} \leftarrow \Lambda'_{s,f} + |H'_{s,f,k}|^2$
8:     **end for**
9:     $\Lambda \leftarrow \Lambda + \Lambda_{s,f}, \Lambda' \leftarrow \Lambda' + \Lambda'_{s,f}$
10:     **for** $n = 1, 2, \cdots, N$ **do**
11:       $z[n] \leftarrow \sum_{k=1}^{N} H_{s,f,k} H'^{*}_{s,f,k} e^{-j\frac{2\pi n(k-1)}{N}}$
12:     **end for**
13:     $n^{\star} = \underset{n=1,2,\cdots,N}{\operatorname{argmax}} |z[n]|$
14:     $\overline{\eta}_{s,f} = z[n^{\star}]$
15:     $\eta \leftarrow \eta + \overline{\eta}_{s,f}$
16:   **end for**
17: **end for**
18: $\gamma[\overline{\mathbf{H}}, \overline{\mathbf{H}'}] \leftarrow \left(\frac{\eta}{\sqrt{\Lambda}\sqrt{\Lambda'}}\right)^2$

---

on the virtual link (s, f). The channel energies and modified cross-correlation on each virtual link are accumulated as shown in steps 9 and 15, respectively. Finally, the TRRS is obtained by step 18. The computation of $\eta_{s,f}$ is approximated by $\overline{\eta}_{s,f} = \max_n \left| \sum_{k=1}^{K} H_{s,f,k} H'^{*}_{s,f,k} e^{-j\frac{2\pi n(k-1)}{N}} \right|$ that takes the same format of a discrete Fourier transform of size N and thus can be computed efficiently by fast Fourier transform. Using a large N in the computations leads to a more accurate approximation of $\eta_{s,f}$.

References

[1] H. Liu, H. Darabi, P. Banerjee, and J. Liu, "Survey of wireless indoor positioning techniques and systems," *IEEE Transactions on Systems, Man, and Cybernetics, Part C*, vol. 37, no. 6, pp. 1067–1080, Nov. 2007.

[2] L. Dimitrios, J. Liu, X. Yang, R. R. Choudhury, V. Handziski, and S. Sen. "A realistic evaluation and comparison of indoor location technologies: Experiences and lessons learned," in *Proceedings of the 14th ACM International Conference on Information Processing in Sensor Networks*, pp. 178–189, 2015.

[3] M. Youssef, A. Youssef, C. Rieger, U. Shankar, and A. Agrawala, "Pinpoint: An asynchronous time-based location determination system," in *Proceedings of 4th ACM International Conference on Mobile Systems, Applications, and Services*, pp. 165–176, 2006.

[4] R. J. Fontana and S. J. Gunderson, "Ultra-wideband precision asset location system," in *Proceedings of the IEEE Conference on Ultra Wideband Systems and Technologies*, pp. 147–150, May 2002.

[5] D. Niculescu and B. Nath, "VOR base stations for indoor 802.11 positioning," in *Proceedings of 10th ACM Annual International Conference on Mobile Computing and Networking*, pp. 58–69, 2004.

[6] J. Xiong and K. Jamieson, "Arraytrack: A fine-grained indoor location system," in *Proceedings of 10th USENIX Symposium on Networked Systems Design and Implementation*, pp. 71–84, 2013.

[7] Y. Zhao, "Standardization of mobile phone positioning for 3G systems," *IEEE Communications Magazine*, vol. 40, no. 7, pp. 108–116, Jul. 2002.

[8] G. Sun, J. Chen, W. Guo, and K. J. R. Liu, "Signal processing techniques in network-aided positioning: A survey of state-of-the-art positioning designs," *IEEE Signal Processing Magazine*, vol. 22, no. 4, pp. 12–23, Jul. 2005.

[9] A. Banerjee, D. Maas, M. Bocca, N. Patwari, and S. Kasera, "Violating privacy through walls by passive monitoring of radio windows," in *Proceedings of the 2014 ACM Conference on Security and Privacy in Wireless & Mobile Networks*, pp. 69–80, 2014.

[10] H. Abdelnasser, M. Youssef, and K. A. Harras, "WiGest: A ubiquitous WiFi-based gesture recognition system," in *Proceedings of the IEEE International Conference on Computer Communications*, pp. 1472–1480, 2015.

[11] J. Xiao, K. Wu, Y. Yi, L. Wang, and L. Ni, "FIMD: Fine-grained device-free motion detection," in *Proceedings of the 18th IEEE International Conference on Parallel and Distributed Systems*, pp. 229–235, Dec. 2012.

[12] F. Adib and D. Katabi, "See through walls with WiFi!" in *Proceedings of the ACM Special Interest Group on Data Communications*, pp. 75–86, 2013.

[13] C. Han, K. Wu, Y. Wang, and L. Ni, "WiFall: Device-free fall detection by wireless networks," in *Proceedings of the IEEE International Conference on Computer Communications*, pp. 271–279, Apr. 2014.

[14] Y. Wang, J. Liu, Y. Chen, M. Gruteser, J. Yang, and H. Liu, "E-eyes: Device-free location-oriented activity identification using fine-grained WiFi signatures," in *Proceedings of the 20th ACM Annual International Conference on Mobile Computing and Networking*, pp. 617–628, 2014.

[15] W. Xi, J. Zhao, X.-Y. Li, K. Zhao, S. Tang, X. Liu, and Z. Jiang, "Electronic frog eye: Counting crowd using WiFi," in *Proceedings of the IEEE International Conference on Computer Communications*, pp. 361–369, Apr. 2014.

[16] W. Wang, A. X. Liu, M. Shahzad, K. Ling, and S. Lu, "Understanding and modeling of WiFi signal based human activity recognition," in *Proceedings of the 21st ACM Annual International Conference on Mobile Computing and Networking*, pp. 65–76, 2015.

[17] Y. Yang and A. Fathy, "Design and implementation of a low-cost real-time ultra-wide band see-through-wall imaging radar system," in *Proceedings of the IEEE/MTT-S International Microwave Symposium*, pp. 1467–1470, Jun. 2007.

[18] F. Adib, Z. Kabelac, D. Katabi, and R. C. Miller, "3D tracking via body radio reflections," in *Proceedings of the 11th USENIX Symposium on Networked Systems Design and Implementation*, pp. 317–329, Apr. 2014.

[19] F. Adib, H. Mao, Z. Kabelac, D. Katabi, and R. C. Miller, "Smart homes that monitor breathing and heart rate," in *Proceedings of the 33rd Annual ACM Conference on Human Factors in Computing Systems*, pp. 837–846, 2015.

[20] H. Abdelnasser, K. A. Harras, and M. Youssef, "Ubibreathe: A ubiquitous non-invasive WiFi-based breathing estimator," in *Proceedings of the 16th ACM International Symposium on Mobile Ad Hoc Networking and Computing*, pp. 277–286, 2015.

[21] J. Liu, Y. Wang, Y. Chen, J. Yang, X. Chen, and J. Cheng, "Tracking vital signs during sleep leveraging off-the-shelf WiFi," in *Proceedings of the 16th ACM International Symposium on Mobile Ad Hoc Networking and Computing*, pp. 267–276, 2015.

[22] B. Wang, Y. Wu, F. Han, Y.-H. Yang, and K. J. R. Liu, "Green wireless communications: A time-reversal paradigm," *IEEE Journal on Selected Areas in Communications*, vol. 29, no. 8, pp. 1698–1710, Sep. 2011.

[23] Y. Jin, J. M. F. Moura, Y. Jiang, D. Stancil, and A. Cepni, "Time reversal detection in clutter: Additional experimental results," *IEEE Transactions on Aerospace and Electronic System*, vol. 47, no. 1, pp. 140–154, Jan. 2011.

[24] J. M. F. Moura and Y. Jin, "Detection by time reversal: Single antenna," *IEEE Transactions on Signal Processing*, vol. 55, no.1, pp. 187–201, Jan. 2007.

[25] Y. Chen, F. Han, Y.-H. Yang, H. Ma, Y. Han, C. Jiang, H. Q. Lai, D. Claffey, Z. Safar, and K. J. R. Liu, "Time-reversal wireless paradigm for green Internet of Things: An overview," *IEEE Internet of Things Journal*, vol. 1, no. 1, pp. 81–98, 2014.

[26] G. Lerosey, J. de Rosny, A. Tourin, A. Derode, G. Montaldo, and M. Fink, "Time reversal of electromagnetic waves," *Physical Review Letters*, vol. 92, p. 193904(3), May 2004.

[27] Q. Xu, C. Jiang, Y. Han, B. Wang, and K. J. R. Liu, "Waveforming: An overview with beamforming," *IEEE Communications Surveys & Tutorials*, vol. 20, no. 1, pp. 132–149, 2018.

[28] Y. Chen, B. Wang, Y. Han, H. Q. Lai, Z. Safar, and K. J. R. Liu, "Why time reversal for future 5G wireless?" *IEEE Signal Processing Magazine*, vol. 33, no. 2, pp. 17–26, Mar. 2016.

[29] Y. Han, Y. Chen, B. Wang, and K. J. R. Liu, "Time-reversal massive multipath effect: A single-antenna massive MIMO solution," *IEEE Transactions on Communications*, vol. 64, no. 8, pp. 3382–3394, Aug. 2016.

[30] F. Zhang, C. Chen, B. Wang, H. Q. Lai, and K. J. R. Liu, "A time-reversal spatial hardening effect for indoor speed estimation, " in *Proceedings of the IEEE International Conference on Acoustics, Speech and Signal Processing*, Mar. 2017.

[31] A. Goldsmith, *Wireless Communications*, New York: Cambridge University Press, 2005.

[32] F. Rusek, D. Persson, B. K. Lau, E. G. Larsson, T. L. Marzetta, O. Edfors, and F. Tufvesson, "Scaling up MIMO: Opportunities and challenges with very large arrays," *IEEE Signal Processing Magazine*, vol. 30, no. 1, pp. 40–60, Jan. 2013.

[33] B. P. Bogert, "Demonstration of delay distortion correction by time-reversal Ttchniques," *IRE Transactions on Communications Systems*, vol. 5, no. 3, pp. 2–7, Dec. 1957.

[34] D. Harasty and A. Oppenheim, "Television signal deghosting by noncausal recursive filtering," in *Proceedings of the IEEE International Conference on Acoustics, Speech, and Signal Processing*, pp. 1778–1781, Apr. 1988.

[35] M. Fink, "Time reversal of ultrasonic fields. I. Basic principles," *IEEE Transactions on Ultrasonics, Ferroelectrics and Frequency Control*, vol. 39, no. 5, pp. 555–566, 1992.

[36] B. Wang, Y. Wu, F. Han, Y. H. Yang, K. J. R. Liu, "Green wireless communications: A time-reversal paradigm," *IEEE Journal of Selected Areas in Communications, special issue on Energy-Efficient Wireless Communications*, vol. 29, no. 8, pp. 1698–1710, Sep. 2011.

[37] N. Guo, B. M. Sadler, and R. C. Qiu, "Reduced-complexity UWB timereversal techniques and experimental results," *IEEE Transactions on Wireless Communications*, vol. 6, no. 12, pp. 4221–4226, Dec. 2007.

[38] F. Han, Y.-H. Yang, B. Wang, Y. Wu, and K. J. R. Liu, "Time-reversal division multiple access over multi-path channels," *IEEE Transactions on Communications*, vol. 60, no. 7, pp. 1953–1965, Jul. 2012.

[39] Z. H. Wu, Y. Han, Y. Chen, and K. J. R. Liu, "A time-reversal paradigm for indoor positioning system," *IEEE Transactions on Vehicular Technology, special issue on Indoor Localization, Tracking, and Mapping with Heterogeneous Technologies*, vol. 64, no. 4, pp. 1331–1339, Apr. 2015.

[40] G. Lerosey, J. de Rosny, A. Tourin, and M. Fink, "Focusing beyond the diffraction limit with far-field time reversal," *Science*, vol. 315, pp. 1120–1122, Feb. 2007.

[41] C. Chen, Y. Han, Y. Chen, and K. J. R. Liu, "Indoor global positioning system with centimeter accuracy using Wi-Fi," *IEEE Signal Processing Magazine*, vol. 33, no. 6, pp. 128–134, Nov. 2016.

[42] M. Speth, S. Fechtel, G. Fock, and H. Meyr, "Optimum receiver design for wireless broadband systems using OFDM Part I," *IEEE Transactions on Communications*, vol. 47, pp. 1668–1677, Nov. 1999.

Part I

Indoor Locationing and Tracking

2 Centimeter-Accuracy Indoor Positioning

In an indoor environment, there commonly exist a large number of multipaths due to rich scatterers. These multipaths make the indoor positioning problem very challenging. The main reason is that most of the transmitted signals are significantly distorted by the multipaths before arriving at the receiver, which causes inaccuracies in the estimation of the positioning features such as the time of arrival (TOA) and the angle of arrival (AOA). On the other hand, the multipath effect can be very constructive when employed in the time-reversal (TR) radio transmission. By utilizing the uniqueness of the multipath profile at each location, TR can create a resonating effect of focusing the energy of the transmitted signal only onto the intended location, which is known as the spatial focusing effect. With this effect, the notion of walls is essentially removed, i.e., there is no notion of walls or obstacles as with a simple time reversal operation a radio wave can go back to the original location. In this chapter, we consider exploiting such a high-resolution focusing effect in the indoor positioning problem. Specifically, we develop a TR indoor positioning system (TRIPS) by utilizing the location-specific characteristic of multipaths. By doing so, we decompose the ill-posed indoor positioning problem into two well-defined subproblems. The first subproblem is to create a database by mapping the physical geographical location with the logical location in the channel impulse response (CIR) space, whereas the second subproblem is to determine the real physical location by matching the estimated CIR with those in the database. To evaluate the performance of the TRIPS, we build a prototype to conduct real experiments. The experimental results show that, with a single AP working in the 5.4-GHz band under the non-line-of-sight (NLOS) condition, the TRIPS can achieve perfect 1 to 2 cm localization accuracy.

2.1 Introduction

With the advancement of communication technology, handheld devices such as mobile phones, tablets, and laptops have become an important and indispensable part of our daily lives. We use them to check emails, to connect to various social networks, and to watch video streaming, just to name a few. Because these devices can provide us with all-day connectivity through wireless communication techniques such as Wi-Fi and Fourth-Generation Long Term Evolution (4G LTE), we carry them with us all the time, due to which it is possible to record and trace our activities by tracking these devices. Specifically, with the sensors installed in the handheld devices, one can gather various kinds of user information that can reveal users' behavior at different locations and times, e.g., users' location, movement, and data usage. Through analyzing these collected pieces of information, service providers can estimate and learn users' behaviors and preferences and, thus, provide user-specific services.

To successfully provide users with the right versatile services, it is crucial for the service provider to know the exact location of users. In the literature, many indoor positioning system (IPS) approaches have been developed, and most of them can be classified into three categories [1]: triangulation, proximity methods, and scene analysis. In triangulation, the terminal device (TD) measures the time of arrival (TOA) [2], time difference of arrival (TDOA) [3], angle of arrival (AOA) [4, 5] of the signals sent from the access point (AP) with known positions and then uses physical principles of wave propagation to calculate the geographical location based on the measurements. Although the concept of triangulation is simple, some special requirements are needed, e.g., precise measurements of TOA and/or AOA, synchronization between the TD and the AP, and specialized apparatus for AP. However, due to the rich scattering characteristic of indoor environment, the measurements are generally not very precise, which leads to poor indoor positioning performance of these triangulation methods.

The second category of IPS algorithms is a proximity method that can provide symbolic relative location information. This kind of algorithm relies on the dense deployment of the infrastructure. When the TD moves in the target area, the TD is considered to be located with the antenna that detects the TD. If multiple antennas can detect the TD, then the TD is simply considered to be located with the antenna that receives the strongest signal. Most of the radio-frequency (RF) identification and the cell identification [6] positioning systems fall into this category. Because the TD will be considered to be colocated with the antenna, this kind of algorithm cannot give precise location information. Moreover, due to the dense deployment of the antennas, the implementation cost is very high.

The third category of IPS algorithms is the scene analysis method, which first collects features of the scene and then matches online measurements with the collected features to estimate the location. Most of the scene analysis-based IPS algorithms make use of the received signal strength (RSS) and/or the channel state information (CSI), while the matching method can be either deterministic or probabilistic [7]. In a deterministic method, the position is determined by finding the minimum distance between the measurements to the database. In [8], it was proposed to first use spatial filtering to reduce the number of reference APs and then use kernel functions as distance measures. A root-mean-square error of 2.71 m was reported using three APs. An RF-based tracking system named RADAR was proposed in [9]. The system uses empirically determined and theoretically computed signal strength for triangulation, and triangulation is done using the signal strength information gathered at multiple locations. A median resolution was reported to be in the range of 2–3 m using three APs. A linear approximation model on the RSS versus the Euclidean distance between the AP and the TD in an anonymous environment without necessary offline training was proposed in [10] and achieves a mean estimation error of 15 m. A compressive sensing scheme was proposed in [11] for localization using the sparsity characteristics in positioning problems with 1.5-m error.

On the other hand, in a probabilistic method, the estimation is based on some probabilistic criteria such as maximum a posteriori (MAP) and maximal likelihood (ML). In [12] and [13], a positioning algorithm based on Wi-Fi RSS was proposed. The RSS information from multiple Wi-Fi APs is collected, and the distribution of the RSS is estimated. During the online positioning phase, the MAP or ML criterion is used to determine the location and achieve a mean error of 40 cm with multiple APs. In [14], the RSS of Wi-Fi and FM signals was used to jointly estimate the cumulative distribution function of RSS for indoor positioning. The smaller variation of FM signals in an indoor environment provides extra information and precision over Wi-Fi-only systems and achieves better room-level accuracy. In addition to the RSS, the CSI has been also used in the literature for positioning. In [15], it was proposed to use the amplitude of channel impulse response (CIR) as the fingerprint for localization. The amplitude of CIR is used as an input to a nonparametric kernel regression method for location estimation. In [16] and [17], it was proposed to utilize the complex CIR as a link signature for location distinction, where the normalized minimal Euclidean distance is adopted as the distance measure. The CSI was proposed to be used in the orthogonal frequency-division multiplexing (OFDM) systems as the fingerprints in the positioning algorithm [18]. Because there are a lot of partitioned channels in an OFDM system, the CSI provides rich information for positioning. In the online phase, the CSI from the TD is matched to the stored database using a MAP algorithm. The authors report a mean accuracy value of 65 cm in a 5-m by 8-m office using three APs.

However, most of the existing IPS algorithms cannot achieve a desired centimeter-level localization accuracy value, particularly for a single AP working in the NLOS condition. The main reason is that it is generally very difficult or even impossible to obtain precise measurements due to the rich scattering indoor environment. Such imprecise measures lead to ambiguity when performing positioning algorithms. To reduce ambiguity, most existing algorithms require more online measurements and/or multiple APs. Different from the existing approaches, in this chapter, we consider a single-AP indoor positioning algorithm that can achieve centimeter-level localization accuracy with single realization of online measurement by utilizing the TR technique. TR technique is known to be able to focus the energy of the transmitted signal only onto the intended location, i.e., the spatial focusing effect. The foundation of spatial focusing is that the CIR in a rich scattering indoor environment is location specific and unique for each location [19], i.e., each CIR corresponds to a physical geographical location. Therefore, by utilizing such a unique location-specific CIR, the TRIPS is able to position the TD by matching the CIR with the geographical location. Because spatial focusing is a half-wavelength focus spot, the TRIPS can achieve a centimeter-level localization accuracy value even with a single AP working in the NLOS condition.

The rest of the chapter is organized as follows. In Section 2.2, we will briefly review the TR technique and describe in details the TRIPS. Then, in Section 2.3, we will discuss the experimental results, including the properties of the TR technique and the performance of the TRIPS.

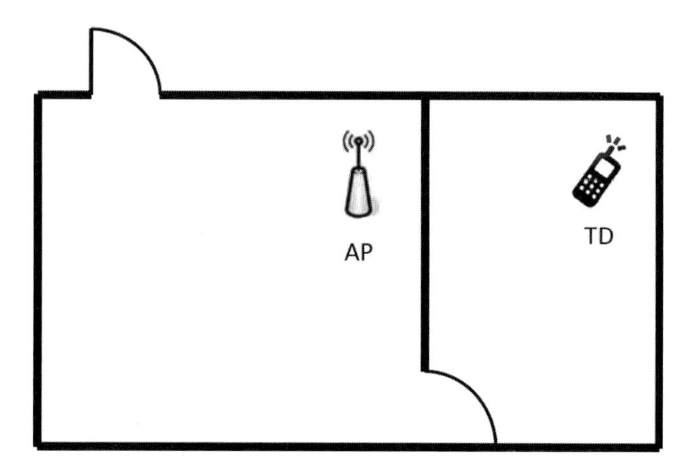

Figure 2.1 System model.

2.2 Time Reversal Indoor Positioning System

As shown in Figure 2.1, we study the indoor localization problem where there is an AP and a TD in an indoor environment. The AP is positioned in an arbitrarily known location, whereas the location of the TD is unknown. The TD transmits some known signals, e.g., fixed pseudorandom sequences, to the AP, and the AP tries to estimate the location of the TD based on the received signals. Due to the multipaths in the indoor environment, the received signal at the AP is significantly distorted [12]. In such a case, it is generally impossible to identify the location purely based on the received signal of a single AP, i.e., the single-AP indoor localization problem is ill posed.

To address this problem, we consider a TRIPS by decomposing the ill-posed problem into two well-defined subproblems. Specifically, in the first subproblem, we build a database offline by mapping the physical geographical locations to the logical locations in the CIR space. Then, in the second subproblem, we match the online estimated CIR of the TD to those in the database to position the TD. In the following sections, we first give a brief introduction of the TR technique and then discuss in detail the TR-based indoor positioning system.

2.2.1 Background of Time Reversal

TR is a technology that can focus the power of the transmitted signal in both time and space domains. The phenomenon of TR was first observed by Zeldovich et al. in 1985 [20]. Later, the TR technique was studied and applied into signal processing by Fink et al. in 1989 [21], followed by several theoretical and experimental works in acoustic and ultrasonic communications, verifying that the transmitted wave energy can be focused at the intended location with high spatial and temporal resolution [22–24]. Due to the fact that TR does not require complicated channel processing and equalization, it was also analyzed, tested, and validated in wireless communications [19, 25–35].

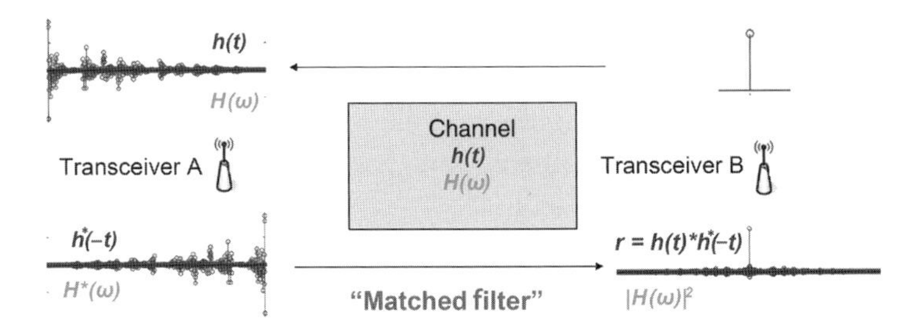

Figure 2.2 The time reversal signal processing principle.

Moreover, with a potential of over an order of magnitude of reduction in power consumption and interference alleviation, as well as the natural capability of supporting heterogeneous TDs and providing an additional security and privacy guarantee, TR technique is shown to be a promising solution for green Internet of Things [36].

Figure 2.2 demonstrates a simple TR communication system [19]. When transceiver A wants to transmit information to transceiver B, transceiver B first sends an impulse signal to transceiver A. This is called the channel probing phase. Then, transceiver A time-reverses (and conjugates if the signal is complex) the received waveform from transceiver B and uses the time-reversed version of waveform to transmit the information back to transceiver B. This second phase is called the TR transmission phase.

The TR technique relies on two basic assumptions, i.e., channel reciprocity and channel stationarity. Channel reciprocity requires the CIRs of the forward and backward links to be highly correlated, whereas channel stationarity requires the CIR to be stationary for at least one probing-and-transmission phase. These two assumptions generally hold in practice, as validated by experiments in [27] and [19]. In [27], an experiment was conducted in a laboratory area and showed that the correlation of CIR between the forward and backward links is as high as 0.98, whereas in [19], it was shown that the multipath channel in a typical office environment does not vary much over time. Specifically, the CIR had a snapshot once every minute for a total of 40 min, where the first 20 snapshots correspond to a stationary environment, the 21st to 30th snapshots correspond to a moderately varying environment, and the last 10 snapshots correspond to a varying environment. The experimental results show that the correlation coefficients between different snapshots are above 0.95 for a stationary environment and above 0.8 for a varying environment.

With the property of the channel reciprocity and stationarity, the re-emitted TR signal will retrace the incoming paths and form a constructive sum of signals at the intended location, resulting in a peak in the signal power distribution over the space, i.e., spatial focusing effect. Because TR utilizes all the multipaths as a matched filter, the transmitted signal will be focused in the time domain, which is referred to as the temporal focusing effect. Moreover, by using the environment as matched filters, the transceiver design complexity can be significantly reduced. In an indoor environment, the wireless multipaths come from the surrounding reflectors. Because the received waveforms from the TD at different locations undergo different reflecting paths and delays, the multipath profile is unique for each location. By utilizing this unique location-specific multipath profile, TR can create the spatial focusing effect at the intended location, i.e., the received signals are added coherently at the intended location but incoherently at any unintended location. As will be discussed in the next section, our algorithm leverages such a special feature to solve the ill-posed single-AP indoor localization problem.

2.2.2 Time Reversal Indoor Positioning Algorithm

Here, we will discuss in detail the TR indoor positioning algorithm. With the spatial focusing effect, we know that the CIR in the TR system is location specific, which means that we can map the physical geographical locations into logical locations in the CIR space where one physical geographical location corresponds to a unique CIR in the TR system. Then, the indoor localization problem becomes a classical classification problem that identifies the class of the TD in the CIR space. Therefore, the TR indoor positioning algorithm contains two phases. The first phase is an offline training phase, where we build a CIR database to map the physical geographical location into the logical location in the CIR space, and the second phase is an online positioning phase where we match the estimated CIR of the TD with the CIR database to localize the TD.

2.2.2.1 Offline Training Phase

In the offline training phase, we are building a CIR database for the online positioning phase. Because the database has a direct consequence to the localization performance, how to build the database is critical to the indoor positioning algorithm. Note that the CIR at different locations will be different if the distance between two locations is larger than the wavelength and may be similar if the distance is smaller than the wavelength. Moreover, the CIR at a certain location may slightly vary over time due to the change of environment. With such an intuition, for each intended location, we obtain a series of CIRs at different time. Specifically, for each intended location p_i, we collect the CIRs information $\mathbf{H}_i$ as follows:

$$\mathbf{H}_i = \{\mathbf{h}_i(t = t_0), \quad \mathbf{h}_i(t = t_1), \ldots, \mathbf{h}_i(t = t_M)\}, \tag{2.1}$$

where $\mathbf{h}_i(t = t_l)$ stands for the estimated CIR information of location p_i at time t_l.

Therefore, the database $\mathbf{D}$ is the collection of all $\mathbf{H}_i^{\prime}s$

$$\mathbf{D} = \{\mathbf{H}_i, \forall i\}, \tag{2.2}$$

2.2.2.2 Online Positioning Phase

In the online positioning phase, we first estimate the CIR information based on the signal received at the AP. Then, our objective is to localize the TD by matching the estimated CIR information with the database using a classification technique. Because the dimension of the information for each location in the database is very high, classification based on the raw CIR information may not work. Therefore, it is necessary to preprocess the CIR information to obtain important features for the classification.

As we have previously discussed, because the received signals undergo different reflecting paths and delays for the receiver at different locations, the CIR can be viewed as a unique location-specific signature. When convolving the time-reversed CIR with the CIR in the database, only that at the intended location will produce a peak, which is known as spatial focusing effect. For the locations other than the intended location, there is no focusing effect. Therefore, we can design a TR-based dimension reduction approach to extract the effective feature for localization. To do so, we first introduce a definition of *TR resonating strength* as follows.

DEFINITION 2.2.1 (Time Reversal Resonating Index)*: The TR resonating strength* $\gamma(\boldsymbol{h}_1, \boldsymbol{h}_2)$*, between two CIRs* $\boldsymbol{h}_1 = [h_1[0], \; h_1[1], \ldots, h_1[L-1]]$ *and* $\boldsymbol{h}_2 = [h_2[0], h_2[1], \ldots, h_2[L-1]]$ *is defined as*

$$\gamma(\boldsymbol{h}_1, \boldsymbol{h}_2) = \frac{\max_i \left|\left(\boldsymbol{h}_1 * \boldsymbol{g}_2\right)[i]\right|^2}{\left(\sum_{i=0}^{L-1} |h_1[i]|^2\right)\left(\sum_{j=0}^{L-1} |g_2[j]|^2\right)}, \tag{2.3}$$

where $\boldsymbol{g}_2 = \left[g_2[0], \; g_2[1], \ldots, g_2[L-1]\right]$ *is defined as the time reversed and conjugated version of* $\boldsymbol{h}_2$ *as follows*

$$g_2[k] = h_2^*[L-1-k], \quad k = 0, 1, \ldots, L-1. \tag{2.4}$$

A close look at (2.3) would reveal that the TR resonating strength is the maximal amplitude of the entries of the cross correlation between two complex CIRs, which is different from the conventional correlation coefficient between two complex CIRs where there is no max operation and the index $[i]$ in (2.3) is replaced with index $[L1]$. The main reason for using the TR resonating strength instead of the conventional correlation coefficient is to increase the robustness for the tolerance of channel estimation error. Note that most of the channel estimation schemes may not be able to perfectly estimate the CIR due to the synchronization error, i.e., a few taps may be added or dropped during the channel estimation process. In such a case, the conventional correlation coefficient without max operation may not reflect the true similarity between two CIRs, whereas the TR resonating strength is able to capture the real similarity and, thus, increase the robustness.

With the definition of TR resonating strength, we are now ready to describe the online positioning phase. Let $\hat{\mathbf{h}}$ be the CIR that we estimate for the TD with unknown location. To match $\hat{\mathbf{h}}$ with the logical locations in the database, we first extract the feature using the TR resonating strength for each location. Specifically, for each location p_i, we compute the maximal TR resonating strength η_i as follows:

$$\eta_i = \max_{\mathbf{h}_i(t=t_j) \in \mathbf{H}_i} \eta(\hat{\mathbf{h}}, \mathbf{h}_i(t = t_j)). \tag{2.5}$$

By computing η_i for all possible locations, i.e., $\mathbf{H}_i \in \mathbf{D}$, we can obtain $\eta_1, \eta_2, \ldots, \eta_N$. Then, the estimated location, $p_{\hat{i}}$, is simply the one that can give the maximal η_i, i.e., $\hat{i}$ can be derived as follows

$$\hat{i} = \arg\max_i \eta_i. \tag{2.6}$$

Although our algorithm is very simple, it can achieve very good localization performance, as we will see in the experiment in the next section.

2.3 Experiments

2.3.1 Experiment Setting

To evaluate the performance of our algorithm, we build a TR system prototype that operates at 5.4-GHz band with a bandwidth of 125 MHz. A snapshot of the radio stations of our prototype is shown in Figure 2.3, where the antenna is attached to a small cart with an RF board and computer installed on the cart. We test the performance of our prototype in a typical office room that is located on the second floor of the Jeong H. Kim Engineering Building at the University of Maryland College Park. The layout of the floor plan of the office room is shown in Figure 2.4(a), where the AP is located at the place with the mark "**AP**" and the TD is located in the smaller office room marked as "**A**." The detailed floor layout of room A is shown in Figure 2.4(b). Notice that with such a setting, the AP is working in the NLOS condition.

2.3.2 Evaluation of TR Properties

Here, we evaluate three important properties of the TR system, namely, channel reciprocity, temporal stationarity, and spatial focusing. Note that channel reciprocity and temporal stationarity are the two underlying assumptions of TR system, whereas spatial focusing is the key feature for the success of the TRIPS.

Figure 2.3 Radio stations of the TR system prototype.

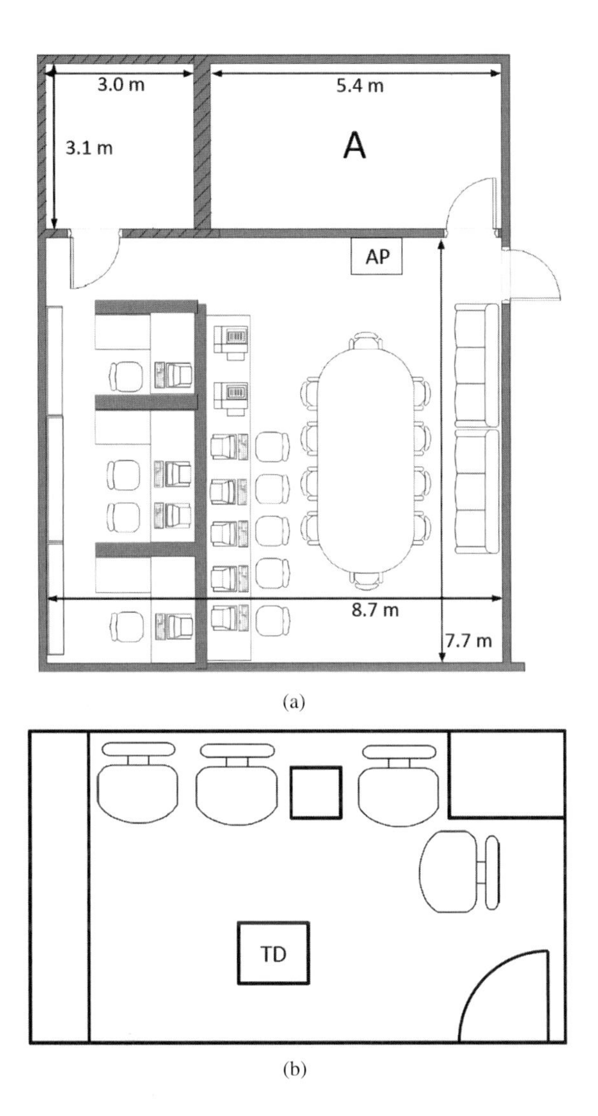

Figure 2.4 (a) Floor plan of the office room where we conduct our experiments; (b) Floor plan of room A.

2.3.2.1 Channel Reciprocity

We explore channel reciprocity by examining the CIR of the forward and backward links between the TD and the AP. Specifically, the TD first transmits a channel probing signal to the AP, and the AP records the CIR of the forward link. Immediately after that, the AP transmits a channel probing signal to the TD, and the TD records the CIR of the backward link. These procedures are repeated 18 times. One CIR realization of forward and backward links is shown in Figure 2.5, where (a) shows the amplitude and phase of the forward channel and (b) shows those of the backward channel. In these figures, we can see that the forward and backward channels are very similar. By computing the

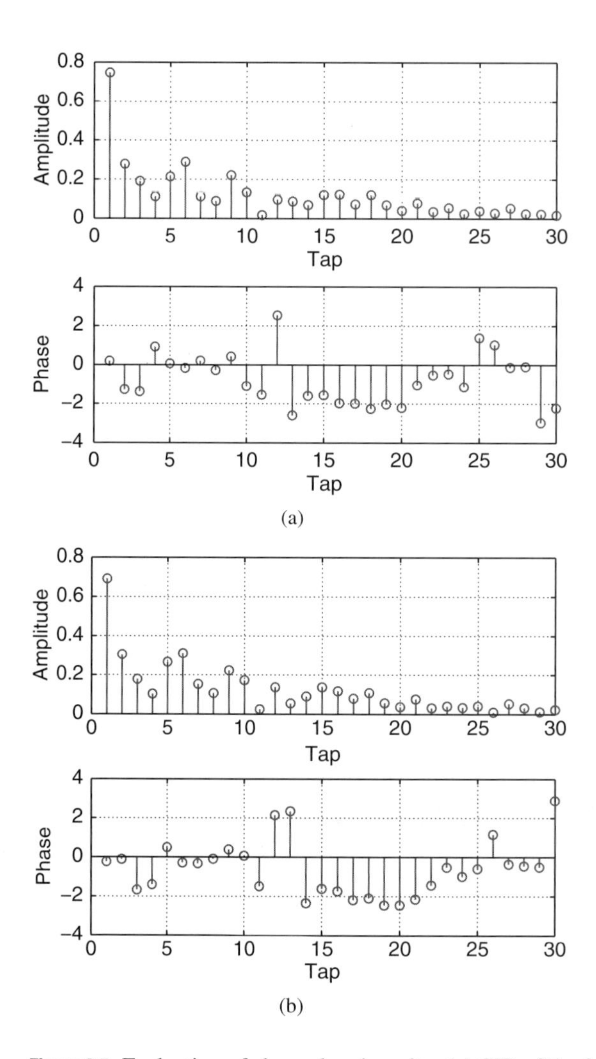

Figure 2.5 Evaluation of channel reciprocity: (a) CIR of the forward link; (b) CIR of the backward link.

correlation between the CIR of the forward link and that of the backward link, as shown in Figure 2.6, we can see that, indeed, the forward and backward channels are highly reciprocal. Figure 2.7 shows the TR resonating strength between any of the 18 forward and backward channel measurements with mean η to be over 0.95. This result shows that the reciprocity is stationary over time.

2.3.2.2 Channel Stationarity

We then evaluate the channel stationarity of the TR system by measuring the CIR of the link from the TD to AP under three different settings: short-interval, long-interval, and dynamic environments with a person walking around. In the short-interval experiment, we measure the CIR repeatedly 30 times, and the duration between two consecutive

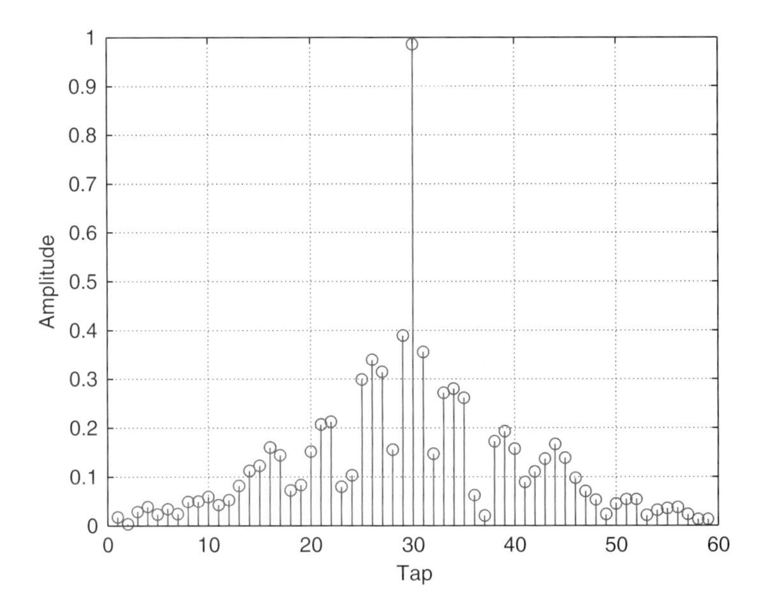

Figure 2.6 Cross correlation between the CIR of forward link and that of the backward link. Note that the center tap is the TR resonating strength between the CIR of forward link and that of the backward link.

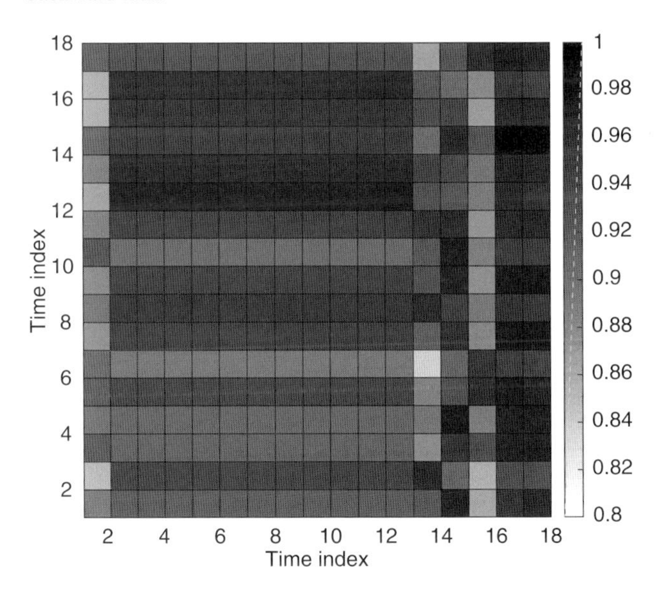

Figure 2.7 TR resonating strength between CIRs of the forward link and those of the backward link.

measurements is 2 min. For the long-interval experiment, we collect a total of 18 CIRs with 1-h interval from 9 A.M. to 5 P.M. over a weekend. Figure 2.8 shows the TR resonating strength η between any two CIRs from all 30 CIRs in the short-interval experiment, and Figure 2.9 shows the TR resonating strength between any two CIRs

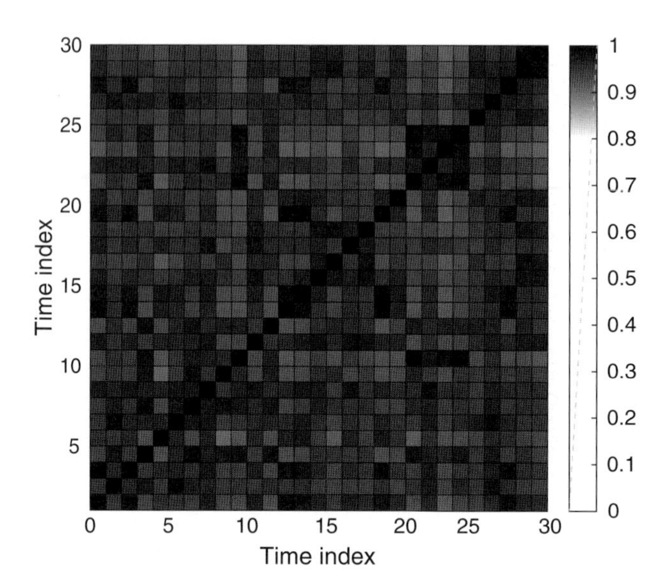

Figure 2.8 Evaluation of short temporal stationarity using the TR resonating strength between any two CIRs from the thirty CIRs of the link between the TD to the AP.

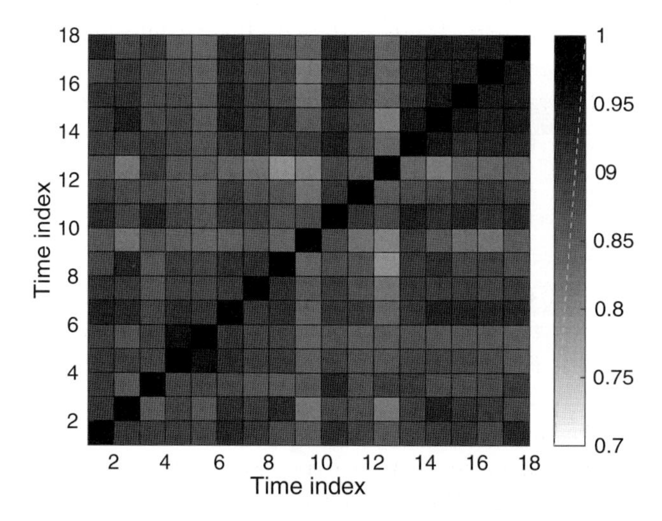

Figure 2.9 Evaluation of long temporal stationarity using the TR resonating strength between any two CIRs from the 18 CIRs collected over a weekend between the AP and TD.

from the 18 CIRs collected in the long-interval experiment. We can see that the CIRs at different time instances are highly correlated for both the short interval and long interval, which means that the channel in an ordinary office does not vary much over time even with long duration. We then investigate the effect of human movement. We collect, every 30 s, the CIRs with a person walking randomly between the AP and the TD. Figure 2.10 shows the TR resonating strength η between the fifteen collected CIRs.

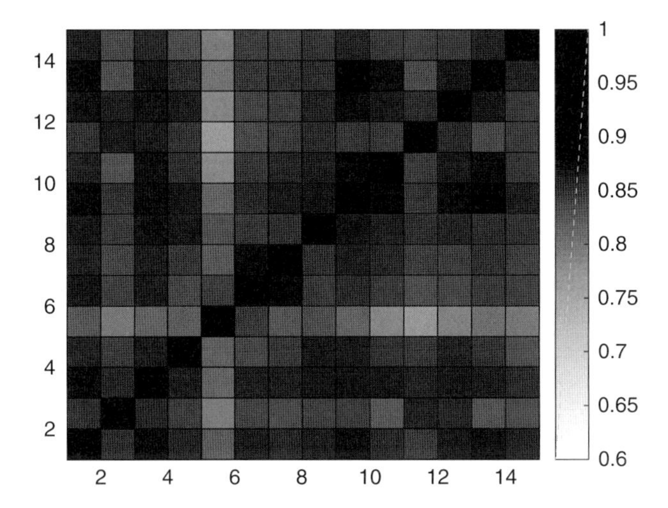

Figure 2.10 Evaluation of channel stationarity under minor environment change using TR resonating strength between any two CIRs collected with a person walking around.

The experimental result shows that, even with a person walking around, the TR resonating strength remains high among all of the collected CIRs. Therefore, the TR positioning system does not require a frequent update of the CIR information. All these results are consistent with the observations in [19], the main reason being that the multipaths come from the refractions and reflections of the indoor environment, which are quite stable, as long as there is no severe disturbance of the environment.

2.3.2.3 Spatial Focusing

As we have previously discussed, the CIR comes from the surrounding scatterers, and such scatterers are generally different for different geographical locations. Therefore, the CIR is location specific and unique for each location. By utilizing such a unique location-specific CIR, TR can focus the transmitted power only to the intended location, which is known as the spatial focusing effect of the TR system. We quantify such a spatial focusing effect using the maximum energy that the TD can harvest from the AP. To evaluate the spatial focusing effect, we conduct experiments by moving the locations of the TD on a 3-D architecture, as shown in Figure 2.11, within a 1-m by 0.9-m area in room A. The grid points are 10 cm apart, which leads to 110 evaluated locations in total.

We collect the CIR of all evaluated locations and compute the focusing gain, which is defined as the square of the TR resonating strength, i.e., η^2, by varying the intended location. The results are shown in Figure 2.12, where we can see that the focusing gain at the intended location is much larger than that at the unintended location, i.e., there exists a very good spatial focusing effect. In Figure 2.12, we also observe some repetitive patterns. Such repetitive behavior is due to the representation of 2-D locations using 1-D index. To better illustrate the spatial focusing effect, we fix the intended location as the center of the test area and show in Figure 2.13 the spatial focusing by directly using the

Figure 2.11 Three-dimensional architecture for moving the locations of the TD.

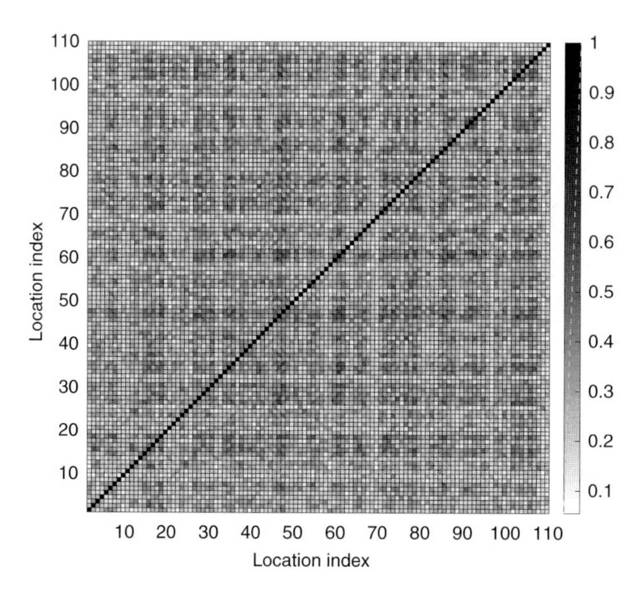

Figure 2.12 Focusing gain, η^2 of all grid points by moving the intended location within the 1-m by 0.9-m area. Every dot in the figure stands for one grid point where two neighboring grids points are 10 cm away from each other. The horizontal/vertical axis is the location index with 1-D representation. Each value in (i, j) represents the focusing gain at location j (location index with 1-D representation) when the intended location is i (location index with 1-D representation).

real geographical locations. Clearly, we can see very good spatial focusing performance. Note that similar results are observed for all other intended locations.

We further evaluate the spatial focusing effect in a finer scale with 1-cm grid spacing, and the results are shown in Figure 2.14. We can see that there is reasonably graceful

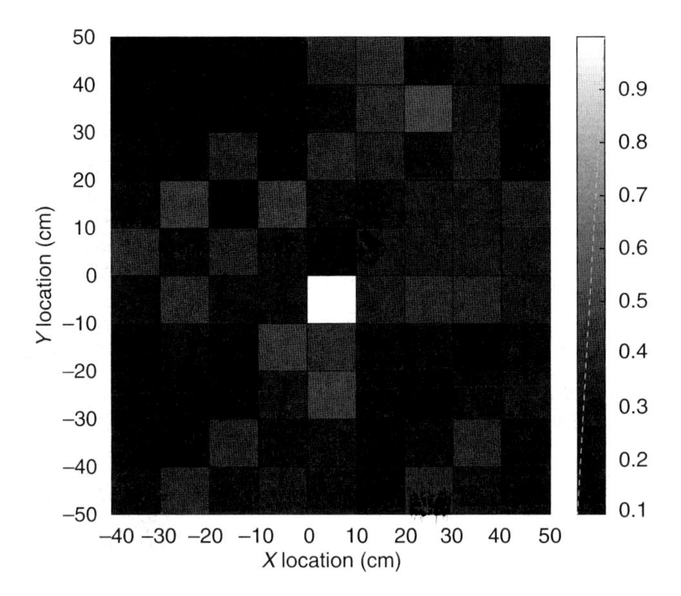

Figure 2.13 Geographic distribution of η^2 with the intended location at the center of the area of interest.

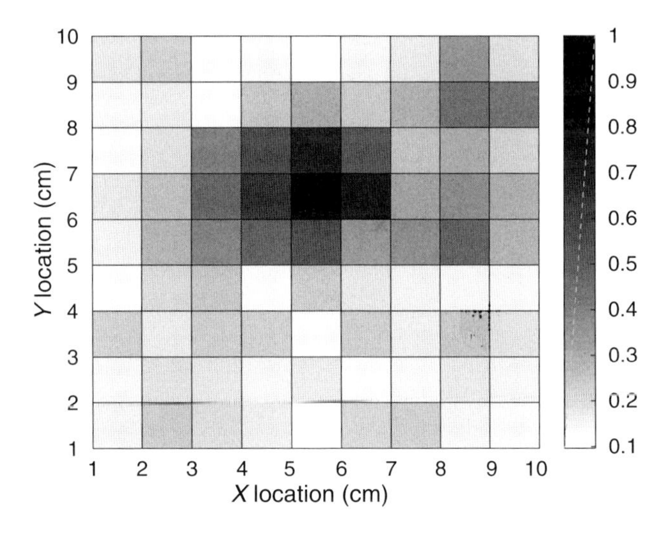

Figure 2.14 Fine-scale geographic distribution of η^2.

degradation in terms of the spatial focusing effect within a 5-cm by 5-cm region, which is consistent with the fact that channels are uncorrelated with a half-wavelength spacing (the wavelength is around 5 cm when the carrier frequency is 5.4 GHz). In such a case, when a user is located between grid points with 10-cm spacing, it may not be localized correctly. Nevertheless, this can be easily solved by asking the user to rotate the device, e.g., smartphone, such that the antenna can cross the 10-cm grid points.

Table 2.1 Localization performance with 10-cm localization accuracy

Number of Trials	3016
Number of Error	0
Error Rate	0%

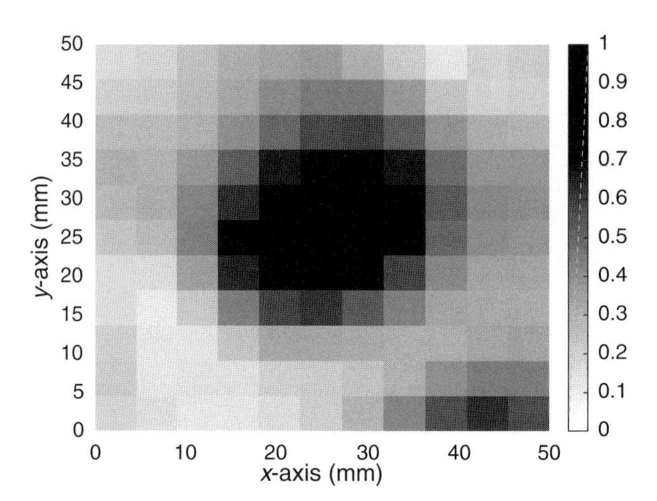

Figure 2.15 TR resonating strength near the intended location with a measurement resolution of 0.5 cm.

In a recent experiment, we refine the measurement resolution to 0.5 cm to study the accuracy. The TR resonating strength near the intended location is shown in Figure 2.15, which demonstrates that the localization accuracy can reach $1 \sim 2$ cm in a NLOS environment.

2.3.3 Localization Performance

From the results in the previous section, we can see that the CIR acts as a signature between the AP and the TD, and it drastically changes, even if the location is only 10 cm away. Here, we will examine the performance of our indoor positioning algorithm.

To evaluate the performance, we use the leave-one-out cross validation. Specifically, we pick each CIR as the test sample and leave the rest as training samples in the database. Then, we perform our algorithm, i.e., the online positioning algorithm, and evaluate the corresponding performance. There are totally 3016 CIRs for the 110 grid points, which leads to a total of 3016 trials. The localization performance is shown in Table 2.1, in which we can see that our indoor localization algorithm gives zero error out of a total 3016 trials, which achieves 100% accuracy with no error in the 1-m by 0.9-m area of interest. Note that this result is achieved with a single AP working in the NLOS condition using one CIR.

2.3.4 Discussions

From the experimental results and discussions, we can see that the TRIPS is an ideal solution to the indoor positioning problem because it can achieve very high localization accuracy with a very simple algorithm and low infrastructure cost summarized as follows.

- From the experimental results, we can see that, with a single AP working in the 5.4-GHz band under the NLOS condition, the TRIPS can achieve perfect centimeter localization accuracy. Such localization accuracy is much better than that of existing state-of-the-art IPSs under the NLOS condition, which typically achieve meter-level localization accuracy. Moreover, the accuracy can be improved if we increase the resolution of the database, which, however, will increase the size of the database and, thus, the complexity of the online positioning algorithm.
- Based on the TR technique, the matching algorithm in our TRIPS is very simple, which just computes the TR resonating strength between the estimated CIR and that in the database. Compared with existing approaches, our method does not require complicated calibrations and matchings.
- Although the localization performance can be further improved with multiple APs, our method only uses a single AP and has already achieved very high localization accuracy under the NLOS condition. Moreover, no special apparatus is needed for the AP. Therefore, the infrastructure cost of our TRIPS is very low.
- The size of the database is determined by three factors, i.e., the room size, the resolution of the grid point, and the number of realizations at each grid point. For a typical room such as room "**A**" shown in Figure 2.4(a), the size is 5.4 m by 3.1 m. Considering a resolution with 10-cm spacing between two neighboring grid points, there are a total of 1,760 grid points. Suppose 20 CIR realizations are collected at each grid point, where the length of the channel L is 30 and where each tap of CIR is represented with 4 bytes (2 bytes for the real part and 2 bytes for the imaginary part). Then, the size of the database is $1{,}760 \times 20 \times 30 \times 4 = 4$ 224, 000 bytes (4.2 MB). Such a database is reasonably small, which can be easily stored with an off-the-shelf storage device. Moreover, all system configurations, including the grid size, the number of realizations, and the channel length L, are all adjustable to fit a specific environment at a desired localization performance.
- The TRIPS is not limited to the 5.4-GHz band. It can be also applied to the ultrawide band with a larger bandwidth, where we expect to achieve much higher localization accuracy.

2.4 Summary

In this chapter, we have considered a TRIPS by exploiting the unique location-specific characteristic of CIR. Specifically, we have addressed the ill-posed single-AP localization problem by decomposing it into two well-defined subproblems. One subproblem is calibration by building a database that maps the physical geographical locations to the logical locations in the CIR space, and the other subproblem is matching the estimated CIR with those in the database. We built a real prototype to evaluate the scheme. Experimental results show that, even only with a single AP under the NLOS condition and a single realization of online measurements, the scheme can still achieve 100% localization accuracy at the scale of 1 to 2 cm. For related references, interested readers can refer to [37].

References

[1] H. Liu, H. Darabi, P. Banerjee, and J. Liu, "Survey of wireless indoor positioning techniques and systems," *IEEE Transactions on Systems, Man, and Cybernetics, Part C: Applications and Reviews,* vol. 37, no. 6, pp. 1067–1080, Nov. 2007.

[2] M. Youssef, A. Youssef, C. Rieger, U. Shankar, and A. Agrawala, "Pinpoint: An asynchronous time-based location determination system," in *Proceedings of the 4th ACM International Conference on Mobile Systems, Applications and Services*, 2006, pp. 165–176. [Online]. Available: http://doi.acm.org/10.1145/1134680.1134698.

[3] R. Fontana and S. Gunderson, "Ultra-wideband precision asset location system," in *Proceedings of the IEEE Conference on Ultra Wideband Systems and Technologies,* May 2002, pp. 147–150.

[4] D. Niculescu and B. Nath, "VOR base stations for indoor 802.11 positioning," in *Proceedings of the 10th Annual ACM International Conference on Mobile Computing and Networking*, 2004, pp. 58–69. [Online]. Available: http://doi.acm.org/10.1145/1023720.1023727.

[5] J. Xiong and K. Jamieson, "ArrayTrack: A fine-grained indoor location system," in *Presented as Part of the 10th USENIX Symposium on Networked Systems Design and Implementation (NSDI 13),* 2013, pp. 71–84. [Online]. Available: www.usenix.org/conference/nsdi13/technical-sessions/presentation/xiong.

[6] Y. Zhao, "Standardization of mobile phone positioning for 3G systems," *IEEE Communications Magazine,* vol. 40, no. 7, pp. 108–116, Jul. 2002.

[7] G. Sun, J. Chen, W. Guo, and K. Liu, "Signal processing techniques in network-aided positioning: A survey of state-of-the-art positioning designs," *IEEE Signal Processing Magazine,* vol. 22, no. 4, pp. 12–23, Jul. 2005.

[8] A. Kushki, K. Plataniotis, and A. Venetsanopoulos, "Kernel-based positioning in wireless local area networks," *IEEE Transactions on Mobile Computing,* vol. 6, no. 6, pp. 689–705, Jun. 2007.

[9] P. Bahl and V. Padmanabhan, "Radar: An in-building RF-based user location and tracking system," in *Proceedings of the Nineteenth Annual Joint Conference of the IEEE Computer and Communications Societies*, vol. 2, 2000, pp. 775–784.

[10] J. Koo and H. Cha, "Localizing WiFi access points using signal strength," *IEEE Communications Letters,* vol. 15, no. 2, pp. 187–189, Feb. 2011.

[11] C. Feng, W. Au, S. Valaee, and Z. Tan, "Received-signal-strength-based indoor positioning using compressive sensing," *IEEE Transactions on Mobile Computing,* vol. 11, no. 12, pp. 1983–1993, Dec. 2012.

[12] M. Youssef, A. Agrawala, and A. Udaya Shankar, "WLAN location determination via clustering and probability distributions," in *Proceedings of the First IEEE International*

Conference on Pervasive Computing and Communications, 2003. (PerCom 2003), pp. 143–150, Mar. 2003.

[13] M. Youssef and A. Agrawala, "The Horus WLAN location determination system," in *Proceedings of the 3rd ACM International Conference on Mobile Systems, Applications, and Services*, 2005, pp. 205–218. [Online]. Available: http://doi.acm.org/10.1145/1067170.1067193.

[14] Y. Chen, D. Lymberopoulos, J. Liu, and B. Priyantha, "Indoor localization using FM signals," *IEEE Transactions on Mobile Computing,* vol. 12, no. 8, pp. 1502–1517, Aug. 2013.

[15] Y. Jin, W.-S. Soh, and W.-C. Wong, "Indoor localization with channel impulse response based fingerprint and nonparametric regression," *IEEE Transactions on Wireless Communications,* vol. 9, no. 3, pp. 1120–1127, Mar. 2010.

[16] N. Patwari and S. K. Kasera, "Robust location distinction using temporal link signatures," in *Proceedings of the 13th Annual ACM International Conference on Mobile Computing and Networking*, 2007, pp. 111–122. [Online]. Available: http://doi.acm.org/10.1145/1287853.1287867.

[17] J. Zhang, M. H. Firooz, N. Patwari, and S. K. Kasera, "Advancing wireless link signatures for location distinction," in *Proceedings of the 14th ACM International Conference on Mobile Computing and Networking*, 2008, pp. 26–37. [Online]. Available: http://doi.acm.org/10.1145/1409944.1409949.

[18] K. Wu, J. Xiao, Y. Yi, D. Chen, X. Luo, and L. Ni, "CSI-based indoor localization," *IEEE Transactions on Parallel and Distributed Systems,* vol. 24, no. 7, pp. 1300–1309, Jul. 2013.

[19] B. Wang, Y. Wu, F. Han, Y.-H. Yang, and K. Liu, "Green wireless communications: A time-reversal paradigm," *IEEE Journal on Selected Areas in Communications,* vol. 29, no. 8, pp. 1698–1710, Sep. 2011.

[20] B. I. Zeldovich, N. F. Pilipetskii, and V. V. Shkunov, *Principles of Phase Conjugation*, Berlin and New York, Springer-Verlag (Springer Series in Optical Sciences. Volume 42), 1985, p. 262.

[21] M. Fink, C. Prada, F. Wu, and D. Cassereau, "Self focusing in inhomogeneous media with time reversal acoustic mirrors," in *IEEE Proceedings of the 1989 Ultrasonics Symposium,* pp. 681–686, vol. 2, Oct. 1989.

[22] M. Fink, "Time reversal of ultrasonic fields. I. Basic principles," *IEEE Transactions on Ultrasonics, Ferroelectrics and Frequency Control,* vol. 39, no. 5, pp. 555–566, Sept. 1992.

[23] A. Derode, P. Roux, and M. Fink, "Robust acoustic time reversal with high-order multiple scattering," *Physical Review Letters*, vol. 75, pp. 4206–4209, Dec. 1995. [Online]. Available: http://link.aps.org/doi/10.1103/PhysRevLett.75.4206.

[24] G. Edelmann, T. Akal, W. Hodgkiss, S. Kim, W. Kuperman, and H. C. Song, "An initial demonstration of underwater acoustic communication using time reversal," *IEEE Journal of Oceanic Engineering,* vol. 27, no. 3, pp. 602–609, Jul. 2002.

[25] B. E. Henty and D. D. Stancil, "Multipath-enabled super-resolution for RF and microwave communication using phase-conjugate arrays," *Physical Review Letters*, vol. 93, p. 243904, Dec. 2004. [Online]. Available: http://link.aps.org/doi/10.1103/PhysRevLett.93.243904.

[26] G. Lerosey, J. de Rosny, A. Tourin, A. Derode, G. Montaldo, and M. Fink, "Time reversal of electromagnetic waves," *Physical Review Letters*, vol. 92, p. 193904, May 2004. [Online]. Available: http://link.aps.org/doi/10.1103/PhysRevLett.92.193904.

[27] R. Qiu, C. Zhou, N. Guo, and J. Zhang, "Time reversal with MISO for ultrawideband communications: Experimental results," *IEEE Antennas and Wireless Propagation Letters*, vol. 5, no. 1, pp. 269–273, Dec. 2006.

[28] G. Lerosey, J. De Rosny, A. Tourin, A. Derode, G. Montaldo, and M. Fink, "Time reversal of electromagnetic waves and telecommunication," *Radio Science*, vol. 40, no. 6, 2005.

[29] G. Lerosey, J. De Rosny, A. Tourin, A. Derode, and M. Fink, "Time reversal of wideband microwaves," *Applied Physics Letters*, vol. 88, no. 15, pp. 154 101, Apr. 2006.

[30] I. Naqvi, G. El Zein, G. Lerosey, J. de Rosny, P. Besnier, A. Tourin, and M. Fink, "Experimental validation of time reversal ultra wide-band communication system for high data rates," *IET Microwaves, Antennas Propagation,* vol. 4, no. 5, pp. 643–650, May 2010.

[31] J. De Rosny, G. Lerosey, and M. Fink, "Theory of electromagnetic time-reversal mirrors," *IEEE Transactions on Antennas and Propagation,* vol. 58, no. 10, pp. 3139–3149, Oct. 2010.

[32] F. Han, Y.-H. Yang, B. Wang, Y. Wu, and K. J. R. Liu, "Time-reversal division multiple access over multi-path channels," *IEEE Transactions on Communications*, vol. 60, no. 7, pp. 1953–1965, Jul. 2012.

[33] Y.-H. Yang, B. Wang, W. Lin, and K. J. R. Liu, "Near-optimal waveform design for sum rate optimization in time-reversal multiuser downlink systems," *IEEE Transactions on Wireless Communications*, vol. 12, no. 1, pp. 346–357, Jan. 2013.

[34] Y. Chen, Y.-H. Yang, F. Han, and K. J. R. Liu, "Time-reversal wideband communications," *IEEE Signal Processing Letters*, vol. 20, no. 12, pp. 1219–1222, Dec. 2013.

[35] F. Han and K. J. R. Liu, "A multiuser TRDMA uplink system with 2D parallel interference cancellation," *IEEE Transactions on Communications*, vol. 62, no. 3, pp. 1011–1022, Mar. 2014.

[36] Y. Chen, F. Han, Y.-H. Yang, H. Ma, Y. Han, C. Jiang, H.-Q. Lai, D. Claffey, Z. Safar, and K. J. R. Liu, "Time-reversal wireless paradigm for green Internet of Things: An overview," *IEEE Internet of Things Journal*, vol. 1, no. 1, pp. 81–98, Feb. 2014.

[37] Z.-H. Wu, Y. Han, Y. Chen, and K. J. R. Liu, "A time-reversal paradigm for indoor positioning system," *IEEE Transactions on Vehicular Technology*, vol. 64, no. 4, pp. 1331–1339, 2015.

3 Multiantenna Approach

Channel frequency response (CFR) is a fine-grained location-specific information in Wi-Fi systems that can be utilized in indoor positioning systems (IPSs). However, CFR-based IPSs can hardly achieve an accuracy in the centimeter level due to the limited bandwidth in Wi-Fi systems. To achieve such accuracy using Wi-Fi devices, we consider an IPS that fully harnesses the spatial diversity in Multiple-Input-Multiple-Output (MIMO) Wi-Fi systems, which leads to a much larger effective bandwidth than the bandwidth of a Wi-Fi channel. The presented IPS obtains CFRs associated with locations-of-interest on multiple antenna links during the training phase. In the positioning phase, the IPS captures instantaneous CFRs from a location to be estimated and compares it with the CFRs acquired in the training phase via the time-reversal resonating strength (TRRS) with residual synchronization errors compensated. Extensive experiment results in an office environment with a measurement resolution of 5 cm demonstrate that, with a single pair of Wi-Fi devices and an effective bandwidth of 321 MHz, the presented IPS achieves detection rates of 99.91% and 100% with false alarm rates of 1.81% and 1.65% under the line-of-sight (LOS) and non-line-of-sight (NLOS) scenarios, respectively. Meanwhile, the presented IPS is robust against environment dynamics. Moreover, experiment results with a measurement resolution of 0.5 cm demonstrate a localization accuracy of $1 \sim 2$ cm in the NLOS scenario.

3.1 Introduction

Wireless indoor positioning systems (IPSs) spawn numerous location-based indoor applications, such as campus-wide localization [1], targeted advertisement in supermarkets [2], and shopping mall navigations [3]. The IPSs often demand an accuracy in the submeter level, which cannot be achieved by the Global Positioning System (GPS) due to the severe attenuation of GPS signals indoors.

The imperative demand on the accuracy leads to an extensive development of IPSs using a wide variety of wireless technologies [4]. Among them, Wi-Fi-based approaches are promising candidates because they are built upon the Wi-Fi networks widely available indoor. Many Wi-Fi-based schemes exploit the location-specific fingerprints characterizing the propagation of electromagnetic waves for indoor spaces to facilitate indoor localization. Examples include the received signal strength indicator (RSSI) [5–7] and channel frequency response (CFR) [8–10]. Each of these IPSs consists of a training phase and a positioning phase. During the training phase, the IPS collects fingerprints associated with multiple locations-of-interest and stores the fingerprints into a database, while in the positioning phase, the IPS determines the location by comparing the instantaneous fingerprint against those stored in the database. Nevertheless, the localization performance of these IPSs is limited by the available bandwidth in Wi-Fi systems, which is 20 or 40 MHz for 802.11*n* Wi-Fi networks. The bandwidth limit introduces severe location ambiguity, which leads to meter-level accuracy on average.

Recently, in [11], Wu et al. present the time-reversal indoor positioning system (TRIPS) to achieve a centimeter-level localization accuracy. TRIPS is a single-antenna IPS that uses channel impulse response (CIR) as the fingerprint. It leverages the time-reversal technique to achieve the high-resolution spatial-temporal focusing effect [12] for localization. With a bandwidth of 125 MHz on 5.4 GHz ISM band, TRIPS achieves a perfect detection rate with zero false alarm within an area of 0.9 m $\times$ 1 m with a measurement resolution of 5 cm. The accuracy can be further driven to 1 cm to 2 cm. Despite its centimeter-level accuracy, TRIPS uses dedicated hardware and requires a large bandwidth to reduce location ambiguity.

Is it possible to achieve the centimeter-level accuracy on a Wi-Fi platform by leveraging the TR technique? The answer is affirmative. In [13, 14], Chen et al. harnesses the frequency diversity of Wi-Fi systems by scanning multiple channels and formulates fingerprints by concatenating CFRs from these channels, leading to a perfect detection rate with zero false alarm with 5 cm measurement resolution. One drawback of this approach is the overhead of frequency hopping in acquiring CFRs from a large number of channels.

In this chapter, we discuss an IPS that leverages the spatial diversity on a multi-antenna Wi-Fi device instead of the frequency diversity to achieve the centimeter-level accuracy. The presented IPS optimally concatenates the available bandwidths of different antenna links to formulate a much larger effective bandwidth. The IPS consists of two phases: the training phase and the positioning phase. During the training phase, the IPS captures CFRs from multiple locations-of-interest and then combines CFRs of different links into location fingerprints. In the positioning phase, the IPS obtains instantaneous CFRs and evaluates the TR focusing effect quantitatively by the time-reversal resonating strength (TRRS) between the instantaneous CFRs and those in the training phase. Realizing that the residual synchronization errors are inevitable in Wi-Fi systems, we develop an algorithm to mitigate the impact of such errors in the computations of TRRS. Finally, the IPS determines the locations based on the TRRS.

We conduct extensive experiments in an office environment using a single pair of off-the-shelf Wi-Fi devices to illustrate that the presented IPS can achieve detection rates of 99.91% and 100%, while triggering negligible false alarm rates of 1.81% and 1.65% under LOS and NLOS scenarios, respectively, for locations with a unit distance of 5 cm. We also show that the presented IPS is robust against environment dynamics. Moreover, experiment results with a unit distance of 0.5 cm demonstrate that 1 $\sim$ 2 cm accuracy is achievable with the presented IPS. To the best of our knowledge, this is the first attempt that achieves 1 $\sim$ 2 cm localization accuracy under NLOS scenarios using a single pair of off-the-shelf Wi-Fi devices leveraging spatial diversity.

The major points of this chapter can be summarized as follows:

- We present the concept of effective bandwidth as a measure of exploitable diversities for localization.
- We discuss a robust algorithm that compensates the inevitable synchronization errors in Wi-Fi transceivers when calculating the TRRS for each Wi-Fi link. Then, it fuses the TRRS from different Wi-Fi links via a weighted average scheme, which significantly reduces the location ambiguity caused by the bandwidth limit in Wi-Fi systems.
- We conduct extensive experiments in a typical office environment with dynamics introduced by human activities as well as the movement of objects such as furniture and doors. The experiment results demonstrate that the proposed IPS achieves centimeter-level accuracy, and it is robust against environment dynamics.

The rest of the chapter is organized as follows. In Section 3.2, we present a literature survey on the indoor localization schemes relevant to the proposed IPS. In Section 3.3, we introduce the TR technique and the channel estimation in MIMO-OFDM Wi-Fi systems. In Section 3.4, we elaborate on the localization algorithm for the IPS. In Section 3.5, we demonstrate the experiment results.

3.2 Related Work

In this section, we sample a few wireless indoor localization systems from the rich literature with high correlation with the proposed IPS. Based on their principles, these schemes can be further categorized into two classes: triangulation and fingerprinting [4].

3.2.1 Triangulation

Triangulation-based schemes utilize the geometric properties to localize the device using several anchors with known coordinates in the space. These schemes can be further classified into *lateration-based* and *angulation-based*.

3.2.1.1 Lateration

Lateration-based schemes measure the distances from the device to at least three anchors to facilitate triangulation. The distance is generally inferred from other information available in the wireless network relevant to the LOS path, such as RSSI, time of arrival (TOA), time difference of arrival (TDOA), and round-trip time-of-flight (RTOF). RSSI-based approaches estimate the distances between device and anchors based on the received signal strength via the free-space path-loss model or its variants, which take the attenuation due to walls and ceilings into consideration. SpotON [15] uses the RFID technology for indoor localization. A SpotON tag measures the intertag distances using the RSSI values associated with the received tag. LANDMARC deploys extra fixed reference tags for calibration to improve the RFID localization accuracy [16].

Nevertheless, the high heterogeneity of indoor spaces give rise to strong NLOS components, which complicates and degrades the efficacy of the path-loss model. Methods based on TOA, TDOA, and RTOF [17–21] suffer in a strong NLOS environment as well because the LOS component can hardly be discerned from the multipath profile. Moreover, the schemes proposed in [19–21] use ultra-wideband transmission to obtain a highly accurate timing resolution and thus require dedicated hardware, which incurs additional cost in deployment.

3.2.1.2 Angulation

Angulation-based schemes calculate the angle of arrival (AOA) of a signal arrived at several anchors and formulate intersections of multiple spheres to pinpoint the device. In [22], Jie Xiong et al. present ArrayTrack, a phased antenna array composed by multiple access points (APs) with their initial phases synchronized by calibration. Each AP is built by stitching two customized WARP radios together with four omnidirectional antennas and is capable of AOA estimation utilizing the widely used MUSIC algorithm [23]. The mean accuracy reaches 107 cm with three APs and can be further improved to 57 cm by introducing an additional antenna for each AP. When all six APs are working simultaneously, the localization accuracy reaches 31 cm. In [24], Gjengset et al. implements the Phaser system with a high resemblance to ArrayTrack, except that two off-the-shelf Wi-Fi cards are stitched together in Phaser to replace the two dedicated WARP radios utilized in ArrayTrack.

One drawback of ArrayTrack and Phaser is that phase calibration must be conducted for each AP to synchronize the initial phases between different antennas whenever the AP is powered on because the oscillators would locked onto unknown random phases. On the other hand, they are only accurate when the device has a direct link to at least one AP and thus multiple APs are needed in general. Last but not least, ArrayTrack requires specialized hardware, and Phaser makes physical modifications to the APs by stitching two of them together using a cable.

3.2.1.3 Lateration Combined with Angulation

More recently, in [25], Kotaru et al. present SpotFi that fuses both lateration and angulation techniques. SpotFi performs spatial smoothing on the correlation matrix of CFRs and applies a two-dimensional (2-D) MUSIC algorithm to estimate the AoA and ToF of different multipath components jointly. Then, SpotFi identifies the LOS path by performing clustering on the assembled estimations from multiple packets. Finally, by combining the AoA estimation with the RSSI values from multiple APs, SpotFi determines the location via solving a nonconvex optimization problem. SpotFi achieves a median accuracy of 1.9 m, 0.8 m, and 0.6 m with three, four, and five APs, respectively, when the LOS path exists between the device and a majority of the APs, and the error increases to 1.6 m when only one or two APs have decent LOS paths to the device. Like ArrayTrack and Phaser, SpotFi relies on the assumption that at least one AP can establish an LOS link with the device. Also, the computation complexity of SpotFi is intense due to the 2-D MUSIC algorithm, clustering, and solving the nonconvex optimization problem.

3.2.2 Fingerprinting

Fingerprinting-based approaches collect location-specific fingerprints at different locations-of-interest in the offline training phase and compares instantaneous fingerprints against those obtained in the training phase in the online positioning phase. In [6], Youssef et al. utilize RSSI from multiple APs as the fingerprints in the training phase and calculate the probability of the candidate locations in the online phase. The achievable accuracy is 1.4 m in 90% of the time. In [8], Sen et al. propose PinLoc that achieves a mean detection rate of 90% across 50 1 m $\times$ 1 m spots with multiple access points (AP) with less than 7% mean false alarm rate. The CFRs collected from each 1 m $\times$ 1 m spot are modeled as random vectors following the distribution of mixture Gaussian. Then, PinLoc uses variational Bayesian inference to partition the CFRs of every 1 m $\times$ 1m spot into clusters with the centroid of each cluster as the representative CFR. Then, with a probabilistic measure, PinLoc evaluates the possibility of the instantaneous CFR in the online phase coming from each of the 1 m $\times$ 1m spots for localization. However, they are unable to reach the centimeter-level accuracy because they find that multiple locations in a room may exhibit identical fingerprints. This can be justified by the fact that only 20 MHz of bandwidth is utilized for CFR fingerprinting, which increases the location ambiguity.

Utilizing the spatial diversity in multiple-input-multiple-output (MIMO) Wi-Fi systems can further bolster the IPS performances. The fine-grained indoor fingerprinting system (FIFS) proposed in [9] formulates the compressed fingerprints by aggregating CFR amplitudes of different antenna links. It achieves an accuracy of 1.1 m under 2 $\times$ 2 MIMO configuration. The accuracy of FIFS is improved to 0.95 m in [10] by incorporating the phase information contained in the CFRs as well. However, centimeter-level accuracy cannot be achieved in the two schemes as the spatial diversity is not fully harnessed in an optimal way.

3.3 Preliminaries

In this section, we briefly introduce the TR technique and the channel estimation schemes in Wi-Fi systems.

3.3.1 Time-Reversal

TR is a signal processing technique that mitigates the phase distortion of a signal filtered by a linear time-invariant (LTI) system. It leverages the fact that the phase distortion can be removed at a particular time instance when the LTI system $h(t)$ is combined with its time-reversed and conjugated counterpart $h^*(-t)$. The development of TR can be dated back to the 1950s when Bogert used TR technique to correct a slow picture transmission system delay distortion [26]. Later, Kormylo et al. utilized TR in the design of zero-phase digital filters, as the signal is processed in both causal and reverse causal direction to remove the phase distortion [27].

The physical channel can be regarded as LTI if it is inhomogeneous and invertible. When such conditions hold, TR technique can refocus the energy of signal waves at a particular spatial location and a specific time, known as the spatial-temporal focusing effect. This effect is verified experimentally in the field of ultrasonics, acoustics, and electromagnetism [12, 28–30]. More recently, TR has been applied to the broadband wireless communication systems [31].

The concept of TR communication can be referred to Chapter 1.

3.3.2 Channel Estimation in MIMO-OFDM

Assume a MIMO-OFDM system with N_t transmitting antennas and N_r receiving antennas, and denote the CIR between transmitting antenna (TX) n_t and receiving antenna (RX) n_r as $h_{n_t,n_r}[\ell]$ where ℓ runs from 0 to $L_{n_t,n_r} - 1$, and L_{n_t,n_r} is the number of multipath components between TX n_t and RX n_r. To facilitate timing/frequency synchronization and channel estimation, TX n_t sends a training sequence $x_{n_t}[n]$ composed by several short training fields (STFs), guard intervals (Gs), and N_t long training fields (LTFs). The nth sample of the received signal at RX n_r can be expressed as

$$y_{n_r}[n] = \sum_{n_t=1}^{N_t} \sum_{l=0}^{L_{n_t,n_r}-1} h_{n_t,n_r}[l] x_{n_t}[n-l] + w_{n_r}[n] , \tag{3.1}$$

where $w_{n_r}[n]$ is the channel noise at RX n_r. Assisted by the STFs, RX n_r detects the starting position of the first LTF. Then, it performs N-point fast Fourier transform (FFT) on all LTFs. After discarding the null subcarriers, the frequency domain representation of $y_{n_r}[n]$ on the kth usable subcarrier associated with the ith LTF takes the form

$$Y_{i,n_r}[u_k] = \sum_{n_t=1}^{N_t} H_{n_t,n_r}[u_k] X_{i,n_t}[u_k] + W_{i,n_r}[u_k] , \tag{3.2}$$

where u_k is the index for the kth subcarrier with $k = 1, 2, \ldots, N_u$, N_u is the number of usable subcarriers, $H_{n_t,n_r}[u_k]$ is the frequency domain representation of $\{h_{n_t,n_r}[\ell]\}_{\ell=0,1,\ldots,L_{n_t,n_r}-1}$ on the kth subcarrier, $X_{i,n_t}[u_k]$ is the frequency domain representation of $x_{n_t}[n]$ with n in the range of the ith LTF on the kth subcarrier, and $W_{i,n_t}[u_k]$ is the frequency domain noise of the ith LTF on the kth subcarrier.

Figure 3.1 illustrates the channel estimation procedure in MIMO-OFDM Wi-Fi systems [32]. The TX transmits LTFs alternatively such that at any given time instance, only one TX is sending the LTF. Thus, $X_{i,n_t}[u_k]$ is expressed by

$$X_{i,n_t}[u_k] = X[u_k] \mathbf{1}(i = n_t) , \tag{3.3}$$

where $\mathbf{1}(x)$ is the indicator function. Substituting $X_{i,n_t}[u_k]$ back into (3.2) gives

$$Y_{i,n_r}[u_k] = \Big[H_{i,n_r}[u_k] X[u_k] + W_{i,n_r}[u_k] \Big] \mathbf{1}(i = n_t) . \tag{3.4}$$

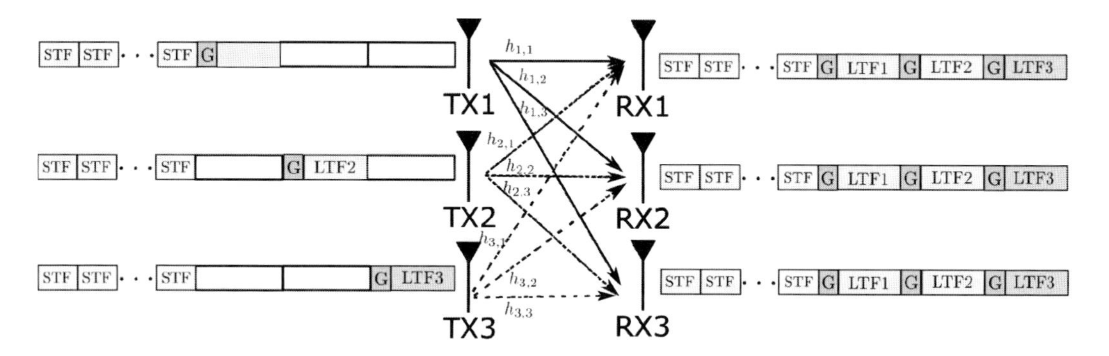

Figure 3.1 Channel estimation in MIMO-OFDM Wi-Fi system.

Assuming that the frequency domain noise follows a Gaussian distribution, the CFR $H_{n_t,n_r}[u_k]$ can be estimated in the least-square sense using

$$\hat{H}_{n_t,n_r}[u_k] = \frac{Y_{n_t,n_r}[u_k]}{X[u_k]} = H_{n_t,n_r}[u_k] + \frac{W_{n_t,n_r}[u_k]}{X[u_k]}. \tag{3.5}$$

Equation (3.5) depicts $H_{n_t,n_r}[u_k]$ in the ideal case. In practice, the Wi-Fi receivers suffer from the carrier frequency offset (CFO), sampling frequency offset (SFO), and symbol timing offset (STO) caused by the mismatches of analog and digital components between the transmitters and receivers. Although Wi-Fi receivers perform synchronization, the residual synchronization errors cannot be neglected. Meanwhile, the phase locked loops (PLLs) at the Wi-Fi receivers produce random common phase offsets (CPO). These additional errors must be considered because they are random in nature and introduce severe uncertainties into the CFRs.

In the presence of the aforementioned synchronization errors, the channel estimation at RX n_r takes the form [33]

$$\begin{aligned}\hat{H}_{n_t,n_r}[u_k] = {} & H_{n_t,n_r}[u_k]e^{j\nu_{n_r}} \\ & \times e^{j2\pi(\alpha_{n_t}(\Delta s_{n_r},\Delta\omega_{n_r})+\epsilon_{n_t}(\Delta\psi_{n_r},\Delta\eta_{n_r})u_k)} + U_{n_t,n_r}[u_k],\end{aligned} \tag{3.6}$$

where $U_{n_t,n_r}[u_k]$ is the estimation noise between TX n_t and RX n_r, ν_{n_r} is the CPO at RX n_r, Δs_{n_r} is the reference absolute time of the detected frame starting point after timing synchronization using STFs, $\Delta\omega_{n_r}$ is the normalized residual SFO at RX n_r given as $\frac{\Delta f_r}{NT_s}$ where Δf_r is the residual CFO at RX n_r, $\Delta\psi_{n_r}$ is the STO at RX n_r, and $\Delta\eta_{n_r}$ is the residual SFO at RX n_r expressed as $\frac{T_s'-T_s}{T_s}$ where T_s and T_s' are the sampling intervals at the TX and RX, respectively. The initial and linear phase distortions are denoted by $\alpha_{n_t}\left(\Delta s_{n_r},\Delta\omega_{n_r}\right)$ and $\epsilon_{n_t}\left(\Delta\psi_{n_r},\Delta\eta_{n_r}\right)$, which take the form

$$\alpha_{n_t}\left(\Delta s_{n_r},\Delta\omega_{n_r}\right) = \frac{\left(\Delta s_{n_r} + (i-1)N_s + N_G + \frac{N}{2}\right)\Delta\omega_{n_r}}{N}, \tag{3.7}$$

$$\epsilon_{n_t}\left(\Delta\psi_{n_r},\Delta\eta_{n_r}\right) = \frac{\Delta\psi_{n_r} + \left((i-1)N_s + N_G + \frac{N}{2}\right)\Delta\eta_{n_r}}{N}. \tag{3.8}$$

Here, N_G is the number of samples of the guard interval, and $N_s = N_G + N$ is the total number of samples of one OFDM block.

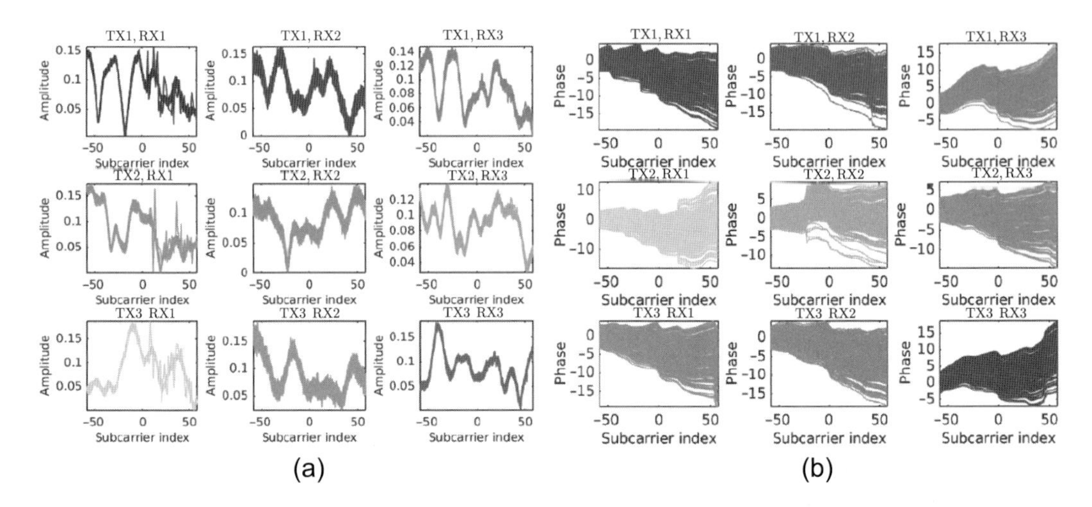

Figure 3.2 A snapshot of CFRs of nine links in a 3 × 3 MIMO-OFDM system collected in 4 seconds. (a) Normalized amplitudes. (b) Unwrapped phases.

To illustrate the impact of phase distortions on CFRs, we show the normalized amplitudes and phases of 200 CFRs captured within 4 seconds in Figures 3.2(a) and 3.2(b). Despite the consistency in the normalized amplitudes, the variations in the phases caused by the aforementioned initial and linear phase distortions differ for different packets and must be compensated.

3.4 Algorithm Design

3.4.1 Calculating TRRS for Each Link

Given two CIRs $\hat{\mathbf{h}} = \left[\hat{h}[0], \hat{h}[1], \ldots, \hat{h}[L-1]\right]^{\mathsf{T}}$ and $\hat{\mathbf{h}}' = \left[\hat{h}'[0], \hat{h}'[1], \ldots, \hat{h}'[L-1]\right]^{\mathsf{T}}$, where $(\cdot)^{\mathsf{T}}$ represents the transpose operator, the TRRS between them is given by [11]

$$\gamma_{\mathsf{CIR}}[\hat{\mathbf{h}}, \hat{\mathbf{h}}'] = \frac{\left(\max_i \left|\left(\hat{\mathbf{h}} * \hat{\mathbf{g}}\right)[i]\right|\right)^2}{\langle \hat{\mathbf{h}}, \hat{\mathbf{h}} \rangle \langle \hat{\mathbf{g}}, \hat{\mathbf{g}} \rangle}, \tag{3.9}$$

where $*$ stands for the linear convolution, $\hat{\mathbf{g}}$ is the time-reversed and conjugate counterpart of $\hat{\mathbf{h}}'$, $\langle \mathbf{x}, \mathbf{y} \rangle$ is the inner product operator between vector $\mathbf{x}$ and $\mathbf{y}$ given as $\mathbf{x}^{\dagger}\mathbf{y}$, and $(\cdot)^{\dagger}$ is the Hermitian operator. In (3.9), taking maximum over all possible i in the calculation of $\gamma_{\mathsf{CIR}}[\hat{\mathbf{h}}, \hat{\mathbf{h}}']$ essentially mitigates the linear phase distortions caused by STO. Meanwhile, taking the absolute value on the numerator as shown in (3.9) eliminates the initial phase distortions. However, the residual SFO is uncompensated.

Because cross correlation in time domain is equivalent to inner product in frequency domain, we redefine the TRRS between two CFRs in the frequency domain by extending (3.9) into MIMO-OFDM systems. Here, we assume two CFRs denoted as $\hat{\mathbf{H}}_{n_t,n_r}$ and $\hat{\mathbf{H}}'_{n_t,n_r}$ between TX n_t and RX n_r. For convenience, we define the *link index* $d = (n_t - 1)N_r + n_r$[1]. Then, $\hat{\mathbf{H}}_d$ is expressed as

[1] For instance, in a 3 × 3 MIMO system, the link between TX antenna 2 and RX antenna 1 is labeled as link 4.

$$\hat{\mathbf{H}}_d = \begin{bmatrix} \hat{H}_d[u_1] & \hat{H}_d[u_2] & \cdots & \hat{H}_d[u_k] & \cdots & \hat{H}_d[u_{N_u}] \end{bmatrix}^{\mathsf{T}}. \quad (3.10)$$

Similar definition applies for $\hat{\mathbf{H}}'_d$. Given $\hat{\mathbf{H}}_d$ and $\hat{\mathbf{H}}'_d$, we define the TRRS on link d as

$$\overline{\phi}_d = \frac{\max_{\epsilon} \left| \sum_{k=1}^{N_u} \hat{H}_d[u_k] \hat{H}'_d[u_k] e^{j\epsilon u_k} \right|^2}{\Lambda_d \Lambda'_d}, \quad (3.11)$$

where

$$\Lambda_d = \langle \mathbf{H}_d, \mathbf{H}_d \rangle,\ \Lambda'_d = \langle \mathbf{H}'_d, \mathbf{H}'_d \rangle \quad (3.12)$$

are the *channel energies* for $\mathbf{H}_d$ and $\mathbf{H}'_d$, respectively. As can be seen from by the numerator of (3.11), the effect of both STO and SFO are alleviated by searching ϵ in the linear term $e^{j\epsilon u_k}$, while the impact of initial phase distortion is totally eliminated by taking the absolute value in (3.11).

Calculating $\overline{\phi}_d$ requires an accurate estimation of ϵ, which can be very inefficient if a brute-force fine-grained search is performed. To obtain ϵ efficiently, we employ an FFT with size N_{ser}, leading to a searching resolution of $2\pi/N_{\mathsf{ser}}$ in the range of $[0, 2\pi)$ for $\overline{\phi}_d$. The algorithm is summarized into Algorithm 2.

The phase correction as shown as Step 8 in Algorithm 2 differs significantly from the phase sanitization scheme in [8]. The phase sanitization scheme performs phase unwrapping on the CFR phases, which is error-prone in the presence of phase noise. Moreover, it totally eliminates the linear phase contained in the unwrapped CFR phases

---

Algorithm 2 Calculating the TRRS $\overline{\phi}_d$ for link d

Input: $\{\hat{H}_d[u_k]\}_{k=1,2,\ldots,N_u}, \{\hat{H}'_d[u_k]\}_{k=1,2,\ldots,N_u}$
Output: $\overline{\phi}_d$

1: Initializing $\Lambda_d = 0$ and $\Lambda'_d = 0$
2: **for** $k = 1, 2, \ldots, N_u$ **do**
3: Calculating $\hat{G}[u_k] = \hat{H}_d[u_k]\hat{H}'^{*}_d[u_k]$
4: Calculating $\Lambda_d = \Lambda_d + \hat{H}_d[u_k]\hat{H}^{*}_d[u_k]$
5: Calculating $\Lambda'_d = \Lambda'_d + \hat{H}'_d[u_k]\hat{H}'^{*}_d[u_k]$
6: **end for**
7: Appending $(N_{\mathsf{ser}} - N_u)$ zeros at the end of $\{\hat{G}[u_k]\}_{k=1,2,\ldots,N_u}$ if $N_{\mathsf{ser}} \geq N_u$. Otherwise, discarding the last $(N_u - N_{\mathsf{ser}})$ entries of $\{\hat{G}[u_k]\}_{k=1,2,\ldots,N_u}$.
8: Performing an N_{ser}-point FFT on $\{\hat{G}[u_k]\}_{k=1,2,\ldots,N_u}$, which leads to $\{g[n]\}_{n=1,2,\ldots,N_{\mathsf{ser}}}$ given as

$$g[n] = \sum_{k=1}^{N_{\mathsf{ser}}} \hat{G}[u_k] e^{-j\frac{2\pi n(k-1)}{N_{\mathsf{ser}}}}. \quad (3.13)$$

9: Calculating $\overline{\phi}_d = \dfrac{\max_{n=1,2,\ldots,N_{\mathsf{ser}}} |g[n]|^2}{\Lambda_d \Lambda'_d}$.
10: **return** $\overline{\phi}_d$

---

via the least-square estimation, which might also remove useful information about the environment as the side effect.[2] On the other hand, the presented phase correction step estimates ϵ by matching two CFRs using FFT and does not perform phase unwrapping. Therefore, the presented method is more robust against noise.

3.4.2 Fusing TR Resonating Strength of Different Links

In a MIMO-OFDM Wi-Fi system, the combined CFR consisting of the CFRs captured from different links can be expressed by

$$\hat{\mathbb{H}} = \left[\hat{\mathbf{H}}_1^{\mathsf{T}} \quad \hat{\mathbf{H}}_2^{\mathsf{T}} \quad \cdots \quad \hat{\mathbf{H}}_d^{\mathsf{T}} \quad \cdots \quad \hat{\mathbf{H}}_D^{\mathsf{T}}\right]^{\mathsf{T}} \tag{3.14}$$

with $\hat{\mathbb{H}}'$ defined similarly, we calculate $\{\overline{\phi}_d\}_{d=1,2,\dots,D}$ and fuse them together into the combined TRRS $\gamma[\hat{\mathbb{H}}, \hat{\mathbb{H}}']$, expressed by

$$\gamma[\hat{\mathbb{H}}, \hat{\mathbb{H}}'] = \left(\sum_{d=1}^{D} \omega_d \sqrt{\overline{\phi}_d}\right)^2, \tag{3.15}$$

where

$$\omega_d = \frac{\sqrt{\Lambda_d \Lambda_d'}}{\sqrt{\sum_{d=1}^{D} \Lambda_d}\sqrt{\sum_{d=1}^{D} \Lambda_d'}} \tag{3.16}$$

is the weight for the dth link. The intuition behind the choice of ω_d lies in that given identical channel noise on different link pairs, those link pairs with higher channel energy products are more robust against noise and thus should be allocated a higher weight in calculating the combined TRRS. The denominator of ω_d scales $\gamma[\hat{\mathbb{H}}, \hat{\mathbb{H}}']$ into the range of $[0, 1]$.

3.4.3 Effective Bandwidth

Because we fully utilize the information contained in $\hat{\mathbb{H}}$ and $\hat{\mathbb{H}}'$ in computing the combined TRRS $\gamma[\hat{\mathbb{H}}, \hat{\mathbb{H}}']$, we achieve an effective bandwidth W_e of

$$W_e = \frac{D N_u B}{N}, \tag{3.17}$$

where B is the bandwidth per link. For 802.11n Wi-Fi systems, B can be as large as 40 MHz. Notice that the effective bandwidth is different from the physical bandwidth allocated to a Wi-Fi channel. In this chapter, the effective bandwidth is used as a metric to quantify the available resources in a fingerprint-based IPS that can be harnessed for localization. A larger effective bandwidth generally leads to a better localization performance in terms of the detection rates and the false alarm rates and thus can provide an insight into the performance of the IPS.

[2] The reflectors in the environment also introduce linear phase shifts into the frequency-domain CFRs.

3.4.4 Localization Using Combined TRRS

The presented IPS consists of a training phase and a positioning phase, which are elaborated in the subsequent part of this section.

3.4.4.1 Training Phase

During the training phase, we collect R CFR realizations from each of the L locations-of-interest. The $L \times R$ CFRs are stored into the CFR database denoted as $\mathbb{D}_{\text{train}}$. The ith column of $\mathbb{D}_{\text{train}}$ is given by $\hat{\mathbb{H}}_i$, with $\hat{\mathbb{H}}_i$ shown as (3.14), and i is the *training index*. Denote the realization index as r and the location index as ℓ, the training index i can be mapped from (r, ℓ) as $i = (\ell - 1)R + r$.

3.4.4.2 Positioning Phase

The problem of determining the device location can be cast into an multi-hypothesis testing problem. More specifically, assume that we collect an instantaneous CFR $\hat{\mathbb{H}}'$ from a location ℓ' to be estimated. Then, we calculate the combined TRRS between each CFR in $\mathbb{D}_{\text{train}}$ and $\hat{\mathbb{H}}'$ shown as (3.15), which leads to $\{\gamma[\hat{\mathbb{H}}_i, \hat{\mathbb{H}}']\}_{i=1,2,\dots,LR}$. After that, we take the maximum of the multiple combined TRRS evaluated at the same training location ℓ but with different realization index r, expressed by

$$\gamma_\ell = \max_{\substack{i=(\ell-1)R+r\\ r=1,2,\dots,R.}} \gamma[\hat{\mathbb{H}}_i, \hat{\mathbb{H}}'], \tag{3.18}$$

Now, we define a total of $L + 1$ hypothesis $\mathcal{H}_0, \mathcal{H}_1, \mathcal{H}_2, \dots, \mathcal{H}_\ell, \dots, \mathcal{H}_L$, where $\mathcal{H}_{\ell,\ell\neq 0}$ stands for the hypothesis that the device is located at location ℓ in the training phase, and $\mathcal{H}_0$ represents the hypothesis that the device is located at an unknown location excluded from the training phase. We determine that $\mathcal{H}_{\ell,\ell\neq 0}$ is true, i.e., the device is located at the ℓth location in the training database, if the following two conditions are satisfied:

$$\gamma_\ell \geq \Gamma, \; \gamma_\ell = \max_{\ell'=1,2,\dots,L} \gamma_{\ell'}, \tag{3.19}$$

where Γ is a threshold in the range of $[0, 1]$. On the other hand, if $\gamma_\ell \leq \Gamma$, $\forall \ell = 1, 2, \dots, L$, we determine that $\mathcal{H}_0$ is true, i.e., we are unable to localize the device because there is no match between the instantaneous CFRs and those in $\mathbb{D}_{\text{train}}$.

3.4.4.3 Configuration of Threshold

The IPS performance is significantly affected by Γ. A well-chosen Γ leads to a high detection rate and incurring negligible false alarm rate. The detection rate, denoted by $P_D(\Gamma)$, characterizes the probability that the IPS successfully determines the correct locations of the device under Γ, while the false alarm rate, denoted as $P_F(\Gamma)$, captures the possibility that the IPS makes incorrect decisions on the device location under Γ.

With a constraint imposed on the detection rate as $P_{D,0}$ and the false alarm rate as $P_{FA,0}$, the IPS learns Γ automatically from CFRs in $\mathbb{D}_{\text{train}}$ in the training phase. First of all, the IPS computes the TRRS matrix $\mathbf{R}$ based on all CFRs in the training database $\mathbb{D}_{\text{train}}$, with the (i, j)th entry of $\mathbf{R}$ given by $\gamma[\mathbb{H}_i, \mathbb{H}_j]$, where $\mathbb{H}_i$ and $\mathbb{H}_j$ are the ith and jth CFR captured in the training phase, respectively. Notice that $[\mathbf{R}]_{i,i} \triangleq 1$. Then, the

IPS evaluates $(P_D(\Gamma), P_{FA}(\Gamma))$ for a variety for Γ, until it finds a specific $\Gamma^\star$ such that $P_D(\Gamma^\star) \geq P_{D,0}$ and $P_{FA}(\Gamma^\star) \leq P_{FA,0}$. Finally, $\Gamma^\star$ is utilized as the threshold in the positioning phase shown in (3.19).

3.5 Experiment Results

3.5.1 Experiment Settings

3.5.1.1 Environment

The experiments are conducted in a typical office in a multistory building. The indoor space is occupied by desks, computers, chairs, and shelves.

3.5.1.2 Devices

We build several prototypes equipped with off-the-shelf Wi-Fi devices. Each Wi-Fi device is equipped with 3 omnidirectional antennas to support 3 × 3 MIMO configuration. Based on functionalities, these Wi-Fi devices can be further classified as APs and Stations (STAs). The center frequency of each AP is configured as 5.24 GHz. The prototype is shown in Figure 3.4.

3.5.1.3 Details of Experiments

We conduct seven experiments in total to assess the performance of the presented IPS with settings illustrated in Figure 3.3. Experiment (Exp.) 1 ∼ 4 are conducted under a measurement resolution of 5 cm to analyze the performance under a static and a dynamic environment with details given below.

Exp. 1 investigates the localization performance of the presented algorithm with a 5 cm resolution. The Wi-Fi devices are placed under the LOS setting, denoted as

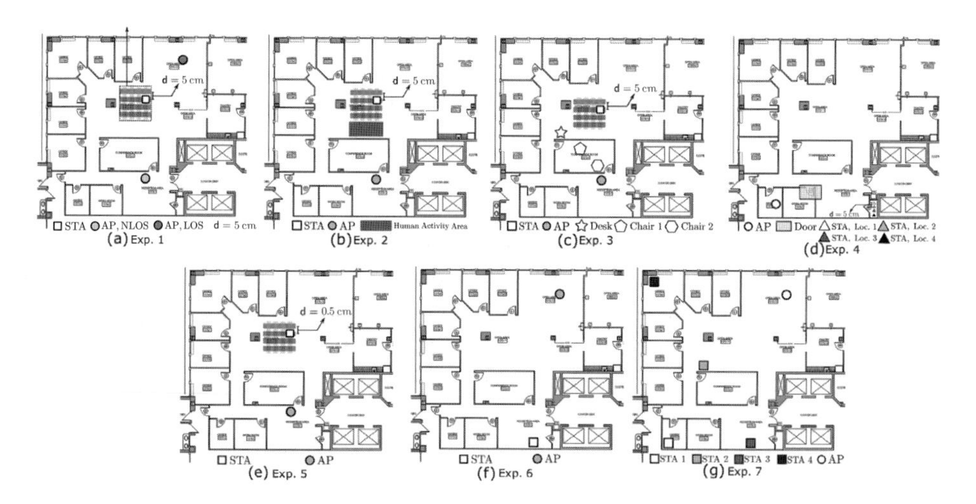

Figure 3.3 Setups for the experiments.

(a)

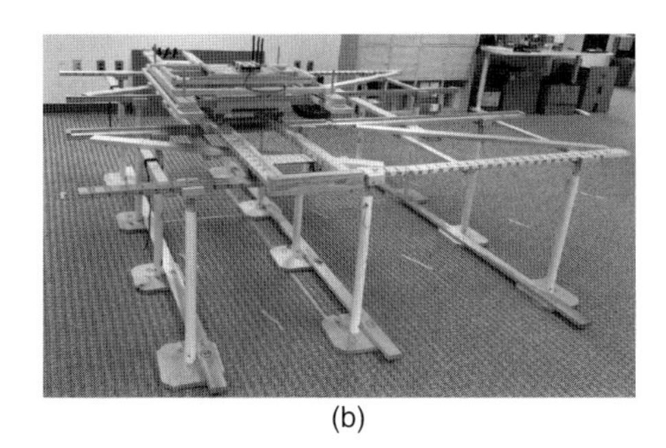

(b)

Figure 3.4 (a) The Wi-Fi prototype for the presented IPS and (b) the measurement structure used in the experiments.

Exp. 1a, as well as the NLOS setting, denoted as Exp. 1b. For each experiment, we measure the CFRs of 100 locations on a measurement structure as shown in Fig 3.4. The measurement resolution is $d = 5$ cm. For each location, 10 CFRs are measured.

Exp. 2 sheds light on the impact of human activities. One participant is asked to walk randomly in the vicinity of the STA on the measurement structure with $d = 5$ cm as the unit distance. The distances between the participant and the STA range from 8 to 10 feet. The AP is placed at the same NLOS position as in Exp. 1b. CFRs from 40 different locations on the structure are collected, with 10 CFRs per location.

Exp. 3 analyzes the localization performance when we introduce environment dynamics via moving the furniture in the office. We measure a total of 5 locations on the measurement structure with a resolution of 5 cm. For each location, we first measure 10 CFRs without furniture movement, followed by another 10 CFRs measured after we move the position of a desk near the measurement structure. Then, we measure 10 CFRs after we move a chair in the conference room and the last 10 CFRs after we move another chair in the conference room.

Exp. 4 studies the impact of door opening/closing on the localization performance. The AP is placed in an office room, with STA located in a closet near the entrance of the office suite. The direct link between the AP and the STA is blocked by two concrete walls. Then, a participant is asked to open and close the door of a room in the middle between the AP and STA. CFRs from 4 positions in the closet are captured with 10 CFRs per location for each door status.

On the other hand, Exp. 5 $\sim$ 7 studies several important aspects of the presented IPS, where

Exp. 5 studies the achievable accuracy of presented IPS. The STA is placed at the same measurement structure with Exp. 1, but with a much finer resolution with $d = 0.5$ cm. CFRs from a total of 400 locations on the grid points of a rectangular area are measured, with 5 CFRs per location.

Exp. 6 studies the effect of the variations in the synchronization parameters. The positions of the AP and the STA are fixed, and we turn on and off the power of the AP to enforce the reinitialization the PLL at the AP. Thus, the synchronization parameters discussed in Section 3.3.2 also changes. The power cycling is repeated 20 rounds with 10 CFRs for each round.

Exp. 7 analyzes the long-term behavior of the IPS. One AP and four STAs are deployed in the office with positions shown in Figure 3.3(g). The CFRs are collected every 10 min from the four STAs. The IPS is kept running for 631 hours (26 days) continuously. For each measurement, we collect 5 CFRs from each STA. In the twenty-six days of measurement, the office is fully occupied by around 10 people during weekdays and occasionally occupied during weekends. Also, the furniture is moved randomly every day.

The effective bandwidths W_e in the experiments are calculated from (3.17) with $N_u = 114$, $N = 128$, and $D = 1, 2, 3, \ldots, 9$, with the maximum W_e as 321 MHz obtained by exploiting all available links under the 3×3 MIMO configuration, e.g., $D = 9$. In the performance evaluations, N_{ser} shown in Algorithm 2 is configured as $1{,}024$.

3.5.2 Metrics for Performance Evaluation

For Exp. 1, 2, 3, and 4, we evaluate the detection rate $P_D(\Gamma^\star)$ and the false alarm rate $P_{FA}(\Gamma^\star)$ under threshold $\Gamma^\star$. More specifically, we choose 5 out of the 10 CFRs of each location randomly and consider them as the CFRs obtained in the training phase for each location-of-interest, which are assembled into the training database $\mathbb{D}_{\text{train}}$. The remaining 5 CFRs of each location are considered as the CFRs obtained in the positioning phase and are arranged into the testing database denoted by $\mathbb{D}_{\text{test}}$.

Using the scheme presented in Section 3.4, we calculate $P_D(\Gamma)$ and $P_{FA}(\Gamma)$ for $\Gamma \in [0, 1]$ based on the TRRS matrix $\mathbf{R}$ calculated from $\mathbb{D}_{\text{train}}$. By comparing $P_D(\Gamma)$ against $P_{FA}(\Gamma)$ under various Γ, we demonstrate the receiver operating characteristic (ROC) curve to highlight the trade-offs between detection and false alarm. Then, we choose the minimum $\Gamma^\star$ that satisfies the objective $P_D(\Gamma^\star) \geq 95\%$, $P_{FA}(\Gamma^\star) \leq 2\%$ as the threshold. Last, we calculate the TRRS matrix $\mathbf{R}'$ from $\mathbb{D}_{\text{test}}$ and evaluate $P'_D(\Gamma^\star)$ and $P'_{FA}(\Gamma^\star)$ based upon $\mathbf{R}'$ and $\Gamma^\star$. To fully utilize the collected CFRs, we repeat this process five times by randomizing the selections of CFRs for the training phase and positioning phase. Finally, we compute the average of $\Gamma^\star$, $P'_D(\Gamma^\star)$, and $P'_{FA}(\Gamma^\star)$, denoted as $\overline{\Gamma^\star}$, $\overline{P}'_D$, and $\overline{P}'_{FA}$, respectively.

In Exp. 5, we illustrate the distribution of the TRRS on the measurement structure. In particular, we assemble the CFRs obtained from the middle point of the 10 cm $\times$ 10 cm rectangular grid into $\mathbb{D}_{\text{train}}$ and keep the CFRs of all locations into $\mathbb{D}_{\text{test}}$. Then, we compute the TRRS matrix $\mathbf{R}$ based on $\mathbb{D}_{\text{train}}$ and $\mathbb{D}_{\text{test}}$.

In Exp. 6, we build $\mathbb{D}_{\text{train}}$ with all CFRs, with $\mathbb{D}_{\text{test}}$ the same as $\mathbb{D}_{\text{train}}$. Thus, the TRRS matrix $\mathbf{R}$ encapsulates the impact of time-varying synchronization parameters on the localization performance.

In Exp. 7, for each STA, we construct $\mathbb{D}_{\text{train}}$ using the five CFRs collected in the first measurement and keep all CFRs measured at different times into $\mathbb{D}_{\text{test}}$. After calculating the TRRS matrix $\mathbf{R}$, we take the column average of $\mathbf{R}$, denoted as $\overline{\mathbf{R}}$, which represents the average TRRS between the CFR in $\mathbb{D}_{\text{train}}$ and those in $\mathbb{D}_{\text{test}}$ of every 10 min. Using $\overline{\mathbf{R}}$, we evaluate the detection rate and false alarm rate.

3.5.3 Performance Evaluation

Exp. 1a: LOS with 5 cm Resolution

In Figure 3.5(a), (b), (c), we demonstrate the TRRS matrix $\mathbf{R}$ with different W_e under the LOS scenario. As we can see, increasing W_e shrinks the off-diagonal components of $\mathbf{R}$. In other words, a larger W_e alleviates the ambiguity among different locations. On the other hand, the TRRS values measured at the same location only degrade slightly and are still close to 1 with a large W_e. The net effect of using a large W_e is a larger margin between the TRRS calculated at the same location and among different locations,

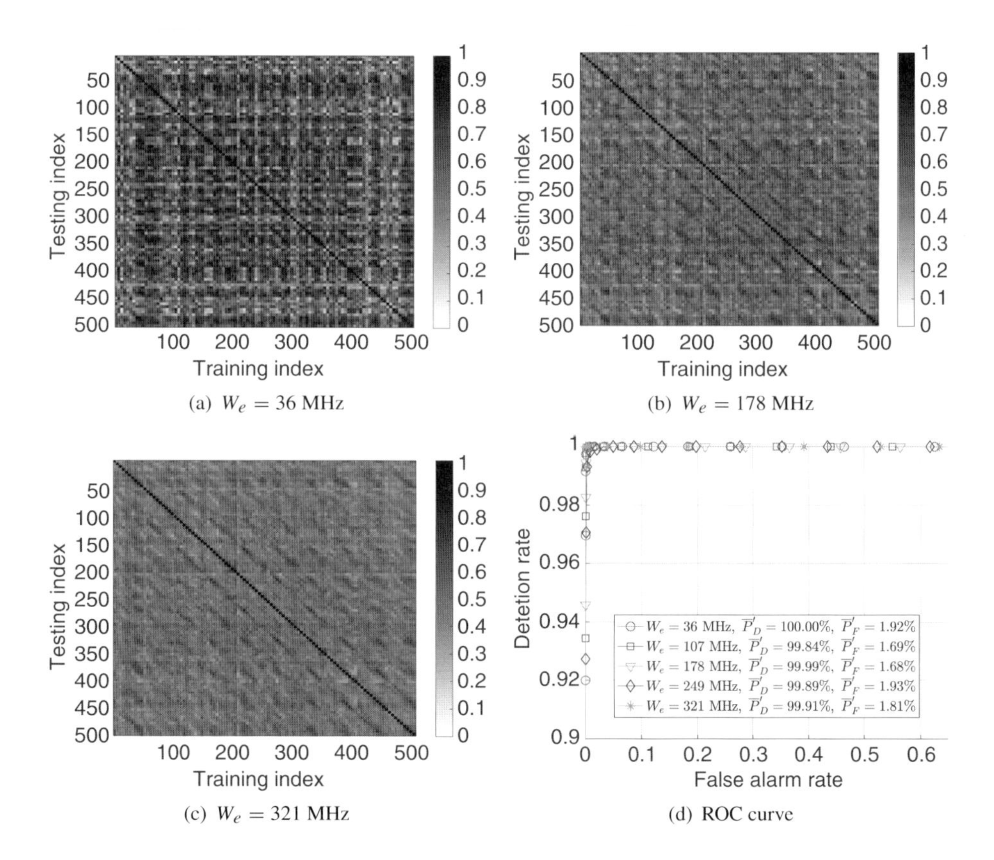

(a) $W_e = 36$ MHz (b) $W_e = 178$ MHz (c) $W_e = 321$ MHz (d) ROC curve

Figure 3.5 Results of Exp. 1a under LOS.

evidencing an enhanced location distinction. The ROC curve in Figure 3.5(d) demonstrates that the IPS achieves nearly perfect localization performance with $\overline{P}'_D \geq 99.84\%$ and $\overline{P}'_{FA} \leq 1.93\%$ under different W_e. When $W_e = 36$ MHz, we can achieve $\overline{P}'_D = 100\%$ and $\overline{P}'_{FA} = 1.92\%$, which implies that there exists ambiguity among locations as shown in Figure 3.5(a) when $W_e = 36$ MHz, we are able to find a good $\overline{\Gamma^\star}$ to distinguish different locations.

However, in general, the threshold $\overline{\Gamma^\star}$ is large when W_e is small. Therefore, under a small W_e, the IPS is highly sensitive to noise and deterioration of CFRs associated with different locations, e.g., when there exists significant environment dynamics due to human or object movement. On the other hand, $\overline{\Gamma^\star}$ is much smaller when W_e is large, which leaves a larger margin for noise and dynamics and thus improves the robustness of the presented IPS. This is justified by Figure 3.7, where we demonstrate $\overline{\Gamma^\star}$ under different W_e for Exp. 1a. It can be seen that a threshold as large as 0.86 is required when $W_e = 36$ MHz, which decreases as W_e increases. When $W_e = 321$ MHz, the threshold drops to 0.63.

Exp. 1b: NLOS with 5 cm Resolution

In Figure 3.6(a), (b), and (c), we show the TRRS matrix $\mathbf{R}$ with different W_e under the NLOS scenario. Comparing with Figure 3.5(a), (b), and (c), we see that the location

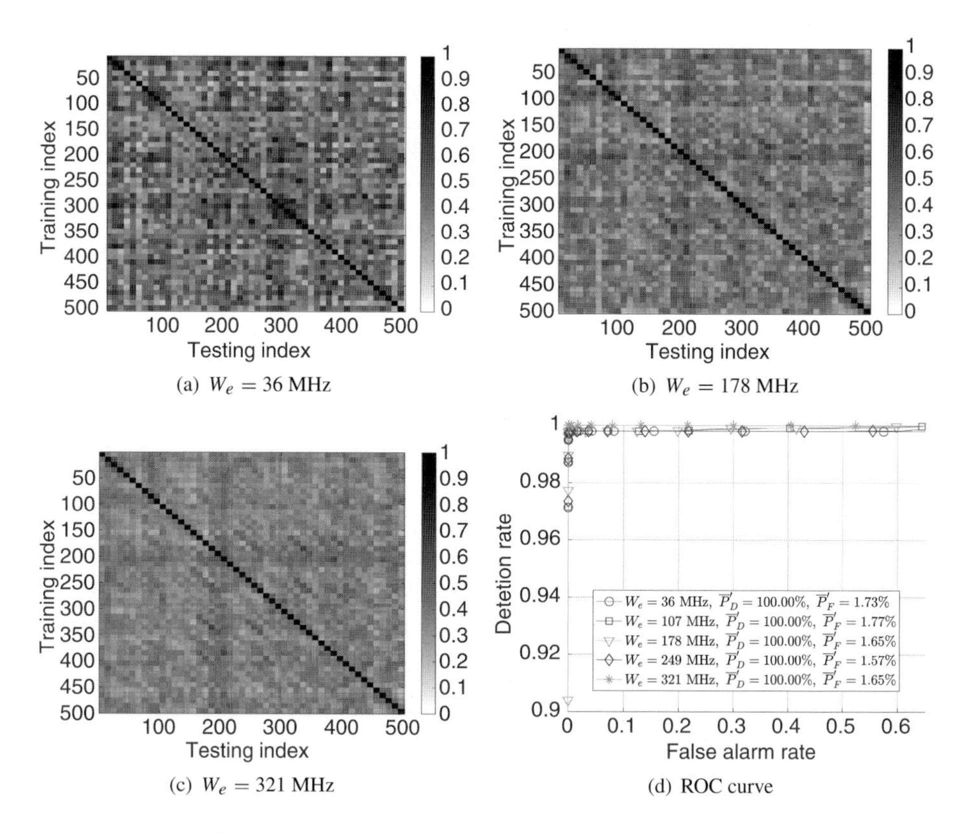

(a) $W_e = 36$ MHz (b) $W_e = 178$ MHz (c) $W_e = 321$ MHz (d) ROC curve

Figure 3.6 Results of Exp. 1b under NLOS.

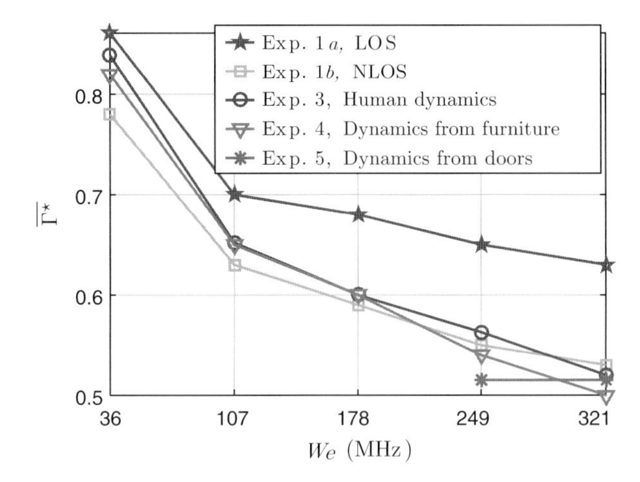

Figure 3.7 $\overline{\Gamma^{\star}}$ under different W_e.

ambiguity is lower for the NLOS scenario than the LOS scenario, indicated by the smaller TRRS values measured between different locations. This can be justified by the fact that the channel energy spreads over more multipath components under NLOS than LOS and provides richer information of the environment. Similar to the results of Exp. 1a, we find that a larger W_e mitigates the location ambiguity and enhances the overall IPS performance, with $\overline{P}'_D = 100\%$ and $\overline{P}'_{FA} = 1.65\%$ when $W_e = 321$ MHz. Additionally, from Figure 3.7, we observe that $\overline{\Gamma^{\star}}$ decreases more rapidly when W_e enlarges than the LOS case, reducing $\overline{\Gamma^{\star}}$ from 0.78 under $W_e = 36$ MHz to 0.53 when $W_e = 321$ MHz.

The negligible false alarm rates in Exp. 1 also imply that the localization error is 0 cm under most cases. In fact, the false alarm rates can be further reduced by increasing $\overline{\Gamma^{\star}}$, leading to a false alarm rate of 0.06% and a detection rate of 99.48% under $\overline{\Gamma^{\star}} = 0.74$ for the LOS case, and 0% false alarm rate and a detection rate of 99.45% under $\overline{\Gamma^{\star}} = 0.71$.

Exp. 2: Effect of Human Activities

Figure 3.8 shows the impact of human activities on the performance of the presented IPS. From Figure 3.8(a), (b), and (c), we find that a large W_e improves the robustness against environment dynamics. Figure 3.8(d) illustrates the ROC curve using different W_e and further verifies that a large W_e can enhance the localization performance, leading to $\overline{P}'_D = 99.88\%$ and $\overline{P}'_{FA} = 1.66\%$ when $W_e = 321$ MHz. As shown in Figure 3.7, when $W_e = 321$ MHz, a threshold of 0.52 suffices to achieve a good performance.

Exp. 3: Impact of Furniture Movement

In Figure 3.9, we show the performance in the presence of furniture movement. Similar to the observations in Exp. 1 and Exp. 2, a larger W_e enhances the robustness against the dynamics from furniture movement and reduces ambiguity among locations. As can be seen from Figure 3.9, location 1 and location 2 are highly correlated implied by a large TRRS value, and the ambiguity between locations 1 and 2 is alleviated when W_e

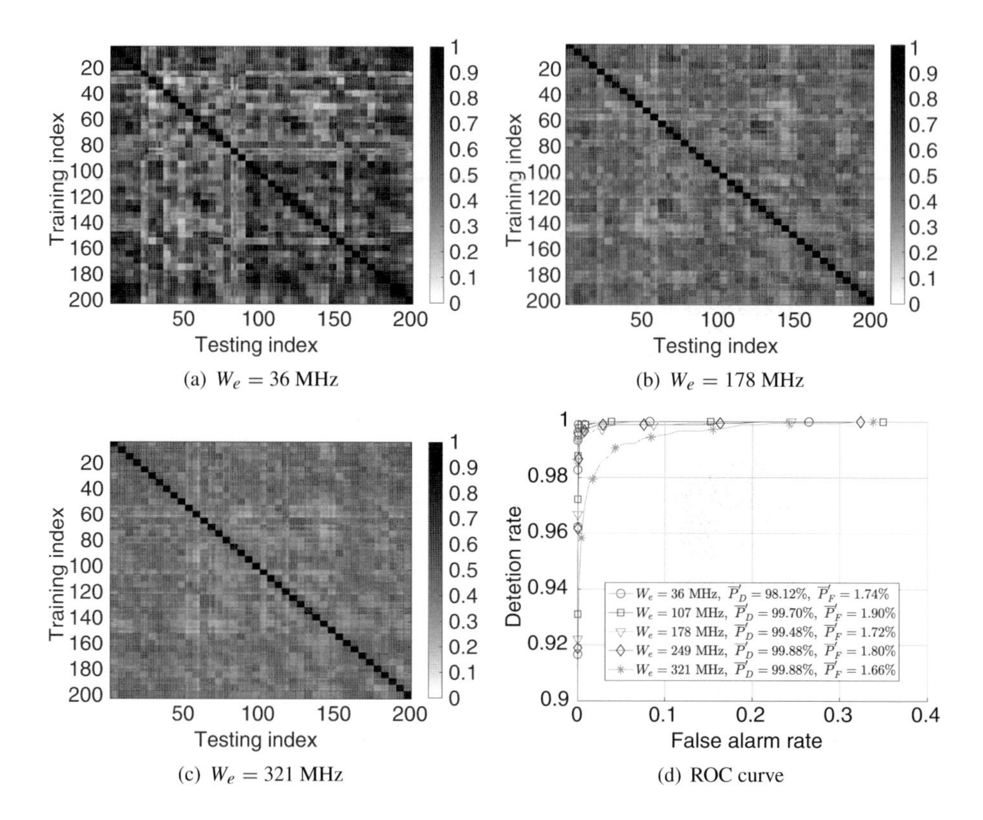

Figure 3.8 Results of Exp. 2 with human dynamics.

increases to 178 MHz and 321 MHz, leading to an improved detection rate and false alarm rate. When $W_e = 321$ MHz, we achieve $\overline{P}'_D = 98.86\%$ and $\overline{P}'_{FA} = 1.95\%$ under a threshold of 0.50 as shown in Figure 3.7.

Also, we notice that the performance does not improve monotonically with W_e. This is because the quality of different links in the multiantenna Wi-Fi system differs due to the discrepancy in their noise and interference levels. Thus, combining multiple links based on the channel energies might not be optimal in this case. This can be solved by calculating the TRRS using criterion robust against noise and interference on different Wi-Fi links.

Exp. 4: Impact of Door

The impact of door status on localization is more severe than the human activities when the door acts as a major reflector in the propagation of electromagnetic waves, and thus its status greatly affects the CFRs.

In Figure 3.10, we illustrate the results under different W_e. Obviously, a large W_e is indispensable in this case because the TRRS values measured at location 1 and 2 degrade to 0.42 and 0.17 under different door status with $W_e = 36$ MHz, and the IPS fails to find a $\Gamma^\star$ to achieve at least 95% detection rate and at most 2% false alarm

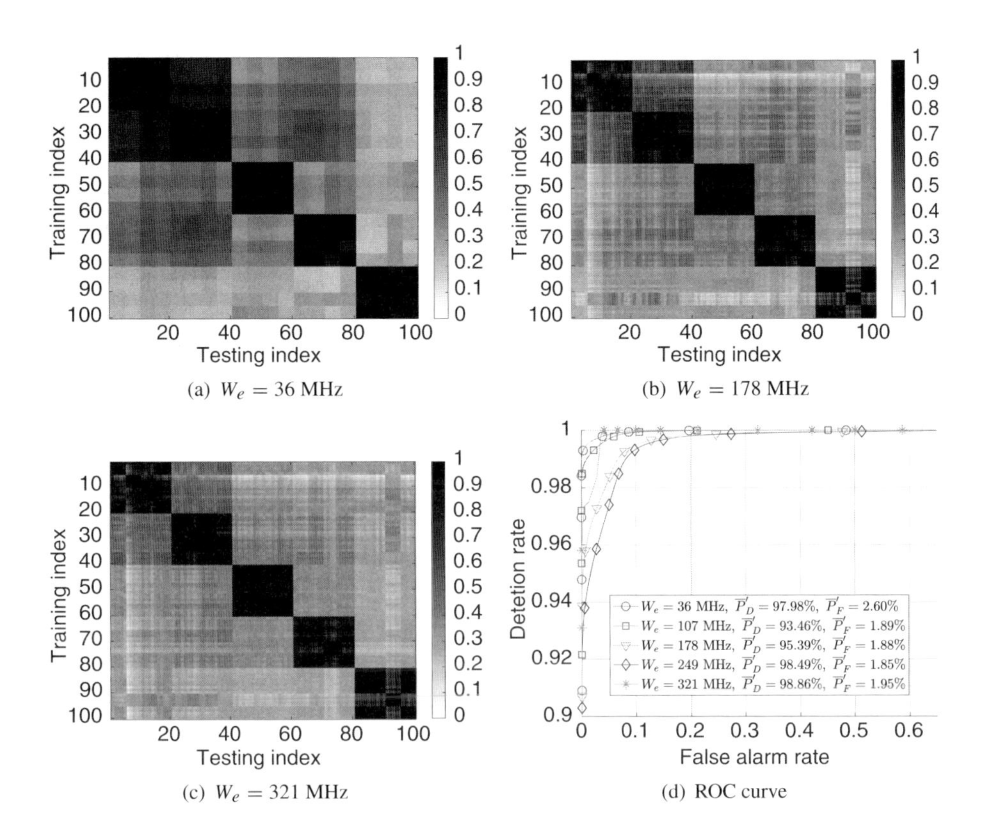

(a) $W_e = 36$ MHz (b) $W_e = 178$ MHz (c) $W_e = 321$ MHz (d) ROC curve

Figure 3.9 Results of Exp. 3 with furniture movement.

rate. On the other hand, increasing W_e increases to 321 MHz partially recovers the similarities of the CFRs collected at the same location with different door status, and we achieve $\overline{P}'_D = 98.39\%$ and $\overline{P}'_{FA} = 1.43\%$. This could be justified by the inherent spatial diversity naturally existing in multiantenna Wi-Fi systems becauase different links can be considered uncorrelated and thus sense the environment from different perspectives. Therefore, even the door affects a majority of the multipath components on some Wi-Fi links; its impact on other Wi-Fi links is much less pronounced. Figure 3.7 shows that the target of more than 95% detection rate and less than 2% false alarm rate is achievable when $W_e \geq 249$ MHz. In conclusion, a large W_e is paramount for the presented IPS when there exists strong environmental dynamics.

Impact of Effective Bandwidth on the CDF of TRRS

We observe from the analysis on the results of Exp. 1, 2, 3, and 4 that the gap between the TRRS measured at the same locations and among different locations enlarges with respect to an increased W_e. To further validate the observation, we draw the cumulative density functions (CDFs) of the TRRS values in Figure 3.11 under various W_e, where the solid lines represent the CDFs with $W_e = 321$ MHz. It shows that the TRRS values among different locations are more concentrated in a region with a small average of

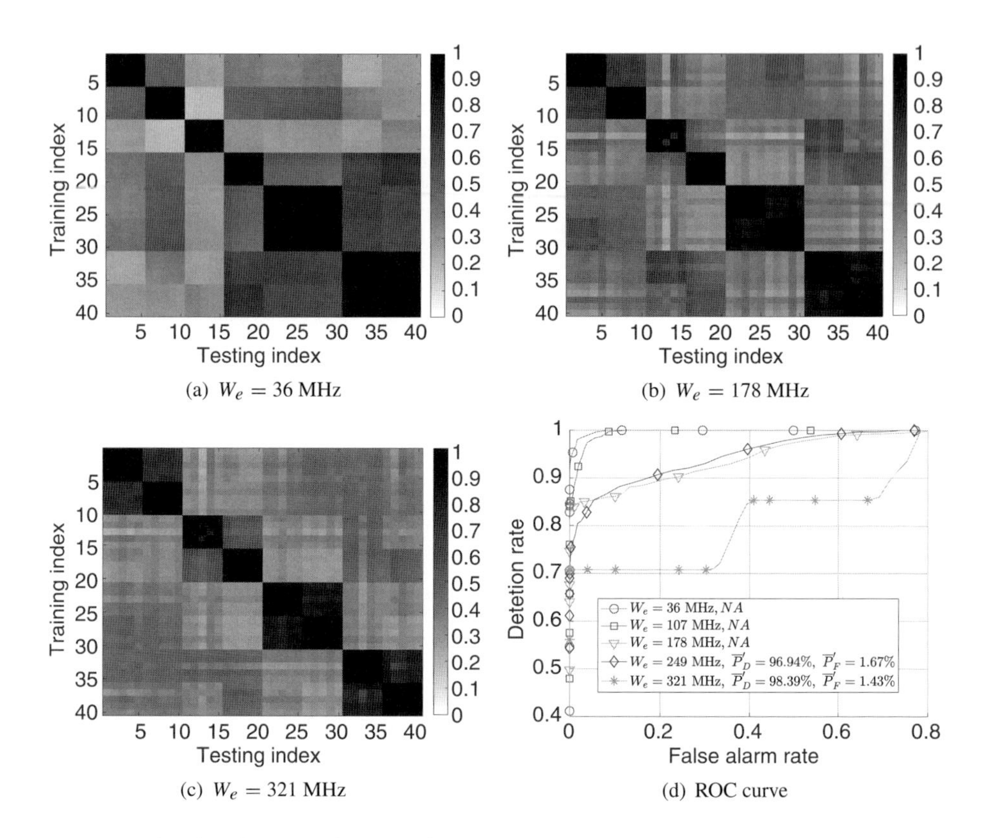

(a) $W_e = 36$ MHz (b) $W_e = 178$ MHz

(c) $W_e = 321$ MHz (d) ROC curve

Figure 3.10 Results of Exp. 4 with door effect.

TRRS when W_e is large, while the TRRS values measured at the same locations are still highly concentrated in a region with an average TRRS close to 1 for Exp. 1a and Exp. 1b. The decrease of the TRRS at the same location is more significant for Exp. 2, 3, and 4. Yet, the degradation is still within the tolerance level as implied by the $\overline{P}'_D$ and $\overline{P}'_{FA}$ performances. Therefore, it is crucial to use a large W_e to enhance both performance and robustness.

Exp. 5**: Results under** 0.5 **cm Measurement Resolution**

Figure 3.12(a), (b), and (c) visualize the TRRS matrix $\mathbf{R}$ calculated under different W_e. We observe that large TRRS values are highly concentrated within a small and uniform circular area with a radius of $1 \sim 2$ cm surrounding the middle point when $W_e \geq 178$ MHz, while the TRRS are more decentralized when $W_e = 36$ MHz.

Figure 3.13 shows the average TRRS decay along different directions calculated using $\mathbf{R}$. A larger W_e accelerates the decay of the TRRS values and improves the location distinction. With a distance larger than 1 cm, the TRRS drops below 0.75. Therefore, with an appropriate threshold, the presented IPS can achieve an accuracy of $1 \sim 2$ cm.

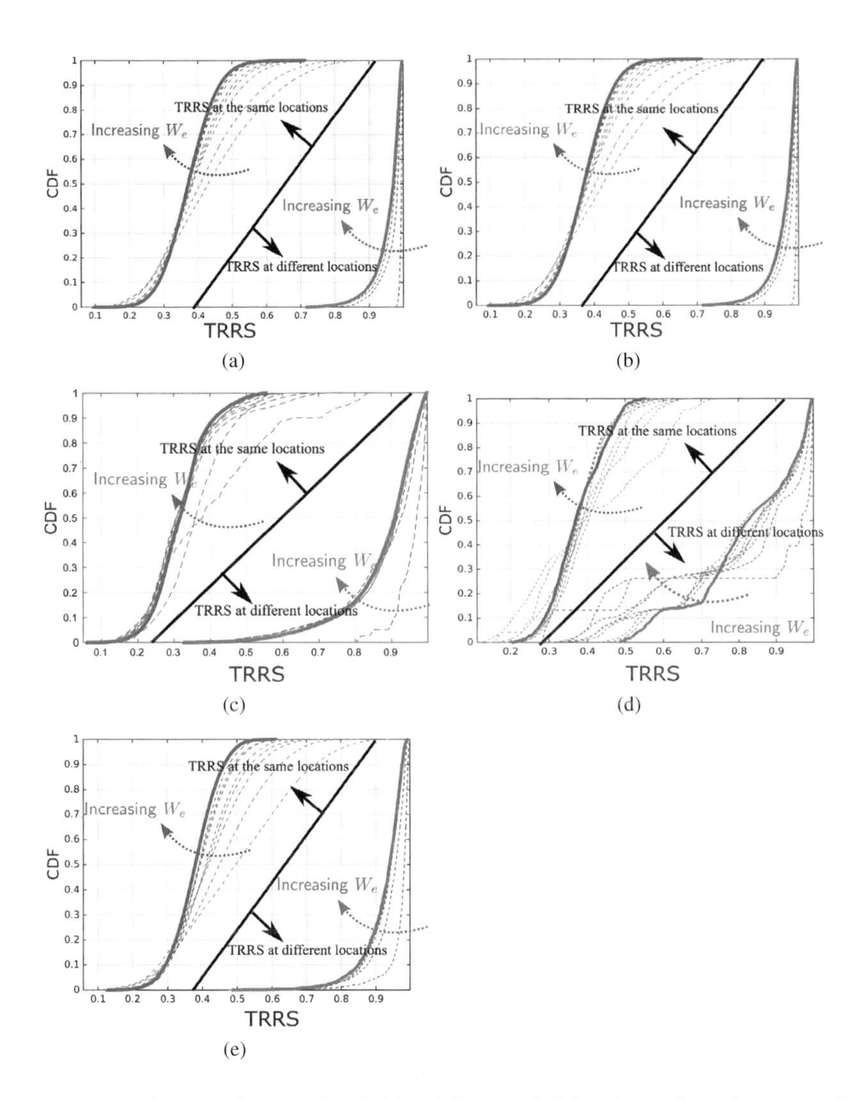

Figure 3.11 Impact of W_e on the TRRS: (a) Exp. 1a, LOS under static environment; (b) Exp. 1b, NLOS under static environment; (c) Exp. 2, dynamic environment with human activities; (d) Exp. 3, dynamic environment with furniture movement; and (e) Exp. 4, dynamic environment with door opening and closing.

Exp. 6: Impact of Power Cycling on Localization Performance

Figure 3.14 shows **R** with different W_e for Exp. 7. Clearly, when $W_e = 36$ MHz, there exists large fluctuation in the TRRS, and the localization performance is deteriorated. The performance loss can be remedied by using $W_e \geq 178$ MHz, which again demonstrates the importance of a large W_e.

Exp. 7: Localization Performance over 26 Days

In Figure 3.15, we sketch the time evolution of the TRRS evaluated at the same STA locations and among different STA locations for different STAs in the 26 days of

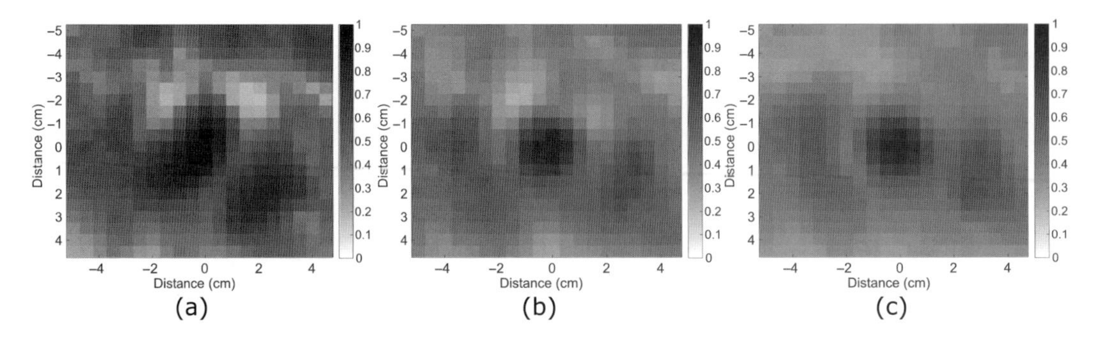

Figure 3.12 Results of Exp. 5 under a measurement resolution of 0.5 cm. (a) $W_e = 36$ MHz, (b) $W_e = 178$ MHz, (c) $W_e = 321$ MHz.

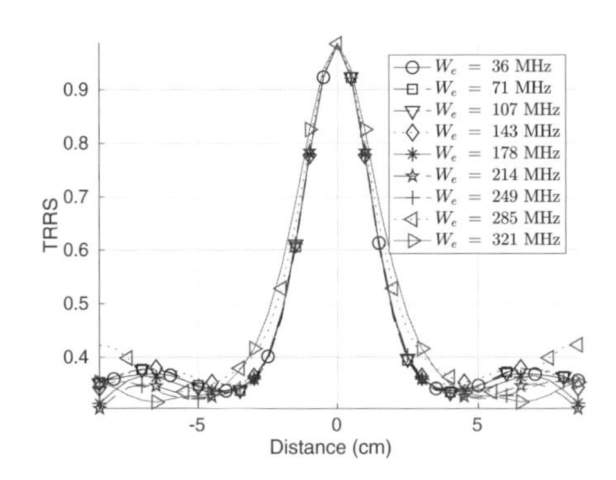

Figure 3.13 Decaying of TRRS with distance in Exp. 5.

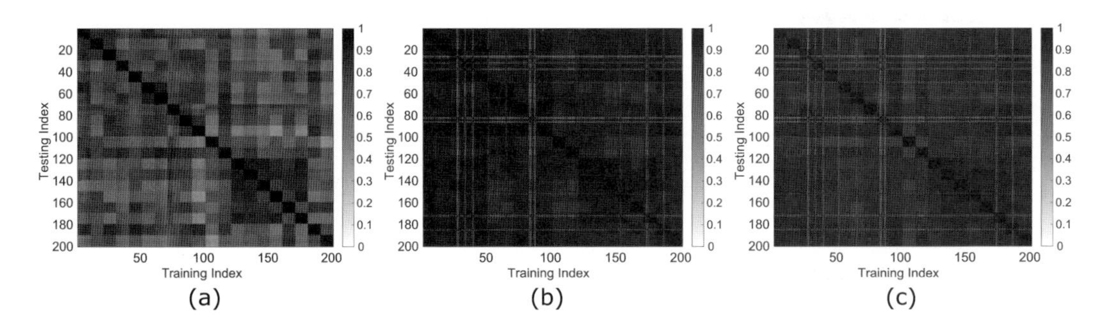

Figure 3.14 Results of Exp. 6. (a) $W_e = 36$ MHz (b) $W_e = 178$ MHz (c) $W_e = 321$ MHz.

measurement. We observe that the TRRS changes with time due to the environment variations. We also find that when $W_e = 321$ MHz, the decay in TRRS is less severe than the case with $W_e = 36$ MHz.

In Figure 3.15(e), (j), and (o), we display the ROC curves for the four STAs under different W_e. Obviously, in comparison with the results of $W_e = 36$ MHz, using a

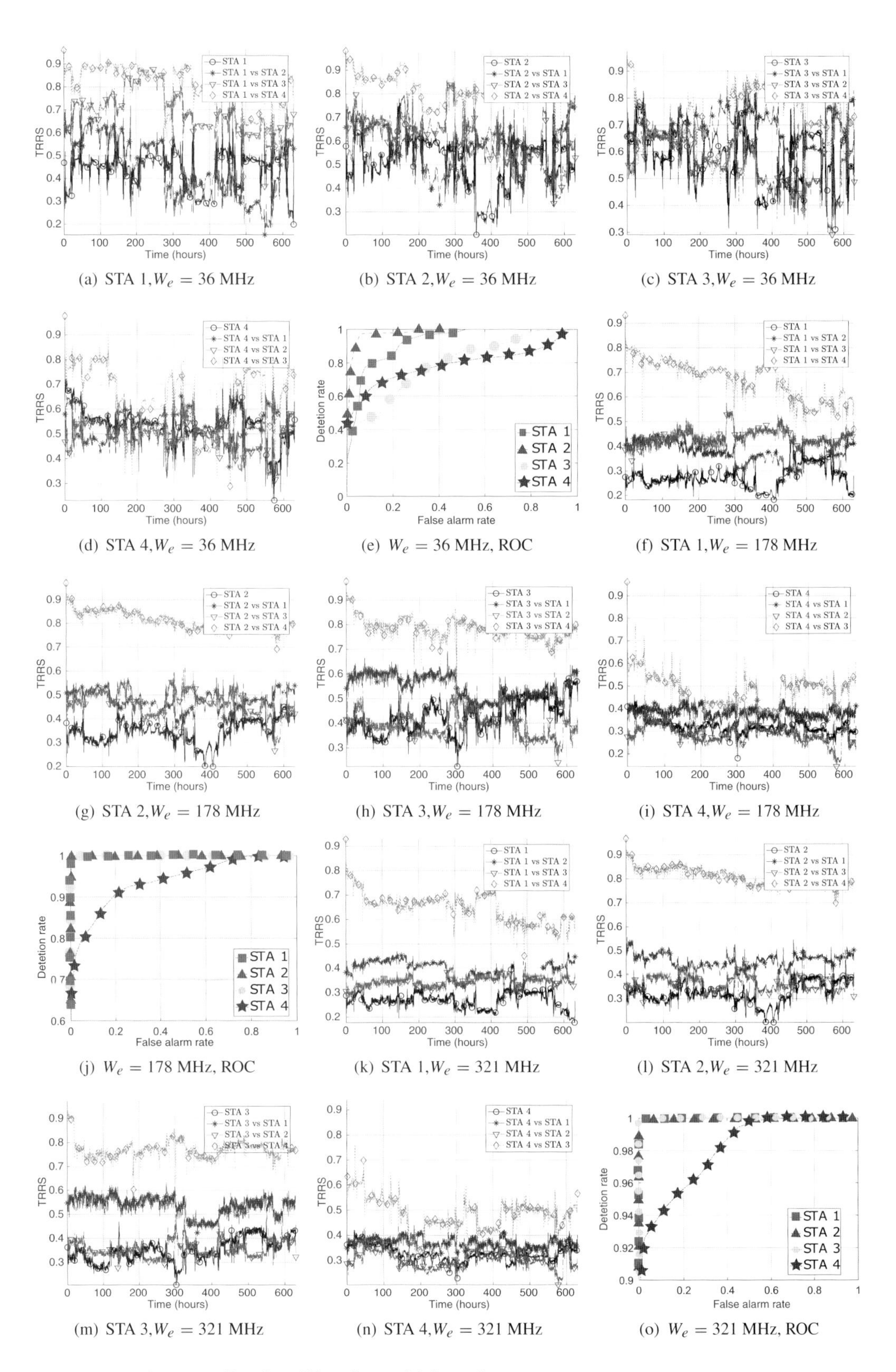

Figure 3.15 Results of Exp. 7 over 26 days of measurement.

Table 3.1 Performances with $\theta = 0.60$ and $W_e = 321$ MHz

	Exp. 1 LOS	Exp. 1 NLOS	Exp. 2 Human activities	Exp. 3 Furniture movement	Exp. 4 Door effect
Detection rate (%)	99.94	100	99.72	96.61	85.00
False alarm rate (%)	3.96	0.14	0.01	0	0

W_e as large as 178 MHz and 321 MHz combats the degradation incurred by inevitable environment changes and improves the IPS performance by a large margin. Therefore, using a large W_e can greatly reduce the overhead introduced by the training phase because it becomes unnecessary to update $\mathbb{D}_{\text{train}}$ frequently.

Although the distances between the STAs in this experiment exceed 5 cm, the results would be very similar if these STAs are placed closer at a centimeter level with a large W_e. This is because the TRRS values calculated between two locations are identical as long as the distance between them exceeds 5 cm as shown in Figures 3.5 and 3.6.

Results under a Universal $\Gamma^\star$

In Exp. 1, 2, 3, and 4, we assume that the presented IPS learns a specific $\Gamma^\star$ from $\mathbb{D}_{\text{train}}$ to achieve $\overline{P}'_D \geq 95\%$ and $\overline{P}'_{FA} \leq 2\%$. In practice, due to the randomness of the environment, we might only be able to roughly find a fixed $\Gamma^\star$ based on a very limited training database without much dynamics, and the performance might degrade consequently. To investigate the performance loss under a fixed $\Gamma^\star$, we configure $\Gamma^\star$ as 0.6 and W_e as 321 MHz.

The performances are summarized in Table 3.1, which shows that except Exp. 4, the presented IPS still achieves a detection rate higher than 96.61% with a false alarm rate smaller than 3.96%. The performance of Exp. 4 degrades but still maintains a detection rate of 85% and a false alarm rate of 0.

In practice, we perform a large number of experiments in a variety of environment. The results reveal that the $1 \sim 2$ cm accuracy is universal instead of limited only to a small area, given that the number of multipath components is sufficient.

3.6 Summary

In this chapter, we discussed a Wi-Fi-based IPS leveraging the TR focusing effect that achieves centimeter-level accuracy for indoor localization. The presented IPS fully utilizes the spatial diversity in MIMO-OFDM Wi-Fi systems to formulate a large effective bandwidth. Extensive experimental results show that with a measurement resolution of 5 cm, the presented IPS achieves true positive rates of 99.91% and 100%, and incurs false-positive rates of 1.81% and 1.65% under the LOS and NLOS scenarios, respectively. Meanwhile, the IPS is robust against the environment dynamics caused by human activities and object movements. Experiment results with a measurement resolution of

0.5 cm demonstrate a localization accuracy of $1 \sim 2$ cm achieved by the presented IPS. For related references, readers can refer to [34].

References

[1] T. J. Gallagher, B. Li, A. G. Dempster, and C. Rizos, "A sector-based campus-wide indoor positioning system," in *2010 IEEE International Conference on Indoor Positioning and Indoor Navigation (IPIN),* 2010, pp. 1–8.

[2] K.-L. Sue, "MAMBO: A mobile advertisement mechanism based on obscure customer's location by RFID," in *2012 IEEE International Conference on Computing, Measurement, Control and Sensor Network (CMCSN),* 2012, pp. 425–428.

[3] L. Wang, W. Liu, N. Jing, and X. Mao, "Simultaneous navigation and pathway mapping with participating sensing," *Wireless Networks*, vol. 21, no. 8, pp. 2727–2745, 2015.

[4] H. Liu, H. Darabi, P. Banerjee, and J. Liu, "Survey of wireless indoor positioning techniques and systems," *IEEE Transactions on Systems, Man, and Cybernetics, Part C: Applications and Reviews,* vol. 37, no. 6, pp. 1067–1080, Nov. 2007.

[5] P. Bahl and V. Padmanabhan, "RADAR: An in-building RF-based user location and tracking system," in *Proceedings of IEEE INFOCOM*, vol. 2, 2000, pp. 775–784.

[6] M. Youssef and A. Agrawala, "The Horus WLAN location determination system," in *Proceedings of the 3rd ACM International Conference on Mobile Systems, Applications, and Services*, 2005, pp. 205–218.

[7] P. Prasithsangaree, P. Krishnamurthy, and P. Chrysanthis, "On indoor position location with wireless LANs," in *IEEE International Symposium on Personal, Indoor and Mobile Radio Communications,* vol. 2, Sep. 2002, pp. 720–724.

[8] S. Sen, B. Radunovic, R. R. Choudhury, and T. Minka, "You are facing the Mona Lisa: Spot localization using PHY layer information," in *Proceedings of the 10th ACM International Conference on Mobile Systems, Applications, and Services*, 2012, pp. 183–196. [Online]. Available: http://doi.acm.org/10.1145/2307636.2307654.

[9] J. Xiao, W. K. S., Y. Yi, and L. Ni, "FIFS: Fine-grained indoor fingerprinting system," in *21st International Conference on Computer Communications and Networks (ICCCN),* Jul. 2012, pp. 1–7.

[10] Y. Chapre, A. Ignjatovic, A. Seneviratne, and S. Jha, "CSI-MIMO: Indoor Wi-Fi fingerprinting system," in *IEEE 30th Conference on Local Computer Networks (LCN),* Sep. 2014, pp. 202–209.

[11] Z.-H. Wu, Y. Han, Y. Chen, and K. J. R. Liu, "A time-reversal paradigm for indoor positioning system," *IEEE Transactions on Vehicular Communications*, vol. 64, no. 4, pp. 1331–1339, Apr. 2015.

[12] B. Wang, Y. Wu, F. Han, Y.-H. Yang, and K. J. R. Liu, "Green wireless communications: A time-reversal paradigm," *IEEE Journal on Selected Areas in Communications*, vol. 29, no. 8, pp. 1698–1710, Sep. 2011.

[13] C. Chen, Y. Chen, H. Q. Lai, Y. Han, and K. J. R. Liu, "High accuracy indoor localization: A WiFi-based approach," in *2016 IEEE International Conference on Acoustics, Speech and Signal Processing (ICASSP),* Mar. 2016, pp. 6245–6249.

[14] C. Chen, Y. Chen, Y. Han, H.-Q. Lai, and K. J. R. Liu, "Achieving centimeter accuracy indoor localization on WiFi platforms: A frequency hopping approach," *IEEE Internet of Things Journal*, vol. 4, no. 1, pp. 111–121, Feb. 2017.

[15] J. Hightower, R. Want, and G. Borriello, "SpotON: An indoor 3D location sensing technology based on RF signal strength," University of Washington, Department of Computer Science and Engineering, Seattle, WA, UW CSE 00-02-02, Feb. 2000.

[16] L. Ni, Y. Liu, Y. C. Lau, and A. Patil, "LANDMARC: Indoor location sensing using active RFID," in *Proceedings of the First IEEE International Conference on Pervasive Computing and Communications (PerCom 2003)*, Mar. 2003, pp. 407–415.

[17] X. Li, K. Pahlavan, M. Latva-aho, and M. Ylianttila, "Comparison of indoor geolocation methods in DSSS and OFDM wireless LAN systems," in *52nd IEEE-VTS Fall Vehicular Technology Conference,* vol. 6, 2000, pp. 3015–3020.

[18] N. Correal, S. Kyperountas, Q. Shi, and M. Welborn, "An UWB relative location system," in *IEEE Conference on Ultra Wideband Systems and Technologies*, Nov. 2003, pp. 394–397.

[19] P. Steggles and S. Gschwind, The Ubisense smart space platform. *Adjunct Proceedings of the Third International Conference on Pervasive Computing*, vol. 191, 73–76, 2005.

[20] B. Campbell, P. Dutta, B. Kempke, Y.-S. Kuo, and P. Pannuto, "DecaWave: Exploring state of the art commercial localization," *Ann Arbor*, vol. 1001, p. 48109.

[21] D. Sapphire, "UWB-based real-time location systems." www.zebra.com/us/en/products/location-technologies/ultra-wideband.html.

[22] J. Xiong and K. Jamieson, "ArrayTrack: A fine-grained indoor location system," in *Proceedings of the 10th USENIX Conference on Networked Systems Design and Implementation*, 2013, pp. 71–84. [Online]. Available: http://dl.acm.org/citation.cfm?id=2482626.2482635.

[23] R. Schmidt, "Multiple emitter location and signal parameter estimation," *IEEE Transactions on Antennas and Propagation*, vol. 34, no. 3, pp. 276–280, Mar. 1986.

[24] J. Gjengset, J. Xiong, G. McPhillips, and K. Jamieson, "Phaser: Enabling phased array signal processing on commodity WiFi access points," in *Proceedings of the 20th Annual ACM International Conference on Mobile Computing and Networking*, 2014, pp. 153–164. [Online]. Available: http://doi.acm.org/10.1145/2639108.2639139.

[25] M. Kotaru, K. Joshi, D. Bharadia, and S. Katti, "SpotFi: Decimeter level localization using WiFi," *SIGCOMM Computer Communication Review*, vol. 45, no. 4, pp. 269–282, Aug. 2015. [Online]. Available: http://doi.acm.org/10.1145/2829988.2787487.

[26] B. Bogert, "Demonstration of delay distortion correction by time-reversal techniques," *IRE Transactions on Communications Systems*, vol. 5, no. 3, pp. 2–7, Dec. 1957.

[27] J. Kormylo and V. Jain, "Two-pass recursive digital filter with zero phase shift," *IEEE Transactions on Acoustics, Speech, and Signal Processing*, vol. 22, no. 5, pp. 384–387, Oct. 1974.

[28] M. Fink and C. Prada, "Acoustic time-reversal mirrors," *Inverse Problems*, vol. 17, no. 1, Feb. 2001.

[29] M. Fink, C. Prada, F. Wu, and D. Cassereau, "Self focusing in inhomogeneous media with time reversal acoustic mirrors," in *Proceedings on the IEEE 1989 Ultrasonics Symposium,* pp. 681–686, vol. 2, Oct. 1989.

[30] C. Dorme, M. Fink, and C. Prada, "Focusing in transmit-receive mode through inhomogeneous media: The matched filter approach," in *Proceedings of the IEEE 1992 Ultrasonics Symposium,* pp. 629–634, vol. 1, Oct. 1992.

[31] F. Han, Y.-H. Yang, B. Wang, Y. Wu, and K. J. R. Liu, "Time-reversal division multiple access over multi-path channels," *IEEE Transactions on Communications*, vol. 60, no. 7, pp. 1953–1965, Jul. 2012.

[32] G. Stuber, J. Barry, S. McLaughlin, Y. Li, M. Ingram, and T. Pratt, "Broadband MIMO-OFDM wireless communications," *Proceedings of the IEEE*, vol. 92, no. 2, pp. 271–294, Feb. 2004.

[33] M. Speth, S. Fechtel, G. Fock, and H. Meyr, "Optimum receiver design for wireless broad-band systems using OFDM – Part I," *IEEE Transactions on Communications*, vol. 47, no. 11, pp. 1668–1677, Nov. 1999.

[34] C. Chen, Y. Chen, Y. Han, H.-Q. Lai, F. Zhang, and K. J. R. Liu, "Achieving centimeter-accuracy indoor localization on WiFi platforms: A multi-antenna approach," *IEEE Internet of Things Journal*, vol. 4, no. 1, pp. 122–134, 2017.

4 Frequency Hopping Approach

Indoor positioning systems (IPSs) are attracting more and more attention from academia and industry recently. Among them, approaches based on Wi-Fi techniques are more favorable because they are built upon the Wi-Fi infrastructures available in most indoor spaces. However, due to the bandwidth limit in mainstream Wi-Fi systems, the indoor positioning system leveraging Wi-Fi can hardly achieve centimeter localization accuracy under strong non-line-of-sight (NLOS) conditions, which is common for indoor environments. In this chapter, to achieve centimeter-level accuracy, we present a Wi-Fi-based indoor positioning system that exploits the frequency diversity via frequency hopping. In the offline phase, the system collects channel frequency responses (CFRs) from multiple channels and from a number of locations-of-interest. Then, the CFRs are postprocessed to mitigate the synchronization errors as well as interference from other Wi-Fi networks. Then, using bandwidth concatenation, the CFRs from multiple channels are combined into location fingerprints, which are stored into a local database. During the online phase, CFRs are formulated into the location fingerprint and is compared against the fingerprints in the database via the time-reversal resonating strength (TRRS). Finally, the IPS determines the location according to the TRRS. Extensive experiment results demonstrate a perfect centimeter-level accuracy in an office environment with strong NLOS using only one pair of single-antenna Wi-Fi devices.

4.1 Introduction

Global Positioning System (GPS) is an outdoor positioning system that provides real-time location information under all weather conditions near the Earth's surface, as long as there exists an unobstructed line-of-sight (LOS) from the device to at least four GPS satellites [1]. On the other hand, accurate indoor localization is highly desirable because nowadays people spend much more time indoors than outdoors. A high-accuracy indoor positioning system (IPS) can enable a wide variety of applications, e.g., providing museum guides to tourists by localizing their exact locations [2], or supplementing users with location information in shopping malls [3]. Unfortunately, the GPS signal can be too weak to be useful in indoor spaces due to the severe attenuation of obstacles as well as scattering in the presence of a large number of reflectors.

Many research efforts have been devoted to the development of accurate and robust IPSs. According to the technologies adopted, these IPSs can be further classified into two classes, i.e., ranging-based and fingerprint-based [4]. For the ranging-based methods, at least three anchors are deployed into the indoor environment to triangulate the device through measuring the relative distances between the device to the anchors. The distances are generally obtained from other measurements, e.g., received signal strength indicator (RSSI), time of arrival (ToA), time of flight (ToF), and angle of arrival (AoA). RSSI-based ranging methods [5–7] utilizes the path-loss model to derive the distance and can typically achieve an accuracy of $1 \sim 3$ m on average under LOS scenarios, while ToA-based ranging methods retrieve the ToA of the first arrived multipath component from the channel impulse response (CIR). To achieve a fine timing resolution, ToA-based methods require a large bandwidth, which is achievable with ultra-wideband (UWB) techniques that lead to an accuracy of $10 \sim 15$ cm in a LOS setting [8, 9]. In [10], Vasisht et al. present a decimeter-level localization using a single Wi-Fi access point. They utilize frequency hopping to acquire the channel frequency response (CFR), a fine-grained information that depicts the propagation of electromagnetic waves and thus portraits the environment with high granularity. Leveraging the non-uniform discrete Fourier Transform (NDFT), they recover the time-domain CIR and use the time delay of the dominant peak of the profile as the ToF measurement. However, in a strong NLOS environment, the dominant peak of CIR does not necessarily characterize the direct path between the Wi-Fi devices, which leads to an increased localization error. The AoA-based schemes proposed in [11] and [12] have the same issue that incurs accuracy degradation in a complicated NLOS indoor environment.

On the other hand, the fingerprint-based approaches harness the naturally existing spatial features associated with different locations, e.g., RSSI, CIR, and CFR. In these schemes, fingerprints of different locations are stored in a database during the offline phase. In the online phase, the fingerprint of the current location is compared against those in the database to estimate the device location. In [13–15], RSSI values from multiple access points (APs) are utilized as the fingerprint, leading to an accuracy of $2 \sim 5$ m. In [16], Wu et al. utilize multi-dimensional scaling to construct a stress-free floorplan as well as its associated fingerprint space containing the RSSI values obtained from locations on the stress-free floorplan for crowdsourcing-based indoor localization. The average error is around 2 m with the maximum error within 8 m. The accuracy can be further improved to $0.95 \sim 1.1$ m by taking CFRs as the fingerprint [17–19]. In [20], Wu et al. obtain CIR fingerprints under a bandwidth of 125 MHz and calculate the time-reversal (TR) resonating strength (TRRS) as the similarity measure among different locations, leading to an accuracy of $1 \sim 2$ cm under NLOS scenarios.

Summarizing the ranging-based and fingerprint-based schemes, we find that

(i) *The accuracy of the ranging-based methods are susceptible to the correctness of the physical rules, e.g., path-loss model, which degrades severely in the complex indoor environment.* The existence of a large number of multipath components and blockage of obstacles in indoor spaces impair the precision of the physical rules.

(ii) *The fingerprint-based methods, which can work under a strong NLOS environment, require a large bandwidth for accurate localization.* Because the maximum bandwidth of the mainstream 802.11n is 40 MHz, IPSs utilizing Wi-Fi techniques cannot resolve enough independent multipath components in the environment, which introduces ambiguities into fingerprints associated with different locations. Thus, the localization performance is degraded. On the other hand, a bandwidth as large as 125 MHz that leads to centimeter accuracy [20] can only be achieved on dedicated hardware and incurs additional costs in deployment.

Is there any approach that can achieve the centimeter localization accuracy using Wi-Fi devices in an NLOS environment? The answer is affirmative. In [21], Chen et al. present an IPS that achieves centimeter accuracy using one pair of single-antenna Wi-Fi devices under strong NLOS conditions using frequency hopping. The IPS obtains CFRs and formulates location fingerprints from multiple Wi-Fi channels in the offline phase and calculates TRRS for localization in the online phase. However, interference from other Wi-Fi networks might corrupt the fingerprint, which is neglected in [21]. To deal with the interference, in this chapter, we introduce an additional step of CFR sifting. Moreover, we utilize CFR averaging to mitigate the impact of channel noise and refine the fingerprint. Additionally, we provide much more detail and analysis on the experiment results. In comparison with most of the existing works that dedicate to mitigate the impact of multipath propagation, the method discussed in this chapter embraces the multipath effect. Moreover, it is infrastructure-free because it is built upon the Wi-Fi networks available in most indoor spaces.

The major points of this chapter can be summarized as follows:

- We develop an IPS that achieves centimeter accuracy in an NLOS environment with one pair of single-antenna Wi-Fi devices. The presented IPS eliminates the impact of interference from other Wi-Fi networks through the process of CFR sifting.
- Leveraging the frequency diversity, we demonstrate that a large effective bandwidth can be achieved on Wi-Fi devices by means of frequency hopping to overcome the issue of location ambiguity issue on traditional Wi-Fi-based approaches.
- We conduct extensive experiments in a typical office environment to show the centimeter accuracy within an area of 20 cm $\times$ 70 cm under strong NLOS conditions.

The rest of the chapter is organized as follows. In Section 4.2, we introduce the TR technique and the channel estimation in Wi-Fi systems. In Section 4.3, we elaborate on the presented localization algorithm. In Section 4.4, we present the frequency hopping mechanism. In Section 4.5, we demonstrate the experiment results in a typical office environment. In Section 4.6, we present some discussions on several aspects of the IPS.

4.2 Preliminaries

In this part, we introduce the background of the TR technique and the channel estimation schemes in Wi-Fi systems.

4.2.1 Time-Reversal

TR is a signal processing technique capable of mitigating the phase distortion of a signal passing a linear time-invariant (LTI) filtering system. It is based upon the fact that when the LTI system $h(t)$ is concatenated with its time-reversed and conjugated version $h^*(-t)$, the phase distortion is completely canceled out at a particular time instance.

A physical medium can be regarded as LTI if it satisfies inhomogeneity and invertibility. When both conditions hold, TR focuses the signal energy at a specific time and at a particular location, known as the spatial-temporal focusing effect. Such focusing effect is observed experimentally in the fields of ultrasonics, and acoustics, as well as electromagnetism [22–25]. Leveraging the focusing effect, TR is applied successfully to the broadband wireless communication systems [26].

The architecture of the TR communication system consisting of two phases, namely, channel probing phase and transmission phase, the details of which can be referred to Chapter 1. In virtue of the high-resolution TR focusing effect, in this chapter, we utilize TR as the signal processing technique to measure the similarity among fingerprints of different locations.

4.2.2 Channel Estimation in Wi-Fi Systems

In a Wi-Fi system adopting the orthogonal frequency-division multiplexing (OFDM) scheme, the transmitted data symbols are spread onto multiple subcarriers to combat against the frequency-selective fading incurred by the multipath effect. Assuming a total of K usable subcarriers and denoting the transmitted frequency domain data symbol on the kth subcarrier with index u_k as X_{u_k}, the received frequency domain signal on subcarrier u_k, denoted by Y_{u_k}, takes the form as [27]

$$Y_{u_k} = H_{u_k} X_{u_k} + W_{u_k}, \quad k = 1, 2, \ldots, K, \tag{4.1}$$

where H_{u_k} is the CFR on subcarrier u_k and W_{u_k} is the complex Gaussian noise on subcarrier u_k. The estimation of H_{u_k} in the least-square sense takes the form

$$\hat{H}_{u_k} = \frac{Y_{u_k}}{X_{u_k}} = H_{u_k} + W'_{u_k}, \quad k = 1, 2, \ldots, K \,, \tag{4.2}$$

where $W'_{u_k} = \frac{W_{u_k}}{X_{u_k}}$ given a priori knowledge of X_{u_k}.

Equation (4.2) is only valid in the absence of synchronization errors, which cannot be neglected in practice. The synchronization errors mainly consist of (i) channel frequency offset (CFO) (ii) sampling frequency offset (SFO), and (iii) symbol timing offset (STO). The CFO, denoted as Δf, is caused by the misalignment of the local oscillators at the transmitter and receiver. Given N samples per OFDM block and a sampling interval of T_s, the normalized CFO ϵ can be written as $\Delta f N T_s$. The SFO, denoted as η, is introduced by the mismatch between the sampling interval at the transmitter and that at the receiver. Given a sampling interval of T_s at the transmitter side and a sampling interval of T_s' at the receiver side, η can be expressed as $(T_s' - T_s)/T_s$. The STO, denoted as Δn_0, is caused by the imperfect timing synchronization at the receiver. These synchronization errors introduce additional phase rotations as well as amplitude attenuation into $\hat{H}_{u_k}$. Although the Wi-Fi receivers perform timing and frequency synchronizations, the residual of these errors cannot be neglected.

Denote the estimated CFR associated with the ith received OFDM symbol on the kth subcarrier as $\hat{H}_i^{u_k}$. In the presence of the residual synchronization errors, $\hat{H}_i^{u_k}$ can be modified from (4.2) into [28]

$$\hat{H}_i^{u_k} = \text{sinc}(\pi(\Delta\epsilon + \Delta\eta u_k))H_{u_k}e^{j2\pi(\beta_i u_k + \alpha_i)} + W'_{i,u_k} \tag{4.3}$$

for $k = 1, 2, \ldots, K$, where

$$\alpha_i = \left(\frac{1}{2} + \frac{iN_s + N_g}{N}\right)\Delta\epsilon \tag{4.4}$$

$$\beta_i = \frac{\Delta n_0}{N} + \left(\frac{1}{2} + \frac{iN_s + N_g}{N}\right)\Delta\eta \tag{4.5}$$

are the initial and linear phase distortions respectively. Here, $\Delta\epsilon$ and $\Delta\eta$ represent the residual errors of the normalized CFO ϵ and the normalized SFO η, respectively. $\text{sinc}(\pi(\Delta\epsilon + \Delta\eta u_k))$ is the amplitude attenuation and can be approximated as 1 given typical values of $\Delta\epsilon$ and $\Delta\eta$. N_g is the length of the cyclic prefix, N_s is the total length of one OFDM frame with length $N + N_g$, and W'_{i,u_k} is the estimation noise on subcarrier u_k for the ith OFDM symbol that can be modeled as complex Gaussian noise [29].

In practice, preambles are utilized at the Wi-Fi receiver to assist synchronization and channel estimation. Figure 4.1 demonstrates the physical layer (PHY) frame structure of 802.11a [30]. Before the transmission of the data payloads, the Wi-Fi transmitter sends preambles composed by short training preambles (STPs), long training preambles (LTPs), and cyclic prefix. The Wi-Fi receiver performs timing and frequency synchronization using the STPs and then compensates the synchronization errors. Because the receiver has the full knowledge of the OFDM symbols of the LTPs, it performs channel estimation based on the LTPs to extract the CFRs, which leads to $\hat{H}_i^{u_k}$ as shown in (4.3).

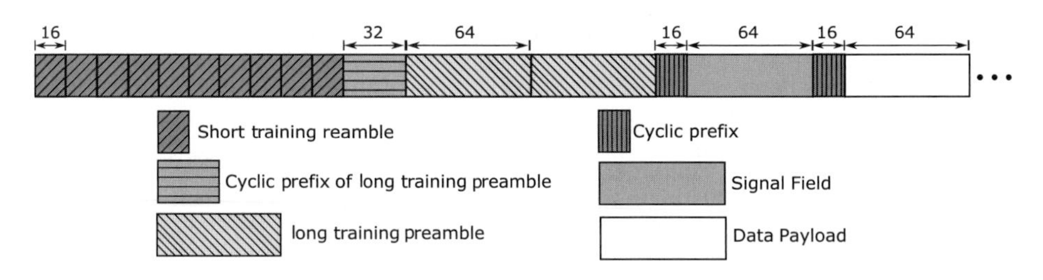

Figure 4.1 The frame structure for 802.11a.

4.3 Algorithm Design

4.3.1 Calculation of TRRS in Frequency Domain

In the presented IPS, the similarity between two locations is measured by the TRRS between their fingerprints. In this section, we provide details of TRRS computation.

Given two time-domain CIRs $\hat{\mathbf{h}}$ and $\hat{\mathbf{h}}'$, with $\hat{\mathbf{h}} = [\hat{h}[0], \hat{h}[1], \dots, \hat{h}[L-1]]^T$ and $\hat{\mathbf{h}}'$ defined similarly, where T is the transpose operator, the TRRS between $\hat{\mathbf{h}}$ and $\hat{\mathbf{h}}'$ is calculated as [20]

$$\gamma_{\mathsf{CIR}}[\hat{\mathbf{h}}, \hat{\mathbf{h}}'] = \frac{\max_i \left| \left(\hat{\mathbf{h}} * \hat{\mathbf{g}} \right) [i] \right|^2}{\langle \hat{\mathbf{h}}, \hat{\mathbf{h}} \rangle \langle \hat{\mathbf{g}}, \hat{\mathbf{g}} \rangle}, \tag{4.6}$$

where $*$ denotes the convolution operator, $\hat{\mathbf{g}}$ is the time-reversed and conjugate version of $\hat{\mathbf{h}}'$, and $\langle \mathbf{x}, \mathbf{y} \rangle$ is the inner product operator between complex vector $\mathbf{x}$ and complex vector $\mathbf{y}$, expressed by $\mathbf{x}^\dagger \mathbf{y}$ where $(\cdot)^\dagger$ is the Hermitian operator. Notice that, the computation of $\gamma_{\mathsf{CIR}}[\hat{\mathbf{h}}, \hat{\mathbf{h}}']$ removes the impact of STO by searching all possible index i across the output of $\left| \left(\hat{\mathbf{h}} * \hat{\mathbf{g}} \right) [i] \right|$. It can be shown that $0 \le \gamma_{\mathsf{CIR}}[\hat{\mathbf{h}}, \hat{\mathbf{h}}'] \le 1$.

Because the convolution in the time domain can be cast to the inner product in the frequency domain [31], the TRRS can be calculated using CFRs, the frequency-domain counterparts of CIRs. Given two CFRs $\hat{\mathbf{H}} = [\hat{H}_{u_1}, \hat{H}_{u_2}, \dots, \hat{H}_{u_K}]^T$ and $\hat{\mathbf{H}}'$ defined similarly and assume that the synchronization errors are mostly mitigated, the TRRS in frequency domain is given by

$$\gamma[\hat{\mathbf{H}}, \hat{\mathbf{H}}'] = \frac{\left| \sum_{k=1}^{K} \hat{H}_{u_k} \hat{H}'_{u_k} \right|^2}{\langle \hat{\mathbf{H}}, \hat{\mathbf{H}} \rangle \langle \hat{\mathbf{H}}', \hat{\mathbf{H}}' \rangle}. \tag{4.7}$$

It is straightforward to prove that $0 \le \gamma[\hat{\mathbf{H}}, \hat{\mathbf{H}}'] \le 1$, and $\gamma[\hat{\mathbf{H}}, \hat{\mathbf{H}}'] = 1$ if and only if $\hat{\mathbf{H}} = C\hat{\mathbf{H}}'$ where $C \neq 0$ is any complex scaling factor. Therefore, the TRRS can be regarded as a measure of similarity between two CFRs.

4.3.2 Indoor Localization Based on TRRS

The presented localization algorithm consists of an offline phase and an online phase. The details of the two phases are illustrated in Figure 4.2 and are elaborated next.

4.3.2.1 Offline Phase

In the offline phase, the CFRs are measured at D channels, denoted by $f_1, f_2, \dots, f_d, \dots, f_D$, and at L locations-of-interest, denoted by $1, 2, \dots, \ell, \dots, L$. Assuming that a total of N_{ℓ, f_d} CFRs are measured from the first and second LTPs at location ℓ and channel f_d, we write the CFR matrix as

$$\hat{\mathbb{H}}_i \left[\ell, f_d \right] = \left[\hat{\mathbf{H}}_{i,1}[\ell, f_d] \cdots \hat{\mathbf{H}}_{i,m}[\ell, f_d] \cdots \hat{\mathbf{H}}_{i,N_{\ell,f_d}}[\ell, f_d] \right], \tag{4.8}$$

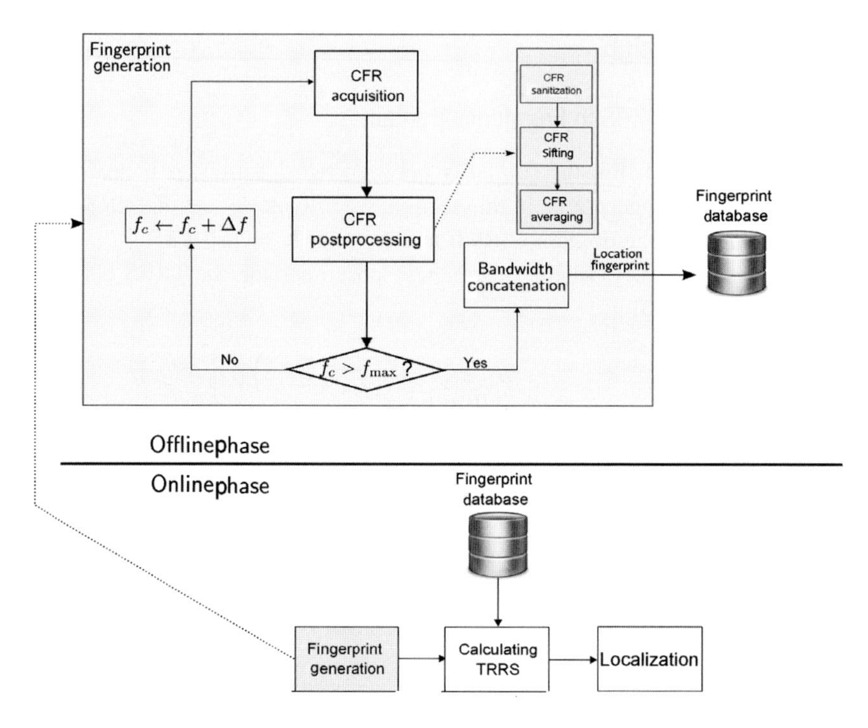

Figure 4.2 Flowchart of the algorithm.

where $m = 1,2,\ldots,N_{\ell,f_d}$ is the realization index, $i \in \{1,2\}$ is the LTP index, and $\hat{\mathbf{H}}_{i,m}[\ell, f_d] = [\hat{H}^{u_1}_{i,m}[\ell, f_d] \cdots \hat{H}^{u_k}_{i,m}[\ell, f_d] \cdots \hat{H}^{u_K}_{i,m}[\ell, f_d]]^T$ with $\hat{H}^{u_1}_{i,m}[\ell, f_d]$ standing for the mth CFR of the ith LTP on subcarrier u_k, and at location ℓ, channel f_d.

The location fingerprint is generated from $\hat{\mathbb{H}}_i\left[\ell, f_d\right]$. The process contains four steps, which are presented next.

1. **CFR Sanitization**
   The captured CFRs must be sanitized to mitigate the impact of initial and linear phase distortions shown in (4.3). First of all, we estimate the residual CFO and SFO from the channel estimation using [32]

$$\Omega^{u_k}_m[\ell, f_d] = \left[\hat{H}^{u_k}_{1,m}[\ell, f_d]\right]^* \times \hat{H}^{u_k}_{2,m}[\ell, f_d]$$
$$= e^{j2\pi \frac{N_s}{N}\phi_{u_k}} |H^{u_k}_{1,m}[\ell, f_d]|^2 + \psi^{u_k}_m[\ell, f_d], \tag{4.9}$$

   where $\phi_k = \Delta\epsilon + \Delta\eta k$ and $\psi^{u_k}_m[\ell, f_d]$ contains all cross terms. Therefore, ϕ_{u_k} can be estimated by

$$\hat{\phi}_{u_k} = \measuredangle\left[\Omega^{u_k}_m[\ell, f_d]\right], \tag{4.10}$$

   where $\measuredangle[X]$ is the angle of X measured in radians. Compensating $\hat{\phi}_{u_k}$ gives

$$\tilde{H}^{u_k}_{i,m}[\ell, f_d] = \hat{H}^{u_k}_{i,m}[\ell, f_d] e^{-j\pi\hat{\phi}_{u_k}} e^{-j2\pi\frac{N_g+(i-1)N_s}{N}\hat{\phi}_{u_k}} \tag{4.11}$$

Substituting (4.11) into (4.8) and writing the updated $\hat{\mathbb{H}}_i\left[\ell, f_d\right]$ in (4.8) as $\tilde{\mathbb{H}}_i\left[\ell, f_d\right]$, we take the average of $\tilde{\mathbb{H}}_1\left[\ell, f_d\right]$ and $\tilde{\mathbb{H}}_2\left[\ell, f_d\right]$ as $\tilde{\mathbb{H}}\left[\ell, f_d\right] = \left(\tilde{\mathbb{H}}_1\left[\ell, f_d\right] + \tilde{\mathbb{H}}_2\left[\ell, f_d\right]\right)/2$.

After the removal of residual CFO and SFO, the STO still remains to be compensated. Write

$$\tilde{\mathbb{H}}\left[\ell, f_d\right] = \left[\tilde{\mathbf{H}}_1[\ell, f_d] \cdots \tilde{\mathbf{H}}_m[\ell, f_d] \cdots \tilde{\mathbf{H}}_{N_{\ell, f_d}}[\ell, f_d]\right], \tag{4.12}$$

where $\tilde{\mathbf{H}}_m[\ell, f_d] = [\tilde{H}_m^{u_1}[\ell, f_d] \cdots \tilde{H}_m^{u_k}[\ell, f_d] \cdots \tilde{H}_m^{u_K}[\ell, f_d]]^T$ is the CFR vector for the mth realization on usable subcarriers after CFO/SFO correction. Denoting $A_m^{u_k}[\ell, f_d] = \measuredangle\left\{\tilde{H}_m^{u_k}[\ell, f_d]\right\}$ as the angle of $\tilde{H}_m^{u_k}[\ell, f_d]$, we perform phase unwrapping on $A_m^{u_k}[\ell, f_d]$ to yield $A_m^{'u_k}[\ell, f_d]$. The slope of $A_m^{'u_k}[\ell, f_d]$ is linear with STO if we disregard the noise and interference. To estimate the slope, we perform a least-square fitting on $A_m^{'u_k}[\ell, f_d]$ expressed by

$$\widehat{\Delta n_0} = \frac{N \sum_{k=1}^{K} [(u_k - \overline{u})]\left[A_m^{'u_k}[\ell, f_d] - \overline{A}\right]}{2\pi \sum_{k=1}^{K} [u_k - \overline{u}]^2}, \tag{4.13}$$

where $\overline{u} = \frac{\sum_{k=1}^{K} u_k}{K}$ and $\overline{A} = \frac{\sum_{k=1}^{K} A_m^{'u_k}[\ell, f_d]}{K}$. Therefore, $\tilde{H}_m^{u_k}[\ell, f_d]$ is compensated as

$$\check{H}_m^{u_k}[\ell, f_d] = \tilde{H}_m^{u_k}[\ell, f_d] e^{-j u_k \mathrm{round}(\widehat{\Delta n_0}) \frac{2\pi}{N}}, \tag{4.14}$$

where round(x) rounds the argument x to its nearest integer. The compensated CFR matrix is denoted by

$$\check{\mathbb{H}}\left[\ell, f_d\right] = \left[\check{\mathbf{H}}_1[\ell, f_d] \cdots \check{\mathbf{H}}_m[\ell, f_d] \cdots \check{\mathbf{H}}_{N_{\ell, f_d}}[\ell, f_d]\right]. \tag{4.15}$$

2. **CFR Sifting**

Due to the presence of other Wi-Fi devices in the environment, some CFR measurements might suffer from interference from nearby Wi-Fi devices or other radio-frequency systems such as Bluetooth and should be excluded from further calculations. The interference introduces random noise onto the CFRs and impairs the CFR qualities. To combat the interference, first, we use $\check{\mathbf{H}}_m[\ell, f_d]$ to calculate the $N_{\ell, f_d} \times N_{\ell, f_d}$ TRRS matrix $\mathbb{R}_{\ell, f_d}$, where $\check{\mathbf{H}}_m[\ell, f_d] = \left[\check{H}_m^{u_1}[\ell, f_d] \cdots \check{H}_m^{u_k}[\ell, f_d] \cdots \check{H}_m^{u_K}[\ell, f_d]\right]^T$ with $\gamma[\cdot, \cdot]$ defined in (4.7). The (i, j)th entry of $\mathbb{R}_{\ell, f_d}$ is

$$\left[\mathbb{R}_{\ell, f_d}\right]_{i,j} = \gamma\left[\check{\mathbf{H}}_i[\ell, f_d], \quad \check{\mathbf{H}}_j[\ell, f_d]\right]. \tag{4.16}$$

Second, we compute the column-wise average of $\mathbb{R}_{\ell, f_d}$ denoted as O_j with $j = 1, 2, \ldots, N_{\ell, f_d}$, given by

$$O_j = \frac{1}{N_{\ell, f_d} - 1} \sum_{\substack{i=1,2,\ldots,N_{\ell, f_d} \\ i \neq j}} \left[\mathbb{R}_{\ell, f_d}\right]_{i,j}. \tag{4.17}$$

Finally, we remove the j'th column of $\check{\mathbb{H}}\left[\ell, f_d\right]$ if $O_{j'} \leq \tau$, where τ is a threshold.

We assume that the number of remaining CFRs after CFR sifting is N'_{ℓ,f_d}, and the corresponding index of the remaining CFRs are $t_1, \ldots, t_m, \ldots, t_{N'_{\ell,f_d}}$.

3. **CFR Averaging**
At location ℓ, for channel f_d, we generate the averaged CFR $\mathbf{S}\left[\ell, f_d\right] = [S^{u_1}_{\ell,f_d} \cdots S^{u_k}_{\ell,f_d} \cdots S^{u_K}_{\ell,f_d}]^T$ with dimension $K \times 1$ as

$$\mathbf{S}\left[\ell, f_d\right] = \frac{1}{N'_{\ell,f_d}} \sum_{m=1}^{N'_{\ell,f_d}} \check{\mathbf{H}}_{t_m}[\ell, f_d] \cdot \mathbf{W}_m , \tag{4.18}$$

where $\cdot$ stands for the element-wise dot product between two vectors. $\mathbf{W}_m$ is a $K \times 1$ vector represented as

$$\mathbf{W}_m = \left[w_m[\ell, f_d] \quad w_m[\ell, f_d] \quad \cdots \quad w_m[\ell, f_d]\right]^T , \tag{4.19}$$

where $w_m[\ell, f_d] = e^{j\left(\measuredangle\left[\check{H}^{u_1}_{t_1}[\ell, f_d]\right] - \measuredangle\left[\check{H}^{u_1}_{t_m}[\ell, f_d]\right]\right)}$. The purpose of introducing $\mathbf{W}_m$ is to match the initial phases of $\check{\mathbf{H}}_{t_m}[\ell, f_d]$ with $m > 1$ to the first realization $\check{\mathbf{H}}_{t_1}[\ell, f_d]$, so that $\check{\mathbf{H}}_{t_m}[\ell, f_d]$ can be accumulated coherently, and the noise variance contained in $\check{\mathbf{H}}_{t_m}[\ell, f_d]$ is reduced by N'_{ℓ,f_d} times consequently.

4. **Bandwidth Concatenation**
At location ℓ, we obtain the fingerprint vector with dimension $DK \times 1$ by concatenating the averaged CFRs from all channels $\{f_d\}_{d=1,2,\ldots,D}$ as

$$\mathbf{G}[\ell] = \left[\mathbf{S}^T\left[\ell, f_1\right] V_1 \cdots \mathbf{S}^T\left[\ell, f_d\right] V_d \cdots \mathbf{S}^T\left[\ell, f_D\right] V_D\right]^T , \tag{4.20}$$

where $V_d = e^{-j\measuredangle\left[S^{u_1}_{\ell,f_d}\right]}$ is introduced to nullify the initial phases of different $\mathbf{S}^T\left[\ell, f_d\right]$.

Figure 4.3 demonstrates an example of the fingerprint generation procedure. As can be observed from Figure 4.3, the CFR postprocessing effectively removes the phase distortions caused by the synchronization errors. The CFR averaging combines different realizations coherently, and the bandwidth concatenation associates the two averaged CFRs into the location fingerprint.

Because we concatenate all available bandwidths from D channels, we achieve a much larger effective bandwidth denoted by $W_e = DW$, where W is the bandwidth per channel.

4.3.2.2 Online Phase

The CFRs from an unknown location are formulated into the location fingerprint in the same manner as described in the offline phase. Assuming that the location fingerprint from the unknown location ℓ' is given by $\mathbf{G}[\ell']$, the TRRS between location ℓ' and location ℓ is computed as $\gamma\left[\mathbf{G}[\ell], \mathbf{G}[\ell']\right]$. Defining $\ell^\star = \underset{\ell=1,2,\ldots,L}{\text{argmax}}\ \gamma\left[\mathbf{G}[\ell], \mathbf{G}[\ell']\right]$, the estimated location $\hat{\ell}'$ takes the form

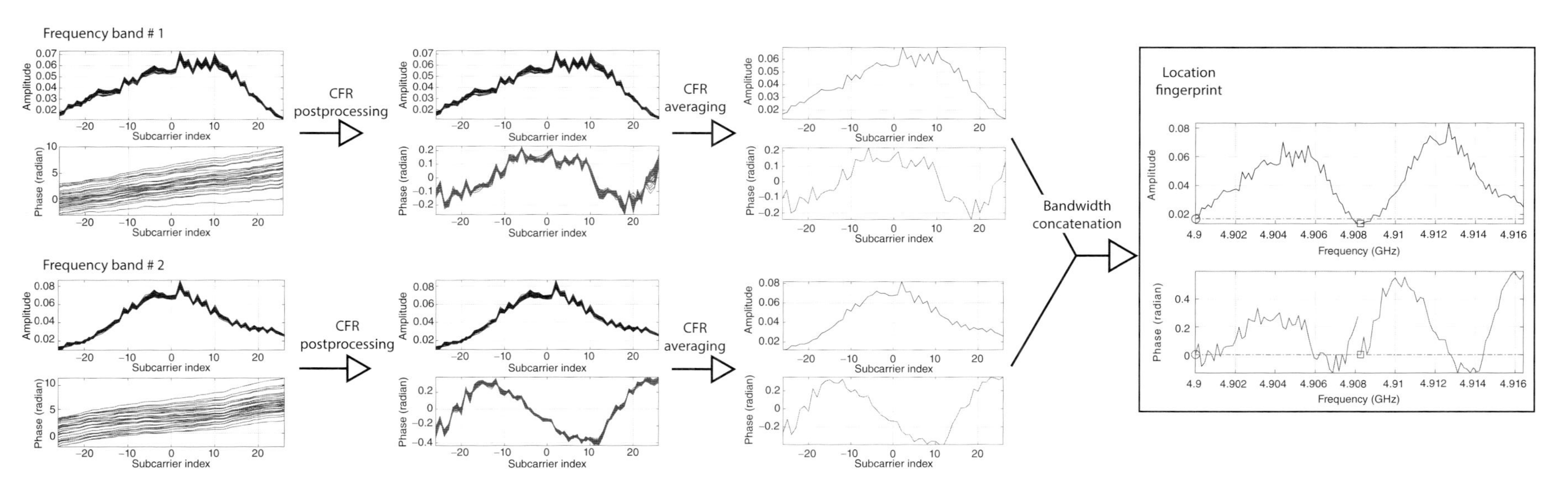

Figure 4.3 An example of CFR postprocessing, channel fingerprint generation, and location fingerprint generation.

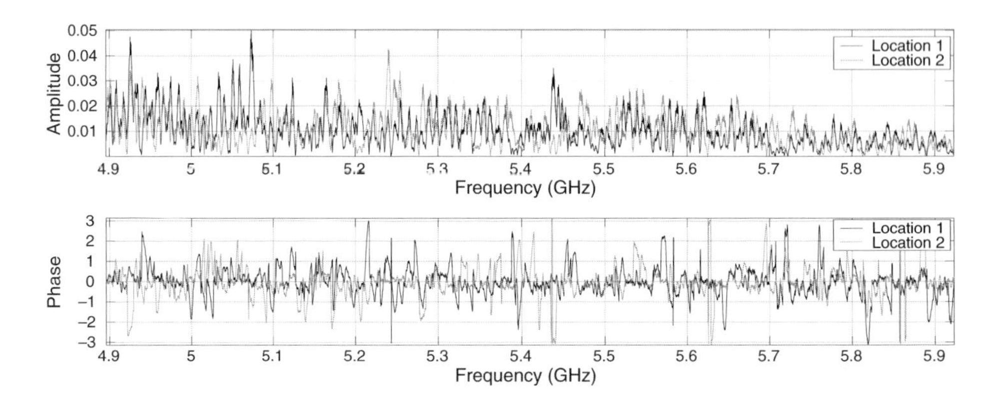

Figure 4.4 A snapshot of location fingerprints after bandwidth concatenation at two different locations.

$$\hat{\ell}' = \begin{cases} \ell^\star, & \text{if } \gamma\left[\mathbf{G}[\ell^\star], \mathbf{G}[\ell']\right] \geq \Gamma \\ 0, & \text{Otherwise}, \end{cases} \tag{4.21}$$

where Γ is a tunable threshold. Notice that, in case of $\gamma\left[\mathbf{G}[\ell^\star], \mathbf{G}[\ell']\right] < \Gamma$, the presented IPS fails to localize the device, and the algorithm returns 0 to imply an *unknown location*.

In Figure 4.4, we show an example of location fingerprints generated at two different locations. For each location, we formulate five location fingerprints. As we can see, the differences among the location fingerprints at the same location are minor, while the differences of location fingerprints between the two different locations are much more pronounced.

4.4 Frequency Hopping Mechanism

In this section, we elaborate on the implementation details of the presented IPS.

4.4.1 CFR Acquisition Using USRPs

We build two Universal Software Radio Peripherals (USRP) N210 [33] into prototypes for localization. Each USRP is equipped with one omnidirectional antenna.

In [34], Bastian et al. develop a Wi-Fi transceiver supporting Wi-Fi standards 802.11a/g/p under the framework of GNU Radio [35]. The proposed WiFi transceiver in [34] extracts the CFRs by the four frequency-domain subcarrier pilots followed by an interpolation to fully recover the CFRs on the 48 usable data subcarriers. However, due to the scarcity of the subcarrier pilots, the estimated CFRs are not accurate enough to provide fine details about the environment to facilitate indoor localization.

To acquire CFRs with high quality, we extend the framework in [34] by including a channel estimator leveraging the two LTPs as shown in the Wi-Fi frame structure in

Figure 4.1. Each LTP is composed by 56 data subcarriers which are known in advance at the receiver side, and the CFRs are extracted using (4.2) in Section 4.2. To reduce the impact of synchronization errors on the CFRs, we estimate and compensate the STO, SFO, and CFO using the STPs as shown in Figure 4.1. The estimated and compensated CFRs are used to equalize the signal field frame, which contains the information of the coding rate as well as the signal constellation of the transmitted OFDM symbols. Then, the receiver decodes the data payloads based on this information.

We also notice that the framework in [34] lacks the mechanism of carrier sense multiple access (CSMA). Therefore, interference from other Wi-Fi devices cannot be avoided. In light of this issue, we only keep those CFRs associated with the data payloads that could be successfully decoded.

4.4.2 Implementing the Frequency Hopping Mechanism

In the presented IPS, frequency hopping is used to acquire CFRs from a multitude of frequency bands. In Figure 4.5, we demonstrate the timing diagram of the mechanism of synchronous frequency hopping with feedback between two devices from the center frequency f_0 to f_1. Here, ACK stands for the acknowledgment frame, and REQ denotes the frequency hopping request frame. Device 2 initializes the procedure by tuning its center frequency at f_0. Then, device 1 starts transmission at f_0 as well to facilitate CFR acquisition on device 2. Assume that the minimum number of CFRs per frequency band is M_{min}. After obtaining M_{min} CFRs at f_0, device 2 sends an ACK frame to device 1, and device 1 feedbacks a REQ frame to device 2. On reception of the REQ frame, device 2 adjusts its center frequency to f_1, and device 1 begins transmission at f_1.

In Figure 4.5, we assume that the two devices perform full-duplex communication, i.e., transmitting signals while listening simultaneously to acquire the ACK and REQ frames. However, in practice, the USRP N210 devices in the presented IPS are half-duplex, i.e., one device cannot perform Wi-Fi transmitting and receiving at the same time. Thus, each device needs to switch between the transmitting mode and the receiving mode in different time slots. Figure 4.6 shows an example of frequency hopping from f_0 to f_1. The details for each time-of-interest denoted as $t_1, t_2, \ldots, t_{12}$ in Figure 4.6 are presented here.

t_0**:** Device 2 (D2) tunes its center frequency to f_0 and stays in the receiving mode.

t_1**:** Device 1 (D1) tunes its center frequency to f_0 and begins data transmission. D2 detects the presence of data transmission and performs channel estimation to extract CFRs from each data frame. Device D2 stays in the receiving mode until the number of CFRs exceeds M_{min}.

t_2**:** D1 switches to receiving mode to determine whether D2 sends an acknowledgment signal (ACK) by encoding the message in the data payloads. Suppose that at this moment, D2 obtains $M' < M_{\text{min}}$ CFRs. Because the number of CFRs is insufficient, D2 still stays in the receiver mode. Notice that, if D2 acquires sufficient CFRs in this stage, D2 would switch to the transmitter mode and send an ACK frame to D1, and the procedure would continue from t_7.

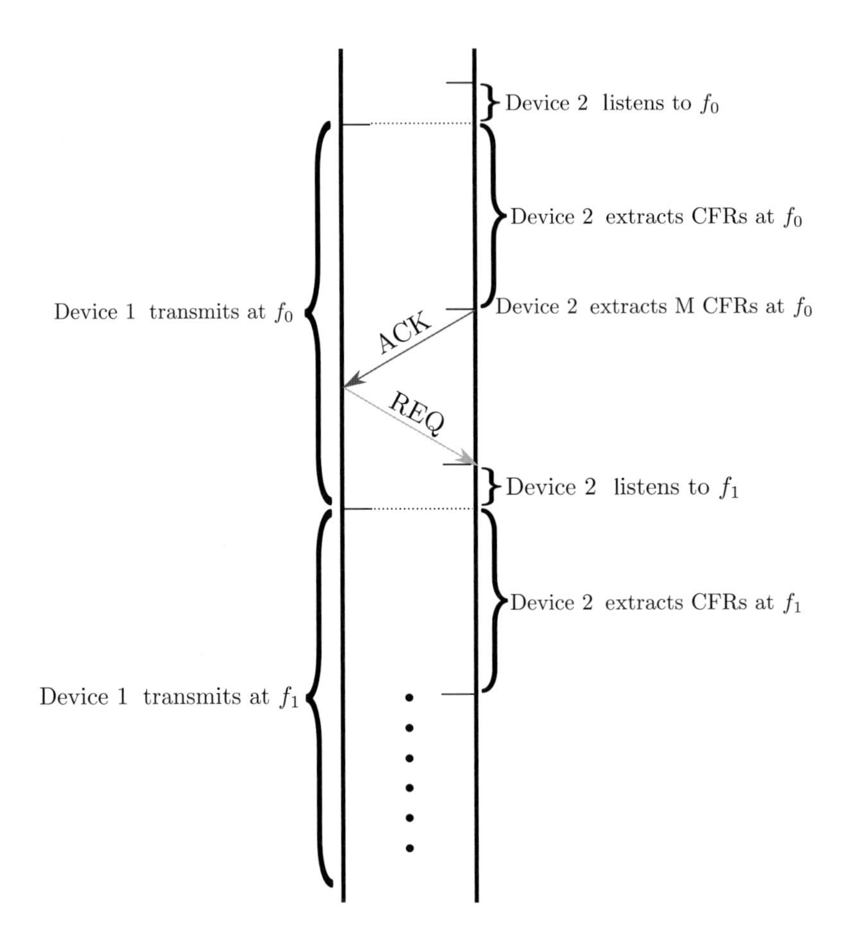

Figure 4.5 Timing diagram of the frequency hopping mechanism.

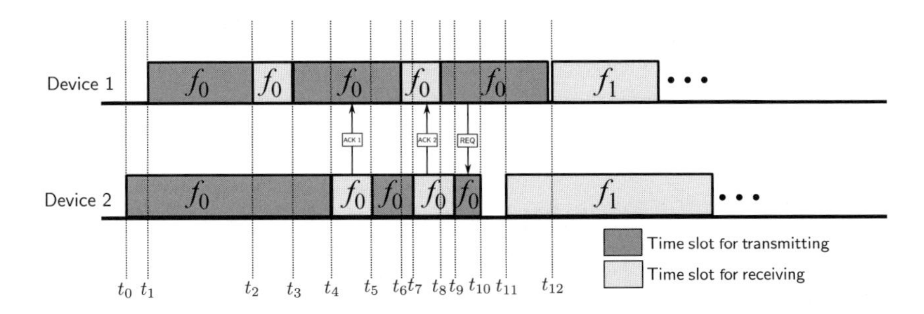

Figure 4.6 The timing diagram for frequency hopping.

t_3**:** D1 does not receive the ACK frame from D2 and thus switches back to the transmitter mode and continues data transmission.

t_4**:** D2 receives the targeted $M_{\min}$ CFRs and switches to the transmitter mode. It then transmits an ACK signal to D1. Nevertheless, because D1 is in transmitter mode, the ACK signal transmission fails.

t_5: D2 switches to receiver mode to decide whether D1 sends a frequency hopping request (REQ), which is encoded into the data payloads. Due to the failure of the ACK signal transmission at t_4, D1 is unable to send the REQ signal.

t_6: D1 switches to the receiver mode again.

t_7: D2 switches to the transmitter mode again and sends out another ACK signal.

t_8: D1 receives the ACK signal and switches to the transmitter mode to send out an REQ. However, Device 2 is still in the transmitter mode and cannot receive the request at this moment.

t_9: D2 switches to the receiver mode and receives the REQ signal because D1 stays in the transmitter mode.

t_{10}: D2 begins the process of tuning its center frequency to f_1.

t_{11}: D2 successfully tunes its center frequency to f_1 and awaits the transmission from D1 at f_1 as well. Because D1 is still transmitting using f_0, D2 is unable to decode the signal.

t_{12}: D1 also tunes its center frequency to f_1 and begins transmission.

The same protocol is repeated until CFRs from all desirable frequency bands are measured.

4.5 Experiment Results

4.5.1 Experiment Settings

Figure 4.7 shows the setups of the experiments with details given here.

4.5.1.1 Environment

The experiments are conducted in a typical office suite composed by a large and a small office room in a multistory building. The two office rooms are blocked by a wall.

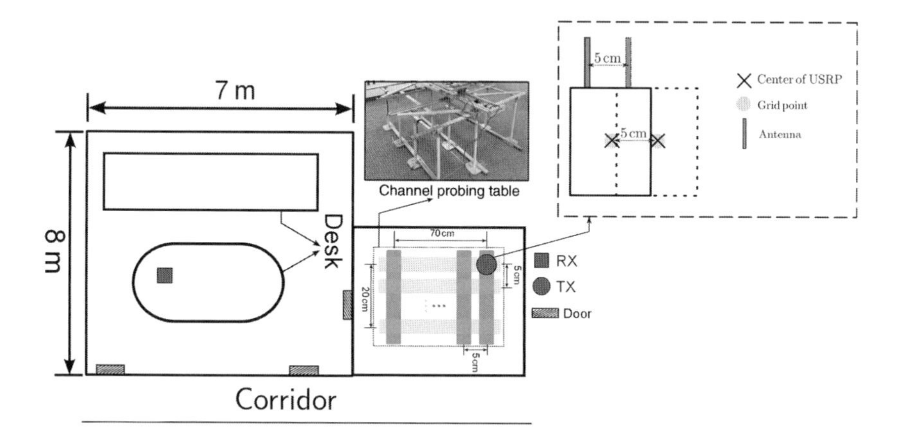

Figure 4.7 Experiment settings.

In addition to the two large desks, the indoor space is filled with other furniture including chairs and computers, which are not shown in Figure 4.7 for brevity.

4.5.1.2 Configurations

Two USRPs are used to obtain the CFRs with bandwidth configured as $W = 10$ MHz. The two USRPs coordinate with each other to perform synchronous frequency hopping using the mechanism discussed in Section 4.4. The step size of frequency hopping is fixed at $\Delta f = 8.28$ MHz.[1] The minimum number of CFRs per channel is set as $M_{\min} = 10$.

4.5.1.3 Details of Measurement

One USRP is placed on the grid points on a measurement platform in the small room as shown in Figure 4.7. The center of the USRP is aligned with the grid point. The distance between two adjacent grid points is 5 cm. The other USRP is placed on the table of the larger room. CFRs from $L = 75$ different grid points are measured within an area of 70 cm $\times$ 20 cm. For each measurement, the two USRPs sweep the frequency band from 4.9 to 5.9 GHz, leading to a total of $D = 124$ times of frequency hopping with a step size of 8.28 MHz. The effective bandwidth W_e is thus 1 GHz. For each of the 75 locations, we formulate $M = 10$ location fingerprints.

4.5.2 Metrics for Performance Evaluation

For the $M = 10$ fingerprints collected at each location, we store the first $M_1 = 5$ CFRs into the fingerprint database in the offline phase, and consider the other $M_2 = 5$ fingerprints as samples collected in the online phase. Denoting the mth location fingerprint formulated at location ℓ as $\mathbf{G}_m[\ell]$, we calculate the TRRS matrix $\mathbb{R}$ with the (i, j)th entry of $\mathbb{R}$ given by $\gamma[\mathbf{G}_m[\ell], \mathbf{G}_n[\ell']]$, where $m = \mathsf{Mod}(i, M_1) + 1$, $\ell = \frac{i-m-1}{M_1} + 1$, $n = \mathsf{Mod}(j, M_2) + 1$, and $\ell' = \frac{j-n-1}{M_2} + 1$. Here, Mod is the modulus operator, i is termed as the training index, and j is termed as the testing index.

We define the entries of $\mathbb{R}$ calculated from CFRs obtained at the same locations as the *diagonal entries* and those calculated using CFRs obtained from different locations as the *off-diagonal entries*. We demonstrate the histograms and cumulative density functions for the diagonal and off-diagonal entries.

Based on $\mathbb{R}$, we evaluate the localization performances using the metrics of the true-positive rate, denoted as P_{TP}, and the false-positive rate, denoted as P_{FP}. P_{TP} is defined as the probability that the IPS localizes the device to its correct location, while P_{FP} captures the probability that the IPS localizes the device to a wrong location, or fails to localize the device.

In the performance evaluation, the CFR sifting parameter τ is set as 0.8.

[1] Considering the null subcarriers at both edges of the Wi-Fi channel spectrum, we adjust the frequency hopping step size such that the entire spectrum can be covered without spectrum holes. Notice that the presented IPS does not require the measured frequency band to be contiguous.

4.5.3 Performance Evaluation

4.5.3.1 TRRS Matrix under Different W_e

Figure 4.8 demonstrates $\mathbb{R}$ with $W_e \in [10, 40, 120, 1,000]$ MHz. We observe that when $W_e = 10$ MHz, there exists many large off-diagonal entries in $\mathbb{R}$, indicating severe ambiguities among different locations. When the total bandwidth W_e increases, the ambiguities among different locations are significantly eliminated, while the TRRSs within the same location are almost unchanged.

4.5.3.2 Distribution of Diagonal and Off-Diagonal Entries under Different W_e

Figure 4.9 visualizes the distribution of the diagonal and off-diagonal entries of $\mathbb{R}$ with different $W_e \in [10, 40, 120, 1,000]$ MHz using histograms. Statistics of the diagonal and off-diagonal entries are shown as well. As we can see, the TRRS values at the same locations are identical with different W_e, implying high stationarity of the presented IPS. On the other hand, the off-diagonal entries are more suppressed and approach a Gaussian-like distribution when W_e increases. We also observe an enlarged gap between the diagonal and off-diagonal entries when W_e increases, indicating a better separability among different locations. The increase of W_e also reduces the variations of diagonal and off-diagonal entries, as shown by the decreasing standard deviations. Moreover,

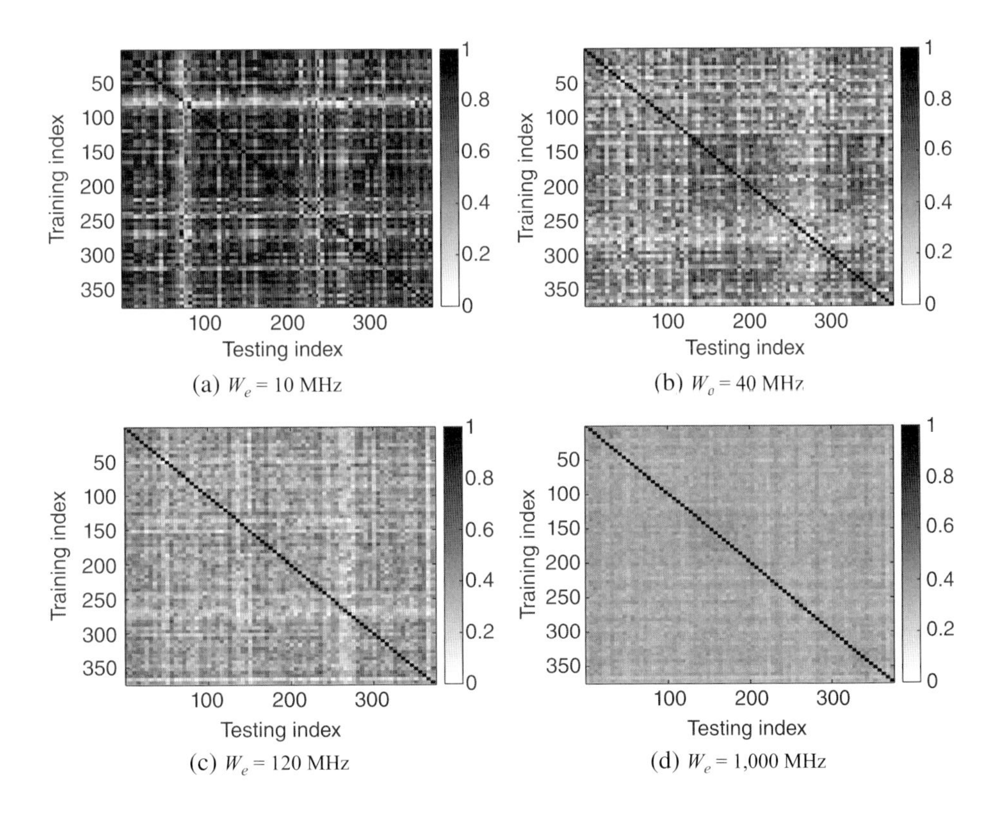

Figure 4.8 TRRS matrix under different W_e.

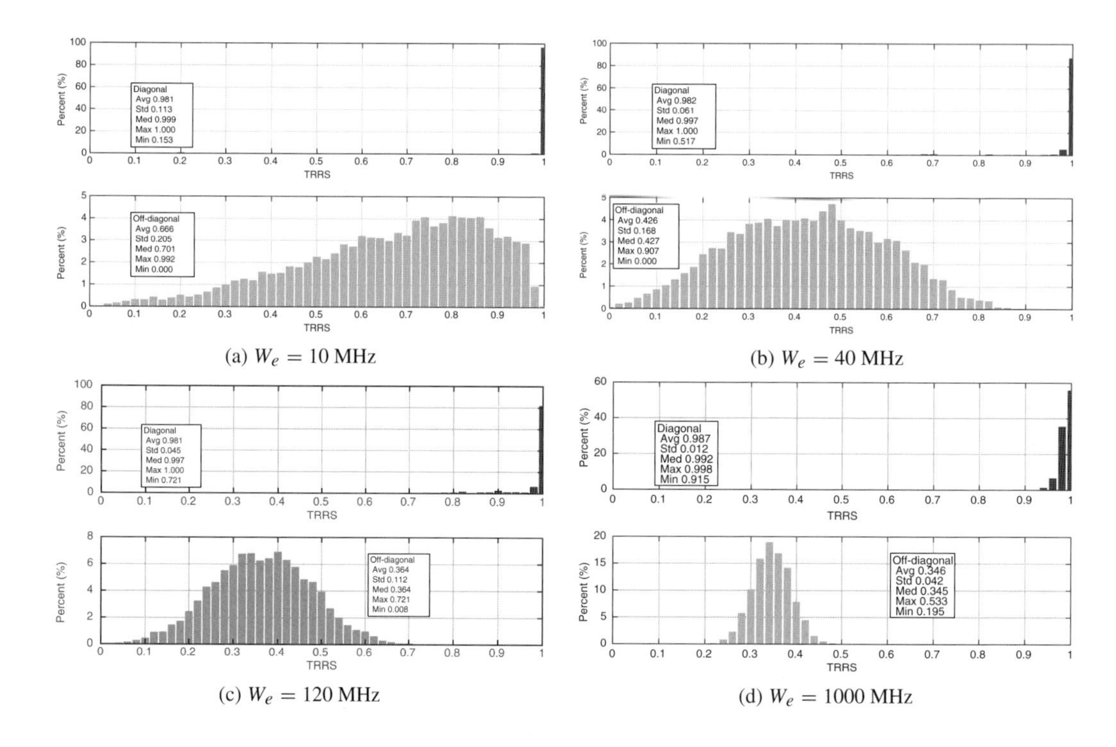

(a) $W_e = 10$ MHz

(b) $W_e = 40$ MHz

(c) $W_e = 120$ MHz

(d) $W_e = 1000$ MHz

Figure 4.9 Histogram of diagonal and off-diagonal entries under different W_e.

a large W_e removes the outliers in the diagonal entries: when $W_e = 10$ MHz, the minimum value of diagonal entries is 0.153, while the minimum value increases to 0.915 when $W_e = 1{,}000$ MHz. Thus, a large W_e improves the robustness of the IPS against outliers.

4.5.3.3 Cumulative Density Functions of Diagonal and Off-Diagonal Entries under Different W_e

In Figure 4.10, we demonstrate the cumulative density functions of diagonal and off-diagonal entries with $W_e \in [10, 20, 40, 80, 120, 300, 500, 1{,}000]$ MHz. As can be seen from the figure, a large W_e reduces the spread of both the diagonal and off-diagonal entries, which agrees with the results shown in Figure 4.9.

4.5.3.4 Mean and Standard Deviation Performances under Different W_e

Figure 4.11 depicts the impact of W_e on the mean and standard deviation performances for both diagonal and off-diagonal entries. The upper and lower bars indicate the $\pm\sigma$ bounds with respect to the average, where σ stands for the standard deviation. We conclude that a large W_e improves the distinction among different locations, but also reduces the variation of the TRRS at the same locations as well as among different locations. In other words, a large W_e makes the IPS performance more stable and predictable.

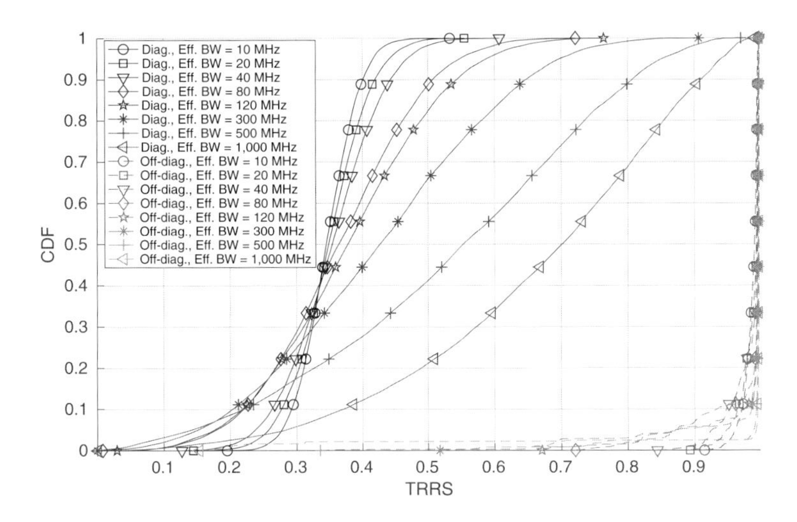

Figure 4.10 Cumulative density functions of diagonal and off-diagonal entries of the TRRS matrix under different W_e.

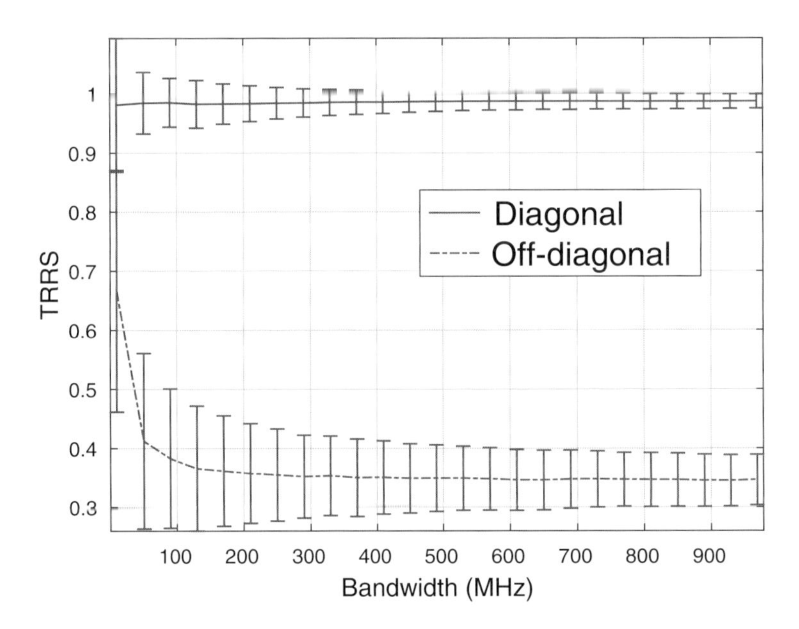

Figure 4.11 Mean and standard deviation of the diagonal and off-diagonal entries of the TRRS matrix under different W_e.

4.5.3.5 Threshold Γ Settings under Different W_e

Figure 4.12 depicts the smallest threshold Γ under $W_e = [20, 60, 100, \ldots, 1{,}000]$ MHz to achieve (i) $P_{\text{TP}} = 100\%$ and $P_{\text{FP}} = 0\%$ and (ii) $P_{\text{TP}} \geq 95\%$ and $P_{\text{FP}} \leq 5\%$. We observe a decreasing in Γ when W_e is larger, which can be justified by the fact that the gap between the diagonal and off-diagonal entries enlarges when W_e becomes larger. When $W_e = 20$ MHz, the IPS fails to achieve $P_{\text{TP}} = 100\%$ and $P_{\text{FP}} = 0\%$.

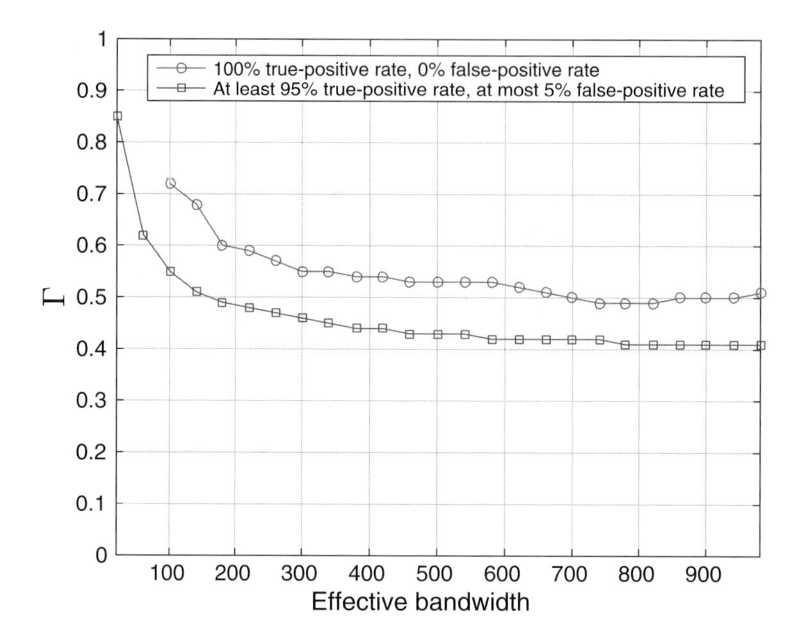

Figure 4.12 Threshold Γ under different W_e to achieve (i) $P_{\mathsf{TP}} = 100\%$ and $P_{\mathsf{FP}} = 0\%$ and (ii) $P_{\mathsf{TP}} \geq 95\%$ and $P_{\mathsf{FP}} \leq 5\%$.

Figure 4.12 also implies that we can achieve a perfect 5 cm localization if Γ is chosen appropriately.

Based on the experiment results, we conclude that a large W_e is imperative for the robustness, stability, and performance of the presented IPS. By formulating the location fingerprint that concatenates multiple channels, the presented IPS achieves a perfect centimeter localization accuracy in an NLOS environment with one pair of single-antenna Wi-Fi devices.

4.6 Discussion

4.6.1 Achievable Localization Accuracy

In Section 4.5, we demonstrate the centimeter-level localization accuracy of the presented IPS with a fine-grained measurement of 5 cm resolution. In a recent experiment, we refine the measurement resolution to 0.5 cm to study the accuracy. The TRRS near the intended location is shown in Figure 4.13 with $W_e = 125$ MHz, which demonstrates that the localization accuracy can reach $1 \sim 2$ cm in an NLOS environment.

4.6.2 Complexity of Fingerprint Collecting

In this chapter, the CFRs are collected in a two-dimensional (2D) space. In practice, localization of an object requires CFR measurement from a three-dimensional (3D)

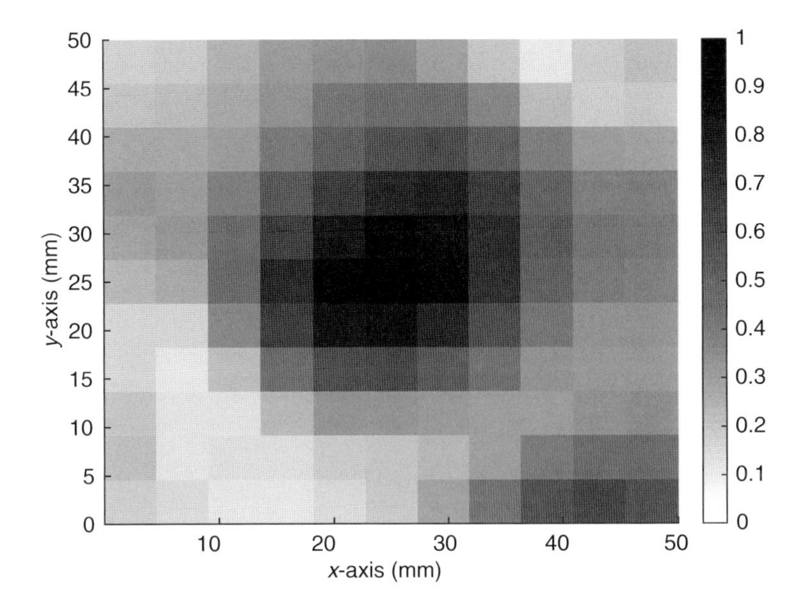

Figure 4.13 TRRS near the intended location with a measurement resolution of 0.5cm.

space with a centimeter-level granularity. In this case, the complexity of CFR measurement can be too high to be practical, especially for a large indoor space.

The burden of measurement can be significantly reduced because we only need to obtain the fingerprints of a limited number of areas that are more critical than the others. For instance, in an office, the main entrance and exit of the office as well as the entrance to some office rooms are of higher importance than the other areas, while in a museum, areas closer to the paintings could be more important. Fine-grained CFR measurements can be confined to these areas-of-interest. On the other hand, the efficiency of measurement can be boosted by automation techniques such as robotics.

4.6.3 Scalability

We notice that most of the calculations in the offline phase and online phase can be interpreted as linear operations. Thus, the computational complexity of the presented IPS scales linearly with the number of location fingerprints stored in the database. As the offline phase can in general tolerate a large delay, the increase in the computational complexity of the offline phase is less significant. On the other hand, the increase of the complexity imposes a challenge to the online phase because the online phase is much more time-sensitive than the offline phase. This issue becomes more severe when a huge number of fingerprints are stored in the database.

To deal with this problem, other information such as the sensory information or the RSSI values can be retrieved to supplement the presented IPS with a coarse position estimation. Then, the presented IPS can choose a subset of the fingerprints from the database that are collected nearby the estimated location to formulate a refined estimation.

4.6.4 Fingerprint Degradation with Time

In an indoor space, movement of small and large objects such as chairs and desks should be expected. These movements slightly change the environment and thus introduce deviations into the CFRs collected in the offline phase. According to our most recent work in [36], we find that a large effective bandwidth can reduce the sensitivity of the fingerprints to the environmental dynamics, which can be achieved by concatenating a large enough number of channels using frequency hopping.

4.6.5 CFR Acquisition on Commercial Wi-Fi Devices

USRPs are used as the prototype to acquire CFRs due to the fact that CFRs are unavailable on most commercial Wi-Fi devices. More recently, the CFRs can be obtained on the off-the-shelf 802.11n device, Intel Wi-Fi Wireless Link 5300, after modification of the firmware and the wireless driver [37]. Currently, we are investigating the IPS performance using the off-the-shelf Wi-Fi devices as well as implementing the frequency hopping mechanism.

4.7 Summary

In this chapter, we presented a Wi-Fi-based IPS that exploits the frequency diversity to achieve centimeter accuracy for indoor localization. The presented IPS fully harnesses the frequency diversity by CFR measurements on multiple channels via frequency hopping. Impacts of synchronization errors and interference are mitigated by CFR sanitization, sifting, and averaging. The averaged CFRs of different channels are then concatenated together into location fingerprints to augment the effective bandwidth. The location fingerprints are stored into a database in the offline phase and are used to calculate the TRRS in the online phase. Finally, the presented IPS determines the location based on the TRRS. Extensive experiment results of measurements on a 1 GHz frequency band demonstrate the centimeter localization accuracy of the presented IPS in a typical office environment with a large effective bandwidth. For related references, interested readers can refer to [38].

References

[1] J. G. McNeff, "The global positioning system," *IEEE Transactions on Microwave Theory and Techniques*, vol. 50, no. 3, pp. 645–652, Mar. 2002.

[2] E. Bruns, B. Brombach, T. Zeidler, and O. Bimber, "Enabling mobile phones to support large-scale museum guidance," *IEEE MultiMedia*, vol. 14, no. 2, pp. 16–25, Apr. 2007.

[3] S. Wang, S. Fidler, and R. Urtasun, "Lost shopping! Monocular localization in large indoor spaces," in *Proceedings of the IEEE International Conference on Computer Vision*, 2015, pp. 2695–2703.

[4] Z. Yang, Z. Zhou, and Y. Liu, "From RSSI to CSI: Indoor localization via channel response," *ACM Computing Surveys (CSUR)*, vol. 46, no. 2, pp. 25–1, 2013.

[5] J. Hightower, R. Want, and G. Borriello, "SpotON: An indoor 3D location sensing technology based on RF signal strength," University of Washington, Department of Computer Science and Engineering, Seattle, WA, UW CSE 00-02-02, Feb. 2000.

[6] L. Ni, Y. Liu, Y. C. Lau, and A. Patil, "LANDMARC: Indoor location sensing using active RFID," in *Proceedings of the First IEEE International Conference on Pervasive Computing and Communications (PerCom)*, pp. 407–415, Mar. 2003.

[7] Q. Zhang, C. H. Foh, B. C. Seet, and A. C. M. Fong, "RSS ranging based Wi-Fi localization for unknown path loss exponent," in *2011 IEEE Global Telecommunications Conference (GLOBECOM 2011)*, pp. 1–5, Dec. 2011.

[8] B. Campbell, P. Dutta, B. Kempke, Y.-S. Kuo, and P. Pannuto, "DecaWave: Exploring state of the art commercial localization," Ann Arbor: University of Michigan Electrical Engineering and Computer Science Department, vol. 1001, p. 48109.

[9] P. Steggles and S. Gschwind, The Ubisense smart space platform, *Adjunct Proceedings of the Third International Conference on Pervasive Computing*, vol. 191, 73–76, 2005.

[10] D. Vasisht, S. Kumar, and D. Katabi, "Decimeter-level localization with a single WiFi access point," in *13th USENIX Symposium on Networked Systems Design and Implementation (NSDI 16)*, pp. 165–178, Mar. 2016. [Online]. Available: www.usenix.org/conference/nsdi16/technical-sessions/presentation/vasisht.

[11] J. Gjengset, J. Xiong, G. McPhillips, and K. Jamieson, "Phaser: Enabling phased array signal processing on commodity WiFi access points," in *Proceedings of the 20th Annual ACM International Conference on Mobile Computing and Networking*, pp. 153–164, 2014. [Online]. Available: http://doi.acm.org/10.1145/2639108.2639139.

[12] J. Xiong and K. Jamieson, "ArrayTrack: A fine-grained indoor location system," in *Proceedings of the 10th USENIX Conference on Networked Systems Design and Implementation*, pp. 71–84, 2013. [Online]. Available: http://dl.acm.org/citation.cfm?id=2482626.2482635.

[13] P. Bahl and V. Padmanabhan, "RADAR: An in building RF-based user location and tracking system," in *Proceedings of the IEEE INFOCOM*, vol. 2, pp. 775–784, 2000.

[14] M. Youssef and A. Agrawala, "The Horus WLAN location determination system," in *Proceedings of the 3rd ACM International Conference on Mobile Systems, Applications, and Services*, pp. 205–218, 2005.

[15] P. Prasithsangaree, P. Krishnamurthy, and P. Chrysanthis, "On indoor position location with wireless LANs," in *The 13th IEEE International Symposium on Personal, Indoor and Mobile Radio Communications,* vol. 2, pp. 720–724, Sep. 2002.

[16] C. Wu, Z. Yang, and Y. Liu, "Smartphones based crowdsourcing for indoor localization," *IEEE Transactions on Mobile Computing*, vol. 14, no. 2, pp. 444–457, Feb. 2015.

[17] S. Sen, B. Radunovic, R. R. Choudhury, and T. Minka, "You are facing the Mona Lisa: Spot localization using PHY layer information," in *Proceedings of the 10th ACM International Conference on Mobile Systems, Applications, and Services*, pp. 183–196, 2012. [Online]. Available: http://doi.acm.org/10.1145/2307636.2307654

[18] J. Xiao, K. Wu, Y. Yi, and L. Ni, "FIFS: Fine-grained indoor fingerprinting system," in *21st International Conference on Computer Communications and Networks (ICCCN)*, pp. 1–7, Jul. 2012.

[19] Y. Chapre, A. Ignjatovic, A. Seneviratne, and S. Jha, "CSI-MIMO: Indoor Wi-Fi fingerprinting system," in *IEEE 39th Conference on Local Computer Networks (LCN)*, pp. 202–209, Sep. 2014.

[20] Z. Wu, Y. Han, Y. Chen, and K. J. R. Liu, "A time-reversal paradigm for indoor positioning system," *IEEE Transactions on Vehicular Communications*, vol. 64, no. 4, pp. 1331–1339, Apr. 2015.

[21] C. Chen, Y. Chen, H. Q. Lai, Y. Han, and K. J. R. Liu, "High accuracy indoor localization: A WiFi-based approach," in *IEEE International Conference on Acoustics, Speech and Signal Processing (ICASSP)*, pp. 6245–6249, Mar. 2016.

[22] B. Wang, Y. Wu, F. Han, Y. Yang, and K. J. R. Liu, "Green wireless communications: A time-reversal paradigm," *IEEE Journal on Selected Areas in Communications*, vol. 29, no. 8, pp. 1698–1710, Sep. 2011.

[23] M. Fink and C. Prada, "Acoustic time-reversal mirrors," *Inverse Problems*, vol. 17, no. 1, Feb. 2001.

[24] M. Fink, C. Prada, F. Wu, and D. Cassereau, "Self focusing in inhomogeneous media with time reversal acoustic mirrors," in *Proceedings of the IEEE Ultrasonics Symposium*, pp. 681–686 vol. 2, Oct. 1989.

[25] C. Dorme, M. Fink, and C. Prada, "Focusing in transmit-receive mode through inhomogeneous media: The matched filter approach," in *Proceedings of the IEEE Ultrasonics Symposium*, pp. 629–634 vol. 1, Oct. 1992.

[26] F. Han, Y.-H. Yang, B. Wang, Y. Wu, and K. J. R. Liu, "Time-reversal division multiple access over multi-path channels," *IEEE Transactions on Communications*, vol. 60, no. 7, pp. 1953–1965, Jul. 2012.

[27] J. Heiskala and J. Terry, *OFDM Wireless LANs: A Theoretical and Practical Guide*. Indianapolis, IN: Sams, 2001.

[28] T.-D. Chiueh and P.-Y. Tsai, *OFDM Baseband Receiver Design for Wireless Communications*. John Wiley and Sons (Asia) Pte Ltd, 2007.

[29] M. Speth, S. Fechtel, G. Fock, and H. Meyr, "Optimum receiver design for wireless broadband systems using OFDM – Part I," *IEEE Transactions on Communications*, vol. 47, no. 11, pp. 1668 –1677, Nov. 1999.

[30] Wireless LAN Working Group, "Supplement to IEEE standard for information technology telecommunications and information exchange between systems: Local and metropolitan area networks, Specific requirements, Part 11: Wireless LAN medium access control (MAC) and physical layer (PHY) specifications: High-Speed physical layer in the 5 GHz band," *IEEE Standard*, 1999.

[31] A. V. Oppenheim, R. W. Schafer, and J. R. Buck, *Discrete-Time Signal Processing* (2nd ed.). Upper Saddle River, NJ: Prentice-Hall, Inc., 1999.

[32] M. Speth, S. Fechtel, G. Fock, and H. Meyr, "Optimum receiver design for OFDM-based broadband transmission II: A case study," *IEEE Transactions on Communications*, vol. 49, no. 4, pp. 571 –578, Apr. 2001.

[33] "Ettus Research LLC," www.ettus.com/.

[34] B. Bloessl, M. Segata, C. Sommer, and F. Dressler, "Decoding IEEE 802.11a/g/p OFDM in Software using GNU Radio," in *19th ACM International Conference on Mobile Computing and Networking (MobiCom 2013), Demo Session*, pp. 159–161, Oct. 2013.

[35] "GNU Radio," http://gnuradio.org/.

[36] C. Chen, Y. Chen, Y. Han, H. Lai, F. Zhang, and K. J. R. Liu, "Achieving centimeter accuracy indoor localization on WiFi platforms: A multi-antenna approach," *IEEE Internet of Things Journal*, vol. 4, no. 1, pp. 122–134, Feb. 2017.

[37] D. Halperin, W. Hu, A. Sheth, and D. Wetherall, "Tool release: Gathering 802.11n traces with channel state information," *ACM SIGCOMM CCR*, vol. 41, no. 1, p. 53, Jan. 2011.

[38] C. Chen, Y. Chen, Y. Han, H.-Q. Lai, and K. J. R. Liu, "Achieving centimeter-accuracy indoor localization on WiFi platforms: A frequency hopping approach," *IEEE Internet of Things Journal*, vol. 4, no. 1, pp. 111–121, 2017.

5 Decimeter-Accuracy Indoor Tracking

With the development of the Internet of Things technology, indoor tracking has become a popular application nowadays, but most existing solutions can only work in line-of-sight scenarios or require regular recalibration. In this chapter, we present WiBall, an accurate and calibration-free indoor tracking system that can work well in non-line-of-sight based on radio signals. WiBall leverages a stationary and location-independent property of the time-reversal focusing effect of radio signals for highly accurate moving distance estimation. Together with the direction estimation based on inertial measurement unit and location correction using the constraints from the floorplan, WiBall is shown to be able to track a moving object with decimeter-level accuracy in different environments. Because WiBall can accommodate a large number of users with only a single pair of devices, it is low cost and easily scalable and can be a promising candidate for future indoor tracking applications.

5.1 Introduction

With the proliferation of the Internet of Things (IoT) applications, Indoor Positioning and Indoor Navigation (IPIN) has received an increasing attention in recent years. Technavio forecasts the global IPIN market to grow to USD 7.8 billion by 2021 [1], and more than ever before, enterprises of all sizes are investing in IPIN technology to support a growing list of applications, including patient tracking in hospitals, asset management for large groceries, workflow automation in large factories, navigation in malls, appliance control, etc.

Although Global Positioning System (GPS) can achieve good accuracy with a low cost in outdoor real-time tracking, such a good balance between the cost and performance has not been realized for indoor tracking yet [2].

In this chapter, we present WiBall, a wireless system for indoor tracking, that can work well in both none-line-of-sight (NLOS) and line-of-sight (LOS) scenarios and is robust to environmental dynamics as well. WiBall estimates the incremental displacement of the device at every moment, and thus, it can track the trace of the device in real time. WiBall adopts a completely new paradigm in the moving distance estimation, which is built on the presented discovered physical phenomenon of radio signals.

In the past, the moving distance estimation could be done by analyzing the output of inertial measurement unit (IMU) that is attached to the moving object. Accelerometer readings are used to detect walking steps and then, the walking distance can be estimated by multiplying the number of steps with the stride length [3]. However, pedestrians often have different stride lengths that may vary up to 40% even at the same speed, and 50% with various speeds of the same person [4]. Thus calibration is required to obtain the average stride lengths for different individuals, which is impractical in real applications and thus has not been widely adopted. The moving distance can also be estimated by analyzing radio signals that are affected by the movement of the device. Various methods have been proposed based on the estimation of the maximum Doppler frequency, such as level crossing rate methods [5], covariance based methods [6, 7], and wavelet-based methods [8]. However, the performance of these estimators is unsatisfactory in practical scenarios. For example, the approach in [7] can only differentiate whether a mobile station moves with a fast speed ($\geq$30 km/h) or with a slow speed ($\leq$5 km/h).

In WiBall, a new scheme for moving distance estimation based on the time-reversal (TR) resonating effect [9, 10] is presented. TR is a fundamental physical resonance phenomenon that allows people to focus the energy of a transmitted signal at an intended focal spot, both in the time and spatial domains, by transmitting the TR waveform. The research of TR can be traced back to the 1950s when it was first utilized to align the phase differences caused by multipath fading during long-distance information transmissions. The TR resonating effect was first observed in a practical underwater propagation environment [11] that the energy of a transmitted signal can be refocused at the intended location because by means of TR the RX recollects multipath copies of a transmitted signal in a coherent matter.

In this chapter, we present a new discovery that the energy distribution of the TR focusing effect exhibits a location-independent property, which is only related to the physical parameters of the transmitted electromagnetic (EM) waves. This is because the number of multipath components (MPC) in indoors is so large that the randomness of the received energy at different locations can be averaged out as a result of the law of large numbers. Based on this location-independent feature, WiBall can estimate the moving distance of the device in a complex indoor environment without requiring any precalibration procedures. To cope with the cumulative errors in distance estimation, WiBall incorporates the constraints imposed by the floorplan of buildings and corrects the cumulative errors whenever a landmark, such as a corner, hallway, door, etc., is met. Combining the improved distance estimator and the map-based error corrector, WiBall is shown to be able to achieve decimeter-level accuracy in real-time tracking regardless of the moving speed and environment.

The rest of the chapter is organized as follows. Section 5.2 introduces the related works. Section 5.3 introduces the system architecture of WiBall followed by a discussion on the TR principle for distance estimation. Section 5.4 presents an IMU-based moving direction estimator and a map-based localization corrector, which can correct the accumulated localization error. Experimental evaluation is shown in Section 5.5.

5.2 Related Works

Existing solutions for indoor tracking can be classified into four categories: vision based, audio based, radio based, and IMU based. Vision-based approaches, such as camera [12], laser [13], infrared [14], etc., suffer from high cost of deployment and equipment, sophisticated calibrations and limited coverage, although a very high accuracy can be achieved. The acoustic-based schemes [15] can only cover a limited range and are not scalable to a large number of users. The performance of radio-based approaches, such as RADAR [16], RFID [17], and UWB localization systems [18, 19], is severely affected by the NLOS multipath propagation, which is unavoidable for a typical indoor environment. The localization accuracy of IMU-based methods [3, 4] is mainly limited by the poor estimation of moving distance and the drifting of the gyroscope.

Due to the wide deployment of Wi-Fi in indoors, various indoor localization systems based on Wi-Fi have been proposed, as summarized in Table 5.1. In general, these works can be classified into two categories: modeling-based approach and fingerprinting-based approach. The features utilized in these approaches can be obtained either from the MAC layer information, e.g., receive signal strength indicator (RSSI) readings and the timestamps of the received packets at the receiver (RX), or from the PHY information, e.g., the channel state information (CSI).

In the modeling-based schemes, either the distance [20–23] or the angle [24–26] between an anchor point and the device can be estimated, and the device can be localized by performing geometrical triangulation. The distance between the anchor point and the device can be estimated from the decay of RSSI [27] or from the time of arrival (ToA) of the transmitted packets, which can be extracted from the timestamps of the received packets [28]. The angle in between can be obtained by examining the features of the CSI received by multiple receive antennas, and then, the angle of arrival (AoA) of the direct path to the target can be calculated. ToA-based methods typically require synchronization between the anchor point and the device and thus are very sensitive to timing offsets [29]; AoA-based methods require an array of phased antennas, which are not readily available in commercial Wi-Fi chips [25]. Recently, a decimeter-level tracking system, Widar, is proposed in [30] and [31]; however, the system can only work in a small area with the constraint of LOS. The main challenges for the

Table 5.1 A brief summary of typical Wi-Fi-based approaches

	Method	Existing solutions
Modeling-based	ToA	CAESAR [20], ToneTrack [28]
	AoA	ArrayTrack [36], SpotFi [24], Phaser [25]
	RSSI	RADAR [22]
	CSI	FILA [37]
Fingerprinting-based	RSSI	Horus [27], Nibble [32]
	CSI	PinLoc [38], TRIPS [35], DeepFi [33]

modeling-based approaches are the blockage and reflection of the transmitted signal because only the signal coming from the direct path between the anchor point and the device is useful for localization.

The fingerprinting-based schemes consist of an offline phase and an online phase. During the offline phase, features associated with different locations are extracted from the Wi-Fi signals and stored in a database; in the online phase, the same features are extracted from the instantaneous Wi-Fi signals and compared with the stored features so as to classify the locations. The features can be obtained either from the vector of RSSIs [27, 32] or the detailed CSI [33–35] from a specific location to all the anchor points in range. A major drawback of the fingerprinting-based approaches lies in that the features they use are susceptible to the dynamics of the environment. For example, the change of furniture or the status of doors may have a severe impact on these features, and the database of the mapped fingerprints needs to be updated before it can be used again. In addition, the computational complexity of the fingerprinting-based approaches scales with the size of the database, and thus they are not feasible for low-latency applications, especially when the number of the collected fingerprints is large.

In sum, the performance of most existing solutions for indoor localization degrades dramatically under NLOS conditions, which are common usage scenarios though. Even with a significant overhead in the manual construction of the database, the fingerprinting-based approaches still fail to achieve a decimeter-level accuracy. Therefore, indoor location-based services are not provided as widespread as expected nowadays, which motivates us to design a highly accurate and robust indoor tracking system even without requiring specialized infrastructure.

5.3 TR Focusing Ball Method for Distance Estimation

In this section, we first introduce the overall system architecture of WiBall and the TR radio system. Then, we derive the analytical normalized energy distribution of the TR focal spot. We show that the normalized energy distribution is location-independent and can be used to estimate distance. Last, we discuss the TR-based distance estimator.

5.3.1 Overview of WiBall

WiBall consists of a transmitter (TX) broadcasting beacon signals periodically to all the RXs being tracked. WiBall estimates the paths that the RX travels, i.e., the location of the RX $\vec{\mathbf{x}}$ at time t_i is estimated as

$$\vec{\mathbf{x}}(t_i) = \vec{\mathbf{x}}(t_{i-1}) + \vec{\Delta}(t_i), \tag{5.1}$$

where $\vec{\mathbf{x}}(t_{i-1})$ represents the location of the RX at the previous time t_{i-1}, and $\Delta(t_i)$ is the incremental displacement. The magnitude of $\vec{\Delta}(t_i)$, denoted as $d(t_i)$, and the angle of $\vec{\Delta}(t_i)$, denoted as $\theta(t_i)$, correspond to the moving distance and the change of moving direction of the RX from t_{i-1} to t_i, respectively. As shown in Figure 5.1, the location of

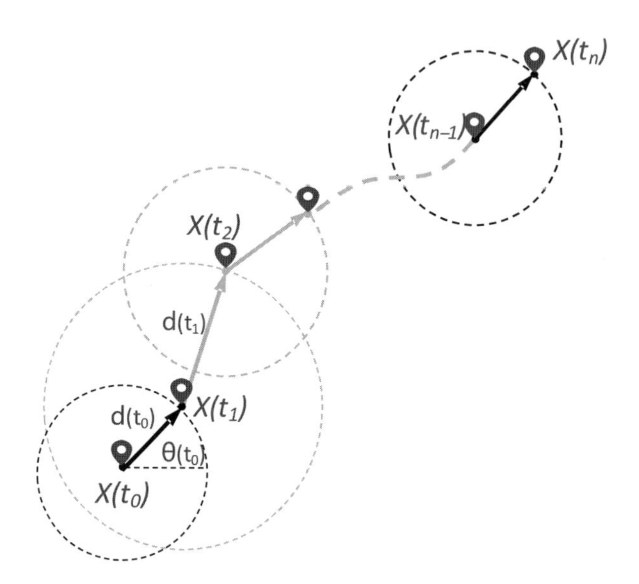

Figure 5.1 Illustration of the tracking procedure.

the RX at time t_n is computed based on the accumulative displacements from time t_0 to t_n and the initial start point $\vec{\mathbf{x}}(t_0)$.

WiBall estimates the moving distance $d(t_i)$ based on the TR resonating effect, which can be obtained from the CSI measurements at the RX. The estimation of $\theta(t_i)$ is based on the angular velocity and gravity direction from IMU, which is a built-in module for most smartphones nowadays.

5.3.2 TR Radio System

Consider a rich-scattering environment, e.g., an indoor or metropolitan area and a wireless transceiver pair, each equipped with a single omnidirectional antenna. Given a large enough bandwidth, the MPCs in a rich-scattering environment can be resolved into multiple taps in discrete-time and let $h(l;\vec{T} \rightarrow \vec{R}_0)$ denote the lth tap of the channel impulse response (CIR) from $\vec{T}$ to $\vec{R}_0$, where $\vec{T}$ and $\vec{R}_0$ denotes the coordinates of the TX and RX, respectively. In the TR transmission scheme, the RX at $\vec{R}_0$ first transmits an impulse, and the TX at $\vec{T}$ captures the CIR from $\vec{R}_0$ to $\vec{T}$. Then the RX at $\vec{T}$ simply transmits back the time-reversed and conjugated version of the captured CIR, i.e., $h^*(-l;\vec{R}_0 \rightarrow \vec{T})$, where $*$ denotes complex conjugation. With channel reciprocity, i.e., the forward and backward channels are identical [35], the received signal at any location $\vec{R}$ when the TR waveform $h^*(-l;\vec{R}_0 \rightarrow \vec{T})$ is transmitted can be written as [39]

$$s(l;\vec{R}) = \sum_{m=0}^{L-1} h(m;\vec{T} \rightarrow \vec{R})h^*(m-l;\vec{R}_0 \rightarrow \vec{T}), \tag{5.2}$$

where L is the number of resolved multipaths in the environment. When $\vec{R} = \vec{R}_0$ and $l = 0$, we have $s(0; \vec{R}) = \sum_{m=0}^{L-1} |h(m, \vec{T} \rightarrow \vec{R}_0)|^2$ with all MPCs added up coherently, i.e., the signal energy is refocused at a particular spatial location at a specific time instance. This phenomenon is termed the TR spatial-temporal resonating effect [40, 41].

To study the TR resonating effect in the spatial domain, we fix time index $l = 0$ and define the TR resonating strength (TRRS) between the CIRs of two locations, $\vec{R}_0$ and $\vec{R}$, as the normalized energy of the received signal when the TR waveform for location $\vec{R}_0$ is transmitted:

$$\eta(\mathbf{h}(\vec{R}_0), \mathbf{h}(\vec{R})) = \left| \frac{s(0; \vec{R})}{\sqrt{\sum\limits_{l_1=0}^{L-1} |h(l_1; \vec{T} \rightarrow \vec{R}_0)|^2} \sqrt{\sum\limits_{l_2=0}^{L-1} |h(l_2; \vec{T} \rightarrow \vec{R})|^2}} \right|^2, \tag{5.3}$$

where we use $\mathbf{h}(\vec{R})$ as an abbreviation of $h(l; \vec{T} \rightarrow \vec{R})$, $l = 0, \ldots, L-1$, when $\vec{T}$ is fixed. Note that the range of TRRS is normalized to be $[0, 1]$ and TRRS is symmetric, i.e., $\eta(\mathbf{h}(\vec{R}_0), \quad \mathbf{h}(\vec{R})) = \eta(\mathbf{h}(\vec{R}), \mathbf{h}(\vec{R}_0))$.

We built a pair of customized TR devices to measure the TRRS at different locations, as shown in Figure 5.2(a). The devices operate at $f_0 = 5.8$ GHz ISM band with 125 MHz bandwidth, and the corresponding wavelength is $\lambda = c/f_0 = 5.17$ cm. The RX is placed on a 5 cm $\times$ 5 cm square area above a channel probing table with 0.5 cm resolution, and the center of the square is set to be the focal spot $\vec{R}_0$. The TRRS distribution around $\vec{R}_0$ in the spatial domain and the normalized received energy at $\vec{R}_0$ in the time domain are shown in Figures 5.2(b) and 5.2(c), respectively. As we can see from the results, the received energy is concentrated around $\vec{R}_0$ both in spatial and time domains almost symmetrically, which shows that a bandwidth of 125 MHz is able to achieve the TR resonating effect in a typical indoor environment.

5.3.3 Energy Distribution of TR Focal Spot

Assume that all the EM waves propagate in a far-field zone, and then each MPC can be approximated by a plane EM wave. For the purpose of illustration, the receive antenna is placed in the origin of the space, and each MPC can be represented by a point in the space whose coordinate is determined by its angle of arrival and propagation distance, e.g., point A, as shown in Figure 5.3, where r stands for the total traveled distance of the MPC, θ denotes the direction of arrival of the MPC, and $G(\omega)$ denotes the power gain with $\omega = (r, \theta)$. In a rich-scattering environment, we can also assume that ω is uniformly distributed in the space, and the total number of MPCs is large. When a vertically polarized antenna is used, only the EM waves with the direction of electric field orthogonal to the horizontal plane are collected. Then, the received signal is just a scalar sum of the electric field of the impinging EM waves along the vertical direction. In the sequel, without loss of generality, we only consider the TRRS distribution in the horizontal plane because its distribution in the vertical plane is similar.

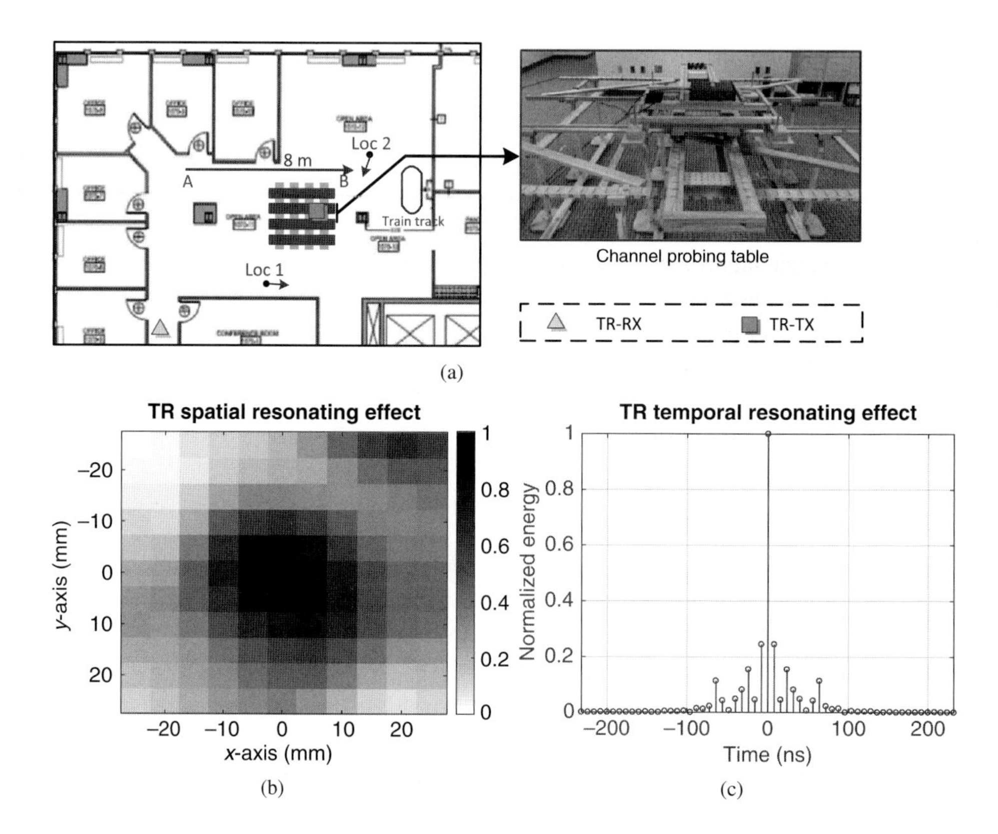

Figure 5.2 (a) TR prototype and the environment of the measurement, (b) the TRRS distribution in the spatial domain, and (c) the normalized energy of the received signal at the focal spot $\vec{R}_0$ in the time domain.

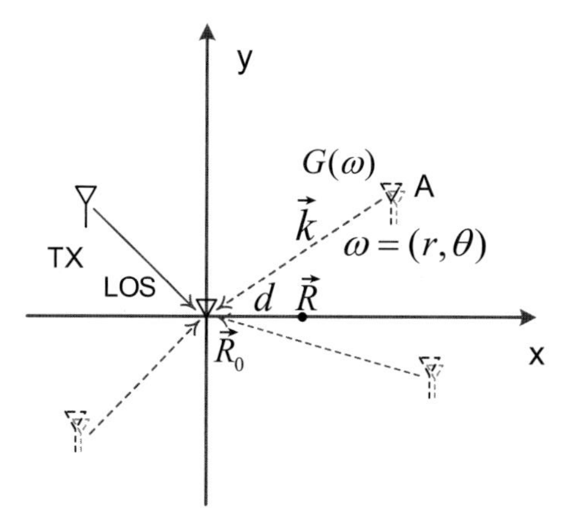

Figure 5.3 Illustration of the polar coordinates in the analysis.

For a system with bandwidth B, two MPCs would be divided into different taps of the measured CIR as long as the difference of their time of arrival is larger than the sampling period $1/B$, that is, any two MPCs with a difference of their traveled distances larger than c/B can be separated. With a sufficiently large system bandwidth B, the distance resolution c/B of the system is so small that all of the MPCs with significant energy can be separated in the spatial domain, i.e., each significant MPC can be represented by a single tap of a measured CIR. Assume that the distribution of the energy of each MPC is uniform in direction θ, i.e., the distribution of $G(\omega)$ is only a function of r. Then the energy of MPCs coming from different directions would be approximately the same when the number of MPCs is large. Mathematically, for any point $\vec{R}$ in a source-free region with constant mean electric and magnetic fields, the channel impulse response when a delta-like pulse is transmitted can be written as [39]

$$h(t;\vec{T} \to \vec{R}) = \sum_{\omega\in\Omega} G(\omega)q(t-\tau(\omega))e^{i(2\pi f_0(t-\tau(\omega))-\phi(\omega)-\vec{k}(\omega)\cdot\vec{R})}, \tag{5.4}$$

where $q(t)$ is the pulse shaper, $\tau(\omega) = r/c$ is the propagation delay of the MPC ω, f_0 is the carrier frequency, Ω is the set of MPCs, $\phi(\omega)$ is the change of phase due to reflections, and $\vec{k}(\omega)$ is the wave vector with amplitude $k = c/f_0$. Accordingly, the lth tap of a sampled CIR at location $\vec{R}$ can be expressed as

$$h(l;\vec{T} \to \vec{R}) = \sum_{\tau(\omega)\in[lT-\frac{T}{2},lT+\frac{T}{2})} G(\omega)q(\Delta\tau(l,\omega))e^{i(2\pi f_0\Delta\tau(l,\omega)-\phi(\omega)-\vec{k}(\omega)\cdot\vec{R})} \tag{5.5}$$

where T is the channel measurement interval and $\Delta\tau(l,\omega) = lT - \tau(\omega)$ for $l = 0, 1, \ldots, L-1$. When the TR waveform $h^*(-l;\vec{R}_0 \to \vec{T})$ is transmitted, the corresponding received signal at the focal spot $\vec{R}_0$ can be written as

$$s(0;\vec{R}) = \sum_{l=1}^{L} \left| \sum_{\tau\in[lT-\frac{T}{2},lT+\frac{T}{2})} G(\omega)q(\Delta\tau(l,\omega))e^{i(2\pi f_0\Delta\tau(l,\omega)-\phi(\omega))} \right|^2. \tag{5.6}$$

Equation (5.6) shows that the MPCs with propagation delays $\tau(\omega) \in [lT - \frac{T}{2}, lT + \frac{T}{2})$ for each l would be merged into one single tap, and the signals coming from different taps would add up coherently while the MPCs within each sampling period T would add up incoherently. It indicates that the larger the bandwidth, the larger the TR focusing gain that can be achieved because more MPCs can be aligned and added up coherently. When the bandwidth is sufficiently large, the received signal at each point $\vec{R}$ can be approximated as

$$s(0;\vec{R}) \approx \sum_{l=1}^{L} |G(\omega)q(\Delta\tau(l,\omega))|^2 e^{-i\vec{k}(\omega)\cdot(\vec{R}-\vec{R}_0)}. \tag{5.7}$$

When a rectangular pulse shaper is used, i.e., $q(t) = 1$ for $t \in [-\frac{T}{2}, \frac{T}{2})$ and $q(t) = 0$ otherwise, under the preceding symmetric scattering assumption, the received signal $s(0; \vec{R})$ can thus be approximated as

$$\begin{aligned} s(0; \vec{R}) &= \sum_{\omega \in \Omega} |G(\omega)|^2 e^{-i\vec{k}\cdot(\vec{R}-\vec{R}_0)} \\ &\approx \int_0^{2\pi} P(\theta) e^{-ikd\cos(\theta)} d\theta \\ &= P J_0(kd), \end{aligned} \tag{5.8}$$

where the coordinate system in Figure 5.3 is used, Ω stands for the set of all significant MPCs, $J_0(x)$ is the 0th-order Bessel function of the first kind, and d is the Euclidean distance between $\vec{R}_0$ and $\vec{R}$. Here we use a continuous integral to approximate the discrete sum, and $P(\theta) = P$ denotes the density of the energy of MPCs coming from direction θ. For $\vec{R} = \vec{R}_0$, it degenerates to the case of $d = 0$, and thus $s(0; \vec{R}_0) \approx P$. Because the denominator of (5.3) is the product of the energy received at two focal spots, it would converge to P^2. At the same time, the numerator is approximately $P^2 J_0^2(kd)$ as discussed earlier. As a result, the TRRS defined in (5.3) can be approximated as

$$\eta(\mathbf{h}(\vec{R}_0), \mathbf{h}(\vec{R})) \approx J_0^2(kd). \tag{5.9}$$

In the following, because the theoretic approximation of the TRRS distribution only depends on the distance between two points, we use $\bar{\eta}(d) = J_0^2(kd)$ to stand for the approximation of TRRS between two points with distance d.

To evaluate the preceding theoretic approximation, we also built a mobile channel probing platform equipped with stepping motors that can control the granularity of the CIR measurements precisely along any predefined direction. Extensive measurements of CIRs from different locations have been collected in the environment shown in Figure 5.2(a). Figure 5.4(b) shows two typical experimental results measured at Location 1 and Location 2 with a separation of 10 m as shown in Figure 5.2(a). The distance d away from each predefined focal spot increases from 0 to 2λ with a resolution of 1 mm. The measured TRRS distribution functions agree with the theoretic approximation quite well in the way that the positions of the peaks and valleys in the measured curves are almost the same as those of the theoretic curve. Although Locations 1 and 2 are far apart, the measured TRRS distribution functions exhibit similar damping pattern when the distance d increases.

We also observe that the measured TRRS distribution functions are far above 0. This is due to the contribution of the direct path between the TR devices. Therefore, the energy density function $P(\theta)$ in (5.8) consists of a term that is symmetric in direction due to NLOS components and another term that is asymmetric in direction due to LOS components. As a result, the TRRS is indeed a superposition of $J_0^2(kd)$ and some unknown function, which is the result of the asymmetric normalized energy distribution of MPCs in certain directions. Because the pattern of $J_0^2(kd)$, embedded in the TRRS distribution function, is location-independent, we can exploit this feature for speed estimation.

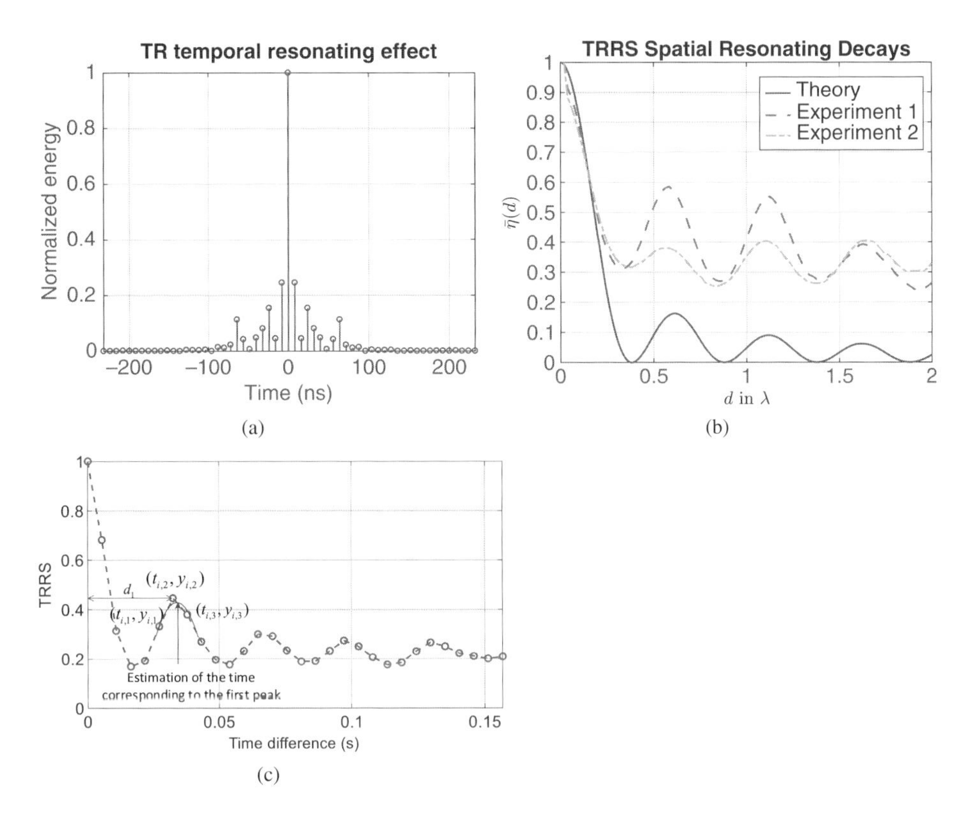

Figure 5.4 The distributions of TRRS. (a) TRRS in time domain. (b) Comparison of the TRRS distribution between experimental results and the theoretical result. Experiments 1 and 2 correspond to Location 1 and Location 2, respectively, in Figure 5.2(a). (c) Illustration of the presented TR-based speed. estimation algorithm.

A numerical simulation using a ray-tracing approach is also implemented to study the impact of bandwidth on TRRS distribution. In the simulation, the carrier frequency of the transmitted signals is set to be 5.8 GHz. Two hundred scatterers are uniformly distributed in a 7.5 m by 7.5 m square area. The reflection coefficient is distributed uniformly and independently in $(0, 1)$ for each scatterer. The distance between the TX and RX is 30 m, and the RX (focal spot) is set to be the center of the square area. Figure 5.5 shows the distributions of TRRS around the focal spot when the system bandwidth is 40 MHz, 125 MHz, and 500 MHz, respectively. As we can see from the results, as the bandwidth increases, the distribution of TRRS in the horizontal plane becomes more deterministic-like and converges to the theoretical approximations.

5.3.4 TR-Based Distance Estimator

Because the shape of the TRRS distribution function $\bar{\eta}(d) \approx J_0^2(kd)$ is only determined by the wave number k, which is independent of specific locations, it can be utilized as an intrinsic ruler to measure distance in the space. Consider that an RX moves

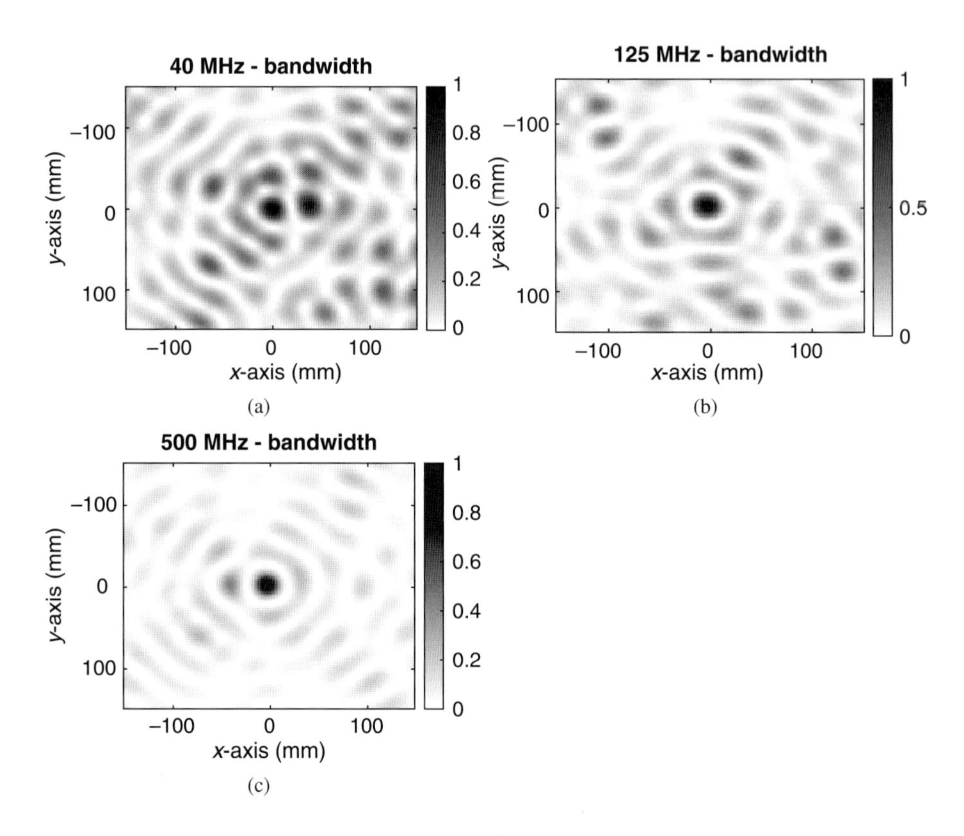

Figure 5.5 Numerical simulations of the distributions of TRRS with varying bandwidth: (a) 40 MHz, (b) 125 MHz, and (c) 500 MHz.

along a straight line with a constant speed v starting from location $\vec{R}_0$, and a TX keeps transmitting the TR waveform corresponding to $\vec{R}_0$ at regular intervals. Then, the TRRS measured at the RX is just a sampled version of $\eta(d)$, which would also exhibit the Bessel-function-like pattern, as illustrated in Figure 5.4(c).

Take the first local peak of $\eta(d)$, for example. The corresponding theoretical distance d_1 is about 0.61λ according to the Bessel-function-like pattern. In order to estimate the moving speed, we only need to estimate how much time $\hat{t}$ it takes for the RX to reach the first local peak starting from point $\vec{R}_0$. We use a quadratic curve to approximate the shape of the first local peak. Combining the knowledge of the timestamps of each CIR measurement, $\hat{t}$ can be estimated by the vertex of the quadratic curve. Therefore, we obtain the speed estimation as $\hat{v} = (0.61\lambda)/\hat{t}$, and then, the moving distance can be calculated by integrating the instantaneous speed over time. One thing to note is that as long as the rate of CIR measurement is fast enough, it is reasonable to assume that the moving speed is constant during the measurement of the TRRS distribution. For example, in Figure 5.4(c) the duration is about 0.16 s.

In practice, the channel is not measured with a uniform interval, and the empirical probability density function (PDF) of the time interval between adjacent channel measurements is shown in Figure 5.6. To overcome the imperfect channel sampling process,

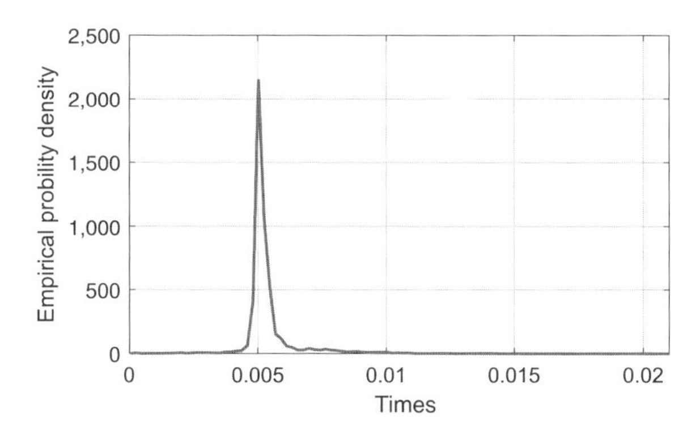

Figure 5.6 Empirical probability density for time intervals between adjacent packets.

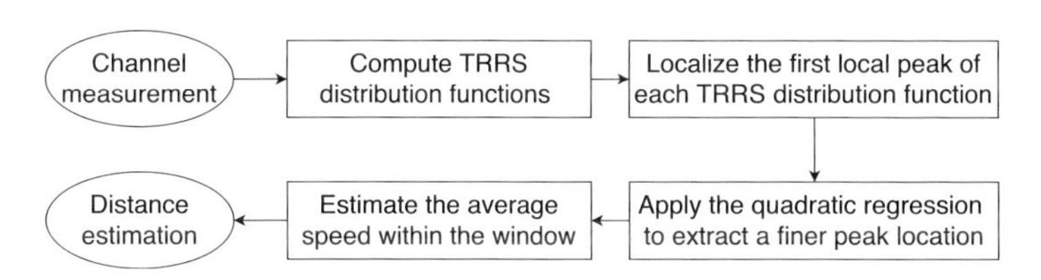

Figure 5.7 Flowchart of the TR-based distance estimator.

we combine multiple realizations of the TRRS distribution function measured at adjacent time slots to increase the accuracy of the estimation of $\hat{t}$. For the ith measurement, first find the data points near the first local peak $(t_{i,j}, y_{i,j}), i = 1, \ldots, N, j = 1, 2, 3$, as shown in Figure 5.4(c), where N is the number of TRRS distribution functions obtained within the window of channel measurements. Then fit the data points with a quadratic regression model $y_{i,j} = \alpha + \beta t_{i,j} + \gamma t_{i,j}^2 + e_{i,j}$, and thus estimation of the elapsed time is $\hat{t} = -\hat{\beta}_{\mathrm{LS}}/(2\hat{\gamma}_{\mathrm{LS}})$, where $\hat{\beta}_{\mathrm{LS}}$ and $\hat{\gamma}_{\mathrm{LS}}$ are the least-square estimators of β and γ, respectively. Different reference points can be used as well, such as the first local valley, the second local peak, and so on, to increase the accuracy of estimation. Therefore, the moving distance at time t can be estimated as $\hat{d}(t) = \hat{v}(t)\Delta t$, where Δt denotes the time duration between the current packet and the previous packet. The procedures of the presented TR-based distance estimator have been summarized in the flowchart shown in Figure 5.7.

Note that besides taking advantage of TR spatial focusing effect, the presented distance estimator also exploits the physical properties of EM waves and thus does not require any precalibration, while the estimator presented in our previous work [40] needs to measure the TRRS spatial decay curve in advance.

5.4 Moving Direction Estimation and Error Correction

In this section, we introduce the other two key components of WiBall: the IMU-based moving direction estimator and the map-based position corrector.

5.4.1 IMU-based Moving Direction Estimator

If the RX is placed in parallel to the horizontal plane, the change of moving direction can be directly measured by the readings of the gyroscope in the z-axis, i.e., $\theta(t_i) = \omega_z(t_{i-1})(t_i - t_{i-1})$, where $\omega_z(t_{i-1})$ denotes the angular velocity of the RX with respect to z-axis in its local coordinate system at time slot t_{i-1}. However, in practice, the angle of the inclination between the RX and the horizontal plane is not zero, as shown in Figure 5.8, and WiBall needs to transform the rotation of the RX into the change of the moving direction in the horizontal plane. Because the direction of the gravity $\vec{\mathbf{g}}/\|\vec{\mathbf{g}}\|$ can be estimated by the linear accelerometer, the rotation of the RX in the horizontal plane, which is orthogonal to the $\vec{\mathbf{g}}/\|\vec{\mathbf{g}}\|$, can be obtained by projecting the angular velocity vector $\vec{\omega} = \omega_x \hat{\mathbf{x}} + \omega_y \hat{\mathbf{y}} + \omega_z \hat{\mathbf{z}}$ with respect to its local coordinate system onto the direction $\vec{\mathbf{g}}/\|\vec{\mathbf{g}}\|$. Therefore, the change of moving direction $\theta(t_i)$ is obtained as

$$\theta(t_i) = \frac{\vec{\omega}^T(t_{i-1})\vec{\mathbf{g}}(t_{i-1})}{\|\vec{\mathbf{g}}(t_{i-1})\|} \cdot (t_i - t_{i-1}), \tag{5.10}$$

where $\vec{\omega}(t_{i-1})$ and $\vec{\mathbf{g}}(t_{i-1})$ denote the vector of angular velocity and the gravity at time t_{i-1}, respectively.

5.4.2 Map-Based Position Corrector

Because WiBall estimates the current location of the RX based on the previous locations, its performance is limited by the cumulative error. However, for typical indoor environments, there are certain constraints in the floorplan that can be utilized as landmarks, and thus, the cumulative errors may be corrected correspondingly as long as a landmark is identified. For example, Figure 5.9 shows a T-shaped corridor, and two possible estimated paths are illustrated in the figure. The moving distance of path #1 is under-

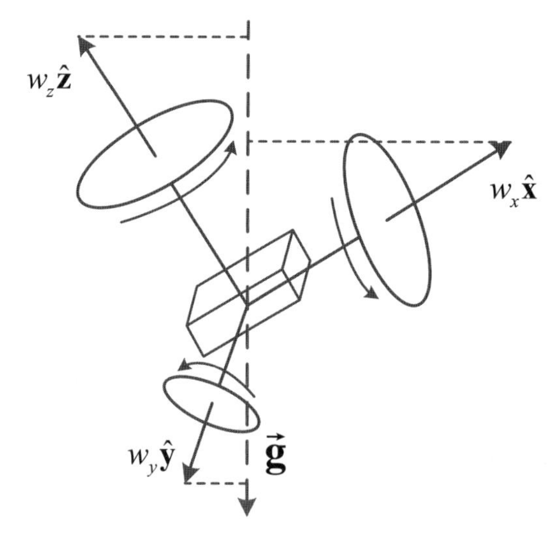

Figure 5.8 Transforming the rotation of RX into the moving direction in horizontal plane.

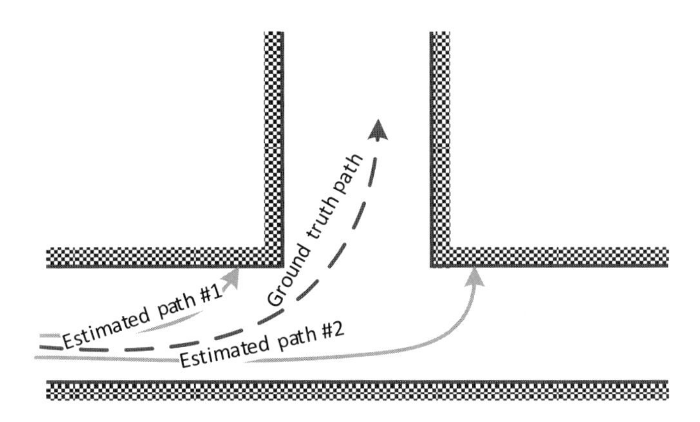

Figure 5.9 Two possible estimated paths and the ground truth path.

estimated and that of path #2 is overestimated, while the dotted line corresponds to the ground truth path. Both of the estimated traces would penetrate the wall in the floorplan if the errors are not corrected, which violates the physical constraints imposed by the structure of the building. In these two cases, a reasoning procedure can be implemented, and WiBall tries to find the most possible path that can be fitted to the floorplan where all the border constraints imposed by the floorplan are satisfied. Therefore, the cumulative errors of both the distance estimations and direction estimations can be corrected when a map-based position correction is implemented.

5.5 Performance Evaluation

To evaluate the performance of WiBall, various experiments are conducted in different indoor environments using the prototype as shown in Figure 5.2(a). In this section, we first evaluate the performance of the TR-based distance estimator. Then, the performance of WiBall in tracking a moving object in two different environments is studied. Last, the impact of packet loss and system window length on the presented system is also discussed.

5.5.1 Evaluations of TR Distance Estimator

The first experiment is to estimate the moving distance of a toy train running on a track. We put one RX on a toy train as shown in Figure 5.10(a) and place one TX about 20 m from the RX with two walls between them. The sampling period between adjacent channel measurement is set to $T = 5$ ms. CIRs are collected continuously when the toy train is running on the track. We also set an anchor point as shown in Figure 5.10(a) on the train track and collect the CIR when the train is at the anchor. The TRRS values between all the measured CIRs and the CIR of the anchor are computed and shown in Figure 5.10(b). The peaks in the dash line indicate that the train passes the anchor

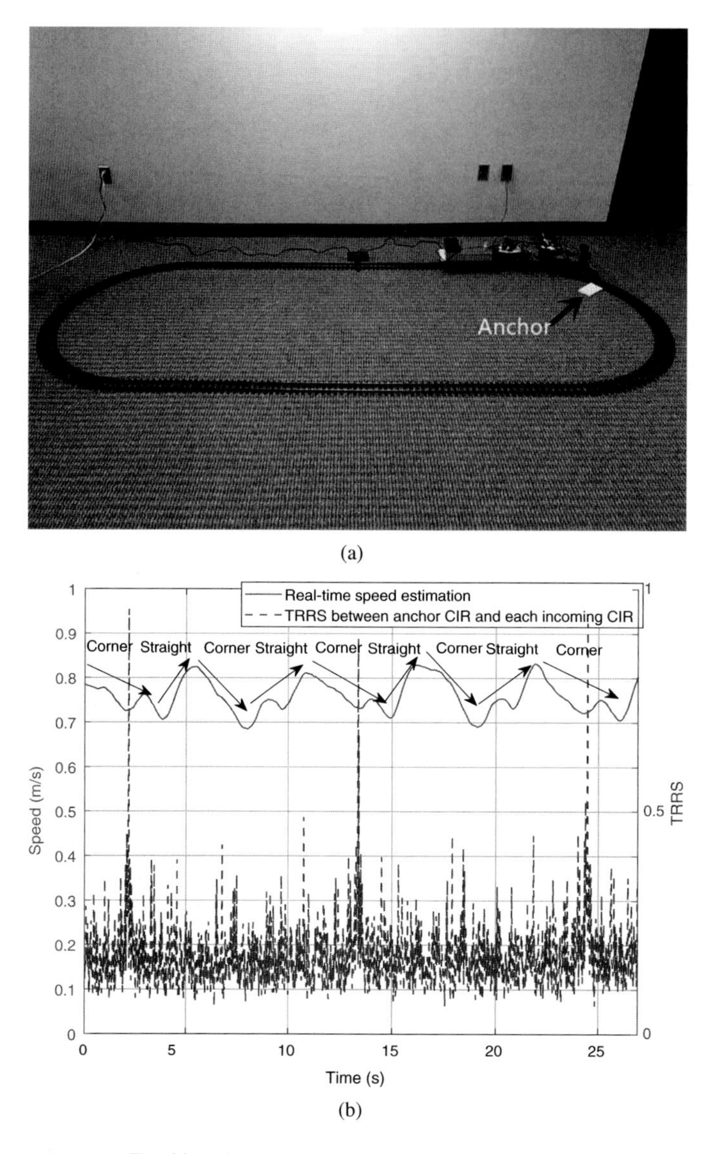

Figure 5.10 Tracking the speed of the toy train: (a) the toy train and the train track used in the experiment and (b) the estimated speed of the toy train over time.

three times. The estimated length of the track for this single loop is 8.12 m and the error is 1.50%, given the actual length of the train track is 8.00 m. The train slows down when it makes turns due to the increased friction and then speeds up in the straight line. This trend is reflected in the speed estimation shown by the solid curve. To show the consistency of the distance estimator over time, we collect the CIRs for 100 laps in total and estimate the track length for each lap separately. The histogram of the estimation results is shown in Figure 5.11. The mean of the estimation error is about 0.02 m and the

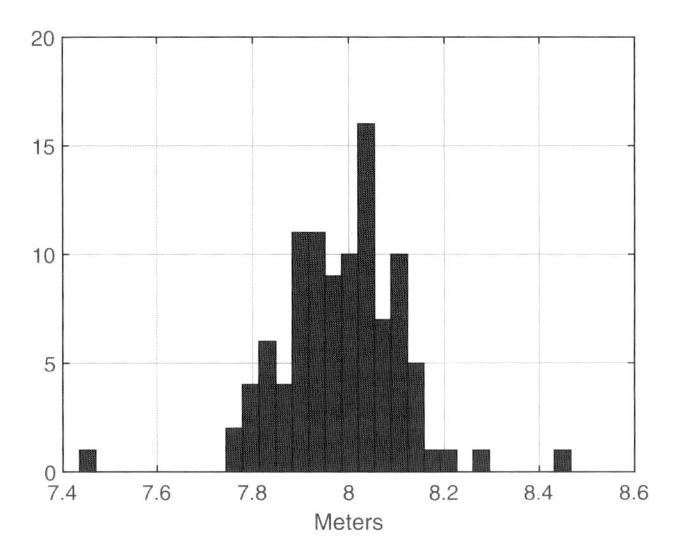

Figure 5.11 The histogram of the estimated track lengths for a total of 100 experiments.

standard error deviation is about 0.13 m, which shows that the estimation is consistent even over a long period.

The second experiment is to estimate the human walking distance. One RX is put on a cart, and one participant pushes the cart along the line from point A to point B shown in Figure 5.2(a) with an approximately constant speed of 1 m/s. To control the walking speed, the participant uses a timer and landmarks placed on the floor during the experiment. In the upper panel of Figure 5.12(a), for each time t, the TRRS values between the CIR measured at time t and those measured at time $t - \Delta t$, where $\Delta t \in (0s, 0.16s]$, are plotted along the vertical axis. As we can see from the figure, when the person moves slowly (e.g., at the beginning or the end of the experiment), the time differences between the local peaks of the measured TRRS distribution along the vertical axis are greater than that when the person moves faster. In addition, for $t \in [0.5\text{s}, 3.5\text{s}]$, the asymmetric part of the density function $P(\theta)$ of the energy of MPCs is more significant compared to the case when $t \in [3\text{s}, 9.5\text{s}]$ and thus the pattern of $J^2(kd)$ is less obvious than the latter one. The bottom panel of Figure 5.12(a) shows the corresponding walking speed estimation. The actual distance is 8 m and the estimated walking distance is 7.65 m, so the corresponding error is 4.4%. The loss of performance is from the blockage of signals by the human body, which reduces the number of significant MPCs.

We further let the participant carry the RX and walk for a distance of 2 m, 4 m, 6 m, 8 m, 10 m, and 12 m, respectively. The ground-truth travel distances are measured by a laser distance meter. For each ground-truth distance, the experiment is repeated 20 times with different paths, and the walking speed does not need to be constant. The results are shown in Figure 5.12(b), where the 5th, 25th, 75th, and 95th percentiles of the estimated distances for each actual distance are plotted from the bottom line to the top line for each

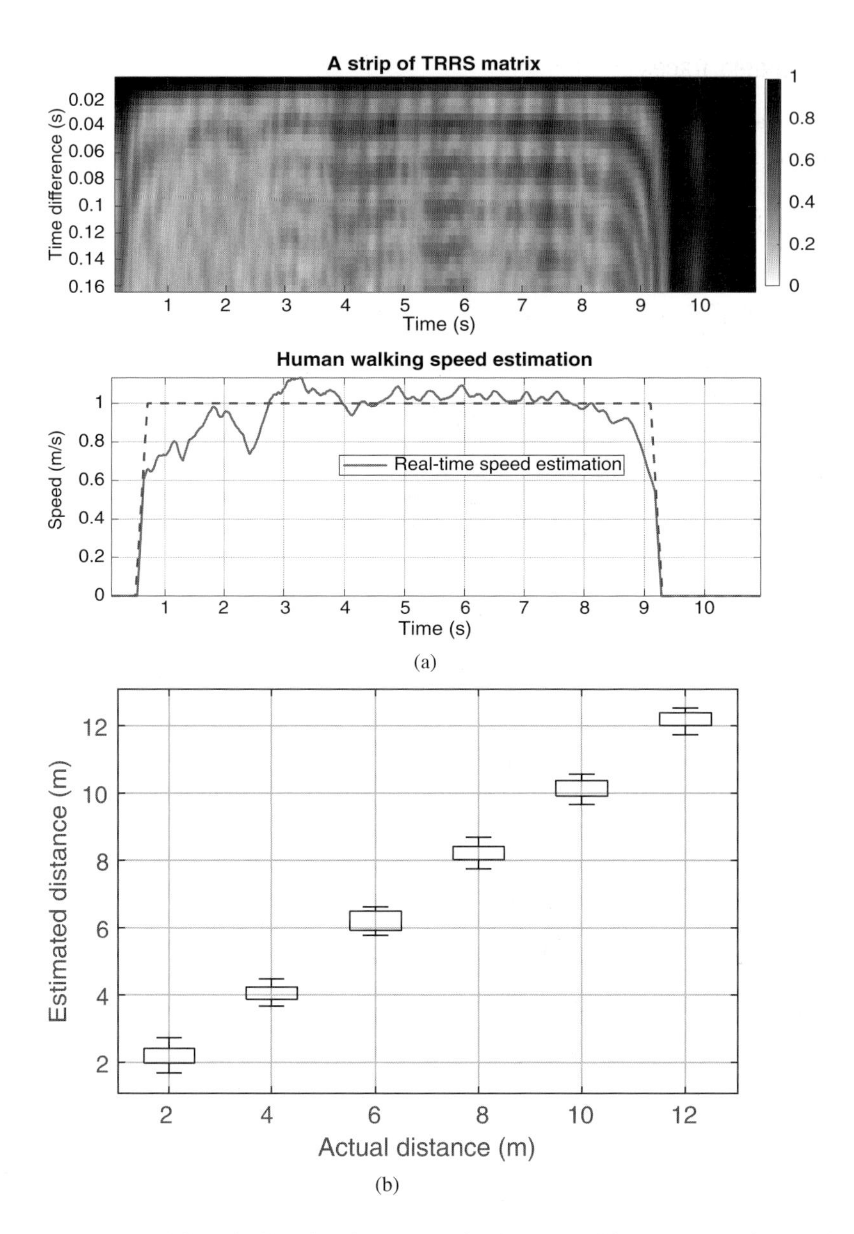

Figure 5.12 (a) Speed estimation for a controlled human walking speed (1m/s). (b) The results for walking distance estimation.

block. We find that when the ground-truth distance is small, the error tends to be large. This is mainly because the participant could introduce additional sources of errors that are uncontrollable, such as not following the path strictly, shaking during walking, and not stopping at the exact point in the end. When the distance is short, the impact of this kind of error can be magnified greatly. However, when the walking distance is large, the impact of the uncontrollable errors on the estimation result is insignificant.

5.5.2 Estimated Traces in Different Environment

We evaluate the performance of indoor tracking using WiBall in two sets of experiments. In the first set of experiments, a participant walks inside a building with a large open space. He carriers the RX with him and walks from Point A on the second floor to Point B on the first floor, as shown in Figure 5.13(a). The TX is placed close to the

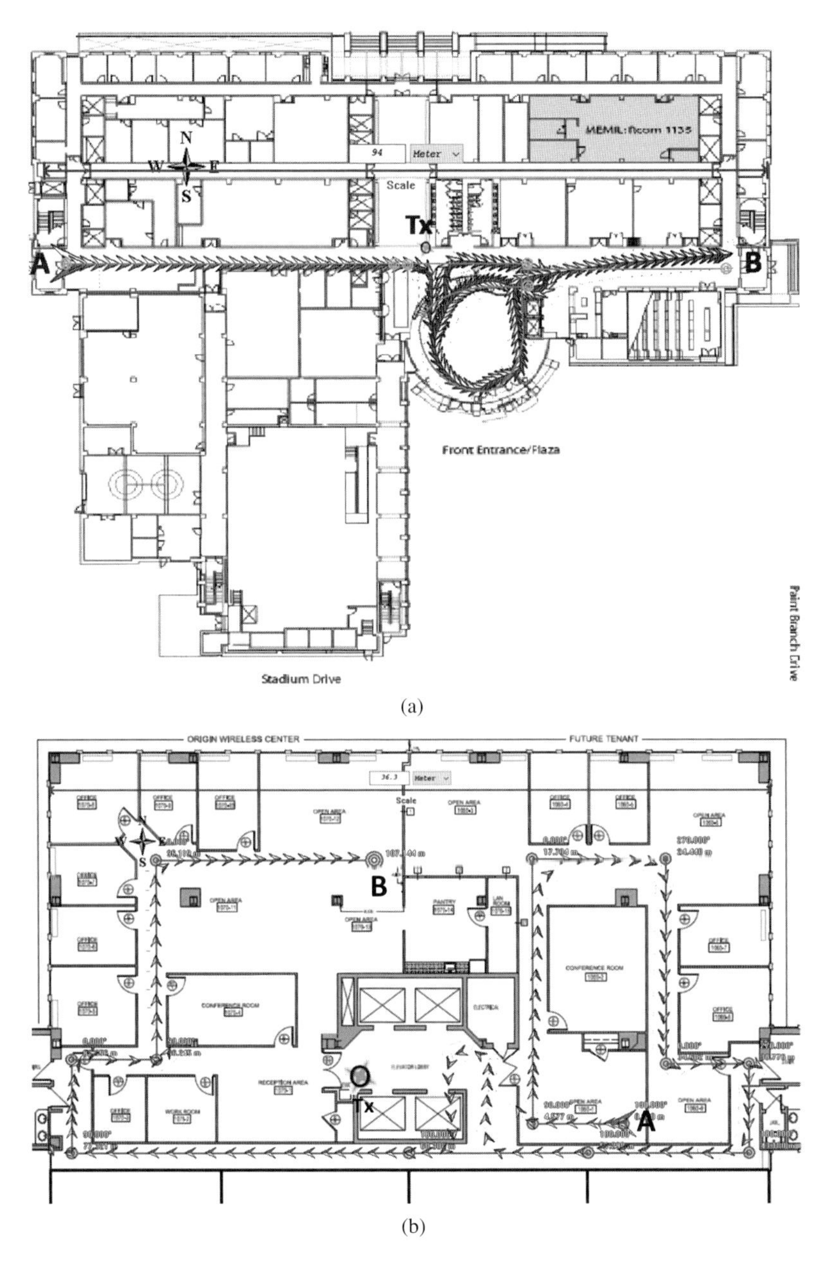

Figure 5.13 Experiment results in different environments: (a) estimated path in a building with a lot of open space and (b) estimated path in an office environment.

middle of the path on the second floor. The dimension of the building is around 94 m × 73 m. Although the moving distance of the first segment of the path is overestimated, the estimated path is corrected when the participant enters the staircase leading to the first floor.

In the second set of experiments, the participant walks inside an office environment. Figure 5.13(b) demonstrates a typical example of the estimated traces in a typical office of a multi-storey. One RX is put on a cart, and the participant pushes the cart along the route from Point A to Point B, as illustrated in the figure. The dimension of the environment is around 36.3 m × 19 m and the placement of the TX is also shown in the figure. As we can see from the figure, the estimated path matches the ground truth path very well because the cumulative errors have been corrected by the constraints from the floorplan.

5.5.3 Statistical Analysis of Localization Error

To evaluate the distribution of the localization errors, extensive experiments have been conducted in the same office environment shown in Figure 5.13(b). The participant pushes the cart with the RX on the cart, following the route as shown in Figure 5.14.

The RX starts from Point A and stops at different locations in the route shown in Figure 5.14. The lengths of the chosen paths are 5, 21, 25, 30, 40, 64, and 69 m, respectively, and the end of each path is marked with two circles. For each specific path, the experiment is repeated for 25 times. The estimation error for different paths has been analyzed through empirical cumulative distribution function (CDF), as shown in Figure 5.15. Based on the results, the median of the estimation error for the selected paths is around 0.33 m, and the 80 percentile of the estimation error is within 1 m. Therefore, WiBall is able to achieve a submeter median error in this complex indoor environment.

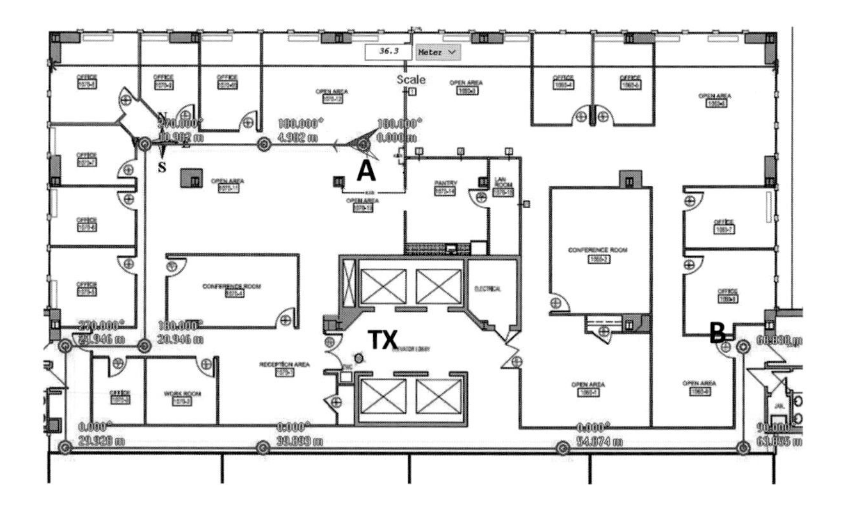

Figure 5.14 The route for the evaluation of statistical errors.

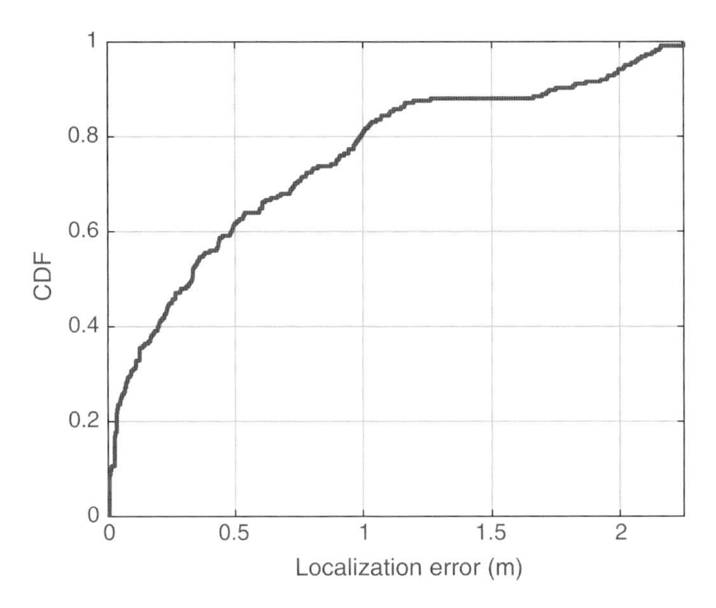

Figure 5.15 Empirical CDF of localization error.

5.5.4 Impact of Packet Loss on Distance Estimation

In the previous experiments, WiBall operates on a vacant band, and the packet loss rate can thus be neglected. However, in practice, the RF interference from other RF devices operated on the same frequency band will increase the packet loss rate. Because WiBall relies on the first peak of the TRRS distribution for distance estimation, enough samples need to be collected to estimate the first peak accurately, and a high packet loss can affect the peak estimation and thus increase distance estimation error.

To study the impact of RF interference, a pair of RF devices is configured to operate in the same frequency band of WiBall to act as an interference source, and we run the tracking system for 100 times with the ground-truth distance being 10 m. When the interfering devices are placed closer to the transmission pair of WiBall, WiBall encounters a higher packet loss rate. Therefore, to obtain various packet loss rates, the interfering devices are placed at different locations during the experiment. The average estimated distance and standard deviation of the estimation under different packet loss rates are shown in Figure 5.16. It is seen that a large packet loss rate would lead to an underestimation of the moving distance and increase the deviations of the estimates.

5.5.5 Impact of Window Length on Distance Estimation

In the following, the impact of window length on the performance of the presented TR-based distance estimator is studied. One implicit assumption of the presented estimator is that the speed of the moving device is constant within the observation window of channel measurements. The length of the observation window should be at least $0.61\lambda/v$, where v is the actual speed of the device. Furthermore, more samples of

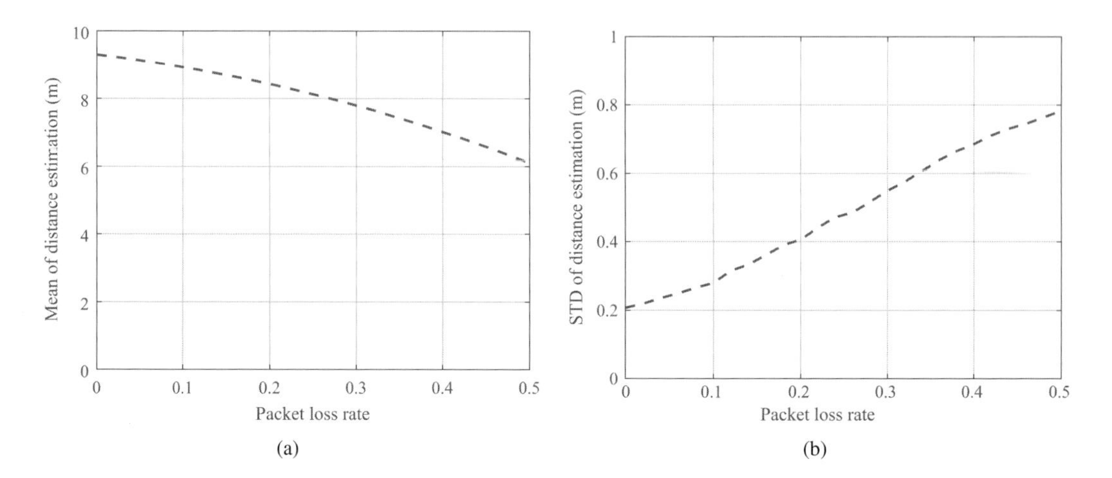

Figure 5.16 The impact of packet loss on the accuracy of distance estimation: (a) sample mean of the distance estimation and (b) standard deviation of the distance estimation.

channel measurements will also improve the accuracy of $\hat{t}$ as described in Section 5.3.4. Therefore, the window length is an important system parameter of WiBall.

In the following experiments, one RX is put on a toy train whose speed can be tuned and remains constant during each experiment. One TX is placed in two different locations: a LOS location where the RX is within the fields of vision of the TX, and a NLOS location where the direct path between the RX and TX is blocked by walls. The TX keeps transmitting packets with a uniform interval of 5 ms. Two experiments are conducted for each location of the TX with two different speeds of the toy train, 0.72 m/s and 0.63 m/s, respectively, and each experiment lasts 10 min. The 5th and 95th percentiles and the sample mean of the estimated speed have been shown in Figure 5.17 with different window lengths. It can be observed that when the window length is smaller than 30 samples, the speed estimates have a bias for the both cases; when the window length is greater or equal to 30 samples, the bias is close to zero, and the range of the estimates becomes stable. In addition, a higher accuracy can be achieved when the TX is placed in the NLOS location.

5.6 Summary

In this chapter, we present WiBall, which offers an accurate, low-cost, calibration-free, and robust solution for INIP in indoor environments without requiring any infrastructure. WiBall leverages the physical properties of the TR focusing ball in radio signals, such as Wi-Fi, LTE, 5G, etc., to estimate the moving distance of an object being monitored and does not need any specialized hardware. It is shown through extensive experiments that WiBall can achieve a decimeter-level accuracy. WiBall is also easily scalable and can accommodate a large number of users with only a single access point

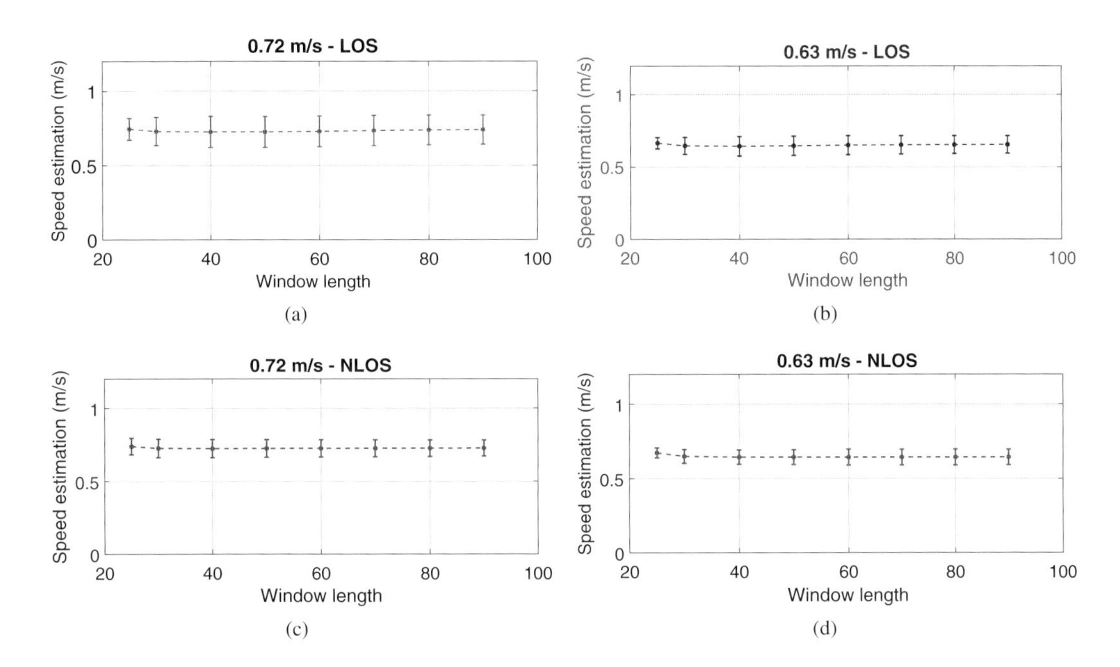

Figure 5.17 Impact of window length in terms of speed estimations.

or TX. Therefore, WiBall can be a very promising candidate for future indoor tracking systems. For related references, readers can refer to [42].

References

[1] "Global Indoor Positioning and Indoor Navigation (IPIN) Market 2017–2021," https://tinyurl.com/ya35sh2z [Accessed: 07-Dec-2017], 2017, [Online].

[2] C. Chen, Y. Chen, Y. Han, H.-Q. Lai, and K. J. R. Liu, "Achieving centimeter-accuracy indoor localization on Wi-Fi platforms: A frequency hopping approach," *IEEE Internet of Things Journal*, vol. 4, no. 1, pp. 111–121, 2017.

[3] H. Wang, S. Sen, A. Elgohary, M. Farid, M. Youssef, and R. R. Choudhury, "No need to war-drive: Unsupervised indoor localization," in *Proceedings of the 10th ACM International Conference on Mobile Systems, Applications, and Services*, pp. 197–210, 2012.

[4] Z. Yang, C. Wu, Z. Zhou, X. Zhang, X. Wang, and Y. Liu, "Mobility increases localizability: A survey on wireless indoor localization using inertial sensors," *ACM Computing Surveys (CSUR)*, vol. 47, no. 3, p. 54, 2015.

[5] G. Park, D. Hong, and C. Kang, "Level crossing rate estimation with Doppler adaptive noise suppression technique in frequency domain," in *IEEE 58th Vehicular Technology Conference,* vol. 2, pp. 1192–1195, 2003.

[6] A. Sampath and J. M. Holtzman, "Estimation of maximum Doppler frequency for handoff decisions," in *43rd IEEE Vehicular Technology Conference,* pp. 859–862, 1993.

[7] C. Xiao, K. D. Mann, and J. C. Olivier, "Mobile speed estimation for TDMA-based hierarchical cellular systems," *IEEE Transactions on Vehicular Technology*, vol. 50, no. 4, pp. 981–991, 2001.

[8] R. Narasimhan and D. C. Cox, "Speed estimation in wireless systems using wavelets," *IEEE Transactions on Communications*, vol. 47, no. 9, pp. 1357–1364, 1999.

[9] G. Lerosey, J. De Rosny, A. Tourin, A. Derode, G. Montaldo, and M. Fink, "Time reversal of electromagnetic waves," *Physical Review Letters*, vol. 92, no. 19, p. 193904, 2004.

[10] B. Wang, Y. Wu, F. Han, Y.-H. Yang, and K. J. R. Liu, "Green wireless communications: A time-reversal paradigm," *IEEE Journal on Selected Areas in Communications*, vol. 29, no. 8, pp. 1698–1710, 2011.

[11] P. Roux, B. Roman, and M. Fink, "Time-reversal in an ultrasonic waveguide," *Applied Physics Letters*, vol. 70, no. 14, pp. 1811–1813, 1997.

[12] M. Werner, M. Kessel, and C. Marouane, "Indoor positioning using smartphone camera," in *2011 IEEE International Conference on Indoor Positioning and Indoor Navigation (IPIN)*, pp. 1–6, 2011.

[13] R. Mautz and S. Tilch, "Survey of optical indoor positioning systems," in *2011 IEEE International Conference on Indoor Positioning and Indoor Navigation (IPIN)*, pp. 1–7, 2011.

[14] E. M. Gorostiza, J. L. Lázaro Galilea, F. J. Meca Meca, D. Salido Monzú, F. Espinosa Zapata, and L. Pallarés Puerto, "Infrared sensor system for mobile-robot positioning in intelligent spaces," *Sensors*, vol. 11, no. 5, pp. 5416–5438, 2011.

[15] I. Rishabh, D. Kimber, and J. Adcock, "Indoor localization using controlled ambient sounds," in *2012 IEEE International Conference on Indoor Positioning and Indoor Navigation (IPIN)*, pp. 1–10, 2012.

[16] D. K. Barton, *Radar System Analysis and Modeling*. Washington, DC: Artech House, 2004, vol. 1.

[17] L. Shangguan, Z. Yang, A. X. Liu, Z. Zhou, and Y. Liu, "STPP: Spatial-temporal phase profiling-based method for relative RFID tag localization," *IEEE/ACM Transactions on Networking (ToN)*, vol. 25, no. 1, pp. 596–609, 2017.

[18] M. Kuhn, C. Zhang, B. Merkl, D. Yang, Y. Wang, M. Mahfouz, and A. Fathy, "High accuracy UWB localization in dense indoor environments," in *2008 IEEE International Conference on Ultra-Wideband*, vol. 2, pp. 129–132, 2008.

[19] J.-Y. Lee and R. A. Scholtz, "Ranging in a dense multipath environment using an UWB radio link," *IEEE Journal on Selected Areas in Communications*, vol. 20, no. 9, pp. 1677–1683, 2002.

[20] D. Giustiniano and S. Mangold, "Caesar: Carrier sense-based ranging in off-the-shelf 802.11 wireless LAN," in *Proceedings of the Seventh ACM Conference on Emerging Networking Experiments and Technologies*, p. 10, 2011.

[21] S. Sen, D. Kim, S. Laroche, K.-H. Kim, and J. Lee, "Bringing CUPID indoor positioning system to practice," in *Proceedings of the 24th International Conference on World Wide Web*, International World Wide Web Conferences Steering Committee, pp. 938–948, 2015.

[22] P. Bahl and V. N. Padmanabhan, "Radar: An in-building RF-based user location and tracking system," in *Proceedings of the IEEE INFOCOM*, vol. 2, pp. 775–784, 2000.

[23] Y. Xie, Z. Li, and M. Li, "Precise power delay profiling with commodity Wi-Fi," in *Proceedings of the 21st Annual ACM International Conference on Mobile Computing and Networking*, pp. 53–64, 2015.

[24] M. Kotaru, K. Joshi, D. Bharadia, and S. Katti, "SpotFi: Decimeter level localization using Wi-Fi," in *ACM SIGCOMM Computer Communication Review*, vol. 45, no. 4, pp. 269–282, 2015.

[25] J. Gjengset, J. Xiong, G. McPhillips, and K. Jamieson, "Phaser: Enabling phased array signal processing on commodity Wi-Fi access points," in *Proceedings of the 20th Annual ACM International Conference on Mobile Computing and Networking*, pp. 153–164, 2014.

[26] S. Sen, J. Lee, K.-H. Kim, and P. Congdon, "Avoiding multipath to revive inbuilding Wi-Fi localization," in *Proceeding of the 11th Annual ACM International Conference on Mobile Systems, Applications, and Services*, pp. 249–262, 2013.

[27] M. Youssef and A. Agrawala, "The Horus WLAN location determination system," in *Proceedings of the 3rd ACM International Conference on Mobile Systems, Applications, and Services*, pp. 205–218, 2005.

[28] J. Xiong, K. Sundaresan, and K. Jamieson, "Tonetrack: Leveraging frequency-agile radios for time-based indoor wireless localization," in *Proceedings of the 21st Annual ACM International Conference on Mobile Computing and Networking*, 2015, pp. 537–549.

[29] S. A. Golden and S. S. Bateman, "Sensor measurements for Wi-Fi location with emphasis on time-of-arrival ranging," *IEEE Transactions on Mobile Computing*, vol. 6, no. 10, 2007.

[30] K. Qian, C. Wu, Z. Yang, Y. Liu, and K. Jamieson, "Widar: Decimeter-level passive tracking via velocity monitoring with commodity Wi-Fi," in *Proceedings of the 18th ACM International Symposium on Mobile Ad Hoc Networking and Computing*, p. 6, 2017.

[31] K. Qian, C. Wu, Y. Zhang, G. Zhang, Z. Yang, and Y. Liu, "Widar2. 0: Passive human tracking with a single Wi-Fi link," *Proceedings of ACM MobiSys*, 2018.

[32] P. Castro, P. Chiu, T. Kremenek, and R. Muntz, "A probabilistic room location service for wireless networked environments," in *International Conference on Ubiquitous Computing*, Springer, Berlin, Heidelberg, pp. 18–34, 2001.

[33] X. Wang, L. Gao, S. Mao, and S. Pandey, "DeepFi: Deep learning for indoor fingerprinting using channel state information," in *IEEE Wireless Communications and Networking Conference (WCNC)*, pp. 1666–1671, 2015.

[34] X. Wang, L. Gao, and S. Mao, "PhaseFi: Phase fingerprinting for indoor localization with a deep learning approach," in *IEEE Global Communications Conference (GLOBECOM)*, pp. 1–6, 2015.

[35] Z.-H. Wu, Y. Han, Y. Chen, and K. J. R. Liu, "A time-reversal paradigm for indoor positioning system," *IEEE Transactions on Vehicular Technology*, vol. 64, no. 4, pp. 1331–1339, 2015.

[36] J. Xiong and K. Jamieson, "ArrayTrack: A fine-grained indoor location system," In Presented as part of the 10th USENIX Symposium on Networked Systems Design and Implementation," pp. 71–84, 2013.

[37] K. Wu, J. Xiao, Y. Yi, D. Chen, X. Luo, and L. M. Ni, "CSI-based indoor localization," *IEEE Transactions on Parallel and Distributed Systems*, vol. 24, no. 7, pp. 1300–1309, 2013.

[38] S. Sen, B. Radunovic, R. Roy Choudhury, and T. Minka, "Precise indoor localization using PHY information," in *Proceedings of the 9th ACM International Conference on Mobile Systems, Applications, and Services*, pp. 413–414, 2011.

[39] H. El-Sallabi, P. Kyritsi, A. Paulraj, and G. Papanicolaou, "Experimental investigation on time reversal precoding for space–time focusing in wireless communications," *IEEE Transactions on Instrumentation and Measurement*, vol. 59, no. 6, pp. 1537–1543, 2010.

[40] F. Zhang, C. Chen, B. Wang, H.-Q. Lai, and K. J. R. Liu, "A time-reversal spatial hardening effect for indoor speed estimation," in *Proceedings of IEEE ICASSP*, pp. 5955–5959, Mar. 2017.

[41] Q. Xu, Y. Chen, and K. R. Liu, "Combating strong–weak spatial–temporal resonances in time-reversal uplinks," *IEEE Transactions on Wireless Communications*, vol. 15, no. 1, pp. 568–580, 2016.

[42] F. Zhang, C. Chen, B. Wang, H.-Q. Lai, Y. Han, and K. J. R. Liu, "WiBall: A time-reversal focusing ball method for decimeter-accuracy indoor tracking," *IEEE Internet of Things Journal*, vol. 5, no. 5, pp. 4031–4041, Oct. 2018.

Part II

Wireless Sensing and Analytics

6 Wireless Events Detection

In this chapter, we present a novel wireless indoor events detection system, TRIEDS. By leveraging the time-reversal (TR) technique to capture the changes of channel state information (CSI) in the indoor environment, TRIEDS enables low-complexity single-antenna devices that operate in the ISM band to perform through-the-wall indoor multiple events detection. The multipath phenomenon denotes that the electromagnetic signals undergo different reflecting and scattering paths in a rich-scattering environment. In TRIEDS, each indoor event is detected by matching the instantaneous CSI to a multipath profile in a training database. To validate the feasibility of TRIEDS and to evaluate the performance, we build a prototype that works on ISM band with carrier frequency being 5.4 GHz and a 125 MHZ bandwidth. Experiments are conducted to detect the states of the indoor wooden doors. Experimental results show that with a single receiver (AP) and transmitter (client), TRIEDS can achieve a detection rate higher than 96.92% and a false alarm rate smaller than 3.08% under either line-of-sight (LOS) or non-LOS transmission.

6.1 Introduction

The past few decades have witnessed the increase in the demand of surveillance systems that aim to capture and to identify unauthorized individuals and events. With the development of technologies, traditional outdoor surveillance systems have become more compact and lower cost. In order to guarantee the security in offices and residences, indoor monitoring systems are now ubiquitous, and their demand is rising both in quality and quantity. For example, they can be designed to guard empty houses and to alarm when break-ins happen.

Currently, most indoor monitor systems basically rely on video recording and require camera deployments in target areas. Techniques in computer vision and image processing are applied on the captured videos to extract information for real-time detection and analysis [1–4]. However, conventional vision-based indoor monitor systems have many limitations. They cannot be installed in places requiring a high level of privacy like restrooms or fitting rooms. Owing to the prevalence of malicious softwares on the Internet, vision-based indoor surveillance systems may lead to more dangers than protections, contradicting their intention. Moreover, vision-based approaches have a fundamental requirement of a line-of-sight (LOS) environment with enough illumination.

On the other hand, sensing with the wireless signals to detect indoor events has gained a lot of attention [5]. By utilizing the fact that the received radio frequency (RF) signals can be altered by the propagation environment, device-free indoor sensing systems are capable of capturing activities in the environment through the changes in received RF signals. Common features of RF signals to identify variations during signal transmission for indoor events detection include the received signal strength (RSS) and channel state information (CSI). Due to its susceptibility to the environmental changes, the RSS indicator (RSSI) has been applied to indicate and further recognize indoor activities [6–9]. Sigg et al. proposed a method that links the patterns of RSSI fluctuation to different human activities [7]. An approach where the direction of human movement was determined according to the RSSI degradation among different receivers was proposed in [8]. Recently, a RSSI-based gesture recognition system was built where seven gestures were identified with percent 56% accuracy [9]. Furthermore, CSI information, including the amplitude and the phase, is now accessible in many commercial devices and has been used for indoor event detection [10–16]. In [10], the first two largest eigenvalues of CSI correlation matrix were viewed as features to determine whether environment is static or dynamic. Abid et al. applied MIMO interference nulling technique to eliminate reflections off static objects and focus on a moving target and used beam steering and smoothed MUSIC algorithm to extract the angle information of targe [11]. Han et al. treated the CSI in the 3×3 MIMO system independently, and the standard deviations of the CSI were combined with SVM for human activity detection [12]. In [13], in order to locate the client with a fixed AP, both the amplitudes of the CSI and the frequency diversity in OFDM spectrum were used to build a model for calculating the distance between the AP and the client. In [14] the histograms of the CSI amplitudes were utilized to distinguish between different human activities. In [15] a coarse relationship between variation in CSI amplitudes and the number of persons present was established. Wang et al. proposed the CARM that leveraged the CSI-speed and CSI-activity models for detection [16]. Moreover, a lip reading system based on Wi-Fi signals was developed where the features of mouth motions were extracted through the discrete wavelet packet decomposition on CSI's amplitudes and classified with the help of dynamic time wrapping [17]. However, most aforementioned CSI-based indoor sensing systems rely on only the amplitudes of the CSI, whereas the phase information is discarded regardless of how informative it is.

Another category of technologies in device-free indoor monitor systems is adopted from radar imaging technology to track targets [18–21]. The radar technique can identify the delays of subnanoseconds in the time-of-flight (ToF) of wireless signals through different paths, by using the ultra-wideband (UWB) sensing. Hence, radar-based systems are capable of separating the reflection from the moving object behind the walls against the reflections from walls or other static objects [18]. However, the UWB transmission is impractical in commercial indoor monitoring systems, because it requires specific hardwares for implementation. Recently, Katabi et al. proposed a new radar-based system to keep track of different ToFs of reflected signals by leveraging a specially designed frequency modulated carrier wave (FMCW) that sweeps over different carrier frequencies [19–21]. However, their techniques consume over 1 GHz bandwidth to sense the environment and only the images of result are obtained from the sensors, which requires further effort to detect the types of indoor events.

The aforementioned device-free systems have limitations in that they either require multiple antennas and dedicated sensors or require LOS transmission environment and ultra-wideband to capture features that can guarantee the accuracy of detection. In contrast, in this chapter, we present a time-reversal (TR) based wireless indoor events detection system, TRIEDS, capable of through-the-wall indoor events detections with only one pair of single-antenna devices. In the wireless transmission, the multipath is the propagation phenomenon that the RF signal reaches the receiving antenna through two or more different paths. TR technique treats each path of the multipath channel in a rich scattering environment as a widely distributed virtual antenna and provides a high-resolution spatial-temporal resonance, commonly known as the focusing effect [22]. In physics, the TR spatial-temporal resonance can be viewed as the result of the resonance of electromagnetic (EM) field in response to the environment. When the propagation environment changes, the involved multipath signal varies correspondingly, and consequently the spatial-temporal resonance also changes.

Taking use of the spatial-temporal resonance, a novel TR-based indoor localization approach, namely TRIPS, was recently proposed in [23]. By exploiting the unique location-specific characteristic of channel impulse response (CIR), TR creates a spatial-temporal resonance that focuses the energy of the transmitted signal only on the intended location. The TRIPS mapped the real physical location to the estimated CSI through the spatial-temporal resonance. The TR indoor locationing system was implemented on a Wi-Fi platform, and the concatenated CSI from a total equivalent bandwidth of 1 GHz has been treated as the location-specific fingerprints [24]. Through NLOS experiments, the Wi-Fi-based TR indoor locationing system achieved a perfect 5 cm precision with a single access point (AP). TR-based indoor locationing system was an active localization system in that it required the object to be located to carry one of the transmitting or receiving device, such that the difference in the TR resonances between different locations of device is large.

Based on a similar principle as TRIPS, we utilize the TR technique to capture the variations in the multipath CSI due to different indoor events, and present TRIEDS for indoor event detection. More specifically, thanks to the nature of TR that captures the variations in the CSI, maps different multipath profiles of indoor events into separate points in the TR space, and compresses the complex-valued features into a real-valued scalar called the spatial-temporal resonance strength, the presented TRIEDS supports simplest detection and classification algorithms with a good performance. Compared with previous works on indoor monitoring systems, which require multiple antennas, dedicated sensors, ultra-wideband transmission, or LOS environment and rely on only the amplitude information in the CSI, TRIEDS introduces a novel and practical solution that can well support through-the-wall detection and only requires low-complexity single-antenna hardware operating in the ISM band. To demonstrate the capability of TRIEDS in detecting indoor events in real office environments, we build a prototype that operates at 5.4 GHz band with a bandwidth of 125 MHz, as shown in Figure 6.1, and conduct extensive experiments in an office on the tenth floor of an sixteen-story

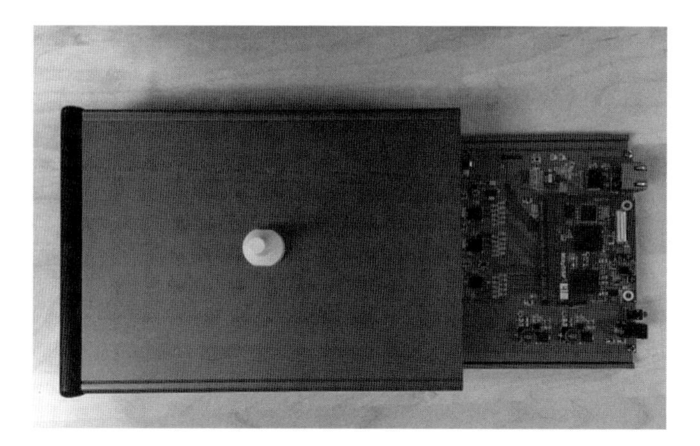

Figure 6.1 Prototype of TRIEDS.

building. During the experiments, we test the capability of TRIEDS of monitoring the states of multiple doors at different locations simultaneously. Using only one pair of single-antenna devices, TRIEDS could achieve perfect detection in LOS scenario and near 100% accuracy in detection when events happen in the absence of LOS path between the transmitter (TX) and the receiver (RX).

The rest of the chapter is organized as follows. In Section 6.2, the system overview for TRIEDS is briefly discussed and an introduction to TR technique is given. The details of how TRIEDS works are studied and analyzed in Section 6.3, consisting of an offline training phase and an online testing phase. Moreover, extensive experiments of TRIEDS in detecting indoor events in real office environments are conducted and the experimental results are investigated in Section 6.4. Based on the results in Section 6.4, we further discuss how the system parameters and human motions will affect the accuracy of TRIEDS.

6.2 TRIEDS Overview

When an EM signal travels over the air in a rich-scattering indoor environment, it encounters reflectors and scatters that alter and attenuate signals differently. Consequently, the received signal at the receiving antenna is a combination of multiple altered copies of the same transmitted signal coming from different paths and suffering different delays. This phenomenon is well known as multipath propagation. In order to detect an indoor event, wireless sensors should be capable of tracking the targets against all other interferences. The previous indoor monitoring work can be categorized into two classes. The first class ignores the multipath effect and only uses a single-valued CSI feature like RSSIs for detection, which leads to the degradation of accuracy to some extent. On the other hand, the second class tries to separate different components in a multipath channel, by means of UWB transmission and specially designed modulated signals.

The previous work either views the multipath as the compromise to the system or separates the components in the multipath CSI by radar-based techniques. As opposed to them, TRIEDS is considered as a novel system that monitors and detects different indoor events by utilizing TR technique.

6.2.1 Time-Reversal Technique

The principle of the TR technique can be referred to Chapter 1. As originally investigated in the phase compensation over telephone line [25], TR technique was then extended to the acoustics [26]. The spatial-temporal resonance of the TR has been proposed as theory and validated through experiments in both acoustic domain and RF domain [27]. In the RF domain, the property of TR spatial-temporal resonances of EM waves have been studied in [28, 29]. Moreover, the TR technique relies on two assumptions, i.e., the channel reciprocity and the channel stationarity. The channel reciprocity demonstrates the phenomenon that the CSI for both the forward and the backward links is highly correlated, whereas the channel stationarity requires that the CSI remains highly correlated during a certain period. Both of the assumptions were validated in [30, 31] and [23], respectively.

In the indoor environment, there exists a large amount of propagation paths for EM signals due to the presence of scatters and reflectors. As long as the indoor propagation environment changes, the received multipath profile varies accordingly. As demonstrated in Figure 6.2, each dot in the CSI logical space represents an indoor event or location, which is uniquely determined by the multipath profile $\mathbf{h}$. By taking a time-reverse and conjugate operation over the multipaths, the corresponding TR signatures $\mathbf{g}$ are generated and the points in the CSI logical space as marked by "A," "B," and "C" are mapped into the TR space as "A'," "B'," and "C'." In the TR space, the similarity between two indoor events or indoor locations is quantified by the strength of TR

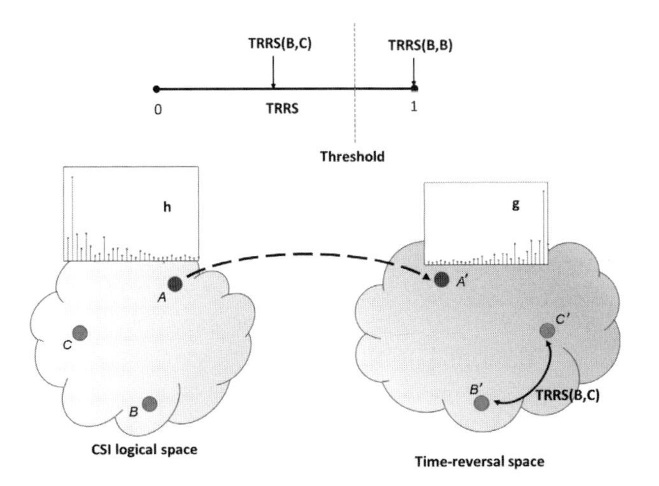

Figure 6.2 Mapping between the CSI logical space and the time-reversal space.

resonances. The definition of TR resonating strength (TRRS) is given in (6.3), where $\mathbf{h}_1$ and $\mathbf{h}_2$ represent the multipath profiles in the CSI logical space, and $\mathbf{g}_2$ is the TR signature in the TR space. The higher the TRRS is, the more similar two points are in the TR space. Similar events defined by a threshold on TRRS will be treated as a single class in TRIEDS. Leveraging the TR technique, a centimeter-level accurate indoor locationing system, named as TRIPS, was proposed in [23]. In TRIPS, each of the indoor physical locations was mapped into a logical location in the TR space and can be easily separated and identified using TRRSs. Taking advantage of the TR space to separate multipath profiles with small differences, TRIEDS is capable of monitoring and detecting different indoor events with a high accuracy.

6.3 System Model

In this section, we present a detailed introduction to the presented TR-based indoor events detection system, TRIEDS. The TRIEDS exploits the intrinsic property of TR technique that the spatial-temporal resonance fuses and compresses the information of the multipath propagation environment. To implement the indoor events detection based on the TR spatial-temporal resonances, TRIEDS consists of two phases: the offline training and the online testing. During the first phase, a training database is built by collecting the signature $\mathbf{g}$ of each indoor events through the TR channel probing phase. After training, in the second phase, TRIEDS estimates the instantaneous multipath CSI $\mathbf{h}$ for current state and makes the prediction according to the signatures in the offline training database by means of the strength of the generated spatial-temporal resonance. The detailed operations are discussed in the following.

6.3.1 Phase 1: Offline Training

As discussed earlier, TRIEDS leverages the unique indoor multipath profile and TR technique to distinguish and detect indoor events. During the offline training phase, we are going to build a database where the multipath profiles of any targets are collected and stored the corresponding TR signatures in the TR space. Unfortunately, due to noise and channel fading, the CSI from a specific state may slightly change over the time. To combat this kind of variations, for each state, we collect several instantaneous CSI samples to build the training set.

Specifically, for each indoor state $S_i \in \mathcal{D}$ with $\mathcal{D}$ being the state set, the corresponding training CSI is estimated and form a $\mathbf{H}_i$ as,

$$\mathbf{H}_i = [\mathbf{h}_{i,t_0}, \quad \mathbf{h}_{i,t_1}, \ldots, \ \mathbf{h}_{i,t_{N-1}}], \tag{6.1}$$

where N is the size of the CSI samples for a training state. $\mathbf{h}_{i,t_j}$ represents the estimated CSI vector of state S_i at time t_j, and $\mathbf{H}_i$ is named as the CSI matrix for state S_i. An example of estimated indoor CSI obtained by the prototype in Figure 6.1 is shown in Figure 6.3, where the total length of the CSI is 30. From Figure 6.3(a), we can find out that there exist at least 10 to 15 significant multipath components.

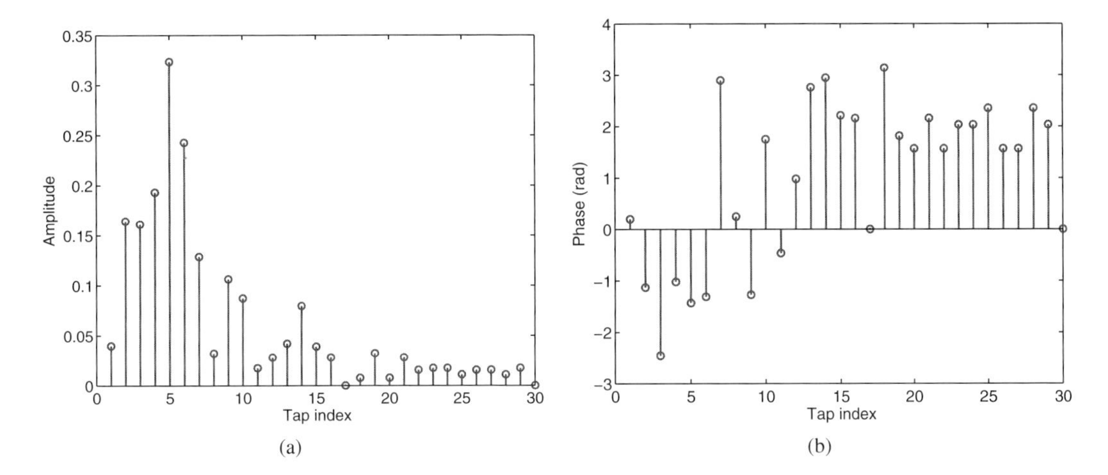

Figure 6.3 An example of indoor CSI: (a) amplitude of CSI and (b) phase of CSI.

The corresponding TR signature matrix $\mathbf{G}_i$ can be obtained by time-reversing the conjugated version of $\mathbf{H}_i$ as:

$$\mathbf{G}_i = [\mathbf{g}_{i,t_0}, \quad \mathbf{g}_{i,t_1}, \ldots, \mathbf{g}_{i,t_{N-1}}], \tag{6.2}$$

where the TR signature $g_{i,t_j}[k] = h^*_{i,t_j}[L-k]$ here, the superscript $*$ on a vector variable represents the conjugate operator. L denotes the length of a CSI vectors, and k denotes the index of taps. Then the training database $\mathcal{G}$ is the collection of $\mathbf{G}_i$'s.

6.3.2 Phase 2: Online Testing

After constructing the training database $\mathcal{G}$, TRIEDS is ready for real-time indoor events detection. The indoor events detection is indeed a classification problem. Our objective is to detect the state of indoor targets through evaluating the similarity between the testing TR signatures and the TR signatures in the training database $\mathcal{G}$. The raw CSI information is complex-valued and of high dimensions, which complicates the detection problem and increases the computational complexity if we directly treat the CSI as the feature. To tackle this problem, by leveraging the TR technique, we are able to naturally compress the dimensions of the CSI vectors through mapping them into the strength of the spatial-temporal resonances. The definition of the strength of the spatial-temporal resonance is given as follows.

Definition: The strength of the spatial-temporal resonance $\mathcal{TR}(\mathbf{h}_1, \mathbf{h}_2)$ between two CSI samples $\mathbf{h}_1$ and $\mathbf{h}_2$ is defined as

$$\mathcal{TR}(\mathbf{h}_1, \mathbf{h}_2) = \left(\frac{\max_i \left|(\mathbf{h}_1 * \mathbf{g}_2)[i]\right|}{\sqrt{\sum_{l=0}^{L-1} |h_1[l]|^2} \sqrt{\sum_{l=0}^{L-1} |h_2[l]|^2}} \right)^2, \tag{6.3}$$

where "$*$" denotes the convolution and $\mathbf{g}_2$ is the TR signature of $\mathbf{h}_2$ as

$$g_2[k] = h_2^*[L-k-1], \quad k = 0, 1, \ldots, L-1. \tag{6.4}$$

When comparing two estimated multipath profiles, they are first mapped into the TR space where each of them is represented as one TR signature. Then the TR spatial-temporal resonating strength is a metric that quantifies the similarity between these two multipath profiles in the mapped TR space. The higher the TRRS is, the more similar two multipath profiles are in the TR space. The resonating strength defined in (6.3) is similar to the definition of cross-correlation coefficient between $\mathbf{h}_1$ and $\mathbf{h}_2$ as the inner product of $\mathbf{h}_1$ and $\mathbf{h}_2^*$, which is equivalent to $(\mathbf{h}_1 * \mathbf{g}_2)[L-1]$. However, the numerator in (6.3) is the maximal absolute value in the convolved sequence. This step is important, in terms of combating any possible synchronization error between two CSI estimations, e.g., the first several taps of CSI may be missed or added in different measurements. Hence, due to its robustness to the synchronization errors in the CSI estimation, the TRRS is capable of capturing all the similarities between multipath CSI samples and increasing the accuracy.

During the online monitoring phase, the receiver keeps matching the current estimated CSI to the TR signatures in $\mathcal{G}$ to find the one that yields the strongest TR spatial-temporal resonance. The TRRS between the unknown testing CSI $\tilde{\mathbf{H}}$ and state S_i is defined as

$$\mathcal{TR}_{S_i}(\tilde{\mathbf{H}}) = \max_{\tilde{\mathbf{h}} \in \tilde{\mathbf{H}}} \max_{\mathbf{h}_i \in \mathbf{H}_i} \mathcal{TR}(\tilde{\mathbf{h}}, \mathbf{h}_i), \tag{6.5}$$

where $\tilde{\mathbf{H}}$ is a group of CSI samples assumed to be drawn from the same state as

$$\tilde{\mathbf{H}} = [\tilde{\mathbf{h}}_{t_0}, \tilde{\mathbf{h}}_{t_1}, \ldots, \tilde{\mathbf{h}}_{i,t_{M-1}}], \tag{6.6}$$

and M is the number of CSI samples in one testing group, similar to the N in the training phase defined in (6.1).

Once we obtain the TRRS for each event, the most possible state for the testing CSI matrix $\tilde{\mathbf{H}}$ can be found by searching for the maximum among $\mathcal{TR}_{S_i}(\tilde{\mathbf{H}})$, $\forall\, i$, as

$$S^* = \arg\max_{S_i \in \mathcal{D}} \mathcal{TR}_{S_i}(\tilde{\mathbf{H}}). \tag{6.7}$$

The superscript $*$ on S denotes the optimal.

Besides finding the most possible state S^* by comparing the TR spatial-temporal resonances, TRIEDS adopts a threshold-trigger mechanism, in order to avoid false alarms introduced by events outside of the state class $\mathcal{D}$. TRIEDS reports a change of states to S^* only if the TRRS $\mathcal{TR}_{S^*}(\tilde{\mathbf{H}})$ reaches a predefined threshold γ.

$$\hat{S} = \begin{cases} S^*, & \text{if } \mathcal{TR}_{S^*}(\tilde{\mathbf{H}}) \geq \gamma, \\ 0, & otherwise, \end{cases} \tag{6.8}$$

where $\hat{S} = 0$ means the state of current environment is not changed, i.e., TRIEDS is not triggered for any trained states in $\mathcal{D}$. According to the aforementioned detection rule, a false alarm for state S_i happens whenever a CSI is detected as $\hat{S} = S_i$ but it is not from state S_i.

Although the algorithm for TRIEDS is simple, the accuracy of indoor events detection is high, and its performance is validated through multiple experiments in the next section.

6.4 Experimental Evaluation

To empirically evaluate the performance of TRIEDS, we conduct several experiments for door states detection in a commercial office environment with different transmitter-receiver locations.

To begin with, a simple LOS experiment for validating the feasibility of TRIEDS is conducted in a controlled environment, with 7 transmitter locations, one receiver location, and two events. Then, the validation is further extended to both LOS and NLOS cases in a controlled office environment with 3 receiver locations, 15 locations for transmitter, and 8 targeted doors made of wood. Meanwhile, experiments are conducted in an uncontrolled indoor environment during normal working hours with people around. Furthermore, the performance of the TRIEDS is also compared with that of the RSS-based indoor monitoring approach, which can be easily extracted from the channel information and classified the using k-nearest neighbor (kNN) method. To further evaluate the accuracy of the TRIEDS in real environments, the performance of TRIEDS with intentional human movements is studied. Last but not least, results of TRIEDS being as a guard system to secure a closed room are discussed.

6.4.1 Experimental Setting

The prototype of the TRIEDS requires one pair of single-antenna transmitter and receiver that work on the ISM band with the carrier frequency being 5.4 GHz and a 125 MHz bandwidth. Moreover, during the experiment, the system runs with a channel probing interval around 20 millisecond (ms). A snapshot of the hardware device for TRIEDS is shown in Figure 6.1 with the antenna installed on the top of the radio box.

The experiments are carried out in the offices at the 10th floor in a commercial building of 16 floors in total. The experimental offices are surrounded by multiple offices and elevators. The detailed setup is shown in the floorplans in Figure 6.4. During the experiments, we are detecting the open/closed states of multiple wooden doors labeled as D1 to D8. The TX-RX locations include both LOS and NLOS transmissions.

In TRIEDS experiments, the receiver and the transmitter are placed on the top of stands at the intended locations, with the height from the ground being 4.3 ft and 3.6 ft, respectively, as shown in Figure 6.5(a) and Figure 6.5(b).

In all the experiments, we choose the number of the training CSI and the testing CSI to be $N = 10$ and $M = 10$ as defined in (6.1) and (6.6).

6.4.2 Feasibility Validation

To begin with, the feasibility for the TRIEDS to detect indoor events is verified in a LOS case where the receiver is placed at the location "D" in Figure 6.4, the transmitter is moving along the seven dots in a vertical line in Figure 6.4 with the dot closest to the targeted door labeled as index "1." Our task is to detect whether the wooden door D3 is closed or open.

The multipath CSI samples for D3 open and closed are obtained through TR channel probing phase and the corresponding TR signatures are stored in the database. In the

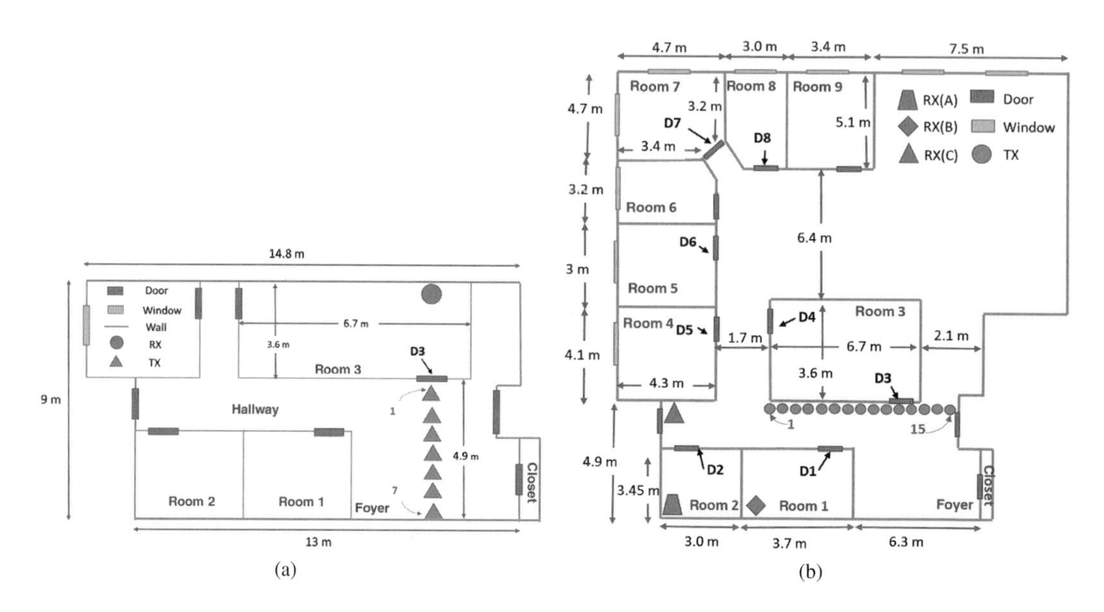

(a) (b)

Figure 6.4 Experimental setting for TRIEDS: (a) setting 1 floor plan and (b) setting 2 floor plan.

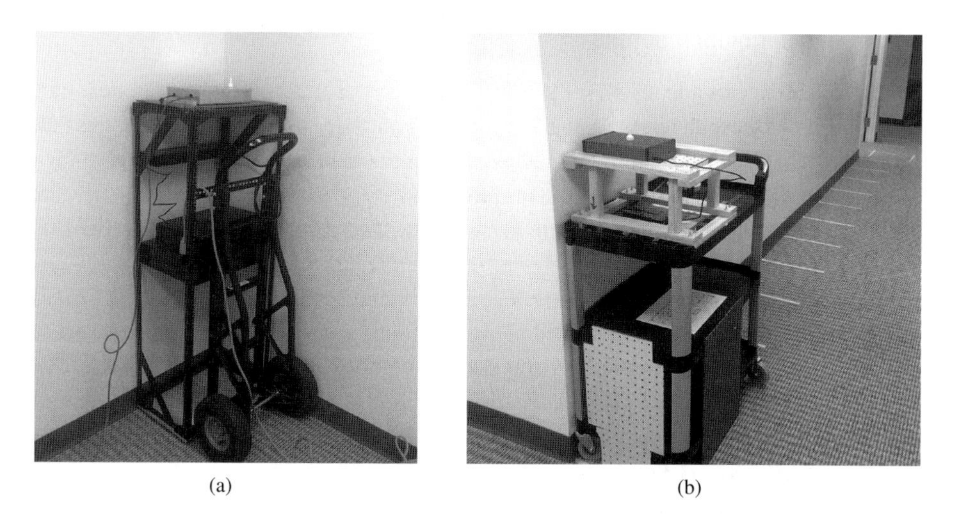

(a) (b)

Figure 6.5 Experiment setting: (a) the receiver and (b) the transmitter.

testing phase, we keep listening to the multipath channel and matching the collected testing CSI to the database for. With any threshold γ smaller than 0.97, we can achieve the perfect detection for all the seven transmitter locations as in Table 6.1.

In this case, the TRIEDS indeed performs a detection for the events on the LOS path between the transmitter and the receiver. Through this simple experiment, we have demonstrated the feasibility of TRIEDS to use the TR spatial-temporal resonance to capture the changes in the indoor multipath environment. Next, the performance of TRIEDS is further evaluated under more complicated changes of the multipath environment and with both LOS and NLOS TX-RX transmissions.

Table 6.1 Performance of the TRIEDS in easy case.

Location index	1	2	3	4	5	6	7
False alarm rate	0	0	0	0	0	0	0
Detection rate	1	1	1	1	1	1	1

6.4.3 Single-Door Monitoring

In this part, the experiments are conducted to understand how locations of the receiver, the transmitter, and the targeted objects affect the performance of TRIEDS. The receiver is placed at location "A," "B," and "C," whereas the transmitter is moving along the 15 locations marked by the circles and separated by 0.5 meters in a horizontal line as shown in Figure 6.4. The objective of TRIEDS is to monitor the states of wooden door D1. During the experiment, for each location and each indoor event, we measure 3,000 samples of the CSI, which lasts about 5 minutes by using our built prototype, leading to a total experimental time to be 10 minutes for each TX-RX location.

Here, the location "A" (LOC A) represent a through-the-wall detection scenario in the absence of a LOS path between the transmitter and the receiver and between the receiver and where the indoor event happens. Under the case when the receiver is at the location "B" (LOC B), there is always a LOS path between the receiver and where the indoor event happens because they are in the same room. However, the LOS path between the transmitter and the receiver disappears regarding most of the possible transmitter locations, and it exists only if the transmitter, the receiver, and the door D1 form a line. However, the transmitter and the receiver always perform LOS transmission when the receiver is at the location "C" (LOC C). Meanwhile, the door D1 to be detected falls outside the LOS link between the transmitter and the receiver.

6.4.3.1 LOC A: NLOS Case

As we discussed earlier, when the receiver is on LOC A, there is no LOS path between the receiver and the transmitter, and the receiver and door D1 are isolated by walls. One example of the multipath CSI for the open and the closed state of door D1 is shown in Figure 6.6. In Figure 6.6 where only the amplitudes of the CSI are plotted, it is clear to observe a change in how the energy is distributed on each tap. In the TRIEDS, not only the amplitude information but also the phase for each tap is taken into consideration by means of the TR spatial-temporal resonance.

From the experiment, with a threshold γ no larger than 0.9, we can achieve a perfect detection rate and zero false alarm rate for all 15 transmitter locations. Hence, we can conclude that TRIEDS is capable of detecting an event in an NLOS environment with through-the-wall detection, and the distance between the receiver and the transmitter has little effect on the performance.

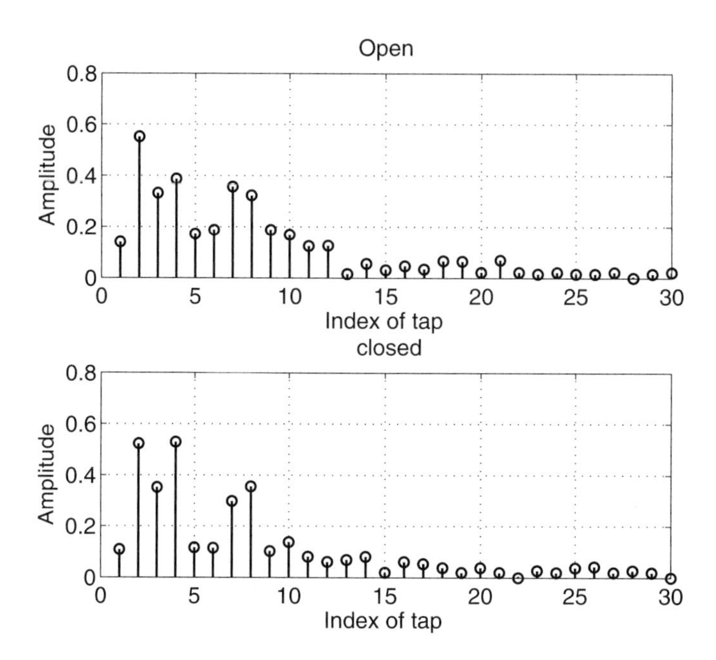

Figure 6.6 Multipath profiles (amplitude part) of door D1 under LOC A.

6.4.3.2 LOC B: LOS and NLOS Case

When the receiver is on LOC B, as the transmitter moving from the location "1" to the location "4" (the fourth dot right to the one marked as "1"), the transmission scenario between the transmitter and the receiver is NLOS due to the absence of a direct LOS link. Then, the transmission scenario become LOS, when the transmitter is on the location "5" to the location "6." When the transmitter moves farther away (i.e., from the dot "7"), there is no LOS path again between the transmitter and the receiver, and the transmission scenario becomes NLOS. In Figures 6.7(a) and 6.7(b), examples of the CSI for each event are plotted to demonstrate the changes in the amplitudes of the multipath profile corresponding to the indoor event.

Considering the accuracy for TRIEDS, with a threshold $\gamma \leq 0.9$, the detection rate for all 15 transmitter locations is higher than 99.9%. Except when the transmitter is at the location "6," the detection probability drops to 95.9%. Nevertheless, the corresponding false alarm rates are all below 0.1%. Because the experiment is carried out in a commercial office building, there exist outside activities that we cannot control but indeed change the multipath CSI to fall out the collected indoor events. So the reason for the detection probability at the sixth location being 95.9% might be the existence of uncontrollable outside activities. For example, the elevator running may greatly change the outside multipath propagation because it is close to the environmental office and is made of metal. Moreover, generally, TRIEDS is robust to the various distances between the transmitter, the receiver, and where the indoor event happens.

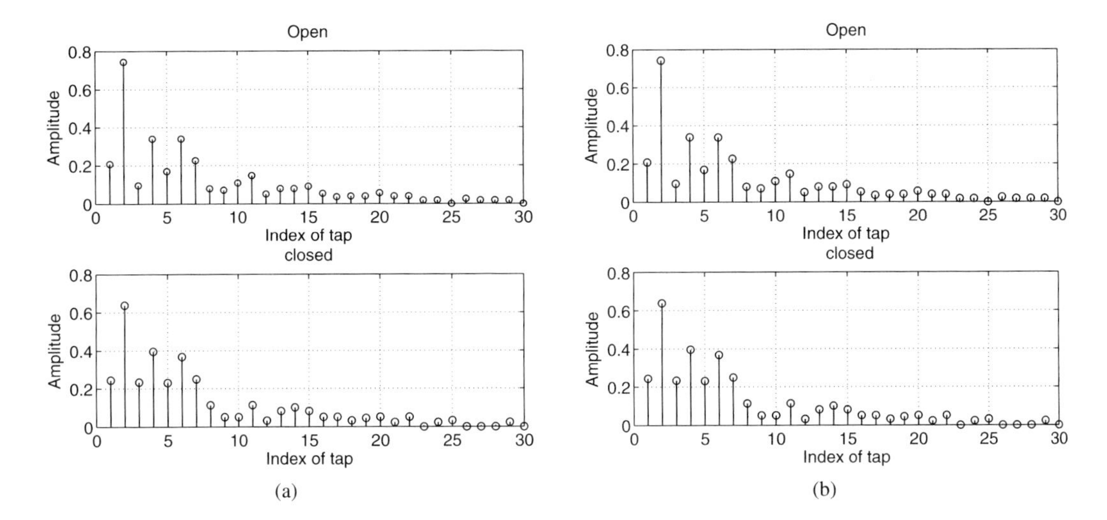

Figure 6.7 Multipath profiles of door D1 under LOC B: (a) multipath profiles (amplitude part) when TX on location "1" (NLOS) and (b) multipath profiles (amplitude part) when TX on location "5" (LOS).

6.4.3.3 LOC C: LOS Case

When the receiver is on LOC C, no matter which circle the transmitter is on, they are transmitting under LOS scenario, which leads to a dominant multipath component that exists in the multipath CSI.

The LOS transmission brings difficulties to indoor events detection when an event locates outside the LOS path between the transmitter and the receiver. The reason for that can be decomposed into two parts. In the first place, in this experiment, the object door D1 is located parallel with the transmission link between the transmitter and the receiver and has little influence to the dominant LOS component in the multipath profile. Second, because more energy is focused on the LOS path dominant in the CSI, the other multipath components that contain the event information are more noise-like and less informative. Hence, as most of the information for the event is buried in the CSI components with only a little energy, it is hard to detect an event happening outside the direct link between the transmitter and the receiver in an LOS-dominant wireless system. This can be shown by an example of the multipath CSI with respect to the open and closed states of door D1 in Figure 6.8, where the dominant path remains the same and contains most of the energy in the CSI.

In the experiment, TRIEDS yields a 100% detection rate and a 0 false alarm rate for all the 15 transmitter locations with the threshold $\gamma \leq 0.93$. The experimental result supports our claim that the TRIEDS can capture even minor changes in the multipath profile by using TR technique.

6.4.4 TRIEDS in Controlled Environments

In the previous sections, we have validated the capability of the presented system of detecting two indoor events with both LOS and NLOS transmission in controlled indoor

Table 6.2 State list for TRIEDS to detect

State index	Description
S_1	All the doors are open.
S_{i+1}	Door Di closed and the others open, $\forall i = 1, 2, \ldots, 8$.

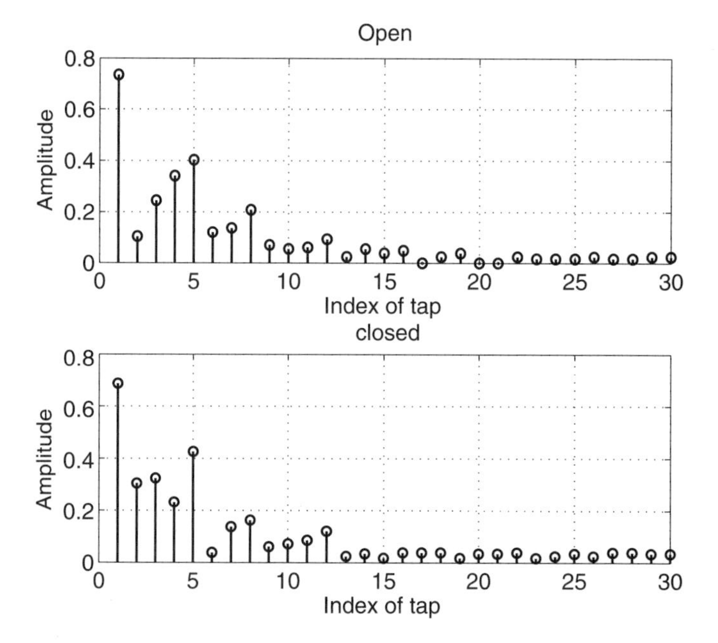

Figure 6.8 Multipath profiles (amplitude part) of door D1 under LOC C.

environments. In this part, we are going to study the performance of TRIEDS in detecting multiple indoor events. Moreover, the performance comparison between the RSSI-based indoor detecting approach and the TRIEDS is further investigated.

In the experiment, the receiver is placed on either LOC B or LOC C, whereas the transmitter moves and stops on every two circles that are separated by 1 meter, named from "axis 1" to "axis 4," respectively. In total, we have two receiver locations and four transmitter locations, i.e., eight TX-RX locations. The objective of TRIEDS is to detect which wooden doors among D1 to D8 is closed versus all other doors are open, as labeled in Figure 6.4. During the experiment, for each TX-RX location and each event, we measure 3,000 CSI samples, which takes approximately 5 min, leading to a total monitoring time of 45 min. Table 6.2 is the state table describing all the indoor events in the experiment.

As we claimed and verified in the single-event detection experiment, the TRIEDS can achieve highly accurate detection performance by utilizing the spatial-temporal resonance to capture changes in the multipath profiles. In this section, we evaluate the

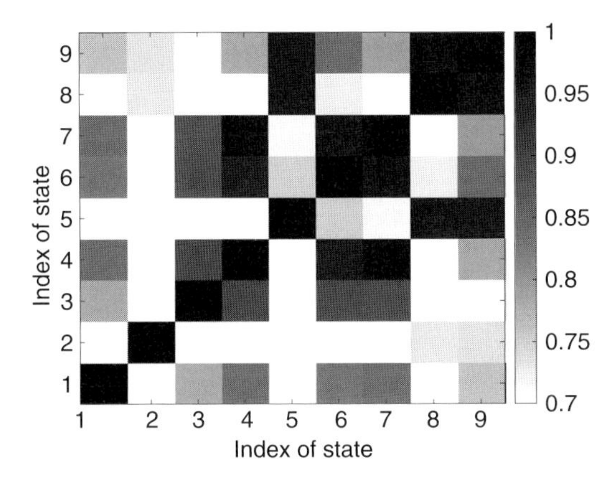

Figure 6.9 Resonance strength map with RX on LOC B and TX on the 1st circle (axis 1).

capability of TRIEDS of detecting multiple events in a controlled indoor environment. The performance analysis for normal office environment during working hours will be discussed in Section 6.4.5.

6.4.4.1 Evaluations on LOC B

To begin with, the performance of TRIEDS when the receiver is on LOC B is studied. In Figure 6.9, we show how the TRRS varies between different events.

Due to the fact that doors D5 and D6 are close to each other whereas they are far away to the receiver and the transmitter, the introduced changes in the multipath profiles of both of them are similar. Consequently, the resonance strength between states S_6 and S_7 is relatively higher than other off-diagonal elements, but it is still smaller than the diagonal ones in Figure 6.9 that represent the in-class resonance strength. Similar phenomenon happens between states S_8 and S_9.

In Figures 6.10 and 6.11, examples of the receiver operating characteristic (ROC) curves for detecting states of indoor doors are plotted for both the TRIEDS system and the conventional RSSI approach. Here, the legend "axis i," $i = 1, 2, 3, 4$, denotes the location of transmitter to be on the $(2 * i - 1)$th green dot in Figure 6.4.

As shown by Figures 6.10 and 6.11, the TRIEDS outperforms the RSSI-based approach in distinguishing between one door that is closed (i.e., S_i, $i \geq 1$) versus all doors that are open (i.e., S_0), by achieving perfect detection and zero false alarm rate. Note that S_9 is the state of door D8, which is blocked from the TX-RX link by a closed office, as an example, Figure 6.10 demonstrates the superiority of TRIEDS in performing a through-the-wall detection. Meanwhile, the performance of the RSSI-based approach degrades as the distance between where the indoor event happens and the TX-RX gets smaller. By leveraging the TR technique, TRIEDS is capable of capturing the changes in a multipath environment in a form of a multidimensional and complex-valued vector with high degree of freedoms, and of distinguishing between different changes in the TR spatial-temporal resonance domain. However, the

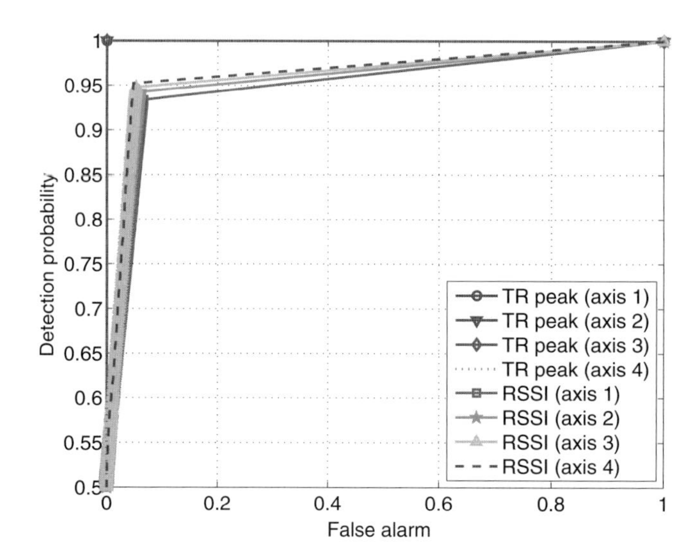

Figure 6.10 ROC curve for distinguishing between S_1 and S_2 under LOC B.

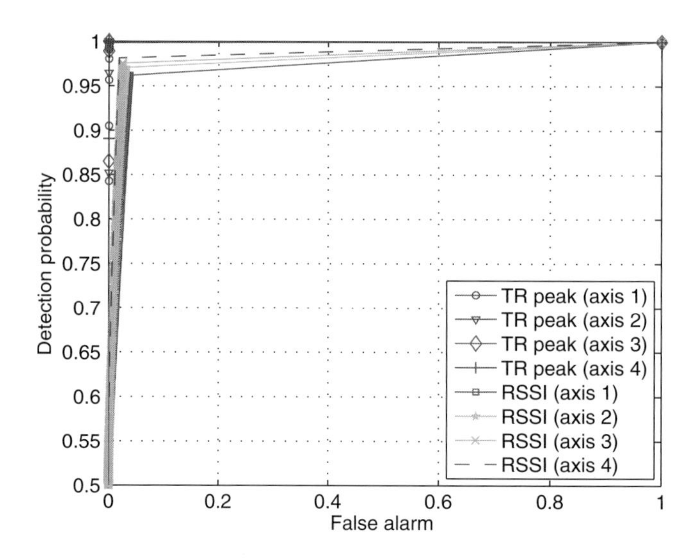

Figure 6.11 ROC curve for distinguishing between S_1 and S_9 under LOC B.

RSSI-based approach tries to monitor the changes in the environment through a real-valued scalar, which due to its dimension loses most of the distinctive information.

Furthermore, the accuracy of detection of TRIEDS improves as the distance between the transmitter and the receiver increases. So does the RSSI-based method. The reason is that when the transmitter and the receiver get far away, more energy will be distributed to the multipath components with longer distance, and thus the sensing system will have a larger coverage. The overall performance obtained by average on all possible events shows that TRIEDS outperforms the RSSI approach in Table 6.3.

Table 6.3 False alarm and detection probability for multievent detection on LOC B in controlled environment

LOC B	Axis 1	Axis 2	Axis 3	Axis 4
Detection rate TRIEDS (%)	99.12	99.5	99.67	99.81
False alarm TRIEDS (%)	0.88	0.5	0.33	0.19
Detection rate RSSI (%)	89.41	91.16	92.07	93.07
False alarm RSSI (%)	10.59	8.84	7.93	6.93

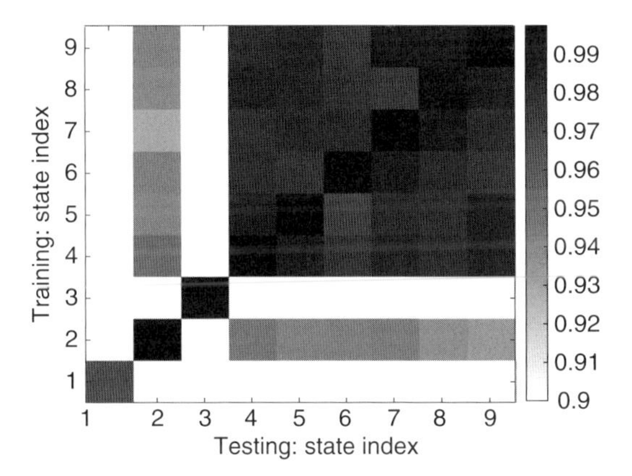

Figure 6.12 Resonance strength map with RX on LOC C and TX on the first circle (axis 1).

6.4.4.2 Evaluations on LOC C

Experiments are further conducted to evaluate the performance of indoor multiple events detection in an LOS transmission scenario by putting the receiver on LOC C. In Figure 6.12, we show the strengths of the TR spatial-temporal resonances between different indoor events. When the receiver and the transmitter transmit in an LOS setting, the CSI is LOS-dominant such that the energy of the multipath profile is concentrated only on a few taps. It makes the coverage of TRIEDS shrink and degrades the performance of TRIEDS, especially when the indoor events happen far from the TX-RX link as shown in Figure 6.12.

Examples of ROC curves to illustrate the detection performance of both TRIEDS and the RSSI-based approach are plotted in Figures 6.13 and 6.14. The performance of the TRIEDS working in an LOS environment is similar to that in an NLOS environment. Generally, TRIEDS achieves a better accuracy for events detection with a lower false alarm rate, compared with the RSSI-based approach. In both scenarios, TRIEDS achieves almost perfect detection performance in differentiating between $S_i, i \geq 1$, and S_0. Moreover, the RSSI method has a better accuracy in the LOS case than that in the NLOS case.

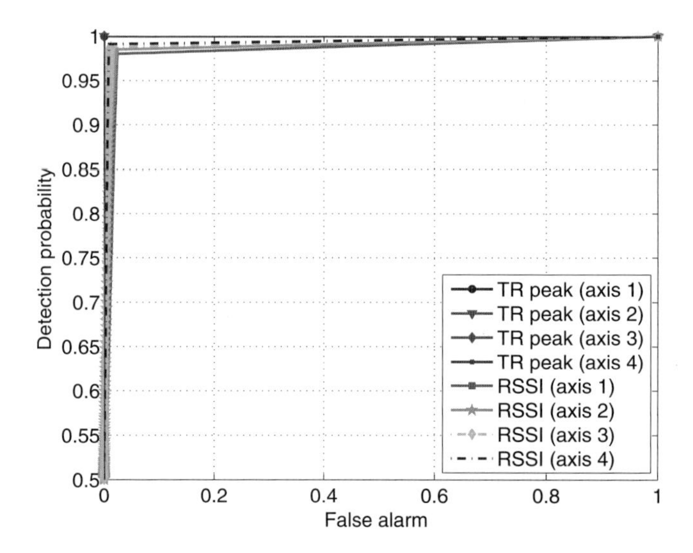

Figure 6.13 ROC curve for distinguishing between S_1 and S_2 under LOC C.

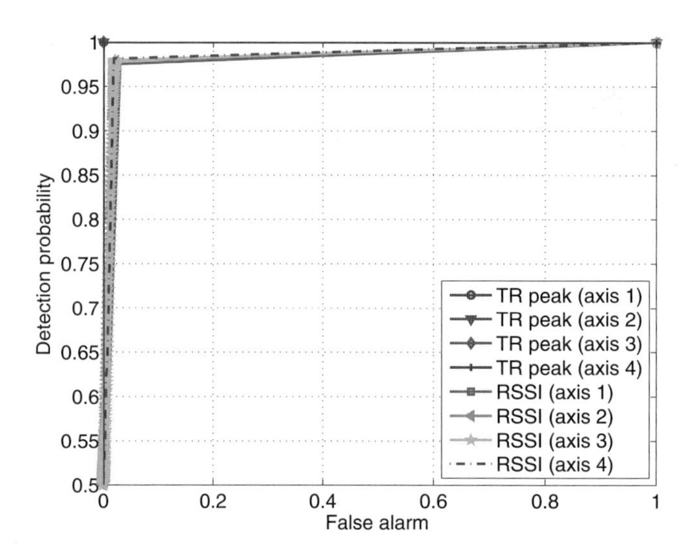

Figure 6.14 ROC curve for distinguishing between S_1 and S_9 under LOC C.

The corresponding overall performance comparison for TRIEDS and the RSSI-based method is shown in Table 6.4. It is obvious that the farther the receiver and the transmitter are separated, the better accuracy TRIEDS achieves. Moreover, compared with Table 6.3, the accuracy of the RSSI-based method improves a lot in the LOS environment, whereas the one of TRIEDS degrades slightly. Moreover, comparing the results in Table 6.3 and Table 6.4, the detection performance for TRIEDS degrades a little when the receiver and the transmitter change the transmission scheme from NLOS to LOS. Because of the dominant LOS path in LOS transmission, the ability to perceive

Table 6.4 False alarm and detection probability for multievent detection on LOC C in controlled environment

LOC C	Axis 1	Axis 2	Axis 3	Axis 4
Detection rate TRIEDS (%)	99.09	99.28	99.31	99.35
False alarm TRIEDS (%)	0.91	0.72	0.69	0.65
Detection rate RSSI (%)	97.24	97.66	97.8	97.88
False alarm RSSI (%)	2.76	2.34	2.2	2.12

Table 6.5 False alarm and detection probability for multievent detection of TRIEDS in normal environment (LOC B)

LOC B	Axis 1	Axis 2	Axis 3	Axis 4
Detection rate TRIEDS (%)	96.92	98.95	99.23	99.4
False alarm TRIEDS (%)	3.08	1.05	0.77	0.6
Detection rate RSSI (%)	92.5	94.16	94.77	95.36
False alarm RSSI (%)	7.5	5.84	5.23	4.64

multipath components that are far away from the direct link degrades, leading to a worse detection accuracy.

6.4.5 TRIEDS in Normal Office Environments

In this section, we repeat the experiments in Section 6.4.4 during working hours in weekdays where approximately ten individuals are working in the experiment area, and all offices surrounding and locating beneath or above the experimental area are occupied with uncontrollable individuals.

The TRIEDS achieves similar accuracy compared with that of the controlled experiment in Section 6.4.4. The overall false alarm and the detection rate for TRIEDS and the RSSI-based approach are shown in the Table 6.5 and Table 6.6.

The results in Tables 6.5 and 6.6 are consistent with the results in Tables 6.3 and 6.4. The performance for TRIEDS is superior to that of the RSSI-based approach, by realizing a better detection rate and a lower false alarm rate. Even in the dynamic environment, the TRIEDS can maintain a detection rate higher than 96.92% and a false alarm smaller than 3.08% under the NLOS case, whereas there is a detection rate higher than 97.89% and a false alarm smaller than 2.11% under the LOS case. Moreover, as the distance between the receiver and the transmitter increases, the accuracy of both

Table 6.6 False alarm and detection probability for multievent detection of TRIEDS in normal environment (LOC C)

LOC C	Axis 1	Axis 2	Axis 3	Axis 4
Detection rate TRIEDS (%)	97.89	98.94	99.18	99.36
False alarm TRIEDS (%)	2.11	1.06	0.82	0.64
Detection rate RSSI (%)	96.73	97.19	97.35	97.43
False alarm RSSI (%)	3.27	2.81	2.65	2.57

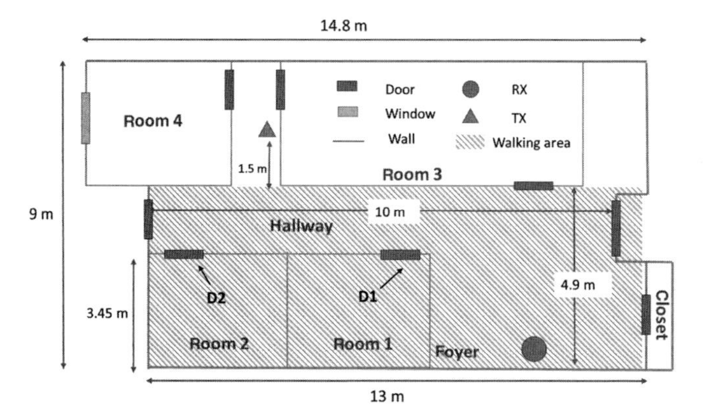

Figure 6.15 Experiment setting for study on human movements.

methods improves. In the comparison of Tables 6.3, 6.4, 6.5, and 6.6, we claim that the TRIEDS has a better tolerance to the environment dynamics.

6.4.6 TRIEDS with Intentional Human Movements

To investigate on the effects that the human movements have on the performance of TRIEDS, we conduct experiments with none, one, and two individuals that keep moving back and forth in the shaded area as Figure 6.15 shows. Meanwhile, the transmitter is put on the triangle and the receiver is on the circle, detecting the states of two adjacent doors labeled as "D1" and "D2." The list of door states is in Table 6.7. For each set of experiments, TRIEDS detects the states of the two doors for 5 min during the normal working hours.

Interference caused by the human movements changes the multipath propagation environment and brings in the variations in the TR spatial-temporal resonances during the monitoring process of TRIEDS. Fortunately, due to the mobility of humans, the introduced interference keeps changing and the duration for each interference is short. To combat the resulted bursted variations in the TRRSs, we adopt the majority vote

Table 6.7 State list for study on human movements

State	00	01	10	11
D1	Open	Open	Closed	Closed
D2	Open	Closed	Open	Closed

Table 6.8 Accuracy comparison of TRIEDS under human movements

Experiment	No HM (%)	One HM (%)	Two HM (%)
No smoothing	97.75	87.25	79.58
With smoothing	98.07	94.37	88.33

method combined with a sliding window to smooth the detection results over time. Supposing we have the previous $K - 1$ outputs S_k^*, $k = t - K + 1, \ldots, t - 1$ and the current result S_t^*, then the decision for time stamp t is made by majority vote over all S_k^*, $k = t - K + 1, \ldots, t$, so on and so forth for all t. K denotes the size of the sliding window for smoothing.

In Table 6.8, we compare the average accuracy over all states for TRIEDS with or without the smoothing algorithm in the absence of human movements (HM), and in the presence of the intentional persistent human movements performed by one individual and two individuals. Here, the length of the sliding window is $K = 20$. First of all, the accuracy of TRIEDS reduces as the number of individuals increases, performing persistent movements near the location of the indoor events to be detected, the transmitter and the receiver. Moreover, the adopted smoothing algorithm improves the robustness of TRIEDS to human movements and enhances the accuracy by 7% to 9% compared with that of the case without smoothing. Meanwhile, during the experiments, we also find that the most vulnerable state is state "00," where all doors are open, such that with human movements TRIEDS is more likely to yield a false alarm than other states. The reason is that as a human moves close to the door location, the human body, viewed as an obstacle at the door location, is similar to a closed wooden door, and hence the changes in the multipath CSI are also similar, especially for D1.

6.4.7 TRIEDS for through-the-Wall Guard

Unlike the previous experiments where we are trying to detect the door states, in this part, TRIBOD is functioning as a through-the-wall guard system. The objective for TRIEDS is to secure a target room through walls and to alarm not only when the door state changes but also when unexpected human movements happen inside the secured room. The system setup is shown in Figure 6.16, where the secured room is shaded.

In this experiment, the transmitter and the receiver of TRIEDS, marked as triangles and circles, are placed in two rooms, respectively, as shown in Figure 6.16. TRIEDS

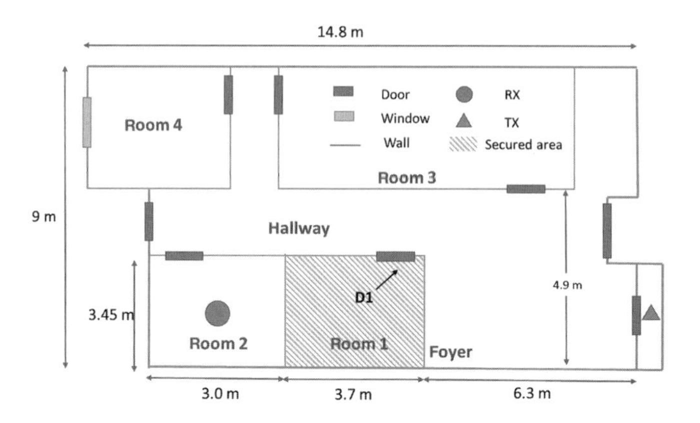

Figure 6.16 Experiment setting for guarding.

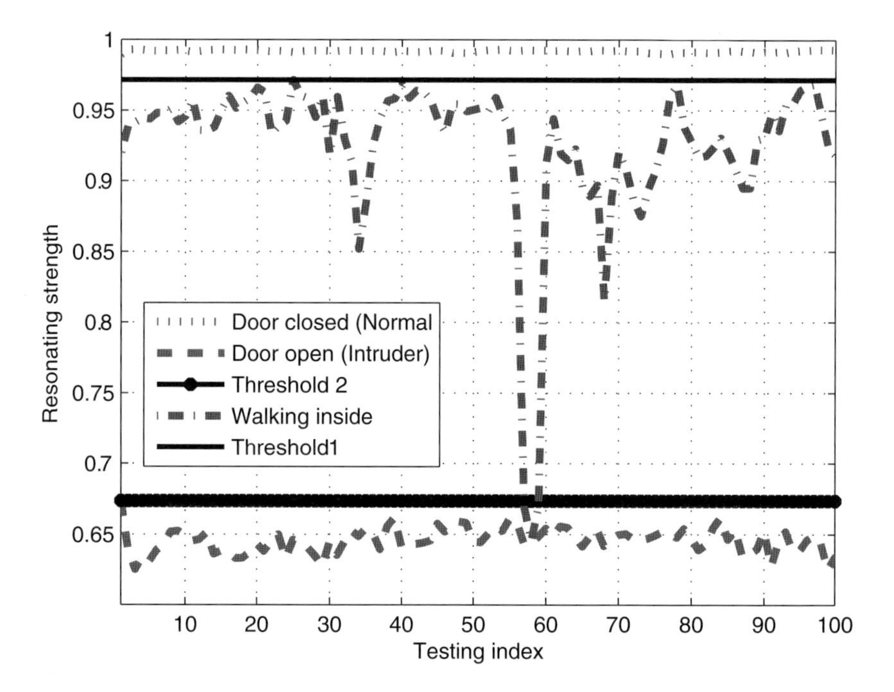

Figure 6.17 Resonating strength of guard system.

is aimed to monitor and secure the room in the middle, which is shaded, and to report as soon as the door of the secured room is opened or someone is walking inside the secured room. TRIEDS only collects the training data for normal state, i.e., door is closed and no one is walking inside the room. The training database consists of 10 samples of the CSI. Once TRIEDS starts monitoring, it will keep sensing the indoor multipath channel profile and compare it with the training database by computing the time reversal resonance strength according to (6.3) and (6.5).

An example is shown in Figure 6.17, where we can see a clear cut between the normal state and the intruder state, and between the normal state and the state where someone is walking inside the room. The *threshold 1* is the threshold for detecting when the indoor states deviates from the normal state, leading to a 100% detection rate and 0 false alarm. Whereas the *threshold 2* is for differentiating between the intruder state (i.e., door is open) and the state when someone is walking inside the secured room with the door closed, based on which TRIEDS only has 3% error by classifying the human activity state as the intruder state. Even with a single-class training dataset, TRIEDS is capable of distinguishing between different events and functioning as an alarm system to secure the rooms through the walls.

6.5 Discussion

6.5.1 Experimental Parameters

(i) *Sampling Frequency:* In this chapter, the sampling frequency of TRIEDS is 50 Hz, i.e., TRIEDS senses the multipath environment every 20 ms. Because usually the changes of door states happen in 1–2 seconds (s), current sampling frequency is enough for capturing binary changes for doors. In order to detect and monitor the entire transition of the changes or other changes happen suddenly, a higher sampling frequency is indispensable.

(ii) *Size of Training and Testing Group:* In the current experiments, we choose both the training group size M in (6.1) and the testing group size N in (6.6) as 10, to address the variations of noise in the CSI estimation. We have studied the performance of TRIEDS with different sizes of training and testing groups. It is found out that with a size greater than 10, the performance does not improve much, but a larger delay for acquiring more CSI samples is introduced. Hence, in this chapter, without sacrificing the time sensitivity of TRIEDS, the size of 10 (i.e., a sensing duration of 0.2s) is adopted.

6.5.2 Impact of Human Movements

TRIEDS utilizes the TR technique to map multipath profiles of indoor events into separate points in the TR space, due to the fact that different indoor events and human movements alter the wireless multipath profiles differently.

In Section 6.4.7, the experimental results of applying TRIEDS in a through-the-wall guard task are discussed. As shown in the Figure 6.17, in most cases, given the door closed event with no human motions, the TRRS of the same event with human motions drops. However, the degradation in the TRRS introduced by human motions is small, whereas the gap between the TRRS of the door closed event and that of the door open event is significantly large. The reason is that due to the small size of human body compared to indoor objects like doors, the human body only alters a small portion of multipath components when moving not close to the transmitter or the receiver, resulting in sparse changes in the amplitude or the phase of a couple of taps in the CSI. Consequently, the point of door closed event with human motions locates at the "proximity" of the point of the static door closed event, i.e., the two points are quite similar measured by the TRRS. They can be viewed as a single cluster given a proper threshold on the TRRS. However, when the human motions are close to the transmitter or the receiver, there is a chance that the altered multipath profile differs a lot from the one of the static indoor events, leading to a great attenuation in the TRRS, and thus a different cluster in the TR space as well as a miss detection in TRIEDS. Moreover, as discussed in Section 6.4.6, the detection accuracy drops compared to the case without intentional motions with intentional human movements. It is because that due to the existence of moving human bodies, the CSI or the multipath profiles in the environment deviate accordingly and keep changing. However, with the help of smoothing over the time domain, the dynamic changes in multipath profiles introduced by human motions can be trimmed out.

6.6 Summary

In this chapter, we presented a novel wireless indoor events detection system, TRIEDS, by leveraging the TR technique to capture changes in the indoor multipath environment. TRIEDS enables low-complexity devices with the single antenna, operating in the ISM band to detect indoor events even through the walls. TRIEDS utilizes the TR spatial-temporal resonances to capture the changes in the EM propagation environment and naturally compresses the high-dimensional features by mapping multipath profiles into the TR space, enabling the implementation of simple and fast detection algorithms. Moreover, we built a real prototype to validate the feasibility and to evaluate the performance of the presented system. According to the experimental results for detecting the states of wooden doors in both controlled and dynamic environments, TRIEDS can achieve a detection rate over 96.92% while maintaining a false alarm rate smaller than 3.08% under both LOS and NLOS transmissions. For related references, interested readers can refer to [32].

References

[1] R. Cucchiara, C. Grana, A. Prati, and R. Vezzani, "Computer vision system for in-house video surveillance," *IEE Proceedings Vision, Image and Signal Processing*, vol. 152, no. 2, pp. 242–249, Apr. 2005.

[2] A. M. Tabar, A. Keshavarz, and H. Aghajan, "Smart home care network using sensor fusion and distributed vision-based reasoning," in *Proceedings of the 4th ACM International Workshop on Video Surveillance and Sensor Networks*, pp. 145–154, 2006.

[3] A. Ghose, K. Chakravarty, A. K. Agrawal, and N. Ahmed, "Unobtrusive indoor surveillance of patients at home using multiple kinect sensors," in *Proceedings of the 11th ACM Conference on Embedded Networked Sensor Systems (SenSys 13)*, pp. 1–2, 2013. [Online]. Available: http://doi.acm.org/10.1145/2517351.2517412.

[4] M. J. Gómez, F. García, D. Martín, A. de la Escalera, and J. M. Armingol, "Intelligent surveillance of indoor environments based on computer vision and 3D point cloud fusion," *Expert Systems with Applications*, vol. 42, no. 21, pp. 8156–8171, 2015.

[5] M. Spadacini, S. Savazzi, M. Nicoli, and S. Nicoli, "Wireless networks for smart surveillance: Technologies, protocol design and experiments," in *Proceedings of IEEE Wireless Communications and Networking Conference Workshops (WCNCW)*, pp. 214–219, 2012.

[6] C. R. R. Sen Souvik and N. Srihari, "SpinLoc: Spin once to know your location," in *Proceedings of the 12th ACM Workshop on Mobile Computing Systems & Applications (HotMobile 12)*, pp. 1–6, 2012. [Online]. Available: http://doi.acm.org/10.1145/2162081.2162099.

[7] S. Sigg, S. Shi, F. Buesching, Y. Ji, and L. Wolf, "Leveraging RF-channel fluctuation for activity recognition: Active and passive systems, continuous and RSSI-based signal features," in *Proceedings of the ACM International Conference on Advances in Mobile Computing & Multimedia (MoMM 13)*, pp. 43–52, 2013. [Online]. Available: http://doi.acm.org/10.1145/2536853.2536873.

[8] A. Banerjee, D. Maas, M. Bocca, N. Patwari, and S. Kasera, "Violating privacy through walls by passive monitoring of radio windows," in *Proceedings of the 2014 ACM Conference on Security and Privacy in Wireless & Mobile Networks (WiSec 14)*, pp. 69–80, 2014. [Online]. Available: http://doi.acm.org/10.1145/2627393.2627418.

[9] H. Abdelnasser, M. Youssef, and K. A. Harras, "WiGest: A ubiquitous WiFi-based gesture recognition system," in *Proceedings of the IEEE Conference on Computer Commununications (INFOCOM)*, pp. 1472–1480, 2015.

[10] J. Xiao, K. Wu, Y. Yi, L. Wang, and L. Ni, "FIMD: Fine-grained device-free motion detection," in *Proceedings of the 18th IEEE International Conference on Parallel and Distributed Systems (ICPADS)*, pp. 229–235, Dec. 2012.

[11] F. Adib and D. Katabi, "See through walls with WiFi!" in *Proceedings of the ACM SIGCOMM*, pp. 75–86, 2013. [Online]. Available: http://doi.acm.org/10.1145/2486001.2486039.

[12] C. Han, K. Wu, Y. Wang, and L. Ni, "WiFall: Device-free fall detection by wireless networks," in *Proceedings of the IEEE International Conference on Computer Communications (INFOCOM)*, pp. 271–279, Apr. 2014.

[13] K. Wu, J. Xiao, Y. Yi, D. Chen, X. Luo, and L. M. Ni, "CSI-based indoor localization," *IEEE Transactions on Parallel and Distributed Systems*, vol. 24, no. 7, pp. 1300–1309, Jul. 2013.

[14] Y. Wang, J. Liu, Y. Chen, M. Gruteser, J. Yang, and H. Liu, "E-eyes: Device-free location-oriented activity identification using fine-grained WiFi signatures," in *Proceedings of the 20th Annual ACM International Conference on Mobile Computing and Networking*, pp. 617–628, 2014.

[15] W. Xi, J. Zhao, X.-Y. Li, K. Zhao, S. Tang, X. Liu, and Z. Jiang, "Electronic frog eye: Counting crowd using WiFi," in *Proceedings of the IEEE International Conference on Computer Communications (INFOCOM)*, pp. 361–369, Apr. 2014.

[16] W. Wang, A. X. Liu, M. Shahzad, K. Ling, and S. Lu, "Understanding and modeling of WiFi signal based human activity recognition," in *Proceedings of the 21st Annual ACM International Conference on Mobile Computing and Networking*, pp. 65–76, 2015.

[17] G. Wang, Y. Zou, Z. Zhou, K. Wu, and L. M. Ni, "We can hear you with Wi-Fi!" in *Proceedings of the 20th Annual ACM International Conference on Mobile Computing and Networking*, pp. 593–604, 2014. [Online]. Available: http://doi.acm.org/10.1145/2639108.2639112.

[18] Y. Yang and A. Fathy, "Design and implementation of a low-cost real-time ultra-wide band see-through-wall imaging radar system," in *Proceedings of the IEEE/MTT-S International Microwave Symposium*, pp. 1467–1470, Jun. 2007.

[19] F. Adib, Z. Kabelac, D. Katabi, and R. C. Miller, "3D tracking via body radio reflections," in *Proceedings of the 11th USENIX Symposium on Networked Systems Design and*

Implementation (NSDI 14), pp. 317–329, Apr. 2014. [Online]. Available: www.usenix.org/conference/nsdi14/technical-sessions/presentation/adib.

[20] F. Adib, C.-Y. Hsu, H. Mao, D. Katabi, and F. Durand, "Capturing the human figure through a wall," *ACM Transactions on Graphics*, vol. 34, no. 6, pp. 1–13, Oct. 2015. [Online]. Available: http://doi.acm.org/10.1145/2816795.2818072.

[21] F. Adib, Z. Kabelac, and D. Katabi, "Multi-person localization via RF body reflections," in *Proceedings of the 12th USENIX Symposium on Networked Systems Design and Implementation (NSDI 15)*, pp. 279–292, May 2015. [Online]. Available: www.usenix.org/conference/nsdi15/technical-sessions/presentation/adib.

[22] Y. Chen, F. Han, Y.-H. Yang, H. Ma, Y. Han, C. Jiang, H.-Q. Lai, D. Claffey, Z. Safar, and K. R. Liu, "Time-reversal wireless paradigm for green internet of things: An overview," *IEEE Internet of Things Journal*, vol. 1, no. 1, pp. 81–98, 2014.

[23] Z.-H. Wu, Y. Han, Y. Chen, and K. Liu, "A time-reversal paradigm for indoor positioning system," *IEEE Transactions on Vehicular Technology*, vol. 64, no. 4, pp. 1331–1339, Apr. 2015.

[24] C. Chen, Y. Chen, K. J. R. Liu, Y. Han, and H.-Q. Lai, "High-accuracy indoor localization: A WiFi-based approach," *The 41st IEEE International Conference on Acoustics, Speech and Signal Processing (ICASSP)*, 2016.

[25] B. Bogert, "Demonstration of delay distortion correction by time-reversal techniques," *IRE Transactions on Communications Systems*, vol. 5, no. 3, pp. 2–7, Dec. 1957.

[26] M. Fink, C. Prada, F. Wu, and D. Cassereau, "Self focusing in inhomogeneous media with time reversal acoustic mirrors," *IEEE Ultrasonics Symposium Proceedings*, pp. 681–686, 1989.

[27] J. de Rosny, G. Lerosey, and M. Fink, "Theory of electromagnetic time-reversal mirrors," *IEEE Transactions on Antennas and Propagation*, vol. 58, no. 10, pp. 3139–3149, 2010.

[28] G. Lerosey, J. De Rosny, A. Tourin, A. Derode, G. Montaldo, and M. Fink, "Time reversal of electromagnetic waves and telecommunication," *Radio Science*, vol. 40, no. 6, pp. 1–10, 2005.

[29] G. Lerosey, J. De Rosny, A. Tourin, A. Derode, and M. Fink, "Time reversal of wideband microwaves," *Applied Physics Letters*, vol. 88, no. 15, p. 154101, 2006.

[30] G. Lerosey, J. De Rosny, A. Tourin, A. Derode, G. Montaldo, and M. Fink, "Time reversal of electromagnetic waves," *Physical Review Letters*, vol. 92, no. 19, p. 193904, 2004.

[31] B. Wang, Y. Wu, F. Han, Y.-H. Yang, and K. Liu, "Green wireless communications: A time-reversal paradigm," *IEEE Journal on Selected Areas in Communications*, vol. 29, no. 8, pp. 1698–1710, 2011.

[32] Q. Xu, Y. Chen, B. Wang, and K. J. R. Liu, "TRIEDS: Wireless events detection through the wall," *IEEE Internet of Things Journal*, vol. 4, no. 3, pp. 723–735, 2017.

7 Statistical Learning for Indoor Monitoring

As embedded in wireless signals, information on an indoor environment is captured during radio propagation, motivating the development of emerging wireless sensing technologies. In this chapter, we discuss a smart radio system that leverages the informative wireless radios to enable intelligent environment and extend human senses to perceive the world. In particular, owing to the time-reversal (TR) technique that captures changes in multipath profiles, the presented TR indoor monitoring system (TRIMS) is capable of monitoring indoor events and detecting motion through walls in real time. A statistic model of intraclass TR resonance strength (TRRS) is developed and treated as the feature for TRIMS. Moreover, a prototype of TRIMS is implemented using commercial Wi-Fi devices with three antennas. We investigate the performance of TRIMS in different single-family houses with normal resident activities. In general, TRIMS can have a perfect detection rate with almost zero false alarm rates for seven target events, whereas during a 2-week experiment TRIMS achieves a detection rate of 95.45% in the indoor multievent monitoring. The presented TRIMS illustrates the potential of smart radio applications in smart homes, thanks to the ubiquitous Wi-Fi.

7.1 Introduction

The development of emerging wireless sensing technologies has enabled a plenty of applications that utilize wireless signals, or more specifically the wireless channel state information (CSI), to perceive and exploit the information hidden in the indoor environment. By deploying wireless transceivers indoors, both macro changes introduced by human activities and moving objects, and micro changes generated by gestures and vital signals, can be extracted from the CSI and recognized through passive wireless sensing.

The feasibility of wireless passive sensing relies on the multipath propagation. Multipath propagation is the phenomenon that a transmitted wireless signal reaches the receiver through different paths after being reflected and scattered by different objects in the indoor environment. A typical indoor multipath environment is demonstrated in Figure 7.1, where the channel between the transmitter (TX) and the receiver (RX) consists of paths affected by: (1) walls, (2) doors of each room, and (3) a moving human. Hence, as long as the states of an indoor environment change, the channel will record it in the CSI, enabling the detection of activities by wireless passive sensing.

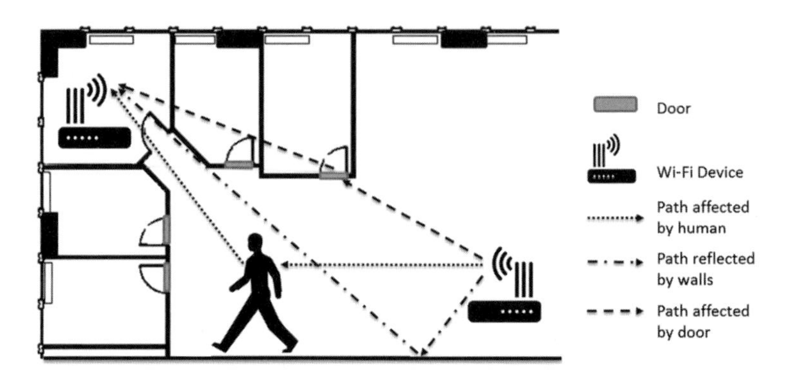

Figure 7.1 Illustration of an indoor multipath environment.

Existing research on wireless passive sensing can be categorized into different groups based on the features extracted from the wireless channel. To begin with, traditional wireless passive sensing systems are mainly based on the received signal strength (RSS) [1–5]. However, due to the fact that the RSS is coarse-grained and can be easily corrupted by multipath effect, RSS-based sensing systems often require a line-of-sight (LOS) transmission, resulting in a limited accuracy in indoor activity detection. In order to improve the accuracy and expand the applicable scenario of traditional wireless passive sensing, a much more informative feature, the CSI, becomes prevalent. Because the CSI is typically of high dimensions, it contains more detailed information and supports fine-grained classification applications, such as human motion detection [6–15], and hand motion recognition [16, 17]. Among most of these works, due to the randomness of phase distortion in the CSI, only amplitude of the CSI was used to detect indoor activities, while ignoring the information in the phase. Later, both the amplitude and phase information of the CSI was utilized in [7] to detect the dynamics of an indoor environment. However, it can only differentiate between the static and dynamic states in a LOS setting and the phase information was sanitized through linear fitting with notable drawbacks. A home intrusion detection system was proposed in [14] which treated the amplitude of the CSI as the feature. However, there is no study on the false alarm rate and long-term performance for the proposed system. Another category of wireless passive sensing techniques relies on the time-of-flight (ToF) of received signals to track the distance changes of reflected moving objects [18–24]. However, in order to extract the fine-grained ToF information, extremely large bandwidths or specially designed frequency-modulated continuous-wave (FMCW) signals are required. Hence, those techniques cannot be implemented on off-the-shelf Wi-Fi devices, and their ability of detecting multiple indoor events has not been studied yet.

Recently, thanks to its capability of capturing the difference between different CSI, time-reversal (TR) technique has been applied to wireless event detection in an indoor environment [25]. Even though the system achieved an accuracy over 96.9% in detecting multiple events by utilizing information in the complex valued CSI, the system required a transmission under 125 MHz bandwidth which cannot be implemented with commodity Wi-Fi. Moreover, it has no experimental results which evaluates the accuracy in motion detection. Meanwhile, it lacks a long-term study on performance in practical use with critical interference introduced by resident activities.

Given the limitations of the aforementioned studies, we are motivated to develop a new indoor monitoring system that not only can fully utilize the information embedded in multipath channels, but can also support simple implementation with commercial Wi-Fi devices while maintaining a high detection accuracy. To achieve this goal, we present TRIMS (abbreviation for the TR-based Indoor Monitoring System), which utilizes both the amplitude and the phase information in the CSI obtained from off-the-shelf Wi-Fi devices and succeeds in monitoring indoor environments in real time under both LOS and NLOS sensing scenarios. In particular, TRIMS is implemented on off-the-shelf Wi-Fi devices that operate around 5.8 GHz with 40 MHz bandwidth, and are capable of both multievent detection and motion detection. Moreover, unlike the aforementioned works that use the strength of TR resonance (TRRS) directly as a similarity score for recognition and localization, TRIMS relies on the statistical behavior of TRRS to differentiate different events. The statistics of TRRS is derived in this chapter and used as features in TRIMS for event detection and motion monitoring. The performance of TRIMS is evaluated through experiments conducted in different single-family houses with resident activities. TRIMS is shown to have high accuracy in monitoring different indoor events and detecting the existence of indoor motion. Furthermore, the accuracy of TRIMS is maintained over 95% during a long-term test lasting for 2 weeks.

The major points of this chapter are summarized following.

- To fully utilize the information in the CSI, both amplitude and phase information is considered. Moreover, we explore the TR technique to capture the difference in the CSI and use the TRRS to quantify the similarity between CSI samples.
- The statistical behavior of intraclass TRRS is first studied in this chapter. The derived statistical model of intraclass TRRS is then served as the feature in the presented smart radio, TRIMS, to differentiate between different indoor events.
- Built upon the theoretic analysis, the smart radio, TRIMS, is presented to monitor indoor environments, recognize different events, and detect the existence of motion in real time. TRIMS is implemented on commodity Wi-Fi devices and evaluated through extensive long-term experiments conducted in real homes with resident activities.

The rest of the chapter is organized as follows. We introduce the theoretical foundation of the smart radio system in Section 7.2. Section 7.3 presents an overview of the TRIMS as well as the details of both the event detector and the motion detector in TRIMS. The performance of TRIMS is studied and evaluated in Section 7.4, where the long-term behavior of TRIMS is also investigated. We briefly discuss the future works as well as the limitations in Section 7.5.

7.2 Preliminaries

In this section, the theoretical foundations of the presented smart radio system, TRIMS, are discussed. We introduce and explain the concept of the TR space where each indoor event is represented by a distinct TR signal. Moreover, we derive the statistics of intraclass TRRS, which later is used as the feature for the event detector in TRIMS.

7.2.1 Time-Reversal Resonance

What is TR technique? In a rich scattering and reflecting environment, the wireless channels are indeed multipath channels that contain the characteristics of an indoor environment. The evolution of TR technique can be dated back to 1957 [26], when it was proposed to compensate the delay distortion in picture transmission. Later, TR technique has been extended to applications in acoustics [27–29] and the electromagnetic (EM) field [30–34]. More recently, TR has been advocated as a novel solution for green wireless communication systems, and the TR signal transmission was introduced in [35].

The TR signal transmission consists of two phases: (1) channel probing phase during which the CSI $h(t)$ between the transmitter and the receiver is estimated at the transmitter, and (2) data transmission phase during which the TR signature $g(t)$ is convolved with data signals and sent out from the transmitter to the receiver, which is the time-reversed and conjugated version of $h(t)$. Through TR signal transmission, a spatial-temporal resonance is produced by fully collecting the energy in the multipath channel and concentrating it at the intended location. In physics, the spatial-temporal resonance is the result of a resonance of electromagnetic (EM) field, in response to the environment. Hence, a strong TR resonance indicates a match between the transmitted TR signature and its propagation channel. In other words, TRRS can be viewed as a similarity measurement between different CSI. TR technique has been utilized in many indoor sensing applications, including indoor locationing [36], indoor human recognition [37], and vital sign monitoring [38].

As shown in Figure 6.2 in the previous chapter, because each multipath profile is uniquely determined by a physical location or an indoor event in the real world, we can use the multipath profile to represent them directly. Moreover, the CSI obtained from Wi-Fi devices is in the frequency domain, i.e., the CSI is in the form of the channel frequency response (CFR). In the time domain where the CSI is represented by the channel impulse response (CIR), the TR signature is the time-reversed and conjugated copy of the CIR $h(t)$, i.e., $g(t) = h^*(-t)$. Hence, in the frequency domain the corresponding TR signature $\mathbf{g}$ of the CFR $\mathbf{h}$ is given by $\mathbf{g} = \mathcal{F}\{g(t)\} = \mathcal{F}\{h^*(-t)\} = \mathbf{h}^*$. With the help of the TR space, the similarity between two physical events or locations associated with different multipath profiles, a.k.a, CFRs, is quantified by TRRS, which is defined as follows.

Definition: The strength of TR spatial-temporal resonance (TRRS) $\mathcal{TR}(\mathbf{h}_1, \mathbf{h}_2)$ between two CFRs $\mathbf{h}_1$ and $\mathbf{h}_2$ is defined as

$$\begin{aligned}\mathcal{TR}(\mathbf{h}_1, \mathbf{h}_2) &= \frac{\left|\sum_k g_1^*[k] g_2[k]\right|^2}{\left(\sum_{l=0}^{L-1} |g_1[l]|^2\right)\left(\sum_{l=0}^{L-1} |g_2[l]|^2\right)} \\ &= \frac{\left|\sum_k h_1[k] h_2^*[k]\right|^2}{\left(\sum_{l=0}^{L-1} |h_1[l]|^2\right)\left(\sum_{l=0}^{L-1} |h_2[l]|^2\right)}\end{aligned} \tag{7.1}$$

where L is the length of the CFR vector, k is the subcarrier index, and $(\cdot)^*$ denotes taking conjugation.

The higher the TRRS is, the more similar two CFRs are. When the TRRS between two CFRs exceeds a certain value, then both of them can be viewed as representing the same physical location or indoor event. In [39, 40], a centimeter-level accurate indoor locationing system was proposed and implemented by mapping indoor physical locations into logical locations in the TR space. TR technique has been applied to indoor passive RF-sensing systems to detect indoor events and identify humans with a high accuracy [25, 37].

In this chapter, by leveraging the information of indoor activities and events embedded in wireless channels, we adopt the TR technique and present an indoor monitoring system that can detect indoor events and human motion in real time with commodity Wi-Fi devices. Unlike the aforementioned works that use the TRRS directly, the presented system relies on the statistics of TRRS to classify different multipath profiles, with the purpose of monitoring indoor environment. The details are discussed in the following.

7.2.2 Statistics of TRRS

Based on the assumption of channel stationarity, if CFRs $\mathbf{h}_0$ and $\mathbf{h}_1$ are captured from the same indoor multipath propagation environment, we can model $\mathbf{h}_1$ as

$$\mathbf{h}_1 = \mathbf{h}_0 + \mathbf{n} \tag{7.2}$$

where $\mathbf{n}$ is the Gaussian noise vector, $\mathbf{n} \sim \mathcal{CN}(\mathbf{0}, \frac{\sigma^2}{L}\mathbb{I})$, and $E\left[\|\mathbf{n}\|^2\right] = \sigma^2$ with $\|\cdot\|_2$ representing the L2-norm of a vector.

Without loss of generality, we assume unit channel gain for $\mathbf{h}_0$, i.e., $\|\mathbf{h}_0\|^2 = 1$. Then, the TRRS defined in Section 7.2.1 between $\mathbf{h}_0$ and $\mathbf{h}_1$ can be calculated as

$$\mathcal{TR}(\mathbf{h}_0, \mathbf{h}_1) = \frac{\left|\sum_k h_0^*[k](h_0[k] + n[k])\right|^2}{\|\mathbf{h}_0\|^2\|\mathbf{h}_0 + \mathbf{n}\|^2} = \frac{\left|1 + \mathbf{h}_0^{\mathrm{H}}\mathbf{n}\right|^2}{\|\mathbf{h}_0 + \mathbf{n}\|^2}, \tag{7.3}$$

where $(\cdot)^{\mathrm{H}}$ denotes the Hermitian operator, i.e., transpose and conjugate.

Based on (7.3), we introduce a new metric γ, and its definition is given by the following.

$$\gamma = 1 - \mathcal{TR}(\mathbf{h}_0, \mathbf{h}_1) = 1 - \frac{\left|1 + \mathbf{h}_0^{\mathrm{H}}\mathbf{n}\right|^2}{\|\mathbf{h}_0 + \mathbf{n}\|^2} = \frac{\|\mathbf{n}\|^2 - \left|\mathbf{h}_0^{\mathrm{H}}\mathbf{n}\right|^2}{\|\mathbf{h}_0 + \mathbf{n}\|^2}. \tag{7.4}$$

According to the Cauchy-Schwartz inequality, we can have $|\mathbf{h}_0^{\mathrm{H}}\mathbf{n}|^2 \leq \|\mathbf{n}\|^2\|\mathbf{h}_0\|^2$, with equality holds if and only if $\mathbf{n}$ is a multiplier of $\mathbf{h}_0$, which is rare to happen because $\mathbf{n}$ is a Gaussian random vector and $\mathbf{h}_0$ is deterministic. Hence, we can assume $\|\mathbf{n}\|^2 > |\mathbf{h}_0^{\mathrm{H}}\mathbf{n}|^2$ given $\|\mathbf{h}_0\|^2 = 1$, leading to $\gamma > 0$.

By taking the logarithm on both sides of (7.4), we have

$$\ln(\gamma) = \ln\left(\|\mathbf{n}\|^2 - |\mathbf{h}_0^{\mathrm{H}}\mathbf{n}|^2\right) - \ln\left(\|\mathbf{h}_0 + \mathbf{n}\|^2\right). \tag{7.5}$$

Let us denote $X = \frac{2L}{\sigma^2}\|\mathbf{n}\|^2$, $Y = \frac{2L}{\sigma^2}|\mathbf{h}_0^{\mathrm{H}}\mathbf{n}|^2$, and $Z = \frac{2L}{\sigma^2}\|\mathbf{h}_0 + \mathbf{n}\|^2$. It is easy to prove that $X \sim \chi^2(2L)$, $Y \sim \chi^2(2)$, and $Z \sim \chi_{2L}^{\prime 2}(\frac{2L}{\sigma^2})$. Here, $\chi^2(k)$ denotes a chi-squared distribution with k degrees of freedom, and $\chi_k^{\prime 2}(\mu)$ represents a noncentral chi-squared distribution with k degrees of freedom and noncentrality parameter μ. By utilizing the statistics of X, Y, and Z, we can have the following properties as

$$
\begin{aligned}
E\big[\|\mathbf{n}\|^2\big] &= \sigma^2, \quad Var\big[\|\mathbf{n}\|^2\big] = \frac{\sigma^4}{L}, \\
E\big[|\mathbf{h}_0^{\mathrm{H}}\mathbf{n}|^2\big] &= \frac{\sigma^2}{L}, \quad Var\big[|\mathbf{h}_0^{\mathrm{H}}\mathbf{n}|^2\big] = \frac{\sigma^4}{L^2} \\
E\big[\|\mathbf{h}_0 + \mathbf{n}\|^2\big] &= 1 + \sigma^2, \quad Var\big[\|\mathbf{h}_0 + \mathbf{n}\|^2\big] = \frac{\sigma^4 + 2\sigma^2}{L},
\end{aligned} \tag{7.6}
$$

where $E[\cdot]$ denotes the expectation and $Var[\cdot]$ represents the variance.

According (7.6), it is reasonable to establish the following approximation as $|\mathbf{h}_0^{\mathrm{H}}\mathbf{n}|^2 \simeq \frac{\sigma^2}{L}$, whose mean square error of approximation is equal to $Var\big[|\mathbf{h}_0^{\mathrm{H}}\mathbf{n}|^2\big] = \frac{\sigma^4}{L^2}$. Considering that in a typical OFDM system σ^4 usually has a magnitude smaller than 10^{-4} after normalization while L^2 is about 10^4, we have $Var\big[|\mathbf{h}_0^{\mathrm{H}}\mathbf{n}|^2\big] = \frac{\sigma^4}{L^2} \to 0$. Then, substituting $|\mathbf{h}_0^{\mathrm{H}}\mathbf{n}|^2$ with $\frac{\sigma^2}{L}$, (7.5) becomes the following.

$$
\begin{aligned}
\ln(\gamma) &\simeq \ln\left(\frac{\sigma^2}{2L}X - \frac{\sigma^2}{L}\right) - \ln\left(\frac{\sigma^2}{2L}Z\right) \\
&= \ln(\sigma^2) + \ln\left(\frac{1}{2L}X - \frac{1}{L}\right) - \ln\left(\frac{\sigma^2}{2L}Z\right).
\end{aligned} \tag{7.7}
$$

Moreover, considering that it is typical to have $L > 100$ and $\sigma^2 < 10^{-2}$ in a real OFDM system, $\frac{1}{2L}X - \frac{1}{L} \to 1$ with a mean square error being $1/L^2 + 1/L$, which approximates to 0. Similarly, it is easy to derive that $\frac{\sigma^2}{2L}Z \to 1$. By utilizing the linear approximation of logarithm, i.e., $\ln(x+1) \simeq x$ when $x \to 0$, along with $\frac{1}{2L}X - \frac{1}{L} \to 1$ and $\frac{\sigma^2}{2L}Z \to 1$, (7.7) can be approximated as follows.

$$
\begin{aligned}
\ln(\gamma) &\simeq \ln(\sigma^2) + \left(\frac{1}{2L}X - \frac{1}{L} - 1\right) - \left(\frac{\sigma^2}{2L}Z - 1\right) \\
&= \ln(\sigma^2) - \frac{1}{L} + \frac{1}{2L}(X - \sigma^2 Z)
\end{aligned} \tag{7.8}
$$

Referring to the definition of X and Z, the last term in (7.8) can be rewritten as

$$
X - \sigma^2 Z = \frac{2L}{\sigma^2}\|\mathbf{n}\|^2 + 2L\|\mathbf{h}_0 + \mathbf{n}\|^2 = \sum_{i=1}^{2L} W_i
$$

where W_i is defined as follows.

$$
W_i = \begin{cases} w_i^2 - (\sqrt{2L}\Re\{h_0[k]\} + \sigma w_i)^2, & \text{if } i = 2k \\ w_i^2 - (\sqrt{2L}\Im\{h_0[k]\} + \sigma w_i)^2, & \text{if } i = 2k - 1. \end{cases} \tag{7.9}
$$

Here, w_i is independent and identically distributed (i.i.d.) with $w_i \sim \mathcal{N}(0,1), \forall i$. $\Re\{\cdot\}$ denotes the function to take the real part of a complex value, while $\Im\{\cdot\}$ for the imaginary part. Given the statistics of w_i, the mean and variance of W_i are derived and listed in (7.10) and (7.11), respectively,

$$E\Big[W_i\Big] = \begin{cases} 1 - 2L\Re\{h_0[k]\}^2 - \sigma^2, & \text{if } i = 2k \\ 1 - 2L\Im\{h_0[k]\}^2 - \sigma^2, & \text{if } i = 2k-1 \end{cases} \tag{7.10}$$

and

$$Var\Big[W_i\Big] = \begin{cases} 2\Big(1 + \sigma^4 + (2L\Re\{h_0[k]\}^2 - 1)\sigma^2\Big), \\ \qquad \text{if } i = 2k \\ 2\Big(1 + \sigma^4 + (2L\Im\{h_0[k]\}^2 - 1)\sigma^2\Big). \\ \qquad \text{if } i = 2k-1 \end{cases} \tag{7.11}$$

Due to the fact that $L > 100$ in typical OFDM system, $\sum_i^{2L} W_i$ will exhibit an asymptotic behavior, according to the *Central Limit Theorem*. Hence we define a new normal-distributed variable S_{2L} as follows.

$$S_{2L} = \frac{\sum_i^{2L} W_i + 2L\sigma^2}{\sqrt{4L(1+\sigma^4)}} \sim \mathcal{N}(0,1). \tag{7.12}$$

After substituting (7.12) into (7.9), we finally get the statistical distribution of γ as follows.

$$\begin{aligned} \ln(\gamma) &\simeq \ln(\sigma^2) - \frac{1}{L} + \frac{1}{2L}\sum_{i=1}^{2L} W_i \\ &= \ln(\sigma^2) - \frac{1}{L} - \sigma^2 + \frac{\sqrt{4L(1+\sigma^4)}}{2L} S_{2L} \\ &\sim \mathcal{N}\left(\ln(\sigma^2) - \frac{1}{L} - \sigma^2, \frac{1+\sigma^4}{L}\right). \end{aligned} \tag{7.13}$$

Hence, the metric γ, i.e., $1 - \mathcal{TR}(\mathbf{h}_0, \mathbf{h}_1)$, follows the log-normal distribution with the location parameter $\mu_{\text{logn}} = \ln(\sigma^2) - \frac{1}{L} - \sigma^2$ and the scale parameter $\sigma_{\text{logn}} = \sqrt{\frac{1+\sigma^4}{L}}$.

The derived statistical model is verified by fitting over real measured CSI samples and CSI samples generated from the model in (7.2), as shown in Figure 7.2. First, we adopt the Kolmogorov–Smirnov test (K–S test) to quantitatively evaluate the accuracy of the derived log-normal distribution model on the real CSI measurements. The score of the K–S test is denoted as D, which measures the difference between the empirical cumulative distribution function (E-CDF) and the log-normal cumulative distribution function (CDF). As depicted by the example in Figures 7.2(a) and 7.2(b), the log-normal distribution fits better over CSI samples captured from real channels, compared with the normal distribution. Moreover, the derived log-normal distribution model is further investigated on simulated CSI samples through studying the mean square errors of parameter estimations against the signal-to-noise radio (SNR), a.k.a., σ^{-1} in dB.

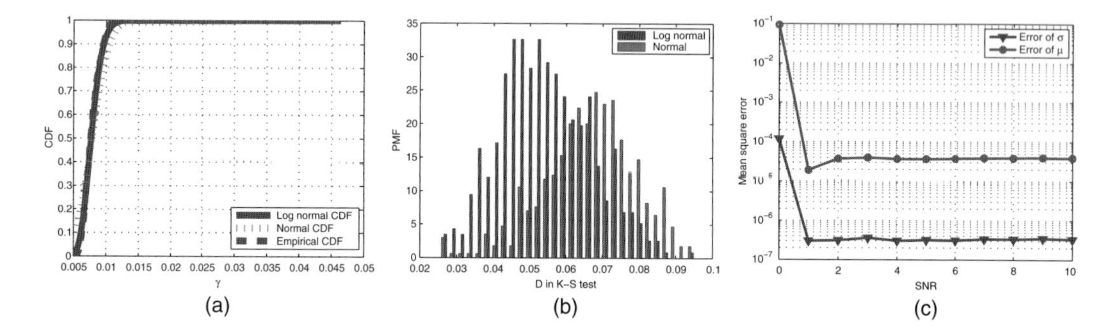

Figure 7.2 Examples for evaluating the derived statistical model. (a) Distribution fitting for 500 real CSI measurements. (b) The histogram of scores of K-S test from 500 real CSI measurements. (c) The mean square error of log-normal parameter estimation for simulated CFRs.

As plotted in Figure 7.2(c), in terms of parameter estimation for the log-normal distribution, the derived model is accurate with almost zero mean square error, especially when SNR is high.

7.3 Design of TRIMS

Intelligent systems have become popular recently, in that with the help of learning they are capable of comprehending an object or even the world in the way humans do. For example, researchers have spent decades on computer vision or machine vision systems that achieve a high-level understanding over digital images and videos that is comparable or even better than the human visual system.

Can Wi-Fi perceive an indoor environment? To answer this question, in this chapter, we discuss an intelligent indoor monitoring system, TRIMS, which enables real-time indoor monitoring with commercial Wi-Fi devices by leveraging TR technique. This novel indoor monitoring system consists of the following components.

(i) Event Detector: With the purpose of perceiving a monitored environment and recognizing specific events, an event detector is included in TRIMS. The presented event detector in TRIMS relies on TR technique to evaluate the difference and similarity between various indoor events. It consists of an offline training phase where the CSI and corresponding statistics of training events are learned and an online monitoring phase where the event detector of TRIMS will report the occurrence of trained events in real time. The details are discussed in Section 7.3.1.

(ii) Motion Detector: TRIMS not only has the functionality of detecting the occurrence of trained events, but it is also capable of detecting dynamics in the environment, i.e., motion inside the protected area. The presented motion detector leverages fluctuations in TRRS values within a time window to indicate environmental dynamics and the sensitivity is auto-adapted for each environment through the training phase. In Section 7.3.2, we will introduce details of the presented motion detector in TRIMS.

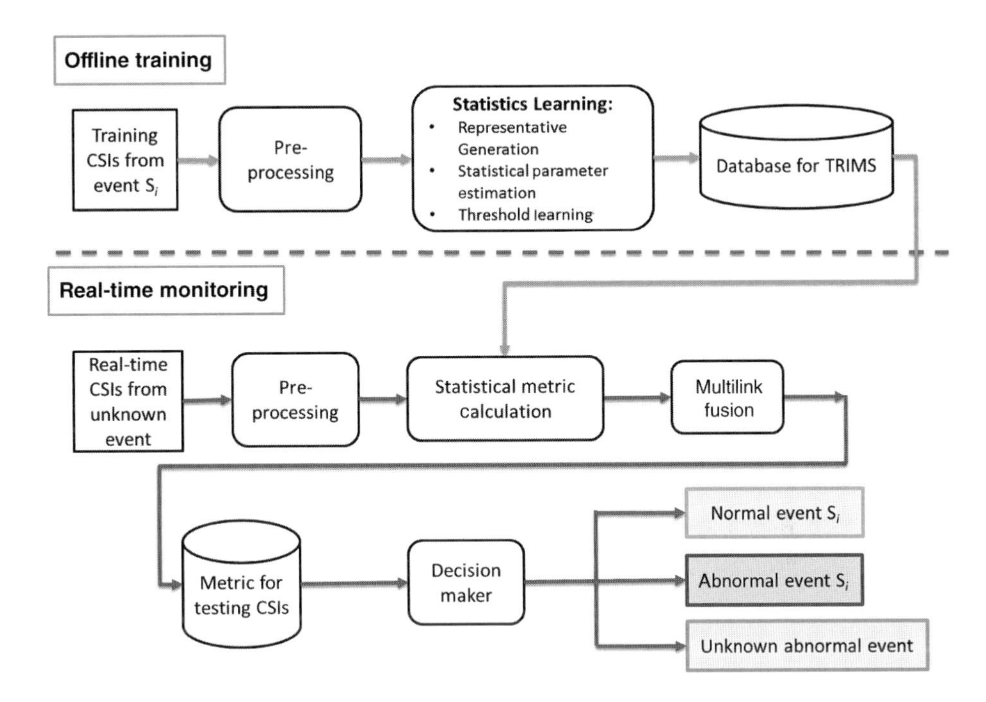

Figure 7.3 Diagram of the presented event detector in TRIMS.

7.3.1 TRIMS: Event Detector

By leveraging the fundamental theories and techniques discussed in Section 7.2, we design a real-time event detection module in TRIMS, utilizing the statistics of TRRS between the CSI as the metric for categorizing indoor environments and recognizing different indoor events. In this section, the details of a statistics-based event detector are introduced, and the diagram illustrating how the event detector works is shown in Figure 7.3. The details are discussed in the following.

7.3.1.1 Offline Training Phase

In the offline training phase, the presented system aims to build a database that stores, for each of the training events, the log-normal statistics of TRRSs between the intraclass CSI and a representative CSI sample.

Specifically, for each indoor event $S_i \in \mathcal{S}$ with $\mathcal{S}$ being the set of indoor events to be monitored, the corresponding CFRs are obtained through channel sounding and estimated at the receiver side as

$$\mathbf{H}_i = \left[\mathbf{h}_i^{(1)}, \mathbf{h}_i^{(2)}, \ldots, \mathbf{h}_i^{(M)}\right], \quad i = 1, 2, \ldots, N, \tag{7.14}$$

where N is the size of $\mathcal{S}$, i.e., the number of events of interest, and M is the number of links between the transmitter and the receiver. Each link represents the channel between a single TX-RX antenna pair. The dimension of $\mathbf{H}_i$ is $L \times M$, with L being the number of active subcarriers in a wireless OFDM system. The statistics of intraclass TRRS is estimated through the following steps.

- **Preprocessing:** A phase sanitization algorithm is applied to compensate all CFRs for phase offsets, which are introduced by carrier frequency offset (CFO), sampling frequency offset (SFO), and symbol timing offset (STO).
- **CSI Representative Generation:** For each link m, a CSI representative is found for every indoor event S_i in the training set. The CSI representative is selected as the one that is most similar to all other CFRs on link m from S_i. In particular, to quantitatively evaluate the similarity, the pair-wise TRRSs on link m between all the CFRs collected for indoor event S_i are calculated first. Then the CSI representative is selected on link m for event S_i as the one that is most similar to the majority of other CSI samples in the same class. $\mathbf{H}_{rep,i}$ is the collection of CSI representatives on all links for event S_i, which is defined as follows:

$$\mathbf{H}_{rep,i} = \left[\mathbf{h}^{(1)}_{rep,i}, \mathbf{h}^{(2)}_{rep,i}, \ldots, \mathbf{h}^{(M)}_{rep,i}\right], \forall i. \tag{7.15}$$

- **Lognormal Parameter Estimation:** Once the CSI representative is selected, the log-normal distribution parameters can be estimated from intra-class TRRSs. For link m and event S_i, the TRRSs between the CSI representative $\mathbf{h}^{(m)}_{rep,i}$ and all other realizations $\mathbf{h}^{(m)}_i(n)$, $\forall\, n$ are calculated using (7.1) and denoted as

$$\mathcal{TR}^{(m)}_i(n) = \mathcal{TR}\left(\mathbf{h}^{(m)}_{rep,i}, \mathbf{h}^{(m)}_i(n)\right), \quad n = 1, 2, \ldots, Z-1, \tag{7.16}$$

where n is the realization index of CFRs collected for event S_i, and Z is the total number of CFRs. Then the log-normal parameters $(\mu^{(m)}_i, \sigma^{(m)}_i)$ of $\gamma = 1 - \mathcal{TR}^{(m)}_i$ for event S_i on link m are estimated by

$$\mu^{(m)}_i = \frac{1}{Z-1}\sum_{n=1}^{Z-1} \ln\left(1 - \mathcal{TR}^{(m)}_i(n)\right) \tag{7.17}$$

$$\sigma^{(m)}_i = \sqrt{Var\left[\ln\left(1 - \mathcal{TR}^{(m)}_i\right)\right]}, \tag{7.18}$$

where $Var[\cdot]$ is the sample variance function.

The training database is built with the collection of CSI representatives and log-normal distribution parameters for all the trained events. All the trained events can be divided into two groups: the normal events group $\mathcal{S}_{normal}$, where no alarm will be sounded when being detected, and the abnormal events group $\mathcal{S}_{abnormal}$, where an alarm will be reported to users when an abnormal event is detected.

- **Threshold Learning:** Based on the knowledge of $\mathcal{H}_{rep}$ and $\mathcal{Q}_{rep}$, the system builds the normal event checker and the abnormal event checker, through which the label of the testing CSI sample is determined in the monitoring phase. To determine whether event testing CSI sample $\mathbf{H}_{test}$ belongs to an event S_i, a score is calculated first as

$$\mathcal{W}_{i,test} = \prod_{m=1}^{M} \mathcal{W}_{i,test}^{(m)} = \prod_{m=1}^{M} F_{(\mu_i^{(m)},\sigma_i^{(m)})}\left(1 - \mathcal{TR}_{i,test}^{(m)}\right), \tag{7.19}$$

where $\mathcal{TR}_{i,test}^{(m)} = \mathcal{TR}\left(\mathbf{h}_{rep,i}^{(m)}, \mathbf{h}_{test}^{(m)}\right)$. $\mathcal{W}_{i,test}^{(m)}$ is the statistical metric on link m of $\mathbf{H}_{test}$ conditioned on event S_i, defined as the value of log-normal cumulative distribution function (CDF) of $1-\mathcal{TR}_{i,test}^{(m)}$ with parameter being $\mu_i^{(m)}$ and $\sigma_i^{(m)}$. The operation $\prod_{m=1}^{M}(\cdot)$ fuses the information among all links. $F_{(\mu,\sigma)}(x)$ represents the CDF of log-normal distribution with parameters (μ, σ) and the variable x.

The smaller the value of $\mathcal{W}_{i,test}^{(m)}$ is, the higher the probability for $\mathbf{H}_{test}$ belonging to event S_i is. Two thresholds, γ_{normal} and $\gamma_{abnormal}$, are required for the normal event checker and the abnormal event checker to define the boundary for the value of metric $\mathcal{W}_{i,j}$. Consequently, when the value of $\mathcal{W}_{i,test}$ falls below the threshold γ_{normal} or $\gamma_{abnormal}$, $\mathbf{H}_{test}$ is viewed as from event S_i. Hence, in order to correctly distinguish different events, both γ_{normal} and $\gamma_{abnormal}$ are carefully learned based on the metrics $\mathcal{W}_{i,test}$, where $\mathbf{H}_{test}$ is replaced by $\mathbf{H}_{rep,j}$ during the training phase. The criteria for choosing γ_{normal} and $\gamma_{abnormal}$ are as follows:

$$\begin{aligned}\gamma_{normal} &= \min_{S_i \in \mathcal{S}_{normal},\ S_j \in \mathcal{S}_{abnormal}} \mathcal{W}_{i,j}\\ \gamma_{abnormal} &= \min_{S_i \in \mathcal{S}_{abnormal},\ S_j \in \mathcal{S},\ S_j \neq S_i} \mathcal{W}_{i,j}.\end{aligned} \tag{7.20}$$

7.3.1.2 Online Monitoring Phase

The statistics-based event detector is designed to identify the real-time indoor events with the knowledge of training database. Once the occurrence of a trained event is detected, the system will decide to sound an alarm based on the characteristics of that event. If an untrained event is detected, the system will also notify the user about the situation. The details are discussed as follows.

During the monitoring phase, the receiver keeps monitoring the environment by collecting the CSI as $\mathbf{H}_{test} = \left[\mathbf{h}_{test}^{(1)}, \mathbf{h}_{test}^{(2)}, \ldots, \mathbf{h}_{test}^{(M)}\right]$.

- **Statistical Metric Calculation:** Because the obtained CSI measurement $\mathbf{H}_{test}$ is corrupted by random phase offsets, a phase sanitization algorithm is applied. After that, for each trained indoor event, the TRRS between the CSI representative and the testing measurement is calculated. Given the TRRSs between the testing CSI sample and trained events, the statistical metric $\mathcal{W}_{i,test}$ between $\mathbf{H}_{test}$ and the trained event S_i is calculated using (7.19).
- **Decision:** The statistical metric $\mathcal{W}_{i,test}^{(m)}$ is a monotonic function of $\mathcal{TR}_{i,test}^{(m)}$, which depicts the similarity between the testing CSI samples and the CSI representative of event S_i. In other words, the more similar two CSI samples are, the smaller the value of $\mathcal{W}_{i,test}$ is. The detailed decision protocol based on $\mathcal{W}_{i,test}$ is described in the following.
  (i) *Step 1 – Normal Event Checker*:
  To begin with, the event detector checks whether the environment is normal, i.e., only one of the normal events in S_{normal} occurs, by the following rule.

$$D_{event} = \begin{cases} \arg\min_{S_i \in \mathcal{S}_{normal}} \mathcal{W}_{i,test}, \\ \quad \text{if } \min_{S_i \in \mathcal{S}_{normal}} \mathcal{W}_{i,test} \leq \gamma_{normal} \\ \text{go to } \textit{Step 2}, \text{ otherwise.} \end{cases} \tag{7.21}$$

(ii) *Step 2 – Abnormal Event Checker*:
In order to determine which trained abnormal event in $S_{abnormal}$ occurs, it follows the rule below:

$$D_{event} = \begin{cases} \arg\min_{S_i \in \mathcal{S}_{abnormal}} \mathcal{W}_{i,test}, \\ \quad \text{if } \min_{S_i \in \mathcal{S}_{abnormal}} \mathcal{W}_{i,test} \leq \gamma_{abnormal} \\ 0, \text{ otherwise,} \end{cases} \tag{7.22}$$

where $D_{event} = 0$ indicates the occurrence of some untrained event.

To summarize, the event detector labels the CSI sample $\mathbf{H}_{test}$ by the following rule:

$$D_{event} = \begin{cases} \arg\min_{S_i \in \mathcal{S}_{normal}} \mathcal{W}_{i,test}, \\ \quad \text{if } \min_{S_i \in \mathcal{S}_{normal}} \mathcal{W}_{i,test} \leq \gamma_{normal} \\ \arg\min_{S_i \in \mathcal{S}_{abnormal}} \mathcal{W}_{i,test}, \\ \quad \text{if } \min_{S_i \in \mathcal{S}_{normal}} \mathcal{W}_{i,test} > \gamma_{normal} \text{ and} \\ \quad \min_{S_i \in \mathcal{S}_{abnormal}} \mathcal{W}_{i,test} \leq \gamma_{abnormal} \\ 0, \quad \text{otherwise.} \end{cases} \tag{7.23}$$

7.3.2 TRIMS: Motion Detector

TRIMS is designed not only to determine which trained indoor event happens, but also to detect if environment has any dynamics by means of a motion detector presented in TRIMS.

Motion always introduces fluctuations in the radio propagation environment, leading to significant changes of TRRSs between CSI samples within a time window. The impact introduced by motion is larger compared to the impacts brought by channel fading and noise, especially when motion happens close to the transmitter or the receiver. In this part, we consider a motion detector that uses the variance of TRRSs between CSI samples within an observation window as the metric to indicate the indoor dynamics. The presented motion detector consists of two phases: an offline training phase and a real-time monitoring phase. The flowchart of the presented motion detector is depicted in Figure 7.4.

7.3.2.1 Phase I. Offline Training

In the training phase, the presented motion detector is trained with the dynamics, measured by the variance of a TRRSs time sequence, under both the static state and

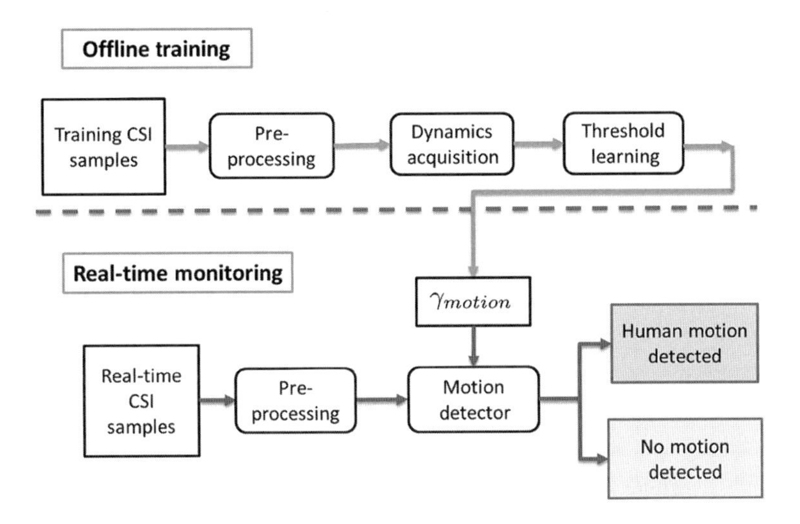

Figure 7.4 Diagram of the motion detector in TRIMS.

the dynamic state with motion in the indoor environment. The detailed steps are listed as follows.

- **Data Acquisition:** First, the state of an indoor environment is divided into two classes: $\mathcal{S}_1$, where the environment is static, and $\mathcal{S}_0$, where there is some motion happening in the monitoring area. The CSI is collected continuously in time for both classes as $\mathbf{H}_i(t) = [\mathbf{h}_i^{(1)}(t), \quad \mathbf{h}_i^{(2)}(t), \quad \ldots, \quad \mathbf{h}_i^{(M)}(t)]$, where $\mathbf{H}_0(t)$ is collected when the environment is static, and $\mathbf{H}_1(t)$ is from the dynamic environment. t is the time instance when the CFR is captured. The phase offset in CFRs is compensated individually and independently before learning the dynamics.
- **Dynamics Acquisition:** After time sequences of CFR measurements under both static state $\mathcal{S}_0$ and dynamic state $\mathcal{S}_1$ are obtained, the environmental dynamics is evaluated by tracking the variance of TRRSs within a time window.

  To study the variance under both states $\mathcal{S}_i$, $i = 0, 1$, a sliding window with length W samples and overlap $W - 1$ is applied on the time sequence of $\mathbf{H}_i(t)$. For example, in a window of length W, CFRs from $\mathbf{H}_i(t_0)$ to $\mathbf{H}_i(t_0 + (W - 1) * T_s)$ are stored, where T_s is the channel probing interval. Within each window, the corresponding TRRS sequence between t_0 and $t_0 + (W - 1) * T_s$ is denoted as $\mathcal{TR}\big(\mathbf{H}_i(t_0), \mathbf{H}_i(t)\big), \quad t_0 \le t \le t_0 + (W - 1) * T_s$, which is calculated as follows.

$$\mathcal{TR}\Big(\mathbf{H}_i(t_0), \mathbf{H}_i(t)\Big) = \frac{\sum_{m=1}^{M} \mathcal{TR}\Big(\mathbf{h}_i^{(m)}(t_0), \mathbf{h}_i^{(m)}(t)\Big)}{M} \tag{7.24}$$

  Then the dynamics within the time window can be quantitatively evaluated by the variance of $\big\{\mathcal{TR}\big(\mathbf{H}_i(t_0), \mathbf{H}_i(t)\big), \quad t_0 \le t \le t_0 + (W - 1) * T_s\big\}$, which is denoted as $\sigma_i(t_0), i = 0, 1$. In order to have a fair and comprehensive analysis, multiple $\sigma_i' s, i = 0, 1$ are captured at different time.

- **Threshold Learning:** After dynamics acquisition, multiple instances of σ_0 and σ_i are obtained, and the threshold γ_{motion} for differentiating between S_0 and S_i is determined by

$$\gamma_{motion} = \begin{cases} \alpha \max_t \sigma_0(t) + (1-\alpha)\overline{\sigma_1(t)}, \\ \qquad \text{if } \max_t \sigma_0(t) \leq \overline{\sigma_1(t)} \\ \max_t \sigma_0(t), \text{ otherwise,} \end{cases} \tag{7.25}$$

where $\overline{\sigma_1(t)}$ denotes the average of multiple $\sigma_1' s$ captured at different time. α, $0 \leq \alpha \leq 1$, is a sensitivity coefficient for motion detections in that the sensitivity of the presented motion detector increases as α decreases.

7.3.2.2 Phase II. Online Monitoring

During the online monitoring phase, the dynamics in the environment are tracked by comparing the variance on real-time TRRSs with γ_{motion} as:

$$D_{motion}(t_0) = \begin{cases} 1, \ \sigma_{test}(t_0) \geq \gamma_{motion}, \\ 0, \ \text{otherwise,} \end{cases} \tag{7.26}$$

where $\sigma_{test}(t_0)$ is the variance on the testing TRRS sample sequence within a window of length W and overlap $W-1$ at time instance t_0. $D_{motion}(t_0) = 1$ indicates the existence of motion, i.e., someone is moving inside the monitoring area, while $D_{motion}(t_0) = 0$ means the environment is static.

7.3.3 TRIMS: Time-Diversity for Smoothing

In a real environment, noise in wireless transmission and outside activities exist and corrupt the estimated CSI, leading to a misdetection or a false alarm in both the event detector and the motion detector of TRIMS. However, by leveraging the fact that these interferences are typically sparse and abrupt, a smoothing method relying on the time diversity is considered in this chapter to address that problem.

The essential idea of the presented time-diversity smoothing algorithm is by applying the majority vote over decisions of each testing CSI sample, assuming that the typical indoor event lasts for a couple of seconds. In both the event detector and the motion detector, decisions will be accepted only if they are consistent along a short time period. The details are as follows.

With the help of a sliding window SW whose length is W and overlap length is O, the decisions $D_{out}(n)$ at time index n is obtained through

$$D_{out}(w) = \text{MV}\big\{D_{in}(1+(w-1)*O), \ldots, D_{in}(W+(w-1)*O)\big\}, \tag{7.27}$$

where $D_{in}(w)$ is the input decision sample at time index w, and $MV\{\cdot\}$ is the operator for taking majority vote. The corresponding time delay introduced by the sliding window SW is in general $(W - O) \times T$, where T is the time interval between consecutive D_{in} samples.

For example, in order to alleviate false alarms introduced by outside activities and imperfect CSI estimation to the presented event detector, a two-level time-diversity smoothing is applied as follows.

(i) *Level I:* A majority vote is applied directly on the raw decisions D_{motion} of each single CSI sample. Given a sliding window SW_1 whose length is W_1 and overlap length is O_1, the decisions of index w, $D_{MV1}(w)$, is obtained from taking a majority vote over $D_{event}(i + (w - 1) * O_1)$, $1 \leq i \leq W_1$.

(ii) *Level II:* A second sliding window, SW_2, is applied on $D_{MV1}(w)$ with length W_2 and overlap O_2. Consequently, the final decision output is $D_{final}(n)$. The system suffers a time delay $(W_2 - O_2) \times (W_1 - O_1) \times T_s$.

7.4 Experimental Results

In order to evaluate the feasibility and the performance of the presented TRIMS in indoor monitoring, we build a prototype on commodity Wi-Fi devices performing 3×3 multiple-input and multiple-output (MIMO) transmission at 5.845 GHz carrier frequency under the IEEE 802.11n standard. According to the IEEE 802.11n standard, both 2.4 GHz band and 5 GHz band support a 40 MHz bandwidth, and the CSI at those two bands should share the same resolution. Therefore, with the obtained CSI, the presented system should achieve a detection performance at 2.4 GHz similar to that from 5.8 GHz. In the prototype, the CSI is extracted from the Qualcomm network interface card (NIC) and composed by a complex-valued matrix for accessible subcarriers on all nine links. With a single pair of devices, we conduct extensive experiments in two real indoor environments: House #1 and House #2 with regular residence activities, whose floor plans are shown in Figures 7.5(a) and 7.5(b). The locations of the transmitter and the receiver are marked in the floor plans.

7.4.1 TRIMS: Event Detector

We start from the performance study of the presented event detector in TRIMS and experiments are conducted in both facilities. In order to learn the statistics of intraclass TRRS, at least 300 realizations of CFRs corresponding to each indoor propagation environment should be collected. Furthermore, the CSI sounding rate is 100 Hz in the training phase, while it becomes 30 Hz in the real-time monitoring phase for the event detector in TRIMS. The two-level time diversity algorithm is applied in the event detector with $W_1 = 15, O_1 = 14, W_2 = 45$, and $O_2 = 15$, considering that 30 CSI samples are collected per second.

Table 7.1 Events of interest in House #1

State Index	Description
e1	All doors are closed.
e2	Front door open.
e3	Back door open.
e4	Bob's room door open.
e5	Study room door open.
e6	Alice's room door open.
e7	Restroom door open.
e8	Window 1 open.
e9	Window 2 open.

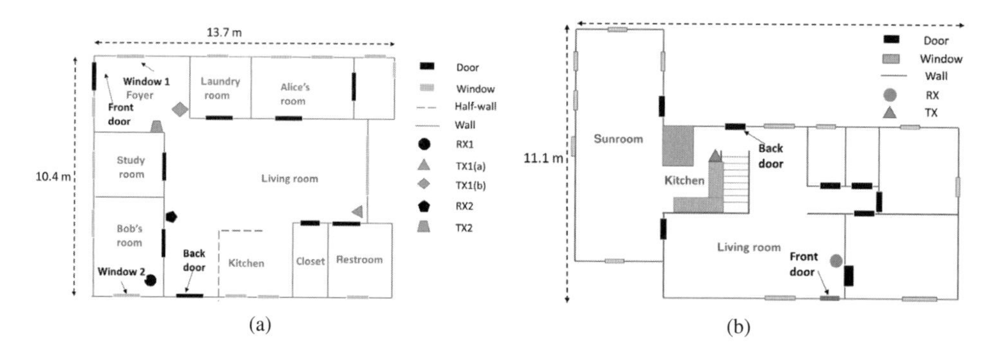

Figure 7.5 Experimental setting for TRIMS. (a) The floorplan of House #1. (b) The floorplan of House #2.

7.4.1.1 Study on Location of TX-RX

As discussed in the previous sections, the presented event detector is aimed at monitoring and detecting indoor events by leveraging the TR technique to capture changes in the CSI. Different events introduce different changes, depending on not only the characteristics of each indoor event but also the distance between the event location to the transceivers. The closer the indoor event is, the larger impact it introduces. Hence, it is crucial to study how the locations of the TX and the RX affect the performance of the presented event detector.

In House #1, we study the impact of TX-RX locations on TRIMS's performance in the event detection, and the events of interest are listed in Table 7.1, while the candidate locations of TX and RX are labeled with "TX_1" and "RX_1" in Figure 7.5(a). The receiver is fixed in the study room while the transmitter is located either in the foyer against a wall or outside the restroom. The performance is evaluated through the receiver operating characteristic (ROC) curve, where the x-axis is the false alarm rate of an event e_i, i.e., the probability of other events being misclassified as e_i, whereas the y-axis is the detection rate of e_i.

As shown in Figure 7.6(a), the presented event detector fails to differentiate between the CSI of e1, e6, and e7, in that the false alarm rates of e1, e6, and e7 are extremely high

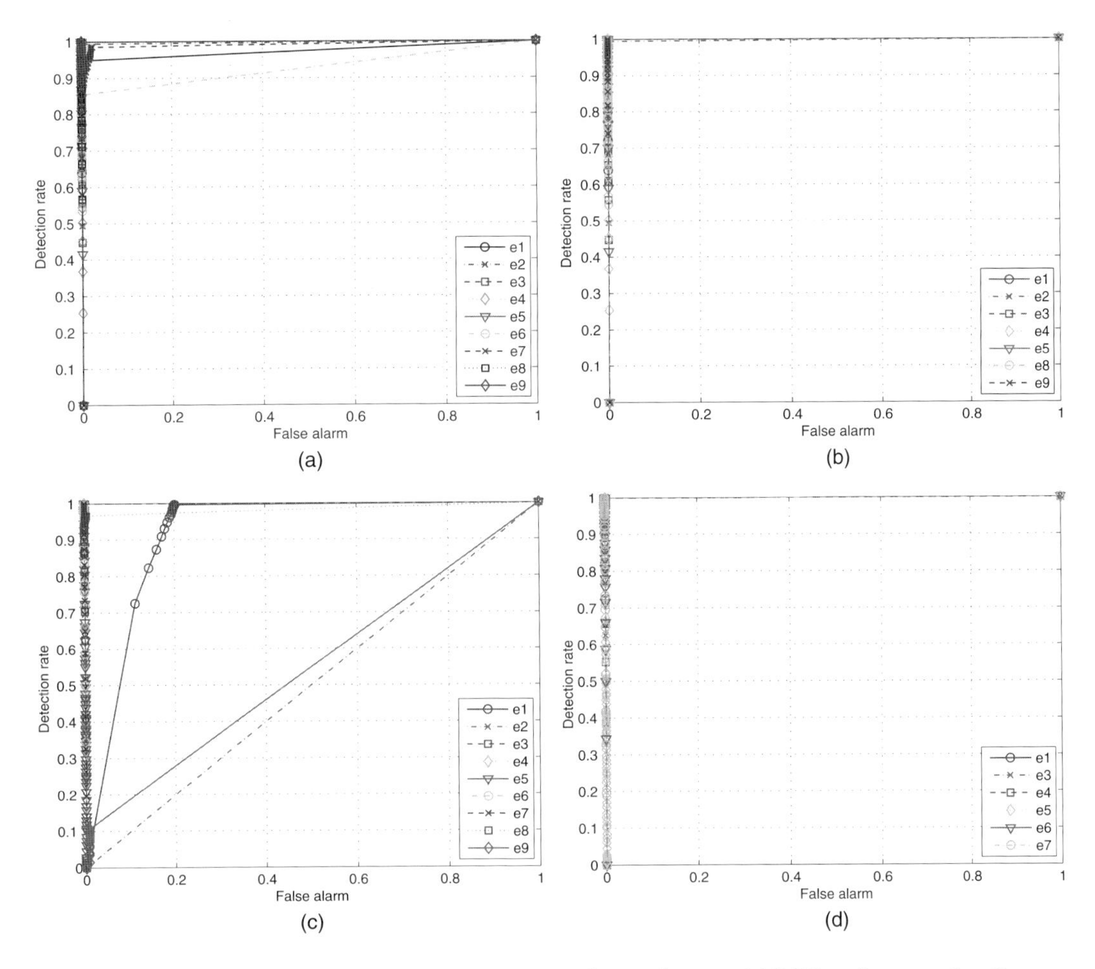

Figure 7.6 ROC performance for the presented event detector. (a) ROC performance for all events (TX in the foyer). (b) ROC performance for targets events (TX in the foyer). (c) ROC performance for all events (TX outside the study room). (d) ROC performance for targets events (TX outside the study room).

under the same detection rate, compared with others. The reason is that the changes in the wireless multipath channel introduced by e6 and e7 are too small for the presented event detector to capture. A possible reason is that events of e6 and e7 are far from the TX and the RX, when both devices are located in the front part of the house. Similarly, in Figure 7.6(c), when the TX is put outside the study room, i.e., in the back part of the house, events of e2 and e8 are too far away while e9 is outside the circle range defined by the line segment between the TX and the RX. Consequently, the presented event detector has an ambiguity over e1, e2, e8, and e9.

Here, we introduce the concept of "target event," to whom the presented event detector has a perfect accuracy, as shown in Figure 7.6(b) and 7.6(d). The target events are those events that satisfy a rule-of-thumb, which says that in order to have it detected, the event should either be close to the TX-RX link or have an LOS path to one of the devices, given the location of the TX and the RX. Under the rule-of-thumb, the target event is able to change the CSI between the TX and the RX in a way that is significant enough. The presented event detector can achieve a perfect ROC performance for target events.

7.4.1.2 Operational Test in House #1

In this part, to further study the performance of TRIMS in the real-time event monitoring, we imitate several intrusion and postman cases with locations of TX and RX being "TX_2" and "RX_2." In the intrusion test, an intruder enters the house from a door and walks inside the house before leaving by the same door. On the other hand, in the postman test, some-one is walking outside the front door of each house to imitate a postman.

Moreover, in this part, the system is only trained for events e1, e2, and e3. In Figure 7.7, the system output is plotted along the time. The y-axis is the output decision, where "Allclosed" indicates e1, "Front" and "Back" represent e2 and e3, respectively, and "Unknown" means untrained events happening. Take Figure 7.7(a) as an example. The presented event detector outputs state 1, i.e., "all doors are closed," during the test. As shown in Figure 7.7(b), during the test, the presented event detector first reports e1 for about 20 s and then detects the occurrence of e2 when the front door is opened at the time index of the 20th s, with a single detection over the untrained event, i.e., the

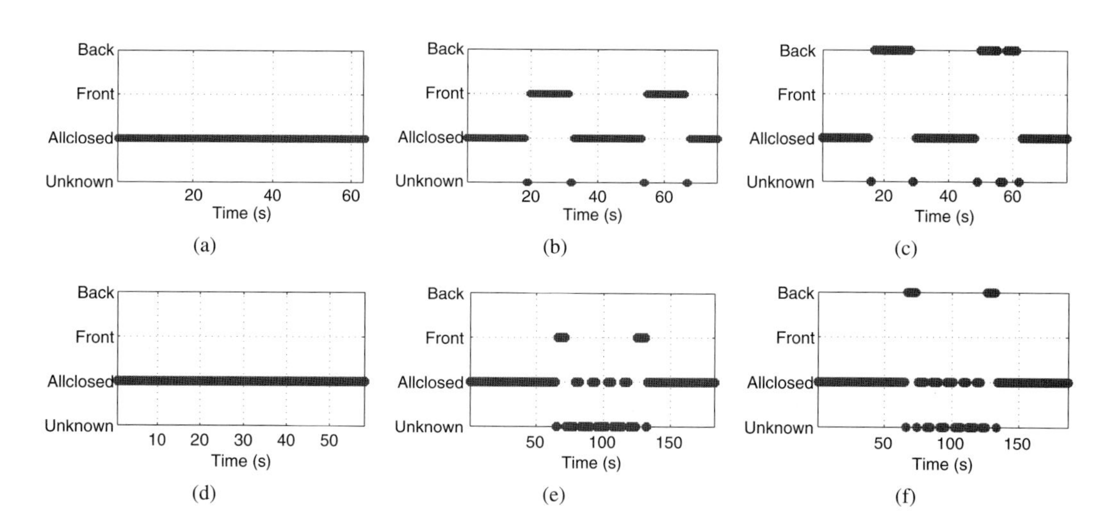

Figure 7.7 Monitoring results of the presented event detector for operational tests in House #1. (a) Test under all doors closed. (b) Test under opening (around the 20th and the 50th s) and closing (around the 30th and the 60th s) front door from outside the house twice. (c) Test under opening (around the 25th and the 55th s) and closing (around the 35th and the 65th s) back door from outside the house twice. (d) Postman test with someone walking outside the front door. (e) Test when an intruder comes in (around the 60th s), walks inside, and leaves (around the 120th s) through front door. (f) Test when an intruder comes in (around the 60th s), walks inside, and leaves (around the 120th s) through back door.

output falls to "Unknown." The system starts to report e1 when the front door is closed at around the 30th s.

Figures 7.7(a), 7.7(b), and 7.7(c) illustrate the ability of the presented event detector to perfectly monitor and detect the trained events in real time when (a) the environment is quiet and all doors are closed, (b) the front door is opened and then closed twice from the outside, and (c) the back door is opened and closed twice from the outside. In Figure 7.7(d), we simulate the postman case where someone wanders outside the front door, close to the target event. The presented event detector shows its robustness to outside activities by reporting no false alarms in the postman case. In the next test, we simulate intrusions made by an intruder through the front door and the back door, and the intruder is required to leave through the same door after walking inside the house for a certain period. As demonstrated in Figures 7.7(e) and 7.7(f), the presented event detector succeeds in capturing the intrusion. Moreover, between the door opening in both figures, the decision of the presented event detector may become "Unknown," which is owing to the interference to the multipath channel brought by human motion inside the house.

7.4.1.3 Long-Term Test in House #1

Furthermore, we conduct a long-term monitoring test for the presented event detector in TRIMS in House #1 for 6 days. The result is compared with that of a commercial home security system whose contact sensors are installed on the front and back doors. During the first 6 days, TRIMS has zero false alarms when the ground truth state of the indoor environment is e1. It detects with 100% accuracy over 21 times of front door opening while it detects 15 out of 18 times of back door opening, i.e., an average accuracy of 92.31%.

The degradation in the accuracy is because the wireless channel keeps fading along the time while the training data for front door opening and back door opening is not updated. Hence, there eventually will be a mismatch between the testing CSI measurements and the training profiles. Considering the channel fading, the presented event detector is designed to have an automatic updating scheme for e1, i.e., it will periodically update the training data of e1 as long as the environment is recognized as in the state of all doors closed by the event detector. The periodic refresh of e1 training metrics is to address the uncontrollable changes in the indoor environment but fails to fully resolve the problem. Due to the difficulty of labeling the testing CSI measurements from door opening in an unsupervised way, in this chapter we do not consider to update the training data for other events automatically.

7.4.1.4 Operational Test in House #2

In this part, the performance of TRIMS in real-time event monitoring is studied in House #2 with the event list being in Table 7.2. Moreover, all the parameters and hardware settings are as the same as the ones in House #1.

To begin with, the presented event detector is set to monitor when (a) the environment is quiet and all doors are closed, (b) someone opens the front door and then closes it from the outside (twice), (c) someone opens the back door and then closes it from

Table 7.2 Events of interest in House #2

State Index	Description
e1	All doors are closed.
e2	Front door open.
e3	Back door open.

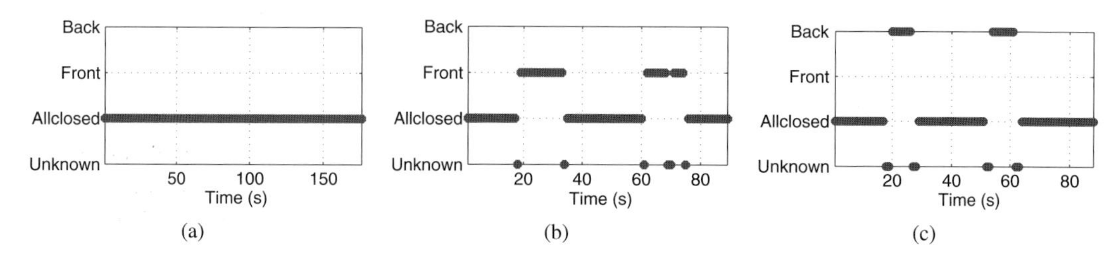

Figure 7.8 Monitoring results of the presented event detector for operational tests in House #2. (a) Test under all doors closed. (b) Test under opening (around the 19th and the 61th s) and closing (around the 36th and the 72th s) front door from the outside the house twice. (c) Test under opening (around the 18th and the 50th s) and closing (around the 28th and the 64th s) back door from the outside the house twice.

the outside (twice). The results are shown in Figure 7.8, where the decision output being "Unknown" means that an untrained event is happening, "Allclosed" indicates that environment is in the all-doors-closed and quiet, and "Front" and "Back" represent front door and back door is opening, respectively. All figures can be interpreted in the same way as those in Figure 7.7. The presented event detector succeeds in capturing the trained events perfectly without false alarms.

7.4.1.5 Long-Term Test in House #2

Furthermore, the long-term behavior of the presented event detector in TRIMS is investigated in House #2 through a test that lasts for 2 weeks. During the long-term test, resident activities are more often than that in House #1, and thus the indoor environment changes every day, which might jeopardize the presented event detector trained in day 1. Every day during the long-term test, the tester performed the same operational test as in Section 7.4.1.4 to evaluate the detection performance of the presented event detector. The system outputs along the time are plotted in Figure 7.9, where the y-axis is the system output with "unknown," "Allclosed," "Front," and "Back" representing the occurrence of untrained events, e1 "all doors are closed," e2 "front door is opened," and e3 "back door is opened," respectively.

As shown in Figures 7.9(a), 7.9(f), and 7.9(k), the presented event detector is good at detecting the trained events with no false alarm during the same day when the system is trained. However, after 1 week or even 2 weeks, with the original training database built on day 1, the presented system fails to detect the trained events and has a high false alarm rate on e2, as shown in Figures 7.9(b), 7.9(d), 7.9(g), 7.9(i), 7.9(l), and 7.9(n).

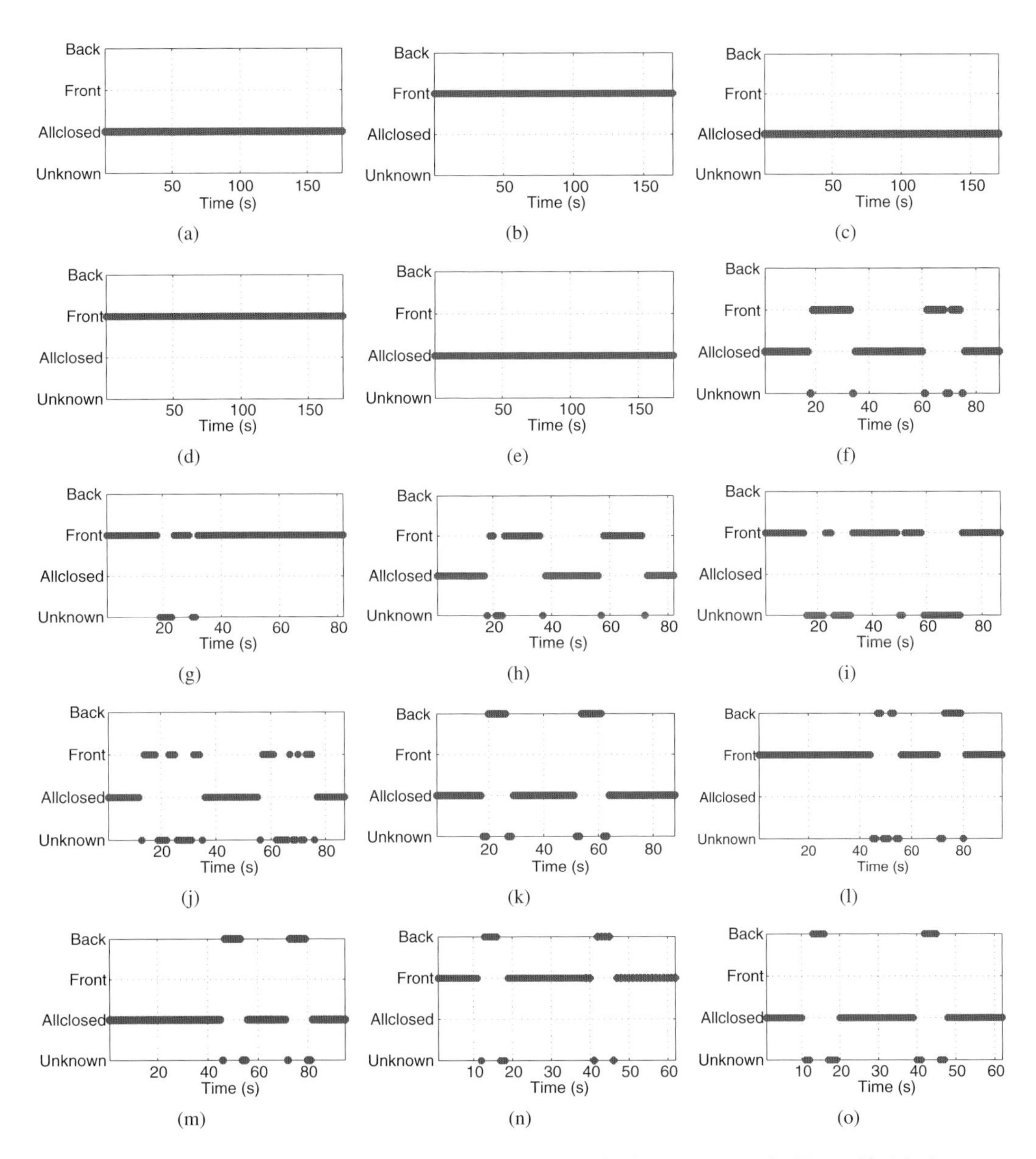

Figure 7.9 Monitoring results of the event detector for long-term tests in House #2. (a) e1 test on day 1. (b) e1 test on day 7 w/o e1 update. (c) e1 test on day 7 w/ e1 update. (d) e1 test on day 14 w/o e1 update. (e) e1 test on day 14 w/ e1 update. (f) e2 test on day 1 (open around the 20th and the 62th s and close around the 34th and the 74th s). (g) e2 test on day 7 w/o e1 update (open around the 20th and the 60th s and close around the 30th and the 70th s). (h) e2 test on day 7 w/ e1 update. (i) e2 test on day 14 w/o e1 update (open around the 18th and the 58th s and close around the 30th and the 70th s). (j) e2 test on day 14 w/ e1 update. (k) e3 test on day 1 (open around the 18th and the 52th s and close around the 24th and the 62th s). (l) e3 test on day 7 w/o e1 update (open around the 46th and the 72th s and close around the 52th and the 80th s). (m) e3 test on day 7 w/ e1 update. (n) e3 test on day 14 w/o e1 update (open around the 12th and the 40th s and close around the 20th and the 48th s). (o) e3 test on day 14 w/ e1 update.

For example, as depicted in Figure 7.9(b), the system keeps reporting "front door is opened," when the ground truth of the indoor state being e1 "all doors are closed." With uncontrolled resident activities, the indoor environment changes result in a different multipath profile not only for e1 but also for e2 and e3. With the help of the auto-update of e1, the presented event detector is able to detect the trained events e2 and e3 during the 2-week experiment with no false alarm. The results are as shown in Figures 7.9(c), 7.9(e), 7.9(h), 7.9(j), 7.9(m), and 7.9(o).

As demonstrated by examples in Figure 7.9, with an automatic and periodic update of the training data for e1, TRIMS can maintain its accuracy in differentiating between and recognizing trained events in a single-family house with normal resident activities during 2 weeks. The monitoring results of TRIMS in the 14-day experiment are compared with the history log provided by a commercial home security system. In general, the presented event detector captures the incidents of e2 and e3, i.e., opening the front or the back door from the outside of the house, with an accuracy being 95.45% while a single misdetection happens on day 13.

7.4.2 TRIMS: Motion Detector

The performance of the motion detector is tested in House #1 with the TX and RX devices located at positions marked by "TX_2" and "RX_2" in Figure 7.5(a). The parameter α defined in (7.25) is set to be either 0.8 or 0.2, while $W = 30$ indicates a 1-s window of continuously collected CSI as defined in (7.24). For the $D_{motion}(t_0)$ in (7.26), we apply a time-diversity smoothing with only one-level majority vote whose $W = 45$ and $O = 15$, to eliminate any possible false alarms due to burst noise or error in the CSI estimation.

During the training phase, the presented motion detector learns the threshold γ_{motion} based on the data from 1-min monitoring data collected under e1 and someone walking in and around the center of the house.

Given a zero false alarm, the detection rates of motion at different locations are listed in Table 7.3. The presented motion detector is intelligent in that it learns and adapts its sensitivity automatically based on the characteristics of the radio propagation environment where it is deployed, through the training phase. The change that motion

Table 7.3 Detection rate for motion at different locations under 0 false alarm rate.

Walking location	Detection rate $\alpha = 0.8(\%)$	Detection rate $\alpha = 0.2(\%)$
Postman	0	0
Foyer	59.18	12.24
Laundry room	100	81.25
Alice's room	3.92	0
Study room	1.92	0
Center of the house	83.33	75.93
Kitchen	48	36
Living room	30.91	0
Restroom	0	0

introduces to the channel is proportional to the amount of the reflected signal energy that is generated by the moving object and collected at the receiver. Hence, by relying on the motion detector, TRIMS succeeds in captures motion inside the house occurring close to the devices or have a LOS path to either the TX or the RX. However, due to the large path loss for EM waves penetrating multiple walls, motion occurring inside Alice's room or the restroom will have no or tiny impacts on the CSI measurements and thus cannot trigger the motion detector. Moreover, a smaller value of α indicates the system being less sensitive and a smaller coverage of monitoring area.

7.5 Discussions

In this section, we are going to discuss some limitations of TRIMS presented in this chapter, along with topics for further extending this chapter.

7.5.1 Retrain of TRIMS

As discussed in the long-term test in Section 7.4.1.3, the system keeps automatically and periodically updating the training data of e1, i.e., the state of all doors are closed, based on its real-time detection results. Due to the difficulty in labeling the testing CSI of door opening in an unsupervised way, the presented auto-updating scheme can only work for e1. As verified by experiments, with that automatic updating scheme, the presented system is robust to normal EM perturbations introduced by noise and slight environmental changes. However, environmental changes affect not only the CSI of state e1, but also that of events e2 and e3. If the environment changes significantly from that when e2 and e3 were trained, the presented system would fail to find a match between the testing CSI and the one in the training database. That is when the system needs retraining, and it can be determined by comparing the very first e1 training data with the current one. Through experiments we found that when the TRRS between the earliest e1 training CSI and the current CSI measured under e1 drops below an empirical threshold of 0.7, the presented system requires retraining over all states.

7.5.2 Monitoring with Multiple Transmitters

In current days, there are more than one device that usually connects to the same Wi-Fi router in an office or at home, which inspires us to extend this chapter by developing TRIMS to accommodate more transmitters. The performance of TRIMS can be improved because the information has more degrees of freedom by means of an increased spatial (device-level) diversity. Moreover, as shown in Sections 7.4.1.1 and 7.4.2, for a single pair of the TX and the RX devices, it has a limited coverage in detecting events and motion. By deploying more transmitters at different locations, the monitoring area will be expanded. However, it requires further study to optimize the performance of the multi-TX TRIMS and will be one of our future work.

7.5.3 Detecting Dynamic Event

In the current event detector of TRIMS, the training database is built upon static CSI measurements collected for each events. Each dynamic event can be decomposed into several intermediate states sampled during its occurrence. Because the intermediate state can be viewed as static, the presented algorithm can be applied to detect the occurrence of its intermediate states. Consequently, the state transition that depicts the occurrence of dynamic events can be captured, and thus dynamic events can also be monitored by the presented system.

7.5.4 Identifying Motion

In this chapter, the presented motion detector in TRIMS manages to detect the incidents of motion. Nevertheless, it is worthwhile to study how to utilize the TR technique to extract the characteristics of a motion with Wi-Fi signals, e.g., the direction and the velocity. The potentiality of extracting motion information and even identifying motion with commercial Wi-Fi devices is beneficial to various applications like elderly assistance and life monitoring.

7.5.5 Potential of TRIMS

The presented TRIMS is not confined by Wi-Fi and can be applied to other wireless technologies as long as CSI with enough resolution can be obtained. The spatial resolution of CSI is determined by the transmission bandwidth of the radio frequency (RF) device. Ultra-wideband (UWB) communication whose bandwidth exceeds 500 MHz can provide CSI of a finer spatial resolution and enable a better discrimination than Wi-Fi does. However, UWB-based indoor monitoring systems require to deploy specially designed RF devices, and the coverage is small. On the other hand, as demonstrated by experiments in this chapter, with the help of TRIMS, commercial Wi-Fi devices with only a 40 MHz bandwidth can support high-accuracy indoor monitoring. Due to the explosive popularity of wireless devices, increasing wireless traffic clogs Wi-Fi and collisions delay the CSI probing with an unknown offset, which introduces difficulty to real-time wireless sensing systems. Taking advantage of the presented smoothing algorithm, TRIMS is robust to nonuniform CSI probing and packages loss. Moreover, thanks to the ubiquitous deployment of Wi-Fi, the presented system is ready and can be easily put into practice for smart home indoor monitoring. In general, the presented system can be integrated with all kinds of wireless technologies where CSI with enough resolution is accessible.

7.6 Summary

In this chapter, we presented a smart radio system, TRIMS, for real-time indoor monitoring, which utilizes TR technique to exploit the information in multipath propagations. Moreover, the statistical behavior of intraclass TRRS is analyzed theoretically. An event detector is built, where different indoor events are differentiated and quantitatively evaluated through TRRS statistics of the associated CSI. Furthermore, a motion detector is designed in TRIMS to detect the existence of dynamics in the environment. The performance of TRIMS is studied through extensive experiments, which are conducted with TRIMS's prototype implemented on a single pair of commodity Wi-Fi devices. Experimental results demonstrate that TRIMS addresses the problem of recognizing different indoor events in real-time. We also evaluate the performance through a 2-week monitoring test in a single-family house with normal resident activities. TRIMS succeeds in achieving high accuracy in long-term indoor monitoring experiments, demonstrating its prominent and promising role in future intelligent Wi-Fi-based low-complexity smart radios. For related references, readers can refer to [41].

References

[1] D. Zhang, J. Ma, Q. Chen, and L. M. Ni, "An RF-based system for tracking transceiver-free objects," in *Fifth Annual IEEE International Conference on Pervasive Computing and Communications (PerCom'07)*, pp. 135–144, Mar. 2007.

[2] S. Sigg, S. Shi, F. Buesching, Y. Ji, and L. Wolf, "Leveraging RF-channel fluctuation for activity recognition: Active and passive systems, continuous and RSSI-based signal features," in *Proceedings of ACM International Conference on Advances in Mobile Computing & Multimedia*, pp. 43:52, 2013.

[3] C. Han, K. Wu, Y. Wang, and L. Ni, "WiFall: Device-free fall detection by wireless networks," in *Proceedings of the International Conference on Computer Communications*, pp. 271–279, Apr. 2014.

[4] Y. Gu, F. Ren, and J. Li, "PAWS: Passive human activity recognition based on WiFi ambient signals," *IEEE Internet of Things Journal*, vol. 3, no. 5, pp. 796–805, Oct. 2016.

[5] H. Abdelnasser, M. Youssef, and K. A. Harras, "WiGest: A ubiquitous WiFi-based gesture recognition system," in *IEEE Conference on Computer Communications (INFOCOM)*, pp. 1472–1480, Apr. 2015.

[6] F. Adib and D. Katabi, "See through walls with WiFi!" in *Proceedings of ACM SIGCOMM*, pp. 75–86, 2013.

[7] K. Qian, C. Wu, Z. Yang, Y. Liu, and Z. Zhou, "PADS: Passive detection of moving targets with dynamic speed using PHY layer information," in *IEEE International Conference on Parallel and Distributed Systems (ICPADS)*, pp. 1–8, Dec. 2014.

[8] Y. Zeng, P. H. Pathak, C. Xu, and P. Mohapatra, "Your AP knows how you move: Fine-grained device motion recognition through WiFi," in *Proceedings of the 1st ACM Workshop on Hot Topics in Wireless*, pp. 49–54, 2014.

[9] A. Banerjee, D. Maas, M. Bocca, N. Patwari, and S. Kasera, "Violating privacy through walls by passive monitoring of radio windows," in *Proceedings of the ACM Conference on Security and Privacy in Wireless & Mobile Networks*, pp. 69–80, 2014.

[10] Y. Wang, J. Liu, Y. Chen, M. Gruteser, J. Yang, and H. Liu, "E-eyes: Device-free location-oriented activity identification using fine-grained WiFi signatures," in *Proceedings of the 20th Annual ACM International Conference on Mobile Computing and Networking*, pp. 617–628, 2014.

[11] C. Wu, Z. Yang, Z. Zhou, X. Liu, Y. Liu, and J. Cao, "Non-invasive detection of moving and stationary human with WiFi," *IEEE Journal on Selected Areas in Communications*, vol. 33, no. 11, pp. 2329–2342, Nov. 2015.

[12] D. Zhang, H. Wang, Y. Wang, and J. Ma, "Anti-fall: A non-intrusive and real-time fall detector leveraging CSI from commodity WiFi devices," in *International Conference on Smart Homes and Health Telematics*, pp. 181–193, 2015.

[13] W. Wang, A. X. Liu, M. Shahzad, K. Ling, and S. Lu, "Understanding and modeling of WiFi signal based human activity recognition," in *Proceedings of the 21st Annual ACM International Conference on Mobile Computing and Networking*, pp. 65–76, 2015.

[14] M. A. A. Al-qaness, F. Li, X. Ma, and G. Liu, "Device-free home intruder detection and alarm system using Wi-Fi channel state information," *International Journal of Future Computer and Communication*, vol. 5, no. 4, p. 180, 2016.

[15] H. Wang, D. Zhang, Y. Wang, J. Ma, Y. Wang, and S. Li, "RT-Fall: A real-time and contactless fall detection system with commodity WiFi devices," *IEEE Transactions on Mobile Computing*, vol. 16, no. 2, pp. 511–526, Feb. 2017.

[16] R. Nandakumar, B. Kellogg, and S. Gollakota, "Wi-Fi gesture recognition on existing devices," *CoRR*, vol. abs/1411.5394, 2014. [Online]. Available: http://arxiv.org/abs/1411.5394.

[17] K. Ali, A. X. Liu, W. Wang, and M. Shahzad, "Keystroke recognition using WiFi signals," in *Proceedings of the 21st Annual ACM International Conference on Mobile Computing and Networking*, pp. 90–102, 2015.

[18] R. M. Narayanan, "Through-wall radar imaging using UWB noise waveforms," *Journal of the Franklin Institute*, vol. 345, no. 6, pp. 659–678, 2008.

[19] G. K. Nanani and M. Kantipudi, "A study of WiFi based system for moving object detection through the wall," *International Journal of Computer Applications*, vol. 79, no. 7, 2013.

[20] D. Huang, R. Nandakumar, and S. Gollakota, "Feasibility and limits of WiFi imaging," in *Proceedings of the 12th ACM Conference on Embedded Network Sensor Systems*, pp. 266–279, 2014.

[21] D. Pastina, F. Colone, T. Martelli, and P. Falcone, "Parasitic exploitation of WiFi signals for indoor radar surveillance," *IEEE Transactions on Vehicular Technology*, vol. 64, no. 4, pp. 1401–1415, Apr. 2015.

[22] F. Adib, Z. Kabelac, D. Katabi, and R. C. Miller, "3D tracking via body radio reflections," in *Proceedings of the 11th USENIX Symposium on Networked Systems Design and Implementation*, pp. 317–329, Apr. 2014.

[23] F. Adib, C.-Y. Hsu, H. Mao, D. Katabi, and F. Durand, "Capturing the human figure through a wall," *ACM Transactions on Graphics*, vol. 34, no. 6, pp. 219, Oct. 2015.

[24] F. Adib, Z. Kabelac, and D. Katabi, "Multi-person localization via RF body reflections," in *Proceedings of the 12th USENIX Symposium on Networked Systems Design and Implementation*, pp. 279–292, May. 2015.

[25] Q. Xu, Y. Chen, B. Wang, and K. J. R. Liu, "TRIEDS: Wireless events detection through the wall," *IEEE Internet of Things Journal*, vol. 4, no. 3, pp. 723–735, Jun. 2017.

[26] B. Bogert, "Demonstration of delay distortion correction by time-reversal techniques," *IRE Transactions on Communications Systems*, vol. 5, no. 3, pp. 2–7, Dec. 1957.

[27] M. Fink, C. Prada, F. Wu, and D. Cassereau, "Self focusing in inhomogeneous media with time reversal acoustic mirrors," *IEEE Ultrasonics Symposium Proceedings*, pp. 681–686, 1989.

[28] M. Fink, "Time reversal of ultrasonic fields. I. Basic principles," *IEEE Transactions on Ultrasonics, Ferroelectrics, and Frequency Control*, vol. 39, no. 5, pp. 555–566, 1992.
[29] F. Wu, J.-L. Thomas, and M. Fink, "Time reversal of ultrasonic fields. II. Experimental results," *IEEE Transactions on Ultrasonics, Ferroelectrics, and Frequency Control*, vol. 39, no. 5, pp. 567–578, 1992.
[30] B. E. Henty and D. D. Stancil, "Multipath-enabled super-resolution for RF and microwave communication using phase-conjugate arrays," *Physical Review Letters*, vol. 93, no. 24, p. 243904, 2004.
[31] G. Lerosey, J. De Rosny, A. Tourin, A. Derode, G. Montaldo, and M. Fink, "Time reversal of electromagnetic waves," *Physical Review Letters*, vol. 92, no. 19, p. 193904, 2004.
[32] "Time reversal of electromagnetic waves and telecommunication," *Radio Science*, vol. 40, no. 6, pp. 1–10, 2005.
[33] G. Lerosey, J. De Rosny, A. Tourin, A. Derode, and M. Fink, "Time reversal of wideband microwaves," *Applied Physics Letters*, vol. 88, no. 15, p. 154101, 2006.
[34] J. de Rosny, G. Lerosey, and M. Fink, "Theory of electromagnetic time-reversal mirrors," *IEEE Transactions on Antennas and Propagation*, vol. 58, no. 10, pp. 3139–3149, 2010.
[35] B. Wang, Y. Wu, F. Han, Y.-H. Yang, and K. J. R. Liu, "Green wireless communications: A time-reversal paradigm," *IEEE Journal on Selected Areas in Communications*, vol. 29, no. 8, pp. 1698–1710, 2011.
[36] C. Chen, Y. Han, Y. Chen, and K. J. R. Liu, "Indoor global positioning system with centimeter accuracy using Wi-Fi," *IEEE Signal Processing Magazine*, vol. 33, no. 6, pp. 128–134, Nov. 2016.
[37] Q. Xu, Y. Chen, B. Wang, and K. J. R. Liu, "Radio biometrics: Human recognition through a wall," *IEEE Transactions on Information Forensics and Security*, vol. 12, no. 5, pp. 1141–1155, May 2017.
[38] C. Chen, Y. Han, Y. Chen, and K. J. R. Liu, "Multi-person breathing rate estimation using time-reversal on WiFi platforms," in *IEEE Global Conference on Signal and Information Processing*, p. 1, Dec. 2016.
[39] Z.-H. Wu, Y. Han, Y. Chen, and K. J. R. Liu, "A time-reversal paradigm for indoor positioning system," *IEEE Transactions on Vehicular Technology*, vol. 64, no. 4, pp. 1331–1339, Apr. 2015.
[40] C. Chen, Y. Chen, K. J. R. Liu, Y. Han, and H.-Q. Lai, "High accuracy indoor localization: A WiFi-based approach," in *IEEE International Conference on Acoustics, Speech and Signal Processing (ICASSP)*, pp. 6245–6249, Mar. 2016.
[41] Q. Xu, Z. Safar, Y. Han, B. Wang, and K. J. R. Liu, "Statistical learning over time-reversal space for indoor monitoring system," *IEEE Internet of Things Journal*, vol. 5, no. 2, pp. 970–983, 2018.

8 Radio Biometrics for Human Recognition

In this chapter, we show the existence of human radio biometrics and present a human identification system that can discriminate individuals even through the walls in a non-line-of-sight (NLOS) condition. Using commodity Wi-Fi devices, the presented system captures the channel state information (CSI) and extracts human radio biometric information from Wi-Fi signals using time-reversal (TR) technique. By leveraging the fact that broadband wireless CSI has a significant number of multipaths, which can be altered by human body interferences, the presented system can recognize individuals in the TR domain without line-of-sight radio. We built a prototype of the TR human identification system using standard Wi-Fi chipsets with 3×3 multiple-input and multiple-output (MIMO) transmission. The performance of the presented system is evaluated and validated through multiple experiments. In general, the TR human identification system achieves an accuracy of 98.78% for identifying about a dozen individuals using a single transmitter and receiver pair. Thanks to the ubiquitousness of Wi-Fi, the presented system shows the promise for future low-cost, low-complexity reliable human identification applications based on radio biometrics.

8.1 Introduction

Nowadays, the capability of performing reliable human identification and recognition has become a crucial requirement in many applications, such as forensics, airport custom check, and bank securities. Current state-of-the-art techniques for human identification rely on the discriminative physiological and behavioral characteristics of humans, known as biometrics.

Biometric recognition refers to the automated recognition of individuals based on their human biological and behavioral characteristics [1, 2]. The well-known biometrics for human recognition include fingerprints, face, iris, and voice. Because biometrics are inherent and distinctive to an individual, biometric traits are widely used in surveillance systems for human identification. Moreover, due to the difficulty for biometrics counterfeit, techniques based on biometrics have clear-cut advantages over traditional security methods such as passwords and signatures in countering the growing security threats and in facilitating personalization and convenience. Even though the current biometrics systems are accurate and can be applied in all environments, all of them require special devices that capture human biometric traits in an extremely line-of-sight

(LOS) environment, i.e., the subject should make contact with the devices. In this chapter, a novel concept of radio biometrics is presented, and accurate human identifications and verifications can be implemented with commercial Wi-Fi devices in a through-the-wall setting.

In [3], researchers studied the relationship between the electromagnetic (EM) absorption of human bodies and the human physical characteristics in the carrier frequency range of 1–15 GHz, in which the body's surface area is found to have a dominant effect on absorption. Moreover, the interaction of EM waves with biological tissue was studied [4], and the dielectric properties of biological tissues were measured in [5, 6]. According to the literature, the wireless propagation around the human body highly depends on the physical characteristic (e.g., height and mass), the total body water volume, the skin condition, and other biological tissues. The human-affected wireless signal under attenuations and alterations, containing the identity information, is defined as *human radio biometrics*. Considering the combination of all the physical characteristics and other biological features that affect the propagation of EM waves around the human body and how variable those features can be among different individuals, the chance for two humans to have the identical combinations is significantly small, no matter how similar those features are.

Even if two have the same height, weight, clothing, and gender, other inherent biological characteristics may be different, resulting in different wireless propagation patterns round the human body. Take the DNA sequence as an example. Even though all humans are 99.5% similar to any other humans, no two humans are genetically identical, which is the key to techniques such as genetic fingerprinting [7]. Because the probability of two individuals to have exactly the same physical and biological characteristics is extremely small, the multipath profiles after human interferences are therefore different among different persons.

Consequently, human radio biometrics, which record how the wireless signal interacts with a human body, are altered accordingly to individuals' biological and physical characteristics and can be viewed as unique among different individuals. One example is that the face recognition has been implemented for many years to distinguish from and recognize different people, thanks to the fact that different individuals have different facial features. Human radio biometrics, which record how RF signals respond to the entire body of a human, including the face, should contain more information than a face and thus become more distinct among humans. In this chapter, the presented TR human identification system utilizes not only the face, but also the entire individual physical characteristic profiles.

In the recent past, a number of attempts have been made to detect and recognize indoor human activities through wireless indoor sensing. Systems have been built to detect indoor human motions based on the variations of CSI [8–10]. In [8], the first two largest eigenvalues of the CSI correlation matrix were viewed as features to determine whether the environment is static or dynamic. The standard deviation of the CSI samples from a 3×3 MIMO system combined were fed into a support vector machine (SVM) to detect human activities such as falling [11]. The received signal strength (RSS) is an indicator for the fluctuation of the wireless channel quality and thus has been applied to recognize indoor human activities [12–15]. Moreover, tracking and recording vital signals using wireless signal has been widely studied [16–19]. Liu et al. proposed a system to track human breathing and heartbeat rate using off-the-shelf Wi-Fi signals [16]. Vital-Radio system was proposed in [18] that monitors vital signs using radar technique to separate different reflections. On the other hand, the recognition of gestures and small hand motions has been implemented using wireless signals [20–23]. Moreover, by sending a specially designed frequency modulated carrier wave (FMCW), which sweeps over different carrier frequencies, Katabi et al. proposed a new radar-based system to keep track of the different time-of-flights (ToFs) of the reflected signals [24–27]. However, as focusing on differentiating between different human movements, e.g., standing, walking, falling down, and small gestures, none of them have addressed the problem of distinguishing one individual from others, who hold the same posture and stand at the same location, by only using Wi-Fi signals in a through-the-wall setting. Recently, in [27], a RF-capture system was presented to image human body contour through the wall. Owing to the distinctiveness of silhouettes, it can differentiate between different individuals by applying image processing and machine learning techniques to the captured human figures. However, to get a high-resolution ToF profile, it requires special devices that can scan over 1 GHz spectrum. Moreover, the computational complexity introduced by the necessary image processing and machine learning algorithms is high. On the contrary, this chapter presents a novel human identification system that aims at distinguishing and identifying different individuals accurately with commercial MIMO Wi-Fi devices of a 40 MHz transmission bandwidth. The presented system supports simple and efficient algorithms to achieve a high-accuracy performance.

To achieve this goal, we utilize the time-reversal (TR) technique to capture the differences between human radio biometrics and to reduce the dimension of features. In an indoor environment, there exists a large amount of reflectors and scatterers. When a wireless signal emitted from the transmitter encounters them, it will travel along different propagation paths with different distances and suffer different fading effects. Consequently, at the receiver the received signal is a combination of the copies of the same transmitted signal through different paths and delays. This phenomenon is well known as the multipath propagation. TR technique takes advantage of the multipath propagation to produce a spatial-temporal resonance effect, the details of which can be referred to Chapter 1. The TR spatial-temporal resonance is generated by fully collecting the energy of the multipath channel and concentrating into a particular location. In physics, the spatial-temporal resonance, which is commonly known as the focusing effect, is the result of a resonance of electromagnetic (EM) field, in response to the environment. This resonance is sensitive to the environment changes, which can be used for capturing the difference in the multipath CSI.

In [28], the concept of TR spatial-temporal resonance has been established as theory and validated through experiments. The TR technique relies on two verified assumptions of channel reciprocity [29, 30] and channel stationarity [31]. Channel reciprocity demonstrates the phenomenon that the CSI of both forward and backward links is highly correlated, whereas channel stationarity establishes that the CSI remains highly correlated during a certain period. A novel TR-based indoor localization approach was first proposed, and a prototype was implemented under a 125 MHz bandwidth, achieving a centimeter accuracy even with a single AP working in non-line-of-sight (NLOS) environments [30]. Recently, in [32], a TR indoor locationing system on a Wi-Fi platform was proposed and built, which utilizes the location-specific fingerprints generated by concatenating the CSI with a total equivalent bandwidth of 1 GHz.

In this chapter, we present a TR human identification system to identify individuals through the walls (i.e., in the absence of any LOS path), based on the human radio biometrics in Wi-Fi signals. To the best of our knowledge, this is the first effort to show and verify the existence of human radio biometrics, which can be found embedding in the wireless channel state information (CSI). Moreover, we present a human recognition system that extracts the unique radio biometrics as features from the CSI for differentiating between people through the wall. We define the term *radio shot* as the procedure to take and record human radio biometrics via Wi-Fi signals. The system consists of two main algorithmic parts: the refinement of human radio biometrics and the TR-based identification. The refinement is designed to remove the common CSI components coming from static objects in the environment and the similarity in the radio biometrics of all participants, and to extract the CSI components that contain distinctive human radio biometrics. In the TR-based identification part, the extracted human radio biometric information is mapped into the TR space, and the similarity between different biometrics is quantified and evaluated using the time-reversal resonance strength (TRRS). The performance of the presented identification system is evaluated, and the accuracy can achieve a 98.78% identification rate when distinguishing between 11 individuals. The detailed study of performance is in Section 8.5.

The major points of this chapter are summarized here.

- We introduce for the first time the concept of human radio biometrics, which account for the wireless signal attenuation and alteration brought by humans. Through experiments, its existence has been verified and its ability for human identification has been illustrated. The procedure to collect human radio biometrics is named as radio shot.
- Due to the fact that the dominant component in the CSI comes from the static environment rather than human body, the human radio biometrics are embedded and buried in the multipath CSI. To boost the identification performance, we design novel algorithms for extracting individual human radio biometrics from the wireless channel information.
- Radio biometrics extracted from the raw CSI are complex-valued and high-dimensional, which complicates the classification problem. To address this problem, we apply the TR technique to fuse and compress the human radio biometrics and to differentiate between radio biometrics of different people, by using the strength of the spatial-temporal resonances.
- For performance evaluation, we build the first prototype that implements the TR human identification system using off-the-shelf Wi-Fi chipsets and test in an indoor office environment during normal working hours with an identification rate as 98.78% in identifying about a dozen of individuals.

This chapter demonstrates the potential of using commercial Wi-Fi signals to capture human radio biometrics for individual identifications.

8.2 TR Human Identification

The presented TR human identification system is capable of capturing human biometrics and identifying different individuals through the walls. The human radio biometrics that are embedded in the CSI contain the Wi-Fi reflections and scattering by human body in the indoor environment. As a result, the human radio biometrics, owing to the differences in human biological metrics, are different among different individuals. Moreover, by leveraging the TR technique, in the presented system, the human radio biometrics can be easily extracted from the CSI for distinguishing between individuals. This procedure is called *radio shot*.

8.2.1 Time-Reversal Space

During the wireless transmission, signals encounter different objects in the environment, and the corresponding propagation path and characteristics change accordingly before arriving at the receiver. As has been demonstrated in Figure 6.2, each dot in the channel state information (CSI) logical space represents a snapshot of the indoor environment, e.g., an indoor location and an indoor event, which can be uniquely determined by the multipath profile $\mathbf{h}$. By taking a time-reverse and conjugate operation over the multipath profile, the corresponding TR signature $\mathbf{g}$ is generated. Consequently, each of the points in the CSI logical space as marked by "A," "B," and "C" is mapped into the TR space as "A'," "B'," and "C'." In the TR space, the similarity between two profiles is quantified by TRRSs. The higher the TRRS is, the more similar two profiles in the TR space are. Similar profiles constrained by a threshold on TRRS can be treated as a single class. Taking advantage of the TR technique and the TR space, a centimeter-level accurate indoor locationing system was proposed in [30], where each of the indoor physical locations is mapped into a logical location in the TR space and can be easily separated and identified using TRRS. The TR-based centimeter-level indoor locationing system was implemented using commercial Wi-Fi chipsets in [32]. By leveraging the TR technique to capture the characteristics of multipath profile at different locations, two locations, even only with a distance of 1 to 2 centimeters, are far away in the TR space and can be easily distinguished.

According to the literature, the wireless propagation around the human body highly depends on the physical characteristic (e.g., height and mass), the total body water volume, the skin condition, and other biological tissues. The human radio biometrics, recording the features in interactions between EM waves and human bodies, are unique among different individuals and are mapped into separate points in the TR space. Hence, the presented system, leveraging the TR technique, is capable of capturing the difference in the multipath profile introduced by different individuals, even when they stand at the same location with the same posture under a through-the-wall setting.

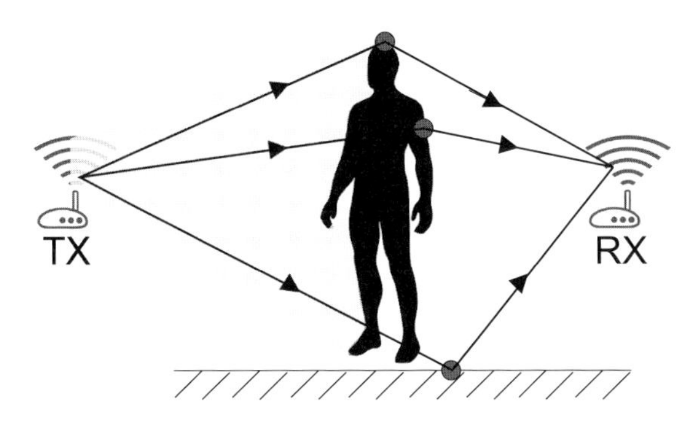

Figure 8.1 RF reflections and scattering.

8.2.2 The Implementation

The system prototype consists of one three-antenna transmitter (TX) and one three-antenna receiver (RX). The CSI samples are obtained from commodity Wi-Fi chips. Moreover, the system is operated at carrier frequency 5.845 GHz with 40 MHz bandwidth. Due to the 3×3 MIMO transmission, each measurement consists of nine pieces of the CSI for each transmitting–receiving antenna pair. Moreover, for each CSI, it contains 114 complex values representing 114 accessible subcarriers in a 40 MHz band.

To the best of our knowledge, the presented system is the first that utilizes commodity WiFi signals for human identification.

8.2.3 The Challenges

Consider the simplified example in Figure 8.1. In an indoor wireless signal propagation environment, the human body acts as a reflector, and the dots represent the reflecting and scattering points due to the human body and other objects. Because the wireless signal reaches the receiving antenna from more than one path, the human radio biometrics are implicitly embedded in the multipath CSI profile. However, the human body may only introduce a few paths to the multipath CSI, and the energy of those paths is small due to the low reflectivity and permittivity, compared with other static objects such as the walls and furniture. As a result, the human radio biometrics, captured through radio shot, are buried by other useless components in the CSI.

Furthermore, due to the fact that the raw CSI obtained from Wi-Fi chips is a 9×114 complex-valued matrix, the resulting raw radio biometrics are of high-dimensional and complex valued, which further complicates the identification and classification problem and increases the computation complexity.

8.2.4 The Solutions

To address the preceding problems, we exploit the TR techniques and consider several postprocessing algorithms to extract the human radio biometrics and magnify the difference among individuals. Specifically, we develop a background subtraction algorithm such that the common information in the CSI can be removed, and the distinctive human radio biometrics are preserved. By leveraging the TR technique, the human radio biometrics in the form of complex-valued matrices are related to the corresponding individual through a real-valued scalar, the TRRS.

The design of the presented time-reversal human identification system exploits the preceding idea and is made up of two key components:

- *Human radio biometrics refinement:* This module extracts the human biometric information from the raw CSI measurement, which is a 9×114 complex-valued matrix. Due to the independency of each link, the background for each link should be calculated and compensated individually. An important consideration is that, for each CSI measurement, it may be corrupted by the sampling frequency offset (SFO) and the symbol timing offset (STO). Hence, before background calculation and compensation, the phase of each CSI measurement should be aligned first. After alignment, based on the assumption that the human radio biometrics only contribute small changes in the multipath, the background can be obtained by taking the average of several CSI measurements.
- *TR-based identification:* Once the 9×114 complex-valued human radio biometric information is refined, this component simplifies the identification problem by reducing the high-dimension complex-valued feature into a real-valued scalar. By leveraging the TR technique, the human radio biometrics are mapped into the TR space, and the TRRS quantifies the differences between different radio biometrics. The detailed methodology will be discussed in Section 14.2.

8.3 System Model

The presented system is built upon the fact that the wireless multipath comes from the environment where the EM signals undergo different reflecting and scattering paths and delays. According to the literature, the wireless propagation around the human body highly depends on individual physical characteristics and conditions of biological tissues. Because it is rare for two individuals to have exactly the same biological physical characteristics, the multipath profiles after human interferences are therefore different among different persons. The human radio biometrics, which record how the wireless signal interacts with a human body, is altered accordingly to individuals' biological physical characteristics and can be viewed as unique among different individuals. Through Wi-Fi sounding, the wireless CSI is collected, as well as the human radio biometrics.

Mathematically, the indoor CSI (a.k.a. Channel frequency response, CFR) for the mth link with the presence of human body can be modeled as the sum of the common CSI component and the human affected component:

$$\mathbf{h}_i^{(m)} = \mathbf{h}_0^{(m)} + \delta\mathbf{h}_i^{(m)}, \quad i = 1,2,\ldots,N, \tag{8.1}$$

where N is the number of individuals to be identified. $\mathbf{h}_i^{(m)}$ is an $L \times 1$ complex-valued vector, which denotes the CSI when the ith individual is inside. L is the number of subcarriers, i.e., the length of the CSI. $\mathbf{h}_0^{(m)}$, defined as the static CSI component, is generated from the static environment in the absence of human, and $\delta\mathbf{h}_i^{(m)}$ denotes the perturbation in the CSI introduced by the ith individual. Here, the $\delta\mathbf{h}_i^{(m)}$ is the raw human radio biometric information of the ith individual embedding in the CSI of the mth link.

At the receiver side, after each channel state sounding, we can collect an $L \times M$ raw CSI matrix for each individual as

$$\mathbf{H}_i = [\mathbf{h}_i^{(1)}, \mathbf{h}_i^{(2)}, \ldots,, \mathbf{h}_i^{(M)}], \; \forall\, i, \tag{8.2}$$

with the corresponding human radio biometric information matrix being

$$\delta\mathbf{H}_i = [\delta\mathbf{h}_i^{(1)}, \delta\mathbf{h}_i^{(2)}, \ldots,, \delta\mathbf{h}_i^{(M)}], \; \forall\, i, \tag{8.3}$$

where M is the number of links between the transmitter and the receiver.

At this point, for human identification and recognition, there are two major problems:

(i) both $\delta\mathbf{H}_i$ and $\mathbf{H}_i$ are $L \times M$ complex-valued matrix. Without appropriate data processing, the classification problem based on the raw data is complex-valued and of high computation complexity.

(ii) Because we have no idea of what $\mathbf{h}_0^{(m)}$ is, it is hard to extract the buried biometric information $\delta\mathbf{H}_i$ directly from the CSI measurement $\mathbf{H}_i$.

To tackle the first problem, we incorporate the TR technique to reduce the data dimension by transforming the feature space into TR spatial-temporal resonance as defined in Section 8.3.1. Furthermore, for the second problem, data postprocessing algorithms are presented to refine the human radio biometrics from the raw CSI information as discussed in Section 8.4.

8.3.1 Time-Reversal Spatial-Temporal Resonance

As discussed in Section 9.1, when transmitting back the TR signature through the corresponding multipath channel, a spatial temporal resonance is generated by fully collecting energy of the multipath channel into a particular location in a rich-scattering indoor environment. The spatial-temporal resonance captures even minor changes in the multipath channel, and it can be utilized to characterize the similarity between two multipath CSI realizations.

The strength of TR spatial-temporal resonance, i.e., the TRRS, in frequency domain is defined as follows.

Definition: The strength of TR spatial-temporal resonance $\mathcal{TR}(\mathbf{h}_1, \mathbf{h}_2)$ in frequency domain between two CFRs $\mathbf{h}_1$ and $\mathbf{h}_2$ is defined as

$$\mathcal{TR}(\mathbf{h}_1, \mathbf{h}_2) = \frac{\max\limits_{\phi} \left| \sum\limits_{k} \mathbf{h}_1[k]\mathbf{g}_2[k]e^{jk\phi} \right|^2}{\left(\sum_{l=0}^{L-1} |\mathbf{h}_1[l]|^2\right)\left(\sum_{l=0}^{L-1} |\mathbf{h}_2[l]|^2\right)}. \tag{8.4}$$

Here, L is the length of CFR and $\mathbf{g}_2$ is the TR signature of $\mathbf{h}_2$ obtained as,

$$\mathbf{g}_2[k] = \mathbf{h}_2^*[k], \quad k = 0, 1, \ldots, L-1. \tag{8.5}$$

Hence, the higher the value of $\mathcal{TR}(\mathbf{h}_1, \mathbf{h}_2)$ is, the more similar $\mathbf{h}_1$ and $\mathbf{h}_2$ are.

For two CSI measurements $\mathbf{H}_i$ and $\mathbf{H}_j$ in a MIMO transmission, we can obtain a $1 \times M$ TRRS vector as

$$[\mathcal{TR}(\mathbf{h}_i^{(1)}, \mathbf{h}_j^{(1)}), \mathcal{TR}(\mathbf{h}_i^{(2)}, \mathbf{h}_j^{(2)}), \ldots,, \mathcal{TR}(\mathbf{h}_i^{(M)}, \mathbf{h}_j^{(M)})].$$

Then, the TRRS between two CSI matrices, $\mathbf{H}_i$ and $\mathbf{H}_j$, is defined as the average of the TRRSs on each of the links,

$$\mathcal{TR}(\mathbf{H}_i, \mathbf{H}_j) = \frac{1}{M} \sum_{m=1}^{M} \mathcal{TR}(\mathbf{h}_i^{(m)}, \mathbf{h}_j^{(m)}). \tag{8.6}$$

We show an example of the TRRS matrices of each link for different CSI measurements captured by commodity Wi-Fi chips in Figure 8.2. Due to the different spatial

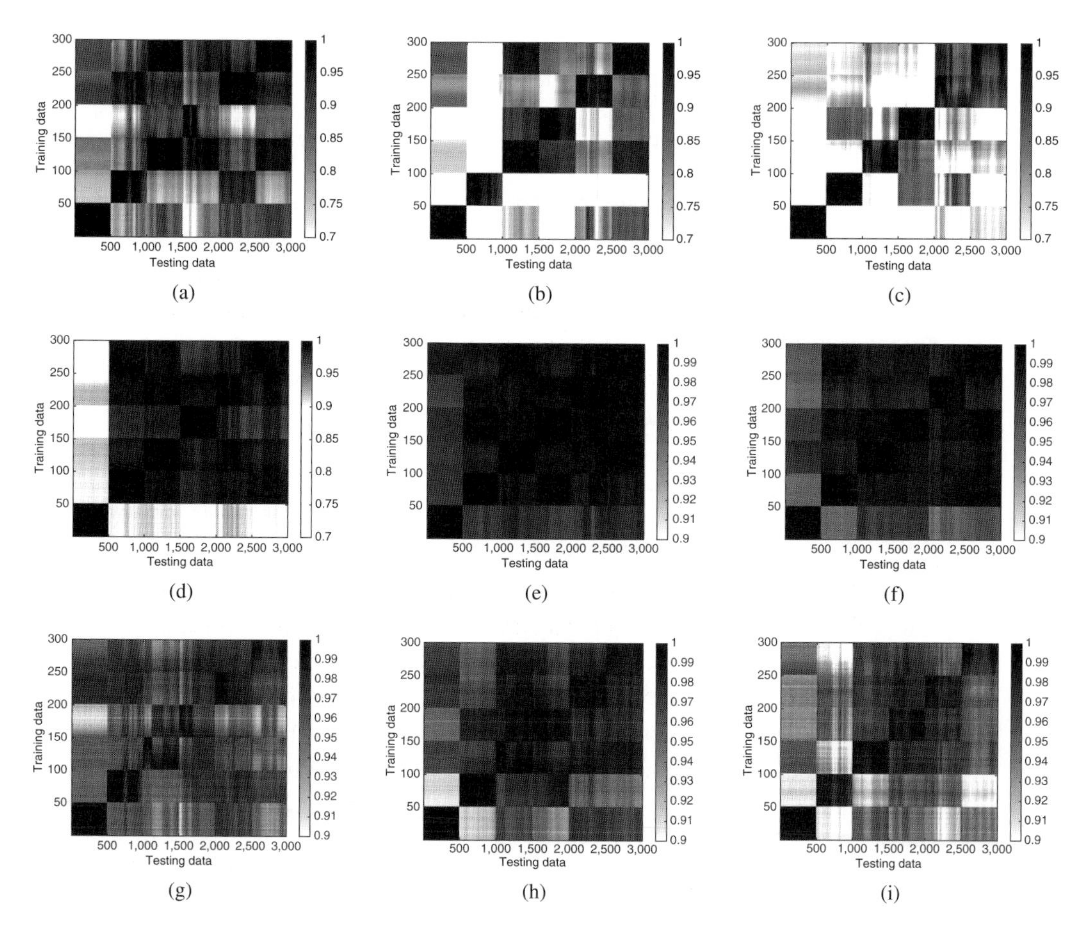

Figure 8.2 TRRS map for each link: (a) Link 1, (b) Link 2, (c) Link 3, (d) Link 4, (e) Link 5, (f) Link 6, (g) Link 7, (h) Link 8, and (i) Link 9.

distributions of each link, how the human body affects the CSI of each link varies. Some links succeed in capturing the human biometric information and show distinct TRRSs between different individuals as in Figure 8.2(c). Some links fail, and the TRRSs between test subjects are similar as shown in Figure 8.2(e).

8.3.2 Identification Methodology

After taking the radio shot, by means of the TR signal processing, the high-dimension complex-valued human radio biometrics embedded in the CSI measurements are mapped into the TR space, and the feature dimension is reduced from $L \times M$ to 1. The human recognition problem can be implemented as a simple multiclass classification problem as following.

For any CSI measurement $\mathbf{H}$, given a training database consisting of the CSI samples of each individual $\mathbf{H}_i$, $\forall\, i$, the predicted individual identity (ID) is obtained based on the TRRS as:

$$\hat{i} = \begin{cases} arg \max\limits_i \mathcal{TR}(\mathbf{H}, \mathbf{H}_i), & \text{if} \max\limits_i \mathcal{TR}(\mathbf{H}, \mathbf{H}_i) \geq \mu, \\ 0, & otherwise, \end{cases} \tag{8.7}$$

where μ is a predefined threshold for triggering the identification, and $\hat{i} = 0$ denotes an unidentified individual.

However, as discussed earlier, the embedded human radio biometric information $\delta\mathbf{H}$ is small compared with other CSI components in measurement $\mathbf{H}$. The resulting TRRS $\mathcal{TR}(\mathbf{H}, \mathbf{H}_i)$ may become quite similar among different samples, and thus the accuracy of identification degrades. In order to improve the identification performance, we need to remove the common components from each CSI measurement and to extract and refine the embedded human biometrics features after taking the radio shot.

8.4 Radio Biometrics Refinement Algorithm

As the presence of human body changes the multipath propagation environment of Wi-Fi signals, the human radio biometrics are implicitly embedded in the CSI measurements. However, owing to the fact that only a few paths are affected by the human body, the human biometrics CSI component for the ith individual in the mth link, $\delta\mathbf{h}_i^{(m)}$, is small in energy, compared with the common CSI component $\mathbf{h}_0^{(m)}$ in (8.1). Without a refinement of the radio biometric information, the common feature $\mathbf{h}_0^{(m)}$ in the CSI dominates in the TRRS in (8.4) and (8.6). Moreover, because there exists similarity between different human bodies, it is inevitable to have resemblances in the human radio biometric information $\delta\mathbf{h}_i^{(m)}$. As a result, even though the spatial-temporal resonance captures the $\delta\mathbf{h}_i^{(m)}$, the difference between the TRRSs for different individuals may become too small to differentiate between people. In this chapter, we discuss postprocessing algorithms to extract the useful human radio biometric information from the CSI, after taking the radio shot.

The process of the human radio biometrics refinement includes the following two steps:

(i) *Phase compensation:* In reality, the estimated CSI can be corrupted by different initial phases of each measurement and different linear phases on each subcarrier due to the time synchronization error. Therefore, in order for the presented system to extract and subtract out correct background CSI components, it is indispensable to compensate for phase errors in all the raw CSI measurements.

(ii) *Background information subtraction:* Note that the CSI is modeled as the sum of static background CSI components and human biometrics CSI components, so the radio biometric information can be extracted by the system through subtracting out the common information in the CSI.

In what follows, we describe each of the algorithms in detail.

8.4.1 Phase Alignment Algorithm

Considering the phase errors, each CSI $\mathbf{h}^{(m)}$ can be mathematically modeled as:

$$\mathbf{h}^{(m)}[k] = \left|\mathbf{h}^{(m)}[k]\right| \exp\Big\{ - j(k\phi_{linear} + \phi_{ini})\Big\}, \quad k = 0, 1, \ldots, L-1, \tag{8.8}$$

where ϕ_{linear} denotes the slope of the linear phase, ϕ_{ini} is the initial phase, and both of them are different for each CSI.

Unfortunately, there is no way to explicitly estimate either ϕ_{linear} or ϕ_{ini}. To address the phase misalignment among the CSI measurements, for each identification task, we pick one CSI measurement in the training database as the reference and align all the other CSI measurements based on this reference.

To begin with, we find the linear phase difference $\delta\phi_{linear}$ between the reference and the other CSI samples. For any given CSI $\mathbf{h}_2$ and reference $\mathbf{h}_1$ from the same link, we can have

$$\delta\phi_{linear} = arg \max_{\phi} \left| \sum_k \mathbf{h}_1[k]\mathbf{h}_2^*[k] \exp\{jk\phi\} \right|. \tag{8.9}$$

To align the linear phase of the CSI $\mathbf{h}_2$ according to the reference, we simply compensate for this difference on each subcarrier through

$$\widehat{\mathbf{h}}_2[k] = \mathbf{h}_2[k] \exp\big\{ - jk\delta\phi_{linear}\big\}, \quad k = 0,\ 1, \ldots, , L-1. \tag{8.10}$$

Once all the linear phase differences of the CSI measurements have been compensated based on the reference, the next step is to cancel the initial phase of the CSI for each link, including the reference. The initial phase is obtained as the phase on the first subcarrier for each CSI $\angle\widehat{\mathbf{h}}[0]$, and can be compensated as

$$\mathbf{h}_{align} = \widehat{\mathbf{h}} \exp\big\{ - j\angle\widehat{\mathbf{h}}[0]\big\}. \tag{8.11}$$

In the following discussion, both the background and the refined human biometric information are extracted from the aligned CSI measurements $\mathbf{h}_{align}$. To simplify notation, we will use $\mathbf{h}$ instead of $\mathbf{h}_{align}$ to denote the aligned CSI in the rest of the chapter.

8.4.2 Background Subtraction Algorithm

In the presented CSI model in (8.1), the radio biometrics $\delta\mathbf{h}_i^{(m)}$ also involves two parts: the common radio biometric information and the distinct radio biometric information. Thus, $\mathbf{h}_i^{(m)}$ can be further decomposed as follows:

$$\mathbf{h}_i^{(m)} = \mathbf{h}_0^{(m)} + \delta\mathbf{h}_{i,ic}^{(m)} + \delta\mathbf{h}_{i,c}^{(m)}, \quad \forall i, m, \tag{8.12}$$

where $\delta\mathbf{h}_i^{(m)} = \delta\mathbf{h}_{i,c}^{(m)} + \delta\mathbf{h}_{i,ic}^{(m)}$. $\delta\mathbf{h}_{i,c}^{(m)}$ denotes the common radio biometric information, which is determined by all the participants in the identification system. Meanwhile, $\delta\mathbf{h}_{i,ic}^{(m)}$ is the corresponding distinct radio biometric information, remaining in the extracted radio biometrics after taking out the common biometric information.

The background CSI components for several CSI measurements of N individuals can be estimated by taking the average over the aligned CSI as:

$$\mathbf{h}_{bg}^{(m)} = \frac{1}{N}\sum_{i=1}^{N} \frac{\mathbf{h}_i^{(m)}}{\left\|\mathbf{h}_i^{(m)}\right\|^2}. \tag{8.13}$$

Then the human radio biometrics for each individual can be extracted through subtracting a scaled version of the background in (8.13) from the original CSI.

$$\widetilde{\mathbf{h}}_i^{(m)} = \mathbf{h}_i^{(m)} - \alpha\mathbf{h}_{bg}^{(m)}, \tag{8.14}$$

where α is the background subtraction factor, $0 \leq \alpha \leq 1$. It cannot be too close to 1 as the remaining CSI will be noise-like. The impact of α is studied in Section 8.5.2.

After obtaining the refined radio biometrics $\widetilde{\mathbf{h}}_i^{(m)}$ for each link, the classification problem based on the TRRS in (8.7) becomes:

$$\hat{i} = \begin{cases} arg\max_i \mathcal{TR}(\widetilde{\mathbf{H}}, \widetilde{\mathbf{H}}_i), & \text{if} \max_i \mathcal{TR}(\widetilde{\mathbf{H}}, \widetilde{\mathbf{H}}_i) \geq \mu, \\ 0, \quad otherwise, \end{cases} \tag{8.15}$$

where $\widetilde{\mathbf{H}}_i$ is the refined radio biometric information matrix for individual i and

$$\widetilde{\mathbf{H}}_i = [\widetilde{\mathbf{h}}_i^{(1)}, \quad \widetilde{\mathbf{h}}_i^{(2)}, \ldots, \widetilde{\mathbf{h}}_i^{(M)}], \; \forall\, i. \tag{8.16}$$

$\widetilde{\mathbf{H}}_i$ is an approximation of the distinctive component in the human radio biometric information matrix $\delta\mathbf{H}_i$ defined in (8.3).

An example is shown in Figure 8.4, where the TRRS $\mathcal{TR}(\mathbf{H}, \mathbf{H}_i)$ before background subtraction is plotted in Figure 8.4(a), while that of $\mathcal{TR}(\widetilde{\mathbf{H}}, \widetilde{\mathbf{H}}_i)$ is in Figure 8.4(b), with the background as the average of all CSI measurements in training database. The comparison between two figures demonstrates that the refinement of human radio biometrics helps to improve the sensitivity of TRRS for differentiating between individuals. The presented background subtraction algorithm suppresses the spatial-temporal resonance of the CSI between different classes while maintaining strong resonance within the same class.

For the presented system, if there are K subjects to be identified, the computational complexities for building the training database and testing are both $O(M \times (K + 1) \times N\log_2 N)$, where M is the number of either the training CSI samples or the testing CSI samples for each subject. N is the search resolution for ϕ in (8.4) and (8.9), where typical values for N are 512 and 1024.

8.5 Performance Evaluation

By leveraging the TR technique to capture human radio biometrics embedded in the CSI of Wi-Fi signals, the presented system is capable of identifying different individuals in real office environments with high accuracy. In this section, the performance of human identification is evaluated. For the presented system, the training, i.e., taking the radio shot, is simple and can be done in seconds.

8.5.1 Experiment Settings

The evaluation experiments are conducted in the office on the 10th floor of a commercial office building with a total of 16 floors. The floor plan of the experiment office

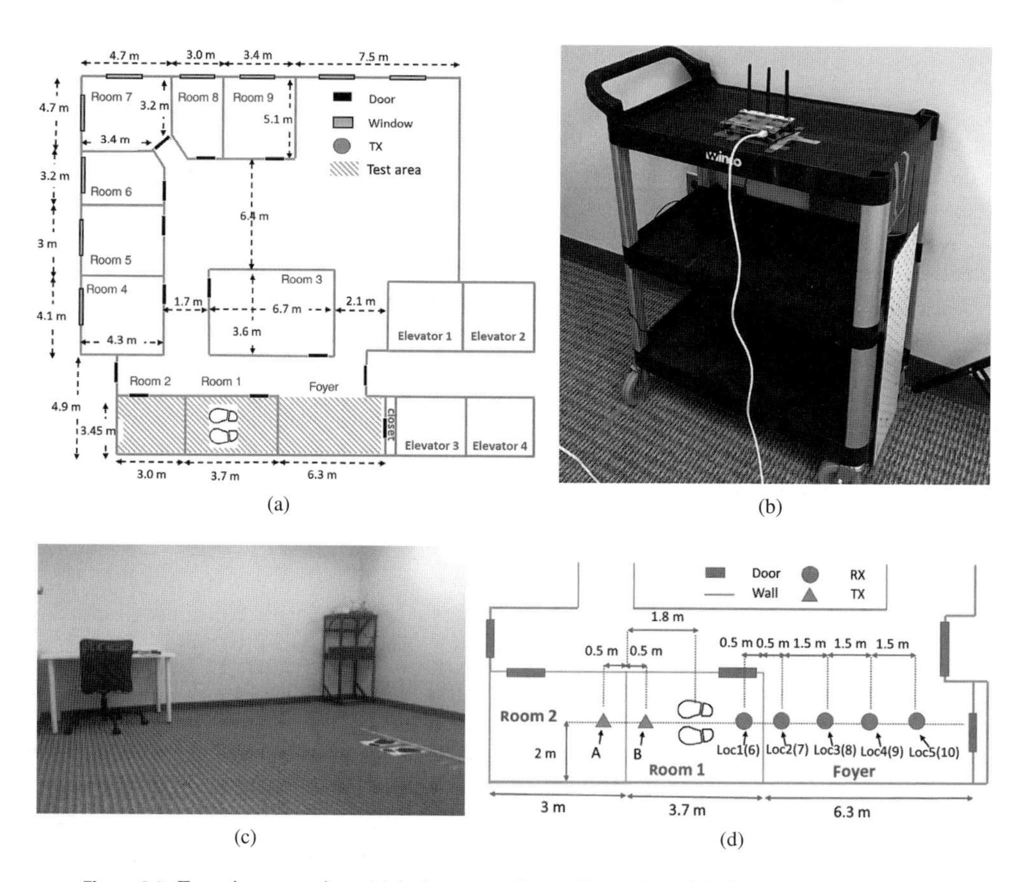

Figure 8.3 Experiment setting: (a) indoor experiment floor plan with dimensions, (b) transmitter or receiver devices, (c) test room configuration, and (d) locations of test subjects and devices.

Table 8.1 Physical characteristics of test subjects in human identification experiment

Test subject	♯1	♯2	♯3	♯4	♯5	♯6	♯7	♯8	♯9	♯10	♯11
Height (cm)	172	164	173	168	176	170	170	172	180	166	155
Weight (kg)	74	53	70	90	90	90	70	69	75	68	45
Gender (M/F)	M	F	M	M	M	M	F	M	M	M	F
Glasses (Y/N)	Y	N	Y	Y	Y	Y	N	Y	Y	Y	N

is shown in Figure 8.3(a). Surrounding the experiment office are four elevators and multiple occupied offices. All the experiments are conducted during normal working hours in weekdays, so that outside the experiment office there are many activities, such as humans walking and elevators running, happening at the same time as the experiments run.

In Figure 8.3(d) the experiment configurations of the transmitter, receiver and individuals are demonstrated. Both Wi-Fi devices are placed on the cart or table with height from the ground being 2.8 ft as shown in Figure 8.3(b). When the transmitter (bot) is on location denoted as "A," the receiver (RX) is placed on the locations denoted from "Loc 1" to "Loc 5." Otherwise when the bot is on location "B," the receiver is on "Loc 6" to "Loc 10," respectively. These 10 TX-RX locations can represent an LOS scenario ("Loc 1"), NLOS scenarios ("Loc 2" to "Loc 6"), and through-the-wall scenarios ("Loc 7" to "Loc 10"). When taking the radio shot, each individual, to be recognized, stands in the room on the point marked by the small footprint, and the door of this room is closed.

Furthermore, in the experiments, we build the training database with 50 CSI measurements for each class, while the size of the testing database for identification is 500 CSI measurements per class. The physical characteristics of test subjects are listed in Table 8.1. The first five subjects participate in experiments in Sections 8.5.2 and 8.5.3, while all the 11 subjects take part in the identification experiment in Section 8.5.4. The second individual is the subject in the verification experiments in Section 8.5.5.

8.5.2 Impact of Background Subtraction

To begin with, we first quantitatively study the impact of the presented background subtraction and biometrics refinement algorithms on human recognition.

As shown by Figure 8.4, after refinement the spatial-temporal resonance between the training and the testing CSI from different classes is suppressed a lot while maintaining a high TRRS for the CSI from the same class. In Table 8.2, the performance matrices for human identification are listed to show the performance improvement after refining the radio biometrics. Each element of the performance matrix is the probability for that the TRRS between the training and the testing classes is higher than the threshold μ. A higher value in the diagonal means a larger chance of correct identifications. However, larger off-diagonal elements indicate higher false alarm rates because it implies that the testing sample may be misclassified to the wrong training class with a higher probability if the testing class has never been included in the training set.

Table 8.2 Performance matrix of individual identification: (a) no background subtraction and (b) after background subtraction with $\alpha = 0.5$

(a)

Training data \ Testing data	Empty room	Human #1	Human #2	Human #3	Human #4	Human #5
Empty room	1	0	0	0	0.4944	0.9613
Human #1	0	1	0	0.5754	0.4138	0
Human #2	0	0	1	1	0.9974	1
Human #3	0	0.3159	1	1	0.1912	0.8869
Human #4	0.8722	0.4951	0.9917	0.5672	1	1
Human #5	0.9781	0	1	0.9999	1	1

(b)

Training data \ Testing data	Empty room	Human #1	Human #2	Human #3	Human #4	Human #5
Empty room	1	0	0	0	0	0
Human #1	0	0.9812	0	0	0	0
Human #2	0	0	0.9972	0.0024	0	0
Human #3	0	0	0	0.9635	0	0
Human #4	0	0	0	0	0.9696	0
Human #5	0	0	0.0011	0	0	0.9842

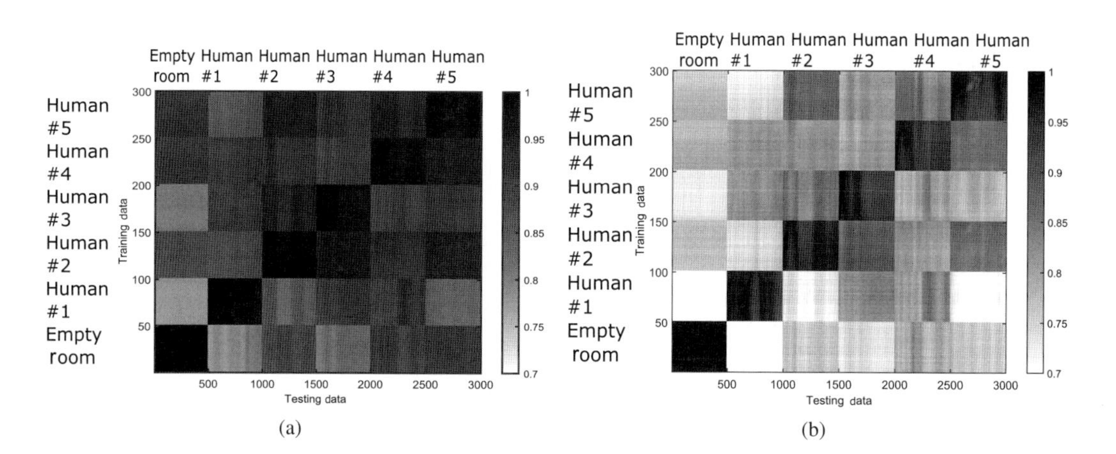

Figure 8.4 Comparison on TRRS maps: (a) no background subtraction and (b) after background subtraction with $\alpha = 0.5$.

Both of the matrices in Table 8.2 have the same threshold $\mu = 0.9$ as defined in (8.7) and (8.15). Without background subtraction, although the diagonal value can reach 100%, the off-diagonal ones can be as high as 99.99% as shown in Table 8.2(a). A high off-diagonal value implies a larger chance to have a false alarm between these particular training and testing classes. Nevertheless, after background subtraction, when using the refined radio biometrics for identification, the largest off-diagonal value drops to 0.24% while maintaining the diagonal elements higher than 96.35%.

8.5.2.1 Background Selection

How to choose the background CSI components is essential for a good radio biometrics refinement. In this part, we study the performance of identification under three schemes: no background subtraction, subtraction with the static environment background, and subtraction with the background consisting of static environment and common radio biometrics. We compare the receiver operating characteristic (ROC) curves in Figure 8.5(a).

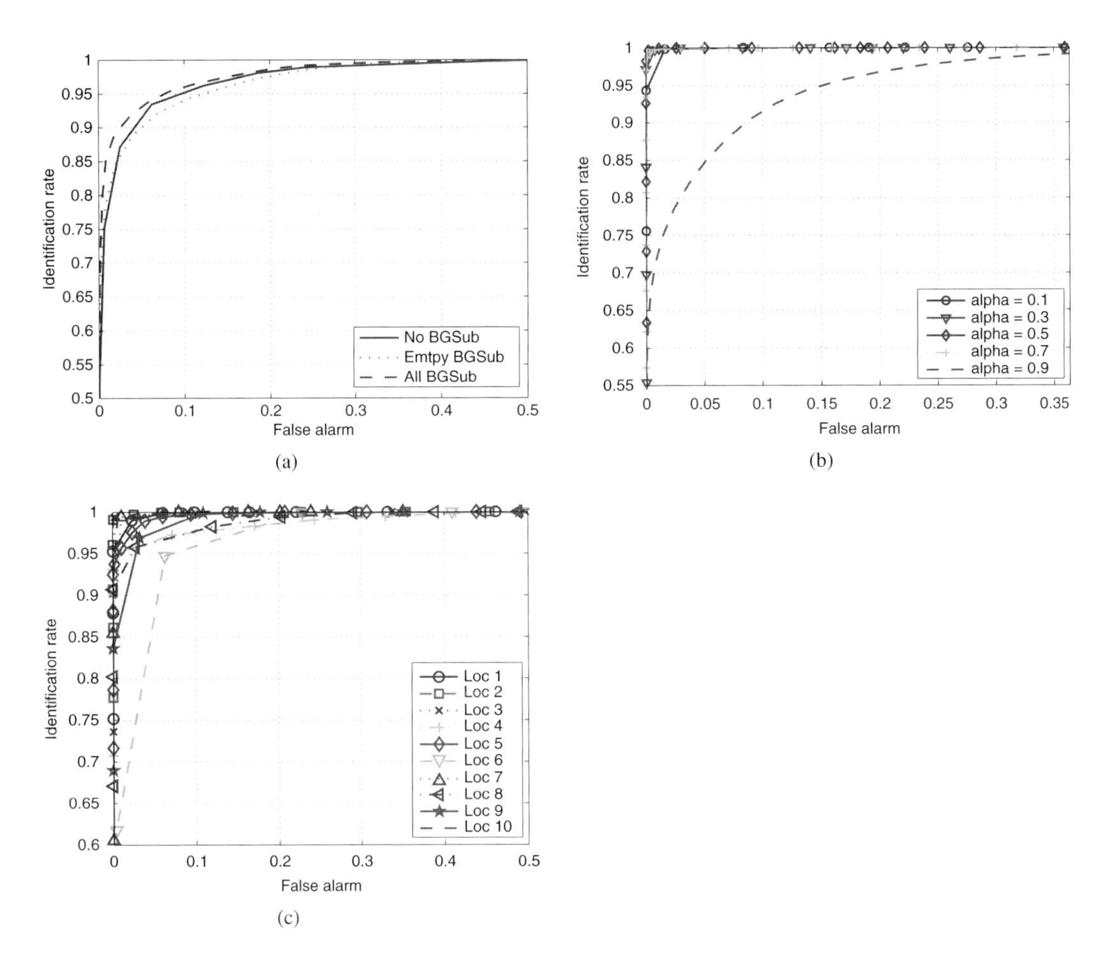

Figure 8.5 Evaluation on ROC curves: (a) different selected backgrounds, (b) different α for Loc 7, and (c) different TX-RX locations.

The ROC curves, which are obtained by averaging the ROC performance measured at all 10 TX-RX locations, show how the identification rate and false alarm rate vary as the decision threshold μ changes. The dashed line denotes the performance when using all the CSI measurements in the training data set as the background (i.e., the background consisting of static environment and common radio biometrics), while the solid line and dotted line represent the case of no background subtraction and subtraction with the static environment background, respectively. Here, the background subtraction factor is $\alpha = 0.5$. The performance of the system using all the training CSI measurements outperforms the others. The reason is that, by taking the average of the CSI samples from all the classes as the background, we effectively eliminate the high correlated and similar component in radio biometrics for different individuals, which is the estimation of $\mathbf{h}_0^{(m)} + \delta\mathbf{h}_{i,c}^{(m)}$ as defined in (8.12), and thus enlarge the difference between the radio biometrics of different people.

8.5.2.2 Optimal Background Subtraction Factor

After we have determined the optimal background, the next question is to find the optimal background subtraction factor α. In Figure 8.5(b), the ROC performance is plotted to evaluate the impact of different α. When $\alpha = 0.9$, the identification performance is the worst because the remaining CSI components after background subtraction is noisy and has little information for human biometrics. Through the experiment, we find $\alpha = 0.5$ is optimal for individual identification. In the rest experiments, we adopt $\alpha = 0.5$ and the all-CSI background scheme.

8.5.3 Impact of TX-RX Locations

Next, we would like to evaluate the impacts of TX-RX configurations on the performance of human identification. "Loc 1" represents LOS scenario where the transmitter, receiver and experiment individual are in the same room. "Loc 2" to "Loc 6" represent the NLOS case where either the transmitter or the receiver is in the same room with the individual, while the other device is placed outside. Moreover, in the through-the-wall scenarios, represented by "Loc 7" to "Loc 10," the individual to be identified is in the room while both the transmitter and the receiver are outside and in different locations.

The identification performance of different scenarios is plotted in Figure 8.5(c). The performance comparison can be summarized from the best to the worst as: Loc 7 > Loc 2 > Loc 3 > Loc 10 > Loc 1 >Loc 5 > Loc 9 > Loc 4 > Loc 8 > Loc 6. There is no direct relation between identification performance and the distance between the transmitter and the receiver. Moreover, the LOS scenario is not the best configuration for human identification. As we discussed, the human radio biometrics are embedded in the multipath CSI. Due to the independency of each path in the multipath CSI, the more paths the CSI contains, the larger number of degrees of freedom it can provide in the embedded human radio biometrics. Consequently, owing to the fact that there are fewer multipath components in the CSI of the LOS scenario, less informative radio biometrics are extracted, which degrades the performance of identification. The results in Figure 8.5(c) also demonstrate the capability of the presented system for through-the-wall human identification, in that no matter which configuration is selected the presented system has a high accuracy.

8.5.3.1 Special Case Study

To better understand the impact of TX-RX locations on the identification capability of the presented system, six examples are investigated and compared in Table 8.3 by using the performance matrices defined at the beginning of Section 8.5.2.

In Table 8.3(a), 8.3(b), and 8.3(c), the performance matrices for LOS case "Loc 1," NLOS case "Loc 6," and the through-the-wall case "Loc 7" with the threshold $\mu = 0.9$ are listed. For "Loc 1," there is no off-diagonal element larger than 0, but the diagonal element for the fifth individual is only 51.59%. This is because in the LOS configuration the human body to be identified is close to both the transmitter and the receiver, which leads to stronger radio biometrics embedded in the CSI. This makes different individuals more distinguishable while making the identification system sensitive and vulnerable to

Table 8.3 Comparison on performance matrices: (a) Loc 1 with threshold $\mu = 0.9$, (b) Loc 6 with threshold $\mu = 0.9$, (c) Loc 7 with threshold $\mu = 0.9$, (d) Loc 1 with minimum diagonal > 0.99, (e) Loc 6 with minimum diagonal > 0.99, and (f) Loc 7 with minimum diagonal > 0.99

(a)

Training data \ Testing data	Empty room	Human #1	Human #2	Human #3	Human #4	Human #5
Empty room	1	0	0	0	0	0
Human #1	0	0.9762	0	0	0	0
Human #2	0	0	0.9887	0	0	0
Human #3	0	0	0	0.9272	0	0
Human #4	0	0	0	0	0.9306	0
Human #5	0	0	0	0	0	0.5159

(b)

Training data \ Testing data	Empty room	Human #1	Human #2	Human #3	Human #4	Human #5
Empty room	1	0	0	0	0	0
Human #1	0	0.9896	0	0	0.0912	0
Human #2	0	0	0.9836	0.8820	0.7128	0.1296
Human #3	0	0	0.9732	0.9969	0.2052	0.3014
Human #4	0	0	0.1190	0	1	0.0174
Human #5	0	0	0.1426	0.3633	0	0.9991

(c)

Training data \ Testing data	Empty room	Human #1	Human #2	Human #3	Human #4	Human #5
Empty room	1	0	0	0	0	0
Human #1	0	0.9812	0	0	0	0
Human #2	0	0	0.9972	0.0024	0	0
Human #3	0	0	0	0.9635	0	0
Human #4	0	0	0	0	0.9696	0
Human #5	0	0	0.0011	0	0	0.9842

(d)

Training data \ Testing data	Empty room	Human #1	Human #2	Human #3	Human #4	Human #5
Empty room	1	0	0	0	0	0
Human #1	0	1	0	0	0	0.0430
Human #2	0	0	1	0.6678	0	0
Human #3	0	0	0.9190	0.9997	0	0
Human #4	0	0	0	0.0004	1	0
Human #5	0	0.1564	0	0	0	0.9977

(e)

Training data \ Testing data	Empty room	Human #1	Human #2	Human #3	Human #4	Human #5
Empty room	1	0	0	0	0	0
Human #1	0	0.9972	0	0	0.4843	0
Human #2	0	0	0.9956	0.9753	0.8852	0.3906
Human #3	0	0	0.9947	0.9990	0.3744	0.6113
Human #4	0	0.0110	0.5126	0.0130	1	0.0771
Human #5	0	0	0.5020	0.6238	0.0048	0.9999

(f)

Training data \ Testing data	Empty room	Human #1	Human #2	Human #3	Human #4	Human #5
Empty room	1	0	0	0	0	0
Human #1	0	0.9966	0	0	0	0
Human #2	0	0	0.9995	0.0443	0	0.0005
Human #3	0	0	0	0.9905	0	0
Human #4	0	0	0	0	0.9936	0
Human #5	0	0	0.0248	0	0	0.9984

small variations on the human body, e.g., the slight inconsistency in poses and standing location of humans. "Loc 6" has the worst performance because its off-diagonal element could reach 97.32%. Meanwhile, the through-the-wall scenario "Loc 7" becomes the most ideal configuration for individual identification in that the minimum diagonal element is higher than 96% and the largest off-diagonal element is only 0.24%.

Similarly, in Table 8.3(d), 8.3(e), and 8.3(f), with the requirement of a minimum diagonal element larger than 99%, the corresponding performance matrices of the aforementioned three cases are shown. To maintain the diagonal values, the identification system has to reduce the threshold μ, which inevitably introduces larger off-diagonal elements and more false alarms. Except for the ideal configuration "Loc 7," the other two examples sacrifice the off-diagonal performance to 91.9% and 99.46%, respectively.

We can conclude that among the 10 TX-RX locations tested in the experiment, "Loc 7" is the optimal configuration for the presented system and is adopted in the following experiments.

8.5.4 Human Identification

From the preceding analysis, we have already observed that the performance of the presented human identification system is influenced by both the background subtraction and the TX-RX configurations. In this part, the performance is evaluated in a large data set of 11 individuals, with optimal background subtraction applied and "Loc 7" TX-RX configuration. The corresponding ROC curve is plotted in Figure 8.6. With a threshold μ being 0.91, the average identification rate is 98.78%, and the average false alarm rate is 9.75%. This is because, when two individuals have similar body contours, the possibility of misclassifying between them increases. However, because not only the contour but also the permittivity and conductivity of body tissue, which is more distinct for different individuals, will affect the Wi-Fi signal propagation that encounters the human body, the accuracy of identification is still high. In the current performance evaluation, the number of participants is 11. We are inviting more people to participate in the experiment and collecting more data for further validation and analysis.

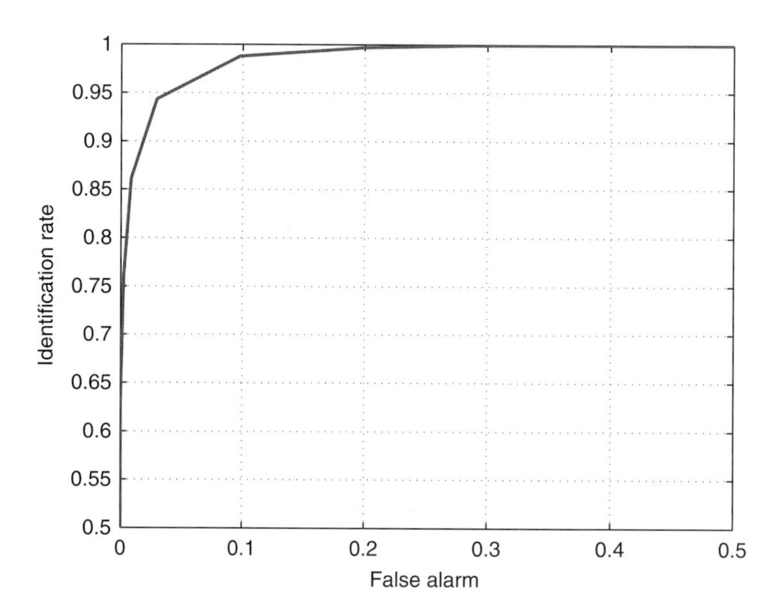

Figure 8.6 ROC curve of identifying 11 individuals.

8.5.5 Individual Verification

In this set of experiments, we study the performance of individual verification using the presented system. Instead of finding the correct identity among several possible ones, the individual verification is to recognize a specific individual with variations in both the human body and the environment.

8.5.5.1 Stationarity over Time

To begin with, the stationarity of human verification performance is discussed. We collect the CSI measurements for both the empty room and with one individual inside twice a day for three consecutive days. The TRRS maps are demonstrated in Figure 8.7. As shown in Figure 8.7(a), if we only use the CSI from the first measurement as the training set, the TRRS within the same class gradually decreases. This leads to a 90.83% identification rate with the threshold $\mu = 0.75$. However, if we update the training set every time after measurement and identification, e.g., for Day 2 morning experiment the training set consists of the CSI from measurements at Day 1 morning and afternoon, the identification rate increases to 97.35%. The details of the verification accuracy are listed in Table 8.4. Hence, to combat the variations over time, the training data set for both identification and verification should be updated regularly.

8.5.5.2 Other Variations

In this experiment, the impact of other types of variations such as wearing a coat or carrying a backpack/laptop on the accuracy of verification is discussed. We consider six classes as listed in the Table 8.5, and the corresponding TRRS map is shown in Figure 8.10.

The detailed verification performance is discussed in Table 8.6, where the relation of the threshold μ and the capability of differentiating between different variations is studied. Here, the training set only contains the CSI from class #1. A low threshold μ

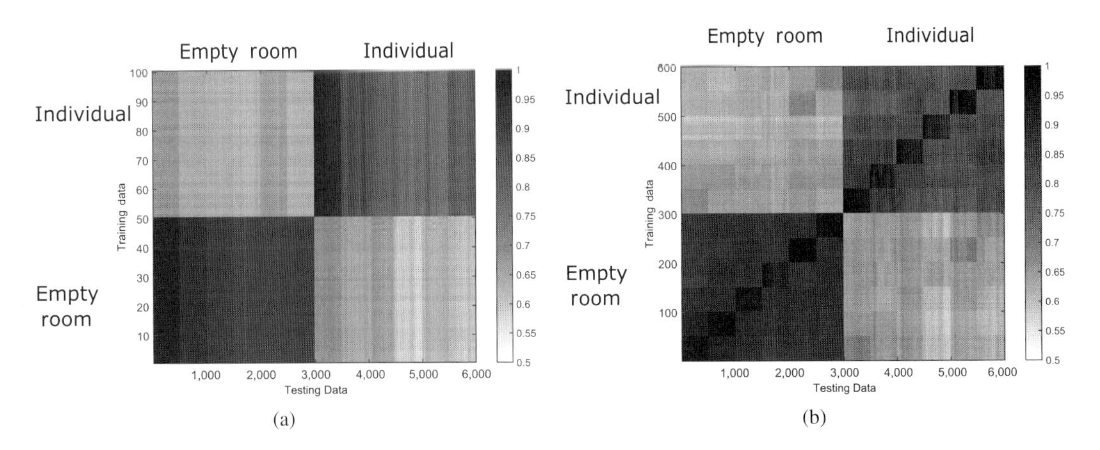

Figure 8.7 Comparison on TRRS map on stationarity: (a) no training database updating and (b) with training database updating.

Table 8.4 Performance matrix for stationarity study

Training data \ Testing data	Day1-AM	Day1-PM	Day2-AM	Day2-PM	Day3-AM	Day3-PM
Day1-AM	1	1	1	0.8522	0.7400	1
Day1-PM	1	1	1	0.9998	0.9856	1
Day2-AM	1	0.9989	1	0.9990	0.9997	1
Day2-PM	1	0.9926	1	1	0.9999	0.9997
Day3-AM	0.88858	0.8005	0.9997	0.9833	1	0.9996
Day3-PM	1	0.9746	0.9998	0.9420	0.9996	1

Table 8.5 List of the six classes of variation

Class	Coat	Backpack	Laptop in the backpack
#1	No	No	No
#2	Yes	No	No
#3	Yes	Yes	No
#4	Yes	Yes	Yes
#5	No	Yes	No
#6	No	Yes	Yes

reduces the sensitivity of the presented system in verification. When the threshold μ increases, it may be able to tell whether the individual is wearing a coat and a backpack, shown by the 0 percentage for class #3 to be misclassified as class #1 in Table 8.6. In terms of the backpack with or without laptop inside, as they are shadowed by the human body, the introduced variations have a relatively small impact on the accuracy of verification.

8.6 Discussion

Through the preceding experiments, the capability of identifying and verifying individuals through-the-wall of the presented TR human identification system has been proved. In this section, the impacts of obstructions and test subjects' postures are evaluated and discussed. The performance of the presented system is further studied by comparing with an RSSI-based identification system, and the current limitation of the presented system is discussed.

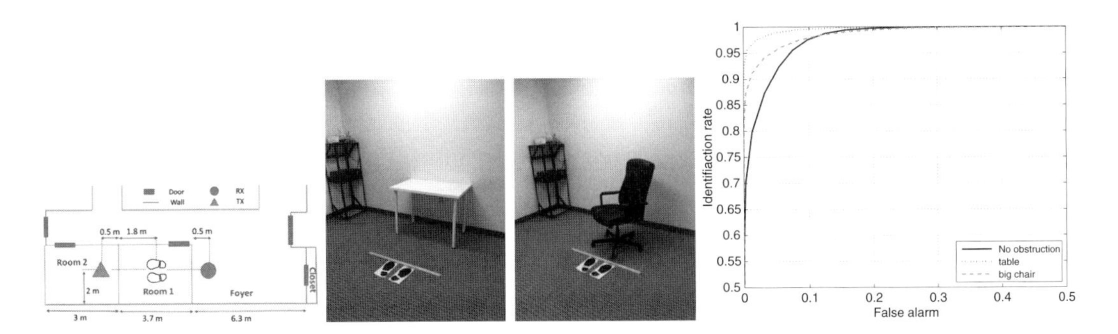

Figure 8.8 Evaluation on impacts of obstructions: (a) test configuration, (b) behind a table, (c) behind a chair, and (d) ROC curves with different obstructions.

8.6.1 Impacts of Obstructions

Experiments are conducted to evaluate and compare the identification accuracy when there is an obstruction in front of and in the same room with the test subject. The office configuration is shown in Figure 8.8(a). The ROC curves for testing under no obstruction, behind a desk as in Figure 8.8(b), and behind a big chair as in Figure 8.8(c) are plotted and compared in Figure 8.8(d). With a similar level of false alarm, the average identification rate for the no-obstruction scenario is 97.57% and the corresponding average false alarm rate is 9.85%. When there is a table in front of the subject against the wall, the average identification rate increases to 99.53% while the average false alarm rate is 8.82%. When a big chair is put in front of the test subject with a very short distance, the system has an average identification rate of 97.44% and an average false alarm rate of 8.43%. When there is an obstruction between the test subject and the transceiver, because of the reflections and penetrations, more copies of the transmitted signal are created, along with more multipath components. If the obstruction does not attenuate the signal much, most of the signals radiated from the obstruction will eventually encounter the test subject. Then more radio biometric information can be captured through the multipath propagation, which helps the identification performance. However, if the obstruction is thick in size and has a large vertical surface which attenuates and blocks most of the incoming signals, there will be fewer multipath components passing through the human body. As a result, less informative radio biometrics are obtained, compared with the no-obstruction case. Furthermore, as demonstrated in this experiment, the existence of furniture as the obstruction does not affect the system much.

However, the multipath profile changes when the obstruction changes, especially when the obstruction locates between the transmitter and the receiver link and in front of the test subject. The TR technique is trying to capture the difference in multipath profile, and of course it will capture the difference introduced by obstruction changes in the meantime. Hence, if an individual is behind a large desk during the training phase and later stands behind a small desk for the testing, the presented system will notice this change in multipath profiles, leading to a mismatch in the training database.

8.6.2 Impacts of Human Postures

Experiments have been conducted to evaluate the effects introduced by human poses. Under the setting in Figure 8.8(a), four participants are asked to stand at the same location and perform five different poses by lifting their arms with different degrees and directions, as shown in Figure 8.9(a). The corresponding ROC curves are shown in Figure 8.9(b).

In the experiment, we select 50 samples for each subject under the first pose as the training set. When the testing samples come from the same pose, the identification rate reaches 97.67% with a false alarm rate being 5.58%. However, as the participants change their poses from the second one to the fifth one, the identification rate drops from 95.66% to 88.06%, 58.83%, and 79.29% with a false alarm rate around 5.6%. The experimental results validate that pose changes will degrade the system performance.

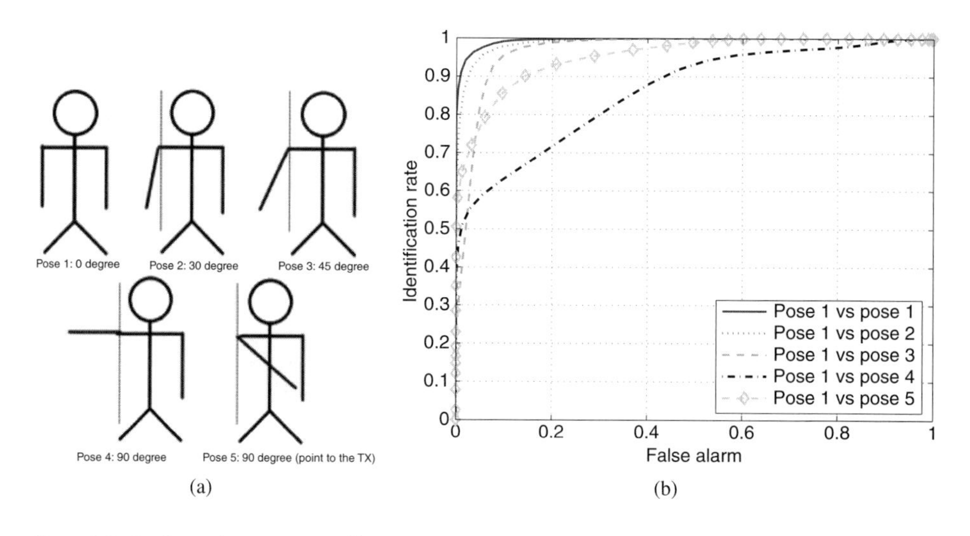

Figure 8.9 Study on human pose effects: (a) test poses and (b) ROC curves with different poses.

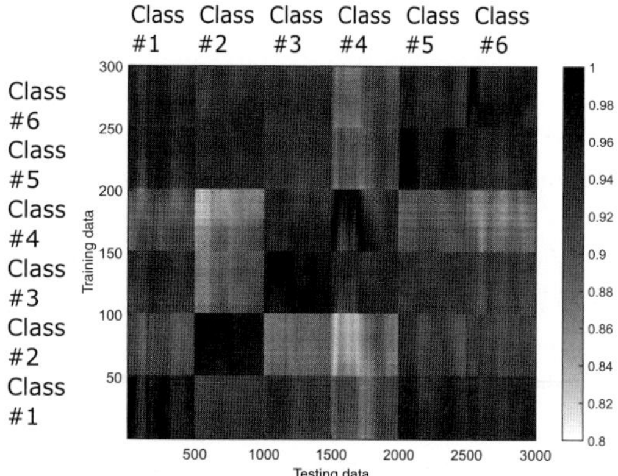

Figure 8.10 TRRS map of variation.

	Threshold 0.92	Threshold 0.9	Threshold 0.85	Threshold 0.84
Class #1	0.9873	0.9994	1	1
Class #2	0.9688	0.9992	1	1
Class #3	0	0.3275	0.9985	1
Class #4	0.4668	0.9756	1	1
Class #5	0.2734	0.9659	1	1
Class #6	0.9720	0.9996	1	1

Table 8.6 Identification rate under variations

The system is robust to slight changes in posture, e.g., from pose 1 to pose 2. However, as shown by the ROC curve of testing over pose 4 data with the pose 1 training in Figure 8.9(b), when the pose alters the propagation environment a lot, the presented TR human identification system fails to find a match in the training database. In the fourth pose, the test subject is asked to lift the left arm to 90 degrees and the direction being perpendicular to the link between the transmitter and the receiver. On the other hand, in the fifth pose, test subjects lift the arm at the same height, but the arm is parallel to the TX-RX link. Compared the result of testing over the fifth pose with that over the fourth pose, it is noticed that the identification accuracy drops more if the pose changes the silhouettes in a manner that is perpendicular to the TX-RX link.

Hence, when poses or standing locations change, the multipath profiles in the TR space for a test subjects might fall out of the "proximity" (range of a high similarity) of his or her self, which results in a reduce in the identification rate. Moreover, a worse situation is that the changed multipath profiles fall into the "proximity" of other test subjects which leads to an increase in the false alarm rate.

8.6.3 Comparison with RSSI-Based Approach

Using the standard WiFi chipsets, besides the CSI, in each measurement we can also obtain a 7×1 RSS vectors, consisting of six RSS values for three receiving antenna in each 20 MHz band and one overall RSS value. Here, we treat each real-valued 7×1 vector as the feature and apply the k nearest neighbors (kNN) classifiers to the measurements.

8.6.3.1 RSSI for Identification

We first test the identification accuracy of the RSSI-based approach on the dataset of 11 individuals.

From the results in Figure 8.11, the RSSI difference between different individuals is small. The false alarm rate is 68.07% and the identification rate is only 31.93%, which is far inferior to the presented identification system.

8.6.3.2 RSSI for Verification

In Figure 8.12, the stationarity is evaluated, and from the plot it is obvious that the RSS value is not stable over time. Without training database update, the identification rate for the individual is only 89.67% with a 10.33% possibility that the individual is misclassified as an empty room. Even with the training database update, the identification rate does not improve due to the instability of the RSS values over time.

Furthermore, in terms of verifying individuals with small variations as listed in Table 8.6, the RSSI-based approach can hardly differentiate between different variations by only using the 7×1 RSS vector as shown in Figure 8.13 and in the confusion matrix of individual verification in Table 8.7. The reason for its insensitivity to small variations is the same as that for its incapability in human identification. The 7×1 RSS vector feature only captures little human radio biometric information and loses the individual discrimination.

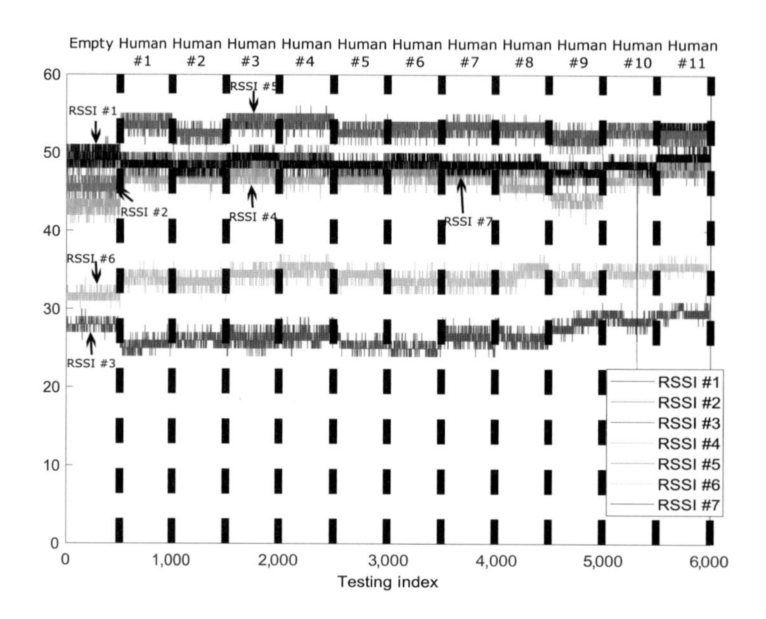

Figure 8.11 RSSI values variation of 11 individuals.

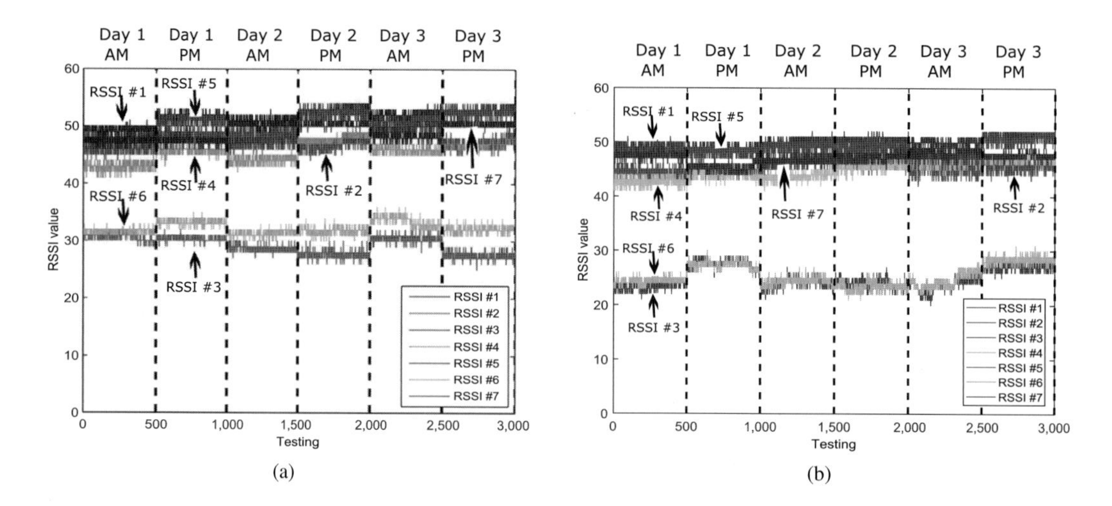

Figure 8.12 RSSI values comparison on stationarity: (a) RSSI for empty room and (b) RSSI with individual present.

Hence, even though the RSSI-based approach is robust to small variations on the human body, it cannot be put into practice for human identification and verification. Moreover, because RSSI is only a real-valued scaler that approximately represents the received signal power, it is less informative, susceptible to noise, and has large intra-class variations that degrade the identification accuracy a lot when the number of test subjects increases. Compared with the RSSI-based approach, the presented TR human

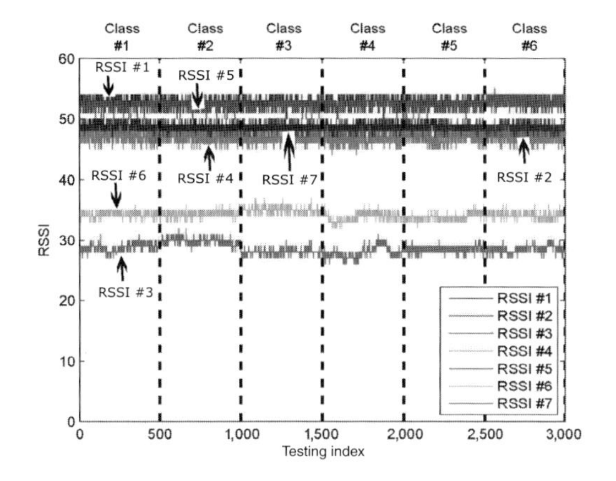

Figure 8.13 RSSI values comparison on variations.

Training data \ Testing data	Class #1	Class #2	Class #3	Class #4	Class #5	Class #6
Class #1	0. 3140	0.3680	0.0840	0.0840	0.1980	0. 2260
Class #2	0. 0220	0. 0280	0. 0020	0	0. 0420	0. 0040
Class #3	0. 0640	0. 0840	0. 2920	0. 1360	0. 0660	0. 0980
Class #4	0. 2340	0. 0220	0. 4600	0. 4360	0. 2940	0. 2600
Class #5	0. 1040	0. 1160	0. 0100	0. 1020	0. 1260	0. 0740
Class #6	0. 2620	0. 3820	0. 1520	0. 2420	0. 2740	0. 3380

Table 8.7 Confusion matrix under RSSI-based approach

identification system succeeds in capturing and extracting the human radio biometric information embedded in the CSI and in distinguishing individuals with high accuracy through-the-wall.

8.6.4 Limitations

At current stage, the presented TR human identification system exhibits some limitations:

(i) The presented system adopts a simple model for human radio biometrics embedded in the CSI as shown in (8.1). As a result, the obtained human radio biometrics $\delta\mathbf{h}$ and the environment component $\mathbf{h}_0$ are correlated. In other words, the human radio biometrics $\delta\mathbf{h}$ is location-dependent, which requires the system to run in an environment consistent over time. Future work includes developing algorithms to separate the human radio biometrics and the outside environment.

(ii) The current system is equipped with only one pair of the transmitter and the receiver, and hence its performance can be improved by deploying more transceiver pairs to capture fine-grained human radio biometrics from different directions simultaneously.

(iii) In the current work, it is difficult to scientifically prove the uniqueness of human radio biometrics, when taking into account the complicated techniques it requires to extract all these biological features from each individual. In the future work, experiments that involve more subjects will be conducted, and techniques that can record other biological features will be utilized to provide more details in human biological characteristics, such as muscle mass index and body temperature. With more detailed information regarding individual biological features besides the common information like height, weight, gender, and clothing, the uniqueness of radio biometrics can be well studied, tested, and verified.

Despite these limitations, we believe the presented TR human identification system should be viewed as a milestone in the development of both human identification systems and wireless sensing systems. For the current system, it can be implemented in the environments that remain stationary most of the time. For example, it can be implemented for identity verification at places like bank vaults to allow the entry of authorized staff. It can also be used in home security systems, functioning as wireless electronic keys in vacation houses. Moreover, the location embedded radio biometrics are helpful in applications that require telling both who the test subject is and where the test subject is. Once the environment-independent radio biometric information is extracted, the presented system can work to identify individuals without being noticed by test subjects and be implemented in applications that require no direct contact with test subjects or where there are obstructions in-between the sensor and the subject.

8.7 Summary

We discuss a TR human identification system, where individuals are distinguished from and identified by the human radio biometrics extracted from the Wi-Fi CSI through the TR technique. Furthermore, the existence of the human radio biometrics, which can be found embedding in the indoor Wi-Fi signal propagation and captured through radio shot, is shown and verified in this chapter. As this new type of biometrics is introduced, it motivates a novel human identification technique relying on wireless sensing with Wi-Fi signals. By leveraging the TR technique to extract radio biometrics, a low-complexity human identification system can be widely implemented without restrictions on the device deployment thanks to the ubiquitousness of Wi-Fi. For related references, interested readers can refer to [33].

References

[1] A. Jain, A. A. Ross, and K. Nandakumar, *Introduction to Biometrics*. Berlin: Springer Science & Business Media, 2011.

[2] A. K. Jain, K. Nandakumar, and A. Ross, "50 years of biometric research: Accomplishments, challenges, and opportunities," *Pattern Recognition Letters*, 2016.

[3] G. Melia, "Electromagnetic absorption by the human body from 1-15 GHz," 2013.

[4] P. Beckmann and A. Spizzichino, *The Scattering of Electromagnetic Waves from Rough Surfaces*, Norwood, MA: Artech House, 1987.

[5] S. Gabriel, R. Lau, and C. Gabriel, "The dielectric properties of biological tissues: II. Measurements in the frequency range 10 Hz to 20 GHz," *Physics in Medicine and Biology*, vol. 41, no. 11, p. 2251, 1996.

[6] "The dielectric properties of biological tissues: III. Parametric models for the dielectric spectrum of tissues," *Physics in Medicine and Biology*, vol. 41, no. 11, p. 2271, 1996.

[7] S. Levy, G. Sutton, P. C. Ng, L. Feuk, A. L. Halpern, B. P. Walenz, N. Axelrod, J. Huang, E. F. Kirkness, G. Denisov et al., "The diploid genome sequence of an individual human," *PLoS Biology*, vol. 5, no. 10, p. e254, 2007.

[8] J. Xiao, K. Wu, Y. Yi, L. Wang, and L. Ni, "FIMD: Fine-grained device-free motion detection," in *Proceedings of the 18th International Conference on Parallel and Distributed Systems*, pp. 229–235, Dec. 2012.

[9] W. Wang, A. X. Liu, M. Shahzad, K. Ling, and S. Lu, "Understanding and modeling of WiFi signal based human activity recognition," in *Proceedings of the 21st Annual ACM International Conference on Mobile Computing and Networking*, pp. 65–76, 2015.

[10] W. Xi, J. Zhao, X.-Y. Li, K. Zhao, S. Tang, X. Liu, and Z. Jiang, "Electronic frog eye: Counting crowd using WiFi," in *Proceedings of the International Conference on Computer Communications*, pp. 361–369, Apr. 2014.

[11] C. Han, K. Wu, Y. Wang, and L. Ni, "WiFall: Device-free fall detection by wireless networks," in *Proceedings of the International Conference on Computer Communications*, pp. 271–279, Apr. 2014.

[12] C. R. R. Sen Souvik and N. Srihari, "SpinLoc: Spin once to know your location," in *Proceedings of the 12th ACM Workshop on Mobile Computing Systems & Applications*, pp. 12:1–12:6, 2012.

[13] S. Sigg, S. Shi, F. Buesching, Y. Ji, and L. Wolf, "Leveraging RF-channel fluctuation for activity recognition: Active and passive systems, continuous and RSSI-based signal features," in *Proceedings of ACM International Conference on Advances in Mobile Computing & Multimedia*, pp. 43:52, 2013.

[14] A. Banerjee, D. Maas, M. Bocca, N. Patwari, and S. Kasera, "Violating privacy through walls by passive monitoring of radio windows," in *Proceedings of the ACM Conference on Security and Privacy in Wireless & Mobile Networks*, pp. 69–80, 2014.

[15] H. Abdelnasser, K. Harras, and M. Youssef, "WiGest demo: A ubiquitous WiFi-based gesture recognition system," in *Proceedings of the International Conference on Computer Communications Workshops*, pp. 17–18, Apr. 2015.

[16] J. Liu, Y. Wang, Y. Chen, J. Yang, X. Chen, and J. Cheng, "Tracking vital signs during sleep leveraging off-the-shelf WiFi," in *Proceedings of the 16th ACM International Symposium on Mobile Ad Hoc Networking and Computing*, pp. 267–276, 2015.

[17] H. Abdelnasser, K. A. Harras, and M. Youssef, "UbiBreathe: A ubiquitous non-invasive WiFi-based breathing estimator," in *Proceedings of the 16th ACM International Symposium on Mobile Ad Hoc Networking and Computing*, pp. 277–286, 2015.

[18] F. Adib, H. Mao, Z. Kabelac, D. Katabi, and R. C. Miller, "Smart homes that monitor breathing and heart rate," in *Proceedings of the 33rd Annual ACM Conference on Human Factors in Computing Systems*, pp. 837–846, 2015.

[19] R. Ravichandran, E. Saba, K. Y. Chen, M. Goel, S. Gupta, and S. N. Patel, "WiBreathe: Estimating respiration rate using wireless signals in natural settings in the home," in *Proceedings of the International Conference on the Pervasive Computing and Communications*, pp. 131–139, Mar. 2015.

[20] Q. Pu, S. Gupta, S. Gollakota, and S. Patel, "Whole-home gesture recognition using wireless signals," in *Proceedings of the 19th Annual ACM International Conference on Mobile Computing & Networking*, pp. 27–38, 2013.

[21] H. Abdelnasser, M. Youssef, and K. A. Harras, "WiGest: A ubiquitous WiFi-based gesture recognition system," in *Proceedings of the IEEE International Conference on Computer Communications*, pp. 1472–1480, 2015.

[22] L. Sun, S. Sen, D. Koutsonikolas, and K.-H. Kim, "WiDraw: Enabling hands-free drawing in the air on commodity WiFi devices," in *Proceedings of the 21st Annual ACM International Conference on Mobile Computing and Networking*, pp. 77–89, 2015.

[23] K. Ali, A. X. Liu, W. Wang, and M. Shahzad, "Keystroke recognition using WiFi signals," in *Proceedings of the 21st Annual International Conference on Mobile Computing and Networking*, pp. 90–102, 2015.

[24] F. Adib and D. Katabi, "See through walls with WiFi!" in *Proceedings of the ACM SIGCOMM*, pp. 75–86, 2013.

[25] F. Adib, Z. Kabelac, D. Katabi, and R. C. Miller, "3D tracking via body radio reflections," in *Proceedings of the 11th USENIX Symposium on Networked Systems Design and Implementation*, pp. 317–329, Apr. 2014.

[26] F. Adib, Z. Kabelac, and D. Katabi, "Multi-person localization via RF body reflections," in *Proceedings of the 12th USENIX Symposium on Networked Systems Design and Implementation*, pp. 279–292, May 2015.

[27] F. Adib, C.-Y. Hsu, H. Mao, D. Katabi, and F. Durand, "Capturing the human figure through a wall," *ACM Transactions on Graphics*, vol. 34, no. 6, pp. 219:1–219:13, Oct. 2015.

[28] J. de Rosny, G. Lerosey, and M. Fink, "Theory of electromagnetic time-reversal mirrors," *IEEE Transactions on Antennas and Propagation*, vol. 58, no. 10, pp. 3139–3149, 2010.

[29] G. Lerosey, J. De Rosny, A. Tourin, A. Derode, G. Montaldo, and M. Fink, "Time reversal of electromagnetic waves," *Physical Review Letters*, vol. 92, no. 19, p. 193904, 2004.

[30] Z.-H. Wu, Y. Han, Y. Chen, and K. J. R. Liu, "A time-reversal paradigm for indoor positioning system," *IEEE Transactions on Vehicular Technology*, vol. 64, no. 4, pp. 1331–1339, Apr. 2015.

[31] B. Wang, Y. Wu, F. Han, Y.-H. Yang, and K. J. R. Liu, "Green wireless communications: A time-reversal paradigm," *IEEE Journal on Selected Areas in Communications*, vol. 29, no. 8, pp. 1698–1710, 2011.

[32] C. Chen, Y. Chen, K. J. R. Liu, Y. Han, and H.-Q. Lai, "High accuracy indoor localization: A WiFi-based approach," in *IEEE International Conference on Acoustics, Speech and Signal Processing (ICASSP)*, pp. 6245–6249, Mar. 2016.

[33] Q. Xu, Y. Chen, B. Wang, and K. J. R. Liu, "Radio biometrics: Human recognition through a wall," *IEEE Transactions on Information Forensics and Security*, vol. 12, no. 5, pp. 1141–1155, 2017.

9 Vital Signs Estimation and Detection

In this chapter, we introduce TR-BREATH, a time-reversal (TR) based contact-free breathing monitoring system. It is capable of breathing detection and multiperson breathing rate estimation within a short period of time using off-the-shelf Wi-Fi devices. The presented system exploits the channel state information (CSI) to capture the miniature variations in the environment caused by breathing. To magnify the CSI variations, TR-BREATH projects CSIs into the TR resonating strength (TRRS) feature space and analyzes the TRRS by the Root-MUSIC and affinity propagation algorithms. Extensive experiment results indoors demonstrate a perfect detection rate of breathing. With only 10 s of measurement, a mean accuracy of 99% can be obtained for single-person breathing rate estimation under the non-line-of-sight (NLOS) scenario. Furthermore, it achieves a mean accuracy of 98.65% in breathing rate estimation for a dozen people under the line-of-sight (LOS) scenario and a mean accuracy of 98.07% in breathing rate estimation of nine people under the NLOS scenario, both with 63 s of measurement. Moreover, TR-BREATH can estimate the number of people with an error around 1. We also demonstrate that TR-BREATH is robust against packet loss and motions. With the prevailing of Wi-Fi, TR-BREATH can be applied for in-home and real-time breathing monitoring.

9.1 Introduction

Breathing rate is an important vital indicator for the health status and predictor of medical conditions [1]. Breathing monitoring is the key technology in the future medical care system. Nevertheless, most conventional breathing monitoring methods are invasive in that they require physical contact with human bodies.

Contact-free breathing monitoring schemes are developed to overcome the drawbacks of conventional schemes for in-home breathing monitoring. Among them, schemes driven by radio frequency (RF) techniques are the most promising candidates due to their abilities to sense breathing in a highly complicated indoor environment by leveraging the propagation of electromagnetic (EM) waves. In terms of techniques, these schemes can be classified into radar-based and Wi-Fi-based. Among the radar-based schemes, Doppler radar is commonly used which measures the frequency shift of the signals caused by the periodic variations of the EM waves reflected from human bodies [2]. Recently, Adib et al. presented a vital sign monitoring system that uses the

Universal Software Radio Peripheral (USRP) as the RF front-end to emulate a frequency modulated continuous radar (FMCW) [3]. However, the requirement of specialized hardware hinders the deployment of these schemes.

On the other hand, Wi-Fi-based schemes are infrastructure-free because they are built upon the existing Wi-Fi networks available indoors. Received signal strength indicator (RSSI) is often used due to its availability on most Wi-Fi devices. In [4], Abdelnasser et al. present UbiBreathe that harnesses RSSI on Wi-Fi devices for breathing estimation. However, UbiBreathe is accurate only when users hold the Wi-Fi devices close to their chests. Another exploitable piece of information on Wi-Fi devices is the channel state information (CSI), a fine-grained information that portraits the EM wave propagation. The scheme proposed by Liu et al. in [5] is one of the first few CSI-based breathing monitoring approaches. Nevertheless, they assume the number of people to be known. Moreover, periodogram is used for spectral analysis that needs a relatively long time for accurate breathing monitoring. In [6], Chen et al. demonstrate the feasibility of high-accuracy multiperson breathing rate estimation using CSIs by leveraging the Root-MUSIC algorithm [7].

In this chapter, we present TR-BREATH, a Wi-Fi-based contact-free breathing monitoring system leveraging time-reversal (TR) technique that detects and monitors multiperson breathing. TR technique is a promising paradigm for future Internet-of-Things applications [8]. The TR technique is utilized for centimeter-level indoor localization [9–12], speed estimation [13], human biometrics [14], and event detection [15]. In this chapter, we demonstrate that TR could also capture the minor but periodic variations embedded in the CSIs.

TR-BREATH measures the CSI variations via the time-reversal resonating strength (TRRS) [16]. The TRRS values are further analyzed by the Root-MUSIC algorithm to produce breathing rate candidates. Then, key statistics are derived based on these candidates to facilitate breathing detection. If breathing is detected, TR-BREATH estimates the multiperson breathing rates via affinity propagation [17], likelihood assignment, and cluster merging. Based on the cluster likelihoods, TR-BREATH could formulate an estimation on the number of people. Also, TR-BREATH makes full use of the sequence numbers in Wi-Fi packets to enhance its robustness against packet loss, which is common in areas with densely deployed Wi-Fi devices.

Extensive experiments in an office environment show that TR-BREATH achieves perfect detection on the existence of breathing within 63 s of measurements. Moreover, with only 10 s of measurements, TR-BREATH achieves 99% accuracy for single-person breathing rate estimation under NLOS. For multiperson breathing monitoring, TR-BREATH achieves a mean accuracy of 98.65% for a dozen people under LOS and 98.07% for nine people under NLOS, both with 63 s of measurement. With the knowledge of the maximum number of people, TR-BREATH can count the people number with an error around 1.

TR-BREATH differs from the prior approaches in the following ways:

- It is infrastructure-free because it utilizes off-the-shelf Wi-Fi devices, while the schemes in [2, 3, 18] require dedicated hardware.

- With the Root-MUSIC algorithm, TR-BREATH can achieve highly accurate breathing rate estimations within 10 s, much shorter than the periodogram schemes used in [3] and [5] so that real-time breathing monitoring is viable.
- It can resolve the breathing rates of nine people breathing concurrently, while in [3] and [5], the authors merely show the results for up to three people.
- It integrates both breathing detection and estimation, while [3, 5, 6] only emphasize breathing rate estimation.
- It can estimate the number of people, which is assumed to be known in advance, in [3, 5, 6].
- It is robust against packet loss in the presence of ambient Wi-Fi traffic, while [3, 5, 6] ignore this practical issue.

The rest of the chapter is organized as follows. Section 9.2 presents an intuitive CSI model that encapsulates the effect of breathing, ambient Wi-Fi traffic, and motions, followed by a brief introduction to the Root-MUSIC algorithm that extracts breathing rates. Section 9.3 elaborates on the algorithm of TR-BREATH. Section 9.4 demonstrates the experiment results for both single-person and multiperson LOS and NLOS scenarios. Section 9.5 demonstrates the performances of TR-BREATH in the presence of a few practical issues.

9.2 Theoretical Foundation

In this section, we first present the CSI model in a static environment without dynamics. Then, we extend the model by considering environmental dynamics, motions, and ambient Wi-Fi traffic. After that, we introduce TRRS as a feature that captures the CSI variations. Finally, we introduce the Root-MUSIC algorithm for breathing rate estimations.

9.2.1 CSI Model without Environmental Dynamics

In the absence of dynamics, the CSI on subcarrier k at time t denoted by $H_k(t)$ can be written as

$$H_k(t) = \sum_{\ell=1}^{L} \zeta_\ell e^{-j2\pi \frac{d_\ell}{\lambda_k}} + e_k(t), \tag{9.1}$$

where $k \in \mathcal{V}$ and $\mathcal{V}$ denote the set of usable subcarriers with a cardinality of V, i.e., V usable subcarriers, L is the total number of multipath components (MPC), ζ_ℓ is the complex gain of MPC ℓ, d_ℓ is the length of MPC ℓ, and λ_k is the wavelength of subcarrier k given by

$$\lambda_k = \frac{c}{f_c + \frac{k}{N_{\mathrm{DFT}} T_s}}, \tag{9.2}$$

where f_c is the carrier frequency, c is the speed of the light, T_s is the sampling interval given as $T_s = \frac{1}{B}$ where B is the baseband bandwidth of the Wi-Fi signals, and N_{DFT} is the size of discrete Fourier Transform (DFT). $e_k(t)$ is the thermal noise on subcarrier k at time t. The MPC gains and delays are time-invariant.

9.2.2 CSI Model with Breathing Impact

With breathing, one or more MPC gains and delays become time-varying. For simplicity, we assume that breathing only affects MPC #1. Then, the gain of MPC #1 takes the form [19]

$$\zeta_1(t) = \zeta_1 \times \left(1 + \frac{\Delta d_1}{d_1} \sin\theta \sin(\frac{2\pi b}{60} t + \phi)\right)^{-\psi} \tag{9.3}$$

where ζ_1 and d_1 are the gain and length for MPC #1 without breathing, Δd_1 is the additional positional displacement of MPC #1 caused by breathing, ψ is the path-loss exponent, θ is the angle between the subject and the impinging EM wave, b is the breathing rate measured in breath-per-minute (BPM), and ϕ is the initial phase of breathing. Given that $d_1 \gg \Delta d_1$, we can approximate $\zeta_1(t)$ with the time-invariant MPC gain ζ_1.

On the other hand, breathing affects the phase of MPC #1 by changing its path length $d_1(t)$ expressed as

$$d_1(t) = d_1 + \Delta d_1 \sin\theta \sin\left(\frac{2\pi b}{60} t + \phi\right). \tag{9.4}$$

Now, $H_k(t)$ takes the form

$$H_k(t) = \zeta_1 e^{-j2\pi \frac{d_1(t)}{\lambda_k}} + \sum_{\ell=2}^{L} \zeta_\ell e^{-j2\pi \frac{d_\ell}{\lambda_k}} + e_k(t), \tag{9.5}$$

which can be further written as

$$\begin{aligned} H_k(t) = {} & \zeta_1 e^{-j2\pi \frac{d_1}{\lambda_k}} e^{-j2\pi \frac{\Delta d_1 \sin\theta \sin(\frac{2\pi b}{60} t + \phi)}{\lambda_k}} \\ & + \sum_{\ell=2}^{L} \zeta_\ell e^{-j2\pi \frac{d_\ell}{\lambda_k}} + e_k(t). \end{aligned} \tag{9.6}$$

The first term on the right-hand side of $H_k(t)$ in (9.6) can be decomposed into an infinite summation according to the Jacobi–Anger expansion [20], as

$$e^{-j2\pi \frac{\Delta d_1 \sin\theta \sin(\frac{2\pi b}{60} t + \phi)}{\lambda_k}} = \sum_{m=-\infty}^{+\infty} (-1)^m J_m(\nu_k) e^{jm \frac{2\pi b}{60} t} e^{jm\phi} \tag{9.7}$$

where $\nu_k = 2\pi \sin\theta \Delta d_1 / \lambda_k$, and $J_m(x)$ is the mth order Bessel function with argument x. It can be seen that in addition to the spectral line at b, there also exists an infinite number of harmonics with spectral lines at mb where m is a nonzero integer.

In practice, $J_m(\nu_k)$ decays quickly for $|m| \geq 2$ given the typical values of ν_k. Thus, (9.7) can be approximated as

$$e^{-j2\pi \frac{\Delta d_1 \sin\theta \sin(\frac{2\pi b}{60}t+\phi)}{\lambda_k}} \approx \sum_{m=-1}^{+1} (-1)^m J_m(\nu_k) e^{jm\frac{2\pi b}{60}t} e^{jm\phi} \tag{9.8}$$

which consists of two spectral lines at $\pm b$ with respect to $m = \pm 1$ as well as a DC component with respect to $m = 0$. Thus, $H_k(t)$ can be expressed as

$$H_k(t) \approx \underbrace{\zeta_1 e^{-j2\pi \frac{d_1}{\lambda_k}} \sum_{m=-1}^{+1} (-1)^m J_m(\nu_k) e^{jm\frac{2\pi b}{60}t} e^{jm\phi}}_{S_k(t)}, + \underbrace{\sum_{\ell=2}^{L} \zeta_\ell e^{-j2\pi \frac{d_\ell}{\lambda_k}}}_{I_k} + e_k(t), \tag{9.9}$$

where $S_k(t)$ stands for the useful signal for breathing monitoring on subcarrier k, and I_k represents the time-invariant part due to the static environment and regarded as the interference. Notice that the dynamic model of $H_k(t)$ shown in (9.9) can be extended easily to the multiperson case.

9.2.3 Impact of Nonidealities on CSIs

In practice, we need to consider two random nonidealities in the CSI model:

- **Random phase distortion** caused by the differences between the local oscillators of the Wi-Fi transmitter and receiver, which consists of an initial and a linear phase distortion.
- **Random amplitude variation** due to the automatic gain control (AGC) in the RF front-end that scales the input voltage into the dynamic range of the analog-to-digital converter (ADC).

With these two nonidealities, $H_k(t)$ in (9.9) should be modified as

$$H_k(t) = \Gamma(t)\,(S_k(t) + I_k)\, e^{j(\omega(t)+\kappa(t)k)} + e_k(t), \tag{9.10}$$

where $\Gamma(t)$ is the real-valued AGC gain at time t, $\omega(t)$ is the initial phase distortion at time t, and $\kappa(t)$ is the linear phase distortion at time t.

9.2.4 Impact of Motions on CSIs

The propagation of EM wave is affected by the motions of the subject under breathing monitoring such as turning heads or bending forward, known as the subject motion, as well as by the motion caused by nearby people and/or objects not under monitoring, known as the ambient motion. Next, we present the CSI model incorporating both effects.

9.2.4.1 Subject Motion

When there exists subject motion, we need to partition time t into two time durations: the time duration without subject motion denoted as $\mathcal{T}_{sm}$, and the time duration with subject motion denoted as $\mathcal{T}_{sm}^c$, which is the complementary of $\mathcal{T}_{sm}$. Then, $S_k(t)$ is modified as

$$S_k(t) = \begin{cases} S_k^0(t) & , t \in \mathcal{T}_{sm} \\ S_k'(t) & , t \in \mathcal{T}_{sm}^c \end{cases}, \tag{9.11}$$

where $S_k^0(t)$ is the original breathing signal and $S_k'(t)$ is a random signal caused by the subject motion.

9.2.4.2 Ambient Motion

In the presence of ambient motion, I_k in (9.10) becomes time-variant, and thus (9.10) should be rewritten as

$$H_k(t) = \Gamma(t) \left(S_k(t) + I_k(t) \right) e^{j(\omega(t) + \kappa(t)k)} + e_k(t) \, . \tag{9.12}$$

The ambient motion $I_k(t)$ can be either periodic or nonperiodic. For instance, $I_k(t)$ can be caused by the breathing of another person in the vicinity of the person under monitoring, or produced by the random motion of nearby people and objects. Clearly, $I_k(t)$ incurs interference into breathing monitoring. In this chapter, we consider the ambient motion as bursty by affecting a portion of the monitoring duration, i.e.,

$$I_k(t) = \begin{cases} I_k^0(t) & , t \in \mathcal{T}_{am} \\ 0 & , t \in \mathcal{T}_{am}^c \end{cases}, \tag{9.13}$$

where $\mathcal{T}_{am}$ is the continuous time duration with ambient motion while $\mathcal{T}_{am}^c$ is the duration without ambient motion, and $I_k^0(t)$ is the original ambient motion signal.

9.2.5 Impact of Ambient Wi-Fi Traffic on CSIs

In reality, CSIs are sampled with a time interval of T_{sp} with an initial time given as t_0. The ith CSI, denoted by $H_k[i]$, is sampled at the ith time instant with a sequence number of $s_i = s_0 + i$. Its reception time is given by $t_i = t_0 + iT_{sp}$, where s_0 is the sequence number of the first CSI sample. Yet, due to the ambient Wi-Fi traffic on the same Wi-Fi channel, packet loss is unavoidable, leading to a sequence number $s_i \neq s_0 + i$ and a reception time $t_i = t_0 + (s_i - s_0)T_{sp} \neq t_0 + iT_{sp}$.

9.2.6 Overall CSI Model

Considering all effects discussed in Sections 9.2.3, 9.2.4, and 9.2.5, the discrete CSI model takes the form of

$$H_k[i] = \Gamma[i] \left(S_k[i] + I_k[i] \right) e^{j(\omega[i] + \kappa[i]k)} + e_k[i] \, , \tag{9.14}$$

where $S_k[i]$ and $I_k[i]$ are the discrete signal and interference of the ith CSI as defined in (9.11) and (9.13), $\Gamma[i]$ and $e_k[i]$ are the discrete AGC gain and thermal noise, and $S_k[i]$ and $I_k[i]$ are given as

$$S_k[i] = \begin{cases} S_k^0[i] & , i \in \mathbb{T}_{sm} \\ S_k' & , i \in \mathbb{T}_{sm}^c \end{cases}, \tag{9.15}$$

$$I_k[i] = \begin{cases} I_k^0[i] & , i \in \mathbb{T}_{am} \\ 0 & , i \in \mathbb{T}_{am}^c \end{cases}, \tag{9.16}$$

where $\mathbb{T}_{sm}$ and $\mathbb{T}_{am}$ are the discrete CSI time index affected by subject motion and ambient motion, respectively, while $\mathbb{T}_{sm}^c$ and $\mathbb{T}_{am}^c$ are the complementary discrete time index sets with respect to $\mathbb{T}_{sm}$ and $\mathbb{T}_{am}$.

9.2.7 Calculating TRRS from CSIs

TRRS is used as a measure of similarity between any two CSIs. Different from the calculation of TRRS in the time domain as shown in [12], we calculate the TRRS between the ith received CSI and the jth received CSI in the frequency domain based on $H_k[i]$ and $H_k[j]$ in (9.14) as follows:

$$\text{TR}\left[\mathbf{H}[i], \mathbf{H}[j]\right] = \frac{\sum_{k\in\mathcal{V}} H_k[i]H_k^*[j]e^{-j(\omega^\star+\kappa^\star k)}}{||\mathbf{H}[i]||_2||\mathbf{H}[j]||_2}, \tag{9.17}$$

where $\mathbf{H}[i] = \{H_k[i]\}_{k\in\mathcal{V}}$ and $||\mathbf{x}||_2$ is the ℓ_2 norm of the vector $\mathbf{x}$. $\omega^\star$ and $\kappa^\star$ in (9.17) are introduced to remove the initial and linear phase distortions, which are given by

$$\kappa^\star = \underset{\kappa}{\text{argmax}} \left| \sum_{k\in\mathcal{V}} H_k[i]H_k^*[j]e^{-j\kappa k} \right| \tag{9.18}$$

$$\omega^\star = \measuredangle \left(\sum_{k\in\mathcal{V}} H_k[i]H_k^*[j]e^{-j\kappa^\star k} \right)^* . \tag{9.19}$$

The denominator of (9.17) normalizes the TRRS so that $\text{TR}\left[\mathbf{H}[i], \mathbf{H}[j]\right] \in [0, 1]$. In other words, the denominator mitigates the impact of the random gains $\Gamma[i]$ and $\Gamma[j]$. $\measuredangle(x)$ is the operator that extracts the phase from the complex argument x.

9.2.8 Extracting Breathing Rates Using Root-MUSIC

Root-MUSIC is a variant to the well-known MUltiple SIgnal Classification (MUSIC) algorithm [21]. It is a super-resolution subspace-based spectral analysis algorithm widely used in signal processing applications [7]. Assuming a total of N CSIs sampled uniformly with an interval of T_{sp}, we can calculate the $N \times N$ TRRS matrix $\mathbf{R}$ based on (9.17), with the (i, j)th element of $\mathbf{R}$ given as $\text{TR}\left[\mathbf{H}[i], \mathbf{H}[j]\right]$.

After calculating $\mathbf{R}$, we perform an eigenvalue decomposition (EVD) on $\mathbf{R}$ to produce

$$\mathbf{R} = \mathbf{U}\Lambda\mathbf{U}^\dagger , \tag{9.20}$$

where † is the transpose and conjugate operator, $\mathbf{U}$ is an $N \times N$ orthonormal matrix such that $\mathbf{U}^{\dagger}\mathbf{U} = \mathbf{I}$ where $\mathbf{I}$ is an $N \times N$ identity matrix, and Λ is an $N \times N$ diagonal matrix with descending real-valued diagonal entries equivalent to the eigenvalues of $\mathbf{R}$.

Second, the orthonormal matrix $\mathbf{U}$ is decomposed into a signal subspace and a noise subspace. The signal subspace, denoted by $\mathbf{U}_s$, consists of the first p columns of $\mathbf{U}$, where $p \leq N - 1$ is the signal subspace dimension. On the other hand, the noise subspace, denoted by $\mathbf{U}_n$, consists of the last $N - p$ columns of $\mathbf{U}$.

Next, we calculate the matrix $\mathbf{Q} = \mathbf{U}_n\mathbf{U}_n^{\dagger}$. Then, we formulate the polynomial $f(z)$ as

$$f(z) = \sum_{m=0}^{N-1}\sum_{n=0}^{N-1} [\mathbf{Q}]_{m,n}\, z^{g_{m,n}}, \tag{9.21}$$

where $[\mathbf{Q}]_{m,n}$ is the (m,n)th element of $\mathbf{Q}$, $z = e^{-j\frac{2\pi b T_{sp}}{60}}$, and $g_{m,n}$ is the *discrete difference function* highlighting the time difference between two CSI samples normalized to T_{sp}, given as

$$g_{m,n} = \begin{cases} s_m - s_n, & \text{Considering Packet Loss} \\ m - n, & \text{Otherwise.} \end{cases} \tag{9.22}$$

Notice that, by using $g_{m,n} = s_m - s_n$, the Root-MUSIC algorithm is robust against Wi-Fi packet loss. Yet, when the ambient Wi-Fi traffic is not severe, setting $g_{m,n}$ as $m - n$ suffices to produce accurate results.

Solving $f(z) = 0$ in (9.21) results in $2N - 2$ complex roots denoted by $\hat{\mathbf{z}} = \{\hat{z}_1, \hat{z}_2, \hat{z}_3, \ldots, \hat{z}_{2N-2}\}$. Because $\mathbf{Q}$ is Hermitian, if $\hat{z}$ is a complex root of $f(z) = 0$, then $1/\hat{z}^*$ is also a complex root of $f(z) = 0$. In other words, the roots of $f(z) = 0$ come in pairs. Considering that only the phase of the complex roots carry the information about the breathing rates, we keep the $N - 1$ complex roots inside the unit circle. Then, we choose p out of the $N - 1$ complex roots closest to the unit circle. The breathing rate estimation can be formulated as

$$\hat{b}_i = 60 \times \frac{\measuredangle \hat{z}_i}{2\pi T_{sp}}, \quad i = 1, 2, \ldots, p\,. \tag{9.23}$$

From (9.12), we find that, while some complex roots are associated with breathing rates, the rest of these complex roots are produced by the motion interference and thermal noise. In particular, the power of the motion interference $I_k(t)$ can be even stronger than the breathing signal $S_k(t)$, e.g., when the motion happens very close to the Wi-Fi devices capturing CSIs. As shown in Section 9.5, as long as the Wi-Fi devices are far away from the motions, the impact of motion can be largely neglected, and most of the complex roots of $f(z)$ in (9.21) are still associated with breathing.

Moreover, we realize that the breathing rates are limited to a finite range $[b_{\min}, b_{\max}]$ because people cannot breathe either too fast or too slow. Thus, we sift the breathing rate estimations $\hat{\mathbf{b}} = [\hat{b}_1, \hat{b}_2, \ldots, \hat{b}_p]$ by discarding those outside the range of $[b_{\min}, b_{\max}]$, which leads to $\tilde{\mathbf{b}} = [\hat{b}_{r_1}, \hat{b}_{r_2}, \ldots, \hat{b}_{r_{p'}}]$, where p' is the number of the remaining complex roots and r_i is the index of the ith remaining estimation.

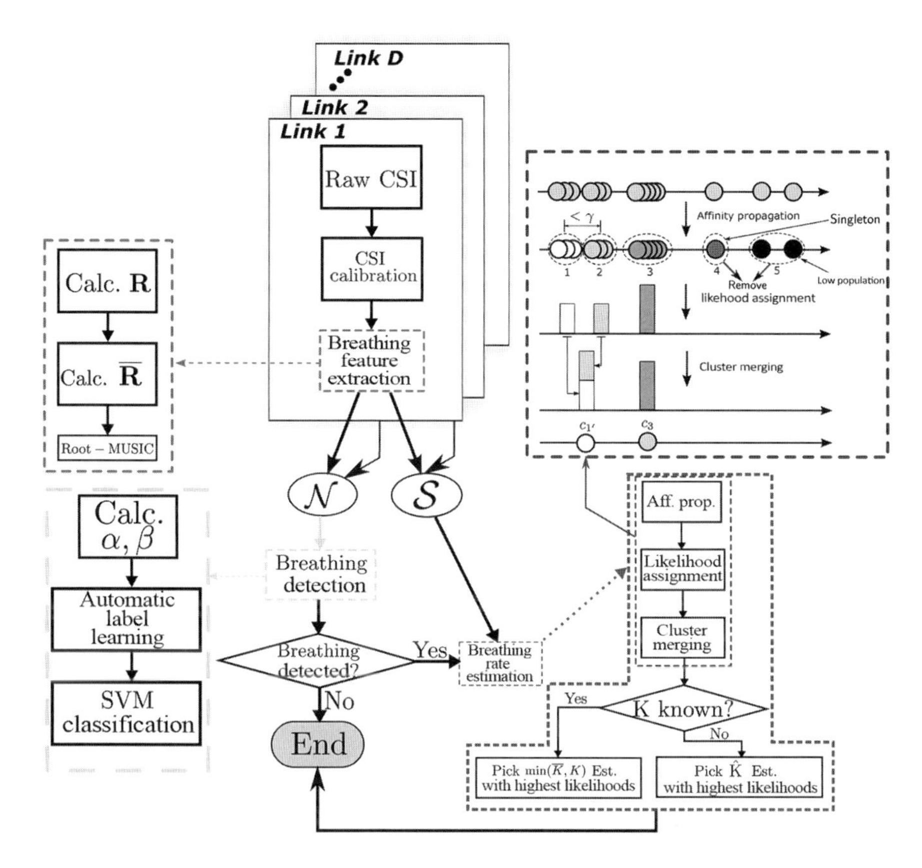

Figure 9.1 Overview of the architecture of TR-BREATH.

9.3 Algorithm

The architecture of TR-BREATH is illustrated in Figure 9.1. We assume the availability of CSIs on a total of D links in a multiantenna Wi-Fi system. In the following parts, we elaborate on the details of the algorithms in TR-BREATH.

9.3.1 CSI Calibration

The miniature and periodic changes in CSIs are masked by the phase distortions caused by residual synchronization errors. To overcome this issue, in the calculation of TRRS, we evaluate $\kappa^\star$ and $\omega^\star$ according to (9.18) and (9.19). This step is executed on all links in parallel.

9.3.2 Breathing Feature Extraction

9.3.2.1 Calculating the TRRS Matrix

Assume that we obtain N CSIs for each link. Because breathing is not strictly stationary in the long run, calculating the $N \times N$ TRRS matrix $\mathbf{R}$ using the calibrated CSIs according to (9.17) is not optimal, which degrades the performance. So, TR-BREATH divides the duration of measurements into multiple blocks, where each block consists of M CSIs where $M \leq N$. Assume that two blocks overlap by P CSIs, TR-BREATH can obtain a total of $B = \lfloor \frac{N-P}{M} \rfloor + 1$ blocks.

For each block, TR-BREATH further partitions the block duration into several overlapping time windows with W CSIs for each, with the CSIs associated with the ith time window given by $\{\mathbf{H}[i], \mathbf{H}[i+1], \ldots, \mathbf{H}[i+W-1]\}$. Two adjacent time windows overlap by 1 CSI.

9.3.2.2 Temporal Smoothing of the TRRS Matrix

To suppress the spurious estimations due to interference and noise, TR-BREATH performs temporal smoothing on the TRRS matrix for each block taking the packet loss into consideration. First, for link d, block b, TR-BREATH parses the sequence numbers for the M CSIs inside that block, denoted as $s_{b(N-P)+1}, s_{b(N-P)+2}, \ldots, s_{b(N-P)+M}$. Then, TR-BREATH calculates the difference M' between the maximum sequence number $s_{\max} = s_{b(N-P)+M}$ and the minimum sequence number $s_{\min} = s_{b(N-P)+1}$. If $M' = s_{\max} - s_{\min} > M$, we infer that $M' - M$ Wi-Fi packets are missing due to ambient Wi-Fi traffic.

Second, TR-BREATH calculates the $M \times M$ TRRS matrix for link d and block b according to (9.17), denoted as $\mathbf{R}_{b,d}$. Then, TR-BREATH forms an extended TRRS matrix $\mathbf{R}'_{b,d}$ with dimension $M' \times M'$. The entries of $\mathbf{R}'_{b,d}$ are initialized with zeros. Then, TR-BREATH fills the (s_i, s_j)th entry of $\mathbf{R}'_{b,d}$ with the (i, j)th element of $\mathbf{R}_{b,d}$. Equivalently speaking, $\mathbf{R}'_{b,d}$ is an interpolated version of $\mathbf{R}_{b,d}$, with entries of zero standing for the index of the missing packets[1]. With a time window size W, TR-BREATH could formulate $Z = M' - W + 1$ time windows in total. Meanwhile, TR-BREATH forms a counting matrix $\mathbf{C}'_{b,d}$ for link d and block b such that

$$[\mathbf{C}'_{b,d}]_{i,j} = \begin{cases} 1, & \text{If } [\mathbf{R}'_{b,d}]_{i,j} > 0 \\ 0, & \text{Otherwise.} \end{cases} \tag{9.24}$$

Next, TR-BREATH partitions $\mathbf{R}'_{b,d}$ into Z square submatrix, with the zth submatrix given by $\mathbf{R}'_{b,d,z}$ composed by the entries of $\mathbf{R}_{b,d}$ from row z to row $z + W - 1$ and column z to column $z + W - 1$. The same operation is performed on $\mathbf{C}'_{b,d}$, leading to Z square submatrix $\{\mathbf{C}'_{b,d,z}\}_{z=1,2,\ldots,Z}$. $\{\mathbf{R}'_{b,d,z}\}_{z=1,2,\ldots,Z}$ and $\{\mathbf{C}'_{b,d,z}\}_{z=1,2,\ldots,Z}$ are accumulated as $\overline{\mathbf{R}'_{b,d}} = \sum_{z=1}^{Z} \mathbf{R}'_{b,d,z}$ and $\overline{\mathbf{C}'_{b,d}} = \sum_{z=1}^{Z} \mathbf{C}'_{b,d,z}$. Also, we replace the sequence numbers with $[1, 2, \ldots, W]$.

Then, we locate and delete the rows and columns of $\overline{\mathbf{R}'_{b,d}}$ and $\overline{\mathbf{C}'_{b,d}}$ with at least one zero, resulting in the matrix $\overline{\mathbf{R}''_{b,d}}$ and $\overline{\mathbf{C}''_{b,d}}$, both with dimension $W' \times W'$ where $W' \leq W$. The deleted index are also removed from the updated sequence numbers in the previous step, leading to the updated sequence numbers $s''_1, s''_2, \ldots, s''_{W'}$.

[1] For example, $\mathbf{R}_{b,d} = \begin{bmatrix} 1 & 0.95 \\ 0.95 & 1 \end{bmatrix}$ and $s_1 = 1, s_2 = 3$. Then, $\mathbf{R}'_{b,d} = \begin{bmatrix} 1 & 0 & 0.95 \\ 0 & 0 & 0 \\ 0.95 & 0 & 1 \end{bmatrix}$.

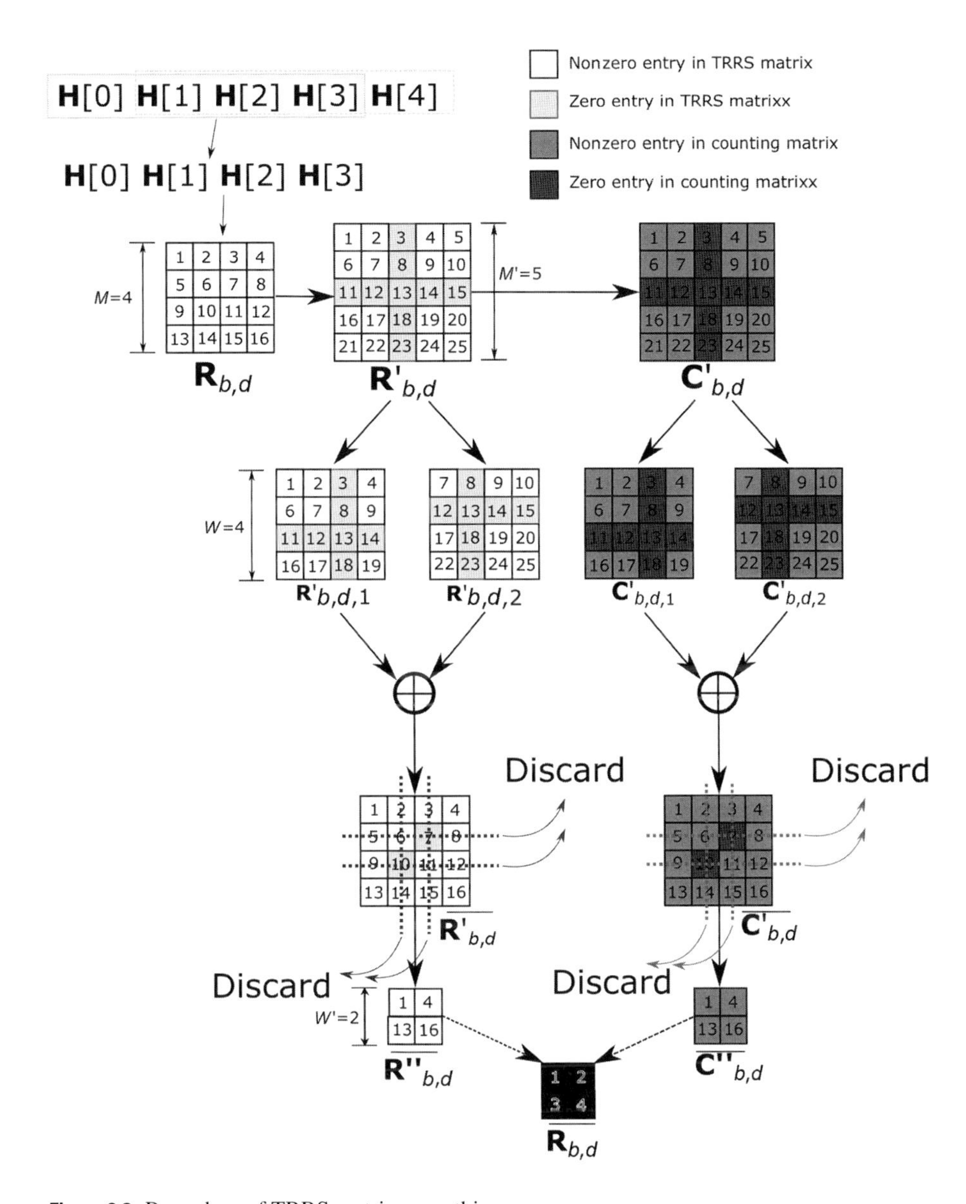

Figure 9.2 Procedure of TRRS matrix smoothing

Finally, we calculate the temporal smoothed matrix $\overline{\mathbf{R}_{b,d}}$ with its (i, j)th element given by $[\overline{\mathbf{R}''_{b,d}}]_{i,j}/[\overline{\mathbf{C}''_{b,d}}]_{i,j}$ for further processing. Figure 9.2 shows an example of generating $\overline{\mathbf{R}_{b,d}}$ under $N = 5, M = 4, M' = 5, W = 4, W' = 2, P = 1$, and $B = 2$. Notice that the parameters indicate the loss of one Wi-Fi packet because $M' - M = 1$.

9.3.2.3 Analysis via Root-MUSIC

The smoothed $W' \times W'$ TRRS matrix $\overline{\mathbf{R}_{b,d}}$ is analyzed via the Root-MUSIC algorithm. An EVD is invoked on $\overline{\mathbf{R}_{b,d}}$, leading to the $W' \times (W' - p)$ noise subspace matrix $\mathbf{U}'_n$ and thus $\mathbf{Q}' = \mathbf{U}'_n(\mathbf{U}'_n)$.† The polynomial is modified as

$$f(z) = \sum_{m=0}^{W'-1} \sum_{n=0}^{W'-1} [\mathbf{Q}']_{m,n} z^{g_{m,n}}. \tag{9.25}$$

where $g_{m,n} = m - n$ if packet loss is not considered, and $g_{m,n} = s''_m - s''_n$ otherwise. Here, p should be set to the maximum possible number of people, e.g., the capacity of a room. When the polynomial in (9.25) cannot produce results in the range $[b_{\min}, b_{\max}]$, we call $f(z) = 0$ insolvable and put an empty solution into a set $\mathcal{N}_{b,d}$. Otherwise, we save the breathing rate candidates $\{\hat{b}_1, \hat{b}_2, \ldots, \hat{b}_{p'}\}$ into a set denoted as $\mathcal{S}_{b,d}$, where p' is the number of candidates after filtering as discussed in Section 9.2.8. After processing all D links, the sets $\{\mathcal{S}_{b,d}\}_{b=1,2,\ldots,B}^{d=1,2,\ldots,D}$ are combined together into $\mathcal{S}$ as $\cup_{d=1}^{D} \cup_{b=1}^{B} \mathcal{S}_{b,d}$ and $\mathcal{N}$ as $\cup_{d=1}^{D} \cup_{b=1}^{B} \mathcal{N}_{b,d}$, where $\cup$ denotes the *set union* operator.

9.3.3 Breathing Detection

Some of the breathing rate candidates generated by the breathing feature extraction might still be noisy estimations caused by interference and/or thermal noise in the CSIs. Therefore, we need to assess how likely these candidates are to be caused by interference and noise. If with high probability, these candidates have no correlation with human breathing, we determine that there are no people breathing. Otherwise, we conclude that breathing is present.

We observe from extensive experiments that the statistics of set $\mathcal{S}$ and set $\mathcal{N}$ are indicator functions of the presence of breathing: In the absence of breathing, it is more likely that the polynomial in (9.25) is insolvable, which yields a large $\mathcal{N}$ and a small $\mathcal{S}$ in terms of their cardinalities, i.e., number of unique set elements. On the contrary, when breathing exists, solving the polynomial in (9.25) would produce many breathing rate candidates, giving rise to a small $\mathcal{N}$ and a large $\mathcal{S}$. We leverage this observation for breathing detection.

9.3.3.1 Calculating α and β

First, we formulate two statistics α and β expressed as

$$\alpha = \frac{\#(\mathcal{N})}{\#(\mathcal{S}) + \#(\mathcal{N})}, \beta = \frac{\#(\mathcal{S})}{BDp}, \tag{9.26}$$

where the denominator of β stands for the total number of possible breathing rate candidates with B blocks, D links, and p estimations per link per time window. The $\#(\cdot)$ denotes the cardinality of a set. α indicates the *insolvability* of (9.25), while β indicates the *diversity* of (9.25). The correlation between (α, β) and the presence of breathing motivates us to develop a detection scheme based on the observed (α, β) values.

9.3.3.2 Automatic Label Learning

TR-BREATH can learn the labels y associated with each (α, β) obtained in the training phase automatically. Write $\boldsymbol{\theta} = (\alpha, \beta)$ for convenience, and by convention, y equals to $+1$ if the associated $\boldsymbol{\theta}$ is measured in the presence of breathing, and y equals to -1 otherwise.

During the training phase, TR-BREATH makes T observations of $\boldsymbol{\theta}$, written as $\{\boldsymbol{\theta}_i\}_{i=1,2,\ldots,T}$. Based on the observations, TR-BREATH extracts the labels $\{\hat{y}_i\}_{i=1,2,\ldots,T}$

using unsupervised label learning consisting of two phases: **(i)** Partition $\{\boldsymbol{\theta}_i\}_{i=1,2,\ldots,T}$ into two classes by invoking k-means clustering [22] with $k = 2$. Denote the centroids of cluster 1 and 2 as $(\hat{\alpha}_1, \hat{\beta}_1)$ and $(\hat{\alpha}_2, \hat{\beta}_2)$, respectively. **(ii)** If $\hat{\alpha}_1 > \hat{\alpha}_2$, label all members of cluster 1 with $\hat{y} = -1$ to indicate that they are observed in the absence of breathing. Then, label the members of cluster 2 with $\hat{y} = +1$. A similar procedure applies to the case of $\hat{\alpha}_1 < \hat{\alpha}_2$. In the rare case that $\hat{\alpha}_1 = \hat{\alpha}_2$, label the elements within the cluster with a larger $\hat{\beta}$ with $\hat{y} = +1$.

9.3.3.3 SVM Classification

Based on $\{\boldsymbol{\theta}_i\}_{i=1,2,\ldots,T}$ and $\{\hat{y}_i\}_{i=1,2,\ldots,T}$, we train a support vector machine (SVM) [23], a widely used binary classifier. SVM returns two weight factors, ω_α and ω_β, as well as a bias ω_b. ω_α and ω_β signify the importance of α and β in breathing detection. After the training phase, given any $\boldsymbol{\theta} = (\alpha, \beta)$, TR-BREATH determines that breathing exists if $\omega_\alpha \alpha + \omega_\beta \beta + \omega_b > 0$ and nonexistent otherwise.

9.3.4 Breathing Rate Estimation

If breathing is detected, TR-BREATH proceeds by formulating multiperson breathing rate estimation.

9.3.4.1 Clustering by Affinity Propagation

The breathing rate candidates in $\mathcal{S}$ are fed into the affinity propagation algorithm [17]. It works by passing the responsibility message to decide which estimations are exemplars and the availability message to determine the membership of an estimation to one of the clusters. Different from k-means [22], affinity propagation does not require the knowledge of the cluster number. Here, we assume that affinity propagation partitions the elements of $\mathcal{S}$ into U clusters.

9.3.4.2 Likelihood Assignment

For each cluster, TR BREATH evaluates its population, variance, and centroid, expressed as p_i, v_i, and c_i. Then, p_i and v_i are normalized as $\overline{p}_i = p_i / \sum_{i=1}^{U} p_i$ and $\overline{v}_i = v_i / \sum_{i=1}^{U} v_i$. The likelihood of cluster i, denoted by l_i, is calculated as

$$l_i = \begin{cases} 0, & (v_i = 0, p_i = 1), \text{or } \overline{p}_i < 2\% \\ \dfrac{e^{\omega_p \overline{p}_i - \omega_v \overline{v}_i - \omega_c c_i}}{\sum_{i=1}^{U} e^{\omega_p \overline{p}_i - \omega_v \overline{v}_i - \omega_c c_i}}, & \text{otherwise} \end{cases}, \tag{9.27}$$

where ω_p, ω_v, and ω_c are positive weighting factors to account for different scales of the corresponding terms. The likelihood assignment in (9.27) incorporates a term related to the cluster centroid c_i. The insight is that a high breathing rate is less likely than a low breathing rate in real life. Meanwhile, high breathing rate candidates are more likely to be caused by the harmonics of breathing rates. Also, (9.27) implies that singletons, i.e., clusters with a single element ($v_i = 0$ and $p_i = 1$), should be assigned with zero likelihoods. Clusters with $\overline{p}_i < 2\%$ are also considered outliers and are eliminated.

9.3.4.3 Cluster Merging

Because the breathing rates are evaluated for each time window and for each link independently, it is likely that breathing rate estimations for the same person differ slightly in a small range. This results in several closely spaced clusters, which should be merged to improve the performance.

To identify the clusters to be merged, we calculate the intercluster distances by calculating the differences in their centroids. Then, we merge clusters with intercluster distance falling below a threshold, known as the merging radius denoted by γ. For example, if $|c_i - c_{i+1}| < \gamma$, then, cluster i and $i+1$ would be merged. Denote the new cluster index as i', the normalized population of cluster i' is given by $\overline{p}_{i'} = \overline{p}_i + \overline{p}_{i+1}$ and the normalized variance $\overline{v}_{i'}$ is recalculated. The centroid of cluster i' is expressed as the weighted average of the merged two clusters, given by $c_{i'} = \frac{\overline{l}_i c_i + \overline{l}_{i+1} c_{i+1}}{\overline{l}_i + \overline{l}_{i+1}}$.

Finally, the likelihood of cluster i' is updated using (9.27). Merging of more than two clusters can be generalized from the aforementioned steps and is omitted here for brevity. The procedures for likelihood assignment and cluster merging are highlighted in Figure 9.1.

Assuming a total of $\overline{K}$ clusters after merging and that the number of people K is known, TR-BREATH directly outputs $K_o = \min(\overline{K}, K)$ centroids with the highest likelihoods as the multiperson breathing rate estimations, i.e., $\hat{b}_i = c_{\mathrm{idx}_i}$, $i = 1, 2, \ldots, K_o$ where idx_i stands for the index of the ith largest likelihood.

9.3.5 Estimating the Number of People

Denote the set $\mathcal{J}$ as $\mathcal{J} = \{j | \sum_{i=1}^{\min(\overline{K}, j)} \overline{l}_{\mathrm{idx}_i} \geq \lambda\}$ where λ is a threshold. In other words, the set $\mathcal{J}$ contains the number of clusters with an accumulated likelihood exceeding λ. When the exact people number is unknown, given the knowledge of the maximum possible number of people, TR-BREATH formulates an estimation $\hat{K}(\lambda)$ given by the minimum element of $\mathcal{J}$ denoted as $\hat{K}(\lambda) = \min(\mathcal{J})$, i.e., the smallest j that satisfies $\sum_{i=1}^{\min(\overline{K}, j)} \overline{l}_{\mathrm{idx}_i} \geq \lambda$.

9.4 Experiment Results

9.4.1 Experiment Setups

9.4.1.1 Environment

We conduct extensive experiments to evaluate the performance of the breathing monitoring system. The experiments are conducted in three different rooms in an office suite with dimensions 5.5 m × 5 m, 8 m × 7 m, and 8 m × 5 m, respectively.

9.4.1.2 Devices

We build one pair of prototypes equipped with off-the-shelf Wi-Fi cards with three omnidirectional antennas to obtain CSIs. Thus, the total number of links D is nine.

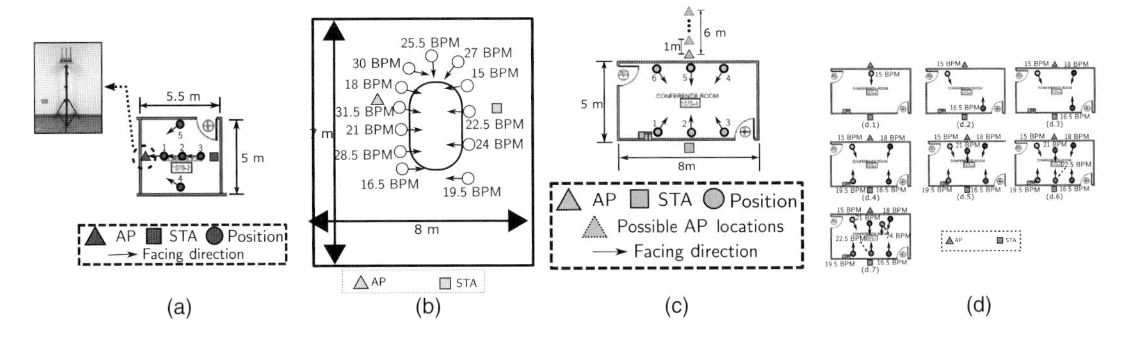

Figure 9.3 Experiment settings: (a) single-person, LOS, (b) multi-person, LOS, (c) single-person, NLOS, and (d) multiperson, NLOS.

One of the prototypes works as the access point (AP), while the other works as the station (STA). The center frequency is configured as 5.765 GHz with a bandwidth of 40 MHz. The transmit power is 20 dBm (100 milliwatts). The set of usable subcarriers $\mathcal{V}$ is given as $\{-58, -57, -56, \ldots, -2, 2, 3, \ldots, 56, 57, 58\}$ with $V = 114$. The size of DFT is $N_{\mathrm{DFT}} = 128$.

9.4.1.3 Placement of Wi-Fi devices

The performance is evaluated in both LOS and NLOS scenarios. For the LOS scenarios, the AP and STA are placed in the same room with people, while for the NLOS scenarios, they are placed outside the room blocked by two walls. The locations of both Wi-Fi devices are marked in Figure 9.3.

9.4.1.4 Participants

A total of 17 different participants were invited. During the experiments, slight movements, e.g., head or limb movements, were allowed.

9.4.1.5 Parameter Settings

The following parameters are used unless otherwise stated:

- Each experiment lasts for 2 min.
- The signal subspace dimension p is configured as 10.
- The merging radius γ is set as 0.5 BPM.
- The range of interest of the breathing rate is from $b_{\min} = 10$ BPM to $b_{\max} = 50$ BPM. This covers the adult breathing rate at rest (10 – 14 BPM), infant breathing rate (37 BPM), and the breathing rate after workout [24, 25].
- The packet rate of Wi-Fi transmission is 10 Hz.[2]

[2] The 10 Hz packet rate agrees with the *beaconing rate* of a commercial Wi-Fi AP, and the packet size containing one CSI measurement is 2.5 KB, resulting in a data rate of 25 KB/s during CSI acquisition. Therefore, the presented system only introduces minor interference to the coexisting Wi-Fi networks on the same Wi-Fi channel.

- The sampling interval T_{sp} is 0.1 s where s stands for second. For notational convenience, we write the time duration of each block measured in seconds as $M_t = MT_{sp}$ and the window size measured in seconds as $W_t = WT_{sp}$. The overlap in terms of seconds between different blocks is $P_t = PT_{sp}$. As default values, we adopt the parameters $M_t = 45$ s, $W_t = 40.5$ s, $P_t = 4.5$ s, and $B = 5$ unless otherwise stated. The total time of CSI measurements T_{tot} is thus $M_t + (B-1) \times P_t$, which equals 63 s.

During the experiments, we only observe $2 \sim 3$ Wi-Fi networks sharing the same Wi-Fi channel with the experimental devices, leading to less than 1% packet loss rate for all experiments. The impact of packet loss can be safely ignored in this case. Therefore, (9.25) reduces to (9.21), and we use $g_{m,n} = m - n$ in (9.22). Meanwhile, M equals to M' as shown in Figure 9.2.

9.4.1.6 Ground-Truths

The performance of the presented monitoring system is evaluated by comparing the breathing rate estimations against the ground-truths. To obtain the ground-truths, we ask each participant to synchronize his/her breathing according to a metronome application on his/her cellphone. After the controlled breathing experiments, we conduct experiments in a more practical setting where the participants are asked to breathe naturally according to their personal habits and count their own breathing rates manually.

9.4.2 Metrics for Performance Evaluation

9.4.2.1 Breathing Detection Rate

The detection performance of the presented system is directly determined by the SVM classification accuracy, which is evaluated by performing K-fold cross-validation on the SVM classifier.

9.4.2.2 Breathing Rate Estimation Accuracy

Assume that K is known in advance with ground-truths given by $\mathbf{b} = [b_1, b_2, \ldots, b_K]$, and the presented system outputs $K_o = \min(\overline{K}, K)$ estimations denoted as $\hat{\mathbf{b}} = [\hat{b}_1, \hat{b}_2, \ldots, \hat{b}_{\overline{K}}]$, the accuracy of estimation is calculated as $\left(1 - \frac{1}{K_o}\sum_{i=1}^{K_o}\left|\frac{\hat{b}_i - b_i}{b_i}\right|\right) \times 100\%$. For instance, the accuracy calculated from $\hat{\mathbf{b}} = [25.1, 29.8]$ BPM and $\mathbf{b} = [25, 30]$ BPM is 99.5%.

9.4.2.3 Average K_o

Still assuming that K is known, the monitoring system outputs $K_o = \min(\overline{K}, K)$ estimations. In this case, there is no penalty if $\overline{K} \geq K$ because the breathing rate estimations are given by the first K estimations with the highest likelihoods. On the other hand, when $\overline{K} < K$, the breathing rates associated with $K - \overline{K}$ people are missing in the estimations. Therefore, the average of K_o, denoted as $\overline{K_o}$, is also an important metric, as $\overline{K_o}$ closer to K indicates that most of the human breathing rates can be resolved by the monitoring system.

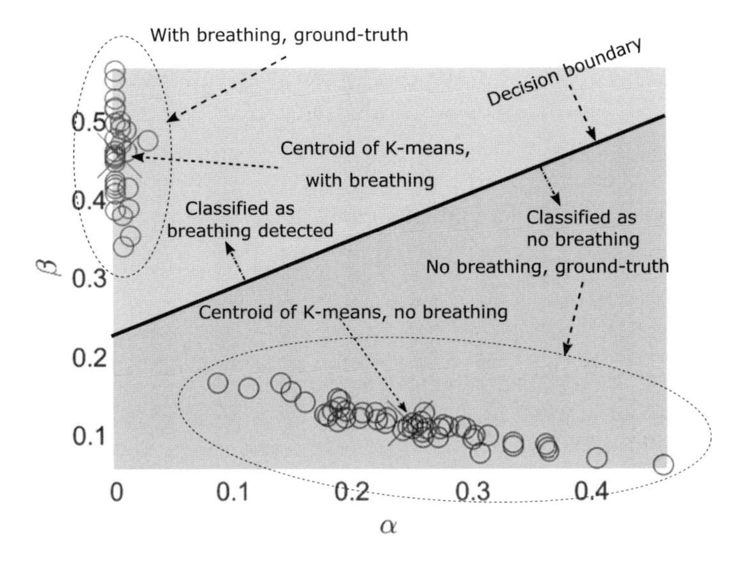

Figure 9.4 Classification performance for breathing detection.

9.4.2.4 Estimation Error of Number of People

When K is unknown, we formulate an estimation on the number of people K via $\hat{K}(\lambda)$, with the performance evaluated by the function $P(\lambda) = \mathrm{E}(|K - \hat{K}(\lambda)|)$, where E stands for the expectation operator.

9.4.3 Breathing Detection Performance

The presented breathing detection scheme determines the existence of breathing based on the output of the SVM algorithm. We use 84 CSI measurements for evaluation, where 32 of them are collected in the presence of at least one person breathing, and 52 measurements are obtained without people breathing in the room. The devices are placed according to the NLOS setting shown in Figure 9.3(c).

In Figure 9.4, we demonstrate the breathing detection performance of the presented system. First of all, we observe that the label $\hat{y}$ can be inferred from (α, β) without errors. Second, we observe that SVM returns a hyperplane that partitions (α, β) perfectly, implying a 100% detection rate. This is further validated by performing K-fold cross-validation on the results, leading to a 100% accuracy for each cross-validation.

9.4.4 Performance of Breathing Rate Estimation

In this part, we evaluate the performance of the presented system based on the ground-truth breathing rates using metronomes.

9.4.4.1 Accuracy under Single-Person LOS Scenario

We ask one participant to sit at five positions as shown in Figure 9.3(a) under the LOS scenario. For each position, the participant breathes at 15 BPM in synchronization to

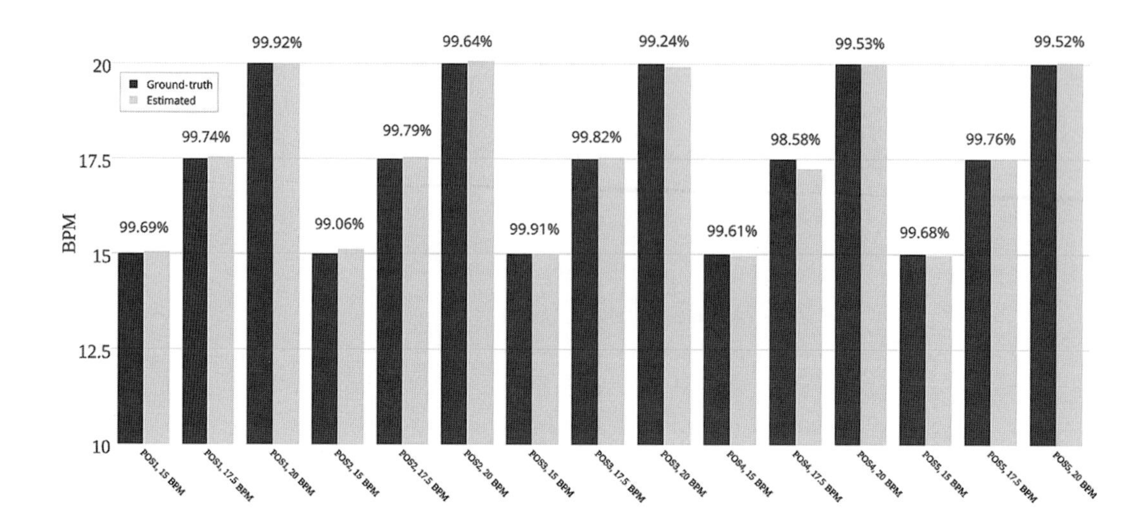

Figure 9.5 Accuracy with single-person breathing under the LOS scenario. $M_t = 45$ s, $W_t = 40.5$ s, $P_t = 4.5$ s, $B = 5$, and $T_{tot} = 63$ s.

the metronome. After that, the participant switches the breathing rate to 17.5 BPM and later 20 BPM. The accuracy performances at the five positions with various breathing rates are depicted in Figure 9.5. For comparison purpose, Figure 9.5 also demonstrates the ground-truths. As can be seen from the figure, the presented system can estimate the breathing rate with an accuracy of 99.56% averaging over all cases. The worst case is when the participant sits at position 4 and breathes at 17.5 BPM with an accuracy of 98.58%, equivalent to an estimation error of ± 0.249 BPM.

9.4.4.2 Accuracy under Multiperson LOS Scenario

A total of 12 people were invited into the conference room as shown in Figure 9.3(b) under the LOS scenario. The details of the position and breathing rate for each participant are displayed in Figure 9.3(b). The normalized population, variance, likelihood, and centroid for each cluster are presented in Figure 9.6. It can be seen that the presented system resolves the breathing rates of nine out of a dozen people with an accuracy of 98.65%.

9.4.4.3 Accuracy under Single-Person NLOS Scenario

One participant was invited into the conference room to breathe with 15 BPM at six different positions, with details shown in Figure 9.3(c). Both Wi-Fi devices are placed outside the conference room. Figure 9.7 shows that a mean accuracy of 98.74% averaging over the six positions is achieved even when the two devices are blocked by two concrete walls of the conference room, which validates the high accuracy under the through-the-wall scenario.

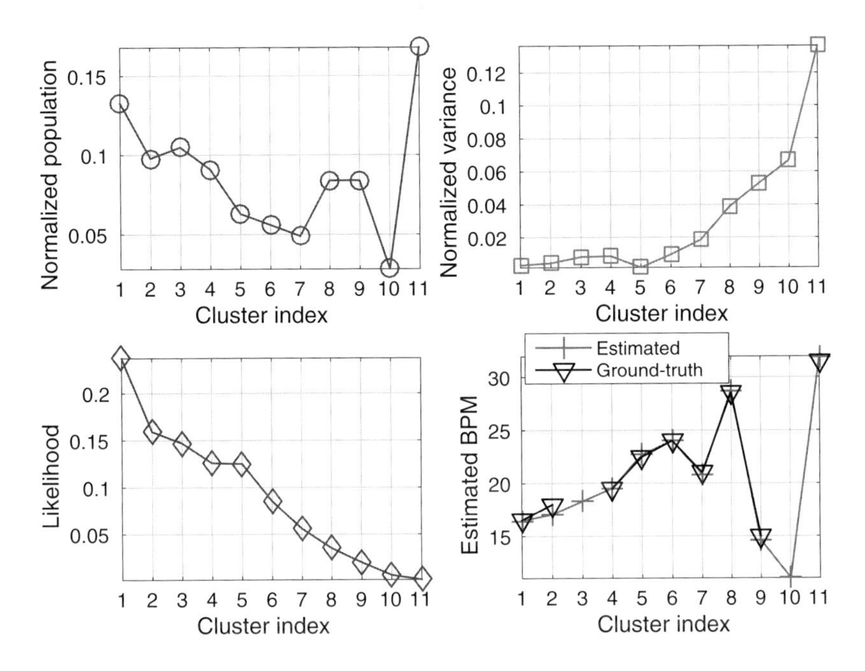

Figure 9.6 Performance of estimating breathing rates of a dozen people under the LOS scenario. $M_t = 45$ s, $W_t = 40.5$ s, $P_t = 4.5$ s, $B = 5$, and $T_{tot} = 63$ s.

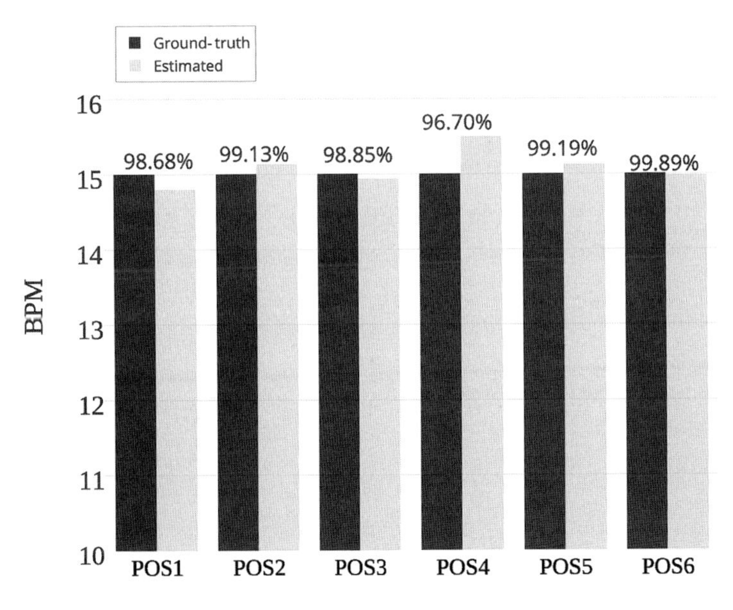

Figure 9.7 Accuracy with single-person breathing under the NLOS scenario. $M_t = 45$ s, $W_t = 40.5$ s, $P_t = 4.5$ s, $B = 5$, and $T_{tot} = 63$ s.

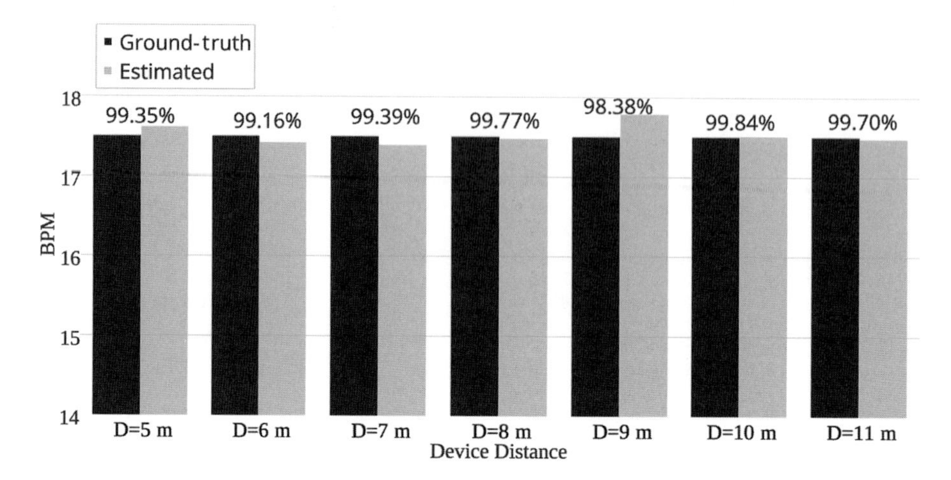

Figure 9.8 Accuracy of breathing rate estimation with various distances. $M_t = 45$ s, $W_t = 40.5$ s, $P_t = 4.5$ s, $B = 5$, and $T_{tot} = 63$ s.

To evaluate the impact of distances between Wi-Fi devices on the performance, we place the AP at six different locations with 1 meter resolution. The participant breathes at 15 BPM in this experiment. The distance between the AP and the STA ranges from 5 meters to 11 meters. As shown in Figure 9.8, the presented scheme achieves more than 98.38% in accuracy, with a mean accuracy of 99.37% averaging over the results of various distances. Even when the device distance reaches 11 meters, the accuracy is maintained at 99.70%, demonstrating the robustness of the presented system under different device distances.

We further evaluate TR-BREATH by reducing M_t to 10 s. Besides, we set $W_t = 9$ s, $P_t = 0.5$ s, and $T_{tot} = 10$ s. The packet rate is increased to 30 Hz. One participant sits at position 1 of Figure 9.3(c) and breathe at 15, 17.5, and 20 BPM, with each breathing rate lasting for 20 s. The total measurement time is 60 s. Figure 9.9 shows that TR-BREATH could track the breathing rate accurately with a mean accuracy of 99%. Therefore, TR-BREATH can provide accurate breathing rates every 10 s for single-person breathing monitoring that fits well to the patient monitoring scenarios.

9.4.4.4 Accuracy under Multiperson NLOS Scenario

We invite up to seven people into one conference room with two devices placed under the NLOS scenario. The positions and breathing rates associated with each person are depicted in Figure 9.3(d). Figure 9.10 summarizes the accuracy performances, which shows that an accuracy of 99.1% when $K = 7$ and a mean accuracy of 97.3% averaging over all seven cases can be achieved.

9.4.4.5 $\overline{K_o}$ under Multiperson NLOS Scenario

Figure 9.11 demonstrates the $\overline{K_o}$ performance for the multiperson NLOS scenario. As we can see, with a various number of people K, $\overline{K_o}$ equals to K, which shows that the presented system could resolve the breathing rates of all people. Combining the results

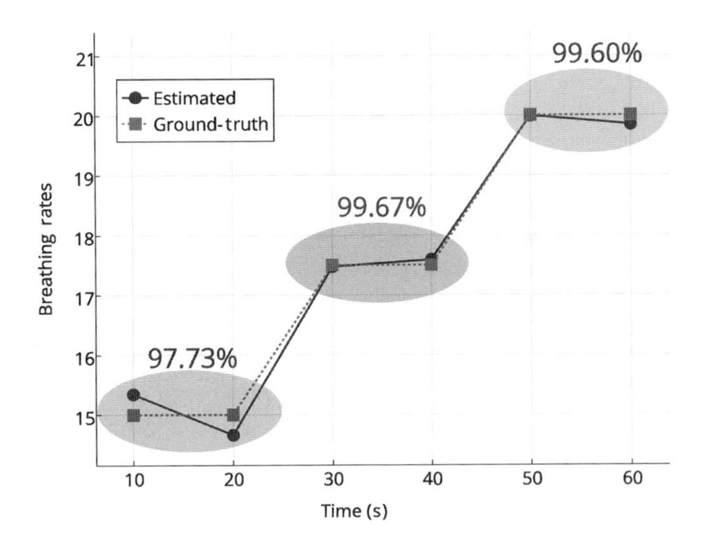

Figure 9.9 Accuracy of breathing rate estimation with 10 s of CSI measurement. $M_t = 10$ s, $W_t = 9$ s, $P_t = 0.5$ s, $B = 1$, and $T_{tot} = 10$ s.

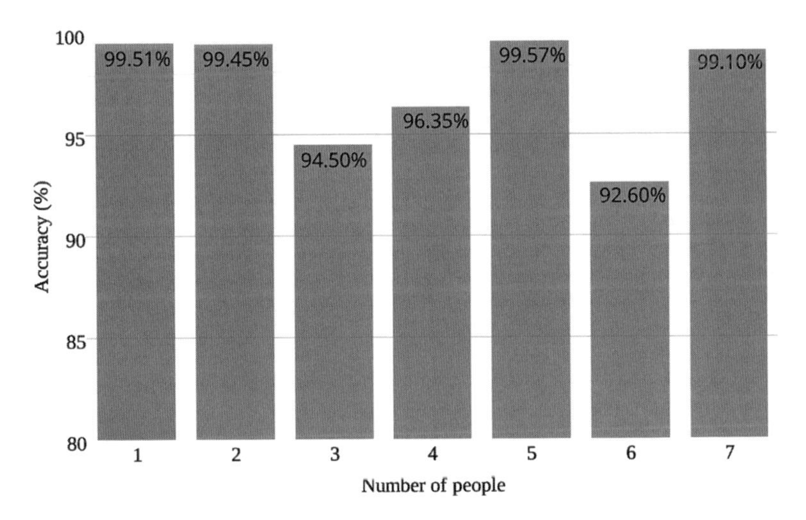

Figure 9.10 Accuracy with multiple people under the NLOS scenario. $M_t = 45$ s, $W_t = 40.5$ s, $P_t = 4.5$ s, $B = 5$, and $T_{tot} = 63$ s.

in Figure 9.10, we conclude that given K people, the presented system resolves the breathing rates of K people with high accuracy.

9.4.5 Performance of Natural Breathing Rate Estimation

In this part, we investigate the performance of the presented system in a more practical setting by asking the participants to breathe naturally. Instead of using the metronomes, the participants were asked to memorize how many breaths they took in a minute.

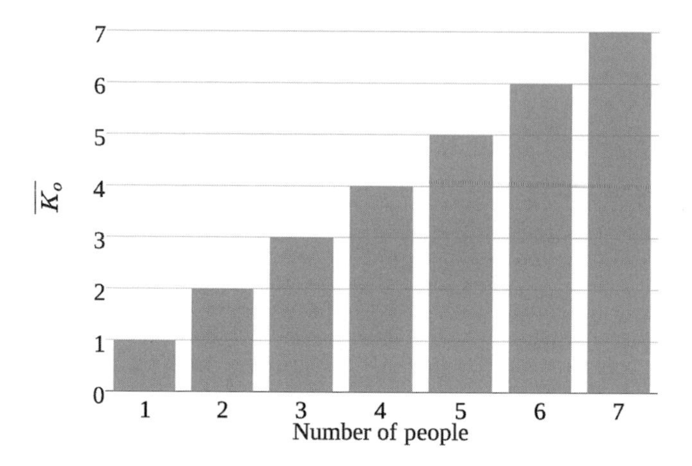

Figure 9.11 $\overline{K_o}$ with multiple people under the NLOS scenario. $M_t = 45$ s, $W_t = 40.5$ s, $P_t = 4.5$ s, $B = 5$, and $T_{tot} = 63$ s.

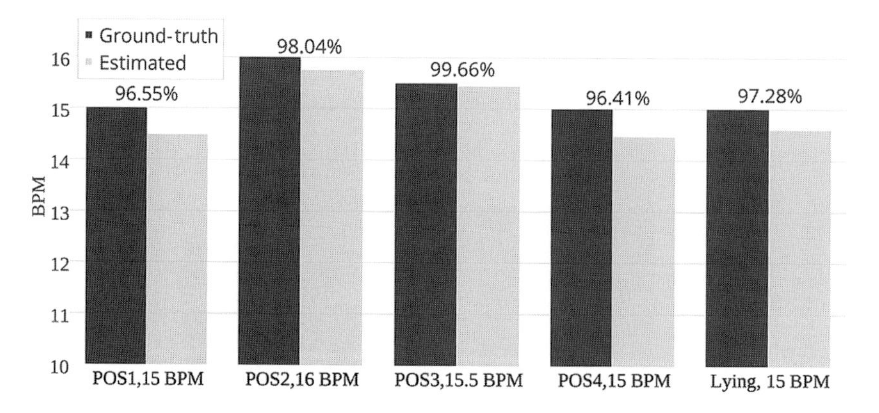

Figure 9.12 Performance of estimating the natural breathing rates of one person under the NLOS scenario. $M_t = 45$ s, $W_t = 40.5$ s, $P_t = 4.5$ s, $T_{tot} = 63$ s.

9.4.5.1 Accuracy under Single-Person NLOS Scenario

One participant is asked to breathe naturally at four different positions in the same conference room as in Figure 9.3(c). Then, the participant lies on the ground and breathes. Figure 9.12 shows that a mean accuracy of 97.0% can be achieved. Moreover, the breathing rate of a person lying on the ground can be estimated accurately, which shows the viability of the presented scheme in monitoring the breathing rate of a sleeping person.

9.4.5.2 Accuracy under Multiperson NLOS Scenario

Nine participants breathe naturally in the conference room shown in Figure 9.3(c). The breathing rates are given as [16, 11.5, 10.5, 12, 13, 15.5, 16.5, 26.5, 12] BPM, where two participants coincide in their breathing rates. Figure 9.13 shows that six out of the eight resolvable breathing rates are obtained with an accuracy of 98.07%.

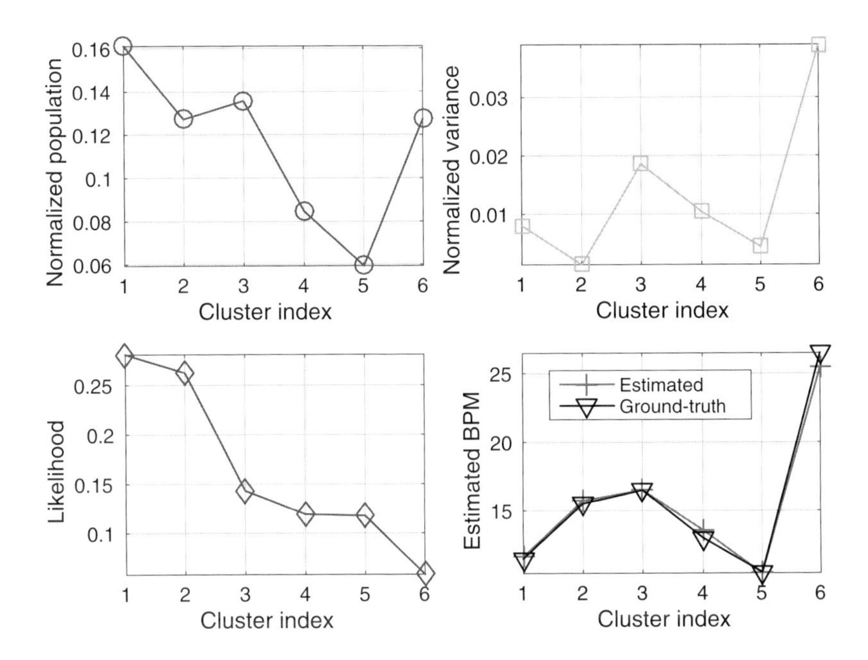

Figure 9.13 Performance of estimating the natural breathing rates of nine people under the NLOS scenario. $M_t = 45\ s$, $W_t = 40.5$ s, $P_t = 4.5$ s, $T_{tot} = 63$ s.

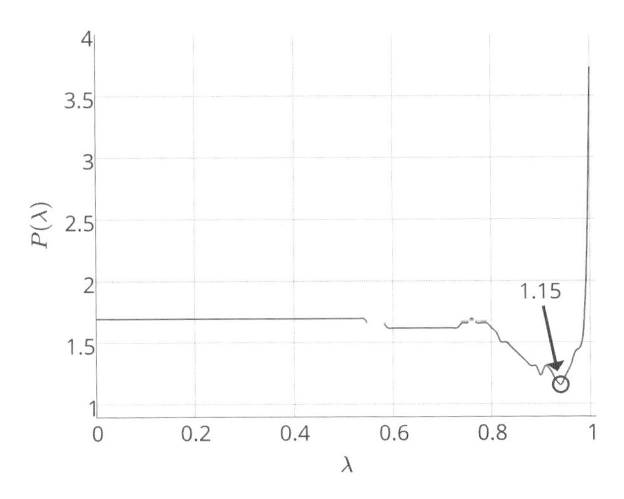

Figure 9.14 Performance of people number estimation. $M_t = 45$ s, $W_t = 40.5$ s, $P_t = 4.5$ s, $B = 5$, and $T_{tot} = 63$ s.

9.4.6 Estimating the Number of People *K*

Figure 9.14 illustrates that the optimal $P(\lambda)$ is 1.15 when $\lambda = 0.88$. Thus, the presented system can estimate the number of people with an error around 1.

9.5 Impact of Various Factors

In this section, we further investigate the performance of TR-BREATH in a more practical application scenario. First of all, we study the performance under the influence of packet loss with various severity. Then, we discuss the effects of motions on TR-BREATH. Finally, we demonstrate the significant improvement of TR-BREATH using both amplitude and phase information compared to the approach using amplitude only in [5]. The parameters are configured to be the same as Section 9.4.1.5 unless otherwise stated.

9.5.1 Impact of Packet Loss

We present the accuracy under the NLOS single-person at position 1 shown in Figure 9.3(c) with different packet loss rates. We consider two packet loss mechanisms, i.e., bursty packet loss and random packet loss. The bursty packet loss is mainly caused by the continuous data transmission among few Wi-Fi devices, which fully jams the medium for a long time. On the other hand, the random packet loss is due to the random access of a large number of nearby Wi-Fi devices, which occupy the medium occasionally.

To emulate packet loss, we intentionally discard collected CSI samples in the experiments. More specifically, for the bursty packet loss, we discard CSI samples within a certain time period, while for the random packet loss, we discard CSI samples with index following a uniform distribution. When the packet loss compensation is enabled, $g_{m,n} = s''_m - s''_n$ is used, otherwise $g_{m,n} = m - n$.

The results with different packet loss rates with the aforementioned two mechanisms are shown in Figure 9.15. We observe that the consequence of random packet loss is much more severe than the bursty packet loss when the packet loss compensation is not enabled. With 10% random packet loss, the accuracy drops to 88.35% from 99.35%. The accuracy further deteriorates to 74.13% and 62.83% with 20% and 30% packet loss, respectively. The advantage of packet loss compensation is obvious because TR-BREATH maintains an accuracy of 99.70% even with 30% packet loss. On the contrary, bursty packet loss does not degrade the accuracy greatly. It can be justified by the fact that most CSIs are still sampled uniformly under this scenario.

9.5.2 Impact of Motion

To study the effect of motion, we perform additional experiments involving ambient motions and subject motions. The experiment settings are shown in Figure 9.16. The participant breathes at 20 BPM.

9.5.2.1 Impact of Ambient Motion

Besides the participant under breathing monitoring, we ask another participant to walk randomly in the eight highlighted areas in Figure 9.16, where S_1 to S_4 stands for the ambient motions in the conference room and S_5 to S_8 in the foyer. We further classify

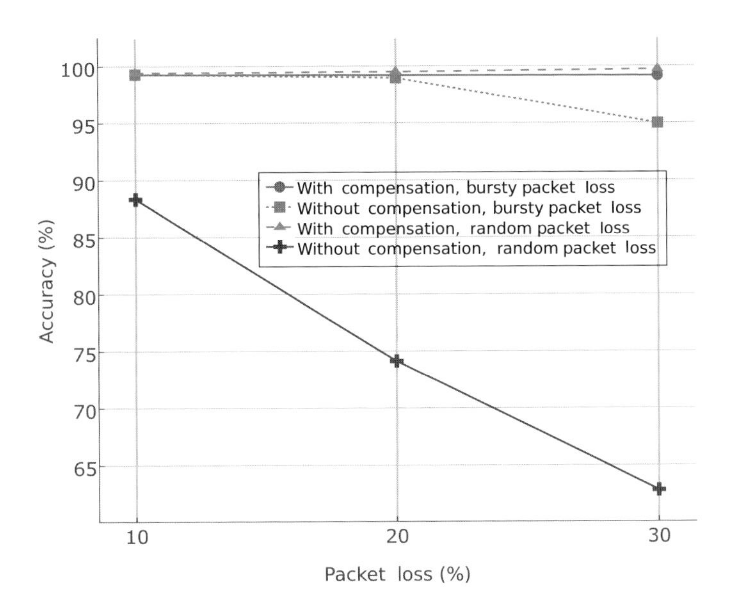

Figure 9.15 Impact of packet loss on accuracy.

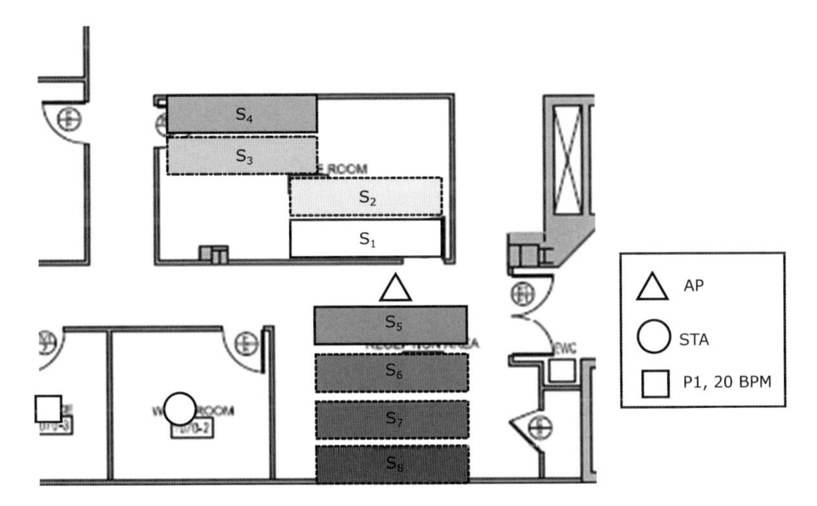

Figure 9.16 Experiment settings for investigation of ambient motions and subject motions.

these areas in terms of their distances to the Wi-Fi AP as very close, close, far, and very far. For instance, S_1 is considered to be very close from the Wi-Fi AP, while S_4 is regarded as very far away from the Wi-Fi AP. Despite the fact that the impact of motion is location-dependent, in general, we find that the motions introduce severe interference into TR-BREATH when they occur within 1 m radius to either the AP or the STA.

The results are depicted in Figure 9.17. Clearly, when the ambient motion occurs very close to the Wi-Fi AP, the accuracy degrades significantly, especially for the case of ambient motions in the foyer area indicated by S_5. When the distance between the motion to the Wi-Fi AP increases, the accuracy is improved. We observe similar results

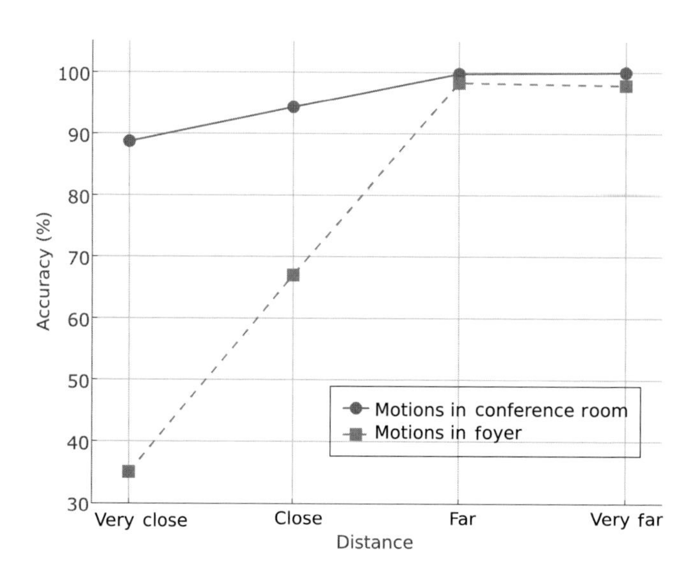

Figure 9.17 Impact of ambient motion on accuracy.

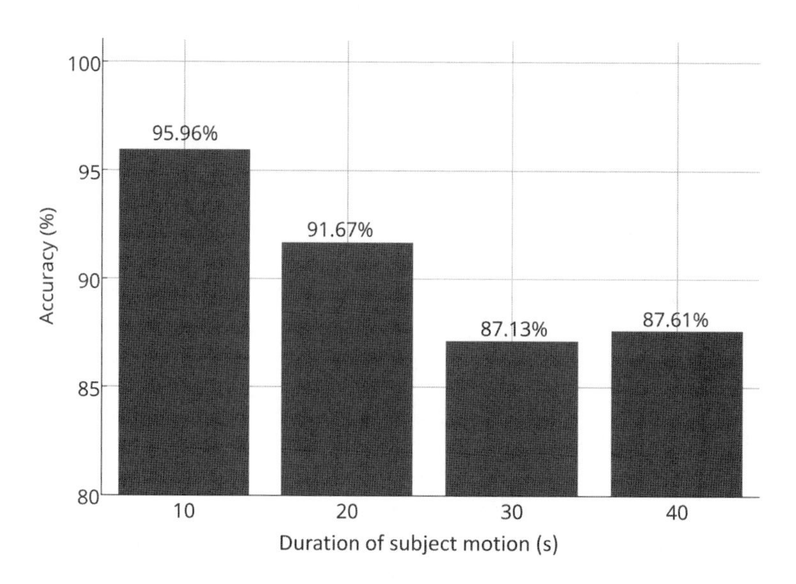

Figure 9.18 Impact of subject motion on accuracy.

when the ambient motion occurs close to the Wi-Fi device. Thus, we conclude that TR-BREATH can tolerate ambient motions as long as both Wi-Fi devices of TR-BREATH are far from these motions.

9.5.2.2 Impact of Subject Motion

In this experiment, we ask the participant under monitoring to move randomly for a certain period of time, and then sit back to the original position as shown in Figure 9.16 to continue breathing. The results are shown in Figure 9.18. We observe that when the

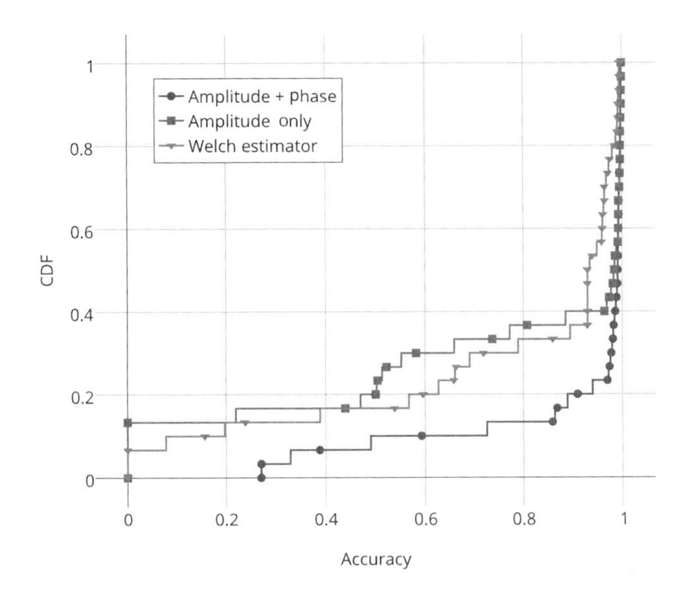

Figure 9.19 Comparison of CDFs among different schemes.

participant only moves for 10 s, the accuracy can be maintained at 95.96%. The accuracy drops to 87.61% when the participant moves for 40 s, corresponding to an error of ±2.48 BPM. This demonstrates that TR-BREATH can tolerate the subject motions given that the participant stays still during most of the time.

9.5.2.3 Impact of using CSI Amplitude Only

Thanks to the additional step of CSI calibration mentioned in Section 9.3, TR-BREATH makes full use of the complex CSIs, which is a major difference from [5], which uses the CSI amplitudes only. In this section, we show that using both CSI amplitudes and phases could improve the performance of TR-BREATH. Furthermore, we replace the Root-MUSIC algorithm with the conventional Welch estimator [26], a widely used nonparametric scheme. For the Welch estimator, we use the CSI amplitude only, which coincides with the spectral analysis scheme used in [5].

We ask the participant to breathe at $[20, 25, 30, 35, 40, 45]$ BPM under the setting of Figure 9.16 without ambient and subject motions. Each experiment lasts for 1 min. The cumulative density functions (CDFs) of the accuracy for this experiment are shown in Figure 9.19. We observe that the results are more concentrated in areas close to 100% accuracy in the complex CSI case, indicating that using complex CSIs outperforms the amplitude-only case and the Welch estimator.

9.6 Summary

In this chapter, we presented TR-BREATH, a contact-free and highly accurate breathing monitoring system leveraging TR for breathing detection and multiperson breathing rate estimations using commercial Wi-Fi devices. The TR resonating strengths are analyzed by the Root-MUSIC algorithm to extract features for breathing detection and breathing rate estimation. Experiment results in a typical indoor environment demonstrate that, with 63 s of measurements, a perfect detection rate can be obtained. Meanwhile, the proposed system can estimate the single-person breathing rate in the NLOS scenario with an accuracy of 99% with only 10 s of measurement. With 63 s of measurement, the presented system achieves a mean accuracy of 98.65% for a dozen people under the LOS scenario and 98.07% for nine people under the NLOS scenario even when the two Wi-Fi devices are blocked by two walls. The presented system can also estimate the number of people with an average error around 1. We also show that TR-BREATH is robust against packet loss and motions in the environment. With the ubiquity of Wi-Fi-enabled mobile devices, TR-BREATH can provide real-time, in-home, and noninvasive breathing monitoring in future medical applications. For related references, interested readers can refer to [27].

References

[1] "Vital signs," www.hopkinsmedicine.org/.

[2] B.-K. Park, S. Yamada, O. Boric-Lubecke, and V. Lubecke, "Single-channel receiver limitations in Doppler radar measurements of periodic motion," in *IEEE Radio and Wireless Symposium*, pp. 99–102, Jan. 2006.

[3] F. Adib, H. Mao, Z. Kabelac, D. Katabi, and R. C. Miller, "Smart homes that monitor breathing and heart rate," in *Proceedings of the 33rd Annual ACM Conference on Human Factors in Computing Systems*, pp. 837–846, 2015. [Online]. Available: http://doi.acm.org/10.1145/2702123.2702200.

[4] H. Abdelnasser, K. A. Harras, and M. Youssef, "UbiBreathe: A ubiquitous non-invasive WiFi-based breathing estimator," in *Proceedings of the 16th ACM International Symposium on Mobile Ad Hoc Networking and Computing*, pp. 277–286, 2015.

[5] J. Liu, Y. Wang, Y. Chen, J. Yang, X. Chen, and J. Cheng, "Tracking vital signs during sleep leveraging off-the-shelf WiFi," in *Proceedings of the 16th ACM International Symposium on Mobile Ad Hoc Networking and Computing*, pp. 267–276, 2015.

[6] C. Chen, Y. Han, Y. Chen, and K. J. R. Liu, "Multi-person breathing rate estimation using time-reversal on WiFi platforms," in *IEEE Global Conference on Signal and Information Processing (GlobalSIP)*, Dec. 2016.

[7] B. D. Rao and K. V. S. Hari, "Performance analysis of root-music," *IEEE Transactions on Acoustics, Speech, and Signal Processing*, vol. 37, no. 12, pp. 1939–1949, Dec. 1989.

[8] Y. Chen, F. Han, Y. H. Yang, H. Ma, Y. Han, C. Jiang, H. Q. Lai, D. Claffey, Z. Safar, and K. J. R. Liu, "Time-reversal wireless paradigm for green Internet of Things: An overview," *IEEE Internet of Things Journal*, vol. 1, no. 1, pp. 81–98, Feb. 2014.

[9] C. Chen, Y. Chen, Y. Han, H. Q. Lai, and K. J. R. Liu, "Achieving centimeter-accuracy indoor localization on WiFi platforms: A frequency hopping approach," *IEEE Internet of Things Journal*, vol. 4, no. 1, pp. 111–121, Feb. 2017.

[10] C. Chen, Y. Chen, Y. Han, H. Q. Lai, F. Zhang, and K. J. R. Liu, "Achieving centimeter-accuracy indoor localization on WiFi platforms: A multi-antenna approach," *IEEE Internet of Things Journal*, vol. 4, no. 1, pp. 122–134, Feb. 2017.

[11] C. Chen, Y. Han, Y. Chen, and K. J. R. Liu, "Indoor global positioning system with centimeter accuracy using Wi-Fi [applications corner]," *IEEE Signal Processing Magazine*, vol. 33, no. 6, pp. 128–134, Nov. 2016.

[12] Z.-H. Wu, Y. Han, Y. Chen, and K. J. R. Liu, "A time-reversal paradigm for indoor positioning system," *IEEE Transactions on Vehicular Communications*, vol. 64, no. 4, pp. 1331–1339, Apr. 2015.

[13] F. Zhang, C. Chen, B. Wang, H.-Q. Lai, and K. J. R. Liu, "A timereversal spatial hardening effect for indoor speed estimation," in *IEEE International Conference on Acoustics, Speech, and Signal Processing (ICASSP)*, Mar. 2017.

[14] Q. Xu, Y. Chen, B. Wang, and K. J. R. Liu, "Radio biometrics: Human recognition through a wall," *IEEE Transactions on Information Forensics and Security*, vol. 12, no. 5, pp. 1141–1155, May 2017.

[15] "TRIEDS: Wireless events detection through the wall," *IEEE Internet of Things Journal*, vol. PP, no. 99, pp. 1–1, 2017.

[16] B. Wang, Y. Wu, F. Han, Y.-H. Yang, and K. J. R. Liu, "Green wireless communications: A time-reversal paradigm," *IEEE Journal on Selected Areas in Communications*, vol. 29, no. 8, pp. 1698–1710, Sep. 2011.

[17] B. J. Frey and D. Dueck, "Clustering by passing messages between data points," *Science*, vol. 315, p. 2007, 2007.

[18] N. V. Rivera, S. Venkatesh, C. Anderson, and R. M. Buehrer, "Multi-target estimation of heart and respiration rates using ultra wideband sensors," in *14th European Signal Processing Conference,* pp. 1–6, Sep. 2006.

[19] A. Goldsmith, *Wireless Communications*. Cambridge University Press, 2005.

[20] M. Abramowitz, *Handbook of Mathematical Functions, With Formulas, Graphs, and Mathematical Tables*. Dover Publications, Incorporated, 1974.

[21] R. Schmidt, "Multiple emitter location and signal parameter estimation," *IEEE Transactions on Antennas and Propagation*, vol. 34, no. 3, pp. 276–280, Mar. 1986.

[22] J. MacQueen, "Some methods for classification and analysis of multivariate observations," in *Proceedings of the Fifth Berkeley Symposium on Mathematical Statistics and Probability, Volume 1: Statistics*, vol. 1, no. 14, pp. 281–297, 1967.

[23] C. J. C. Burges, "A tutorial on support vector machines for pattern recognition," *Data Mining and Knowledge Discovery*, vol. 2, no. 2, pp. 121–167, Jun. 1998.

[24] J. F. Murray, *The Normal Lung: The Basis for Diagnosis and Treatment of Pulmonary Disease*, WB Saunders Company, 1976.

[25] M. Kearon, E. Summers, N. Jones, E. Campbell, and K. Killian, "Breathing during prolonged exercise in humans," *The Journal of Physiology*, vol. 442, p. 477, 1991.

[26] P. Welch, "The use of fast Fourier transform for the estimation of power spectra: A method based on time averaging over short, modified periodograms," *IEEE Transactions on Audio and Electroacoustics*, vol. 15, no. 2, pp. 70–73, Jun. 1967.

[27] C. Chen, Y. Han, Y. Chen, H.-Q. Lai, F. Zhang, B. Wang, and K. J. R. Liu, "TR-BREATH: Time-reversal breathing rate estimation and detection," *IEEE Transactions on Biomedical Engineering*, vol. 65, no. 3, pp. 489–501, 2018.

10 Wireless Motion Detection

Motion detection as a key component in modern security systems has received an increasing attention recently, but most existing solutions require special installation and calibration and only have a limited coverage. In this chapter, we discuss WiDetect, a highly accurate, calibration-free, and low-complexity wireless motion detector. By exploiting the statistical theory of electromagnetic waves, we establish a link between the autocorrelation function of the physical layer channel state information (CSI) and motion in the environment. Temporal, frequency, and spatial diversity are also exploited to further improve the robustness and accuracy of WiDetect. Extensive experiments conducted in several facilities show that WiDetect can achieve similar detection performance compared to a commercial home security system, while with much larger coverage and lower cost.

10.1 Introduction

Motion detection plays a vital role in modern security systems. However, popular approaches that rely on video, infrared, RFID, UWB, etc. all require specialized hardware deployment and have their own limitations in practical applications. For example, the vision-based schemes [1] can only perform motion monitoring in areas covered by camera, and in addition they introduce privacy issues. The infrared-based motion sensors are especially sensitive to thermal radiation, leading to a high false alarm rate.

Recently, WiFi has been considered in wireless sensing due to its deployment flexibility, large coverage, and cost efficiency. RASID [2] exploits the fluctuations of the receive signal strength indicator (RSSI) to detect the presence of humans indoors, based on the dissimilarity in RSSI distribution in a static environment. E-eyes [3] follows a similar idea but uses CSI instead of RSSI as the metric. PILOT [4] decomposes the CSI amplitude correlation matrix using singular value decomposition (SVD) and monitors the variations of the singular vectors along time. Similarly, CARM [5] tracks the variance of the second singular vector to detect motion. WiDar [6] computes cross-correlation among different subcarriers and uses the increase in the correlation between adjacent subcarriers as an indicator of motion. Table 10.1 summarizes the performance of most existing approaches, where the second column shows the false-negative and false-positive rate, and the third column shows whether calibration is needed. As can be

Table 10.1 Related works

Reference	FN/FP	Cal.
RASID [2]	3.8%/4.7%	Yes
PILOT [4]	10.0%/10.0%	Yes
E-eyes [3]	10.0%/1.0%	Yes
Omni-PHD [7]	8.0%/7.0%	Yes
DeMan [8]	5.93%/1.45%	Yes
CARM [5]	2.0%/1.4 times per hour	Yes
SIED [9]	0%/6.4%(slow motion)	Yes

seen, all of them rely on some kind of calibration before use, such as storing the features of normal states, or fine-tuning of parameters, which is not robust to the environmental dynamics and not easy to use for ordinary users. Also, their performance in terms of coverage, accuracy, and computational complexity is quite far from meeting the requirement of real applications.

To address these challenges, in this chapter, we present WiDetect, a highly accurate and robust Wi-Fi-based motion detector that can cover a large area and is easy to use. We first characterize the impact of motion on the autocorrelation function (ACF) of the received channel power response using statistical theory of electromagnetic (EM) waves. Then, we define a *motion statistic* to measure the likelihood of the presence of motion. To improve the accuracy of detection, WiDetect combines all the motion statistics obtained from multiple subcarriers, and the impact of the number of CSI measurements and the number of available subcarriers on WiDetect is quantified. We conduct extensive experiments in an office and a single-family home, where four PIRs are deployed for comparison. Experiment results show that WiDetect is able to detect human motion in a large area while maintaining a negligible false alarm rate.

The rest of this chapter is organized as follows. Section 10.2 presents a statistical modeling of the CSI measurements based on statistical theory of EM waves. The detailed design of WiDetect is presented in Section 10.3, and experimental evaluation is discussed in Section 10.4.

10.2 Statistical Modeling of CSI Measurements

In this section, we discuss the theoretical basis of WiDetect.

10.2.1 CSI Measurement

Consider a pair of WiFi devices deployed in an indoor environment, and the transmitter (Tx) keeps transmitting signals to the receiver (Rx). Let $X(t, f)$ and $Y(t, f)$ be the transmitted and received signals over a subcarrier with frequency f at time t. Then, the CSI on the subcarrier with frequency f at time t is $H(t, f) = \frac{Y(t,f)}{X(t,f)}$ [10], which is a complex number and can be obtained from the PHY layer of commercial Wi-Fi. However, in practice, the estimated $H(t, f)$ often suffers from severe phase distortions [11, 12], so in this chapter we only use the magnitude of $H(t, f)$ and define the power response of the CSI $G(t, f)$ as follows,

$$G(t, f) \triangleq |H(t, f)|^2 = \mu(t, f) + \varepsilon(t, f), \tag{10.1}$$

where $\mu(t, f)$ denotes the part contributed by the propagations of the EM waves, and $\varepsilon(t, f)$ denotes the measurement noise. Let $\mathcal{F}$ denote the set of available subcarriers. For any given subcarrier $f \in \mathcal{F}$, $\varepsilon(t, f)$ can be shown through experiment measurements to be an additive white Gaussian noise, i.e., $\varepsilon(t, f) \sim \mathcal{N}(0, \sigma^2(f))$, and $\varepsilon(t_1, f_1)$ and $\varepsilon(t_2, f_2)$ are independent for any two different subcarriers $f_1 \neq f_2$ or any two different time slots $t_1 \neq t_2$.

10.2.2 Modeling of the Signal Term

Radio propagation in a building interior is in general very difficult to analyze because the EM waves can be absorbed and scattered by walls, doors, windows, moving objects, etc. However, buildings and rooms can be viewed as reverberation cavities in that they exhibit internal multipath propagations. Hence, we refer to a statistical modeling instead of a deterministic one and apply the statistical theory of EM fields developed for reverberation cavities to analyze the statistical properties of the signal term $\mu(t, f)$.

Consider a rich-scattering environment as illustrated in Figure 10.1, which is typical for indoor spaces. The scatterers are assumed to be diffusive and can reflect the impinging EM waves toward all directions. A pair of Tx and Rx are deployed in the environment, both equipped with omnidirectional antennas, and the Tx emits a continuous EM wave via its antennas, which is received by the Rx, and the corresponding received electric field is denoted as $\vec{E}_{Rx}(t, f)$. Actually, $\mu(t, f)$ measures the power of $\vec{E}_{Rx}$, i.e., $\mu(t, f) = \|\vec{E}_{Rx}(t, f)\|^2$, where $\|\cdot\|^2$ denotes the Euclidean norm. Within a sufficiently short period, $\vec{E}_{Rx}(t, f)$ can be decomposed into two parts as $\vec{E}_{Rx}(t, f) \approx \vec{E}_s(f) + \sum_{i \in \Omega_d} \vec{E}_i(t, f)$, where $\vec{E}_s(f)$ and $\vec{E}_i(t, f)$ denote the components contributed

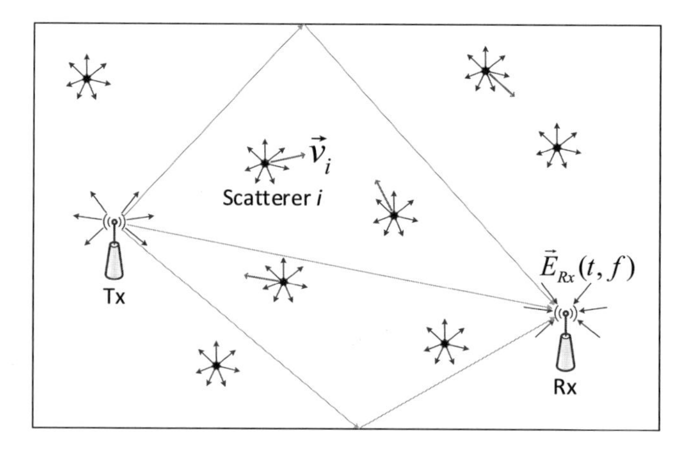

Figure 10.1 Propagation of radio signals in scattering environment.

by all the static scatterers and the ith dynamic scatterer, respectively, and Ω_d denotes the set of dynamic scatterers in the environment. When the environment is static, Ω_d is empty. The intuition behind the decomposition is that each scatterer can be treated as a "virtual antenna" diffusing the received EM waves in all directions, and then these EM waves add up together at the receive antenna after bouncing off the walls, ceilings, furniture, windows, etc. of the building.

Let v_i denote the speed of the ith moving scatterer, and $\vec{E}_i(t, f)$ is expanded in the orthogonal basis as $\vec{E}_i(t, f) = E_{i,x}(t, f)\hat{x} + E_{i,y}(t, f)\hat{y} + E_{i,z}(t, f)\hat{z}$, where $E_{i,u}(t, f)$ denotes the linear component of $\vec{E}_i(t, f)$ along the direction $\hat{u}$, $u \in \{x, y, z\}$, and $\hat{z}$ points to the moving direction of the scatterer. Then, under certain common assumptions on the homogeneity of scattering for reverberation cavities [13], the ACF for each linear component of $\vec{E}_i(t, f)$ can be derived in closed forms as

$$\rho_{E_{i,x}}(\tau, f) = \rho_{E_{i,y}}(\tau, f) = \frac{3}{2}\left[\frac{\sin(kv_i\tau)}{kv_i\tau} - \frac{1}{(kv_i\tau)^2}\left(\frac{\sin(kv_i\tau)}{kv_i\tau} - \cos(kv_i\tau)\right)\right], \quad (10.2)$$

$$\rho_{E_{i,z}}(\tau, f) = \frac{3}{(kv_i\tau)^2}\left[\frac{\sin(kv_i\tau)}{kv_i\tau} - \cos(kv_i\tau)\right], \quad (10.3)$$

where k is the wave number of the transmitted signal, and τ denotes the time lag. Denote $E_i^2(f)$ as the radiation power of the ith scatterer, $E_d^2(f)$ as the variance of $\mu(t, f)$, and assume that $\vec{E}_{i_1}(t, f)$ and $\vec{E}_{i_2}(t, f)$ are statistically uncorrelated for $\forall i_1 \neq i_2$, the ACF of $\mu(t, f)$ can be approximated as

$$\rho_\mu(\tau, f) \approx \frac{1}{E_d^2(f)} \sum_{u\in\{x,y,z\}} \left(\sum_{i\in\Omega_d} \frac{2E_{s,u}^2(f)E_i^2(f)}{3} \rho_{E_{i,u}}(\tau, f) + \sum_{\substack{i_1, i_2 \in \Omega_d \\ i_1 \geq i_2}} \frac{E_{i_1}^2(f)E_{i_2}^2(f)}{9} \rho_{E_{i_1,u}}(\tau, f)\rho_{E_{i_2,u}}(\tau, f) \right). \quad (10.4)$$

An important observation is that when $\tau \to 0$, $\rho_\mu(\tau, f) \to 1$.

10.2.3 Modeling of the CSI Power Response

As $\mu(t, f)$ is due to the propagations of EM waves and $\varepsilon(t, f)$ is due to the imperfect measurements of CSI, it can be shown through experimental results that $\mu(t, f)$ and $\varepsilon(t, f)$ are uncorrelated with each other, i.e., $\mathrm{cov}(\mu(t_1, f), \varepsilon(t_2, f)) = 0$, for $\forall t_1, t_2$, Therefore, the auto-covariance function of $G(t, f)$ can be expressed as

$$\begin{aligned}\gamma_G(\tau, f) &\triangleq \mathrm{cov}\Big(\mu(t, f)+\varepsilon(t, f), \mu(t-\tau, f)+\varepsilon(t-\tau, f)\Big) \\ &= E_d^2(f)\rho_\mu(\tau, f)+\sigma^2(f)\delta(\tau),\end{aligned} \quad (10.5)$$

where $\delta(\cdot)$ is Dirac delta function. The corresponding ACF of $G(t, f)$ can thus be expressed as

$$\rho_G(\tau, f) = \frac{E_d^2(f)}{E_d^2(f) + \sigma^2(f)} \rho_\mu(\tau, f), \tag{10.6}$$

where $\tau \neq 0$. When there exists motion and $\tau \to 0$, with the knowledge of $\rho_\mu(\tau, f) \to 1$, we know $\rho_G(\tau, f) \to \frac{E_d^2(f)}{E_d^2(f)+\sigma^2(f)} > 0$; when there is no motion and $\tau \to 0$, we have $\rho_G(\tau, f) = 0$ because $E_d^2(f) = 0$. Therefore, $\lim_{\tau \to 0} \rho_G(\tau, f)$ is a good indicator of the presence of motion, which is only determined by $E_d^2(f)$ incurred by motion and the power of the measurement noise $\sigma^2(f)$. We will exploit this important observation in the following design of WiDetect.

10.3 Design of WiDetect

In this section, we discuss the motion statistics and the detection rule and analyze the performance of WiDetect.

10.3.1 Motion Statistics

In practice, $\lim_{\tau \to 0} \rho_G(\tau, f)$ cannot be measured directly because $\tau \to 0$ is difficult to achieve due to finite channel sampling rate F_s. Instead, we use the quantity $\rho_G(\tau = \frac{1}{F_s}, f)$ as an approximation as long as F_s is large enough. Then, we define the *motion statistic* from the CSI power response $G(t, f)$ as the sample ACF of $G(t, f)$,

$$\hat{\phi}(f) = \frac{\hat{\gamma}_G\left(\tau = \frac{1}{F_s}, f\right)}{\hat{\gamma}_G(\tau = 0, f)}, \tag{10.7}$$

where $\hat{\gamma}_G(\tau, f)$ denotes the sample auto-covariance function of $G(t, f)$ [14]. When there is no motion, according to the large sample theory [14], the distribution of $\hat{\phi}(f)$ will converge to an asymptotically normal (AN) distribution with mean $-\frac{1}{T}$ and variance $\frac{1}{T}$ as T approaches infinity, i.e., $\hat{\phi}(f) \sim \mathcal{AN}(-\frac{1}{T}, \frac{1}{T})$ as $T \to \infty$, with T as the number of samples. In addition, $\hat{\phi}(f_1)$ and $\hat{\phi}(f_2)$ are i.i.d. for $\forall f_1 \neq f_2$. When there exists motion, $\hat{\phi}(f)$ will converge to a positive constant $\frac{E_d^2(f)}{E_d^2(f)+\sigma^2(f)}$ as $F_s \to \infty$ and $T \to \infty$.

10.3.2 Detection Rule

In order to improve the reliability of WiDetect, the motion statistics obtained from all the available subcarriers can be combined together. In this chapter we define the aggregated motion statistics as the average of all the individual motion statistics, i.e., $\hat{\psi} = \frac{1}{F} \sum_{f \in \mathcal{F}} \hat{\phi}(f)$. We know that when there is no motion, $\hat{\phi}(f)$ converges to an AN distribution, and $\hat{\phi}(f_1)$ and $\hat{\phi}(f_2)$ are i.i.d. for $\forall f_1 \neq f_2$. Therefore, the distribution of

$\hat{\psi}$ can be approximated as $\hat{\psi} \sim \mathcal{AN}(-\frac{1}{T}, \frac{1}{FT})$. Because the variance of $\hat{\psi}$ is inversely proportional to the number of samples T and the number of subcarriers F, increasing T and F will improve the detection performance.

According to the preceding analysis, a simple detection rule is proposed: *WiDetect detects motion only if* $\hat{\psi} \geq \eta$. Given a preset threshold η, the probability of false alarm can be approximated as

$$P(\hat{\psi} \geq \eta) \approx Q\left(\sqrt{FT}\left(\eta + \frac{1}{T}\right)\right), \tag{10.8}$$

where $Q(\cdot)$ denotes the tail probability of the standard normal distribution, i.e., $Q(x) = \frac{1}{2\pi}\int_x^{\infty} \exp(-\frac{u^2}{2})\,du$.

10.4 Experimental Evaluation

To evaluate the performance of WiDetect, a prototype based on a pair of commercial Wi-Fi devices is built to detect human motion in two different environments as shown in Figure 10.2. The carrier frequency is set to 5.805 GHz, and the channel sampling rate is 30 Hz. Each Wi-Fi device is equipped with three omnidirectional antennas, and each antenna-pair link has a total of 114 subcarriers. To avoid the correlations among adjacent subcarriers, we take one subcarrier from every two adjacent ones and only use 58 subcarriers for each link, considering the fact that the CSI of DC subcarrier is not accessible.

10.4.1 Validation of the Theoretical Analysis

We first verify the theoretical analysis described in Section 10.3. The Tx and Rx of WiDetect are placed in a typical office environment as shown in Figure 10.2(a). One subject first walks around in the conference room for 30 min, and then walks in the area outside the conference room but within the square $ABCD$ for another 30 min, during the entire period of which the CSI data is collected. We also collect a set of 1-hour CSI data when the environment is static.

We calculate the false alarm probability using the experimental CSI data and compare with the theoretical false alarm probability according to 10.8, and the comparison is shown in Figure 10.3(a) for different sample sizes T and varying η. The theoretical curves match well with the experimental ones when η is greater, and the gap at smaller η is due to the correlation among different subcarriers, which we assume not existing in the theoretical analysis. In addition, the ROC curves in Figure 10.3(b) show that the performance of WiDetect improves as T increases.

10.4.2 Coverage Test

In this experiment, to test the coverage of WiDetect, one subject walks in different regions of a single-family house as shown in Figure 10.2(b), and the positions of the

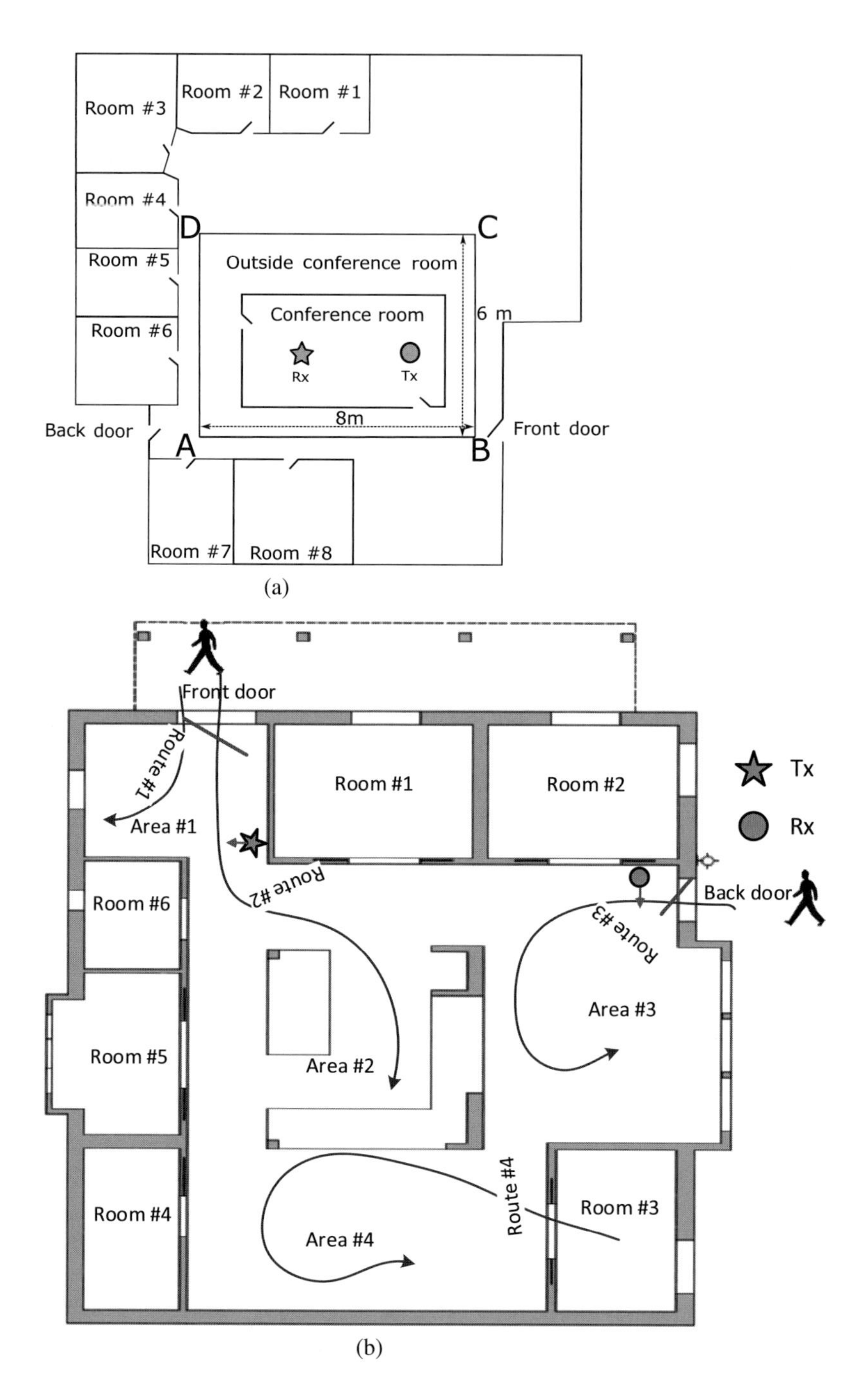

Figure 10.2 Floorplans of two different environments: (a) a typical office and (b) a typical single-family house.

Table 10.2 Detection index (DI) for different regions

Region	R. #1	R. #2	R. #3	R. #4	R. #5
DI	0.52	0.22	0	0	0
Region	R. #6	A. #1	A. #2	A. #3	A. #4
DI	1	0.90	0.93	0.75	0.95

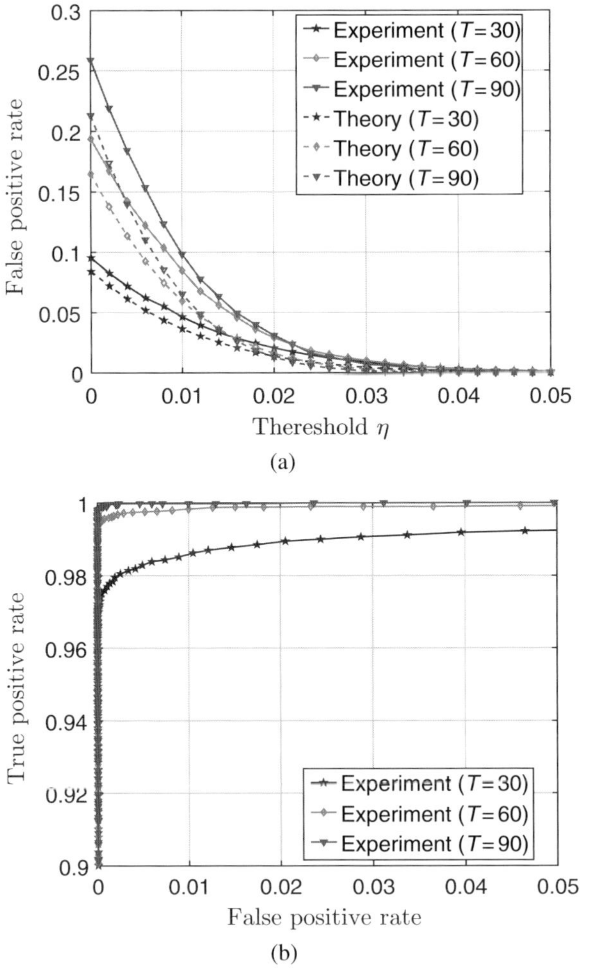

Figure 10.3 The performance curves of WiDetect for one link: (a) false positive rate ($F = 58$) and (b) ROC for varying T.

Tx and Rx are also indicated in the floorplan. We define the detection index (DI) of a region as the ratio between the duration when motion is detected and the total time when motion is present in that region. The results are summarized in Table 10.2. The motion occurring in Rooms #3–#5 cannot be detected because they are far away from the transmission devices. In some regions such as Rooms #1 and #2, motion is not detected all the time. However, as long as there is at least one motion detected along the subject's moving trajectory, the presence of that moving subject can be detected.

10.4.3 Intrusion Test

In this experiment, one subject tries to "break" into the house following four different routes as indicated in Figure 10.2(b), and then leaves the house following the same route. The subject spends about 1 min in the house for each route. The detection index for the four routes are shown in Table 10.3. The results show that the presence of the "intruder" can be detected most of the time for all the routes.

10.4.4 Long-Term Test

To evaluate the false alarm rate, we run WiDetect in the same single-family house for 1 week and compare with the detection that deploys four PIRs in different areas of

Table 10.3 Detection index (DI) for different routes

Route	#1	#2	#3	#4
DI	0.90	0.98	0.83	1

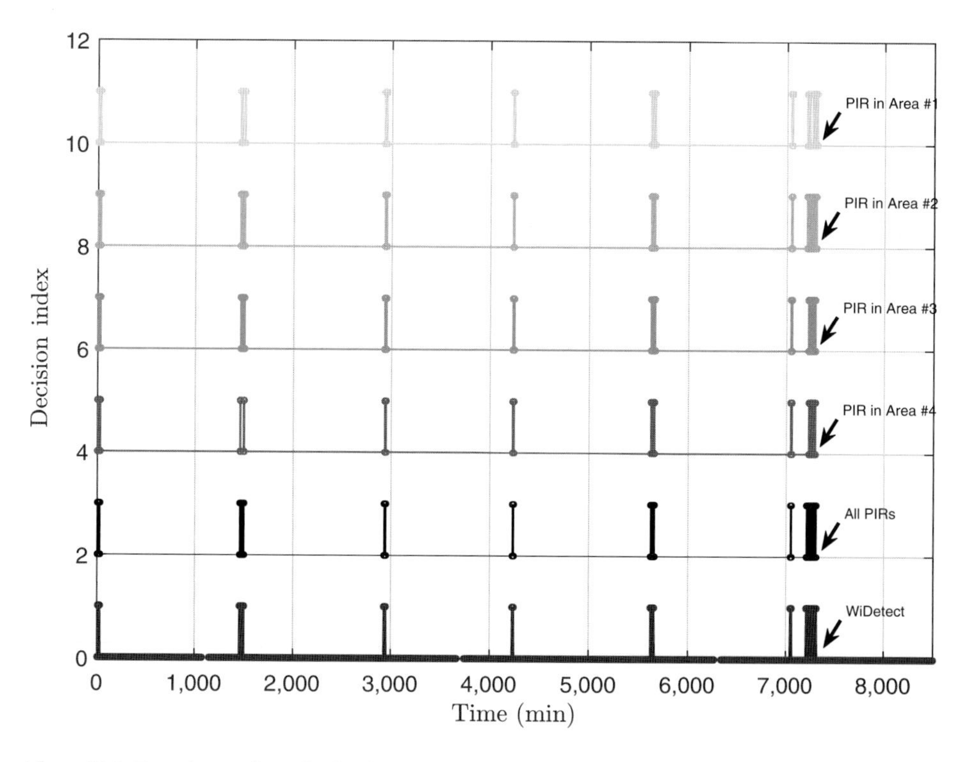

Figure 10.4 Experimental results for long-term test compared with PIRs.

the house. The detection results for both WiDetect and the four PIRs are shown in Figure 10.4, where an even decision index (0, 2, 4, 6, 8, 10) indicates that no motion is detected. The results show that WiDetect can achieve comparable detection performance as the PIRs while having a much larger coverage.

10.5 Summary

In this chapter, we presented WiDetect, a highly accurate and calibration-free motion detection system leveraging CSI of a wireless channel. Extensive experiments show its superiority over existing motion detection approaches. Due to its large coverage, robustness, low cost, and low computational complexity, WiDetect is a very promising candidate for indoor motion detection applications. For related references, interested readers can refer to [15].

References

[1] L. Wang, G. Zhao, L. Cheng, and M. Pietikäinen, *Machine Learning for Vision-Based Motion Analysis: Theory and Techniques*, Berlin: Springer, 2010.

[2] A. E. Kosba, A. Saeed, and M. Youssef, "RASID: A robust WLAN device-free passive motion detection system," in *Proceedings of IEEE International Conferentce on Pervasive Computing and Communications*, pp. 180–189, 2012.

[3] Y. Wang, J. Liu, Y. Chen, M. Gruteser, J. Yang, and H. Liu, "E-eyes: Device-free location-oriented activity identification using fine-grained WiFi signatures," in *Proceedings of the 20th Annual ACM International Conference on Mobile Computing & Networking*. pp. 617–628, ACM, 2014.

[4] J. Xiao, K. Wu, Y. Yi, L. Wang, and L. M. Ni, "Pilot: Passive device-free indoor localization using channel state information," in *Proceedings of the IEEE International Conference on Distributed Computing Systems (ICDCS)*, pp. 236–245, 2013.

[5] W. Wang, A. X. Liu, M. Shahzad, K. Ling, and S. Lu, "Understanding and modeling of WiFi signal based human activity recognition," in *Proceedings of the 21st Annual ACM International Conference on Mobile Computing & Networking*. pp. 65–76, 2015.

[6] K. Qian, C. Wu, Z. Yang, Y. Liu, and K. Jamieson, "WiDar: Decimeter-level passive tracking via velocity monitoring with commodity Wi-Fi," in *Proceedings of the 18th ACM International Symposium on Mobile Ad Hoc Networking and Computing*, p. 6, 2017.

[7] Z. Zhou, Z. Yang, C. Wu, L. Shangguan, and Y. Liu, "Omnidirectional coverage for device-free passive human detection," *IEEE Transactions on Parallel and Distributed Systems*, vol. 25, no. 7, pp. 1819–1829, 2014.

[8] C. Wu, Z. Yang, Z. Zhou, X. Liu, Y. Liu, and J. Cao, "Non-invasive detection of moving and stationary human with WiFi," *IEEE Journal on Selected Areas in Communications*, vol. 33, no. 11, pp. 2329–2342, 2015.

[9] J. Lv, W. Yang, L. Gong, D. Man, and X. Du, "Robust WLAN-based indoor fine-grained intrusion detection," in *Proceedings of the IEEE Global Communications Conference (GLOBECOM)*, pp. 1–6, 2016.

[10] T.-D. Chiueh, P.-Y. Tsai, and I.-W. Lai, *Baseband Receiver Design for Wireless MIMO-OFDM Communications*, Hoboken, NJ: John Wiley & Sons, 2012.

[11] C. Chen, Y. Chen, Y. Han, H. Q. Lai, F. Zhang, and K. J. R. Liu, "Achieving centimeter-accuracy indoor localization on WiFi platforms: A multi-antenna approach," *IEEE Internet of Things Journal*, vol. 4, no. 1, pp. 122–134, Feb. 2017.

[12] S. Sen, B. Radunovic, R. R. Choudhury, and T. Minka, "You are facing the Mona Lisa: Spot localization using PHY layer information," in *Proceedings of the 10th ACM International Conference on Mobile Systems, Applications, and Services*. pp. 183–196, 2012.

[13] D. A. Hill, *Electromagnetic Fields in Cavities: Deterministic and Statistical Theories*, vol. 35, Hoboken, NJ: John Wiley & Sons, 2009.

[14] G. E.P Box, G. M. Jenkins, G. C. Reinsel, and G. M. Ljung, *Time Series Analysis: Forecasting and Control*, Hoboken, NJ: John Wiley & Sons, 2015.

[15] F. Zhang, C. Chen, B. Wang, H.-Q. Lai, Y. Han, and K. J. R. Liu, "WiDetect: A robust and low-complexity wireless motion detector," in *IEEE International Conference on Acoustics, Speech and Signal Processing (ICASSP)*, pp. 6398–6402, 2018.

11 Device-Free Speed Estimation

Due to the severe multipath effect, no satisfactory device-free methods have ever been found for indoor speed estimation problems, especially in non-line-of-sight scenarios, where the direct path between the source and observer is blocked. In this chapter, we present WiSpeed, a universal low-complexity indoor speed estimation system leveraging radio signals, such as commercial Wi-Fi, LTE, 5G, etc., which can work in both device-free and device-based situations. By exploiting the statistical theory of electromagnetic waves, we establish a link between the autocorrelation function of the physical layer channel state information and the speed of a moving object, which lays the foundation of WiSpeed. WiSpeed differs from the other schemes requiring strong line-of-sight conditions between the source and observer in that it embraces the rich-scattering environment typical for indoors to facilitate highly accurate speed estimation. Moreover, as a calibration-free system, WiSpeed saves the users' efforts from large-scale training and fine-tuning of system parameters. In addition, WiSpeed could extract the stride length as well as detect abnormal activities such as falling down, a major threat to seniors that leads to a large number of fatalities every year. Extensive experiments show that WiSpeed achieves a mean absolute percentage error of 4.85% for device-free human walking speed estimation and 4.62% for device-based speed estimation, and a detection rate of 95% without false alarms for fall detection.

11.1 Introduction

As people are spending more and more their time indoors nowadays, understanding their daily indoor activities will become a necessity for future life. Because the speed of the human body is one of the key physical parameters that can characterize the types of human activities, speed estimation of human motions is a critical module in human activity monitoring systems. Compared with traditional wearable sensor-based approaches, device-free speed estimation is more promising due to its better user experience, which can be applied in a wide variety of applications, such as smart homes [1], health care [2], fitness tracking [3], and entertainment.

Nevertheless, indoor device-free speed estimation is very challenging, mainly due to the severe multipath propagations of signals and the blockage between the monitoring devices and the objects under monitoring. Conventional approaches of motion sensing require specialized devices, ranging from RADAR, SONAR, laser, to camera. Among them, the vision-based schemes [4] can only perform motion monitoring in their fields of vision with performance degradation in dim light conditions. Also, they introduce privacy issues. Meanwhile, the speed estimation produced by RADAR or SONAR [5] varies for different moving directions, mainly because of the fact that the speed estimation is derived from the Doppler shift, which is relevant to the moving direction of an object. Also, the multipath propagations of indoor spaces further undermine the efficacy of RADAR and SONAR.

More recently, WiGait [6] and WiDar [7] have been proposed to measure gait velocity and stride length in indoor environments using radio signals. However, WiGait uses specialized hardware to send Frequency Modulated Carrier Wave (FMCW) probing signals, and it requires a bandwidth as large as 1.69 GHz to resolve the multipath components. On the other hand, WiDar can only work well under a strong line-of-sight (LOS) condition and a dense deployment of Wi-Fi devices because its performance relies heavily on the accuracy of ray tracing/geometry techniques.

In this chapter, we present WiSpeed, a robust universal speed estimator for human motions in a rich-scattering indoor environment, which can estimate the speed of a moving object under either the device-free or device-based condition. WiSpeed is actually a fundamental principle that requires no specific hardware, as it can simply utilize only a single pair of commercial off-the-shelf Wi-Fi devices. First, we characterize the impact of motions on the autocorrelation function (ACF) of the received electric field of electromagnetic (EM) waves using the statistical theory of EM waves. However, the received electric field is a vector, and it cannot be easily measured. Therefore, we further derive the relation between the ACF of the power of the received electric field and the speed of motions because the electric field power is directly measurable on commercial Wi-Fi devices [8]. By analyzing different components of the ACF, we find that the first local peak of the ACF differential contains the crucial information of speed of motions, and we present a novel peak identification algorithm to extract the speed. Furthermore, the number of steps and the stride length can be estimated as a by-product of the speed estimation. In addition, fall can be detected from the patterns of the speed estimation.

To assess the performance of WiSpeed, we conduct extensive experiments in two scenarios, namely, human walking monitoring and human fall detection. For human walking monitoring, the accuracy of WiSpeed is evaluated by comparing the estimated walking distances with the ground-truths. Experimental results show that WiSpeed achieves a mean absolute percentage error (MAPE) of 4.85% for the case when the human does not carry the device and a MAPE of 4.62% for the case when the subject carries the device. In addition, WiSpeed can extract the stride lengths and estimates the number of steps from the pattern of the speed estimation under the device-free setting. In terms of human fall detection, WiSpeed is able to differentiate falls from other normal activities, such as sitting down, standing up, picking up items, and walking. The average detection rate is 95% with no false alarms. To the best of our knowledge, WiSpeed is the first device-free/device-based wireless speed estimator for motions that achieves high estimation accuracy, high detection rate, low deployment cost, large coverage, low computational complexity, and privacy preservation at the same time.

Because Wi-Fi infrastructure is readily available for most indoor spaces, WiSpeed is a low-cost solution that can be deployed widely. WiSpeed would enable a large number of important indoor applications such as

(i) *Indoor fitness tracking:* More and more people become aware of their physical conditions and are thus interested in acknowledging their amount of exercise on a daily basis. WiSpeed can assess a person's exercise amount by the estimation of the number of steps through the patterns of the speed estimation. With the assistance of WiSpeed, people can obtain their exercise amount and evaluate their personal fitness conditions without any wearable sensors attached to their bodies.
(ii) *Indoor navigation:* Although outdoor real-time tracking has been successfully solved by GPS, indoor tracking has still left an open problem up to now. Dead reckoning based approach is among the existing popular techniques for indoor navigation, which is based upon measurements of speed and direction of movement to compute the position starting from a reference point. However, the accuracy is mainly limited by the inertial measurement unit (IMU) based moving distance estimation. Because WiSpeed can also measure the speed of a moving Wi-Fi device, the accuracy of distance estimation module in dead reckoning-based systems can be improved dramatically by incorporating WiSpeed.
(iii) *Fall detection:* Real-time speed monitoring for human motions is important to the seniors who live alone in their homes, as the system can detect falls that impose major threats to their lives.
(iv) *Home surveillance:* WiSpeed can play a vital role in the home security system because WiSpeed can distinguish between an intruder and the owner's pet through their different patterns of moving speed and inform the owner as well as law enforcement immediately.

The rest of the chapter is organized as follows. Section 11.2 summarizes the related works about human activity recognition using Wi-Fi signals. Section 11.3 introduces the statistical theory of EM waves in cavities and its extensions for wireless motion sensing. Section 11.4 presents the basic principles of WiSpeed, and Section 11.5 shows the detailed designs of WiSpeed. Experimental evaluation is shown in Section 11.6. Section 11.7 discusses the parameter selections and the computational complexity of WiSpeed.

11.2 Related Works

Existing works on device-free motion sensing techniques using commercial Wi-Fi include gesture recognition [9–13], human activity recognition [14–16], motion tracing [17, 18], passive localization [7, 19], vital signal estimation [20], indoor event detection [21], and so on. These approaches are built upon the phenomenon that human motions inevitably distort the Wi-Fi signal and can be recorded by Wi-Fi receivers for further analysis. In terms of the principles, these works can be divided into two categories: learning based and ray-tracing based. Details of the two categories are elaborated here.

Learning-based: These schemes consist of two phases, namely, an offline phase, and an online phase. During the offline phase, features associated with different human activities are extracted from the Wi-Fi signals and stored in a database; in the online phase, the same set of features are extracted from the instantaneous Wi-Fi signals and compared with the stored features so as to classify the human activities. The features can be obtained either from CSI or the Received Signal Strength Indicator (RSSI), a readily available but low granularity information encapsulating the received power of Wi-Fi signals. For example, E-eyes [14] utilizes histograms of the amplitudes of CSI to recognize daily activities such as washing dishes and brushing teeth. CARM [15] exploits features from the spectral components of CSI dynamics to differentiate human activities. WiGest [9] exploits the features of RSSI variations for gesture recognition.

A major drawback of the learning-based approach lies in that these works utilize the speed of motion to identify different activities, but they only obtain features related to speed instead of directly measuring the speed. One example is the Doppler shift, as it is determined by not only the speed of motion but also the reflection angle from the object as well. These features are thus susceptible to external factors, such as changes in the environment, the heterogeneity in human subjects, the changes of device locations, etc., which might violate their underlying assumption of the reproducibility of the features in the offline and online phases.

Ray-tracing based: Based on the adopted techniques, they can be classified into multipath-avoidance and multipath-attenuation. The multipath-avoidance schemes track the multipath components only reflected by a human body and avoid the other multipath components. Either a high temporal resolution [22] or a “virtual” phased antenna array is used [18], such that the multipath components relevant to motions can be discerned in the time domain or in the spatial domain from those irrelevant to motions. The drawback of these approaches is the requirement of dedicated hardware, such as USRP, WARP [23], etc., to achieve a fine-grained temporal and spatial resolution, which is unavailable on Wi-Fi devices.[1]

In the multipath-attenuation schemes, the impact of multipath components is attenuated by placing the Wi-Fi devices in the close vicinity of the monitored subjects, so that the majority of the multipath components are affected by the subject [7, 10, 17]. The drawback is the requirement of a very strong LOS working condition, which limits their deployment in practice.

WiSpeed differs from the state-of-the-arts in literature in the following ways:

- WiSpeed embraces multipath propagations indoors and can survive and thrive under severe non-line-of-sight (NLOS) conditions, instead of getting rid of the multipath effect [7, 10, 18, 22].

[1] On commercial main-stream 802.11ac Wi-Fi devices, the maximum bandwidth is 160 MHz, much smaller than the 1.69 GHz bandwidth in WiTrack. Meanwhile, commercial Wi-Fi devices with multiple antennas cannot work as a (virtual) phased antenna array out-of-box before carefully tuning the phase differences among the RF front-ends.

- WiSpeed exploits the physical features of EM waves associated with the speed of motion and estimates the speed of motion without detouring. As the physical features hold for different indoor environments and human subjects, WiSpeed can perform well disregarding the changes of environment and subjects, and it is free from any kind of training or calibration.
- WiSpeed enjoys its advantage in a lower computational complexity in comparison with other approaches because costly operations such as principal component analysis (PCA), discrete wavelet transform (DWT), and short-time Fourier transform (STFT) [7, 11, 15] are not required.
- WiSpeed is a low-cost solution because it only deploys a single pair of commercial Wi-Fi devices, while [6, 7, 12, 17, 22] need either specialized hardware or multiple pairs of Wi-Fi devices.

11.3 Statistical Theory of EM Waves for Wireless Motion Sensing

In this section, we first decompose the received electric field at the Rx into different components and then, the statistical behavior of each component is analyzed under certain statistical assumptions.

11.3.1 Decomposition of the Received Electric Field

To provide an insight into the impact of motions on the EM waves, we consider a rich-scattering environment as illustrated in Figure 11.1(a), which is typical for indoor spaces. The scatterers are assumed to be diffusive and can reflect the impinging EM waves toward all directions. A transmitter (Tx) and a receiver (Rx) are deployed in the environment, both equipped with omnidirectional antennas. The Tx emits a continuous EM wave via its antennas, which is received by the Rx. In an indoor environment or a reverberating chamber, the EM waves are usually approximated as plane waves, which can be fully characterized by their electric fields. Let $\vec{E}_{Rx}(t, f)$ denote the electric field received by the receiver at time t, where f is the frequency of the transmitted EM wave. In order to analyze the behavior of the received electric field, we decompose $\vec{E}_{Rx}(t, f)$ into a sum of electric fields contributed by different scatterers based on the superposition principle of electric fields

$$\vec{E}_{Rx}(t, f) = \sum_{i \in \Omega_s(t)} \vec{E}_i(t, f) + \sum_{j \in \Omega_d(t)} \vec{E}_j(t, f), \tag{11.1}$$

where $\Omega_s(t)$ and $\Omega_d(t)$ denote the set of static scatterers and dynamic (moving) scatterers, respectively, and $\vec{E}_i(t, f)$ denotes the part of the received electric field scattered by the ith scatterer. The intuition behind the decomposition is that each scatterer can be treated as a "virtual antenna" diffusing the received EM waves in all directions, and then these EM waves add up together at the receive antenna after bouncing off the walls, ceilings, windows, etc. of the building. When the transmit antenna is static, it can be considered to be a "special" static scatterer, i.e., $Tx \in \Omega_s(t)$; when it is moving, it can be classified in the set of dynamic scatterers, i.e., $Tx \in \Omega_d(t)$. The power of $\vec{E}_{Tx}(t,f)$ dominates that of electric fields scattered by scatterers.

Within a sufficiently short period, it is reasonable to assume that both the sets $\Omega_s(t)$, $\Omega_d(t)$ and the electric fields $\vec{E}_i(t,f)$, $i \in \Omega_s(t)$ change slowly in time. Then, we have the following approximation:

$$\vec{E}_{Rx}(t,f) \approx \vec{E}_s(f) + \sum_{j \in \Omega_d} \vec{E}_j(t,f), \tag{11.2}$$

where $\vec{E}_s(f) \approx \sum_{i \in \Omega_s(t)} \vec{E}_i(t,f)$.

11.3.2 Statistical Behaviors of the Received Electric Field

As is known from the channel reciprocity, EM waves traveling in both directions will undergo the same physical perturbations (i.e., reflection, refraction, diffraction, etc.). Therefore, if the receiver were transmitting EM waves, all the scatterers would receive the same electric fields as they contribute to $\vec{E}_{Rx}(t,f)$, as shown in Figure 11.1(b). Therefore, in order to understand the properties of $\vec{E}_{Rx}(t,f)$, we only need to analyze its individual components $\vec{E}_i(t,f)$, which is equal to the received electric field by the ith scatterer as if the Rx were transmitting. Then, $\vec{E}_i(t,f)$ can be interpreted as an integral of plane waves over all direction angles, as shown in Figure 11.2. For each incoming plane wave with direction angle $\Theta = (\alpha,\beta)$, where α and β denote the elevation and azimuth angles, respectively, let $\vec{k}$ denote its vector wavenumber and let $\vec{F}(\Theta)$ stand for its angular spectrum, which characterizes the electric field of the wave. The vector wavenumber $\vec{k}$ is given by $-k(\hat{x}\sin(\alpha)\cos(\beta) + \hat{y}\sin(\alpha)\sin(\beta) + \hat{z}\cos(\alpha))$ where the corresponding free-space wavenumber is $k = \frac{2\pi f}{c}$ and c is the speed of light. The angular spectrum $\vec{F}(\Theta)$ can be written as $\vec{F}(\Theta) = F_\alpha(\Theta)\hat{\alpha} + F_\beta(\Theta)\hat{\beta}$, where $F_\alpha(\Theta)$, $F_\beta(\Theta)$ are complex numbers, and $\hat{\alpha}$, $\hat{\beta}$ are unit vectors that are orthogonal to each other and to $\vec{k}$. If the speed of the ith scatterer is v_i, then $\vec{E}_i(t,f)$ can be represented as

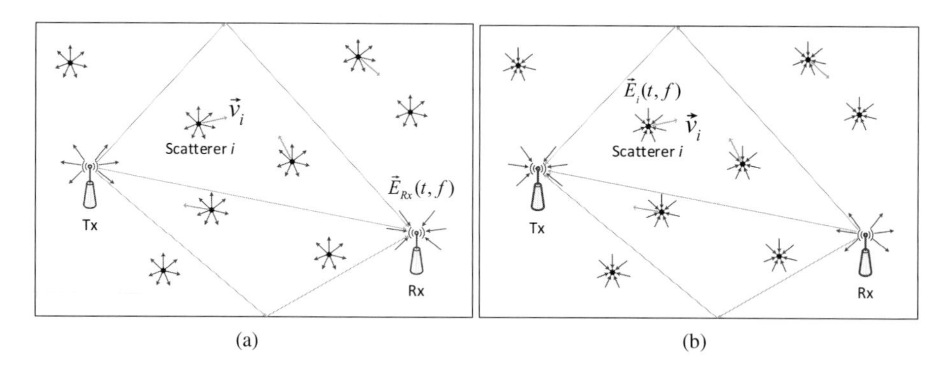

Figure 11.1 Illustration of wave propagation : (a) propagation of radio signals in rich scattering environment and (b) Understanding $\vec{E}_i(t,f)$, $i \in \Omega_d(t)$ using channel reciprocity.

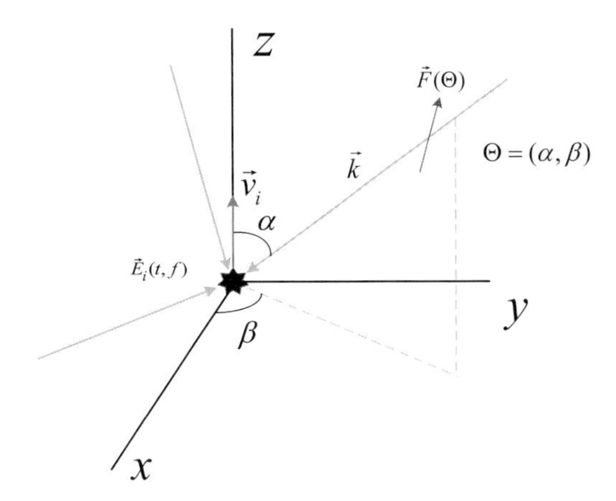

Figure 11.2 Plane wave component $\vec{F}(\Theta)$ of the electric field with vector wavenumber $\vec{k}$.

$$\vec{E}_i(t, f) = \int_0^{2\pi} \int_0^{\pi} \vec{F}(\Theta) \exp(-j\vec{k}\cdot \vec{v}_i t) \sin(\alpha)\, \mathrm{d}\alpha\, \mathrm{d}\beta, \tag{11.3}$$

where the z-axis is aligned with the moving direction of scatterer i, as illustrated in Figure 11.2, and time dependence $\exp(-j2\pi f t)$ is suppressed because it does not affect any results that will be derived later. The angular spectrum $\vec{F}(\Theta)$ could be either deterministic or random. The electric field in (11.3) satisfies Maxwell's equations because each plane-wave component satisfies Maxwell's equations [24].

Radio propagation in a building interior is in general very difficult to analyze because that the EM waves can be absorbed and scattered by walls, doors, windows, moving objects, etc. However, buildings and rooms can be viewed as reverberation cavities in that they exhibit internal multipath propagations. Hence, we refer to a statistical modeling instead of a deterministic one and apply the statistical theory of EM fields developed for reverberation cavities to analyze the statistical properties of $\vec{E}_i(t, f)$. We assume that $\vec{E}_i(t, f)$ is a superposition of a large number of plane waves with uniformly distributed arrival directions, polarizations, and phases, which can well capture the properties of the wave functions of reverberation cavities [24]. Therefore, we take $\vec{F}(\Theta)$ to be a random variable, and the corresponding statistical assumptions on $\vec{F}(\Theta)$ are summarized as follows:

Assumption 11.3.1 *For $\forall\Theta$, $F_\alpha(\Theta)$ and $F_\beta(\Theta)$ are both circularly symmetric Gaussian random variables [25] with the same variance, and they are statistically independent.*

Assumption 11.3.2 *For each dynamic scatterer, the angular spectrum components arriving from different directions are uncorrelated.*

Assumption 11.3.3 *For any two dynamic scatterers i_1, $i_2 \in \Omega_d$, $\vec{E}_{i_1}(t_1, f)$ and $\vec{E}_{i_2}(t_2, f)$ are uncorrelated, for $\forall t_1$, t_2.*

Assumption 11.3.1 is due to the fact that the angular spectrum is a result of many rays or bounces with random phases, and thus it can be assumed that each orthogonal component of $\vec{F}(\Theta)$ tends to be Gaussian under the Central Limit Theorem. Assumption 11.3.2 is because that the angular spectrum components corresponding to different directions have taken very different multiple scattering paths, and they can thus be assumed to be uncorrelated with each other. Assumption 11.3.3 results from the fact that the channel responses of two locations separated by at least half wavelength are statistically uncorrelated [26, 27], and the electric fields contributed by different scatterers can thus be assumed to be uncorrelated.

Under these three assumptions, $\vec{E}_i(t,f), \forall i \in \Omega_d$ can be approximated as a stationary process in time. Define the temporal ACF of an electric field $\vec{E}(t,f)$ as

$$\rho_{\vec{E}}(\tau,f) = \frac{\langle \vec{E}(0,f), \vec{E}(\tau,f)\rangle}{\sqrt{\langle|\vec{E}(0,f)|^2\rangle\langle|\vec{E}(\tau,f)|^2\rangle}}, \tag{11.4}$$

where τ is the time lag, $\langle\ \rangle$ stands for the ensemble average over all realizations, $\langle\vec{X},\vec{Y}\rangle$ denotes the inner product of $\vec{X}$ and $\vec{Y}$, i.e., $\langle\vec{X},\vec{Y}\rangle \triangleq \langle\vec{X}\cdot\vec{Y}^*\rangle$ and * is the operator of complex conjugate and $\cdot$ is dot product, and $|\vec{E}(t,f)|^2$ denotes the square of the absolute value of the electric field. Because $\vec{E}(t,f)$ is assumed to be a stationary process, the denominator of (11.4) degenerates to $E^2(f)$, which stands for the power of the electric field, i.e., $E^2(f) = \langle|\vec{E}(t,f)|^2\rangle, \forall t$, and the ACF is merely a normalized counterpart of the auto-covariance function.

For the ith scatterer with moving velocity $\vec{v}_i$, $\langle\vec{E}_i(0,f)\cdot\vec{E}_i^*(\tau,f)\rangle$ can be derived as [24]

$$\begin{aligned}\langle\vec{E}_i(0,f)\cdot\vec{E}_i^*(\tau,f)\rangle &= \int_{4\pi}\int_{4\pi}\langle\vec{F}(\Theta_1)\cdot\vec{F}(\Theta_2)\rangle\exp(j\vec{k}_2\cdot\vec{v}_i\tau)\,\mathrm{d}\Theta_1\,\mathrm{d}\Theta_2\\ &= \frac{E_i^2(f)}{4\pi}\int_{4\pi}\exp(jkv_i\tau\cos(\alpha_2))\mathrm{d}\Theta_2\\ &= E_i^2(f)\frac{\sin(kv_i\tau)}{kv_i\tau},\end{aligned} \tag{11.5}$$

where we define $\int_{4\pi} \triangleq \int_0^{2\pi}\int_0^{\pi}$ and $\mathrm{d}\Theta \triangleq \sin(\alpha)\,\mathrm{d}\alpha\,\mathrm{d}\beta$, and $E_i^2(f)$ is the power of $\vec{E}_i(t,f)$. With Assumption 11.3.3, the auto-covariance function of $\vec{E}_{Rx}(t,f)$ can be written as

$$\left\langle(\vec{E}_{Rx}(0,f)-\vec{E}_s(f))\cdot(\vec{E}_{Rx}^*(\tau,f)-\vec{E}_s^*(f))\right\rangle = \sum_{i\in\Omega_d}E_i^2(f)\frac{\sin(kv_i\tau)}{kv_i\tau}, \tag{11.6}$$

and the corresponding ACF can thus be derived as

$$\rho_{\vec{E}_{Rx}}(\tau,f) = \frac{1}{\sum_{j\in\Omega_d}E_j^2(f)}\sum_{i\in\Omega_d}E_i^2(f)\frac{\sin(kv_i\tau)}{kv_i\tau}. \tag{11.7}$$

From (11.7), the ACF of $\vec{E}_{Rx}$ is actually a combination of the ACF of each moving scatterer weighted by their radiation power, and the moving direction of each dynamic scatterer does not play a role in the ACF. The importance of (11.7) lies in the fact that the speed information of the dynamic scatterers is actually embedded in the ACF of the received electric field.

11.4 Theoretical Foundation of WiSpeed

In Section 11.3, we have derived the ACF of the received electric field at the Rx, which depends on the speed of the dynamic scatterers. If all or most of the dynamic scatterers move at the same speed v, then the right-hand side of (11.7) would degenerate to $\rho_{\vec{E}_{Rx}}(\tau, f) = \frac{\sin(kv\tau)}{kv\tau}$, and it becomes very simple to estimate the common speed from the ACF. However, it is not easy to directly measure the electric field at the Rx and analyze its ACF. Instead, the power of the electric field can be viewed as equivalent to the power of the channel response that can be measured by commercial Wi-Fi devices. In this section, we will discuss the principle of WiSpeed that utilizes the ACF of the CSI power response for speed estimation.

Without loss of generality, we use the channel response of OFDM-based Wi-Fi systems as an example. Let $X(t, f)$ and $Y(t, f)$ be the transmitted and received signals over a subcarrier with frequency f at time t. Then, the least-square estimator of the CSI for the subcarrier with frequency f measured at time t is $H(t, f) = \frac{Y(t,f)}{X(t,f)}$ [28]. In practice, the obtained estimation of the CSI suffers from synchronization errors, which mainly consists of channel frequency offset (CFO), sampling frequency offset (SFO), and symbol timing offset (STO) [26]. Although the Wi-Fi receivers perform timing and frequency synchronization, the residual of these errors cannot be neglected. However, the impact of synchronization errors on the amplitude of CSI is insignificant, and thus WiSpeed only exploits the amplitude information of the measured CSI.

We define the power response $G(t, f)$ as the square of the magnitude of CSI, which takes the form

$$G(t, f) \triangleq |H(t, f)|^2 = \|\vec{E}_{Rx}(t, f)\|^2 + \varepsilon(t, f), \tag{11.8}$$

where $\|\vec{E}\|^2$ denotes the total power of $\vec{E}$, and $\varepsilon(t, f)$ is assumed to be an additive noise due to the imperfect measurement of CSI.

The noise $\varepsilon(t, f)$ can be assumed to follow a normal distribution. To prove this, we collect a set of one-hour CSI data in a static indoor environment with the channel sampling rate $F_s = 30\,\text{Hz}$. The Q-Q plot of the normalized $G(t, f)$ and standard normal distribution for a given subcarrier is shown in Figure 11.3(a), which shows that the distribution of the noise is very close to a normal distribution. To verify the whiteness of the noise, we also study the ACF of $G(t, f)$ that can be defined as [29] $\rho_G(\tau, f) = \frac{\gamma_G(\tau,f)}{\gamma_G(0,f)}$, where $\gamma_G(\tau, f)$ denotes the auto-covariance function, i.e., $\gamma_G(\tau, f) \triangleq \mathrm{cov}(G(t, f), G(t - \tau, f))$. In practice, sample auto-covariance function $\hat{\gamma}_G(\tau, f)$ is used instead. If $\varepsilon(t, f)$ is white noise, the sample ACF $\hat{\rho}_G(\tau, f)$, for $\forall \tau \neq 0$, can be approximated by a normal random variable with zero mean and standard deviation $\sigma_{\hat{\rho_G}(\tau,f)} = \frac{1}{\sqrt{T}}$. Figure 11.3(b) shows the sample ACF of $G(t, f)$ when 2,000 samples on the first subcarrier are used. As we can see from the figure, all the taps of the sample ACF are within the interval of $\pm 2\sigma_{\hat{\rho_G}(\tau,f)}$, and thus, it can be assumed that $\varepsilon(t, f)$ is an additive white Gaussian noise, i.e., $\varepsilon(t, f) \sim \mathcal{N}(0, \sigma^2(f))$.

In the previous analysis in Section 11.3, we assume that the Tx transmits continuous EM waves, but in practice the transmission time is limited. For example, in IEEE

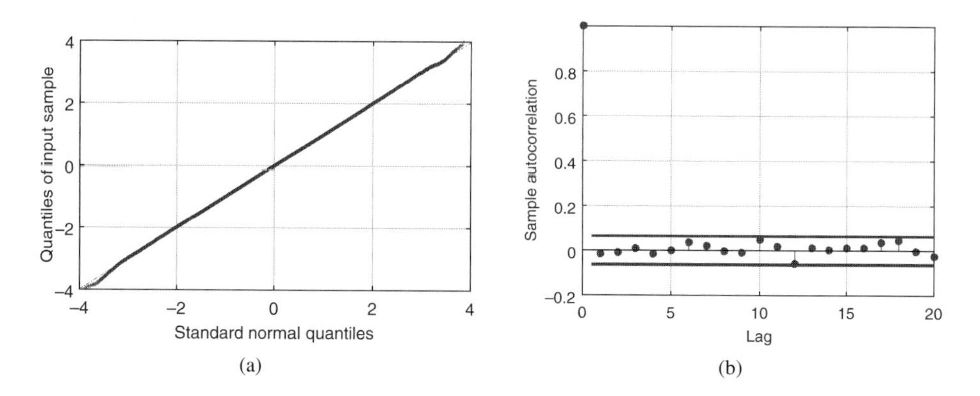

Figure 11.3 (a) The Q-Q plot and (b) sample ACF of a typical CSI power response.

802.11n Wi-Fi systems operated in 5 GHz frequency band with 40 MHz bandwidth channels, a standard Wi-Fi symbol is 4 μs, composed of a 3.2 μs useful symbol duration and a 0.8 μs guard interval. According to [30], for most office buildings, the delay spread is within the range of 40 to 70 ns, which is much smaller than the duration of a standard Wi-Fi symbol. Therefore, we can assume continuous waves are transmitted in Wi-Fi systems.

Based on the above assumptions and (11.2), (11.8) can be approximated as

$$
\begin{aligned}
G(t,f) &\approx \|\vec{E}_s(f) + \sum_{i\in\Omega_d} \vec{E}_i(t,f)\|^2 + \varepsilon(t,f) \\
&= \left\| \sum_{u\in\{x,y,z\}} \left(E_{su}(f)\hat{u} + \sum_{i\in\Omega_d} E_{iu}(t,f)\hat{u} \right) \right\|^2 + \varepsilon(t,f) \\
&= \sum_{u\in\{x,y,z\}} \left| E_{su}(f) + \sum_{i\in\Omega_d} E_{iu}(t,f) \right|^2 + \varepsilon(t,f) \\
&= \sum_{u\in\{x,y,z\}} \left(|E_{su}(f)|^2 + 2\mathrm{Re}\left\{ E^*_{su}(f) \sum_{i\in\Omega_d} E_{iu}(t,f) \right\} \right. \\
&\quad \left. + \left| \sum_{i\in\Omega_d} E_{iu}(t,f) \right|^2 \right), + \varepsilon(t,f),
\end{aligned} \tag{11.9}
$$

where $\hat{x}$, $\hat{y}$ and $\hat{z}$ are unit vectors orthogonal to each other as shown in Figure 11.2, $\mathrm{Re}\{\cdot\}$ denotes the operation of taking the real part of a complex number, and E_{iu} denotes the component of $\vec{E}_i$ in the u-axis direction, for $\forall u \in \{x, y, z\}$. Then, the auto-covariance function of $G(t, f)$ can be derived as

$$
\begin{aligned}
\gamma_G(\tau,f) &= \mathrm{cov}\left(G(t,f), G(t-\tau,f)\right) \\
&\approx \sum_{u\in\{x,y,z\}} \left(2|E_{su}(f)|^2 \sum_{i\in\Omega_d} \mathrm{cov}(E_{iu}(t,f), E_{iu}(t-\tau,f)) \right.
\end{aligned}
$$

$$+ \sum_{\substack{i_1, i_2 \in \Omega_d \\ i_1 \geq i_2}} \operatorname{cov}(E_{i_1 u}(t, f), E_{i_1 u}(t-\tau, f)) \cdot$$

$$\left. \operatorname{cov}(E_{i_2 u}(t, f), E_{i_2 u}(t-\tau, f)) \right) + \delta(\tau)\sigma^2(f), \tag{11.10}$$

where Assumptions 11.3.1–11.3.3 and (11.3) are applied to simplify the expression, and the detailed derivations can be found in the appendix.

According to the relation between the autocovariance and autocorrelation, $\gamma_G(\tau, f)$ can be rewritten in the forms of ACFs of each scatterer as

$$\gamma_G(\tau, f) \approx \sum_{u \in \{x,y,z\}} \left(\sum_{i \in \Omega_d} \frac{2|E_{su}(f)|^2 E_i^2(f)}{3} \rho_{E_{iu}}(\tau, f) \right.$$

$$\left. + \sum_{\substack{i_1, i_2 \in \Omega_d \\ i_1 \geq i_2}} \frac{E_{i_1}^2(f) E_{i_2}^2(f)}{9} \rho_{E_{i_1 u}}(\tau, f) \rho_{E_{i_2 u}}(\tau, f) \right) + \delta(\tau)\sigma^2(f), \tag{11.11}$$

where the right-hand side is obtained by using the relation $E_{iu}^2(f) = \frac{E_i^2(f)}{3}$, $\forall u \in \{x, y, z\}$, $\forall i \in \Omega_d$ [24]. The corresponding ACF $\rho_G(\tau, f)$ of $G(t, f)$ is thus obtained by $\rho_G(\tau, f) = \frac{\gamma_G(\tau, f)}{\gamma_G(0, f)}$, where $\gamma_G(\tau, 0)$ can be obtained by plugging $\rho_{E_{iu}}(0, f) = 1$ into (11.11). When the moving directions of all the dynamic scatterers are approximately the same, then we can choose the z-axis aligned with the common moving direction. Then, the closed forms of $\rho_{E_{iu}}(\tau, f)$, $\forall u \in \{x, y, z\}$, are derived under Assumptions 11.3.1–11.3.2 [24], i.e., for $\forall i \in \Omega_d$,

$$\rho_{E_{ix}}(\tau, f) = \rho_{E_{iy}}(\tau, f) = \frac{3}{2}\left[\frac{\sin(kv_i\tau)}{kv_i\tau} - \frac{1}{(kv_i\tau)^2} \left(\frac{\sin(kv_i\tau)}{kv_i\tau} - \cos(kv_i\tau) \right) \right], \tag{11.12}$$

$$\rho_{E_{iz}}(\tau, f) = \frac{3}{(kv_i\tau)^2} \left[\frac{\sin(kv_i\tau)}{kv_i\tau} - \cos(kv_i\tau) \right]. \tag{11.13}$$

The theoretical spatial ACFs are shown in Figure 11.4(a), where $d \triangleq v_i\tau$. As we can see from Figure 11.4(a), the magnitudes of all the ACFs decay with oscillations as the distance d increases.

For a Wi-Fi system with a bandwidth of 40 MHz and a carrier frequency of 5.805 GHz, the difference in the wavenumber k of each subcarrier can be neglected, e.g., $k_{\max} = 122.00$ and $k_{\min} = 121.16$. Then, we can assume $\rho(\tau, f) \approx \rho(\tau)$, $\forall f$. Thus, we can improve the sample ACF by averaging across all subcarriers, i.e., $\hat{\rho}_G(\tau) \triangleq \frac{1}{F}\sum_{f \in \mathcal{F}} \hat{\rho}_G(\tau, f)$, where $\mathcal{F}$ denotes the set of all the available subcarriers, and F is the total number of subcarriers. When all the dynamic scatterers have the same speed, i.e., $v_i = v$ for $\forall i \in \Omega_d$, which is the case for monitoring the motion

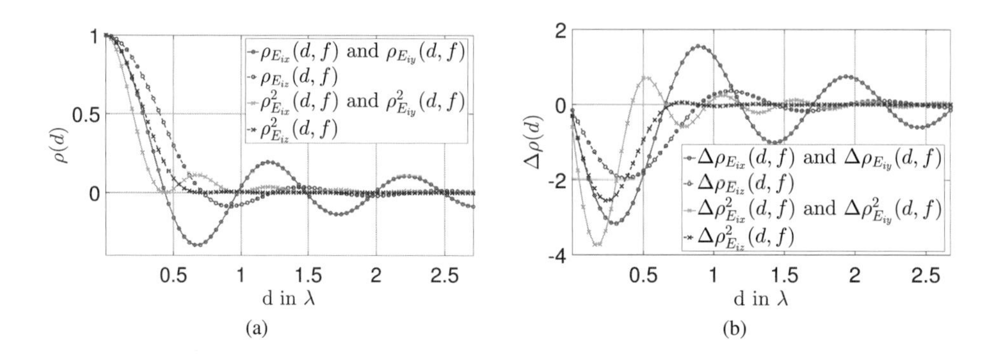

Figure 11.4 (a) Theoretical spatial ACFs and (b) Diff. of spatial ACFs.

for a single human subject, by defining the substitutions $E_{su}^2 \triangleq \frac{2}{F}\sum_{f\in\mathcal{F}}|E_{su}(f)|^2$, $E_d^2 \triangleq \frac{1}{3F}\sum_{i\in\Omega_d}\sum_{f\in\mathcal{F}}E_i^2(f)$, $\hat{\rho}_G(\tau)$ can be further approximated as (for $\tau \neq 0$)

$$\hat{\rho}_G(\tau) \approx C \sum_{u\in\{x,y,z\}} \left(E_d^2 \hat{\rho}^2_{E_{iu}}(\tau) + E_{su}^2 \hat{\rho}_{E_{iu}}(\tau) \right), \tag{11.14}$$

where C is a scaling factor, and the variance of each subcarrier is assumed to be close to each other.

From (11.14), we observe that $\rho_G(\tau)$ is a weighted combination of $\rho_{E_{iu}}(\tau)$ and $\rho^2_{E_{iu}}(\tau)$, $\forall u \in \{x, y, z\}$. The left-hand side of (11.14) can be estimated from CSI, and the speed is embedded in each term on the right-hand side. If we can separate one term from the others on the right-hand side of (11.14), then the speed can be estimated.

Taking the differential of all the theoretical spatial ACFs as shown in Figure 11.4(b) where we use the notation $\Delta\rho(\tau)$ to denote $\frac{\mathrm{d}\rho(\tau)}{\mathrm{d}\tau}$, we find that although the ACFs of different components of the received EM waves are superimposed, the first local peak of $\Delta\rho^2_{E_{iu}}(\tau)$, $\forall u \in \{x, y\}$, happens to be the first local peak of $\Delta\rho_G(\tau)$ as well. Therefore, the component $\rho^2_{E_{iu}}(\tau)$ can be recognized from $\rho_G(\tau)$, and the speed information can thus be obtained by localizing the first local peak of $\Delta\hat{\rho}_G(\tau)$, which is the most important feature that WiSpeed extracts from the noisy CSI measurements.

To verify (11.14), we build a prototype of WiSpeed with commercial Wi-Fi devices. The configurations of the prototype are summarized as follows: both Wi-Fi devices operate on WLAN channel 161 with a center frequency of $f_c = 5.805$ GHz, and the bandwidth is 40 MHz; the Tx is equipped with a commercial Wi-Fi chip and two omnidirectional antennas, while the Rx is equipped with three omnidirectional antennas and uses Intel Ultimate N Wi-Fi Link 5300 with modified firmware and driver [8]. The Tx sends sounding frames with a channel sampling rate F_s of 1500 Hz, and CSI is obtained at the Rx. The transmission power is configured as 20 dBm.

All experiments in this chapter are conducted in a typical indoor office environment as shown in Figure 11.5. In each experiment, the LOS path between the Tx and the Rx is blocked by at least one wall, resulting in a severe NLOS condition. More specifically, we investigate two cases:

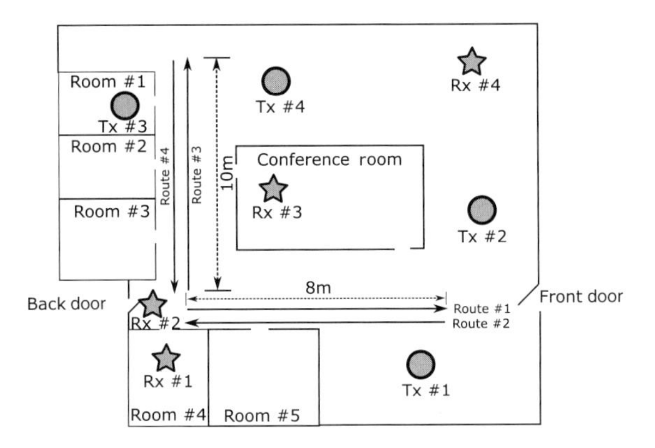

Figure 11.5 Experimental settings in a typical office environment with different Tx/Rx locations and walking routes.

(i) **The Tx is in motion, and the Rx remains static:** The Tx is attached to a cart, and the Rx is placed at Location Rx #1 as shown in Figure 11.5. The cart is pushed forward at an almost constant speed along Route #1 marked in Figure 11.5 from $t = 3.7$ s to $t = 14.3$ s.

(ii) **Both the Tx and the Rx remain static, and a person passes by:** the Tx and Rx are placed at Location Tx #1 and Rx #1, respectively. A person walks along Route #1 at a speed similar to Case (1) from $t = 4.9$ s to $t = 16.2$ s.

Because the theoretical approximations are only valid under the short duration assumption, we set the maximum time lag τ as 0.2 s. In both cases, we compute the sample ACF $\hat{\rho}_G(\tau)$ every 0.05 s.

Figure 11.6 demonstrates the sample ACFs for the two cases. In particular, Figure 11.6(a) visualizes the sample ACF corresponding to a snapshot of Figure 11.6(e) for different subcarriers given a fixed time t with the time lag $\tau \in [0, 0.2s]$, and Figure 11.6(c) shows the average ACF $\hat{\rho}_G(\tau)$, which is much less noisy compared with individual $\hat{\rho}_G(\tau, f)$. In this case, the Tx can be regarded as a moving scatterer with a dominant radiation power compared with the other scatterers, giving rise to the dominance of $E_d^2 \rho^2_{E_{iu}}(\tau)$, $u \in \{x, y, z\}$ over the other components in (11.14). Additionally, $\rho^2_{E_{iz}}(\tau)$ decays much faster than $\rho^2_{E_{ix}}(\tau)$ and $\rho^2_{E_{iy}}(\tau)$, and $\rho^2_{E_{ix}}(\tau) = \rho^2_{E_{iy}}(\tau)$. Thus, a similar pattern between $\hat{\rho}_G(\tau)$ and $\rho^2_{E_{ix}}(\tau)$ ($\rho^2_{E_{iy}}(\tau)$) can be observed with a common and dominant component $\frac{\sin^2(kv\tau)}{(kv\tau)^2}$, where v is the speed of the cart and the person. The experimental result illustrated in Figure 11.6(c) matches well with the theoretical analysis in the sense that only the component $\rho^2_{E_{ix}}(\tau)$ dominates the obtained ACF estimation, and the impacts of the other components can be neglected.

Similarly, for Case (2), Figure 11.6(b) shows the sample ACF $\hat{\rho}_G(\tau, f)$ for different subcarriers, and Figure 11.6(d) shows the average sample ACF $\hat{\rho}_G(\tau)$, which is a snapshot of Figure 11.6(f) given a fixed time t with the time lag $\tau = [0, 0.2s]$. Clearly, the pattern of the component $\rho^2_{E_{iu}}(\tau)$, $u \in \{x, y\}$, in the sample ACF is much less pronounced than Case (1) shown in Figure 11.6(c) and Figure 11.6(e). This can be

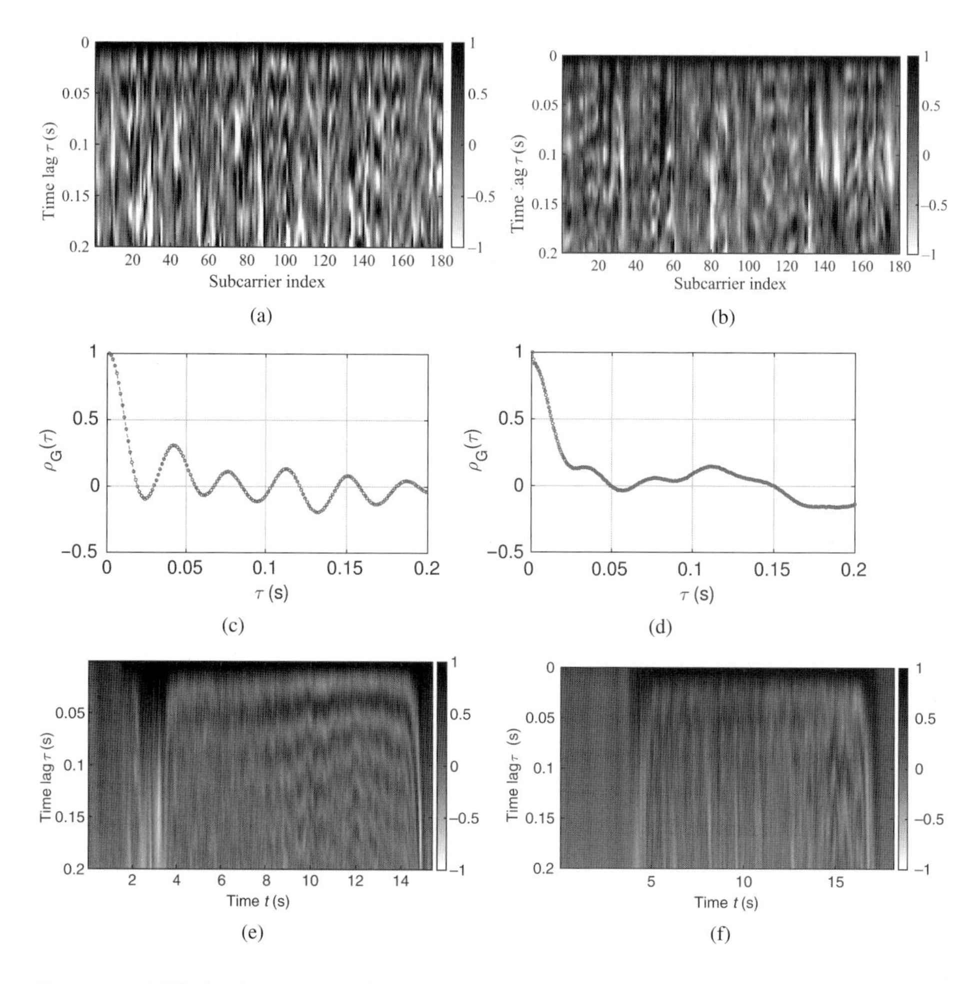

Figure 11.6 ACFs for the two scenarios: (a) ACF measured by different subcarriers for a moving Tx, (b) ACF measured by different subcarriers for a walking human, (c) snapshot of ACF for a moving Tx, (d) snapshot of ACF for a walking human, (e) ACF matrix for a moving Tx, and (f) ACF matrix for a walking human.

justified by the fact that the radiation power E_d^2 is much smaller than that in Case (1), as the set of dynamic scatterers only consists of different parts of a human body in mobility. Consequently, the shape of $\hat{\rho}_G(\tau)$ resembles more closely to $\rho_{E_{iu}}(\tau), \forall u \in \{x, y, z\}$ with a dominant component $\frac{\sin(kv\tau)}{kv\tau}$. Note that the component $\frac{\sin(kv\tau)}{kv\tau}$ oscillates two times slower than the component $\frac{\sin^2(kv\tau)}{(kv\tau)^2}$ does. From Figure 11.6(d), we can observe that the obtained ACF is a result of a weighted sum of these two components. We also observe that the slow-varying trend of the ACF follows the shape of the component $\frac{\sin(kv\tau)}{kv\tau}$ and the component $\frac{\sin^2(kv\tau)}{(kv\tau)^2}$ is only embedded in the trend, the weight of $\frac{\sin(kv\tau)}{kv\tau}$ should be larger than that of $\frac{\sin^2(kv\tau)}{(kv\tau)^2}$. Note that the embedded component $\frac{\sin^2(kv\tau)}{(kv\tau)^2}$ has a similar pattern compared with Case (1) because the moving speeds in the two experiments are similar to each other.

11.5 Key Components of WiSpeed

Based on the theoretical results derived in Section 11.4, we present WiSpeed, which integrates three modules: moving speed estimator, acceleration estimator, and gait cycle estimator. The moving speed estimator is the core module of WiSpeed, while the other two extract useful features from the moving speed estimator to detect falling down and to estimate the gait cycle of a walking person.

11.5.1 Moving Speed Estimator

WiSpeed estimates the moving speed of the subject by calculating the sample ACF $\Delta\hat{\rho}_G(\tau)$ from CSI measurements, localizing the first local peak of $\Delta\hat{\rho}_G(\tau)$, and mapping the peak location to the speed estimation. Because in general, the sample ACF $\Delta\hat{\rho}_G(\tau)$ is noisy as can be seen in Figure 11.6(e) and Figure 11.6(f), we develop a novel robust local peak identification algorithm based on the idea of local regression [31] to reliably detect the location of the first local peak of $\Delta\hat{\rho}_G(\tau)$.

For notational convenience, write the discrete signal for local peak detection as $y[n]$, and our goal is to identify the local peaks in $y[n]$. First of all, we apply a moving window with length $2L + 1$ to $y[n]$, where L is chosen to be comparable with the width of the desired local peaks. Then, for each window with its center located at n, we verify if any potential local peak exists within the window by performing a linear regression and a quadratic regression to the data inside the window, separately. Let SSE denote the sum of squared errors for the quadratic regression and SSE_r denote that for the linear regression. If there is no local peak within the given window, the ratio $\alpha[n] \triangleq \frac{(\text{SSE}_r - \text{SSE})/(3-2)}{\text{SSE}/(2\text{L}+1-3)}$ can be interpreted as a measure of the likelihood of the presence of a peak within the window, and it has a central F-distribution with 1 and $2(L - 1)$ degrees of freedom, under certain assumptions [32]. We choose a potential window with the center point n only when $\alpha[n]$ is larger than a preset threshold η, which is determined by the desired probability of finding a false peak, and $\alpha[n]$ should also be larger than its neighborhoods $\alpha[n - L], \ldots, \alpha[n + L]$. When L is small enough and there exists only one local peak within the window, the location of the local peak can be directly obtained from the fitted quadratic curve.

We use a numerical example in the following to verify the effectiveness of the presented local peak identification algorithm. Let $y(t) = \cos(2\pi f_1 t + 0.2\pi) + \cos(2\pi f_2 t + 0.3\pi) + n(t)$, where we set $f_1 = 1\,\text{Hz}$, $f_2 = 2.5\,\text{Hz}$, and $n(t) \sim \mathcal{N}(0, \sigma^2)$ is additive white Gaussian noise with zero mean and variance σ^2. The signal $y(t)$ is sampled at a rate of 100 Hz from time $t = 0\,\text{s}$ to $t = 1\,\text{s}$. When the noise is absent, the true locations of the two local peaks are $t_1 \approx 0.331\,\text{s}$ and $t_2 \approx 0.760\,\text{s}$, and the estimates of our local peak identification algorithm are $\hat{t}_1 \approx 0.327\,\text{s}$ and $\hat{t}_2 \approx 0.763\,\text{s}$, as shown in Figure 11.7(a). When the noise is present and σ is set to 0.2, the estimates are $\hat{t}_1 \approx 0.336\,\text{s}$ and $\hat{t}_2 \approx 0.762\,\text{s}$, as shown in Figure 11.7(b). As we can see from the results, the estimated locations of the local peaks are very close to those of the actual peaks, even when the signal is corrupted with the noise, which shows the effectiveness of the presented local peak identification algorithm.

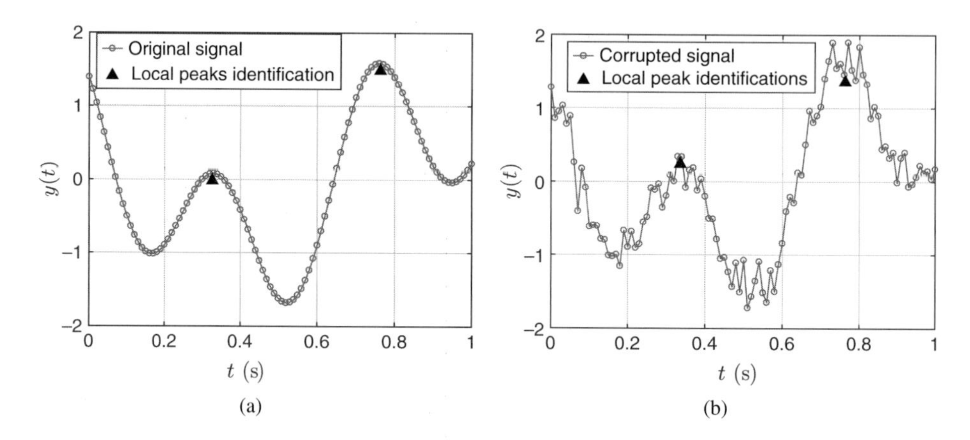

Figure 11.7 An illustration of the peak identification algorithm: (a) original signal and its estimated local peaks and (b) corrupted signal and its estimated local peaks.

Then, the speed of the moving object can be estimated as $\hat{v} = \frac{d_1}{\hat{\tau}}$, where d_1 is the distance between the first local peak of $\Delta\rho^2_{E_{ix}}(d)$ and the origin, and $\hat{\tau}$ is the location of the first local peak of $\Delta\hat{\rho}_G(\tau)$. The distance d_1 can be obtained by solving the equation

$$\frac{\partial^2}{\partial d^2}\rho^2_{E_{ix}}(d, f) = 0, \tag{11.15}$$

where $\rho_{E_{ix}}(d, f)$ denotes the theoretical spatial ACF as shown in Figure 11.4(a). As (11.15) does not have a closed-form solution, we evaluate the second smallest root of (11.15) numerically, which leads to about 0.54λ. A median filter is then applied to the speed estimates to remove the outliers. The presented speed estimator is summarized in Algorithm 3.

Algorithm 3 The presented speed estimator

Input: T consecutive CSI measurements before time t: $H(s, f)$, $s = t - \frac{T-1}{F_s}, \ldots, t - \frac{1}{F_s}, t$, and $f \in \mathcal{F}$;

Output: Speed estimation at t: $\hat{v}(t)$.

1: Calculate the CSI power response: $G(s, f) \leftarrow |H(s, f)|^2$;

2: Calculate the ACF of each subcarrier f: $\hat{\rho}_G(\tau, f) \leftarrow \frac{1}{T}\sum_{s=t-\frac{T-1}{F_s}+\tau}^{t} \left(G(s-\tau, f) - \bar{G}(f)\right)\left(G(s, f) - \bar{G}(f)\right)$, where $\bar{G}(f)$ is the sample mean;

3: Aggregate ACF across all the subcarriers: $\hat{\rho}_G(\tau) \leftarrow \frac{1}{F}\sum_{f\in\mathcal{F}}\hat{\rho}_G(\tau, f)$;

4: Calculate the differential ACF: $\Delta\hat{\rho}_G(\tau) \leftarrow \hat{\rho}_G(\tau) - \hat{\rho}_G(\tau - \frac{1}{F_s})$;

5: Apply the presented peak identification algorithm to estimate the location of the first local peak of $\Delta\hat{\rho}_G(\tau)$: $\hat{\tau}$;

6: Speed estimation at time t: $\hat{v}(t) \leftarrow \frac{0.54\lambda}{\hat{\tau}}$.

11.5.2 Acceleration Estimator

Acceleration can be calculated from $\hat{v}$ obtained in Section 11.5.1. One intuitive method of acceleration estimation is to take the difference of two adjacent speed estimates and then divide the difference of the speeds by the difference of their measurement time. However, this scheme is not robust as it is likely to magnify the estimation noise. Instead, we leverage the fact that the acceleration values can be approximated as a piecewise linear function as long as there are enough speed estimates within a short duration. ℓ_1 trend filter produces trend estimates that are smooth in the sense of being piecewise linear [33] and is well suited to our purpose. Thus, we adopt an ℓ_1 trend filter to extract the piecewise linear trend embedded in the speed estimation and then estimate the accelerations by taking the differential of the smoothed speed estimation.

Mathematically, let $\hat{v}[n]$ denote $\hat{v}(n\Delta T)$, where ΔT is the interval between two estimates, and let $\tilde{v}[n]$ denote the smoothed one. Then, $\tilde{v}[n]$ is obtained by solving the following unconstrained optimization problem:

$$\min_{\tilde{v}[n],\forall n} \sum_{n=1}^{N} \left(\tilde{v}[n]-\hat{v}[n]\right)^2+\lambda\sum_{n=2}^{N-1}\left|\tilde{v}[n-1]-2\tilde{v}[n]+\tilde{v}[n+1]\right|, \tag{11.16}$$

where $\lambda \geq 0$ is the regularization parameter used to control the trade-off between smoothness of $\tilde{v}[n]$ and the size of the residual $|\tilde{v}[n] - \hat{v}[n]|$, and N denotes the size of the speed estimates that need to be smoothed. Then, we obtain the acceleration estimation as $\hat{a}[n] = \frac{(\tilde{v}[n]-\tilde{v}[n-1])}{\Delta T}$. As shown in [33], the complexity of the ℓ_1 filter grows linearly with the length of the data N and can be calculated in real time on most platforms.

11.5.3 Gait Cycle Estimator

When the estimated speed is within a certain range, e.g., from $1\,m/s$ to $2\,m/s$, and the acceleration estimates are small, then WiSpeed starts to estimate the corresponding gait cycle. In fact, the process for walking a single step can be decomposed into three stages: lifting one leg off the ground, using the lifted leg to contact with the ground and pushing the body forward, and keeping still for a short period of time before the next step. The same procedure is repeated until the destination is reached.

In terms of speed, one cycle of walking consists of an acceleration stage followed by a deceleration stage. WiSpeed leverages the periodic pattern of speed changes for gait cycle estimation. More specifically, WiSpeed extracts the local peaks in the speed estimates corresponding to the moments with the largest speeds. To achieve peak localization, we use the persistence-based scheme presented in [34] to formulate multiple pairs of local maximum and local minimum, and the locations of the local maximum are considered as the peak locations. The time interval between every two adjacent peaks is computed as a gait cycle. Meanwhile, the moving distance between every two adjacent peaks is calculated as the estimation of the stride length.

11.6 Experimental Results

In this section, we first introduce the indoor environment and system setups of the experiments. Then, the performance of WiSpeed is evaluated in two applications: human walking monitoring and human fall detection.

11.6.1 Environment

We conducted extensive experiments in a typical office environment, with the floorplan shown in Figure 11.5. The indoor space is occupied by desks, computers, shelves, chairs, and household appliances. The same Wi-Fi devices as introduced in Section 11.4 are used during the experiments.

11.6.2 Experimental Settings

Two sets of experiments were performed. In the first set of experiments, we studied the performance of WiSpeed in estimating the human walking speed. For device-free scenarios, it shows that the number of steps and stride length can also be estimated besides the walking speed. Estimation accuracy is used as the metric that compares the estimated walking distances with the ground-truth distances because measuring walking distance is much easier and more accurate than measuring the speed directly. Different routes and locations of the devices are tested and the details of experiment setup are summarized in Tables 11.1 and 11.2. In the second set of experiments, we investigated the performance of WiSpeed as a human activity monitoring scheme. Two participants were asked to perform different activities, including standing up, sitting down, picking up things from the ground, walking, and falling down.

11.6.3 Human Walking Monitoring

Figure 11.8 visualizes one of the experimental results under Setting #1 of Route #1, i.e., both the Tx and Rx are static, and one experimenter walks along the specified route. Figure 11.8(a) to (c) show three snapshots of estimated ACFs at different time instances marked in Figure 11.8(d). From Figure 11.8, we can conclude that although the ACFs

Table 11.1 Experimental settings for device-free human walking monitoring

Setting \ Config.	Tx loc.	Rx loc.	Route index
Setting #1	Tx #1	Rx #1	Route #1/#2
Setting #2	Tx #1	Rx #2	Route #1/#2
Setting #3	Tx #2	Rx #1	Route #1/#2
Setting #4	Tx #3	Rx #2	Route #3/#4
Setting #5	Tx #4	Rx #2	Route #3/#4
Setting #6	Tx #3	Rx #3	Route #3/#4

Table 11.2 Experimental settings for device-based speed monitoring

Setting \ Config.	Tx loc.	Rx loc.	Route index
Setting #7	moving	Rx #1	Route #1/#2
Setting #8	moving	Rx #4	Route #1/#2
Setting #9	moving	Rx #1	Route #3/#4
Setting #10	moving	Rx #4	Route #3/#4

are very different, the locations of the first local peak of $\Delta\hat{\rho}_G(\tau)$ are highly consistent as long as the ACFs are calculated under similar walking speeds.

Figure 11.8(d) shows the results of walking speed estimation for the experiment, and we can see a very clear pattern of walking due to the acceleration and deceleration. The corresponding stride length estimation is shown in Figure 11.8(e). The estimated walking distance is 8.46 m, and it is within 5.75% of the ground-truth distance of 8 m. On the other hand, the average stride length is 0.7 m and is very close to the average walking stride length of the participants.

Figure 11.9 shows two typical speed estimation results both under Setting #7 of Route #1 where the Tx is attached to a cart and one experimenter pushes the cart along the specified route. The cart moves at different speeds for these two realizations, and Figures 11.9(a) and 11.9(b) show the corresponding speed estimates, respectively. As we can see from the estimated speed patterns, there are no periodic patterns like the device-free walking speed estimates as in Figure 11.8(d). This is because when the Tx is moving, the energy of the EM waves reflected by the human body is dominated by that radiated by the transmit antennas, and WiSpeed can only estimate the speed of moving antennas. The estimated moving distance for the case that Tx moves at a higher speed is 8.26 m and the other one is 8.16 m, where the ground-truth distance is 8 m. Note that the speed estimators, presented in [35] and [36], can also obtain similar results under the same condition; however, they cannot work for device-free scenarios.

Figure 11.10 summarizes the accuracy of the 200 experiments of human walking speed estimation. More specifically, Figure 11.10(a) shows the error distribution for Settings #1 to #6, and Figure 11.10(b) demonstrates the corresponding error distribution for Routes #1 to #4; Figure 11.10(c) shows the error distribution for Settings #7 to #10, and Figure 11.10(d) demonstrates the corresponding error distribution for Routes #1 to #4. The bottom and top error bars stand for the 5% percentiles and 95% percentiles of the estimates, respectively, and the middle of point is the sample mean of the estimates. The ground-truths for Routes #1 to #4 are shown in Figure 11.5. From the results, we find that (i) WiSpeed performs consistently for different Tx/Rx locations, routes, subjects, and walking speeds, indicating the robustness of WiSpeed under various scenarios, and (ii) WiSpeed tends to overestimate the moving distances under device-free settings. This is because we use the route distances as baselines and ignore the displacement of the subjects in the direction of gravity. Because WiSpeed measures the absolute moving

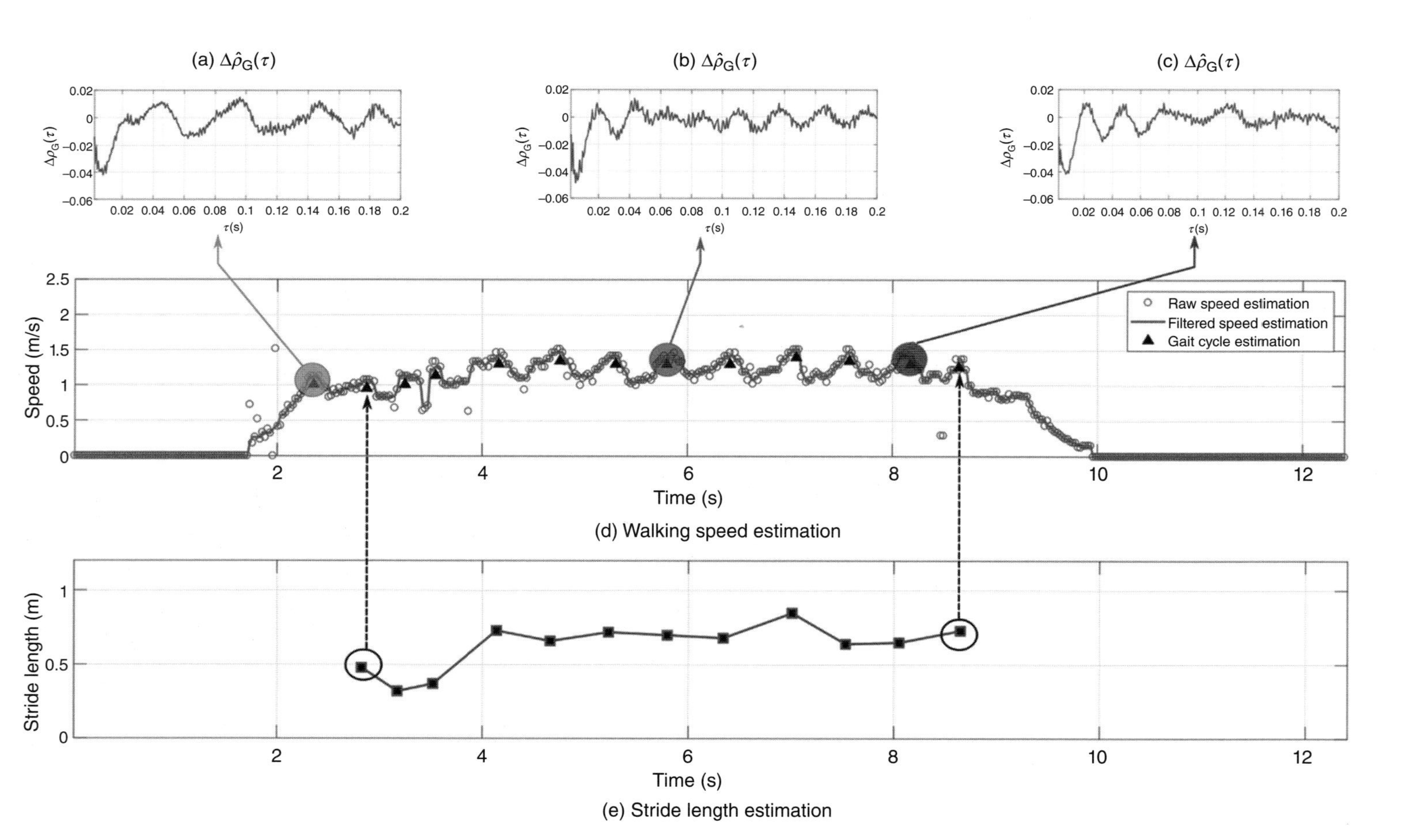

Figure 11.8 Experimental results for human walking monitoring under Setting #1 and Route #1.

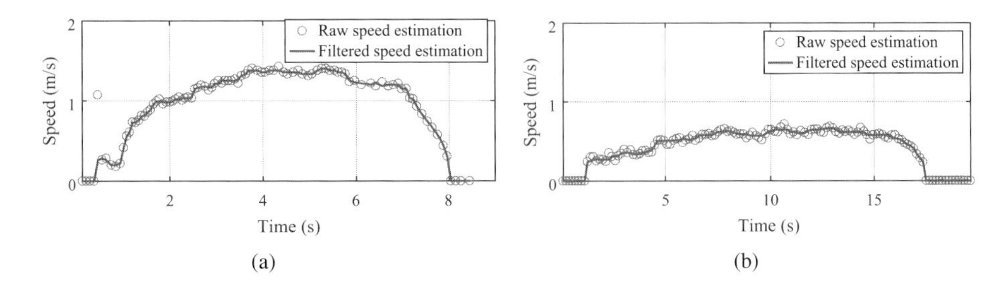

Figure 11.9 Speed estimation for a moving Tx: (a) Tx moves at a higher speed and (b) Tx moves at a lower speed.

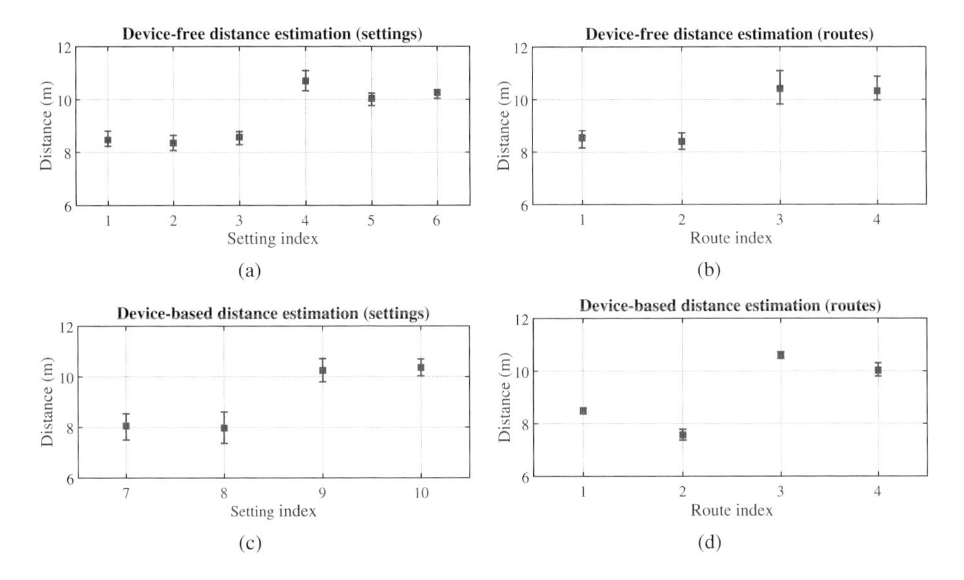

Figure 11.10 Error distribution of distance estimates under different conditions: (a) Settings #1 to #6, (b) Routes #1 to #4, (c) Settings #7 to #10, and Routes #1 to #4.

distance of the subject in the coverage area, the motion in the gravity direction would introduce a bias into the distance estimation.

In summary, WiSpeed achieves a MAPE of 4.85% for device-free human walking speed estimation and 4.62% for device-based speed estimation, which outperforms the existing approaches, even with only a single pair of Wi-Fi devices and in severe NLOS conditions. Note that WiDar [7] can achieve a median speed error of 13%, however, they require multiple pairs of Wi-Fi devices and strong line-of-sight operating conditions, i.e., the object being tracked should be within the fields of vision of both the transmitters and receivers.

11.6.4 Human Fall Detection

In this section, we show that WiSpeed can differentiate falling down from other normal daily activities. We collect a total of five sets of data: (i) falling to the ground, (ii)

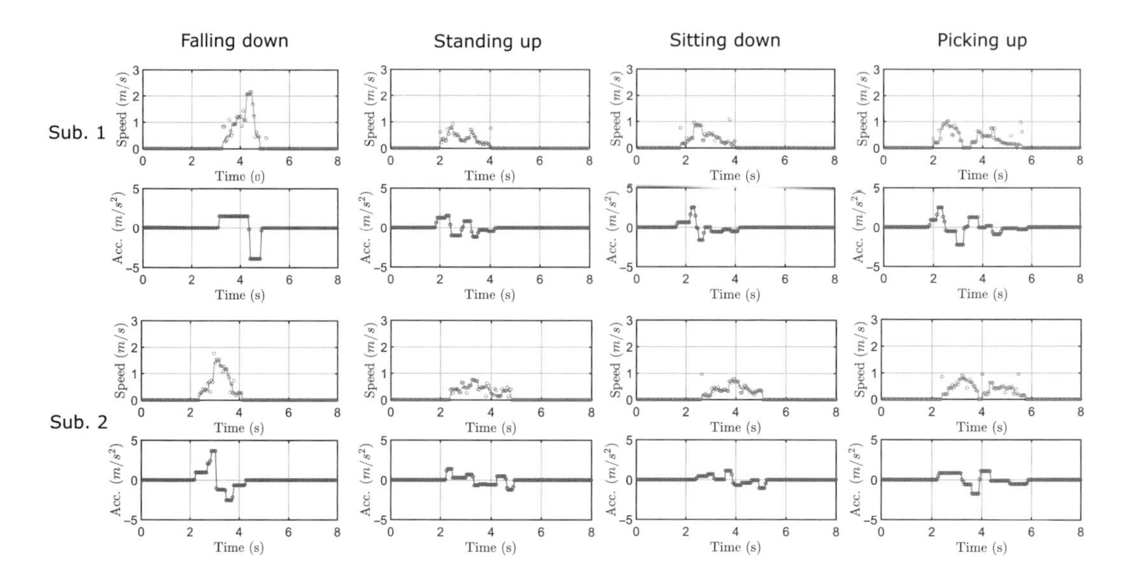

Figure 11.11 Speed and Acceleration for different activities and subjects.

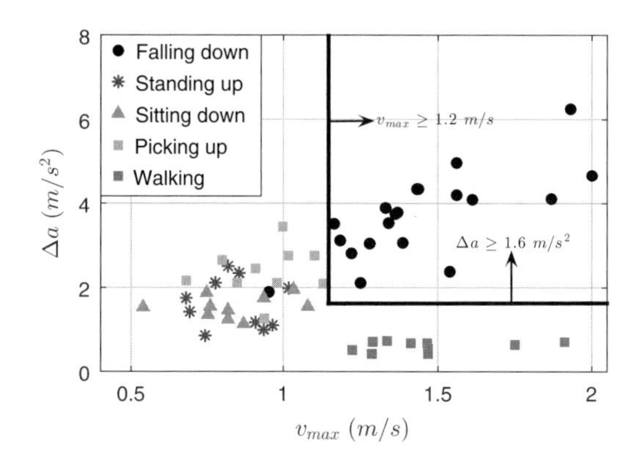

Figure 11.12 Distribution of the two metrics for all the activities.

standing up from a chair, (iii) sitting down on a chair, (iv) bowing and picking up items from the ground, and (v) walking inside the room. Each experiment lasts for 8 s. We collect 20 datasets of the falling down activity from two subjects and 10 datasets for each of the other four activities from the same two subjects. The experiments are conducted in Room #5, and the Wi-Fi Tx and Rx are placed at Location Tx #1 and Rx #2 as shown in Figure 11.5. Figure 11.11 shows a snapshot of speed and acceleration estimation results for different activities and subjects.

Realizing that the duration of a real-world falling down can be as short as 0.5 s and the human body would experience a sudden acceleration and then a deceleration [37], we consider two metrics for falling down detection: (i) the maximum change in acceleration within 0.5 s, denoted as Δa, and (ii) the maximum speed during the period of the maximum change of acceleration, written as $v_{\max}$. Figure 11.12 shows the distribution of

$(\Delta a, v_{\max})$ of all activities from the two subjects. Obviously, by setting two thresholds: $\Delta a \geq 1.6\,m/s^2$ and $v_{\max} \geq 1.2\,m/s$, WiSpeed could differentiate falls from the other four activities except one outlier, leading to a detection rate of 95% and zero false alarm, while [14] requires machine learning techniques. This is because WiSpeed extracts the most important physical features for activity classification, namely, the speed and the change of acceleration, while [14] infers these two physical values indirectly.

11.7 Discussion

In this section, we discuss the system parameter selections for different applications and their impact on the computational complexity of WiSpeed, and the behavior of WiSpeed when multiple objects are present.

11.7.1 Tracking a Fast Moving Object

In order to track a fast speed-varying object, we adopt the following equation with a reduced number of samples to calculate the sample auto-covariance function:

$$\hat{\gamma}_G(\tau, f) = \frac{1}{M} \sum_{t=T-M+1}^{T} \left(G(t-\tau,f) - \bar{G}(f)\right)\left(G(t,f) - \bar{G}(f)\right), \tag{11.17}$$

where T is the length of the window, M is the number of samples for averaging, and $\bar{G}(f)$ is the sample average. Equation (11.17) shows that to estimate a moving subject with speed v, WiSpeed requires a time window with a duration $T_0 = \frac{0.54\lambda}{v} + \frac{M}{F_s}$ s. Essentially, WiSpeed captures the average speed of motion in a period of time rather than the instantaneous moving speed. For instance, with $v = 1.3\,m/s$, $F_s = 1500$ Hz, $f_c = 5.805$ GHz, and $M = 100$, T_0 is around 0.12 s. In case that the speed changes significantly within a duration of T_0, the performance of WiSpeed would degrade. To track the speed of a fast-varying moving subject, a smaller T_0 is desirable, which can be achieved by increasing the channel sampling rate F_s or increasing the carrier frequency to reduce the wavelength λ.

11.7.2 Computational Complexity

The main computational complexity of WiSpeed comes from the estimation of the overall ACF $\hat{\rho}_G(\tau)$, giving rise to a total of FMT_0F_s multiplications where F is the number of available subcarriers. For motions with slow-varying speeds such as walking and standing up, a lower channel sampling rate suffices, which could reduce the complexity. For example, in our experiments of human walking speed estimation and human fall detection, $F_s = 1500$ Hz, $f_c = 5.805$ GHz, $F = 180$, and $M = 100$, the total number of multiplications for WiSpeed to produce one output is around three million. This leads to a computational time of 80.4 ms on a desktop with an Intel Core i7-7500U processor and 16GB memory, which is short enough for real-time applications.

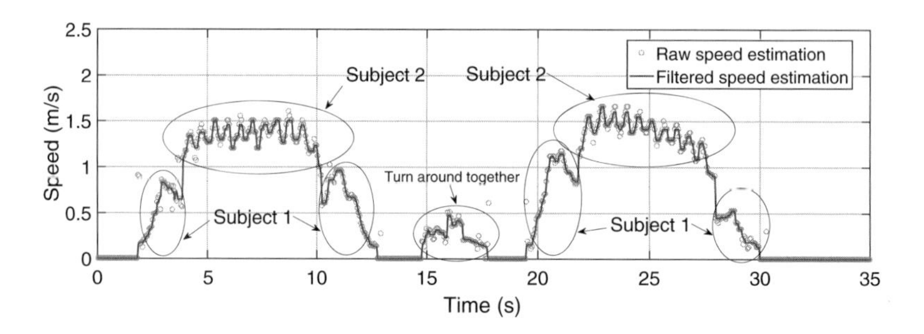

Figure 11.13 Two subjects walking in the environment.

11.7.3 Impact of Multiple Moving Objects

WiSpeed is designed to estimate the speed of a single moving object in the environment. If there exist multiple moving objects within the coverage of WiSpeed, WiSpeed would capture the highest speed among the objects. This is because WiSpeed uses the first local peak of the obtained ACF differential to estimate the speed and the component of ACF contributed by the object with the highest moving speed has the closest peak to the origin.

An experiment is conducted to illustrate the conjecture. Under Setting #4 as described in Section 11.6.2, two subjects first walk along Route #3, and then they turn around at the same time and repeat the process along Route #4. For each Route, Subject #1 walks with a lower speed and starts to walk earlier than Subject #2, and Subject #2 walks with a higher speed and stops earlier than Subject #1. Figure 11.13 shows that WiSpeed first captures the walking speed of Subject #1 while Subject #2 keeps static, and then it captures the speed of Subject #2 when the speed of Subject #2 exceeds that of Subject #1.

One potential solution for detecting the speeds of multiple moving objects is to deploy multiple transmission pairs of WiSpeed. The coverage of each pair can be tuned by varying the distance between the transmitter and receiver. The environment can thus be divided into multiple small regions, and it is reasonable to assume that there is only a single person within each small region.

11.8 Summary

In this chapter, we presented WiSpeed, a universal indoor speed estimation system for human motions leveraging commercial Wi-Fi, which can estimate the speed of a moving object under either device-free or device-based conditions. WiSpeed is built upon the statistical theory of EM waves, which quantifies the impact of human motions on EM waves for indoor environments. We conducted extensive experiments in a typical indoor environment, which demonstrates that WiSpeed can achieve a MAPE of 4.85% for device-free human walking speed monitoring and a MAPE of 4.62% for device-based speed estimation. Meanwhile, it achieves an average detection rate of 95% with no false alarms for human fall detection. Due to its large coverage, robustness, low cost, and low computational complexity, WiSpeed is a very promising candidate for indoor passive human activity monitoring systems. For related references, readers can refer to [38].

Appendix: Derivation of (11.10)

First, we can rewrite $G(t, f)$ as

$$G(t, f) = \sum_{u\in\{x,y,z\}} G_u(t, f) + \varepsilon(t, f), \tag{11.18}$$

where $G_u(t, f) \triangleq |E_{su}(f)|^2 + 2\mathrm{Re}\left\{E^*_{su}(f)\sum_{i\in\Omega_d} E_{iu}(t, f)\right\} + \left|\sum_{i\in\Omega_d} E_{iu}(t, f)\right|^2$. Then, the covariance of $G(t, f)$ can be written as

$$\begin{aligned}\gamma_G(\tau, f) &= \mathrm{cov}\big(G(t, f), G(t-\tau, f)\big)\\ &= \sum_{u\in\{x,y,z\}} \mathrm{cov}\big(G_u(t,f), G_u(t-\tau, f)\big) + \mathrm{cov}\big(\varepsilon(t, f), \varepsilon(t-\tau, f)\big)\\ &= \sum_{u\in\{x,y,z\}} \mathrm{cov}\big(G_u(t, f), G_u(t-\tau, f)\big) + \delta(\tau)\sigma^2(f),\end{aligned} \tag{11.19}$$

which is due to Assumptions 11.3.2–11.3.3 and the assumptions of the noise term. Thus, in the following, we only need to focus on the term $\gamma_{G_u}(\tau, f) \triangleq \mathrm{cov}\big(G_u(t, f), G_u(t-\tau, f)\big)$, that is, for $\forall u \in \{x, y, z\}$, we have the equation (11.20).

$$\begin{aligned}\gamma_{G_u}(\tau, f) &= \Big\langle G_u(t, f) - \langle G_u(t, f)\rangle, G_u(t-\tau, f) - \langle G_u(t-\tau, f)\rangle\Big\rangle\\ &= \Big\langle \underbrace{2\mathrm{Re}\{E^*_{su}(f)\sum_{i\in\Omega_d} E_{iu}(t, f)\}}_{\mathcal{A}_1} + \underbrace{\Big(|\sum_{i\in\Omega_d} E_{iu}(t, f)|^2 - \langle|\sum_{i\in\Omega_d} E_{iu}(t, f)|^2\rangle\Big)}_{\mathcal{A}_2},\\ &\quad \underbrace{2\mathrm{Re}\{E^*_{su}(f)\sum_{i\in\Omega_d} E_{iu}(t-\tau, f)\}}_{\mathcal{A}_3} + \underbrace{\Big(|\sum_{i\in\Omega_d} E_{iu}(t-\tau, f)|^2 - \langle|\sum_{i\in\Omega_d} E_{iu}(t-\tau, f)|^2\rangle\Big)}_{\mathcal{A}_4}\Big\rangle.\end{aligned} \tag{11.20}$$

We begin with the term $\langle\mathcal{A}_1, \mathcal{A}_3\rangle$. For notational convenience, define $E_{iu}(t, f) \triangleq a_i(t) + jb_i(t)$ and $E_{su}(f) \triangleq u + jv$, for $\forall i \in \Omega_d$, $\forall u \in \{x, y, z\}$, and a_i, b_i, u, v are all real.

Then, we have

$$\begin{aligned}
\left\langle \mathcal{A}_1, \mathcal{A}_3 \right\rangle &= 4\left\langle u \sum_{i \in \Omega_d} a_i(t) + v \sum_{i \in \Omega_d} b_i(t), u \sum_{i \in \Omega_d} a_i(t-\tau) + v \sum_{i \in \Omega_d} b_i(t-\tau) \right\rangle \\
&= 4u^2 \sum_{i \in \Omega_d} \left\langle a_i(t), a_i(t-\tau) \right\rangle + 4v^2 \sum_{i \in \Omega_d} \left\langle b_i(t), b_i(t-\tau) \right\rangle \\
&= 4(u^2 + v^2) \sum_{i \in \Omega_d} \left\langle a_i(t), a_i(t-\tau) \right\rangle, \qquad (11.21)
\end{aligned}$$

where we apply the assumption that the real and imaginary parts of the electric field have the same statistical behaviors. At the same time, we have

$$\begin{aligned}
\operatorname{cov}(E_{iu}(t,f), E_{iu}(t-\tau,f)) &= \left\langle E_{iu}(t,f), E_{iu}(t-\tau,f) \right\rangle \\
&= \left\langle a_i(t), a_i(t-\tau) \right\rangle + \left\langle b_i(t), b_i(t-\tau) \right\rangle \\
&= 2\left\langle a_i(t), a_i(t-\tau) \right\rangle. \qquad (11.22)
\end{aligned}$$

Thus, we have

$$\left\langle \mathcal{A}_1, \mathcal{A}_3 \right\rangle = 2|E_{su}(f)|^2 \sum_{i \in \Omega_d} \operatorname{cov}\big(E_{iu}(t,f), E_{iu}(t-\tau,f)\big). \qquad (11.23)$$

Next, we derive the term $\langle \mathcal{A}_1, \mathcal{A}_4 \rangle$ as shown in (11.24).

$$\begin{aligned}
\left\langle \mathcal{A}_1, \mathcal{A}_4 \right\rangle &= 2\left\langle u \sum_{i \in \Omega_d} a_i(t) + v \sum_{i \in \Omega_d} b_i(t), \left(\sum_{i \in \Omega_d} a_i(t-\tau) \right)^2 \right. \\
&\quad \left. + \left(\sum_{i \in \Omega_d} b_i(t-\tau) \right)^2 - \left\langle \left| \sum_{i \in \Omega_d} E_{iu}(t-\tau,f) \right|^2 \right\rangle \right\rangle \\
&= 2\left\langle u \sum_{i \in \Omega_d} a_i(t) + v \sum_{i \in \Omega_d} b_i(t), \left(\sum_{i \in \Omega_d} a_i(t-\tau) \right)^2 + \left(\sum_{i \in \Omega_d} b_i(t-\tau) \right)^2 \right\rangle \\
&= 2u \sum_{i \in \Omega_d} \left\langle a_i(t), a_i^2(t-\tau) \right\rangle + 2v \sum_{i \in \Omega_d} \left\langle b_i(t), b_i^2(t-\tau) \right\rangle. \qquad (11.24)
\end{aligned}$$

According to the integral representation of the electric field in (11.3), we have

$$|E_{iu}(t,f)|^2 = \iint_{4\pi} F_{iu}(\Theta_1) F_{iu}^*(\Theta_2) \exp(-j(\vec{k}(\Theta_1) - \vec{k}(\Theta_2)) \cdot \vec{v}_i t) \mathrm{d}\Theta_1 \mathrm{d}\Theta_2, \quad (11.25)$$

and thus, the covariance between $E_{iu}(t, f)$ and $|E_{iu}(t-\tau, f)|^2$ can be expressed as

$$\begin{aligned}
&\operatorname{cov}(E_{iu}(t, f), |E_{iu}(t-\tau, f)|^2) \\
&= \left\langle E_{iu}(t, f) - \langle E_{iu}(t, f)\rangle, |E_{iu}(t-\tau, f)|^2 - \langle |E_{iu}(t-\tau, f)|^2\rangle \right\rangle \\
&= \left\langle E_{iu}(t, f), |E_{iu}(t-\tau, f)|^2 \right\rangle \\
&= \iiint_{4\pi} \left\langle F_{iu}(\Theta_1), F_{iu}(\Theta_{21}) F_{iu}^*(\Theta_{22}) \right\rangle \exp(-j\vec{k}(\Theta_1)\cdot \vec{v}_i t) \\
&\qquad \exp(-j(\vec{k}(\Theta_{21}) - \vec{k}(\Theta_{22}))\cdot \vec{v}_i (t-\tau))\,\mathrm{d}\Theta_1\,\mathrm{d}\Theta_{21}\,\mathrm{d}\Theta_{22} \\
&= \int_{4\pi} \left\langle F_{iu}(\Theta_1), |F_{iu}(\Theta_1)|^2 \right\rangle \exp(-j\vec{k}(\Theta_1)\cdot \vec{v}_i t)\,\mathrm{d}\Theta_1 \\
&= \int_{4\pi} \left(\left\langle \mathrm{Re}\left\{ F_{iu}(\Theta_1) \right\}, \mathrm{Re}\left\{ F_{iu}(\Theta_1) \right\}^2 \right\rangle \right. \\
&\qquad \left. + j\left\langle \mathrm{Im}\left\{ F_{iu}(\Theta_1) \right\}, \mathrm{Im}\left\{ F_{iu}(\Theta_1) \right\}^2 \right\rangle \right) \exp(-j\vec{k}(\Theta_1)\cdot \vec{v}_i t)\mathrm{d}\Theta_1 \\
&= 0,
\end{aligned} \tag{11.26}$$

because $\langle X^3\rangle \equiv 0$ for any Gaussian random variable with zero mean. At the same time, we have

$$\left\langle E_{iu}(t, f), |E_{iu}(t-\tau, f)|^2 \right\rangle = \left\langle a_i(t), a_i^2(t-\tau) \right\rangle + j\left\langle b_i(t), b_i^2(t-\tau) \right\rangle, \tag{11.27}$$

and thus we have $\langle a_i(t), a_i^2(t-\tau)\rangle = 0$. Plugging this result in (11.24), we can obtain

$$\left\langle \mathcal{A}_1, \mathcal{A}_4 \right\rangle = 0. \tag{11.28}$$

Similarly, we can also derive that $\langle \mathcal{A}_2, \mathcal{A}_3\rangle = 0$. At last, we derive the term $\langle \mathcal{A}_2, \mathcal{A}_4\rangle$ as shown in (11.29).

$$\begin{aligned}
\left\langle \mathcal{A}_2, \mathcal{A}_4 \right\rangle = \operatorname{cov}\Bigg(&\left(\sum_{i\in\Omega_d} a_i(t) \right)^2 \\
&+ \left(\sum_{i\in\Omega_d} b_i(t) \right)^2, \left(\sum_{i\in\Omega_d} a_i(t-\tau) \right)^2 + \left(\sum_{i\in\Omega_d} b_i(t-\tau) \right)^2 \Bigg)
\end{aligned}$$

$$= \operatorname{cov}\left(\left(\sum_{i\in\Omega_d} a_i(t)\right)^2, \left(\sum_{i\in\Omega_d} a_i(t-\tau)\right)^2\right)$$
$$+ \operatorname{cov}\left(\left(\sum_{i\in\Omega_d} b_i(t)\right)^2, \left(\sum_{i\in\Omega_d} b_i(t-\tau)\right)^2\right)$$
$$= 2 \sum_{i_1, i_2 \in \Omega_d} \operatorname{cov}\Big(a_{i1}(t)a_{i2}(t), a_{i1}(t-\tau)a_{i2}(t-\tau)\Big)$$
$$= 2 \sum_{i\in\Omega_d} \operatorname{cov}\Big(a_i^2(t), a_i^2(t-\tau)\Big)$$
$$+ 2 \sum_{\substack{i_1, i_2 \in \Omega_d \\ i_1 \neq i_2}} \operatorname{cov}\Big(a_{i1}(t)a_{i2}(t), a_{i1}(t-\tau)a_{i2}(t-\tau)\Big). \tag{11.29}$$

Because for any two Gaussian random variables, X and Y, with zero mean, the expectations can be evaluated by using the following relationship [39]:

$$\langle X^2Y^2\rangle = \langle X^2\rangle\langle Y^2\rangle + 2\langle XY\rangle^2, \tag{11.30}$$

then, we have, $\forall i \in \Omega_d$,

$$\begin{aligned}
\operatorname{cov}(a_i^2(t), a_i^2(t-\tau)) &= \Big\langle a_i^2(t) - \langle a_i^2(t)\rangle, a_i^2(t-\tau) - \langle a_i^2(t-\tau)\rangle\Big\rangle \\
&= \langle a_i^2(t), a_i^2(t-\tau)\rangle - \langle a_i^2(t)\rangle\langle a_i^2(t-\tau)\rangle \\
&= 2\langle a_i(t), a_i(t-\tau)\rangle^2 \\
&= \frac{1}{2}\operatorname{cov}\big(E_{iu}(t,f), E_{iu}(t-\tau,f)\big)^2.
\end{aligned} \tag{11.31}$$

For $i_1, i_2 \in \Omega_d$ and $i_1 \neq i_2$, we have

$$\begin{aligned}
\operatorname{cov}\big(a_{i1}(t)a_{i2}(t), a_{i1}(t-\tau)a_{i2}(t-\tau)\big) &= \Big\langle a_{i1}(t)a_{i2}(t), a_{i1}(t-\tau)a_{i2}(t-\tau)\Big\rangle \\
&= \Big\langle a_{i1}(t)a_{i1}(t-\tau), a_{i2}(t)a_{i2}(t-\tau)\Big\rangle \\
&= \Big\langle a_{i1}(t), a_{i1}(t-\tau)\Big\rangle\Big\langle a_{i2}(t), a_{i2}(t-\tau)\Big\rangle \\
&= \frac{1}{4}\operatorname{cov}\big(E_{i_1u}(t,f), E_{i_1u}(t-\tau,f)\big) \\
&\quad \operatorname{cov}\big(E_{i_2u}(t,f), E_{i_2u}(t-\tau,f)\big).
\end{aligned} \tag{11.32}$$

Therefore, $\Big\langle \mathcal{A}_2, \mathcal{A}_4\Big\rangle$ can be derived as

$$\Big\langle \mathcal{A}_2, \mathcal{A}_4\Big\rangle = \sum_{\substack{i_1, i_2 \in \Omega_d \\ i_1 \geq i_2}} \operatorname{cov}\big(E_{i_1u}(t,f), E_{i_1u}(t-\tau,f)\big)\operatorname{cov}\big(E_{i_2u}(t,f), E_{i_2u}(t-\tau,f)\big). \tag{11.33}$$

Finally, we can obtain the result shown in (11.10).

References

[1] M. Khan, B. N. Silva, and K. Han, "Internet of things based energy aware smart home control system," *IEEE Access*, vol. 4, pp. 7556–7566, 2016.

[2] S. Pinto, J. Cabral, and T. Gomes, "We-care: An IoT-based health care system for elderly people," in *IEEE International Conference on Industrial Technology (ICIT)*, pp. 1378–1383, Mar. 2017.

[3] S. E. Schaefer, C. C. Ching, H. Breen, and J. B. German, "Wearing, thinking, and moving: Testing the feasibility of fitness tracking with urban youth," *American Journal of Health Education*, vol. 47, no. 1, pp. 8–16, 2016.

[4] L. Wang, G. Zhao, L. Cheng, and M. Pietikäinen, *Machine Learning for Vision-Based Motion Analysis: Theory and Techniques*. Berlin: Springer, 2010.

[5] S. Z. Gurbuz, C. Clemente, A. Balleri, and J. J. Soraghan, "Micro-Doppler-based in-home aided and unaided walking recognition with multiple radar and sonar systems," *IET Radar, Sonar and Navigation*, vol. 11, no. 1, pp. 107–115, 2017.

[6] C.-Y. Hsu, Y. Liu, Z. Kabelac, R. Hristov, D. Katabi, and C. Liu, "Extracting gait velocity and stride length from surrounding radio signals," in *Proceedings of the CHI Conference on Human Factors in Computing Systems*, pp. 2116–2126, ACM, 2017.

[7] K. Qian, C. Wu, Z. Yang, Y. Liu, and K. Jamieson, "WiDar: Decimeter-level passive tracking via velocity monitoring with commodity Wi-Fi," in *Proceedings of the 18th ACM International Symposium on Mobile Ad Hoc Networking and Computing*, p. 6, ACM, 2017.

[8] D. Halperin, W. Hu, A. Sheth, and D. Wetherall, "Tool release: Gathering 802.11n traces with channel state information," *SIGCOMM Computer Communication Review*, vol. 41, pp. 53–53, Jan. 2011.

[9] H. Abdelnasser, M. Youssef, and K. A. Harras, "WiGest: A ubiquitous WiFi-based gesture recognition system," in *Proceedings of IEEE INFOCOM*, pp. 1472–1480, Apr. 2015.

[10] K. Qian, C. Wu, Z. Zhou, Y. Zheng, Z. Yang, and Y. Liu, "Inferring motion direction using commodity Wi-Fi for interactive exergames," in *Proceedings of CHI Conference on Human Factors in Computing Systems*, pp. 1961–1972, ACM, 2017.

[11] K. Ali, A. X. Liu, W. Wang, and M. Shahzad, "Keystroke recognition using WiFi signals," in *Proceedings of the 21st Annual International Conference on Mobile Computing & Networking*, pp. 90–102, ACM, 2015.

[12] Q. Pu, S. Gupta, S. Gollakota, and S. Patel, "Whole-home gesture recognition using wireless signals," in *Proceedings of the 19th Annual International Conference on Mobile Computing & Networking*, pp. 27–38, ACM, 2013.

[13] G. Wang, Y. Zou, Z. Zhou, K. Wu, and L. M. Ni, "We can hear you with Wi-Fi!," *IEEE Transactions on Mobile Computing*, vol. 15, pp. 2907–2920, Nov. 2016.

[14] Y. Wang, J. Liu, Y. Chen, M. Gruteser, J. Yang, and H. Liu, "E-eyes: Device-free location-oriented activity identification using fine-grained WiFi signatures," in *Proceedings of the 20th Annual International Conference on Mobile Computing & Networking*, pp. 617–628, ACM, 2014.

[15] W. Wang, A. X. Liu, M. Shahzad, K. Ling, and S. Lu, "Understanding and modeling of WiFi signal based human activity recognition," in *Proceedings of the 21st Annual International Conference on Mobile Computing & Networking*, pp. 65–76, ACM, 2015.

[16] Y. Wang, K. Wu, and L. M. Ni, "WiFall: Device-free fall detection by wireless networks," *IEEE Transactions on Mobile Computing*, vol. 16, pp. 581–594, Feb. 2017.

[17] L. Sun, S. Sen, D. Koutsonikolas, and K.-H. Kim, "WiDraw: Enabling hands-free drawing in the air on commodity WiFi devices," in *Proceedings of the 21st Annual International Conference on Mobile Computing & Networking*, pp. 77–89, ACM, 2015.

[18] F. Adib and D. Katabi, "See through walls with WiFi!," *SIGCOMM Computer Communication Review*, vol. 43, pp. 75–86, Aug. 2013.

[19] M. Seifeldin, A. Saeed, A. E. Kosba, A. El-Keyi, and M. Youssef, "Nuzzer: A large-scale device-free passive localization system for wireless environments," *IEEE Transactions on Mobile Computing*, vol. 12, pp. 1321–1334, Jul. 2013.

[20] C. Chen, B. Wang, Y. Han, Y. Chen, F. Zhang, H.Q. Lai, and K. J. R. Liu, "TR-BREATH: Time-reversal breathing rate estimation and detection," *IEEE Transactions on Biomedical Engineering*, vol. 65, no. 3, pp. 489–501, Mar. 2018.

[21] Q. Xu, Y. Chen, B. Wang, and K. J. R. Liu, "TRIEDS: Wireless events detection through the wall," *IEEE Internet of Things Journal*, vol. 4, pp. 723–735, Jun. 2017.

[22] F. Adib, Z. Kabelac, D. Katabi, and R. C. Miller, "3D tracking via body radio reflections," in *11th USENIX Symposium on Networked Systems Design and Implementation*, pp. 317–329, USENIX Association, 2014.

[23] P. Murphy, A. Sabharwal, and B. Aazhang, "Design of warp: A flexible wireless open-access research platform," in *Proceedings of EUSIPCO*, pp. 53–54, 2006.

[24] D. A. Hill, *Electromagnetic Fields in Cavities: Deterministic and Statistical Theories*, vol. 35. Hoboken, NJ: John Wiley & Sons, 2009.

[25] D. Tse and P. Viswanath, *Fundamentals of Wireless Communication*. Cambridge, UK: Cambridge University Press, 2005.

[26] C. Chen, Y. Chen, Y. Han, H. Q. Lai, F. Zhang, and K. J. R. Liu, "Achieving centimeter-accuracy indoor localization on WiFi platforms: A multi-antenna approach," *IEEE Internet of Things Journal*, vol. 4, pp. 122–134, Feb. 2017.

[27] Z.-H. Wu, Y. Han, Y. Chen, and K. R. Liu, "A time-reversal paradigm for indoor positioning system," *IEEE Transactions on Vehicular Technology*, vol. 64, no. 4, pp. 1331–1339, 2015.

[28] T.-D. Chiueh, P.-Y. Tsai, and I.-W. Lai, *Baseband Receiver Design for Wireless MIMO-OFDM Communications*. Hoboken, NJ: John Wiley & Sons, 2012.

[29] R. H. Shumway and D. S. Stoffer, *Time Series Analysis and Its Applications with R Examples*, Berlin: Springer, 2006.

[30] R. Van Nee, "Delay spread requirements for wireless networks in the 2.4 GHz and 5 GHzi bands," *IEEE*, vol. 802, pp. 802–822, 1997.

[31] W. S. Cleveland, "Robust locally weighted regression and smoothing scatterplots," *Journal of the American Statistical Association*, vol. 74, no. 368, pp. 829–836, 1979.

[32] H. Scheffe, *The Analysis of Variance*, vol. 72. New York: John Wiley & Sons, 1999.

[33] S.-J. Kim, K. Koh, S. Boyd, and D. Gorinevsky, "l1 trend filtering," *SIAM Review*, vol. 51, no. 2, pp. 339–360, 2009.

[34] Y. Kozlov and T. Weinkauf, "Persistence1D: Extracting and filtering minima and maxima of 1d functions," http://people.mpi-inf.mpg.de/weinkauf/notes/persistence1d.html, pp. 11–01, 2015.

[35] F. Zhang, C. Chen, B. Wang, H. Q. Lai, and K. J. R. Liu, "A time-reversal spatial hardening effect for indoor speed estimation," in *Proceedings of IEEE ICASSP*, pp. 5955–5959, Mar. 2017.

[36] F. Zhang, C. Chen, B. Wang, and K. J. Liu, "WiBall: A time-reversal focusing ball method for indoor tracking," *IEEE Internet of Things Journal*, vol. 5, no. 5, pp. 4031–4041, Oct. 2018.

[37] F. Bagalà, C. Becker, A. Cappello, L. Chiari, K. Aminian, J. M. Hausdorff, W. Zijlstra, and J. Klenk, "Evaluation of accelerometer-based fall detection algorithms on real-world falls," *PloS One*, vol. 7, no. 5, p. e37062, 2012.

[38] F. Zhang, C. Chen, B. Wang, and K. J. R. Liu, "WiSpeed: A statistical electromagnetic approach for device-free indoor speed estimation," *IEEE Internet of Things Journal*, vol. 5, no. 3, pp. 2163–2177, 2018.

[39] A. Papoulis and U. Pillai, *Probability, Random Variables, and Stochastic Processes*. New York: McGraw-Hill, 2002.

Part III

Wireless Power Transfer and Energy Efficiency

12 Time-Reversal for Energy Efficiency

Green wireless communications have received considerable attention recently in hopes of finding novel solutions to improve energy efficiency for the ubiquity of wireless applications. In this chapter, we argue and show that the time-reversal (TR) signal transmission is an ideal paradigm for green wireless communications because of its inherent nature to fully harvest energy from the surrounding environment by exploiting the multipath propagation to recollect *all* the signal energy that would have otherwise been lost in most existing communication paradigms. A green wireless technology must ensure low energy consumption and low radio pollution to others than the intended user. In this chapter, we show through theoretical analysis, numerical simulations, and experiment measurements that TR wireless communications, compared to the conventional direct transmission using a Rake receiver, reveals significant transmission power reduction, achieves high interference alleviation ratio, and exhibits large multipath diversity gain. As such it is an ideal paradigm for the development of green wireless systems. The theoretical analysis and numerical simulations show an order of magnitude improvement in terms of transmit power reduction and interference alleviation. Experimental measurements in a typical indoor environment also demonstrate that the transmit power with TR-based transmission can be as low as 20% of that without TR, and the average radio interference (thus radio pollution) even in a nearby area can be up to 6 dB lower. A strong time correlation is found to be maintained in the multipath channel even when the environment is varying, which indicates high bandwidth efficiency can be achieved in TR radio communications.

12.1 Introduction

In recent years, with the explosive growth of the wireless communication industry in terms of network infrastructures, network users, and various new applications, the energy consumption of wireless networks and devices is experiencing a dramatic increase. Because of ubiquity of wireless applications, such an increasing energy consumption not only results in a high operational cost and an urgent demand for battery/energy capacity to wireless communications operators, but also causes a more severe electromagnetic (EM) pollution to the global environment. Therefore, an emerging concept of "Green Communications" has received considerable attention

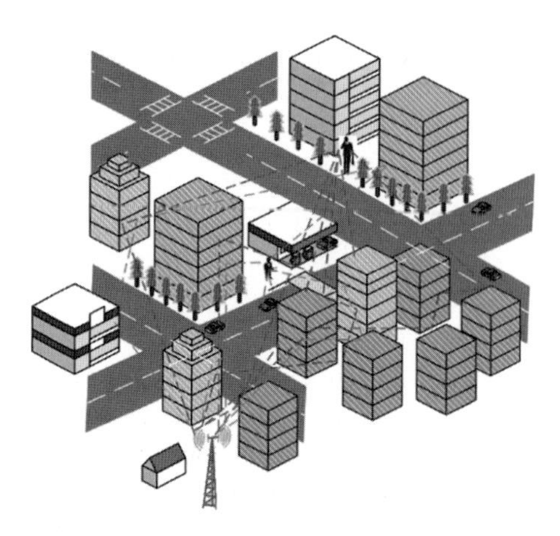

Figure 12.1 Illustration of a typical urban multipath environment.

in hopes of finding novel solutions to improve energy efficiency, relieve/reduce radio pollution to unintended users, and maintain/improve performance metrics.

In this chapter, we argue and show that the time-reversal (TR) signal transmission is an ideal paradigm for green wireless communications because of its inherent nature to fully harvest energy from the surrounding environment by exploiting the multipath propagation, as shown in Figure 12.1, to recollect *all* the signal energy that would have otherwise been lost in most existing communication paradigms. To qualify as a green wireless technology, one must meet two basic requirements: one is low energy consumption (environmental concerns) and the other is low radio pollution to others (health concerns) besides the intended transmitter and receiver. We will illustrate in this chapter that the TR paradigm not only meets the preceding two criteria but also exhibits a very high multipath diversity gain, as well as preserving high bandwidth efficiency due to high channel correlation in practice.

TR wireless communication has been known for some time; however, its applications have been mainly considered as a specialty use for extreme multipath environment. Therefore, not much development and interest can be seen beyond defense applications. The history of applying TR to communication systems dates back to early 1990s. In TR communications, when transceiver A wants to transmit information to transceiver B, transceiver B first has to send a delta-like pilot pulse that propagates through a scattering and multipath environment, and the signals are received by transceiver A; then, transceiver A simply transmits the TR signals back through the same channel to transceiver B. By utilizing channel reciprocity, TR essentially leverages the multipath channel as a matched filter, i.e., treats the environment as a facilitating matched filter computing machine and focuses the wave at the receiver in both space and time domains. As such one can readily see the low-complexity nature of TR communications.

Experiments on TR in acoustics and ultrasound domains [1–4] have shown that acoustic energy can be refocused on the source with very high resolution, and the focusing effect in real propagation environments was further validated by underwater acoustics experiments in the ocean [5–7]. Because TR can make full use of multipath propagation and also requires no complicated channel measurements and estimation, it has been also studied in wireless communication systems. Spatial and temporal focusing properties of EM signal transmission with TR have been demonstrated in [8–10] by taking measurements in radio frequency (RF) communications. A TR-based interference canceller to mitigate the effect of clutter was presented in [11], and target detection in a highly cluttered environment using TR was investigated in [12, 13].

Leveraging from the spatial and temporal focusing effect, in this chapter, we show that the TR technique is indeed an ideal green wireless communication paradigm that can efficiently harvest energy from the environment. We first derive the theoretical transmission power reduction and interference alleviation of the TR-based transmission compared to direct transmission with a Rake receiver. Our theoretical analysis and simulations show that a potential of over an order of magnitude of power reduction and interference alleviation can be achieved. We also investigate the multipath diversity gain of the TR-based transmission, in which we demonstrate a very high multipath diversity gain exhibiting in a TR system. In essence, TR transmission treats each multipath as a virtual antenna and makes full use of *all* the multipaths.

Experimental results obtained from measurements in a real RF multipath environment are shown to demonstrate the great potential of TR-based transmission as an energy-efficient green wireless communication paradigm. It is found that in a typical indoor multipath environment, in order to achieve the same receiver performance, TR-based transmission only costs as low as 20% of the transmission power needed in direct transmission; moreover, the average interference can be up to 6 dB lower than that caused by direct transmission when the interfered receiver is only 1 m away from the intended receiver. It is also shown from channel measurements in different time epochs that a static indoor multipath environment is strongly time-correlated; therefore, there is no need for the receiver to keep sending pilot pulses to the transmitter, and the spectral efficiency can be much higher than a typically achieved value of 50%. We also performed extensive numerical simulation to validate the theoretical derivation.

The rest of this chapter is organized as follows. In Section 12.2, we introduce the system model and multipath channel model. In Section 12.3, we investigate the performance of the TR-based transmission in terms of power reduction, interference alleviation, and multipath diversity gain. Simulation studies are presented in Section 12.4, and experimental results obtained from practical indoor multipath channels are shown in Section 12.5. We briefly present a few prospective applications of the TR-based technology based on its focusing effect in Section 12.6, which suggest that TR-based communication is a promising direction in addition to power efficiency and low interference pollution.

12.2 System Model

In this chapter, we consider a slow fading wireless channel with a large delay spread. The channel impulse response (CIR) at time k between the transmitter and the receiver in discrete time domain is modeled as

$$h[k] = \sum_{l=0}^{L-1} h_l \delta[k-l], \tag{12.1}$$

where h_l is the complex amplitude of lth tap of the CIR, and L is the number of channel taps. Because we assume that the channel is slow fading, the channel taps will not vary during the observation time. To gain some insight into the TR system while keeping the model analytically tractable, the CIRs associated with different receivers at different locations are assumed to be independent, e.g., when the receivers are very far apart. Furthermore, we assume independence among the taps of each CIR, i.e., the paths of each CIR are uncorrelated. Each $h[l]$ is a circular symmetric complex Gaussian (CSCG) random variable with zero mean and

$$E[|h[l]|^2] = e^{-\frac{lT_S}{\sigma_T}}, \tag{12.2}$$

where T_S is the sampling period of this system such that $1/T_S$ equals the system bandwidth B, and σ_T is the delay spread [14] of the channel.

A TR-based communication system is very simple. For example, a base station tries to transmit some information to an end user. Prior to the transmission, the end user has to send out a delta-like pilot pulse which propagates to the base station through a multipath channel, where the base station keeps a record of the received waveform. Then, the base station time-reverses the received waveform, and uses the normalized time-reversed conjugate signals as a basic waveform, i.e.,

$$g[k] = h^*[L-1-k] \Bigg/ \sqrt{\sum_{l=0}^{L-1} |h[l]|^2}\,, \quad k = 0, 1, \ldots, L-1. \tag{12.3}$$

In the preceding equation, we ignore the noise term to simplify derivation.[1] Thanks to the channel reciprocity, the multipath channel forms a *natural matched filter* to the basic waveform $g[k], \quad k = 0, 1, \ldots, L-1$, and hence a peak is expected at the receiver.

The base station loads the data stream on the basic waveform and transmits the signal into the wireless channel. Usually the baud rate is much lower than the sampling rate, and the ratio of the sampling rate to the baud rate is also known as the rate back-off factor D [10]. Mathematically, if a sequence of information symbols is denoted by $\{X[k]\}$ and assumed to be i.i.d. complex random variables with zero mean and a variance of P, the transmitted signal into the wireless channels can be expressed as

$$S[k] = \left(X^{[D]} * g\right)[k], \tag{12.4}$$

where $X^{[D]}[k]$ is an up-sampled sequence of $X[k]$,

$$X^{[D]}[k] = \begin{cases} X[k/D], & \text{if } k \bmod D = 0, \\ 0, & \text{if } k \bmod D \neq 0. \end{cases} \tag{12.5}$$

[1] By sending a large number of channel training sequences from the receiver, the noise term is diminishing asymptotically.

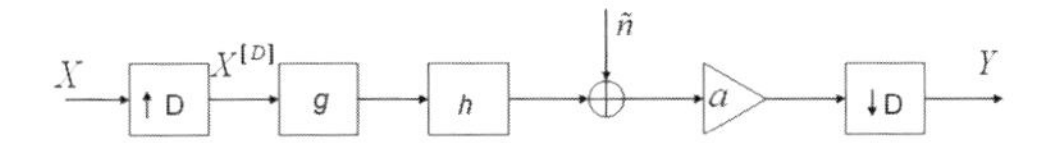

Figure 12.2 The block diagram of a TR-based communication system.

The signal received at the receiver is the convolution of $\{S[k]\}$ and $\{h[k]\}$, plus additive white Gaussian noise (AWGN) $\{\tilde{n}_i[k]\}$ with zero-mean and variance σ^2. The receiver simply performs a one-tap gain adjustment to the received signal, i.e., multiplying a coefficient a, and then down-samples it with the same factor D. The signal before down-sampling can be represented as

$$Y^{[D]}[k] = a\left(X^{[D]} * g * h\right)[k] + a\tilde{n}[k]. \tag{12.6}$$

Accordingly, the down-sampled signal $Y[k]$ is as follows (for simplicity, $L-1$ is assumed to be a multiple of D)

$$Y[k] = a \sum_{l=0}^{(2L-2)/D} (h * g)[Dl]X[k-l] + an[k], \tag{12.7}$$

where

$$(h * g)[k] = \frac{\sum_{l=0}^{L-1} h[l]h^*[L-1-k+l]}{\sqrt{\sum_{l=0}^{L-1} |h[l]|^2}}, \tag{12.8}$$

with $k = 0, 1, \cdots, 2L-2$, and $n[k] = \tilde{n}[Dk]$, a white Gaussian additive noise with zero mean and variance σ^2. The block diagram of a TR-based communication system is summarized in Figure 12.2, and we can see that both the transmitter and receiver are of very low complexity.

12.3 Performance Analysis

In this part, we compare the performance of the TR system to that of conventional direct transmission with Rake receivers, and evaluate several performance metrics, including the transmit power, in order to achieve the same signal-to-interference-and-noise ratio (SINR) and the interference caused to unintended receivers. Finally, we analyze the multipath gain of the TR system.

12.3.1 Power Reduction

Note that in (12.8), when $k = L-1$, it corresponds to the maximum-power central peak of the autocorrelation function of the CIR, i.e.

$$(h * g)[L-1] = \sqrt{\sum_{l=0}^{L-1} |h[l]|^2}. \tag{12.9}$$

Subject to the one-tap constraint, the receiver is designed to estimate $X[k - \frac{L-1}{D}]$ solely based on the observation of $Y[k]$. Then, the remaining components of $Y[k]$ can be further categorized into intersymbol interference (ISI) and noise, as shown in the following

$$\begin{aligned} Y[k] = {} & a(h * g)[L-1]X\left[k - \frac{L-1}{D}\right] && (Signal) \\ & + a \sum_{\substack{l=0 \\ l \neq (L-1)/D}}^{(2L-2)/D} (h * g)[Dl]X[k-l] && (ISI) \\ & + an[k] && (Noise) \end{aligned} \tag{12.10}$$

Given a specific realization of the random CIRs, from (12.10), one can calculate the signal power P_{Sig} as[2]

$$\begin{aligned} P_{Sig} &= E_X\left[\left|(h * g)[L-1]X[k - \frac{L-1}{D}]\right|^2\right] \\ &= P\,|(h * g)[L-1]|^2 = P\left(\sum_{l=0}^{L-1} |h[l]|^2\right), \end{aligned} \tag{12.11}$$

where $E_X[\cdot]$ represents the expectation over X. Similarly, the ISI can be derived as

$$\begin{aligned} P_{ISI} &= E_X\left[\left|\sum_{\substack{l=0 \\ l \neq (L-1)/D}}^{(2L-2)/D} (h * g)[Dl]X[k-l]\right|^2\right] \\ &= P \sum_{\substack{l=0 \\ l \neq (L-1)/D}}^{(2L-2)/D} |(h * g)[Dl]|^2. \end{aligned} \tag{12.12}$$

As D increases, the ISI term P_{ISI} will gradually decrease. In the regime where D is such a large positive number that $P_{ISI} \to 0$, we can focus on the signal-to-noise ratio (SNR) only:

$$\text{SNR} = \frac{P_{Sig}}{\sigma^2}. \tag{12.13}$$

Without using the TR-based transmission, we can express the received signal of direct transmission as

$$Y^{DT}[k] = (X * h)[k] + n[k] = \sum_{l=0}^{L-1} h[l]X[k-l] + n[k], \tag{12.14}$$

[2] Note that the one-tap gain a does not affect the effective SNR (or SINR), so we consider it as $a = 1$ in the subsequent analysis unless otherwise mentioned.

where the superscript "DT" represents "direct transmission," and the AWGN $n[k]$ has a zero mean and a variance σ^2. Using a Rake receiver with L_R fingers, the received signal power[3] can be expressed as [15]

$$P_{Sig}^{DT} = P^{DT} \sum_{l=0}^{L_R-1} |h_{(l)}|^2, \tag{12.15}$$

where P^{DT} denotes the transmit power of direct transmission, and $h_{(l)}$'s, $l = 0, 1, \ldots, L_R - 1$, represent the L_R channel taps with the L_R largest immediate tap gains.

In order for the TR system and the direct transmission to have the same performance, i.e., $\mathsf{SNR}_{TR} = \mathsf{SNR}_{DT}$, we must have

$$P_{Sig} = P_{Sig}^{DT}. \tag{12.16}$$

Then, we can express the ratio of the transmission power of the two schemes as

$$r_P = \frac{P}{P^{DT}} = \frac{\sum_{l=0}^{L_R-1} |h_{(l)}|^2}{\sum_{l=0}^{L-1} |h[l]|^2}, \tag{12.17}$$

and the ratio of the expected transmission power needed for TR and direct transmission can be expressed as

$$\tau_P = \frac{E[P]}{E[P^{DT}]} = \frac{E\left[\sum_{l=0}^{L_R-1} |h_{(l)}|^2\right]}{\sum_{l=0}^{L-1} E[|h[l]|^2]}. \tag{12.18}$$

In order to derive the numerator of (12.18), one needs to analyze the order statistics of the $|h[l]|^2$'s. However, because the $|h[l]|^2$'s are not identically distributed and it is also unknown which L_R out of all the $|h[l]|^2$'s are the L_R largest channel taps, it is very difficult to obtain the closed-form expression of the numerator in (12.18). Therefore, we will first assume that the $|h[l]|^2$'s are identically and independently distributed (i.i.d.) and derive the numerator of (12.18). Then we will calibrate the results for nonidentically distributed $|h[l]|^2$'s.

Before we start our analysis, let us first introduce the concept of *quantile* [16] in order statistics. Denote $F(z)$ as the distribution function for a continuous random variable.

Definition 12.3.1 *Suppose that $F(z)$ is continuous and strictly increasing when $0 < F(z) < 1$. For $0 < q < 1$, the* q-quantile *of $F(z)$ is a number z_q such that $F(z_q) = q$. If F^{-1} represents the inverse of $F(z)$, then $z_q = F^{-1}(q)$.*

Now, let us suppose that the $h[l]$'s are i.i.d random variables; then, $|h[l]|^2$ are also i.i.d. Denote $Z_l \triangleq |h[l]|^2$ for short, and the numerator in (12.18) can be approximated by the sample mean, i.e.,

$$E\left[\sum_{l=0}^{L_R-1} Z_{(l)}\right] \approx \lim_{n\to\infty} \frac{1}{n} \sum_{i=1}^{n} \left[\sum_{l=0}^{L_R-1} z_{(l)}^i\right], \tag{12.19}$$

[3] We assume that rate back-off factor D for direct transmission is also large enough so that the ISI is negligible.

where the superscript i denotes the ith trial, and $z^i_{(0)} \geq z^i_{(1)} \geq \cdots \geq z^i_{(L_R-1)} \geq \cdots \geq z^i_{(L-1)}$ represents the ordered descending realization of the Z_l's in the ith trial.

Because the Z_l's are now supposed to be i.i.d. and the relation between L_R and L generally satisfies $L \gg L_R$, we can further approximate (12.19) as

$$E\left[\sum_{l=0}^{L_R-1} Z_{(l)}\right] \approx \lim_{n\to\infty} \frac{1}{n} \sum_{l=0}^{nL_R-1} z_{(l)}, \tag{12.20}$$

where the $z_{(l)}$'s, $l = 0, \ldots, nL_R - 1$, represent the nL_R largest realizations among the nL realizations of the random variable Z_l. Because $z_{(nL_R-1)}$, the smallest realization among the $z_{(l)}$'s, $l = 0, \ldots, nL_R - 1$, is no less than $nL - nL_R$ out of the total nL realizations, (12.20) can be approximated as

$$E\left[\sum_{l=0}^{L_R-1} Z_{(l)}\right] \approx L_R E\left[Z_l | Z_l \geq z_{l,q}\right], \tag{12.21}$$

where $z_{l,q}$ is the q-quantile of Z_l's distribution $F_{Z_l}(z)$, with $q = \frac{nL-nL_R}{nL} = \frac{L-L_R}{L}$.

However, the Z_l's are not identically distributed, so we need to calibrate the results obtained in (12.21). An upper bound of $E\left[\sum_{l=0}^{L_R-1} |h_{(l)}|^2\right]$ can be obtained by substituting the largest quantile in (12.21), i.e.,

$$E\left[\sum_{l=0}^{L_R-1} Z_{(l)}\right] \leq L_R E\left[Z_{(0)} | Z_{(0)} \geq z_{(0),q}\right], \tag{12.22}$$

and an approximation can be expressed as

$$E\left[\sum_{l=0}^{L_R-1} Z_{(l)}\right] \approx \sum_{l=0}^{L_R-1} E\left[Z_{(l)} | Z_{(l)} \geq z_{(l),q}\right], \tag{12.23}$$

where $z_{(0),q} \geq \cdots \geq z_{(L_R-1),q} \geq \cdots \geq z_{(L-1),q}$, and $Z_{(0)}, \ldots, Z_{(L_R-1)}$ are corresponding random variables.

As defined earlier in Section 12.2, $h[l]$ is a CSCG random variable with $E[|h[l]|^2] = e^{-\frac{lT_s}{\sigma_T}}$. Denote $\sigma_l^2 \triangleq e^{-\frac{lT_s}{\sigma_T}}$, then $\frac{|h[l]|^2}{\sigma_l^2/2} \sim \chi^2(k)$, with $k = 2$. In the special case of $k = 2$, a $\chi^2(k)$ distribution is equivalent to an exponential distribution Exp(λ) with $\lambda = \frac{1}{2}$. After some mathematical derivation, we can get the distribution function of Z_l as

$$F_{Z_l}(z) = \begin{cases} 1 - e^{-\frac{z}{\sigma_l^2}}, & z \geq 0, \\ 0, & z < 0. \end{cases} \tag{12.24}$$

Therefore, Z_l is also exponentially distributed, with mean $E[Z_l] = \sigma_l^2 = e^{-\frac{lT_s}{\sigma_T}}$. Solving the inverse function of $F_{Z_l}(z)$ and substituting $q = \frac{L-L_R}{L}$ yields the q-quantile of Z_l

$$z_{l,q} = -\sigma_l^2 \ln(1-q) = e^{-\frac{lT_s}{\sigma_T}} \ln\left(\frac{L}{L_R}\right). \tag{12.25}$$

Considering the approximation in (12.23), $z_{(l),q}$ is the $(l+1)$th largest q-quantile corresponding to $z_{l,q}$ in (12.25), and $Z_{(l)}$ corresponds to $Z_l \sim \text{Exp}(1/\sigma_l^2)$, we can get

$$E\left[Z_{(l)}|Z_{(l)} \geq z_{(l),q}\right] = \left(1 + \ln\left(\frac{L}{L_R}\right)\right) e^{-\frac{lT_s}{\sigma_T}}. \tag{12.26}$$

Then, the numerator of (12.18) can approximated as

$$E\left[\sum_{l=0}^{L_R-1} |h_{(l)}|^2\right] \approx \left(1 + \ln\left(\frac{L}{L_R}\right)\right) \sum_{l=0}^{L_R-1} e^{-\frac{lT_s}{\sigma_T}}, \tag{12.27}$$

and the upper bound in (12.22) becomes

$$E\left[\sum_{l=0}^{L_R-1} |h_{(l)}|^2\right] \leq L_R \left(1 + \ln\left(\frac{L}{L_R}\right)\right). \tag{12.28}$$

Note that for the $|h[l]|^2$'s, $l = 0, 1, \ldots, L-1$, when l is very large, $E[|h[l]|^2] = e^{-\frac{lT_s}{\sigma_T}}$ may become very small if $lT_s \gg \sigma_T$, and the $|h[l]|^2$'s with very small mean values can be negligible compared to those $|h[l]|^2$'s with large mean values. Therefore, to make the upper bound tight and the approximation more precise, we only keep the significant paths whose expected gain is larger than a predetermined parameter[4] ϵ, i.e., $E[|h[l]|^2] = e^{-\frac{lT_s}{\sigma_T}} \geq \epsilon$. The index of the last significant path is $L_c = \lceil \frac{\sigma_T}{T_s} \ln(\epsilon^{-1}) \rceil$, while the rest paths indexed $l = L_c + 1, L_c + 2, \ldots, L-1$ are neglected in the approximation. Replacing L with L_c in (12.27) and (12.28), and substituting them back into (12.18), we get approximate τ_P as

$$\tau_P \approx \left(1 + \ln\left(\frac{L_c}{L_R}\right)\right) \frac{1 - e^{-L_R T_s/\sigma_T}}{1 - e^{-LT_s/\sigma_T}}, \tag{12.29}$$

with an upper bounded

$$\tau_P \leq L_R \left(1 + \ln\left(\frac{L_c}{L_R}\right)\right) \frac{1 - e^{-T_s/\sigma_T}}{1 - e^{-LT_s/\sigma_T}}. \tag{12.30}$$

Because the number of taps of the CIR is in general much greater than the number of fingers of a Rake receiver, we usually have $1 - e^{-L_R T_s/\sigma_T} \ll 1 - e^{-LT_s/\sigma_T}$, and $L_R(1 - e^{-T_s/\sigma_T}) \ll 1 - e^{-LT_s/\sigma_T}$. Thus, the ratio of the power needed for a TR system to achieve the same performance as direct transmission is much less than 1. With a typical number of fingers $L_R = 4$ (for example, 3GPP2 recommends the Rake receiver shall provide a minimum of four fingers for the CDMA 2,000 system [17]) and the channel length L = 200, the value of (12.29) is about 0.1, which implies an order of magnitude reduction in power consumption. According to our experiment and simulation results with typical parameters setting, the energy needed for a TR-based transmission can be as low as 20% of that needed for a direct transmission with Rake receivers. When the rate back-off factor D is not large, both TR system and the direct

[4] Choices of different ϵ values will affect the approximation, e.g., a greater value of ϵ may tighten the upper bound. In this chapter, we fix $\epsilon = 10^{-3}$, a properly chosen value after trial-and-error, but how to choose a good ϵ is beyond the scope of the chapter.

transmission face the ISI problem. Although it is difficult to analyze accurately, it has been shown that [18] the temporal focusing effects of TR can significantly reduce the presence of ISI by reducing the channel delay spread. Thus, we expect a similar or even higher level of power reduction can be achieved. Therefore, TR is expected to achieve a much better power efficiency than direct transmission.

12.3.2 Interference Alleviation

In this section, we will compare the interference that a transmitter causes to an unintended receiver using TR-based transmission to that using direct transmission. Assume the CIR between the transmitter to the unintended victim receiver is

$$h_1[k] = \sum_{l=0}^{L-1} h_{1,l}\delta[k-l], \tag{12.31}$$

with $h_1[l]$ being the lth tap of the CIR and L the length of the CIR. Each $h_1[l]$ has the same distribution as $h[l]$, i.e., a circular symmetric complex Gaussian random variable with a zero mean and a variance $e^{\frac{-lT_S}{\sigma_T}}$, but they are assumed to be independent due to the location difference.

Then, we can express the received signal from the transmitter at the victim receiver with the TR-based transmission as

$$\begin{aligned} Y_1[k] = {} & a(h_1 * g)[L-1]X\left[k - \frac{L-1}{D}\right] && (Signal) \\ & + a \sum_{\substack{l=0 \\ l \neq (L-1)/D}}^{(2L-2)/D} (h_1 * g)[Dl]X[k-l] && (ISI) \\ & + a n_1[k] && (Noise) \end{aligned} \tag{12.32}$$

For simplicity, we still omit the ISI term by assuming that D is a large positive number, then the interference perceived by the victim receiver is equal to the signal power of $Y_1[k]$, i.e.,

$$I^{TR} = P|(h_1 * g)[L-1]|^2 = P\frac{\left|\sum_{l=0}^{L-1} h_1[l]h^*[l]\right|^2}{\sum_{l=0}^{L-1}|h[l]|^2}. \tag{12.33}$$

With direct transmission, the received signal perceived by the victim receiver can be written as

$$Y_1^{DT}[k] = (h_1 * X)[k] + n_1[k] = \sum_{l=0}^{L-1} h_1[l]X[k-l] + n_1[k]. \tag{12.34}$$

Then, the interference at the un-intended receiver can be expressed as

$$I^{DT} = E_X\left[\left|\sum_{l=0}^{L-1} h_1[l]X[k-l]\right|^2\right] = P^{DT}\sum_{l=0}^{L-1}|h_1[l]|^2, \tag{12.35}$$

and we can obtain the ratio of the interference caused by the two schemes as

$$r_I = \frac{I^{TR}}{I^{DT}}. \tag{12.36}$$

Define

$$\tau_I = \frac{E[I^{TR}]}{E[I^{DT}]} \tag{12.37}$$

as the ratio of the expected interference caused by TR and direct transmission. Substituting (12.33) and (12.35) into (12.37) and taking expectation with respect to h and h_1, we can approximate τ_I as

$$\begin{aligned}
\tau_I &\approx \tau_P \frac{E\left[\left|\sum_{l=0}^{L-1} h_1[l]h^*[l]\right|^2\right]}{E\left[\left(\sum_{l=0}^{L-1}|h[l]|^2\right)\left(\sum_{l=0}^{L-1}|h_1[l]|^2\right)\right]}\\
&= \tau_P \frac{\sum_{l=0}^{L-1}\left(E\left[|h[l]|^2\right]\right)^2}{E\left[\sum_{l=0}^{L-1}|h[l]|^2\right]\cdot E\left[\sum_{l=0}^{L-1}|h_1[l]|^2\right]}\\
&= \tau_P \frac{\sum_{l=0}^{L-1} e^{-\frac{2lT_s}{\sigma_T}}}{\left(\sum_{l=0}^{L-1} e^{-\frac{lT_s}{\sigma_T}}\right)^2}\\
&= \tau_P \frac{1+e^{-\frac{LT_s}{\sigma_T}}}{1-e^{-\frac{LT_s}{\sigma_T}}}\cdot\frac{1-e^{-\frac{T_s}{\sigma_T}}}{1+e^{-\frac{T_s}{\sigma_T}}},
\end{aligned} \tag{12.38}$$

where the second equality holds because $h[l]$ and $h_1[l]$ are i.i.d. random variables, and $h[l]$ and $h_1[k]$ are independent for $l \neq k$. Note that the ratio of the expected transmission power τ_P should be chosen according to (12.18) in order to maintain the same performance.

In general, the observation time LT_s satisfies $LT_s \gg \sigma_T$, and the sampling period T_S is much smaller than the delay spread σ_T, and thus we know that τ_I is much less than 1. According to our simulation results with typical parameters, under the ideal assumption that channel responses of two different locations are completely independent, interference could be made 20 dB lower by using the TR-based system. Even for a practical environment where correlation between channel responses does exist, our experiment measurements show that the interference alleviation can be up to 6 dB when the victim receiver is only 1 m away from the intended receiver. Therefore, the interference caused to an unintended receiver with the TR-based transmission is greatly reduced compared to direct transmission.

12.3.3 Multipath Gain of TR

Because TR can utilize the multipaths as virtual multi-antennas, the multiple paths can provide spatial diversity. In this section, we briefly talk about the maximum achievable diversity order of TR transmission.

We first consider a binary phase-shift keying (BPSK) signaling with amplitude $\sqrt{P}$, i.e., $X[k] = \pm\sqrt{P}$. By omitting the ISI term, the error probability of detecting X is

$$Q\left(\sqrt{\frac{P_{Sig}}{\sigma^2/2}}\right) = Q\left(\sqrt{2\left(\sum_{l=0}^{L-1} |h[l]|^2\right)\mathsf{SNR}}\right), \tag{12.39}$$

where $\mathsf{SNR} = P/\sigma^2$ is the signal-to-noise ratio per symbol time, and $Q(\cdot)$ is the complementary cumulative distribution function of an $N(0, 1)$ random variable. By averaging over the random tap gain h and following similar analysis as in [19], we can express the overall error probability as

$$\begin{aligned} p_e &\le \prod_{l=0}^{L-1}\left(1 + \mathsf{SNR} \cdot e^{-\frac{lT_s}{\sigma_T}}\right)^{-1} \\ &\le \prod_{l=0}^{L-1}\left(\mathsf{SNR} \cdot e^{-\frac{lT_s}{\sigma_T}}\right)^{-1} \\ &= \left(\prod_{l=0}^{L-1} e^{\frac{lT_s}{\sigma_T}}\right)(\mathsf{SNR})^{-L} \\ &= e^{\frac{L(L-1)T_s}{2\sigma_T}}(\mathsf{SNR})^{-L}. \end{aligned} \tag{12.40}$$

Thus, the maximum achievable diversity of TR-filtering is L. Similar conclusions can be drawn when other modulation schemes are used, such as quadrature amplitude modulation (QAM) and M-ary phase-shift keying (PSK). For example, if an M-QAM is used, the error probability of a symbol for a fixed channel can be represented by

$$4KQ\left(\sqrt{b_{\text{QAM}}\left(\sum_{l=0}^{L-1} |h[l]|^2\right)\mathsf{SNR}}\right) - 4K^2Q^2\left(\sqrt{b_{\text{QAM}}\left(\sum_{l=0}^{L-1} |h[l]|^2\right)\mathsf{SNR}}\right), \tag{12.41}$$

where $K = 1 - 1/\sqrt{M}$ and $b_{\text{QAM}} = 3/(M-1)$ [20]. Note that the second term can be dropped because we are interested in the upper bound, and similar derivation can be applied to show that the error probability is asymptotically proportional to $(\mathsf{SNR})^{-L}$.

We have assumed that multipaths on different channel taps are independent, and there are L independent multipaths in total, which account for the diversity order of L. In practice, however, it is possible that some multipath components on nearby channel taps are correlated, and there are possibly some channel taps on which no multipaths fall in. In that case, we only consider those independent multipaths, and according to our analysis, the diversity order of a TR system should be equal to the number of independent multipaths.

12.4 Simulation Results

In this part, we present some simulation results about the performance of TR transmission and justify the theoretical results derived in Section 12.3. Simulation results shown in this section are obtained by choosing $\sigma_T = 125T_s$ in the system model. We are interested in the impact of L_R (number of fingers of the Rake receiver) and L (number of channel taps) on the system performance. Because 3GPP2 recommends the Rake receiver shall provide a minimum of four fingers for the CDMA 2,000 system [17], and too many fingers may result in unaffordable complexity, we believe comparing the TR-based transmission with a Rake receiver who has four to eight fingers is a relative fair comparison.

In Figure 12.3, we compare r_P approximated in (12.29) (denoted by "theory") with the value of $E[r_P]$ by averaging r_P over 5,000 channel realizations. L_R is varied

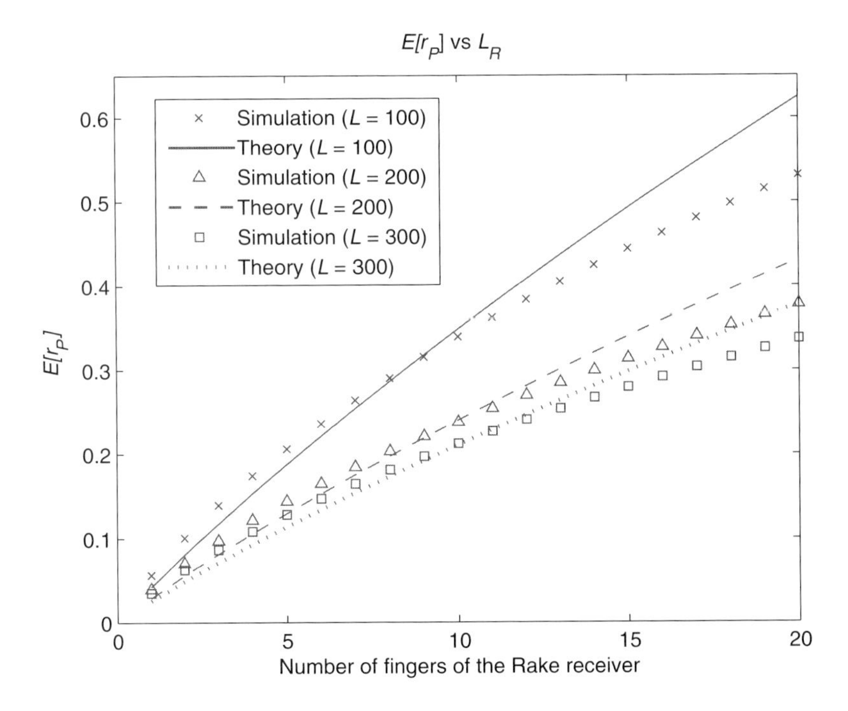

Figure 12.3 The expected ratio of energy needed for a TR-based communication system compared with an L_R-finger Rake receiver.

from 1 to 20, and L is chosen from $\{100, 200, 300\}$. We can see that, as an analytical approximation of $E[r_P]$, τ_P matches simulation results very well in a wide range of L_R ($1 \le L_R < 15$). When there are fewer fingers in the Rake receiver, direct transmission can only get a worse equalization. Thus, in order to have the same receiver performance, direct transmission costs an increasing amount of transmission power compared to TR, and TR becomes more energy-efficient than direct transmission, reflected by a decreasing $E[r_P]$. In addition, TR can benefit more from a richer multipath environment, as shown by the decrease in $E[r_P]$ when L increases from 100 to 300.

In Figure 12.4, we compare τ_I with $E[r_I]$, which is obtained by averaging r_I over 5,000 realizations. We can see that the τ_I also matches the simulation results $E[r_I]$ very well. Under the system model defined in Section 12.2, the interference caused by TR is 22 dB to 38 dB lower than the interference of direct transmission, depending on different choices of L_R and L. Under a normal parameter setting, e.g., $L = 200$ and $L_R = 6$, the interference of TR is about 30 dB lower, which indicates TR signal transmission can greatly reduce the interference and is thus much "greener."

To simplify the analysis of τ_P and τ_I, we have assumed D is so large that the ISI becomes negligible in Section 12.3. In order to better understand the impact of the parameter D on the transmit power reduction and the interference alleviation, we use simulations to demonstrate $E[r_P]$ and $E[r_I]$ where the ISI cannot be neglected in the received signal for both direct transmission and TR-based transmission. In Figure 12.5,

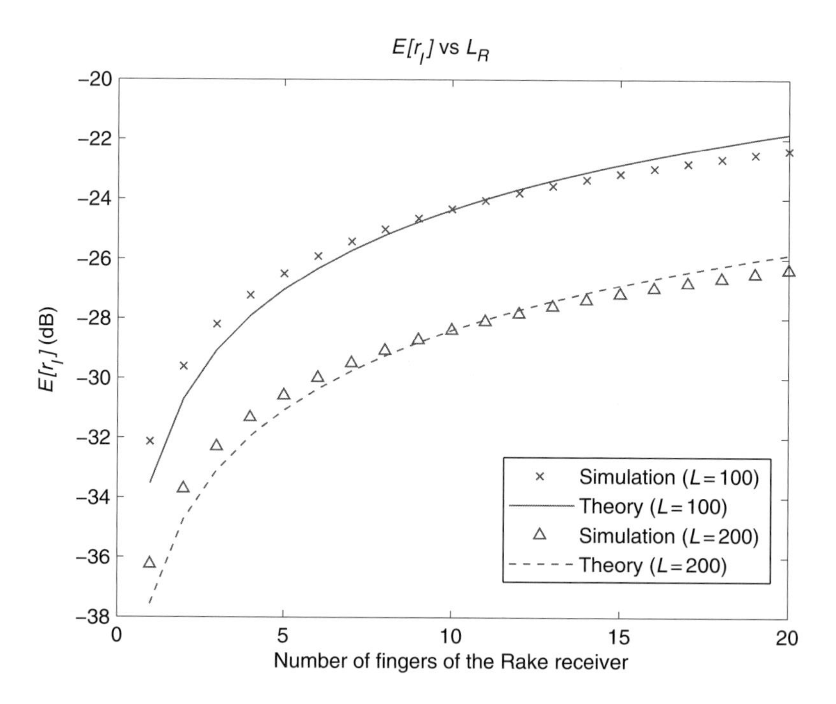

Figure 12.4 The expected interference alleviation of a TR-based communication system compared with an L_R-finger Rake receiver.

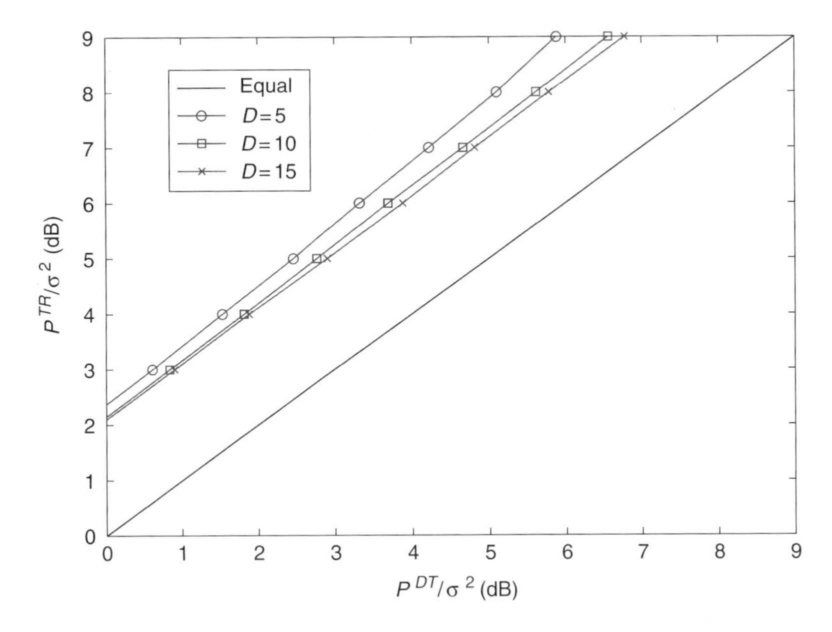

Figure 12.5 Expected transmit power needed for a TR-based system vs an L_R-finger Rake receiver (ISI nonnegligible).

we show the ratio between the transmit signal power required by the two schemes against the noise power, in order to achieve the same received SINR performance. For illustration purposes, we choose $L_R = 6$ fingers and $L = 21$ channel taps. The value of factor D is chosen from $\{5, 10, 15\}$ to represent very large, medium, and small ISI, respectively. The line with legend "equal" is used to represent the benchmark $P^{DT} = P^{TR}$ for comparing the transmit power of the two schemes. We see from Figure 12.5 that in order to achieve the same receiver performance, direct transmission usually requires 2~3 dB higher transmit power than TR-based transmission. In Figure 12.6, we show the interference power comparison when the transmit power of the two schemes follows the relation shown in Figure 12.5. We can see that the interference at a victim receiver caused by TR-based transmission is around 13 dB lower than that caused by direct transmission, when D varies in $[1, 15]$. This clearly shows that the capability of power reduction and interference alleviation of TR-based transmission remains even if we want to transmit the signals with a higher data rate, i.e., a smaller D.

In Section 12.2, in order to make the performance analysis tractable, we have assumed a specific channel model as defined in (12.2). In order to have a more comprehensive comparison on the performance of TR-based transmission and direct transmission, we also conduct numerical simulations under practical channel models. Although 3GPP channel model is a prevailing channel model, it does not fit in the presented TR-based scheme because 3GPP channel models only apply to narrow-band systems, while the TR-based scheme requires a frequency bandwidth of at least several hundred MHz so as to have plenty of multipath components. As will be seen in the next section, the

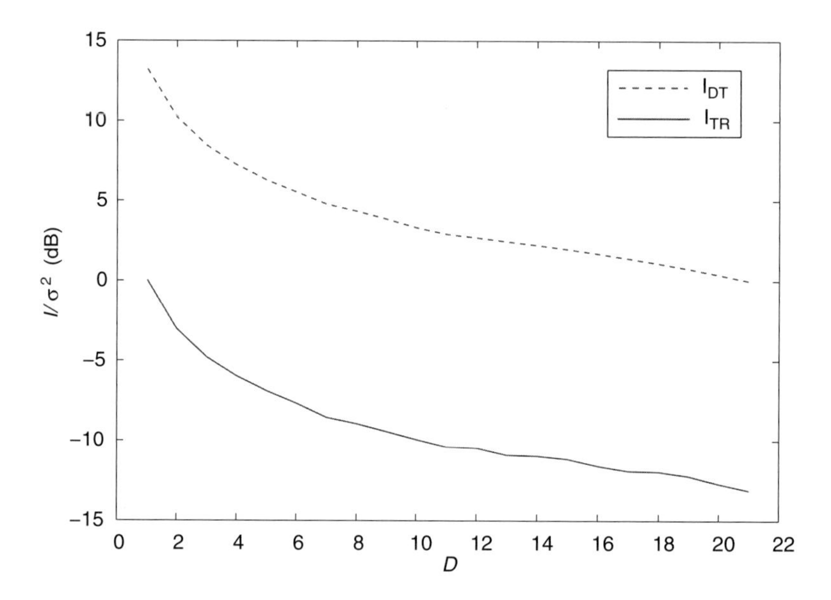

Figure 12.6 Expected interference alleviation of a TR-based system vs an L_R-finger Rake receiver (ISI nonnegligible).

bandwidth in the experimental measurements actually spans from 490 MHz to 870 MHz. Due to this reason, we chose the IEEE 802.15.4a channel model [21], which is a standard model for wideband transmissions, and simulated both the indoor LOS scenario with L $\sim$ 100 taps and the outdoor NLOS scenario with L $\sim$ 500 taps. The simulation results are shown in Figures 12.7 and 12.8, where the x-axis denotes the number of fingers of the Rake receiver varying from 1 to 20, and the y-axis denotes the expected power reduction and interference alleviation, respectively. As can be seen from these figures, compared to direct transmission using a six-finger Rake receiver, TR-based transmission only needs 62% transmit power while reducing the interference by 23 dB in an indoor environment, and for outdoor, TR only needs 48% transmit power while reducing the interference by 27 dB. These clearly show the advantage of TR-based scheme over direct transmission in a practical wireless channel.

Finally, in Figure 12.9, we show the multipath gain of TR, where the channel length is chosen[5] as $L = 5$, and the rate back-off factor is $D = 5$. We can see that in the high SNR regime, the diversity order of TR is around 5, which equals L and thus justifies the derivation in Section 12.3.3.

[5] Although the real channel length is generally much longer than the chosen parameter, computers cannot afford the simulation using real channel length that requires 10^L channel realizations to get an error bit. Therefore, we choose a much shorter multipath channel just for illustration purposes.

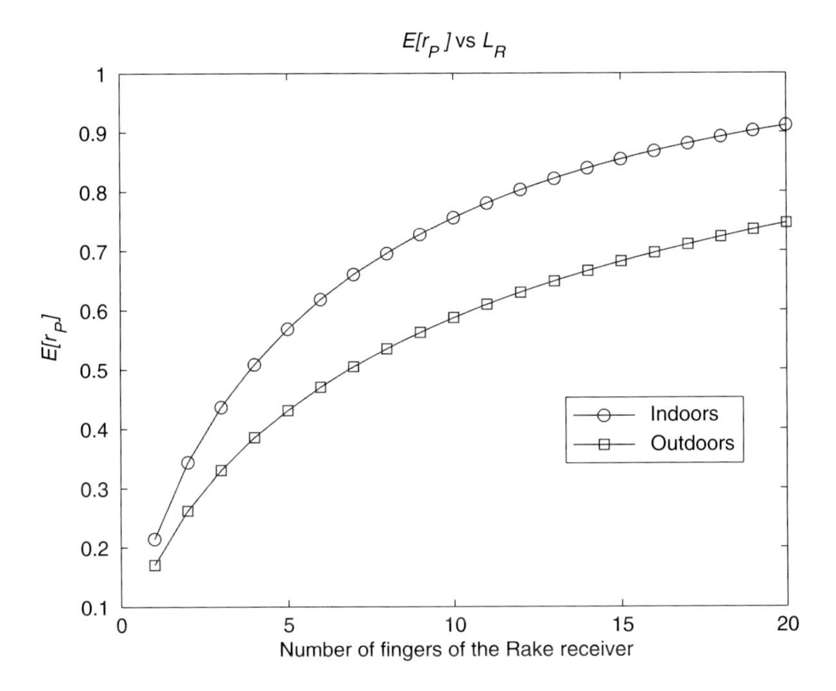

Figure 12.7 Expected ratio of energy needed for a TR-based system vs direct transmission (IEEE 802.15.4a channel model).

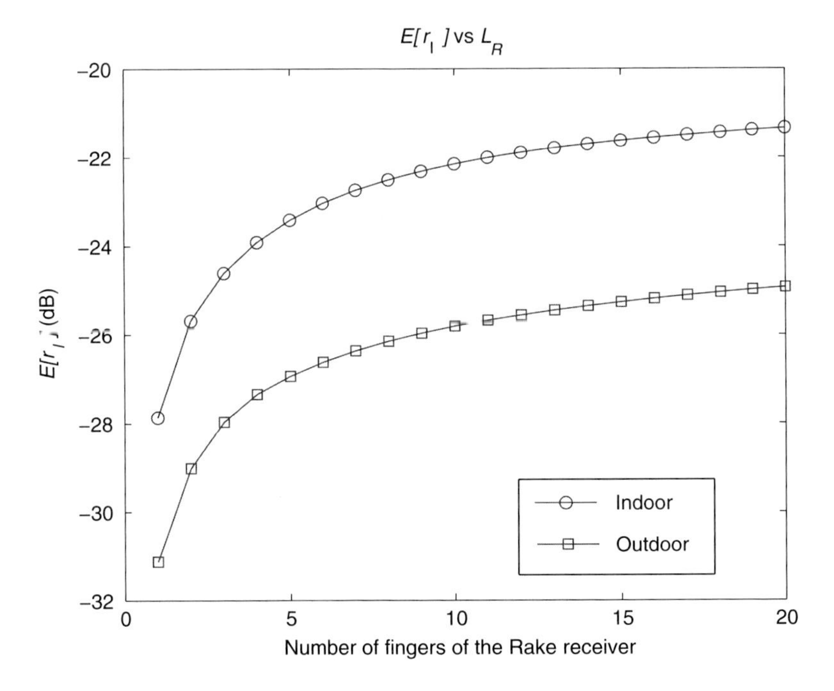

Figure 12.8 Expected interference alleviation for a TR-based system vs direct transmission (IEEE 802.15.4a channel model).

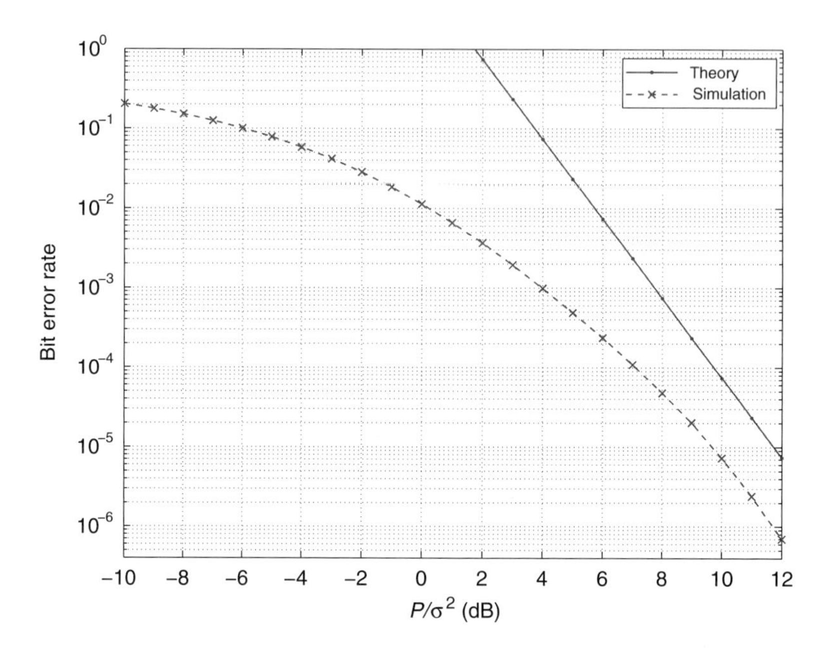

Figure 12.9 Illustration of the diversity order using the bit error rate (BER) curve.

12.5 Experimental Measurements

In this part, we demonstrate some experimental measurements taken in practical multipath channels. The tested signal bandwidth spans from 490 MHz to 870 MHz, centered at the carrier frequency 680 MHz. Two measurement sites are considered, an office room and a corridor, both of which are located on the second floor of the J. H. Kim Engineering Building at the University of Maryland. The layouts of the two sites are given in Figure 12.10, where transceiver A transmits time-reversed signals to transceiver B, and electromagnetic waves are reflected by walls, ceiling/floor, and other objects in the surrounding area. We fixed the location of transceiver A, whereas moving transceiver B in a rectangular area (the length is about four wave lengths) in the experiment.

12.5.1 Channel Impulse Response

In Figure 12.11, we show the amplitude of the channel impulse response (CIR) in the two tested sites. Due to the plentiful reflections by the walls of the small room, there are more paths (larger delay spread) for the office environment than in the corridor. Moreover, the amplitude also decays more slowly in the office environment because the signal waveforms are bounced back and forth and thus last longer in time. In Figure 12.11(c), we show the normalized magnitude of the received signals using the TR transmission in the corridor. We see clearly that TR can compress a substantial portion of signal power into very few taps, i.e., has the temporal focusing effect.

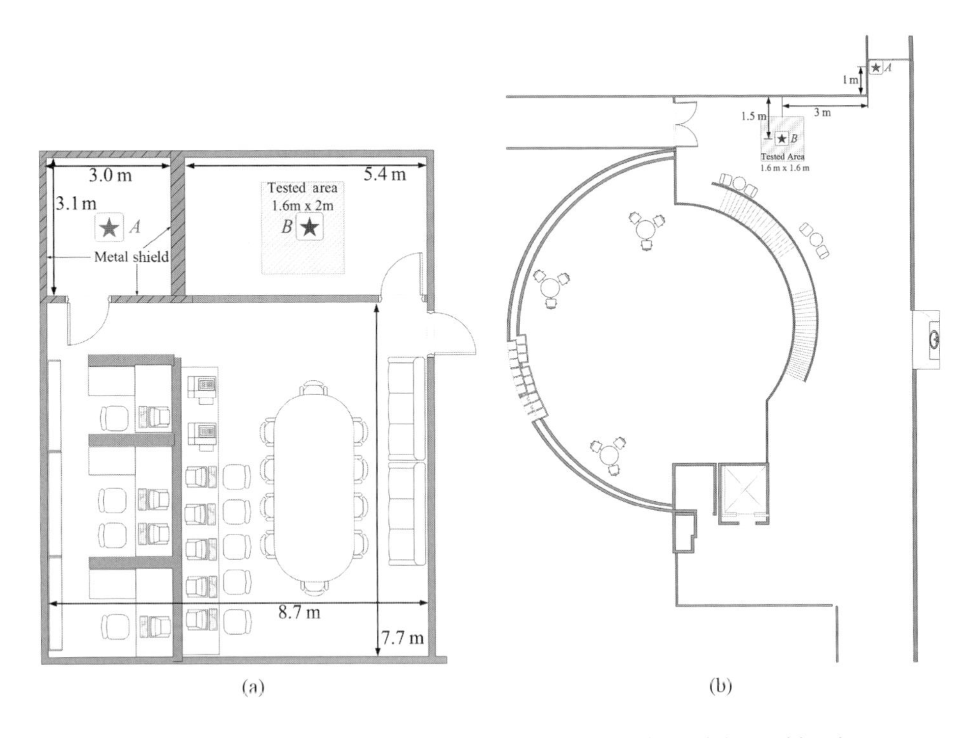

Figure 12.10 Floor plan and the layout of the test sites: (a) office site and (b) corridor site.

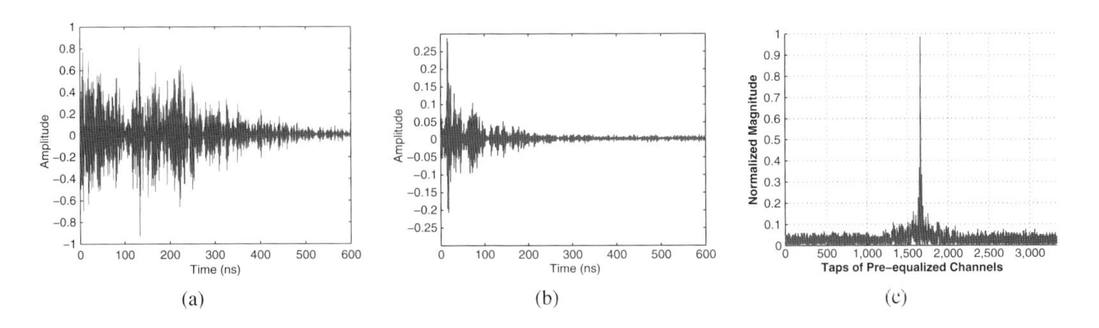

Figure 12.11 Channel impulse responses and temporal focusing effect obtained from experiments: (a) office, (b) CIR (corridor), and (c) temporal focusing effect (corridor).

12.5.2 Power Reduction

Due to the temporal focusing effect, TR can utilize the multipaths as multiple antennas to harvest energy from the environment. By varying the number of fingers of a Rake receiver for direct transmission, we show the ratio of the transmission power of a TR system over direct transmission in Figure 12.12. We can see that in order to achieve the same receiver performance, TR only costs as low as 30% of the transmission power of direct transmission, given that a Rake receiver usually has less ten than fingers, for both the office and the corridor. When the Rake receiver has six fingers, the ratio of

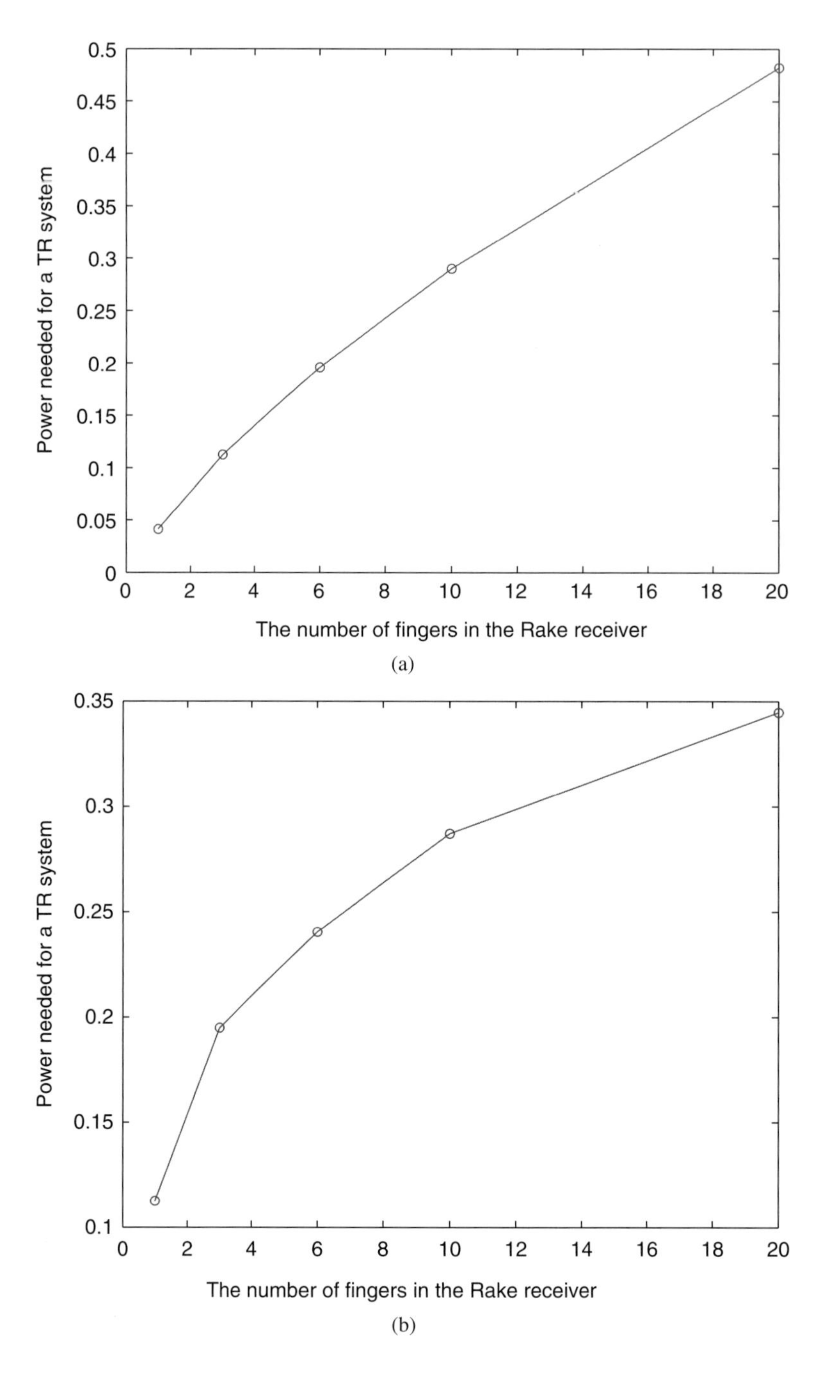

Figure 12.12 Power reduction by the TR-based transmission obtained by the experiment measurements in (a) r_I office and (b) r_I corridor.

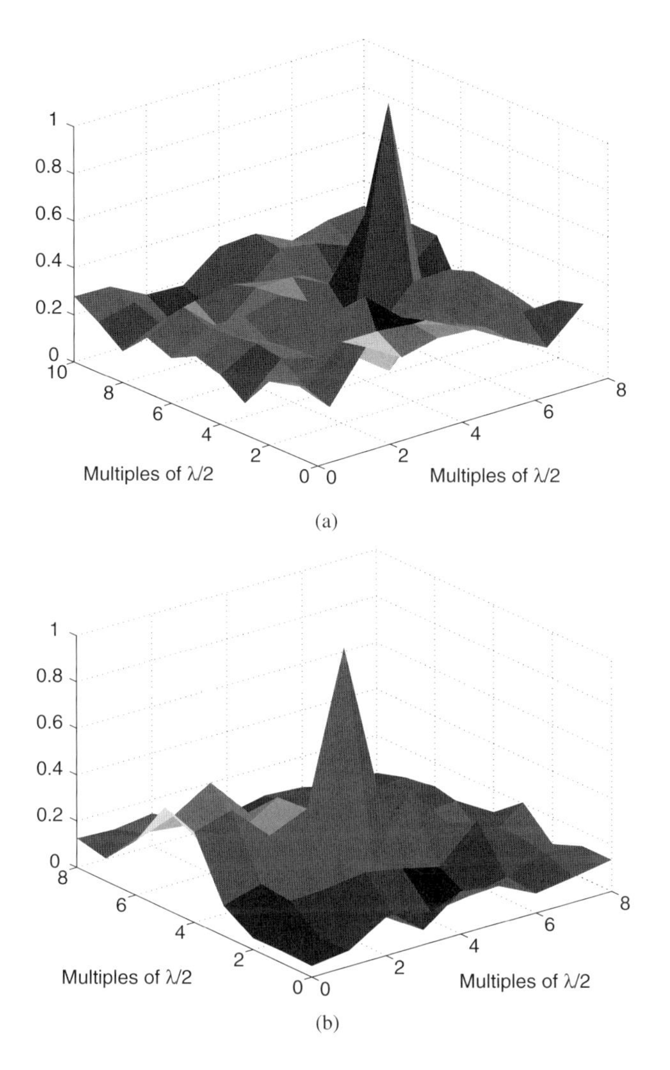

Figure 12.13 Spatial focusing effect of the TR-based transmission from the experiment measurements; received signal power in (a) office and (b) corridor.

power reduces to 20% for the office and 24% for the corridor. This shows that TR can achieve highly energy-efficient communication without requiring much complexity for the transmitter and the receiver. It is worth noting that the experimental measurement shown here in Figure 12.12 has a similar trend as the case of $L = 200$ in Figure 12.3.

12.5.3 Interference Alleviation

Besides energy-efficiency due to the temporal focusing effect, the time-reversed waves can also retrace the incoming paths, resulting in a spiky spatial signal power distribution focused at the intended receiver. This indicates by using TR, a transmitter will cause little interference to an un-intended receiver. In this part, we demonstrate the spacial focusing effect of TR and the resulting interference alleviation. In the experiment, we used the time-reversed CIR associated with the intended receiver as a basic waveform to load data streams and moved the receive antenna by a step size of $\lambda/2$, where λ is the wave length corresponding to carrier frequency 680 MHz.

The received signal power distribution (normalized by the peak power) in the spatial domain is shown in Figure 12.13. We see that the spike is centered at the intended receiver located at point (6, 6) for office measurements and (4, 4) for corridor measurements, whereas the received signal power in the other locations is only 20% to 30% of the signal power in the intended location. Therefore, it is highly possible that the interference leakage caused by a transmitter using the TR-based transmission will be much smaller than that without using TR. We show the interference ratio r_I between TR and direct transmission in Figure 12.14, assuming that the power ratio r_P corresponds

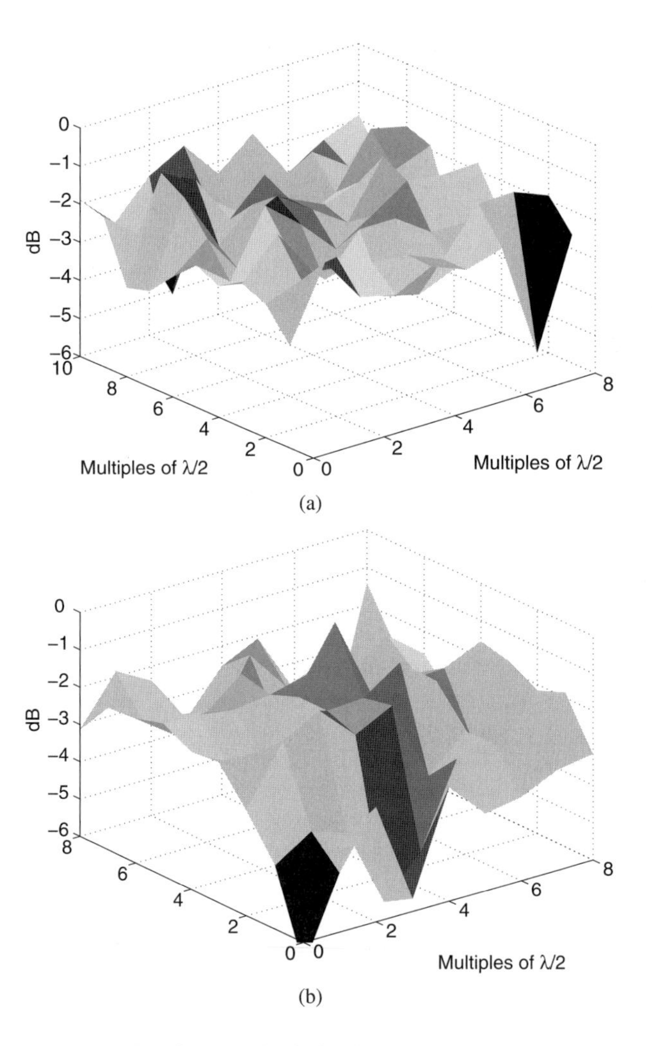

Figure 12.14 Interference alleviation by the TR-based transmission obtained by the experiment measurements: (a) r_P (office) and (b) r_P (corridor).

to a six-finger Rake receiver. We can see that on average the interference caused by TR transmission is 3 dB lower than the interference of direct transmission.

One may notice that the interference alleviation shown here is not as good as that shown in Figure 12.4. The reason is that our system model assumes ideal channel independence among different transmission pairs, e.g., when they are very far apart in space. Thus, the interference shown by the simulation results is much lower than the results obtained by measurements, where the channels are actually not perfectly independent but correlated. However, as shown in Figure 12.14, when the unintended receiver is 2λ (less than 1 m in our experiment) away from the intended receiver, the least interference caused by TR transmission can be as low as 6 dB lower, and we can expect even less severe interference when the unintended receiver is farther away. This result demonstrates that TR transmission has a high resolution of spatial selectivity and low pollution to the surrounding environment, which makes it a perfect candidate paradigm for future green wireless communications.

Furthermore, we can observe that the corridor site has better interference alleviation results than the office site. Because the office is a more enclosed environment where waves resonate between walls and lots of objects, the energy dissipates much slower, and the interference is relatively high. Hence, we can expect further reduction in interference if the communications take place in the outdoor environment, which is an open space.

12.5.4 Spectral Efficiency

A prerequisite of TR transmission is that the transmitter needs to use the time-reversed channel response as the basic waveform to load data. If the channel is fast fading, then the receiver needs to continuously transmit short pilot pulses to the transmitter so that the transmitter can get immediate CIR. In the worst case, the receiver needs to send pilot pulses before every transmission attempt of the transmitter, leading to a spectral efficiency of 50%. In this part, we use experiment results to show that the multipath channel of an office environment is actually not changing a lot. In this experiment, we measured the channel every 1 min, and a total of 40 channel snapshots were taken and stored. In the first 20 min, the testing environment was kept static; in the following 10 min, one experimenter walked randomly around the receive antenna (about 1.5 m–3 m away); in the last 10 min, the experimenter walked very close to the antenna (within 1.5 m). In other words, snapshots 1–20 correspond to a static environment, snapshots 21–30 correspond to a moderately varying environment, and snapshot 31–40 correspond to a varying environment.

We calculated the correlation coefficient between different snapshots to gain an idea of how the channel impulse response varies. Figure 12.15 illustrates the correlation matrix for this experiment, where each grid represents the correlation between two snapshots, whose indices are given by the x- and y-coordinates. Most correlation coefficients between static snapshots (1–20) are above 0.95, which implies that the channel responses are strongly correlated when the testing environment is static. When the experimenter moved around the antenna, some rays might be blocked, and additional reflection paths might be introduced. Therefore, the channel response will vary from

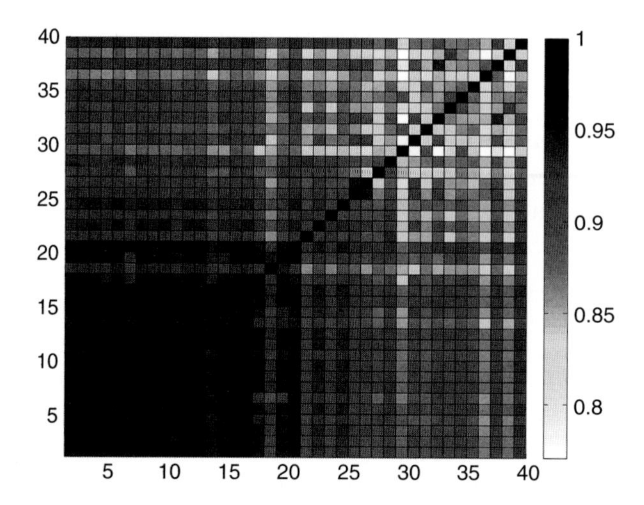

Figure 12.15 Correlation of channel responses at different time epochs.

its baseline, i.e., the static response. From our experiment, although the correlation drops when there are human activities near the antenna (snapshots 21–30) and becomes even weaker when the experimenter is very close to the antenna (snapshots 31–40), most of the coefficients are still higher than 0.8. This suggests that good correlation is maintained even if the environment is varying, and the achievable spectral efficiency will be much higher than 50%.

12.6 Time-Reversal Division Multiplexing and Security

Due to its special features and focusing effect, the TR-based communications will spark a series of unique wireless applications, in addition to the low-power low-interference green communications. In this section, we briefly introduce two prospective applications based on the TR communication technology.

12.6.1 Time-Reversal Division Multiplexing

In a multiuser system, different users have to find a way to share the wireless media. Traditional approaches include time division multiplexing (TDM), frequency division multiplexing (FDM), and code division multiplexing (CDM). The recent advance in multi-input multi-output (MIMO) has brought in a new multiplexing scheme named spatial division multiplexing (SDM), where different users can be distinguished by their channel response vectors due to the equipment of multiple antennas. In a rich scattering environment, because different users have different unique multipath profiles that depend on their physical locations and TR transmission treats each path like a virtual antenna, it is possible to utilize multipath profiles as a way to distinguish different users, which may facilitate the multiplexing. Therefore, a new TR-based multiplexing scheme, time-reversal division multiplexing (TRDM) for a multiuser downlink system can be developed [22].

The TRDM exploits the nature of the multipath environment, utilizes the location-specific signatures between the base station and multiple users to separate intended signals, and thus achieves satisfying performance. Furthermore, the TRDM approach will make possible numerous applications that require accurately locating the receiver, e.g., automatic inventory management in a warehouse, and wireless mailbox where a server could deliver information to a specific office in a building.

12.6.2 Time-Reversal Based Security

Secret communications have been of critical interest for quite a long time. Because of the fast technology evolution, a malicious attacker may easily find some low-cost radio equipment or easily modify the existing equipment to enable a potential intrusion. Moreover, wireless networks are extremely vulnerable to malicious attacks due to the broadcasting nature of wireless transmission and often a distributed network structure. As a result, traditional security measures may become insufficient to protect wireless networks. Therefore, TR-based communications can be exploited to enhance system security based on the unique location-specific multipath profile.

In a rich scattering wireless environment, multiple paths are formed by numerous surrounding reflectors. For receivers at different locations, the received waveforms undergo different reflecting paths and delays, and hence the multipath profile can be viewed as a *unique location-specific signature*. As this information is only available to the transmitter and the intended receiver, it is very difficult for other unauthorized users to infer or forge such a signature. It has been shown in [23] that even when the eavesdroppers are close to the target receiver, the received signal strength is much lower at the eavesdroppers than at the target receiver in an indoor application because the received signals are added incoherently at the eavesdroppers. The security based on multipath profiles is twofold: first, the multipath profile can be used to derive a symmetric key for the transmitter–receiver pair, which protects the secret information from malicious users; second, the transmitter can employ the TR-based transmission to hide the information from eavesdroppers, thanks to the spatial focusing effect.

The scheme is somehow like the direct sequence spread spectrum (DSSS) based secret communications. In DSSS communications, the energy of an original data stream is spread to a much wide spectrum band by using a pseudo-random sequence, and the signal is hidden below the noise floor. It is only those who know the pseudo-random sequence that could recover the original sequence from the noise-like signals. However, if the pseudo-random sequence has been leaked to a malicious user, that user is also capable of decoding the secret message. Nevertheless, for TR-based security, this would no longer be a problem because the underlying spreading sequence is not a fixed choice but instead a location-specific signature. For the intended receiver, the multipath channel automatically serves as a decipher that recovers the original data sent by the transmitter; for all other ineligible users at different locations, the signal that propagates to their receivers would be noise-like and probably is hidden below the noise floor. Therefore, malicious users are unable to recover the secret message because the security is inherent in the physical layer.

12.7 Summary

In this chapter, we argued and showed that a TR-based transmission system is an ideal candidate for green wireless communications. By receiving pilot pulses from the receiver and sending back the reversed waveforms, the transmitter can focus energy at the receiver in both spatial and temporal domains with high resolution, and thus harvest energy from the environment and cause less interference to other receivers. We have investigated the system performance, including power reduction, interference alleviation, and multipath diversity gain. The results show that the TR system has a potential of over an order of magnitude of reduction in power consumption and interference alleviation, as well as a very high multipath diversity gain. Both numerical simulations and experimental measurements have shown that TR-based transmission can greatly reduce transmission power consumption and inter-user interference. Moreover, strong channel correlation is also demonstrated, showing that TR can achieve green wireless communication with high spectral efficiency even in a time-varying environment. For related references, readers can refer to [24].

References

[1] M. Fink, C. Prada, F. Wu, and D. Cassereau, "Self focusing in inhomogeneous media with time reversal acoustic mirrors," *IEEE Ultrasonics Symposium*, vol. 1, pp. 681–686, 1989.

[2] C. Prada, F. Wu, and M. Fink, "The iterative time reversal mirror: A solution to self-focusing in the pulse echo mode," *Journal of the Acoustic Society of America*, vol. 90, pp. 1119–1129, 1991.

[3] M. Fink, "Time reversal of ultrasonic fields. Part I: Basic principles," *IEEE Transactions on Ultrasonic, Ferroelectronic, and Frequency Control*, vol. 39, no. 5, pp. 555–566, Sep. 1992.

[4] C. Dorme and M. Fink, "Focusing in transmit-receive mode through inhomogeneous media: The time reversal matched filter approach," *Journal of the Acoustic Society of America*, vol. 98, no. 2, part. 1, pp. 1155–1162, Aug. 1995.

[5] W. A. Kuperman, W. S. Hodgkiss, and H. C. Song, "Phase conjugation in the ocean: Experimental demonstration of an acoustic time-reversal mirror," *J. Acoustic Society of America*, vol. 103, no. 1, pp. 25–40, Jan. 1998.

[6] H. C. Song, W. A. Kuperman, W. S. Hodgkiss, T. Akal, and C. Ferla, "Iterative time reversal in the ocean," *Journal of the Acoustic Society of America*, vol. 105, no. 6, pp. 3176–3184, Jun. 1999.

[7] D. Rouseff, D. R. Jackson, W. L. Fox, C. D. Jones, J. A. Ritcey, and D. R. Dowling, "Underwater acoustic communication by passive-phase conjugation: Theory and experimental results," *IEEE Journal of Oceanic Engineering*, vol. 26, pp. 821–831, 2001.

[8] B. E. Henty and D. D. Stancil, "Multipath enabled super-resolution for RF and microwave communication using phase-conjugate arrays," *Physical Review Letters*, vol. 93, no. 24, pp. 243904(4), Dec. 2004.

[9] G. Lerosey, J. de Rosny, A. Tourin, A. Derode, G. Montaldo, and M. Fink, "Time reversal of electromagnetic waves," *Physical Review Letters*, vol. 92, pp. 193904(3), May 2004.

[10] M. Emami, M. Vu, J. Hansen, A. J. Paulraj, and G. Papanicolaou, "Matched filtering with rate back-off for low complexity communications in very large delay spread channels,"

Proceedings of the 38th Asilomar Conference on Signals, Systems and Computers, vol. 1, pp. 218–222, Nov. 2004.

[11] Y. Jin and J. M. F. Moura, "Time reversal imaging by adaptive interference canceling," *IEEE Transactions on Signal Processing*, vol. 56, no. 1, pp. 233–247, Jan. 2008.

[12] J. M. F. Moura and Y. Jin, "Detection by time reversal: Single antenna," *IEEE Transactions on Signal Processing*, vol. 55, no. 1, pp. 187–201, 2007.

[13] Y. Jin and J. M. F. Moura, "Time reversal detection using antenna arrays," *IEEE Transactions on Signal Processing*, vol. 57, no. 4, pp. 1396–1414, Apr. 2009.

[14] A. J. Goldsmith, *Wireless Communication*, New York: Cambridge University, 2005.

[15] K. Cheun, "Performance of direct-sequence spread-spectrum RAKE receivers with random spreading sequences," *IEEE Transactions on Communications*, vol. 45, no. 9, pp. 1130–1143, Sep. 1997.

[16] A. M. Law, *Simulation Modeling and Analysis*, 4th ed., New York: McGraw-Hill, 2007.

[17] 3GPP2, *Physical Layer Standard for CDMA2000 Spread Spectrum Systems*, Rev-E, Jun. 2010.

[18] P. Blomgren, P. Kyritsi, A. Kim, and G. Papanicolaou, "Spatial focusing and intersymbol interference in multiple-input-single-output time reversal communication systems," *IEEE Journal of Oceanic Engineering*, vol. 33, no. 3, pp. 341–355, Jul. 2008.

[19] D. Tse and P. Viswanath, *Fundamentals of Wireless Communication*, Cambridge, UK: Cambridge University Press, 2005.

[20] M. K. Simon and M. S. Alouini, "A unified approach to the performance analysis of digital communication over generalized fading channels," *Proceedings of the IEEE*, vol. 86, no. 9, pp. 1860–1877, Sep. 1998.

[21] A. F. Molisch, B. Kannan, D. Cassioli, C. C. Chong, S. Emami, A. Fort, J. Karedal, J. Kunisch, H. Schantz, U. Schuster, and K. Siwiak, "IEEE 802.15.4a channel model – final report," *IEEE 802.15-04-0662-00-004a*, Nov. 2004.

[22] F. Han, Y. H. Yang, B. Wang, Y. Wu, and K. J. R. Liu, "Time-reversal division multiple access in multi-path channels," *IEEE Global Telecommunications Conference (GLOBECOM 2011)*, pp. 1–5, 2011.

[23] X. Zhou, P. Eggers, P. Kyritsi, J. Andersen, G. Pedersen, and J. Nilsen, "Spatial focusing and interference reduction using MISO time reversal in an indoor application," *IEEE Workshop on Statistical Signal Processing (SSP 2007)*, pp. 307–311, 2007.

[24] B. Wang, Y. Wu, F. Han, Y. H. Yang, and K. J. R. Liu, "Green wireless communications: A Time-reversal paradigm," *IEEE Journal of Selected Areas in Communications*, vol. 29, no. 8, pp. 1698–1710, Sep. 2011.

13 Power Waveforming

This chapter explores the idea of time-reversal (TR) technology in wireless power transfer to present a new wireless power transfer paradigm termed as power waveforming (PW), where a transmitter engages in delivering wireless power to an intended receiver by fully utilizing all the available multipaths that serve as virtual antennas. Two power transfer-oriented waveforms, energy waveform and single-tone waveform, are discussed for PW power transfer systems, both of which are no longer time-reversal in essence. The former is designed to maximize the received power, while the latter is a low-complexity alternative with small or even no performance degradation. We theoretically analyze the power transfer gain of the presented power transfer system over the direct transmission scheme, which can achieve about 6 dB gain, under various channel power delay profiles and show that the PW is an ideal paradigm for wireless power transfer because of its inherent ability to recollect all the power that is possible to harvest from the surrounding environment. In addition, the outage performances in harvesting power of the PW system and the conventional multiple-input multiple-output (MIMO) system are derived. It reveals that the PW system can achieve the same outage performance of the MIMO system as long as the number of resolvable multipaths is sufficiently large. Simulation results validate the analytical findings, and experimental results demonstrate the effectiveness of the presented PW technique.

13.1 Introduction

The advent of Internet of Things (IoT) era is anticipated to facilitate ubiquitous wireless connections of many devices, enabling not only data collection from the surrounding environments but also data exchange and interaction with other devices. Unlike the traditional wireless communications that are primarily limited by the availability of spectrum resource, wireless devices in the future are further subject to an energy-hungry problem due to the explosive growth of wireless data services [1]. In particular, such an embarrassment is unavoidable when wireless devices are untethered to a power grid and can only be supplied by batteries with limited capacity [2]. In order to prolong the network lifetime, one immediate solution is to frequently replace batteries before the battery is exhausted, but unfortunately, this strategy is inconvenient, costly, and dangerous for some emerging wireless applications, e.g., sensor networks in monitoring toxic substances.

Recently, energy harvesting has attracted a lot of attention in realizing self-sustainable wireless communications with perpetual power supplies [3, 4]. Being equipped with a rechargeable battery, a wireless device is solely powered by the scavenged energy from the natural environment such as solar, wind, motion, vibration, and radio waves. While the ambient energy sources are environmentally friendly, the random, uncontrollable, and unpredictable characteristics make it difficult to ensure the quality-of-service (QoS) in wireless communications. For example, the intensity of sunlight is affected by the time of the day, the current weather, the seasonal weather and pattern, and the surrounding environment, to name but a few [5].

Alternatively, wireless power transfer using electromagnetic radiation has been recognized as an effective and viable technology to provide reliable energy sources for dedicated low-powered wireless devices, e.g., sensors and radio-frequency identification (RFID) [6]. A comprehensive survey of wireless charging technologies, along with the progress in standardization and the recent advances in network applications, has been studied in [7]. The conventional wireless power transfer technologies include electromagnetic induction coupling, magnetic resonant coupling, and radio frequency (RF) signals [8]. The energy transfer efficiency of the first two technologies is higher than 80%, whereas they are only suitable for short-distance energy transfer applications within a distance of a wavelength. On the other hand, electromagnetic radiation, which appears in the form of RF signals, can propagate up to tens of meters, and the power can be extracted at receive antennas through rectifier circuits; however, the energy transfer efficiency is relatively low due to the severe propagation loss. As compared with the ambient energy sources, the main advantage of wireless power transfer is that a dedicated electromagnetic radiation source is capable of delivering an on-demand energy supply. In this chapter, we will focus on the investigation of an RF-signal-based power transfer system because it is expected to play an important role in unplugged wireless applications in the near future.

The design of the RF signal-based power transfer systems faces two essential challenges. First, the required power sensitivity of an energy receiver is much higher than that for an information receiver, e.g., −50 dBm for information receivers and −10 dBm for energy receivers [9]. Second, the power density of wireless signals at receive antennas is degraded by multipath channels, shadowing effect, and large-scale path loss, and only a small fraction of energy emitted from the transmitter can be harvested at the receiver, resulting in an energy scarcity problem. By harnessing spatial degree-of-freedom of multiple antennas, beamforming techniques have been widely applied to combat the severe signal power loss over distance and to enhance the energy transfer efficiency. Specifically, in wireless power transfer, multiple transmit antennas facilitate focusing a sharp energy beam at an intended receiver, while multiple receive antennas enlarge the effective aperture area.

Various energy beamforming schemes have been studied in the literature [9, 11–16]. In [10], the authors emphasized the performance enhancement by employing multi-antenna techniques, and a tradeoff between wireless information and power transfer was investigated under two cases: a limited feedback multi-antenna technique and a large-scale multiple-input multiple-output (MIMO) technique. In [11], a new network architecture was proposed to enable mobile recharging, where a cellular system is overlaid with randomly deployed power stations, which could radiate power isotropically or directionally toward mobile users by beamforming. In [12], the authors considered wireless-powered communications with a multi-antenna access point, where users' data transmission in the uplink fully hinges on the wireless power from the access point in the downlink. The authors in [13] designed energy beamforming for wirelessly charging multiple RFID tags, and both channel-training energy and energy allocation weights were jointly optimized. In [14], a distributed energy beamforming scheme was developed for reaching a compromise between simultaneous wireless information and power transfer in two-way relay channels, and superimposed energy and information-bearing signals were proposed to enhance the achievable sum rate. A MIMO energy transfer system was considered in [15] in Rician fading channels with the joint optimization of channel acquisition and transmit beamforming, while an energy beamforming scheme was investigated in [16] for a multiuser MIMO energy transfer system with one-bit channel feedback. Some existing works applied a massive antenna array for wireless power transfer [17–24]. In [17], an outage probability for which the energy harvested by a node is less than the energy spent on uplink pilots was analyzed with a huge number of transmit antennas. The works in [18] and [24] focused on a MIMO system equipped with a large-scale antenna array and examined the impact of the number of antennas on the energy transfer performance.

However, there are two drawbacks when utilizing multiple antennas for wireless power transfer. First, additional RF chains increase the implementation cost. Second, for indoor rich scattering environments, which are the most desired application scenarios for wireless energy transfer, the beamforming schemes may not work appropriately because the line-of-sight (LOS) link may be blocked by impenetrable objects or attenuated by penetration loss.

Motivated by the preceding discussions, three fundamental and interesting questions are raised: (1) Can/how a wireless power transfer system wirelessly charge a remote device in non-line-of-sight (NLOS) environments? (2) Can/how wireless power transfer be enabled by a low-complexity system, e.g., a single antenna? (3) Can/how the multipath signals, if they exist, act as a useful resource for wireless power transfer? Generally speaking, the efficiency of wireless power transfer using omnidirectional antennas in the NLOS channels could be very poor because the scatters in the environments will disperse the radiated power to multiple replicas of transmitted signals. In this regard, it is possible to sustain the receiver with a better harvested power level if the power on each multipath can be constructively recollected. While the wireless power transfer has been extensively investigated in the recent literature, the aforementioned questions have not been fully addressed and remain to be answered, and the use of multipaths brings a new research opportunity in designing the wireless power transfer technologies.

Thanks to its ability to make full use of multipath propagation, time-reversal (TR) transmission in a rich-scattering environment has been recognized as an ideal paradigm for low-complexity single-carrier broadband wireless communications [20, 21].

The transmission consists of two phases. During the first phase, a receiver sends an ideal impulse signal to a transmitter for probing the channel impulse response (CIR) of the link. With channel reciprocity, the transmitter simply sends a time-reversed conjugate waveform according to the CIR, which is also called a basic TR waveform, in order to leverage the multipath channel as a cost-free matched filter and to refocus the signal power at the receiver. This phenomenon is commonly referred to as a spatial-temporal focusing effect [22] because it concentrates the signal power at a particular time instant and an intended spatial location. Despite this focusing advantage, the large delay spread causes severe intersymbol interference (ISI), thus waste of energy, especially when the data rate is high. To compensate for this vulnerability, a new waveform, referred to as MaxSINR waveform, was designed in [23] to maximize the signal-to-interference plus noise ratio (SINR) at the receiver. However, none of these previous works attempts to design a wireless power transfer system by taking good advantage of multipath propagation to sustain the received power at an energy receiver. Also the performance of such a wireless power transfer system has not been investigated.

In this chapter, we discuss the new concept of power transfer waveform design generalized from the TR system, called power waveforming (PW), that utilizes all available multipaths as virtual antennas to recollect all power that can be possibly harvested for power transfer and provide a quantitative performance analysis of the PW power transfer system. Specifically, the major points of this chapter are summarized as follows:

- Most of the existing works studied wireless power transfer problems in flat fading channels, e.g., massive MIMO systems in [24], multiple-input single-output (MISO) broadcasting systems in [25], and joint design of training and power transfer in [15]. Some recent works, like [26–28], considered orthogonal frequency division multiplexing (OFDM)-based wireless power transfer in frequency selective fading channels. However, the first two designs mainly focus on the optimization of power and sub-band allocation for simultaneous information and wireless power transfer, and the sub-bands are assumed to be independent of each other. The third design also optimizes the net harvested energy over independent sub-bands for MISO systems. To the best of our knowledge, this chapter is the first one to comprehensively study the use/impact of multipaths on the wireless power transfer. An optimal energy waveform is discussed for the PW power transfer system to maximize the received power at the receiver side by constructively recollecting the dispersed signal power in the multipath channel.
- We also discover a single-tone waveform, which only concentrates its full waveform power on the principle frequency component with the largest amplitude. As compared with the energy waveform, the single-tone waveform has much lower computational complexity in implementation, whereas it can achieve a comparable (near optimal) power transfer performance. We show that the single-tone waveform is exactly the same as the optimal energy waveform, if the transmitted signal is J-periodic, where J is an integer ratio between the number of multipaths and the backoff factor.

- A rigorous performance analysis for the single-tone waveform is conducted, and several quantitative findings are provided. We first define a power transfer gain, which is a ratio between the average power transfer performances of the single-tone waveform and the direct transmission, and this gain can serve as a lower bound for the gain of the optimal energy waveform over the direct transmission. A finite-integral expression of the power transfer gain is derived under general power delay profiles. Furthermore, we obtain a closed-form expression for the power transfer gain in uniform power delay profiles (UPD) as well as a performance lower bound in triangular power delay profiles (TPD).
- The outage performances in harvesting power of the presented PW power transfer system and the conventional MIMO power transfer system with transmit beamforming are theoretically derived. We consider a new notion of power transfer outage to measure the performance of wireless power transfer, which is different from the traditional definition adopted in information decoding. Specifically, a power transfer system is in outage if the harvested power is not larger than a preset threshold. Our analysis shows that the PW system offers great promise in achieving a comparable performance to the MIMO system if the number of resolvable multipaths is sufficiently large. The presented PW power transfer system, however, only requires a single transmit and receive antenna. For example, a PW system with six paths and 24 paths can achieve the same outage probability of 0.9 as an MIMO system with $(N_T, N_R) = (2, 1)$ and $(3, 1)$, respectively, where N_T and N_R are the numbers of transmit and receive antennas.
- Extensive computer simulations are performed to validate the theoretical results in various channel models, including ultra-wideband (UWB) channels in LOS and NLOS office environments, UPD, TPD, and exponential decay power delay (EPD) channel profiles, etc. Several waveforms, including direct transmission, energy waveform, single-tone waveform, basic TR waveform, and MaxSINR waveform, are simulated for performance comparisons. It is found that the power transfer performance is dependent on the extent of the multipath effect, and the power transfer gain provided by the single-tone waveform over the direct transmission is approximately given by 6 dB under practical multipath conditions. In addition, real experiments are conducted to verify the performance of the presented waveforms in LOS and NLOS experimental settings in indoor environments, and it is shown that the PW technique can improve the wireless power transfer efficiency by about 400 percent $\sim$800 percent (6 dB$\sim$ 9 dB) with a single antenna only, as compared with direct transmission.

The following notations are used throughout this chapter. The uppercase and lowercase boldface letters denote matrices and vectors, respectively. The notations $(\cdot)^T$, $(\cdot)^\dagger$, $(\cdot)^*$ and $(\cdot)^{-1}$ stand for transpose, conjugate transpose, element-wise conjugate, and inverse operation, respectively. The matrix $\mathbf{I}_N$ represents an $N \times N$ identity matrix. The notations $\mathbb{E}[\cdot]$ and $\|\mathbf{x}\|_2$ take expectation and the Euclidean norm of a vector $\mathbf{x}$. The operator max (x, y) takes the maximum value between x and y, while the operator min (x, y) finds the minimum value.

The rest of this chapter is organized as follows. In Section 13.2, we introduce the system model for the PW systems, generalized from the TR systems, with multipath channels. Two power transfer waveforms are presented in Section 13.3. Section 13.4 is devoted to analyzing the power transfer gain under different channel power delay profiles. The outage performances of the PW system and the conventional MIMO system are analyzed and compared in Section 13.5. Simulation and numerical results are presented in Section 13.6, and experimental results are given in Section 13.7.

13.2 System Model

In this chapter, we consider an L-tap wireless fading channel, and the channel is assumed to be quasi-static over the observation time. The CIR between the transmitter and the receiver can be modeled as

$$h[n] = \sum_{l=0}^{L-1} h_l \delta[n-l], \quad n = 0, \ldots, L-1, \tag{13.1}$$

where $\delta[n]$ is the Kronecker delta function, and h_l is the channel gain of the lth path, which is a circularly symmetric complex Gaussian random variable with zero mean and variance $\mathbb{E}[|h_l|^2] = \rho_l$, for $l = 0, \ldots, L-1$. Without loss of generality, we assume that the total channel power is one, i.e., $\sum_{l=0}^{L-1} \rho_l = 1$. For simplicity, it is further assumed that the paths of CIR are uncorrelated with each other, i.e., $\mathbb{E}\left[h_i h_j^*\right] = 0$, for $i \neq j$. Besides, the in-phase and quadrature components of each path are also uncorrelated with each other and contain identical power. Note that in practice, the number of resolvable paths increases as the system bandwidth becomes wider, and the number of resolvable paths will reach a limit when the system bandwidth is sufficiently large, which is the maximal number of resolvable paths in the wireless environments.

We first review the signal processing procedures in the TR systems, where a focusing effect is created at the intended receiver by exploiting the multipath propagation to recollect all the possibly available path energy from the wireless environment [20]. To achieve this, the receiver first sends a pilot signal with a delta-like autocorrelation function to measure the CIR at the transmitter side during a channel probing phase. With the channel reciprocity assumption, the transmitter then forms a waveform $g[n]$ to send data symbols according to the channel state information (CSI). Typically, the baud rate is much lower than the sampling rate. Let $\{v_D[n]\}_{n=0}^{(N-1)D}$ be the up-sampling signal of data symbols $\{v[n]\}_{n=0}^{N-1}$, where N is the number of total transmitted data symbols, given by

$$v_D[n] = \begin{cases} v\left[n/D\right], & n \bmod D = 0\,; \\ 0, & \text{otherwise}\,, \end{cases} \tag{13.2}$$

where D is a rate back-off factor, which is a ratio of the sampling rate and the baud rate, and $\mathbb{E}\left[|v[n]|^2\right] = P$. Hence, the transmitted signal $\{s[n]\}_{n=0}^{L-1+(N-1)D}$ after waveform mapping can be expressed as[1]

$$s[n] = (v_D * g)[n] . \tag{13.3}$$

Furthermore, the signal received at the receiver, $\{y_D[n]\}_{n=0}^{2L-2+(N-1)D}$, is given as

$$\begin{aligned} y_D[n] &= (h * s)[n] + z[n] = (f * v_D)[n] + z[n] \\ &= \sum_{l=0}^{2L-2} f[l]\, v_D[n-l] + z[n] , \end{aligned} \tag{13.4}$$

where $f[n] = (h * g)[n]$ is defined as an equivalent impulse response, for $n = 0, \ldots, 2L - 2$, and $z[n]$ is additive complex white Gaussian noise with zero mean and variance σ_z^2. A basic TR waveform is a time-reversed conjugate version of the CIR, given as [20]

$$g_{TR}[n] = \frac{h^*[L-1-n]}{\sqrt{\sum_{l=0}^{L-1} |h[n]|^2}}, \quad n = 0, \ldots, L-1 . \tag{13.5}$$

From (13.4), it is found that the wireless channel naturally performs the operation of matched filtering with respect to $g_{TR}[n]$, and a peak is observed in the received signal $y_D[n]$ for data detection. To further suppress the ISI, a waveform, $g_{SINR}[n]$, is designed in [23] to maximize the SINR. It is also worth mentioning that if there is no particular design for the waveform, i.e., $g_{DT}[n] = \delta[n]$, the considered system model is degenerated to the conventional direct transmission scheme. The waveforms $g_{TR}[n]$, $g_{SINR}[n]$ and $g_{DT}[n]$, however, are not eligible for power transfer because they are mainly designed from the perspective of information transfer, and the ISI due to the multipath effect is not appropriately utilized as a green source of energy. The PW power transfer system follows the same signal processing procedures and models presented in (13.2) to (13.4), but new waveforms are designed for the purpose of wireless power transfer, and $v[n]$ is defined as a noninformation-bearing sequence. For simplicity, we assume that this sequence is random in order to evaluate the wireless power transfer performance in an average sense over distinct random sequences. In the following section, we aim at finding the power transfer waveforms for the PW systems.

[1] The PW schemes are implemented digitally. Before it is radiated, the discrete-time signal is converted into a continuous-time signal with common signal processing at the transmitter, e.g., pulse-shaping filters, digital-to-analog converters, up-converters. For simplicity, equivalent baseband representation is considered here.

13.3 Power Transfer Waveform Designs

13.3.1 Optimal Energy Waveform

By substituting (13.2) into (13.4) and applying change of variables, it yields

$$
\begin{aligned}
y_D[n] &= \sum_{l=\lceil (n-2L+2)/D \rceil}^{\lfloor n/D \rfloor} f\,[n-lD]\,v\,[l] + z\,[n] \\
&= \sum_{l=\lceil (n-2L+2)/D \rceil}^{\lfloor n/D \rfloor} \sum_{m=0}^{L-1} g\,[m]\,h\,[n-lD-m] v\,[l] + z\,[n] \ ,
\end{aligned}
\tag{13.6}
$$

where $\lceil \cdot \rceil$ and $\lfloor \cdot \rfloor$ are the ceiling and floor functions, respectively. Our design goal is to find an optimal energy waveform to maximize the average power transferred to the receiver side, given by

$$
\mathbf{g}_E = \arg \max_{\|\mathbf{g}\|_2^2=1} \left(\lim_{N\to\infty} \frac{1}{2L-1+(N-1)\,D} \sum_{n=0}^{2L-2+(N-1)D} \mathbb{E}\left[|y_D[n]|^2 \right] \right), \tag{13.7}
$$

where $\mathbf{g} = \left[g\,[0]\,, \ldots, g\,[L-1] \right]^T$ is the waveform vector. Let $\mathbf{y}_p = [y_D[L-1+(p-1)D+1], \ldots, y_D[L-1+pD]]^T$ be the pth segment of the received signal $y_D[n]$, and $\mathbf{H}$ be a $(2L-1) \times L$ Toeplitz channel matrix with the column vector $[h\,[0]\,, \ldots, h\,[L-1]\,, 0, \ldots, 0]^T$ as its first column. We can rewrite (13.6) in a compact matrix form as follows:

$$
\mathbf{y}_p = \sum_{q=\lceil p-1-(L-2)/D \rceil}^{\lfloor p+(L-1)/D \rfloor} \mathbf{H}_{(q-p)} \mathbf{g} \cdot v\left[q\right] + \mathbf{z}_p \ , \tag{13.8}
$$

where $\mathbf{z}_p$ represents the noise term contained in the pth segment, and the matrix $\mathbf{H}_j$ is defined as

$$
\mathbf{H}_j = \left[\mathbf{O}^T_{(a_l-L+(j+1)D)\times L}, \mathbf{B}^T, \mathbf{O}^T_{(L-1-jD-a_u)\times L} \right]^T , \tag{13.9}
$$

where $a_l = \max\left(0, \left(L-(j+1)\,D\right)\right)$, $a_u = \min\left(2L-2, \left(L-1-jD\right)\right)$, and $\mathbf{B}$ is a submatrix of the Toeplitz channel matrix $\mathbf{H}$, containing the entries from the a_l^{th} row to the a_u^{th} row of $\mathbf{H}$. Without loss of generality, we assume that the number of channel taps is finite, and the time duration of the transmitted signals is sufficiently larger, i.e., $N \gg L$. From (13.7), as N goes to infinity, the average power transfer maximization problem is equivalent to maximizing the sum power of the pth segment:

$$
\mathbf{g}_E = \arg \max_{\|\mathbf{g}\|_2^2=1} \mathbb{E}\left[\left\| \mathbf{y}_p \right\|_2^2 \right] , \tag{13.10}
$$

where the total power of the designed waveform is subject to one.

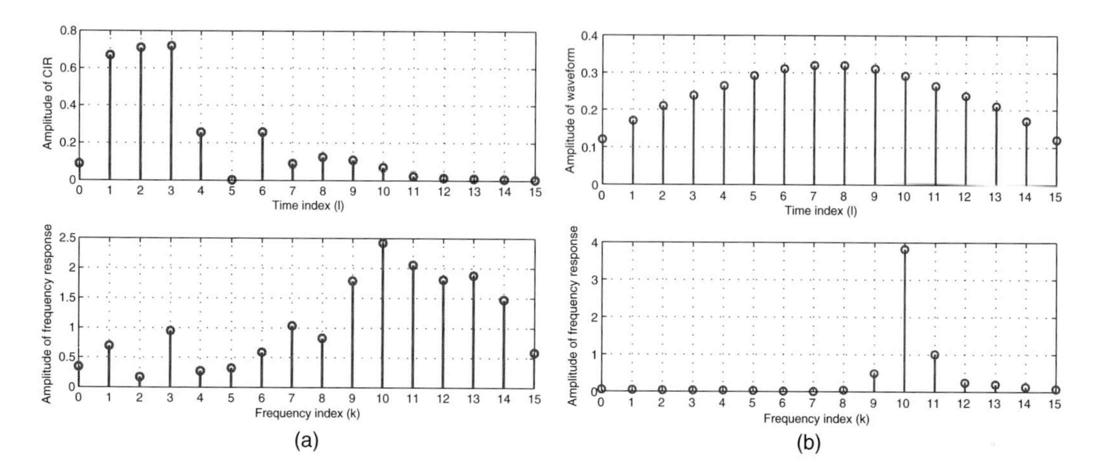

Figure 13.1 An example of the optimal energy waveform for UWB channels in NLOS office environments: (a) UWB channels and (b) optimal energy waveform.

THEOREM 13.3.1 *If the signals $v[n]$ are uncorrelated, i.e., $\mathbb{E}[v[i]v^*[j]] = 0$, $\forall i \neq j$, the optimal energy waveform is given by*

$$\mathbf{g}_E = \mathbf{u}_1 , \tag{13.11}$$

where $\mathbf{u}_1$ is the principal eigenvector of the matrix $\sum_{q=\lceil -1-(L-2)/D \rceil}^{\lfloor (L-1)/D \rfloor} \mathbf{H}_q^\dagger \mathbf{H}_q$.

PROOF. The proof is provided in Appendix A. ▲

An example of the optimal energy waveform for UWB channels is demonstrated in Figure 13.1 under NLOS environments [29]. Figure 13.1(a) shows the amplitude of the CIR as well as the corresponding channel frequency response (CFR). Figure 13.1(b) sketches the amplitude of the optimal energy waveform and its frequency response, which is the discrete Fourier transform (DFT) of the time-domain waveform $\mathbf{g}_E$. A closer look at this figure reveals that the energy waveform tends to concentrate its waveform power on those frequency components ($k = 9$, 10, and 11) with the largest amplitude of the CFR. Interestingly, each tap of the optimal energy waveform has comparable gain in order to spread out the energy of the transmitted signals over the entire time duration, which is different from other waveform designs as depicted in Figure 13.2.

13.3.2 Single-Tone Waveform

Define $\mathbf{F}$ as a DFT matrix, whose (m,n)th entry is given by $\frac{1}{\sqrt{L}} e^{-j2\pi mn/L}$, for $m, n = 0, \ldots, L-1$, and the CFR of the CIR $\mathbf{h} = [h[0], \ldots, h[L-1]]^T$ can be calculated as $\zeta = \mathbf{F}\mathbf{h} = [H[0], \ldots, H[L-1]]^T$. Motivated by the observation in Figure 13.1, a single-tone waveform $\mathbf{g}_{ST} = [g_{ST}[0], \ldots, g_{ST}[L-1]]^T$ is discussed as follows:

$$g_{ST}[n] = \frac{1}{\sqrt{L}} e^{-j\frac{2\pi k_{\max} n}{L}}, \quad n = 0, \ldots, L-1 , \tag{13.12}$$

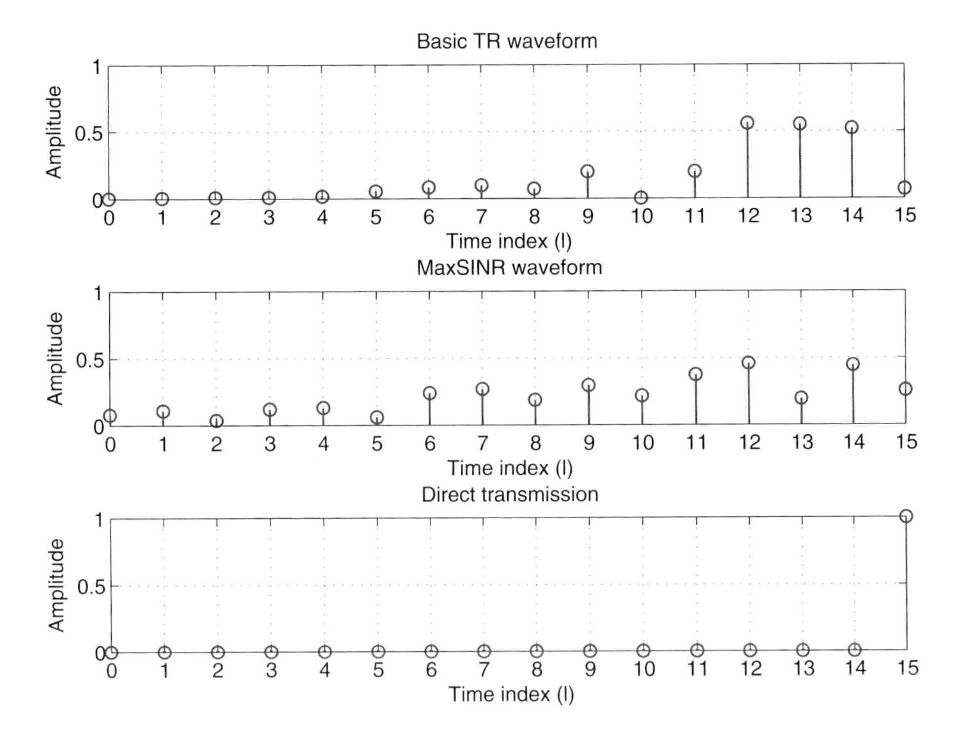

Figure 13.2 An example of different waveform designs, including basic TR waveform, MaxSINR waveform, and direct transmission, for UWB channels in NLOS office environments.

where $k_{\max} = \arg\max_k |H[k]|^2$ denotes the principal frequency component. Remember that the optimal energy waveform is designed by maximizing the average received power in time domain; hence, it is superior to any other waveforms, and the harvested power with the single-tone waveform can serve as a performance lower bound for the maximum received power that can be achieved by the optimal energy waveform. In the following, we provide a theorem to link the relationship between the simple single-tone waveform and the optimal energy waveform.

Theorem 13.3.2 *Assume $L = J \cdot D$, where J is a positive integer number. If the signal $v[n]$ is periodic with the period J, the single-tone waveform is the optimal energy waveform for* (13.10).

Proof. The proof is provided in Appendix B. ▲

Because the waveform has a finite length, the single-tone waveform is not the optimal for achieving the maximum average received power if the signal $v[n]$ is not periodic. The preceding theorem provides a condition for the single-tone waveform to achieve the maximum power transfer value at the receiver. The intuition behind this theorem can be explained as follows. When the transmitted signal is J-periodic, the convolution relation among the transmitted signal $v[n]$, the wireless channel $h[n]$, and the waveform $g[n]$ in (13.6) becomes the circular convolution; hence, the optimal design of the waveform is to concentrate the total waveform power on the frequency component

Table 13.1 Complexity ratio between optimal energy and single-tone waveforms ($D = 1$)

L	5	10	15	20	25	30	35
Complexity ratio	32	90	173	278	404	550	716

with the largest amplitude gain in the frequency domain. In general cases, the energy waveform is the optimal design for wireless power transfer, whereas the simulation results in Section 13.6 will demonstrate that the single-tone waveform is near optimal with a minor performance loss. When it comes to the implementation cost, one can find from (13.11) and (13.12) that the single-tone waveform has an advantage of low computational complexity over the energy waveform. We evaluate the complexity of the two waveforms in terms of the required number of complex multiplications. To be explicit, the calculation of the energy waveform in (13.11) involves a sum of matrix–matrix products and a principal eigenvector of a matrix, leading to a complexity of $L^3 + \left(\left\lfloor \frac{L-1}{D} \right\rfloor - \left\lceil -1 - \frac{L-2}{D} \right\rceil + 1\right) DL^2$. On the other hand, the single-tone waveform in (13.12) requires a DFT and element-wise products to select the main frequency component of the CFR, having a complexity of $L \log_2 L + L$. Taking $D = 1$ as an example, the complexity ratio between the energy waveform, and the single-tone waveform is approximately given by $\frac{3L^2}{\log_2 L}$ when $L \gg 1$, and some numerical results for different values of L are listed in Table 13.1.

13.4 Performance Analysis

Because the performance analysis of the optimal energy waveform is almost intractable, we alternatively analyze the power transfer gain of the presented single-tone scheme in PW systems as compared with the direct transmission scheme. With loss of generality, we focus on the scenario when the transmitted signals is J-periodic because the single-tone waveform in this case can deliver the maximum received power at the receiver side as the optimal energy waveform. By doing so, the analysis result can help us capture the fundamental limit of the power transfer in multipath environments.

From (13.29) to (13.31), for a given multipath channel, the average power transfer for the single-tone waveform $\mathbf{g}_{ST}$ at the receiver side can be expressed as

$$P_{ST} = \mathbb{E}\left[\left\|\mathbf{y}_P\right\|_2^2\right] = P \cdot \mathbf{g}_{ST}^{\dagger}\mathbf{C}^{\dagger}\mathbf{C}\mathbf{g}_{ST} \,. \tag{13.13}$$

Here we ignore the noise term in performance analysis because the noise power is commonly much smaller than the desired received power.[2] By substituting $\mathbf{g}_{ST} = \mathbf{F}^{\dagger}\mathbf{e}_{k_{\max}}$ into P_{ST} and applying the property $\mathbf{FCF}^{\dagger} = \sqrt{L} \cdot Diag\left(\boldsymbol{\zeta}\right)$, we get

[2] In practice, the power sensitivity of an energy harvesting receiver is required to be larger than -10 dBm for efficient RF-DC conversion [4]. Hence, the received signal strength for wireless power transfer must be much higher than the noise power level, e.g., -94 dBm for a bandwidth of 100 MHz.

$$P_{ST} = P\mathbf{e}_{k_{\max}}^{\dagger}\mathbf{F}\mathbf{C}^{\dagger}\mathbf{C}\mathbf{F}^{\dagger}\mathbf{e}_{k_{\max}} = P \cdot L|H[k_{\max}]|^2 . \tag{13.14}$$

On the other hand, because the direct transmission waveform is expressed as $\mathbf{g}_{DT} = \mathbf{e}_0$, the power transfer performance of the direct transmission scheme for a given channel, $P_{DT} = P \cdot \mathbf{g}_{DT}^{\dagger}\mathbf{C}^{\dagger}\mathbf{C}\mathbf{g}_{DT}$, can be expanded as

$$P_{DT} = P\mathbf{e}_0^{\dagger}\mathbf{F}^{\dagger}\mathbf{F}\mathbf{C}^{\dagger}\mathbf{C}\mathbf{F}^{\dagger}\mathbf{F}\mathbf{e}_0 = P \cdot \sum_{k=0}^{L-1} |H[k]|^2 . \tag{13.15}$$

Definition 13.4.1 *The power transfer gain G is a ratio between the average power transfer of the single-tone waveform and that of the direct transmission scheme in wireless channels, i.e., $G = \mathbb{E}[P_{ST}]/\mathbb{E}[P_{DT}]$.*

Specifically, from (13.14) and (13.15), it leads to

$$G = L \cdot \mathbb{E}\left[|H[k_{\max}]|^2\right] , \tag{13.16}$$

where $\mathbb{E}\left[\sum_{k=0}^{L-1} |H[k]|^2\right] = \sum_{l=0}^{L-1} \rho_l = 1$ according to (13.1). Therefore, it is found that the power transfer gain depends on the frequency selection diversity gain over the correlated multivariate Rayleigh random variables $\varepsilon_k = |H[k]|^2$, for $k = 0, \ldots, L-1$. Intuitively, a larger power transfer gain can be obtained, if the multipath channel is more frequency selective.

Before we begin the performance analysis of the exact power transfer gain, the following theorem regarding the characteristic function of the multivariate random variables ε_k under a given channel power delay profile is first provided.

Theorem 13.4.1 *Let $\boldsymbol{\omega} = \left[\omega_0, \ldots, \omega_{L-1}\right]^T$. For a multipath channel with the power delay profile $\boldsymbol{\rho} = \left[\rho_0, \ldots, \rho_{L-1}\right]^T$ in (13.1), the characteristic function of $\boldsymbol{\varepsilon} = \left[\varepsilon_0, \ldots, \varepsilon_{L-1}\right]^T$ is*

$$\Psi_{\varepsilon}(j\boldsymbol{\omega}) = \frac{1}{\det\left(\mathbf{I}_L - 2j \cdot Diag(\boldsymbol{\omega})\,\Sigma\right)} , \tag{13.17}$$

where $j = \sqrt{-1}$, $\Sigma = \mathbf{F}^{-1} Diag(\boldsymbol{\lambda})\,\mathbf{F}$, $\boldsymbol{\lambda} = \left[\lambda_0, \ldots, \lambda_{L-1}\right]^T$, $\lambda_l = \frac{1}{4}\left(\rho_l + \rho_{((L-l)\%L)}\right)$, for $l = 0, \ldots, L-1$, and the notation $(y\%L)$ takes the modulo operation over y with respect to L.

Proof. The proof is provided in Appendix C. ▲ With Theorem 13.4.1, a finite-integral expression for the power transfer gain is theoretically derived in the following theorem.

Theorem 13.4.2 *For a multipath channel with the power delay profile $\boldsymbol{\rho} = [\rho_0, \ldots, \rho_{L-1}]^T$, the power transfer gain G is*

$$G = L \int_0^{\infty} \left(1 - \frac{1}{(2\pi)^L} \int_{-\infty}^{\infty} \cdots \int_{-\infty}^{\infty} \Psi_{\varepsilon}(j\boldsymbol{\omega}) \times \prod_{l=0}^{L-1} \left(\frac{1 - e^{-j\omega_l x}}{j\omega_l}\right) d\omega_0 \cdots d\omega_{L-1}\right) dx . \tag{13.18}$$

Proof. The proof is provided in Appendix D. ▲

While the power transfer gain G is analyzed in Theorem 13.4.2, it is difficult to analytically express the power transfer gain in a closed-form for general power delay profiles as the involved multiple integrals prohibit further simplification. Instead, to get more insight into the effect of the parameters in power delay profiles, e.g., the number of multipaths, on the power transfer gain while keeping the analysis analytically tractable, we consider two power delay profiles: an UPD profile and a TPD profile, which are defined in the following.

Definition 13.4.2 *A power delay profile is called UPD with L channel paths, if* $\rho_l = \frac{1}{L}$, *for* $l = 0, \ldots, L-1$.

Definition 13.4.3 *A power delay profile is called TPD with L channel paths, if* $\rho_l = \frac{2(1-L\rho_0)}{(L-1)L} l + \rho_0$, *for* $l = 0, \ldots, L-1$, *where* $\frac{1}{L} \leq \rho_0 \leq \frac{2}{L}$.

Notice that in Definition 13.4.3, the TPD channel profile is derived under the assumption that the channel path power is linearly decayed with respect to the path delay. When the path power ρ_0 and the number of multipaths L are specified, the TPD profile can be determined with the unity constraint of the total channel power, i.e., $\sum_{l=0}^{L-1} \rho_l = 1$. Although the EPD profile is a common channel model in studying the performance, it is worth mentioning that the TPD profile could be a good approximation for the EPD profile, especially when the tails of the profile are insignificant. Hence, the TPD profile could be a good alternative to the investigation of the inherent power transfer gain in wireless channels owing to the difficulty in performance analysis by exploiting the EPD profile.

Theorem 13.4.3 *In a wireless channel with the UPD channel profile, the power transfer gain G is*

$$G(L) = \sum_{n=1}^{L} \frac{1}{n}. \tag{13.19}$$

Proof. The proof is provided in Appendix E. ▲

This theorem gives two immediate remarks in the following.

Remark 13.4.1 *If* $L = 1$, *the power transfer gain* $G(1) = 1$.

In other words, both the direct transmission scheme and the PW system with the single-tone waveform have the same power transfer performance if there is only one LOS path in the wireless environments between the transmitter and the receiver.

Remark 13.4.2 *In the UPD channel profile, the power transfer gain* $G(L)$ *is monotonically nondecreasing with the number of multipaths L. Moreover, when the number of multipaths is increased from* $L-1$ *to* L, *the gap of* $G(L) - G(L-1)$ *is given by* $\frac{1}{L}$.

From this remark, we can make three important observations for understanding the interaction between the power transfer gain and the number of multipaths. First, the power transfer gain is monotonically increasing with the number of multipaths. From the system point of view, the larger value of L means that either the bandwidth is wider or the environment becomes more scattering. Second, the improvement on the power transfer gain becomes gradually small as the number of resolvable channel paths is enlarged. Third, the result is analogous to the signal-to-noise power ratio (SNR) increase in the selection combining diversity scheme of a single-input multiple-output (SIMO) system when the number of the receive antennas is increased by one [33]. This is because the single-tone waveform design in the PW system has a similar mathematical model to an SIMO system with antenna selection.

Now we turn to the performance analysis of the power transfer gain in wireless channels with the TPD channel profile. Because it is very tough to deal with the multiple integrals in this case, a performance lower bound is offered in the following theorem.

Theorem 13.4.4 *In a wireless channel with the TPD channel profile, the power transfer gain G for $L \geq 2$ is lower bounded by*[3]

$$G \geq \frac{L\rho_0 - 1}{L-1} + \frac{L(1-\rho_0)}{L-1} \sum_{n=1}^{L} \binom{L}{n} (-1)^{n+1} \frac{1}{n} . \tag{13.20}$$

Proof. The proof is provided in Appendix F. ▲

We can further simplify the lower bound formula by substituting $\rho_0 = \frac{a}{L}$, for $1 \leq a \leq 2$, into (13.20), yielding

$$G(L) \geq 1 + \frac{L-a}{L-1} \sum_{n=2}^{L} \frac{1}{n} . \tag{13.21}$$

Remember that the parameter a reflects the decline rate of the TPD channel profile, which becomes flatter as the value of a is closer to one. Some interesting remarks are then summarized here. First, for a fixed number of multipaths L, the smaller the value of a is, the larger the lower bound is obtained. Second, for a fixed value of a, the lower bound is monotonically increased with the number of multipaths because the function $\frac{L-a}{L-1}$ in (13.21) is an increasing function of L, for $1 \leq a \leq 2$. Third, the lower bound is tight when $a = 1$ or $L = a = 2$ or $L \to \infty$. The first case is because the TPD profile is degenerated to the UPD profile. For the second case, the TPD profile is degenerated to a single-path channel according to Definition 13.4.3, i.e., $\rho_0 = 1$ and $\rho_1 = 0$, and thus $G(2) = 1$. The third case can be verified from the proof in Appendix F that the variance of $h_{0,2}$ approaches zero when $L \to \infty$.

13.5 Comparisons between PW Systems and MIMO Systems

In this section, we study the outage performances of the power transfer for the PW systems and the conventional MIMO systems as well as their performance comparisons. It is addressed that the definition of the outage probability is different from the traditional definition used for evaluating the information transfer performance, e.g., SINR and sum rate. The considered outage probability is defined in the following.

[3] For the case of $L = 1$, the power transfer gain is equal to one.

DEFINITION 13.5.1 *A power transfer system is said to be in outage if the harvesting power P_{EH} is smaller than or equal to a preset threshold x, and the outage probability in harvesting power is given as* $\Pr(P_{EH} \leq x)$.

First, an $N_T \times N_R$ MIMO system model is introduced, where N_T and N_R are the numbers of transmit and receive antennas, respectively. We assume that the CSI is perfectly available at the transmitter side, and a transmit beamforming technique is applied for wireless power transfer. Hence, the received signal at the receiver can be expressed as

$$\tilde{\mathbf{y}} = \tilde{\mathbf{H}}\tilde{\mathbf{g}}v + \mathbf{z}\,, \tag{13.22}$$

where v is the transmitted signal with the power value P, i.e., $\mathbb{E}\left[|v|^2\right] = P$, $\tilde{\mathbf{g}}$ is the beamforming vector, $\tilde{\mathbf{H}}$ is an $N_R \times N_T$ identically, independently distributed (i.i.d.) MIMO channel matrix, each entry of which is assumed to be a zero-mean complex Gaussian random variable with the variance one for a fair comparison. Besides, the norm of the beamforming vector is subject to one, i.e., $\left\|\tilde{\mathbf{g}}\right\|_2 = 1$. By ignoring the noise term, which accounts for an insignificant amount of the entire received signals in reality, the received power is given by $\mathbb{E}\left[\tilde{\mathbf{y}}^\dagger\tilde{\mathbf{y}}\right] = P \cdot \tilde{\mathbf{g}}^\dagger\tilde{\mathbf{H}}^\dagger\tilde{\mathbf{H}}\tilde{\mathbf{g}}$. The power transfer can be maximized by choosing the principal eigenvector of the matrix $\tilde{\mathbf{H}}^\dagger\tilde{\mathbf{H}}$, and the power transfer performance for a given channel is computed as

$$P_{MIMO} = P \cdot \lambda_{\max}\left(\mathbf{H}^\dagger\mathbf{H}\right) = P \cdot \lambda_{\max}\left(\mathbf{H}\mathbf{H}^\dagger\right), \tag{13.23}$$

where $\lambda_{\max}(\cdot)$ represents the maximum eigenvalue. Notice that the matrix $\mathbf{H}\mathbf{H}^\dagger$ is of the complex Wishart distribution. By directly applying Corollary 2 in [34], the outage probability of the MIMO power transfer can be derived in (13.24),

$$\Pr(P_{MIMO} \leq x) = \frac{\det(\mathbf{B}(x))}{\prod_{k=1}^{\min(N_T,N_R)} \Gamma(\max(N_T,N_R) - k + 1) \cdot \Gamma(\min(N_T,N_R) - k + 1)}. \tag{13.24}$$

where $\Gamma(\cdot)$ is the gamma function, $\mathbf{B}(x)$ is a matrix, whose (i,j)th element is given by $\gamma\left(\max(N_T,N_R) - \min(N_T,N_R) + i + j + 1, \frac{x}{P}\right)$, for $i, j = 0, 1, \ldots, \min(N_T,N_R) - 1$, and $\gamma(\cdot,\cdot)$ is the lower incomplete gamma function as specified in [35].

From (13.41) and (13.44), it is known that $F_{\varepsilon_{\max}}(x) = \Pr\left(|H[k_{\max}]|^2 \leq x\right) = \left(1 - e^{-Lx}\right)^L$ when the UPD channel profile is considered. Using (13.14), we then obtain the outage performance of the single-tone waveform under the UPD channel profile:

$$\Pr(P_{ST} \leq x) = \left(1 - e^{-x/P}\right)^L. \tag{13.25}$$

It is interesting to compare the outage performances of the PW system and the conventional MIMO system under the same transmit power value P and examine how many multipaths are necessary for the PW system to achieve a comparable outage performance as that provided by the MIMO system.[4] In this aspect, we give the following definition:

DEFINITION 13.5.2 *A PW power transfer system with $L^\star$ resolvable paths is comparable to an $N_T \times N_R$ MIMO power transfer system at an outage probability* $\Pr(P_{MIMO} \le x) = P_{out}$, *if $L^\star$ is the smallest positive integer that satisfies* $\Pr(P_{ST} \le x) \le P_{out}$.

By using the derived outage probabilities in (13.24) and (13.25), it can be shown that the condition $\Pr(P_{ST} \le x) \le \Pr(P_{MIMO} \le x)$ yields

$$L \ge \frac{1}{\log\left(1 - e^{-x/P}\right)} \left(\log \det (\mathrm{B}(x)) - \sum_{k=1}^{\min(N_T, N_R)} \log\left(\Gamma\left(\max\left(N_T, N_R\right) - k + 1\right)\right) - \sum_{k=1}^{\min(N_T, N_R)} \log\left(\Gamma\left(\min\left(N_T, N_R\right) - k + 1\right)\right) \right). \tag{13.26}$$

For a PW power transfer system to achieve the same outage probability of a MIMO power transfer system, the required number of resolvable paths $L^\star$ must be equal to the smallest integer not less than the value on the right-hand side of (13.26).

13.6 Simulation Results and Discussions

In this section, we present the performances of the PW wireless power transfer with different designed waveforms, including the energy, single-tone, direct transmission, basic TR, and MaxSINR waveforms, by computer simulation, and justify the theoretical findings in Sections 13.4 and 13.5. The system bandwidth is set as 125 MHz, i.e., the sampling period $T_S = 8$ ns. Here, a large bandwidth is required to digitally resolve the naturally existing multipaths for the systems, on which the PW is designed in the digital domain. In other words, without a large bandwidth, the equivalent CIR to the system is likely to be a single tap even though a lot of multipaths exist. However, our finding shows that the optimal energy or single-tone waveforms selectively concentrate the waveform power over a few couple of frequency components or a single frequency component. In practice, the PW system could be possibly operated on a selective portion of the industrial, scientific and medical (ISM) bands on which the presented waveform concentrates its power, and the remaining bandwidth could be reserved for conventional wireless information transmission. The rate back-off factor D is chosen to be one. The transmitted signal $v[l]$ is assumed to be random and aperiodic, and the SNR value is

[4] The PW and MIMO schemes could be jointly adopted to further boost the system performance by reaping both beamforming and frequency diversity gains.

15 dB, unless otherwise stated. In addition to the UPD and TPD channel profiles, an EPD channel profile is also included in the simulation:

$$\rho_l = c_\rho \cdot e^{-\frac{lT_S}{\sigma_T}}, \quad l = 0, \ldots, L-1, \tag{13.27}$$

where σ_T is the delay spread of the channel, and c_ρ is a constant such that the summation of ρ_l for all l is equal to one. In a typical indoor environment, the delay spread is on the order of hundreds of nanoseconds, and the total number of resolvable multipaths is around several dozens with respect to $T_S = 8$ ns. If the sampling rate is increased, we could possibly resolve more multipaths, but it is eventually limited by naturally existing multipaths in wireless environments. In the simulation, the average transmit power and the average channel power of different profiles are assumed to be one, i.e., $P = 1$ and $\sum_{l=0}^{L-1} \rho_l = 1$, and we normalize the large-scale path loss effect as our focus is to examine the impact of multipaths on the performance of different waveforms. Accordingly, the average received signal power of the direct transmission scheme is a constant value, given by 0 dB if the noise is excluded, over the number of multipaths. Besides, we simulate the system performance in a Saleh-Valenzuela (SV) channel model [29]. This model is popularly adopted in IEEE 802.15.4a standard and suitable for the wideband applications with a frequency range from 2 to 6 GHz, covering indoor residential, office, outdoor, industrial, and open outdoor environments.

Figure 13.3 shows the power transfer performance versus the number of multipaths for different waveforms under the EPD channel profiles, along with the corresponding SINR. The delay spread σ_T is set to 160 ns. We can observe that the power transfer performance is monotonically increasing with the number of resolvable paths, except

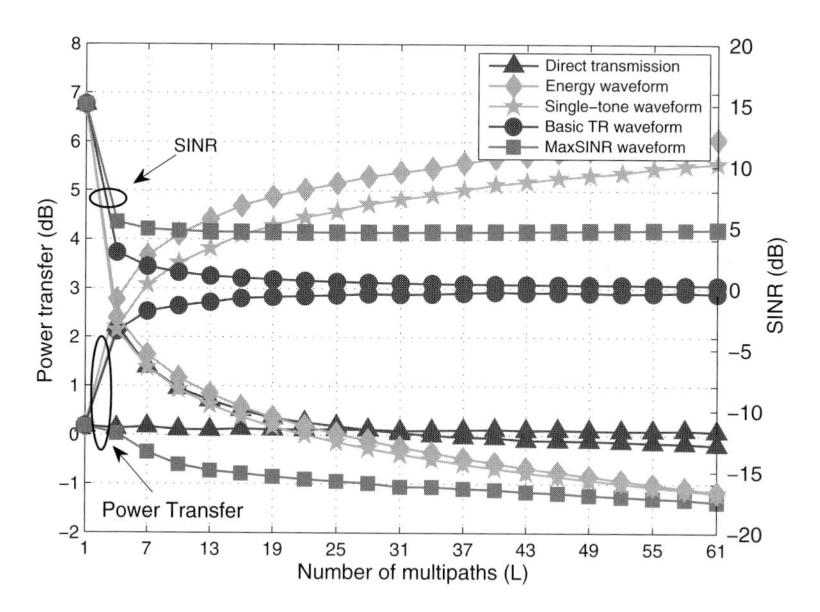

Figure 13.3 Power transfer and SINR performances versus number of multipaths under the EPD channel profiles for different waveforms.

for the direct transmission and the MaxSINR waveforms. Actually, the power transfer performance of the direct transmission keeps constant, while that of the MaxSINR waveform is degraded as L increases. For $L = 61$, the energy waveform and the basic TR waveform outperform the direct transmission by about 6 dB and 3 dB, respectively. Moreover, the energy waveform is superior to the basic TR waveform with a performance gap of 3 dB at $L = 61$. As compared with the energy waveform, the single-tone waveform only incurs a slight performance degradation of no more than 0.5 dB. No matter what waveforms are applied, the SINR performance is degraded as the number of multipaths is increased because the resultant ISI becomes more severe. The SINR performance, in general, has an opposite behavior to that found in the power transfer performance, and it is shown that both the energy waveform and the single-tone waveform exhibit a significantly poorer SINR performance than the basic TR waveform and the MaxSINR waveform. Particularly, the performance difference is enlarged as L increases. This is because the waveforms designed from the power transfer maximization perspective aim at utilizing the multipath propagation effect to spread out the energy over the entire time duration, thereby raising the ISI level.

Figure 13.4 shows the histograms of the power transfer performance for different waveforms in UWB channels with LOS and NLOS office environments. It is obvious from Figure 13.4(a) that for the LOS setting, the power transfer performance of the optimal energy waveform is slightly better than that of the single-tone waveform in terms of the mean values in histograms. Both waveforms have better capability of recollecting the energy of all the possibly available multipaths at the receiver side than the other three waveforms. Again, the MaxSINR waveform has the worst power transfer performance among all the waveforms because it is designed to maximize the received SINR from the information transmission perspective. A similar performance trend can be observed in Figure 13.4(b) in the NLOS environments. However, it is found that the harvested energy per sample in the NLOS environments is larger than that in the LOS environments.

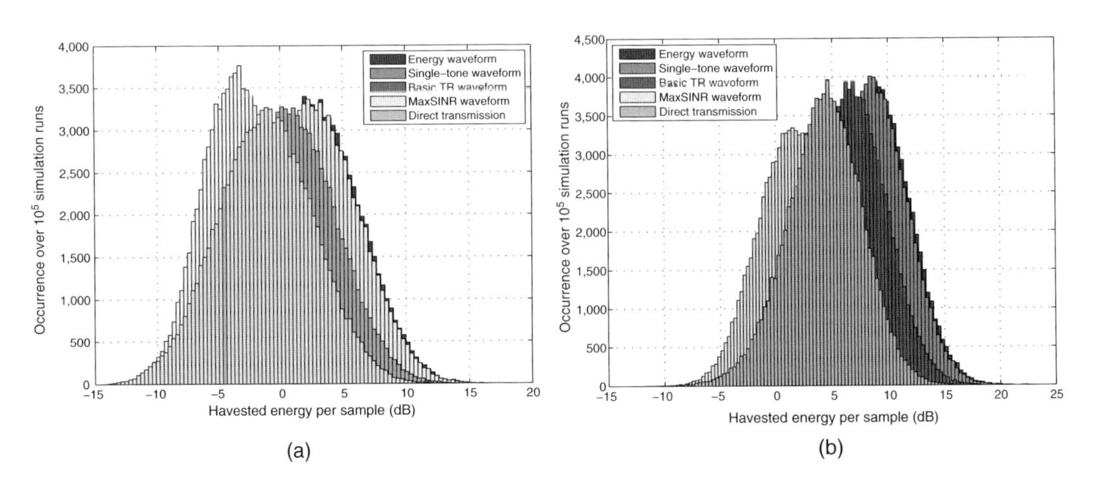

Figure 13.4 Histograms of power transfer for different waveforms in UWB channels with (a) LOS and (b) NLOS environments.

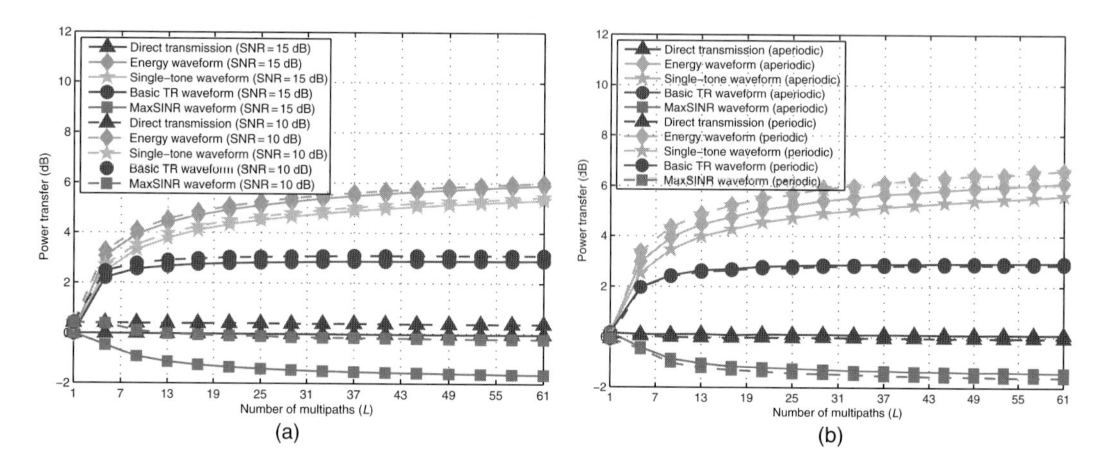

Figure 13.5 Power transfer performance versus number of multipaths for different waveforms: (a) UPD channel profile and (b) TPD channel profile with $\rho_0 = \frac{2}{L}$.

Figures 13.5(a) and 13.5(b) show the power transfer performance for different waveforms versus the number of multipaths under the UPD and TPD channel profiles, respectively. The parameter ρ_0 in the TPD channel profile is set as $\frac{2}{L}$. In Figure 13.5(a), it is found that all waveforms, except for the MaxSINR waveform, follow a similar performance trend when SNR is decreased from 15 dB to 10 dB. This is because the design of the MaxSINR waveform makes a trade-off between the ISI and noise effect, while the design of the other four waveforms is irrelevant to the noise power. In Figure 13.5(b), it reveals that the performance of the two presented waveforms with periodic signals is slightly better than that with aperiodic signals.

In Figure 13.6, the power transfer performance of the single-tone waveform with periodic random transmitted signals is simulated and compared under the channel profiles, EPD, UPD, and TPD. The parameter σ_T in the EPD channel profile could be 10 ns, 40 ns, and 80 ns, while the parameter ρ_0 in the TPD profile is given by $\frac{2}{L}$. For the EPD channel profile, it exhibits that the power transfer performance can be potentially improved as the delay spread σ_T increases, and the performance gets close to that in the UPD channel profile as σ_T is larger than 80 ns. It is possible to roughly infer the power transfer performance in the EPD channel profile by using the TPD channel profile and counting the effective number of multipaths. For example, let $\bar{L}$ be an effective number of dominant paths of a channel profile, which is defined as $\sum_{l=0}^{\bar{L}-1} \rho_l \Big/ \sum_{l=0}^{L-1} \rho_l \geq 99\%$, where $\bar{L}$ is a minimum integer satisfying the preceding inequality. Accordingly, for $L = 61$, the effect number of paths for the EPD channel profile with $\sigma_T = 10$ ns can be calculated as $\bar{L} = 6$. Interestingly, the attainable performance in this case is comparable to that in the TPD channel profile with $L = 6$.

The validity of the analytic power transfer gain for the single-tone waveform with periodic random transmitted signals in Theorems 13.4.3 and 13.4.4 is shown in Figure 13.7. One can observe that for the UPD channel profile, the analytic result is exactly the same as the simulation result. On the other hand, for the TPD channel profile,

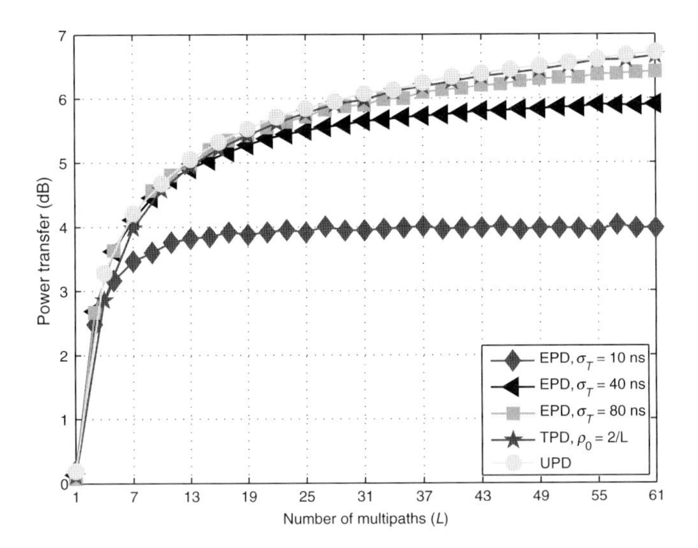

Figure 13.6 Comparisons of power transfer performance of the single-tone waveform with periodic random transmitted signals under different channel profiles.

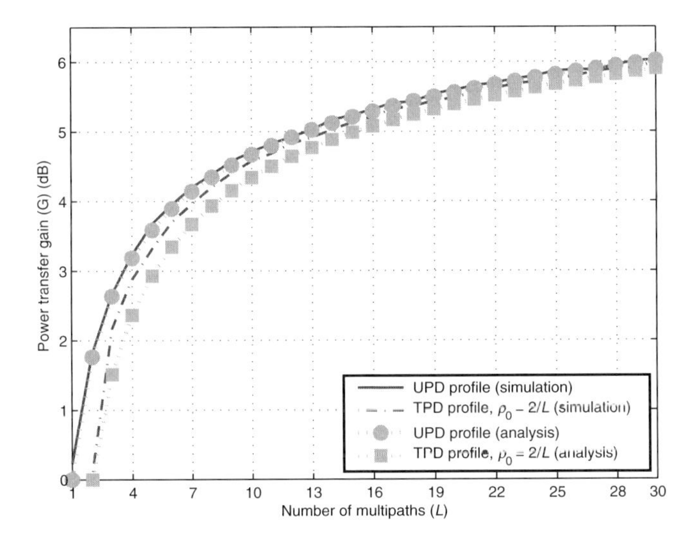

Figure 13.7 Simulation and analytic results for the single-tone waveform with periodic random transmitted signals in UPD and TPD channel profiles.

the analytic result can serve as a good performance lower bound for the corresponding simulation result, and the bound becomes tight when the number of multipaths is sufficiently large. Hence, the provided analytic expressions are useful to predict the characteristic of the power transfer gains in multipath environments.

In Figure 13.8, the analytical outage performances of the PW power transfer system and the MIMO power transfer system, derived in (13.24) and (13.25), are validated and compared by computer simulation, where the analytical results are plotted as solid

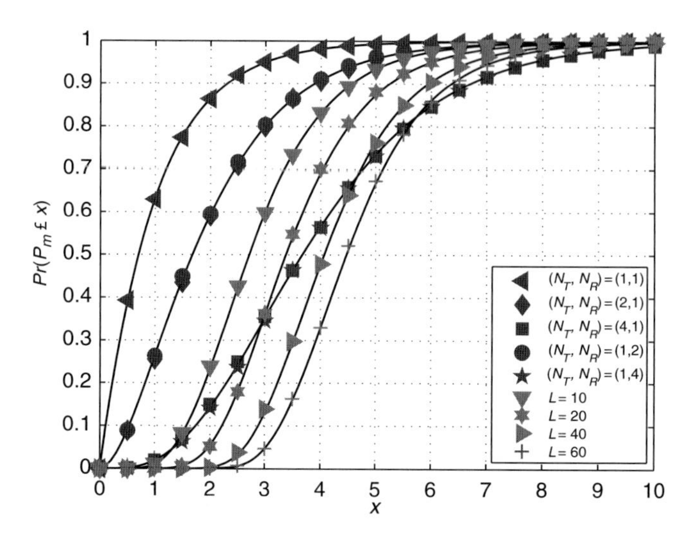

Figure 13.8 Comparisons between the theoretical outage performance and simulation results of the MIMO power transfer system and the PW power transfer system ($P = 1$).

curves, and the corresponding simulation results are marked by different shapes. The subscript m of P_m on the y-axis could be "ST" or "MIMO." Without loss of generality, the transmit power is set as $P = 1$. One can find that there is a perfect match between the analytical and the simulation results. For the MIMO power transfer system, we can make the following observations. On one hand, the outage probability for $N_R = 1$ can be dramatically reduced by increasing the number of transmit antennas due to the transmit beamforming gain. On the other hand, for $N_T = 1$, the outage performance can be enhanced by an increased number of receive antennas. Actually, the outage performance of a MIMO system with $(N_T, N_R) = (N_1, N_2)$ is identical to that of a system with $(N_T, N_R) = (N_2, N_1)$, which can also be justified by using the outage probability formula in (13.24). For the PW power transfer system using the single-tone waveform, it is worth pointing out that the outage performance can be significantly enhanced when the number of resolvable multipaths is increased, while the improvement is less pronounced as L goes beyond 60. Table 13.2 summarizes the comparable setting values between $L^\star$ in the PW power transfer system and (N_T, N_R) in the conventional MIMO system for various outage probabilities P_{out} by using (13.26). We can find that the presented PW power transfer system, which only requires a single antenna in implementation, can support a comparable performance as that using multi-antenna techniques. For instance, if the number of resolvable multipaths is six, i.e., $L^\star = 6$, a PW power transfer system can achieve the same outage performance of $P_{out} = 0.9$ as 1×2 or 2×1 MIMO power transfer systems. When $L^\star$ is increased up to 24 or 43, which can be realized by enlarging the system bandwidth in practice, the presented PW power transfer system is comparable to 3×1 or 2×2 MIMO systems at the outage probability of $P_{out} = 0.9$, respectively. It is worth pointing out that the performance of the PW power transfer system can also be enhanced by incorporating multiple antennas.

Table 13.2 Comparable setting values between PW power transfer systems and conventional MIMO power transfer systems

	$L^\star$	$N_T = 1$	$N_T = 2$	$N_T = 3$	$N_T = 4$
	$P_{out} = 0.7$	2	5	15	44
$N_R = 1$	$P_{out} = 0.8$	2	5	17	61
	$P_{out} = 0.9$	2	6	24	86
	$P_{out} = 0.7$	5	24	107	478
$N_R = 2$	$P_{out} = 0.8$	5	30	149	666
	$P_{out} = 0.9$	6	43	233	1153

13.7 Experimental Results and Discussions

The experimental study for the PW wireless power transfer systems is conducted to demonstrate the performance of the presented PW methods and to explore the essential wireless power transfer characteristics in real wireless scattering environments, involving both the large-scale and small-scale fading effects. Our experiment platform is based on the transmitter/receiver developed by Origin Wireless Inc. The system is operated in the ISM band with the signal bandwidth 125 MHz, centered at the carrier frequency 5 GHz. The measurement is performed in the office environments in Origin Wireless Inc. A transmitter sends signals to a receiver by utilizing different kinds of waveforms in LOS and NLOS fashions, and the electromagnetic waves in the air are reflected by plentiful objectives in the surrounding areas.

The detailed floor plan and the layout of the experiments are shown in Figure 13.9. The total space of the office environment in the experiment is around 882 m^2. For the layout of the experiment in the LOS testing environment, the transmitter and the receiver are randomly deployed in a rectangular shadowed area in an LOS manner, i.e., there is no blocking between them. The wireless power transfer distance is set from 1 meter to 6 meters with an increment of 1 meter, and the power transfer performance is measured at eight random locations (L1–L8) for each distance value. For the layout of the experiment in the NLOS testing environments, five office rooms (R1–R5) are considered, and the transmitter and the receiver are located outside and inside the room, respectively, just as an example sketched for the experiment in the room R2. For each room, three random locations (L1–L3) are adopted for the receiver in the experiments, and the distance between the transmitter and the receiver is set about 2 to 3 meters.

The photograph of the measurement setup and the detailed block diagram, including one transmitter, one receiver, one power splitter, one power meter, one camera, and two laptops, are shown in Figures 13.10(a) and 13.10(b), respectively. As mentioned earlier, the experiment of the PW wireless power transfer systems is mainly comprised of two phases. During the channel probing phase, the receiver sends channel probing signals to the transmitter for the purpose of CIR acquisition [36]. A channel estimator is applied to the received channel probing signals at the transmitter side for the estimation of the CIR [37]. During the power transfer phase, by assuming that the channel is reciprocal,

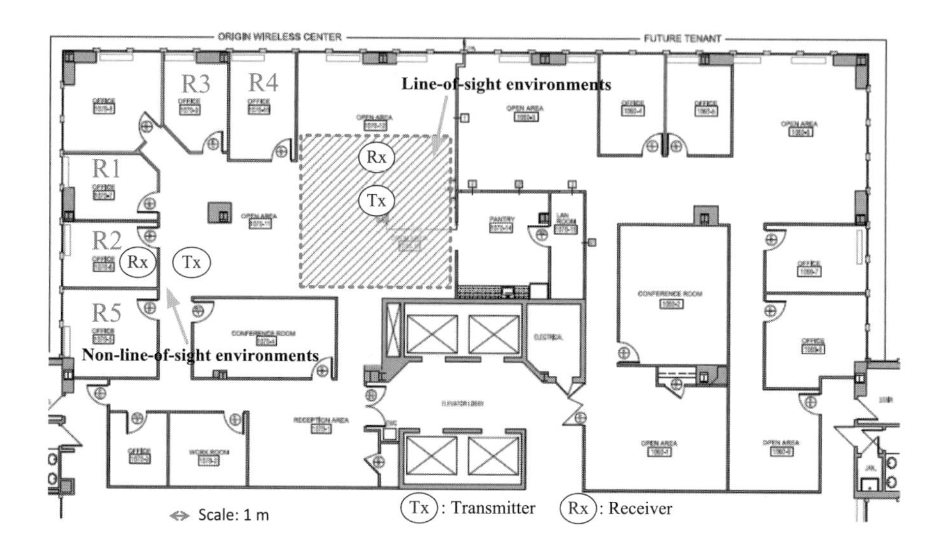

Figure 13.9 Floor plan and the layout of the experiments in the LOS and NLOS testing environments (circles of Tx and Rx indicate the locations of the transmitter and the receiver, respectively).

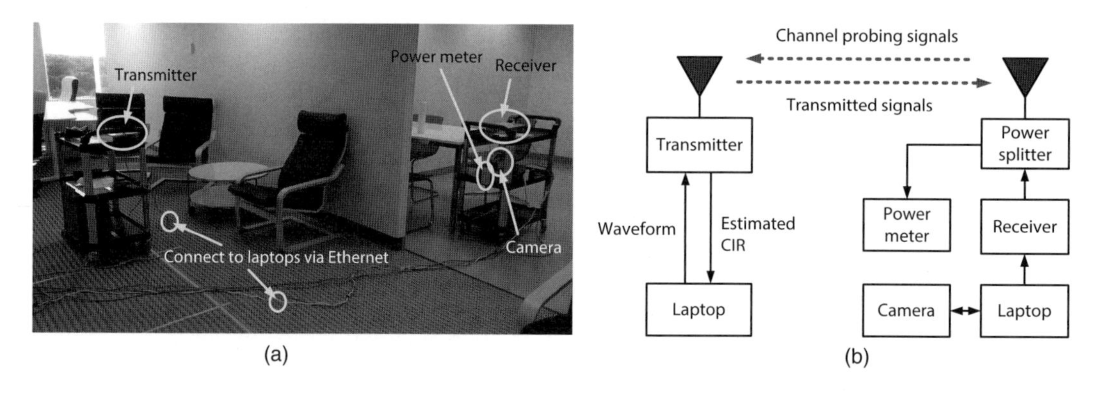

Figure 13.10 Wireless power transfer measurement setup. (a) Photograph of the experiment setup for the transmitter and the receiver on two carts. (b) Detailed block diagram of the experiment setup.

the laptop connected to the transmitter first computes the digital waveforms according to the estimated CIR, and the transmitter then emits the signals shaped by the designed waveforms via an omnidirectional antenna. In addition to the two PW methods, the other three waveforms developed primarily for information or focusing perspectives are also carried out in the experiments for performance comparisons. The maximum transmit power at the transmitter's radio frequency-front end is 20 dBm. At the receiver, a power meter CORNET ED85EXS, which is linked to the antenna via a power splitter ZX10-2-71+, is used to measure the received power at the radio frequency-front end, and a digital camera is set up to record the reported values on the power meter.

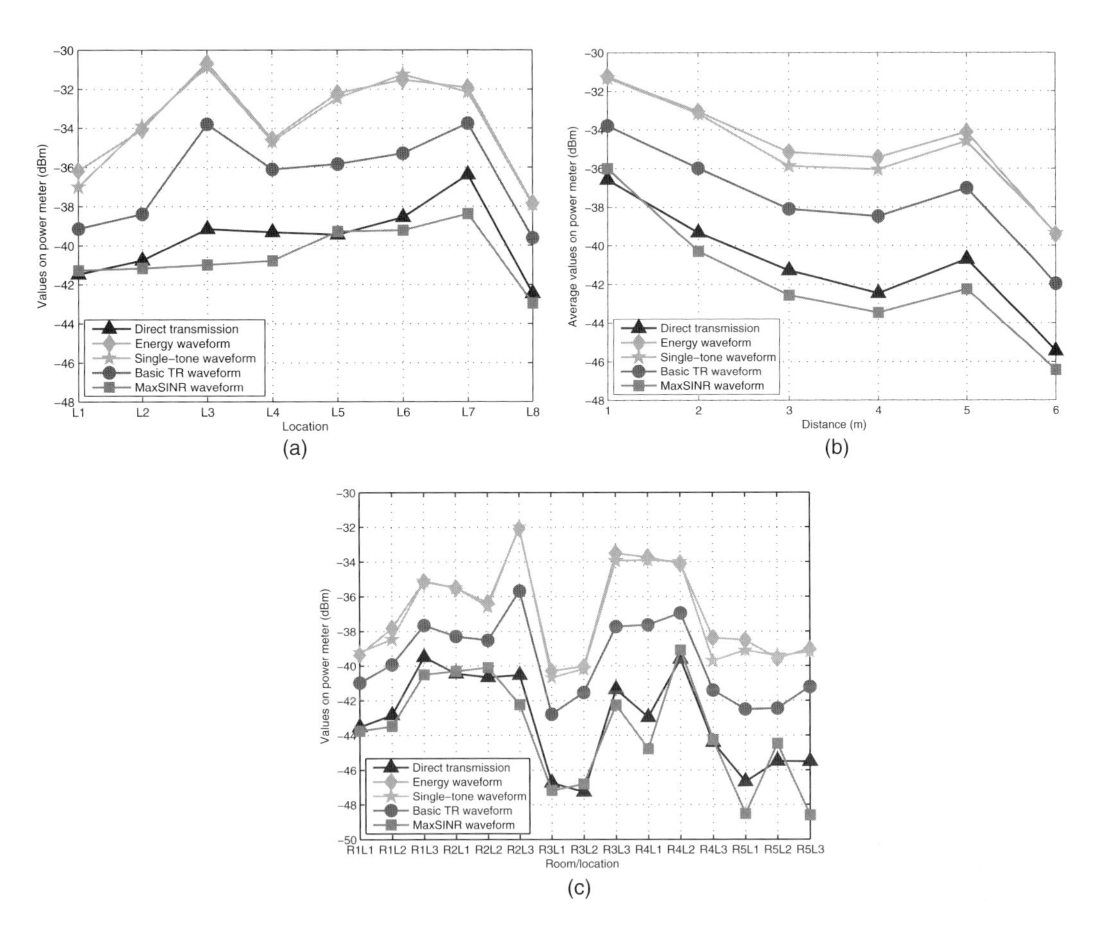

Figure 13.11 PW wireless power transfer performance in LOS and NLOS environments. (a) Measurement results in the LOS environments for various locations at a distance of 2 meters. (b) Average measurement results in the LOS environments as a function of the separation between the transmitter and the receiver. (c) Measurement results in the NLOS environments for various rooms and locations.

The sensitivity range of the power meter is from −55 dBm up to 0 dBm. Finally, all the Wi-Fi access points in the offices are turned off during the experiment.

Figure 13.11 shows the PW wireless power transfer performance for different waveforms in the LOS and NLOS testing environments. For a fair comparison, the average transmit power of the different waveforms is normalized in such a way that the average transmit power is the same as that with the direct transmission waveform. The power transfer performance for the eight measured locations in the LOS setting at the distance of 2 meters is shown in Figure 13.11(a). Due to the multipath effect, the power transfer performance, in terms of the observed power values on the power meter, varies for the different locations, even though the distance is merely two meters and the setting is LOS. As compared with the direct transmission waveform, the performance improvement achieved by the optimal energy waveform ranges from 4 dB to 8 dB. Taking an example,

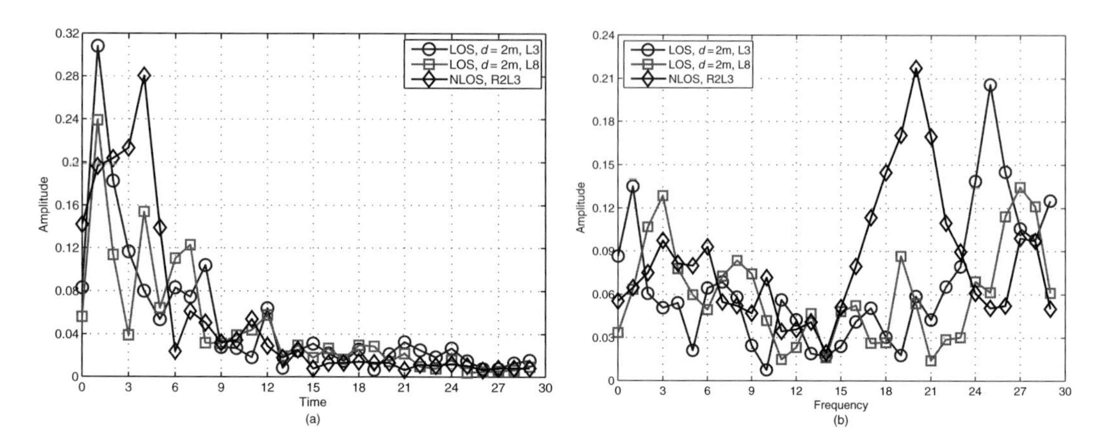

Figure 13.12 Channel probing results in the LOS environments (d = 2 meters, and locations L3 and L8) and the NLOS environments (location R2L3). (a) The amplitude of CIRs versus time index. (b) The amplitude of CFRs versus frequency index.

the improvement values at the locations L8 and L3 are, respectively, given as 4 dB and 8 dB, and one can observe from Figure 13.12 that there are sufficient multipaths available even at such a short distance. Furthermore, the performances of the optimal energy and the single-tone waveforms are comparable, and the basic TR waveform is worse than the optimal energy waveform by about 2 dB~4 dB. Again, the power transfer performance of the MaxSINR waveform is worse in the experiment because it is designed from the information transfer viewpoint.

Figure 13.11(b) demonstrates the average power transfer performance versus the distance in the LOS setting, for which the performance for each distance value is averaged over the eight random locations. We can find that for most of the cases, the average measured power value on the power meter is decreased when the distance becomes large due to the path loss effect. In addition, the optimal energy waveform is superior to the direct transmission waveform by about 6 dB in terms of the received power. Compared with the energy waveform, the single-tone waveform can achieve a near-optimal performance, but requires less computational complexity in implementation. It also exhibits that the performance of the TR waveform is about 3 dB worse than that of the energy waveform. Note that there is a jump at the distance of 5 meters in terms of the performance of the received power. This is because in our testing environments, the wireless channels at the distance of 5 meters often appear to be more dispersed (i.e., more multipaths are collectable), as compared with the wireless channels at the distance of 3 or 4 meters. In general, the PW scheme can provide a larger performance gain, if the multipath effect is more severe. Hence, the average received power value at the distance of 5 meters is larger than those measured at the distance of 3 or 4 meters, even though the radio waves may propagate over a relatively longer distance and suffer a larger path loss.

Figure 13.11(c) shows the power measurement results for various rooms and locations in the NLOS testing environments. We can see that the performance gain of the optimal energy waveform over the direct transmission is larger than 6 dB for most of the locations, e.g., R2L3 and R3L1. It is worth pointing out that this performance gain in the NLOS environments is in general larger than that obtained in the LOS because the NLOS channel can provide a larger frequency selection diversity gain (see Figure 13.12). Also, as compared with the direct transmission, the performance gain of using the optimal energy waveform (or the single-tone waveform) could be significantly different even in the same room but different locations; for example, the achievable performance gain is 9 dB at the location R4L1, whereas it is approximately given by 6 dB at the location R4L2. Overall, the single-tone waveform can still achieve a comparable (near-optimal) performance to the optimal energy waveform even in the NLOS environments. Finally, the performance gap between the energy waveform and the basic TR waveform ranges from 2 dB to 4 dB.

13.8 Summary

In this chapter, a wireless power transfer system incorporating the PW technology was investigated to gather multipath power by fully utilizing the multipath propagation effect. Two power transfer-oriented waveforms, energy waveform and single-tone waveform, were discussed with the goal of maximizing the average received signal power at the receiver. The energy waveform is optimal in the sense of maximum power delivery, while the single-tone waveform is a low-complexity alternative at the expense of small performance degradation. In essence, the single-tone waveform could be just as good as the optimal one if a J-periodic condition for the transmitted signals is satisfied. The power transfer gain of the single-tone waveform over the direct transmission scheme was also theoretically analyzed under various channel power delay profiles. Besides, a comparable relationship between the presented PW power transfer system and the conventional MIMO system is characterized through an outage performance analysis. It is conceived that the PW system can achieve a comparable performance to the MIMO system as long as the number of digitally resolvable multipaths at the transmitter is sufficiently large. As such, the PW technology is a practical and flexible solution to power transfer because the evolution of high-speed analog-to-digital converters (ADCs) allows the transmitter to digitally resolve as many as possible multipaths from wireless channels. On the contrary, the hardware implementation cost of the MIMO solution increases with the number of antennas, a key factor to the performance improvement. Extensive computer simulation was conducted to validate the analysis results, and it was further revealed that the presented energy waveform and the single-tone waveform outperform other waveforms. Generally, the two presented waveforms can provide a performance gain of about 6 dB over the direct transmission scheme. Real experiments were also implemented to justify the achievable performance gains. The results give a quantitative assessment of the impact of the number of resolvable multipaths on the power transfer performance. For related references, interested readers can refer to [38].

Appendix A: Proof of Theorem 13.3.1

Without loss of generality, we consider the zeroth segment of the received signal, i.e., $p = 0$. By applying (13.8) into (13.10) and using the uncorrelated property of $v[n]$, the power transfer maximization problem becomes

$$\mathbf{g}_E = \arg \max_{\|\mathbf{g}\|_2^2=1} \mathbf{g}^\dagger \left(\sum_{q=\lceil -1-(L-2)/D \rceil}^{\lfloor (L-1)/D \rfloor} \mathbf{H}_q^\dagger \mathbf{H}_q \right) \mathbf{g} \,. \tag{13.28}$$

Hence, the optimal energy waveform is given in (13.11). By substituting (13.9) into (13.28), we can also find that the optimal energy waveforms are the same for different values of D.

Appendix B: Proof of Theorem 13.3.2

Because $L = J \cdot D$, we get $p - 1 - J + \lceil 2/D \rceil \leq q \leq p + J - 1$ in the summation of (13.8). As the signal is J-periodic, i.e., $v[n] = v[n + J]$, for any n, the received signal in (13.8) can be simplified as

$$\mathbf{y}_p = \begin{cases} \mathbf{H}_0 \mathbf{g} \cdot v[p] + \sum_{q=1}^{J-1} \left(\mathbf{H}_q + \mathbf{H}_{(q-J)}\right) \mathbf{g} \cdot v[p+q] + \mathbf{z}_p, & D = 1\,; \\ \sum_{q=0}^{J-1} \left(\mathbf{H}_q + \mathbf{H}_{(q-J)}\right) \mathbf{g} \cdot v[p+q] + \mathbf{z}_p, & D \geq 2\,. \end{cases} \tag{13.29}$$

Let $\mathbf{C}$ be a circulant matrix of size $L \times L$, specified by the column vector of the CIR $\mathbf{h}$, and it is straightforward to verify from (13.9) that the matrix $\mathbf{C}$ and $\mathbf{H}_j$ has the following relationship:

$$\mathbf{C} = \begin{cases} \left[\mathbf{H}_{(J-1)}^T, \mathbf{H}_{(J-2)}^T, \ldots, \mathbf{H}_1^T, \mathbf{H}_0^T\right]^T \\ \quad + \left[\mathbf{H}_{-1}^T, \mathbf{H}_{-2}^T, \ldots, \mathbf{H}_{(1-J)}^T, \mathbf{O}_{1\times L}^T\right]^T, & D = 1\,; \\ \left[\mathbf{H}_{(J-1)}^T, \mathbf{H}_{(J-2)}^T, \ldots, \mathbf{H}_0^T\right]^T \\ \quad + \left[\mathbf{H}_{-1}^T, \mathbf{H}_{-2}^T, \ldots, \mathbf{H}_{-J}^T\right]^T, & D \geq 2\,. \end{cases} \tag{13.30}$$

By applying (13.29) and the uncorrelated property of $v[p+q]$, for $q = 0, \ldots, J-1$, the power transfer maximization problem in (13.10) becomes

$$\mathbf{g}_E = \arg \max_{\|\mathbf{g}\|_2^2=1} \mathbf{g}^\dagger \mathbf{C}^\dagger \mathbf{C} \mathbf{g} \,. \tag{13.31}$$

The circulant matrix $\mathbf{C}$ can be diagonalized as $\mathbf{F}\mathbf{C}\mathbf{F}^\dagger = \sqrt{L} \cdot Diag\,(\mathbf{F}\mathbf{h}) = \sqrt{L} \cdot Diag\,(\boldsymbol{\zeta})$. As a result, the optimal energy waveform must satisfy $\mathbf{F}\mathbf{g}_E = \mathbf{e}_{k_{\max}}$, where $\mathbf{e}_i$ is the ith column of the identity matrix $\mathbf{I}_L$, for $i = 0, \ldots, L-1$. Hence, $\mathbf{g}_E = \mathbf{F}^\dagger \mathbf{e}_{k_{\max}}$ is exactly the same as the single-tone waveform $\mathbf{g}_{ST}$ given in (13.12).

Appendix C: Proof of Theorem 13.4.1

Let $\mathbf{h}_I$ and $\mathbf{h}_Q$ be the real and imaginary parts of the CIR $\mathbf{h}$. In addition, we define $\mathbf{F}_I$ and $\mathbf{F}_Q$ as the real and imaginary parts of the DFT matrix $\mathbf{F}$. Because $\zeta = \mathbf{Fh}$, its real and imaginary parts can be expanded as $\zeta_I = \mathbf{F}_I\mathbf{h}_I - \mathbf{F}_Q\mathbf{h}_Q$ and $\zeta_Q = \mathbf{F}_I\mathbf{h}_Q + \mathbf{F}_Q\mathbf{h}_I$, respectively. Therefore, it can be derived that

$$\begin{aligned} \mathbb{E}\left[\zeta_I\zeta_I^T\right] &= \mathbb{E}\left[\zeta_Q\zeta_Q^T\right] \triangleq \Sigma \\ &= \frac{1}{2}\left(\mathbf{F}_I\, Diag\left(\rho\right)\mathbf{F}_I^T + \mathbf{F}_Q\, Diag\left(\rho\right)\mathbf{F}_Q^T\right)\ ; \end{aligned} \tag{13.32}$$

$$\mathbb{E}\left[\zeta_I\zeta_Q^T\right] = \mathbb{E}\left[\zeta_Q\zeta_I^T\right] = \mathbf{O}\,, \tag{13.33}$$

where the assumption of $\mathbb{E}\left[\mathbf{h}_I\mathbf{h}_I^T\right] = \mathbb{E}\left[\mathbf{h}_Q\mathbf{h}_Q^T\right] = \frac{1}{2}Diag\left(\rho\right)$ and $\mathbb{E}\left[\mathbf{h}_I\mathbf{h}_Q^T\right] = \mathbb{E}\left[\mathbf{h}_Q\mathbf{h}_I^T\right] = \mathbf{O}$ in (13.1) is applied. Due to (13.32) and (13.33), the characteristic function of $\boldsymbol{\varepsilon}$ has a form as shown in (13.17) [30], and the matrix Σ can be explicitly calculated in the following.

First, it is shown from (13.32) that the matrix Σ is circulant because the (k,m)th entry of Σ is given by

$$\Sigma\left[k,m\right] = \frac{1}{2L}\sum_{l=0}^{L-1}\rho_l\cos\left(-\frac{2\pi\left(k-m\right)l}{L}\right), \quad k,m = 0,\ldots,L-1\,. \tag{13.34}$$

Hence, by using $\mathbf{F}$, we can diagonalize the matrix into $\Sigma = \mathbf{F}^{-1}Diag\left(\boldsymbol{\lambda}\right)\mathbf{F}$, where the vector $\boldsymbol{\lambda} = \sqrt{L}\mathbf{Fq}$, and $\mathbf{q} = \frac{1}{2L}\sum_{l=0}^{L-1}\left[\rho_l, \rho_l\cos\left(\frac{2\pi 1 l}{L}\right), \ldots, \rho_l\cos\left(\frac{2\pi(L-1)l}{L}\right)\right]^T$ is the first column of the matrix Σ, i.e., setting $m = 0$ in (13.34) [31]. We then rewrite $\mathbf{q}$ into a compact matrix form:

$$\mathbf{q} = \mathbf{M}\rho\,, \tag{13.35}$$

where the matrix $\mathbf{M}$ is defined in (13.36) at the top of the next page.

$$\mathbf{M} = \frac{1}{4L}\begin{bmatrix} 2 & 2 & \cdots & 2 \\ 2 & e^{j2\pi 1\cdot 1/L} + e^{j2\pi 1\cdot(L-1)/L} & \cdots & e^{j2\pi 1\cdot(L-1)/L} + e^{j2\pi 1\cdot 1/L} \\ \vdots & \vdots & \ddots & \vdots \\ 2 & e^{j2\pi(L-1)\cdot 1/L} + e^{j2\pi(L-1)\cdot(L-1)/L} & \cdots & e^{j2\pi(L-1)\cdot(L-1)/L} + e^{j2\pi(L-1)\cdot 1/L} \end{bmatrix}. \tag{13.36}$$

By applying (13.35) and (13.36), the (k,m)th entry of the matrix $\tilde{\mathbf{M}} = \sqrt{L}\mathbf{FM}$ can be computed as

$$\tilde{M}[k,m] = \begin{cases} \frac{1}{2}, & m=0\,;k=0\,; \\ \frac{1}{4}, & m \neq 0\,;k=L-m \neq m\,; \\ \frac{1}{2}, & m \neq 0\,;k=m=\frac{L}{2}\,; \\ 0, & \text{othcrwise}\,, \end{cases} \tag{13.37}$$

if L is even, and

$$\tilde{M}[k,m] = \begin{cases} \frac{1}{2}, & m=0\,;k=0\,; \\ \frac{1}{4}, & m \neq 0\,;k=L-m \neq m\,; \\ 0, & \text{otherwise}\,, \end{cases} \tag{13.38}$$

if L is odd. Because $\boldsymbol{\lambda} = \sqrt{L}\mathbf{F}\mathbf{q} = \tilde{\mathbf{M}}\boldsymbol{\rho}$, we finally get $\lambda_l = \frac{1}{4}\left(\rho_l + \rho_{((L-l)\%L)}\right)$.

Appendix D: Proof of Theorem 13.4.2

The cumulative distribution function (CDF) of $\boldsymbol{\varepsilon}$ can be expressed in terms of $\Psi_{\boldsymbol{\varepsilon}}(j\boldsymbol{\omega})$ as follows [32]:

$$F_{\boldsymbol{\varepsilon}}(\mathbf{x}) = \frac{1}{(2\pi)^L} \int_{-\infty}^{\infty} \cdots \int_{-\infty}^{\infty} \Psi_{\boldsymbol{\varepsilon}}(j\boldsymbol{\omega}) \times \prod_{l=0}^{L-1} \left(\frac{1-e^{-j\omega_l x_l}}{j\omega_l} \right) d\omega_0 \cdots d\omega_{L-1}\,, \tag{13.39}$$

where $\mathbf{x} = \left[x_0, \ldots, x_{L-1}\right]^T$. Let $\varepsilon_{\max} = \max_k \varepsilon_k$. The CDF of $\varepsilon_{\max}$ is then obtained by letting $x_0 = \cdots = x_{L-1} = x$ in (13.39), and we have

$$F_{\varepsilon_{\max}}(x) = \frac{1}{(2\pi)^L} \int_{-\infty}^{\infty} \cdots \int_{-\infty}^{\infty} \Psi_{\boldsymbol{\varepsilon}}(j\boldsymbol{\omega}) \times \prod_{l=0}^{L-1} \left(\frac{1-e^{-j\omega_l x}}{j\omega_l} \right) d\omega_0 \cdots d\omega_{L-1}\,. \tag{13.40}$$

Because $\varepsilon_{\max}$ is a nonnegative random variable, the mean of $\varepsilon_{\max}$ can be directly computed through its CDF, which follows essentially from integration by parts:

$$\mathbb{E}[\varepsilon_{\max}] = \int_0^{\infty} \left(1 - F_{\varepsilon_{\max}}(x)\right) dx\,. \tag{13.41}$$

From (13.16), (13.40), and (13.41), the power transfer gain is finally derived in (13.18).

Appendix E: Proof of Theorem 13.4.3

For the UPD channel profile, we have $\rho_l = \frac{1}{L}$, for $l = 0, \ldots, L-1$. From (13.17) in Theorem 13.4.1, it is then shown that $\lambda_l = \frac{1}{2L}$ and the characteristic function is reduced to

$$\Psi_\varepsilon(j\boldsymbol{\omega}) = \left(\det\left(\mathbf{I}_L - j\frac{1}{L}\cdot Diag(\boldsymbol{\omega})\,\mathbf{F}^{-1}\mathbf{F}\right)\right)^{-1} = \prod_{l=0}^{L-1}\frac{L}{L-j\omega_l}, \quad (13.42)$$

where $\mathbf{F}^{-1}\mathbf{F} = \mathbf{I}_L$ is applied in the second equality. Moreover, considering an exponential random variable x with the probability density function $\mu e^{-\mu x}$, for $\mu > 0$, its characteristic function $\Psi_X(j\omega) = \frac{\mu}{\mu - j\omega}$ and the corresponding CDF $F_X(x) = 1 - e^{-\mu x}$ have the following relationship:

$$F_X(x) = \frac{1}{2\pi}\int_{-\infty}^{\infty}\Psi_X(j\omega)\frac{1-e^{-j\omega x}}{j\omega}d\omega = \frac{1}{2\pi}\int_{-\infty}^{\infty}\frac{\mu}{\mu - j\omega}\times\frac{1-e^{-j\omega x}}{j\omega}d\omega = 1 - e^{-\mu x}. \quad (13.43)$$

By applying (13.43) into (13.18), the power transfer gain can be derived in (13.44) at the top of the next page.

$$\begin{aligned} G &= L\int_0^{\infty}\left(1 - \frac{1}{(2\pi)^L}\int_{-\infty}^{\infty}\cdots\int_{-\infty}^{\infty}\prod_{l=0}^{L-1}\left(\frac{L}{L-j\omega_l}\cdot\frac{1-e^{-j\omega_l x}}{j\omega_l}\right)d\omega_0\cdots d\omega_{L-1}\right)dx \\ &= L\int_0^{\infty}\left(1 - \frac{1}{(2\pi)^{L-1}}\int_{-\infty}^{\infty}\cdots\int_{-\infty}^{\infty}\prod_{l=1}^{L-1}\left(\frac{L}{L-j\omega_l}\cdot\frac{1-e^{-j\omega_l x}}{j\omega_l}\right)\right. \\ &\qquad \left.\cdot\left(\frac{1}{2\pi}\int_{-\infty}^{\infty}\frac{L}{L-j\omega_0}\cdot\frac{1-e^{-j\omega_0 x}}{j\omega_0}d\omega_0\right)d\omega_1\cdots d\omega_{L-1}\right)dx \\ &= L\int_0^{\infty}\left(1 - \left(1-e^{-Lx}\right)^L\right)dx. \end{aligned} \quad (13.44)$$

Using the binomial expansion, we can get

$$\left(1-e^{-Lx}\right)^L = \sum_{n=0}^{L}\binom{L}{n}(-1)^n e^{-nLx}. \quad (13.45)$$

From (13.44) and (13.45), it therefore implies that

$$G = L \sum_{n=1}^{L} \binom{L}{n} (-1)^{n+1} \int_{0}^{\infty} e^{-nLx} dx$$
$$= \sum_{n=1}^{L} \binom{L}{n} (-1)^{n+1} \frac{1}{n}, \tag{13.46}$$

where $\binom{L}{n} = \frac{L!}{n!(L-n)!}$ is a binomial coefficient, and $(\cdot)!$ is a factorial function. To further inspect the impact of the number of multipaths on the performance, we denote the power transfer gain as a function of L, given by $G(L)$. From (13.46), it can be verified that

$$G(L) - G(L-1) = -\frac{1}{L}\left(\sum_{n=0}^{L} \frac{L!}{n!(L-n)!}(-1)^n - 1\right). \tag{13.47}$$

From the binomial expansion, it yields $\sum_{n=0}^{L} \frac{L!}{n!(L-n)!}(-1)^n = (1 + (-1))^L = 0$. Thus, it can be shown that $G(L) - G(L-1) = \frac{1}{L} \geq 0$. Finally, because $G(1) = 1$ according to (13.46), it is straightforward to get (13.19).

Appendix F: Proof of Theorem 13.4.4

The main idea of the proof is to decompose the zeroth channel path of (13.1), h_0, into two independent random variables:

$$h_0 = h_{0,1} + h_{0,2}, \tag{13.48}$$

where $h_{0,1}$ and $h_{0,2}$ are zero-mean complex Gaussian random variables with variance $\frac{1-\rho_0}{L-1}$ and $\frac{L\rho_0-1}{L-1}$, respectively. By utilizing (13.16), the power transfer gain is given as

$$G = L \cdot \mathbb{E}\left[\max_k \left|\frac{1}{\sqrt{L}} h_{0,2} + \tilde{H}[k]\right|^2\right], \tag{13.49}$$

where $\tilde{H}[k] = \frac{1}{\sqrt{L}}\left(h_{0,1} + \sum_{l=1}^{L-1} h_l e^{-j2\pi kl/L}\right)$. Let $\tilde{k}_{\max} = \arg\max_k \left|\tilde{H}[k]\right|^2$. Then, a lower bound can be derived as

$$G \geq L \cdot \mathbb{E}\left[\left|\frac{1}{\sqrt{L}} h_{0,2} + \tilde{H}\left[\tilde{k}_{\max}\right]\right|^2\right]. \tag{13.50}$$

Because $h_{0,2}$ is independent of $\tilde{H}[k]$, we have

$$G \geq \mathbb{E}\left[\left|h_{0,2}\right|^2\right] + L \cdot \mathbb{E}\left[\left|\tilde{H}\left[\tilde{k}_{\max}\right]\right|^2\right]$$
$$= \frac{L\rho_0 - 1}{L-1} + L \cdot \mathbb{E}\left[\left|\tilde{H}\left[\tilde{k}_{\max}\right]\right|^2\right]. \tag{13.51}$$

The remaining problem is to compute the expectation value of $\left|\tilde{H}\left[\tilde{k}_{\max}\right]\right|^2$. Let us define $\tilde{\varepsilon} = \left[\tilde{\varepsilon}_0, \ldots, \tilde{\varepsilon}_{L-1}\right]^T$, where $\tilde{\varepsilon}_k = \left|\tilde{H}\left[k\right]\right|^2$. Similar to the derivation in (13.42) to (13.46), the characteristic function for $\tilde{\varepsilon}$ can be expressed as

$$\Psi_{\tilde{\varepsilon}}(j!) = \left(\det\left(\mathbf{I}_L - j\frac{1-\rho_0}{L-1}\cdot Diag\,(!)\right)\right)^{-1} = \prod_{l=0}^{L-1}\frac{L-1}{L-1-j\left(1-\rho_0\right)\omega_l}. \tag{13.52}$$

Furthermore, we get

$$\mathbb{E}\left[\left|\tilde{H}\left[\tilde{k}_{\max}\right]\right|^2\right] = \int_0^{\infty}\left(1-\left(1-e^{-\frac{L-1}{1-\rho_0}x}\right)^L\right)dx = \frac{1-\rho_0}{L-1}\sum_{n=1}^{L}\binom{L}{n}(-1)^{n+1}\frac{1}{n}. \tag{13.53}$$

By combining (13.51) with (13.53), the result is finally proved.

References

[1] A. Luigi, I. Antonio, and M. Giacomo, "The internet of things: A survey," *Computer Networks*, vol. 54, no. 15, pp. 2787–2805, Oct. 2010.

[2] C. Han, T. Harrold, S. Armour, I. Krikidis, S. Videv, P. Grant, H. Haas, J. Thompson, I. Ku, C.-X. Wang, T.A. Le, M. Nakhai, J. Zhang, and L. Hanzo, "Green radio: Radio techniques to enable energy-efficient networks," *IEEE Communications Magazine*, vol. 49, no. 6, pp. 46–54, Jun. 2011.

[3] S. Sudevalayam and P. Kulkarni, "Energy harvesting sensor nodes: Survey and implications," *IEEE Communications Surveys & Tutorials*, vol. 13, no. 3, pp. 443–461, Third Quarter 2011.

[4] M.-L. Ku, W. Li, Y. Chen, and K. J. Ray Liu, "Advances in energy harvesting communications: Past, present, and future challenges," *IEEE Communications Surveys & Tutorials*, vol. 18, no. 2, pp. 1384–1412, Second Quarter 2016.

[5] M. L. Ku, Y. Chen, and K. J. R. Liu, "Data-driven stochastic models and policies for energy harvesting sensor communications," *IEEE Journal on Selected Areas in Communications*, vol. 33, no. 8, pp. 1505–1520, Aug. 2015.

[6] R. J. M. Vullers, R. V. Schaijk, H. J. Visser, J. Penders, and C. V. Hoof, "Energy harvesting for autonomous wireless sensor networks," *IEEE Solid-State Circuits Magazine*, vol. 2, no. 2, pp. 29–38, Spring 2010.

[7] X. Lu, P. Wang, D. Niyato, D. Kim, and Z. Han, "Wireless charging technologies: Fundamentals, standards, and network applications," *IEEE Communications Surveys & Tutorials*, 2015.

[8] N. B. Carvalho, A. Georgiadis, A. Costanzo, H. Rogier, A. Collado, J. A. Garcia, S. Lucyszyn, P. Mezzanotte, J. Kracek, D. Masotti, A. J. S. Boaventura, M. de las Nieves Ruiz Lavin, M. Pinuela, D. C. Yates, P. D. Mitcheson, M. Mazanek, and V. Pankrac, "Wireless

power transmission: R&D activities within Europe," *IEEE Transactions on Microwave Theory and Techniques*, vol. 62, no. 4, pp. 1031–1045, Apr. 2014.

[9] S. Kim, R. Vyas, J. Bito, K. Niotaki, A. Collado, A. Georgiadis, and M. M. Tentzeris, "Ambient RF energy-harvesting technologies for self-sustainable standalone wireless sensor platforms," *Proceedings of the IEEE*, vol. 102, no. 11, pp. 1649–1666, Nov. 2014.

[10] X. Chen, Z. Zhang, H.-H. Chen, and H. Zhang, "Enhancing wireless information and power transfer by exploiting multi-antenna techniques," *IEEE Communications Magazine*, vol. 53, no. 4, pp. 133–141, Apr. 2015.

[11] K. Huang and V. K. N. Lau, "Enabling wireless power transfer in cellular networks: Architecture, modeling and deployment," *IEEE Transactions on Wireless Communications*, vol. 13, no. 2, pp. 902–912, Feb. 2014.

[12] L. Liu, R. Zhang, and K.-C. Chua, "Multi-antenna wireless powered communication with energy beamforming," *IEEE Transactions on Communications*, vol. 62, no. 12, pp. 4349–4361, Dec. 2014.

[13] G. Yang, C. K. Ho, and Y. L. Guan, "Multi-antenna wireless energy transfer for backscatter communication systems," *IEEE Journal on Selected Areas in Communications*, vol. 33, no. 12, pp. 2974–2987, Dec. 2015.

[14] Z. Fang, X. Yuan, and X. Wang, "Distributed energy beamforming for simultaneous wireless information and power transfer in the two-way relay channel," *IEEE Signal Processing Letters*, vol. 22, no. 6, pp. 656–660, Jun. 2015.

[15] Y. Zeng and R. Zhang, "Optimized training design for wireless energy transfer," *IEEE Transactions on Communications*, vol. 63, no. 2, pp. 536–550, Feb. 2015.

[16] J. Xu and R. Zhang, "Energy beamforming with one-bit feedback," *IEEE Transactions on Signal Processing*, vol. 62, no. 20, pp. 5370–5381, Oct. 2014.

[17] S. Kashyap, E. Bjornson, and E. G. Larsson, "Can wireless power transfer benefit from large transmitter arrays?," *IEEE Wireless Power Transfer Conference*, pp. 1–3, 2015.

[18] X. Chen, X. Wang, and X. Chen, "Energy-efficient optimization for wireless information and power transfer in large-scale MIMO systems employing energy beamforming," *IEEE Wireless Communications Letters*, vol. 2, no. 6, pp. 667–670, Dec. 2013.

[19] G. Yang, C. Keong Ho, R. Zhang, and Y. L. Guan, "Throughput optimization for massive MIMO systems powered by wireless energy transfer," *IEEE Journal on Selected Areas in Communications*, vol. 33, no. 8, pp. 1640–1650, Aug. 2015.

[20] B. Wang, Y. Wu, F. Han, Y.-H. Yang, and K. J. R. Liu, "Green wireless communications: A time-reversal paradigm," *IEEE Journal on Selected Areas in Communications*, vol. 29, no. 8, pp. 1698–1710, Sep. 2011.

[21] Y. Chen, Y.-H. Yang, F. Han, and K. J. R. Liu, "Time-reversal wideband communications," *IEEE Signal Processing Letters*, vol. 20, no. 12, pp. 1219–1222, Dec. 2013.

[22] Y. Chen, B. Wang, Y. Han, H. Q. Lai, Z. Safar, and K. J. R. Liu, "Why time-reversal for future 5G wireless?," *IEEE Signal Processing Magazine*, vol. 33, no. 2, pp. 17–26, Mar. 2016.

[23] Y.-H. Yang, B. Wang, W. S. Lin, and K. J. R. Liu, "Near-optimal waveform design for sum rate optimization in time-reversal multiuser downlink systems," *IEEE Transactions on Wireless Communications*, vol. 12, no. 1, pp. 346–357, Jan. 2013.

[24] G. Yang, C. K. Ho, R. Zhang, and Y. L. Guan, "Throughput optimization for massive MIMO systems powered by wireless energy transfer," *IEEE Journal on Selected Areas in Communications*, vol. 33, no. 8, pp. 1640–1650, Aug. 2015.

[25] S. Luo, J. Xu, T. J. Lim, and R. Zhang, "Capacity region of MISO broadcast channel for simultaneous wireless information and power transfer," *IEEE Transactions on Communications*, vol. 62, no. 10, pp. 3856–3868, Oct. 2015.

[26] X. Zhou, C. K. Ho, and R. Zhang, "Wireless power meets energy harvesting: a joint energy allocation approach in OFDM-based system," *IEEE Transactions on Wireless Communications*, vol. 15, no. 5, pp. 3481–3491, May 2016.

[27] X. Zhou, R. Zhang, and C. K. Ho, "Wireless information and power transfer in multiuser OFDM systems," *IEEE Transactions on Wireless Communications*, vol. 13, no. 4, pp. 2282–2294, Apr. 2014.

[28] Y. Zeng and R. Zhang, "Optimized training for net energy maximization in multi-antenna wireless energy transfer over frequency-selective channel," *IEEE Transactions on Communications*, vol. 63, no. 6, pp. 2360–2373, Jun. 2015.

[29] A. A. M. Saleh and R. A. Valenzuela, "A statistical model for indoor multipath propagation," *IEEE Journal on Selected Areas in Communications*, vol. 5, no. 2, pp. 128–137, Feb. 1987.

[30] R. K. Mallik, "On multivariate Rayleigh and exponential distributions," *IEEE Transactions on Information Theory*, vol. 49, no. 6, pp. 1499–1515, Jun. 2003.

[31] R. A. Horn and C. R. Johnson, *Matrix Analysis*, New York: Cambridge University Press, 1985.

[32] Q. T. Zhang and H. G. Lu "A general analytical approach to multi-branch selection combining over various spatially correlated fading channels," *IEEE Transactions on Communications*, vol. 50, no. 7, pp. 1066–1073, Jul. 2002.

[33] N. Kong and L. B. Milstein "Average SNR of a generalized diversity selection combining scheme," *IEEE Communications Letters*, vol. 3, no. 3, pp. 57–59, Mar. 1999.

[34] M. Kang and M.-S. Alouini, "Largest eigenvalue of complex Wishart matrices and performance analysis of MIMO MRC systems," *IEEE Journal on Selected Areas in Communications*, vol. 21, no. 3, pp. 418–426, Apr. 2003.

[35] M. Abramowitz and I. A. Stegun, *Handbook of Mathematical Functions with Formulas, Graphs, and Mathematical Tables*, 9th ed. New York: Dover, 1970.

[36] S. Budišin "Golay complementary sequences are superior to PN sequences," *IEEE International Conference on Systems Engineering*, pp. 101–104, 1992.

[37] B. M. Popovic, "Efficient Golay correlator," *Electronics Letters*, vol. 35, no. 17, pp. 1427–1428, Aug. 1999.

[38] M. L. Ku, Y. Han, H. Q. Lai, Y. Chen, and K. J. R. Liu, "Power waveforming: Wireless power transfer beyond time-reversal," *IEEE Transactions on Signal Processing*, vol. 64, no. 22, pp. 5819–5834, Nov. 2016.

14 Joint Power Waveforming and Beamforming

In the Internet of Things (IoT), wireless devices need more easily accessible energy resources, which motivates the development of wireless power transfer (WPT) using radio frequency (RF) signals. Beamforming technique has been widely adopted by using multiple transmit antennas to form a sharp energy beam toward an intended receiver. However, few of them have considered the potential gain of multipath propagation. In this chapter, we consider a joint power waveforming and beamfoming in the time domain for WPT, in which the waveforms on multiple transmit antennas driven by a common reference signal are designed to maximize the gain of energy delivery efficiency. We consider both nonperiodic and periodic reference signals and present low-complexity waveforms that can achieve near-optimal performance. It is found that the energy delivery efficiency gain of the proposed approach increases with the waveform length until saturation. We theoretically analyze the outage probability of the proposed approach under a uniform power delay (UPD) channel profile, which quantifies the impact of the number of antennas and multipaths. Simulations are performed to validate the theoretic analysis and the effectiveness of the proposed joint power waveforming and beamforming approach.

14.1 Introduction

In the era of Internet of Things (IoT), it is anticipated that low-power wireless devices, e.g., home appliances, security sensors, smart meters, are widely deployed as fundamental building blocks to support numerous wireless data applications [1, 2]. These wireless devices are often untethered to a power grid and merely rely on equipped batteries for the operation over a long period of time. In addition to the availability of spectrum resources, which is a common issue for the conventional wireless networks, wireless devices in the future are thus constrained by the limited battery capacity due to massive wireless data services. For IoT applications, wireless devices are sometimes eager for more energy resources, rather than spectrum resources, because low-rate data may be continuously and perpetually reported from devices to devices. Besides, the deployment of nodes in poisonous and even unreachable environments does not allow for frequent battery replacement when the battery is exhausted. In recent years, energy harvesting techniques, by which wireless devices are able to harvest and store energy by using rechargeable batteries, have been emerged as an effective way for avoiding the cost and the challenge of battery replacement [3].

In general, energy harvesting nodes can scavenge energy from either ambient energy sources, e.g., solar, wind, and vibration, or dedicated energy sources, e.g., power stations, to prolong the network lifetime [4]. While ambient energy sources are environmentally friendly, the main drawback of these energy sources lies in that the uncertain nature in time, location and weather conditions makes it difficult to guarantee the quality-of-services (QoS) of wireless communications. On the other hand, dedicated energy emitted by power stations are attractive alternatives to supply on-demand energy for wirelessly replenishing devices and fully controllable to meet the QoS requirements, even though an additional cost is incurred with the deployment of power stations, in comparison with the ambient energy sources [5]. The rapid proliferation of energy harvesting communications has urged the development of wireless power transfer (WPT) using electromagnetic radiation, which has been recognized as a promising solution to sustain energy for rechargeable low-power wireless devices over the air. Electromagnetic induction coupling, magnetic resonant coupling, and radio frequency (RF) signals are three common WPT approaches [6]. As compared with the former two approaches, which only afford a limited working range of a few centimeters, the adoption of a RF signal as a medium is capable of wirelessly charging nodes over a long distance (up to several meters) [7]. Considering its potential in far-end WPT applications, we will concentrate on the investigation of RF signal-based WPT systems in this chapter.

A major concern for designing the RF signal-based power transfer systems is the low power transfer efficiency due to the severe radio wave propagation loss, including multipath, shadowing, and large-scale path loss over distance. According to the law of energy conservation, only a small portion of energy radiated from a transmitter can be harvested at a receiver in reality, yielding an energy scarcity problem. Moreover, unlike the conventional information receivers, energy receivers require much higher receiver sensitivity to convert RF signals into DC power via rectifier circuits, which makes the design even more challenging, e.g., -50 dBm for information receivers and -10 dBm for energy receivers [8].

To overcome the aforementioned problem, beamforming techniques have been widely adopted to achieve efficient power transfer by synthesizing a sharp energy beam toward an intended receiver through multiple transmit antennas. Because the transmitter is capable of concentrating the radiated energy toward a particular direction, the power intensity is significantly magnified in the direction of the intended receiver. Various research efforts have been made to improve the WPT performance along this line [9–24]. The performance enhancement with multi-antenna configurations was investigated in [9], and a performance trade-off between wireless information and power transfer was theoretically analyzed. The authors in [10] compared isotropic and directional beamforming antennas in randomly deployed power stations, based on a stochastic geometry model, for charging mobile terminals wirelessly. In [11], energy beamforming was designed for powering RF identification (RFID) tags. In [12], an adaptive WPT scheme was considered for a large-scale sensor network by adapting energy beams to charge the nearby sensors, and stochastic geometry was utilized to derive the distribution of aggregated received power and the active probability of sensors. Massive antennas were applied in [13] and [14] for WPT, and the effect of the number of antennas on the performance was comprehensively analyzed. Moreover, the probability of outage in wireless energy transfer was analyzed in [15] for a base station using massive antenna arrays over flat fading channels. The authors in [16] studied energy beamforming in wireless-powered cellular networks, where downlink energy was used to sustain users during the uplink transmission. The work [17] extended [16] by deriving the asymptotically optimal power allocation for energy beamforming. For broadband wireless systems with multiple transmit antennas, the wideband channel is partitioned into several frequency sub-bands, and the signals on the sub-bands are optimized in the frequency domain to maximize the harvested power at the receiver in [18–23]. An overview on multiband WPT that exploits both beamforming and frequency diversity gains was provided in [20], while a power control problem was cast in [21] for realizing simultaneous information and power transfer in the downlink. The cost of channel training for narrow-band WPT was concerned in [22–24]. The authors in [22] studied channel training methods for distributed energy beamforming. In [23], beamforming, along with the time and energy costs for reverse-link channel training, was designed by exploiting the channel reciprocity. In [24], the preamble length and power allocation for channel estimation were optimized for energy beamforming.

More recently, power waveforming (PW) techniques have been first proposed in [25] to improve the WPT efficiency by exploiting multipath signals. Considering the fact that the radiated power is dispersed into multiple replicas of transmitted signals by the obstacles in wireless environments, waveforms were designed in [25] for the PW systems to constructively recollect the power of all possibly available multipaths at an energy receiver. However, the scheme only considers a single-antenna PW system and subject to the case when the number of multipaths is equal to the waveform length. In this chapter, we attempt to develop a joint power waveforming and beamforming design for WPT, in which the waveforms on multiple transmit antennas, driven by a reference signal, are proposed to maximize an energy delivery efficiency gain, which is defined as a ratio of the harvested energy at the receiver and the energy expenditure at the transmitter.

Most of the existing works studied the multi-antenna WPT by means of beam-forming techniques over frequency flat fading channels or narrow-band transmissions [9–17, 22–24, 26, 27]. Based on orthogonal frequency division multiplexing (OFDM), some recent works such as [18, 21, 28] investigated broadband WPT systems over frequency selective fading channels. However, the work [28] only addressed the problem with a single transmit antenna. For multi-antenna OFDM-based WPT systems, the work [21] focused on a joint power and sub-band allocation problem and bypassed beamforming designs, while the work [18] maximized the net harvested energy. In these two works, the sub-bands were assumed to be independent of each other, and no attempts have been made to appropriately utilize the multipath effect of wireless channels.

To the best of our knowledge, joint power waveforming and beamforming for WPT is first proposed in this chapter to reap both the antenna and multipath gains in wireless channels, which is different from the narrow-band transmission. While OFDM could be one possible solution for WPT, the proposed system in this chapter can be deemed as a *time-domain* waveform design approach, in which the total wideband channel is considered as a whole in the design of the time-domain reference signals and waveforms without relying on the assumption of the orthogonality of sub-bands such as in [18–23].[1] The major points of this chapter are summarized as follows:

- This chapter is the first attempt to formulate the WPT problem by combining power waveforming and beamforming and jointly optimize the reference signal and the waveforms to maximize the energy delivery efficiency gain. Two kinds of reference signals are considered in our study: non-periodic and periodic.
- For the nonperiodic transmission of the reference signals, the problem is iteratively solved by alternating between the optimization of the reference signal and the waveforms, and eigenvalue decomposition is performed at each iteration run. Based on some observations, a simple rule of thumb is found to initialize the reference signal, achieving a better performance than a randomly initialized reference signal. It is shown by simulation results that compared to conventional WPT using beamforming scheme, the proposed approach has a 20–30 dB improvement in energy delivery efficiency.
- In the case of the periodic reference signals, a rigorous analysis for the structure of the optimal waveforms is conducted for the proposed multi-antenna PW systems, and several quantitative findings are provided. In this case, low-complexity single-tone waveforms, which merely concentrate the entire waveform power on a single frequency component, are proven to be the optimal choice when the waveform length is a multiple of the period of a given reference signal. The optimal power allocation and tone selection are also discussed. With the optimal waveforms, we further show that the energy delivery efficiency gain can be linearly increased by prolonging the waveform length when the noise power is sufficiently small.
- In addition, when the period of the reference signal is larger than or equal to the length of multipath channels, we prove that the optimal reference signal with periodic transmission is a complex sinusoidal signal, which is determined by selecting a frequency component with the largest sum-power of channel frequency responses over multiple transmit antennas.
- With the optimal reference signal and waveform designs, we theoretically analyze the outage probability of the average harvested energy as well as the average energy delivery efficiency gain under a uniform power delay (UPD) channel profile. Specifically, an upper bound for the outage performance and a lower bound for the delivery efficiency are obtained in closed forms, which helps us to quantify the impact of the numbers of multipaths and antennas on the WPT performance. Finally, extensive computer simulations are carried out to validate the theoretical results and to demonstrate the effectiveness of the proposed multi-antenna PW schemes in various channel models.

[1] The CIR is explicitly utilized in the time-domain design, which is capable of constructively accumulating all the possibly available multipath power at the receiver in both spatial and temporal domains. The phenomenon in the temporal domain is the so-called multipath gains.

The following notations are adopted throughout this chapter. The uppercase and lowercase boldface letters denote matrices and vectors, respectively. The notations $(\cdot)^T$, $(\cdot)^\dagger$, $(\cdot)^*$, $((\cdot))_N$, and $\star$ stand for transpose, conjugate transpose, element-wise conjugate, modulo-N, and convolution operation, respectively. The matrices $\mathbf{I}_N$ and $\mathbf{F}_N$ represent an $N \times N$ identity matrix and an N-point discrete Fourier transform (DFT) matrix, respectively, and the (k,n)th entry of the DFT matrix is given by $\frac{1}{\sqrt{N}}e^{-j\frac{2\pi kn}{N}}$. The notations $\mathbf{1}_N$, $\mathbf{0}_N$, and $\mathbf{e}_k$ are used to express an all-one vector, an all-zero vector, and the kth column of the identity matrix $\mathbf{I}_N$, respectively. The Kronecker product and Hadamard product are denoted by $\otimes$ and $\odot$, respectively. The notation $\mathbb{E}[\cdot]$ takes expectation, while $\|\mathbf{x}\|_2$ finds the Euclidean norm of a vector $\mathbf{x}$. The matrix $Diag\,[\mathbf{x}]$ represents a diagonal matrix with $\mathbf{x}$ as its diagonal entries. The operators $\zeta_1\,[\mathbf{A}]$ and $\lambda_{\max}\,[\mathbf{A}]$ take the principal eigenvector and eigenvalue of the matrix $\mathbf{A}$, respectively.

The rest of this chapter is organized as follows. In Section 14.2, we introduce the system model for the proposed multi-antenna PW system in multipath channels. The waveform designs for nonperiodic and periodic transmissions of reference signals, along with the optimization of reference signals, are presented Section 14.3. Section 14.4 is devoted to analyze the performance of the multi-antenna PW system, including the outage probability of the average harvested energy and the average energy delivery efficiency gain. Section 14.5 presents simulation and numerical results.

14.2 System Model

The proposed multi-antenna PW system is illustrated in Figure 14.1, where the transmitter is equipped with M antennas, and the receiver has a single antenna. We assume that the wireless channel between each transmit and receive antenna is composed of L taps and quasi-static during the observation time. The channel impulse response (CIR) between the mth transmit antenna and the receiver is modeled as

$$h_m\,[n] = \sum_{l=0}^{L-1} h_{m,l}\delta\,[n-l], \quad n = 0, \ldots, L-1\,, \tag{14.1}$$

where $\delta\,[n]$ is the Kronecker delta function, and $h_{m,l}$'s are identically and independently distributed (i.i.d.) complex Gaussian with zero mean and variance $\rho_{m,l}$, for $l = 0, \ldots, L-1$. Without loss of generality, the total channel power of each channel link is normalized to one, i.e., $\sum_{l=0}^{L-1} \rho_{m,l} = 1$, for $m = 0, \ldots, M-1$.

In Figure 14.1, the implementation of the presented multi-antenna PW system consists of two transmission phases: channel probing phase and power transfer phase. During the first phase, the receiver first sends a pilot sounding signal with a delta-like autocorrelation function to each transmit antenna for estimating the corresponding CIR [29]. The CIR can be estimated very accurately through Golay sequences and the correlation method in [30]. In practice, the number of digitally resolvable multipaths in (14.1), which naturally exist in the wireless environments, at the transmitter increases as the system bandwidth becomes wider, and the number of resolvable multipaths will reach

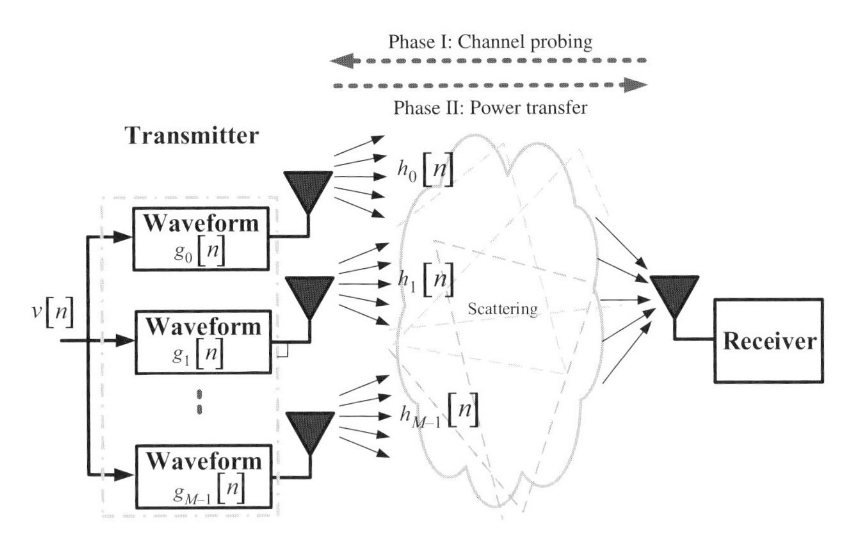

Figure 14.1 The presented multi-antenna PW system with two transmission phases: channel probing and power transfer

an upper limit when the system bandwidth is sufficiently large [31]. With the channel reciprocity and quasi-static assumptions, the transmitter computes a waveform $g_m[n]$ for each transmit antenna based on the estimated CIR during the second phase for the WPT purpose, for $m = 0, \ldots, M-1$ and $n = 0, \ldots, N_g - 1$, and N_g is the length of the waveform. Let $v[n]$ be a reference signal of length N_v, for $n = 0, \ldots, N_v - 1$, which acts as the energy-bearing signal for continuously supplying energy. The multipath channels will disperse the radiated power to multiple replicas of the transmitted signals, resulting in the so-called intersymbol interference (ISI) effect, which causes not only the magnitude and phase variation but also the delay spread of the transmitted signals. By leveraging this effect, the waveforms are designed to constructively accumulate all the possibly available multipath power at the receiver. Hence, the reference signal after the waveform embedding at the mth transmit antenna can be formulated as[2]

$$s_m[n] = \sqrt{P_v}\,(v \star g_m)[n], \quad n = 0, \ldots, N_g + N_v - 2, \tag{14.2}$$

where we assume that the total waveform power is equal to one, i.e., $\sum_{m=0}^{M-1} \sum_{n=0}^{N_g-1} |g_m[n]|^2 = 1$, the average reference signal power is given by $\frac{1}{N_v} \sum_{n=0}^{N_v-1} |v[n]|^2 = 1$, and P_v is the average transmit power. Accordingly, the received signal at the receiver side is given as

[2] Another architecture is to apply different reference signals for the transmit antennas, or equivalently, to combine the reference signal and waveform at each antenna as one set of variables. This is just a special case of Figure 14.1 by setting $N_v = 1$ and $v[0] = 1$ and treating $\mathbf{g}_m$ as the variable to be optimized. Based on our numerical simulations, this architecture performs worse than that adopted in Figure 14.1.

$$y[n] = \sum_{m=0}^{M-1} (s_m \star h_m)[n] + z[n] \quad (14.3)$$
$$= \sqrt{P_v} \sum_{m=0}^{M-1} (v \star g_m \star h_m)[n] + z[n],$$
$$n = 0, \ldots, N_g + N_v + L - 3,$$

where $z[n]$ is additive complex white Gaussian noise at the receiver side with zero mean and variance σ_z^2.

14.3 Power Transfer Waveform and Reference Signal Designs

In this section, we attempt to design the reference signal as well as the waveform at each transmit antenna for the presented multi-antenna PW system in order to maximize the energy delivery efficiency gain. While we only focus on a block transmission of $v[n]$ with the time duration of N_v in (14.3), the aforementioned WPT procedures can be repeated continuously with the different or the same reference signals, i.e., $v[n]$ could be nonperiodic or periodic. The optimal designs of these two scenarios are investigated in the following.

Definition 14.3.1 *An energy delivery efficiency gain is defined as a ratio of the total harvested energy at the receiver and the total energy expenditure of the reference signal at the transmitter.*[3]

14.3.1 Waveform Design with Nonperiodic Reference Signals

Let us first define $\mathbf{y} = \left[y[0], \ldots, y\left[N_g + N_v + L - 3\right]\right]^T$, $\mathbf{v} = [v[0], \ldots, v[N_v - 1]]^T$ and $\mathbf{h}_m = [h_m[0], \ldots, h_m[L-1]]^T$. By rewriting (14.3) into a compact matrix-vector form, it gives

$$\mathbf{y} = \sqrt{P_v} \sum_{m=0}^{M-1} \mathbf{V}\mathbf{H}_m \mathbf{g}_m + \mathbf{z} = \sqrt{P_v}\Phi\mathbf{g} + \mathbf{z}, \quad (14.4)$$

where $\mathbf{g}_m = \left[g_m[0], \ldots, g_m\left[N_g - 1\right]\right]^T$, $\mathbf{z} = \left[z[0], \ldots, z\left[N_g + N_v + L - 3\right]\right]^T$, $\mathbf{H}_m$ is a Toeplitz matrix of size $\left(N_g + L - 1\right) \times N_g$ with the vector $\left[\mathbf{h}_m^T, \mathbf{0}^T\right]^T$ as its first column, and $\mathbf{V}$ is a Toeplitz matrix of size $\left(N_g + N_v + L - 2\right) \times \left(N_g + L - 1\right)$ with the vector $\left[\mathbf{v}^T, \mathbf{0}^T\right]^T$ as its first column. Besides, we define $\mathbf{g} = \left[\mathbf{g}_0^T, \ldots, \mathbf{g}_{M-1}^T\right]^T$ and $\Phi = \mathbf{V}\left[\mathbf{H}_0, \ldots, \mathbf{H}_{M-1}\right]$. From (14.4), the energy delivery efficiency gain is computed as

[3] Here we assume that the large-scale path loss is normalized and mainly address the gain that can be provided by the joint power waveforming and beamforming. The exact energy delivery efficiency is equal to the energy delivery efficiency gain minus the path loss (in dB).

$$E_G = \frac{1}{N_v P_v} \mathbb{E}\left[\|\mathbf{y}\|_2^2 \right] \\ = \frac{1}{N_v P_v} \left(P_v \mathbf{g}^\dagger \Phi^\dagger \Phi \mathbf{g} + \left(N_g + N_v + L - 2 \right) \cdot \sigma_z^2 \right) . \tag{14.5}$$

The maximization problem for the energy delivery efficiency gain is then formulated as

$$\begin{aligned} (\mathbf{P1}) : &\max_{\mathbf{g}, \mathbf{v}} \mathbf{g}^\dagger \Phi^\dagger \Phi \mathbf{g} \\ \text{s.t.} \quad &(C.1) \quad \|\mathbf{g}\|_2^2 = 1 ; \\ &(C.2) \quad \|\mathbf{v}\|_2^2 = N_v . \end{aligned} \tag{14.6}$$

The joint design problem, however, is nonconvex, which cannot be directly solved in its current form. To make the problem tractable, we present an iterative method to handle the problem based on alternating optimization, in which the reference signal and the waveform are optimized and updated alternatively. For a given reference signal $\mathbf{v}$, the optimization problem for the waveform design is equivalent to an eigenvalue maximization problem; that is, the optimal waveform is given by

$$\hat{\mathbf{g}} = \zeta_1 \left[\mathbf{\Phi}^\dagger \mathbf{\Phi} \right] . \tag{14.7}$$

On the other hand, we can rewrite (14.3) as

$$\mathbf{y} = \sqrt{P_v} \sum_{m=0}^{M-1} \bar{\mathbf{H}}_m \mathbf{G}_m \mathbf{v} + \mathbf{z} = \sqrt{P_v} \bar{\mathbf{\Phi}} \mathbf{v} + \mathbf{z} , \tag{14.8}$$

where $\mathbf{G}_m$ is a Toeplitz matrix of size $\left(N_g + N_v - 1\right) \times N_v$ with the vector $\left[\mathbf{g}_m^T, \mathbf{0}^T\right]^T$ as its first column, $\bar{\mathbf{H}}_m$ is a Toeplitz matrix of size $\left(N_g + N_v + L - 2\right) \times \left(N_g + N_v - 1\right)$ with the vector $\left[\mathbf{h}_m^T, \mathbf{0}^T\right]^T$ as its first column, and $\bar{\mathbf{\Phi}} = \sum_{m=0}^{M-1} \bar{\mathbf{H}}_m \mathbf{G}_m$. As a result, for a given waveform $\mathbf{g}$, the optimization problem for designing the reference signal in (14.6) can be expressed as

$$\begin{aligned} (\mathbf{P2}) : &\max_{\mathbf{v}} \mathbf{v}^\dagger \bar{\mathbf{\Phi}}^\dagger \bar{\mathbf{\Phi}} \mathbf{v} \\ \text{s.t.} \quad &(C.1) \quad \|\mathbf{v}\|_2^2 = N_v . \end{aligned} \tag{14.9}$$

Accordingly, the optimal reference signal is given by

$$\hat{\mathbf{v}} = \sqrt{N_v} \cdot \zeta_1 \left[\bar{\Phi}^\dagger \bar{\mathbf{\Phi}} \right] . \tag{14.10}$$

An iterative algorithm for jointly optimizing the reference signal and the waveform is summarized in Table 14.1, where the reference signal $\mathbf{v}$ and the waveform $\mathbf{g}$ are alternatively updated based on the latest value obtained at iteration. The procedures are repeated until a stopping criterion is met. The stopping criterion is to check whether $\frac{1}{N_v} \left\| \mathbf{v}^{(i)} - \mathbf{v}^{(i-1)} \right\|_2^2 \leq \varepsilon$, where ε is a sufficiently small threshold, or the iteration number reaches a predefined limit $I_{\max}$. The convergence is analyzed as follows. Let $\Theta\left(\mathbf{v}, \mathbf{g}\right)$ be the optimal objective value of the problems (**P1**) or (**P2**). The

Table 14.1 Iterative algorithm for finding the reference signal and waveform

1: Set the iteration number $i = 0$ and the maximum allowable iteration number $I_{\max}$;
2: Initialize the reference signal $\mathbf{v}^{(i)}$;
3: **repeat**
4: For the given $\mathbf{v}^{(i)}$, compute the optimal waveform $\mathbf{g}^{(i)}$ using (14.7) ;
5: For the given $\mathbf{g}^{(i)}$, compute the optimal reference signal $\mathbf{v}^{(i+1)}$ using (14.10) ;
6: Set $i \leftarrow i + 1$;
7: **until** $\frac{1}{N_v} \left\| \mathbf{v}^{(i)} - \mathbf{v}^{(i-1)} \right\|_2^2 \leq \varepsilon$ or $i \geq I_{\max}$.

sequence $\left\{\Theta\left(\mathbf{v}^{(i)}, \mathbf{g}^{(i)}\right)\right\}$ is increasing and bounded upper because $\Theta\left(\mathbf{v}^{(i)}, \mathbf{g}^{(i)}\right) \leq \Theta\left(\mathbf{v}^{(i+1)}, \mathbf{g}^{(i)}\right) \leq \Theta\left(\mathbf{v}^{(i+1)}, \mathbf{g}^{(i+1)}\right)$. Therefore, the convergence of the algorithm is guaranteed. In addition, we evaluate the complexity of the presented algorithm in terms of the required number of complex multiplications per iteration. Assume that the complexity of finding the principal eigenvector of an $N \times N$ matrix is N^3. The calculation of (14.7) and (14.10) involves the matrix–matrix products and the principal eigenvector of a matrix, and the total complexity per iteration is no more than $N_{\max}^3 \left(M^3 + 3M^2 + 12M + 4\right)$. From (14.5) and (14.6), the achievable energy delivery efficiency gain is thus given by

$$E_G = \frac{1}{N_v P_v} \left(P_v \lambda_{\max} \left[\mathbf{\Phi}^{(i)\dagger} \mathbf{\Phi}^{(i)} \right] + \left(N_g + N_v + L - 2\right) \sigma_z^2 \right), \tag{14.11}$$

where $\mathbf{\Phi}^{(i)} = \mathbf{V}^{(i)} \left[\mathbf{H}_0, \ldots, \mathbf{H}_{M-1}\right]$, and the matrix $\mathbf{V}^{(i)}$ is obtained by substituting $\mathbf{v}^{(i)}$ into $\mathbf{V}$.

While the reference signal can be easily initialized by randomly generated complex binary signals, it is of great importance to carefully initialize the reference signal in order to achieve better WPT performance once the algorithm gets converged. To get more insight into determining a suitable initial value of $\mathbf{v}^{(0)}$, an example of the optimal waveform in the time domain, along with its relationship to the frequency-domain representations of $v[n]$, $g_m[n]$, and $h_m[n]$, is illustrated in Figure 14.2 for a given reference signal, where $M = 2$, $N_g = 100$, $N_v = 50$, and $P_v = 1$. Here, a Saleh–Valenzuela (SV) channel model in [32] is adopted to generate the multipath channels. An $N_{\max}$-point DFT is performed for the spectrum analysis, where $N_{\max} = \max\left\{N_g, N_v, L\right\}$. Because the signals $v[n]$, $g_m[n]$, and $h_m[n]$ have different lengths, they are zero-padded when applying the $N_{\max}$-point DFT. The frequency-domain representations of $v[n]$, $g_m[n]$, and $h_m[n]$ are denoted as $\tilde{v}[k]$, $\tilde{g}_m[k]$ and $\tilde{h}_m[k]$, respectively, for $k = 0, \ldots, N_{\max} - 1$. Furthermore, we define $\tilde{q}[k] = |\tilde{v}[k]|^2 \sum_{m=0}^{M-1} \left|\tilde{h}_m[k]\right|^2$, for $k = 0, \ldots, N_{\max} - 1$. Figure 14.2(a) exemplifies the optimal waveform with respect to a randomly generated reference signal $v[n]$, whose time-domain and frequency-domain representations are given in Figures 14.2(b) and 14.2(c), respectively. The optimal waveforms for the two transmit antennas in the frequency domain are shown in Figures 14.2(d) and 14.2(e) respectively, and we can make two interesting observations. First, the waveforms both concentrate its allocated power on a peak frequency tone $k = 94$ with the largest value

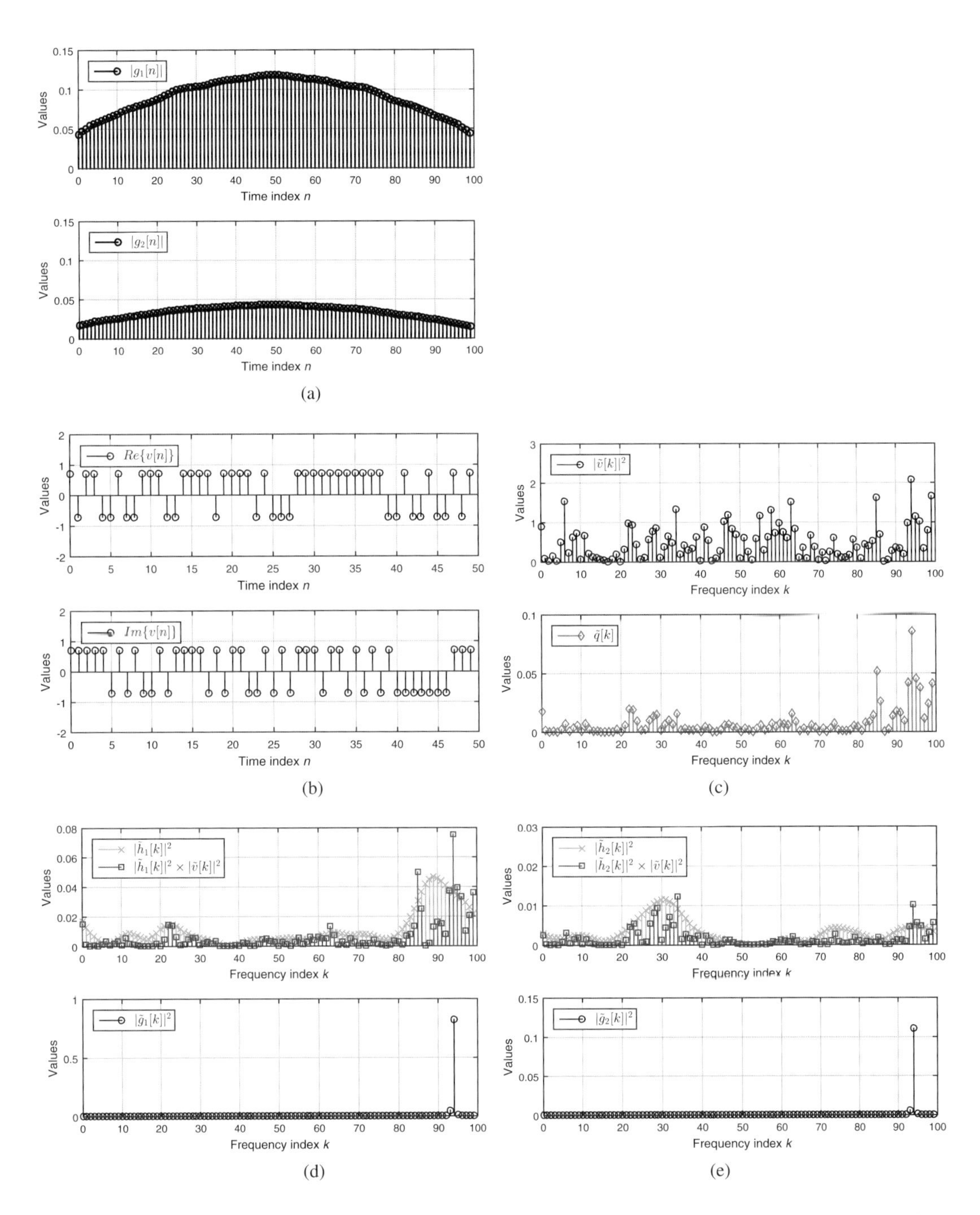

Figure 14.2 An example of the optimal waveform in the time domain and the frequency domain, along with its relationship to $|\tilde{h}_m[k]|^2$, $|\tilde{v}[k]|^2$ and $\tilde{q}[k]$, for a given reference signal ($M = 2$, $N_g = 100$, $N_v = 50$, and $P_v = 1$). (a) The values of $|g_1[n]|$ and $|g_2[n]|$; (b) Real and imaginary parts of $v[n]$; (c) The values of $|\tilde{v}[k]|^2$ and $\tilde{q}[k]$; (d) The values of $|\tilde{h}_1[k]|^2$, $|\tilde{h}_1[k]|^2|\tilde{v}[k]|^2$ and $|\tilde{g}_1[k]|^2$; (e) The values of $|\tilde{h}_2[k]|^2$, $|\tilde{h}_2[k]|^2|\tilde{v}[k]|^2$ and $|\tilde{g}_2[k]|^2$

of $\tilde{q}[k]$. Second, at different antennas, the larger the value of $|\tilde{h}_m[k]|^2|\tilde{v}[k]|^2$ at the peak frequency tone $k = 94$, the more power the waveform is allocated.

From the first observation, the waveforms bear resemblance to single tones with the same frequency, and thus, the received power is in a specific formula. This motivates the initialization format of $\mathbf{v}^{(0)}$ by condensing the power spectrum of the initial reference signal into a frequency tone with the largest value of the summation of the channel power over all the antennas, i.e., $\sum_{m=0}^{M-1} |\tilde{h}_m[k]|^2$, in order to maximize the peak value of $\tilde{q}[k]$. By doing so, the nth entry of the initial reference signal $\mathbf{v}^{(0)}$ is given as

$$v^{(0)}[n] = e^{j\frac{2\pi k_{\max} n}{N_{\max}}}, \quad n = 0, \ldots, N_v - 1, \tag{14.12}$$

where $k_{\max} = \arg \max_{k=0,\ldots,N_{\max}-1} \sum_{m=0}^{M-1} |\tilde{h}_m[k]|^2$. Another reason is that because $\mathbf{g}_m$ is unknown at the initialization stage, by assuming that $\mathbf{g}_m = \mathbf{e}_1$, for all m, the received signal is reduced to $\mathbf{y} = \sqrt{P_v}\sum_{m=0}^{M-1} \bar{\mathbf{H}}_m \mathbf{v} + \mathbf{z}$. From the frequency-domain viewpoint, the received signal is the multiplication between $\tilde{v}[k]$ and $\tilde{h}_m[k]$, and hence, the reference signal in (14.12) is a good initial choice. It is noted that the presented initial reference signal is equivalent to a complex sinusoidal signal truncated by a rectangular window to a length of N_v; hence, its frequency-domain representation is essentially a sinc function centered at the $k_{\max}^{th}$ frequency tone.

14.3.2 Waveform Design with Periodic Reference Signals

In this section, we investigate the optimal waveform design when the reference signal $v[n]$ is periodically transmitted over time, i.e., $v[n] = v[n + N_v]$. From (14.3), because $v[n]$ is N_v-periodic, the received signal at the receiver side is also N_v-periodic, given by

$$\mathbf{y}_c = \sqrt{P_v}\sum_{m=0}^{M-1} \mathbf{R}\mathbf{H}_m \mathbf{g}_m + \mathbf{z}_c, \tag{14.13}$$

where $\mathbf{y}_c = \left[y[0], \ldots, y[N_v - 1]\right]^T$, $\mathbf{z}_c = [z[0], \ldots, z[N_v - 1]]^T$, $\mathbf{R}$ is a generalized circulant matrix of size $N_v \times \left(N_g + L - 1\right)$, whose jth column is the cyclic permutation of the vector $\mathbf{v}$ with an offset $((j))_{N_v}$, for $j = 0, \ldots, N_g + L - 2$. By defining $\mathbf{\Phi}_c = \mathbf{R}\left[\mathbf{H}_0, \ldots, \mathbf{H}_{M-1}\right]$, we can rewrite (14.13) into a compact matrix-vector form:

$$\mathbf{y}_c = \sqrt{P_v}\mathbf{\Phi}_c \mathbf{g} + \mathbf{z}. \tag{14.14}$$

According to Definition 14.3.1, if the transmission time is sufficiently large, the energy delivery efficiency gain can be approximated as

$$E_{c,G} = \frac{1}{N_v P_v}\mathbb{E}\left[\|\mathbf{y}_c\|_2^2\right] = \frac{1}{N_v P_v}\left(P_v \mathbf{g}^\dagger \mathbf{\Phi}_c^\dagger \mathbf{\Phi}_c \mathbf{g} + N_v \sigma_z^2\right). \tag{14.15}$$

Under a given reference signal, the optimal waveform for maximizing the energy delivery efficiency gain in (14.15) can be computed as

$$\hat{\mathbf{g}}_c = \zeta_1\left[\mathbf{\Phi}_c^\dagger \mathbf{\Phi}_c\right], \tag{14.16}$$

where $\hat{\mathbf{g}}_c = \left[\hat{\mathbf{g}}_{c,0}, \ldots, \hat{\mathbf{g}}_{c,M-1}\right]^T$. Similar to the previous section, the reference signal and waveforms in the case of periodic transmission can be jointly optimized by following similar iterative steps in Table 14.1. Due to the limited space, the detailed derivation of the optimal reference signal for given waveforms is omitted here.

According to (14.14), let $\mathbf{\Phi}_{c,m} = \mathbf{R}\mathbf{H}_m$ be the mth submatrix of $\mathbf{\Phi}_c$, which is also a generalized circulant matrix of size $N_v \times N_g$. We define the first column of the matrix $\Phi_{c,m}$ as ϕ_m, for $m = 0, \ldots, M-1$, and its frequency representation as $\tilde{\phi}_m = \left[\tilde{\phi}_m[0], \ldots, \tilde{\phi}_m[N_v - 1]\right]^T = \mathbf{F}_{N_v}\phi_m$. From (14.15), the energy delivery efficiency gain is upper bounded by

$$\begin{aligned} E_{c,G} &\le \frac{1}{N_v P_v}\left(P_v \|\mathbf{\Phi}_c\|_2^2 + N_v \sigma_z^2\right) \\ &= \frac{1}{N_v P_v}\left(P_v \lambda_{\max}\left[\mathbf{\Phi}_c^\dagger \mathbf{\Phi}_c\right] + N_v \sigma_z^2\right) \\ &= \frac{1}{N_v P_v}\left(P_v \lambda_{\max}\left[\mathbf{\Phi}_c \mathbf{\Phi}_c^\dagger\right] + N_v \sigma_z^2\right) \\ &= \frac{1}{N_v P_v}\left(P_v \lambda_{\max}\left[\sum_{m=0}^{M-1} \mathbf{\Phi}_{c,m} \mathbf{\Phi}_{c,m}^\dagger\right] + N_v \sigma_z^2\right), \end{aligned} \tag{14.17}$$

where the relationship of $\|\mathbf{\Phi}_c \mathbf{g}\|_2^2 \le \|\mathbf{\Phi}_c\|_2^2 \|\mathbf{g}\|_2^2$ and $\|\mathbf{g}\|_2^2 = 1$ is applied to the first inequality. Let $\bar{\Phi}_{c,m}$ be an $N_v \times N_v$ circulant matrix with ϕ_m as its first column, and we have $\mathbf{F}_{N_v} \bar{\Phi}_{c,m} \mathbf{F}_{N_v}^\dagger = \sqrt{N_v} Diag[\tilde{\phi}_m]$. Because $N_g = QN_v$, it can be shown from (14.13) that $\mathbf{\Phi}_{c,m} = \mathbf{R}\mathbf{H}_m = \mathbf{1}_Q^T \otimes \bar{\Phi}_{c,m}$, yielding $\mathbf{F}_{N_v} \mathbf{\Phi}_{c,m} (\mathbf{I}_Q \otimes \mathbf{F}_{N_v})^\dagger = \sqrt{N_v}(\mathbf{1}_Q^T \otimes Diag[\tilde{\phi}_m])$. By using the property of $\mathbf{F}_{N_v} \mathbf{F}_{N_v}^\dagger = \mathbf{F}_{N_v}^\dagger \mathbf{F}_{N_v} = \mathbf{I}_{N_v}$ and inserting the DFT matrix into (14.17), one can diagonalize the matrix $\mathbf{\Phi}_{c,m}$ as shown in (14.19). Because $\lambda_{\max}\left[\sum_{m=0}^{M-1} Diag[\tilde{\phi}_m] \cdot Diag[\tilde{\phi}_m]^\dagger\right] = \sum_{m=0}^{M-1} \left|\tilde{\phi}_m[k_{c,\max}]\right|^2$, the energy delivery efficiency gain in (14.19) is finally bounded by

$$E_{c,G} \le Q \cdot \sum_{m=0}^{M-1} \left|\tilde{\phi}_m\left[k_{c,\max}\right]\right|^2 + \frac{\sigma_z^2}{P_v}, \tag{14.18}$$

where $k_{c,\max} = \arg \max_{k=0,\ldots,N_v-1} \sum_{m=0}^{M-1} \left|\tilde{\phi}_m[k]\right|^2$.

$$\begin{aligned} E_{c,G} &\le \frac{1}{N_v P_v}\left(P_v \lambda_{\max}\left[\sum_{m=0}^{M-1} \mathbf{F}_{N_v} \mathbf{\Phi}_{c,m} (\mathbf{I}_Q \otimes \mathbf{F}_{N_v})^\dagger (\mathbf{I}_Q \otimes \mathbf{F}_{N_v}) \mathbf{\Phi}_{c,m}^\dagger \mathbf{F}_{N_v}^\dagger\right] + N_v \sigma_z^2\right) \\ &= \frac{1}{N_v P_v}\left(N_v Q P_v \cdot \lambda_{\max}\left[\sum_{m=0}^{M-1} Diag\left[\tilde{\phi}_m\right] \cdot Diag\left[\tilde{\phi}_m\right]^\dagger\right] + N_v \sigma_z^2\right). \end{aligned} \tag{14.19}$$

Following we are interested in discovering the structure of the optimal waveform and reference signal. In fact, it can be proved that with the periodic transmission of a reference signal, the optimal waveform for each transmit antenna in (14.16) is endowed with a simple single-tone structure, if the lengths of the waveform and the reference signal are appropriately designed. The following theorem is then provided.

Theorem 14.3.1 *Let $N_g = Q \cdot N_v$, where Q is a positive integer. For any reference signal* $\mathbf{v}$*, the optimal waveforms $\hat{\mathbf{g}}_{c,m}$ at transmit antennas are single-tone waveforms, given by (14.21).*

$$\begin{aligned} E_{c,G} &= \frac{1}{N_v P_v}\left(P_v\left\|\sum_{m=0}^{M-1}\mathbf{\Phi}_{c,m}\mathbf{g}_{c,m}\right\|_2^2 + N_v\sigma_z^2\right) \\ &= \frac{1}{N_v P_v}\left(P_v\left\|\sum_{m=0}^{M-1}\mathbf{\Phi}_{c,m}\frac{\tilde{\phi}_m^*\left[k_{c,\max}\right]}{\sqrt{\sum_{m=0}^{M-1}\left|\tilde{\phi}_m\left[k_{c,\max}\right]\right|^2}}\left(\frac{1}{\sqrt{Q}}\mathbf{1}_Q\otimes\left(\mathbf{F}_{N_v}^{\dagger}\mathbf{e}_{k_{c,\max}}\right)\right)\right\|_2^2\right. \\ &\quad \left. + N_v\sigma_z^2\right) \\ &= \frac{1}{N_v P_v}\left(P_v\left\|\mathbf{F}_{N_v}\sum_{m=0}^{M-1}\mathbf{\Phi}_{c,m}\frac{\tilde{\phi}_m^*\left[k_{c,\max}\right]}{\sqrt{\sum_{m=0}^{M-1}\left|\tilde{\phi}_m\left[k_{c,\max}\right]\right|^2}}\left(\frac{1}{\sqrt{Q}}\mathbf{1}_Q\otimes\left(\mathbf{F}_{N_v}^{\dagger}\mathbf{e}_{k_{c,\max}}\right)\right)\right\|_2^2\right. \\ &\quad \left. + N_v\sigma_z^2\right), \end{aligned} \tag{14.20}$$

$$\begin{aligned} \hat{\mathbf{g}}_{c,m} &= \frac{\tilde{\phi}_m^*\left[k_{c,\max}\right]}{\sqrt{\sum_{m=0}^{M-1}\left|\tilde{\phi}_m\left[k_{c,\max}\right]\right|^2}}\left(\frac{1}{\sqrt{Q}}\mathbf{1}_Q\otimes\left(\mathbf{F}_{N_v}^{\dagger}\mathbf{e}_{k_{c,\max}}\right)\right), \\ m &= 0,\ldots,M-1, \end{aligned} \tag{14.21}$$

with $k_{c,\max} = \arg\max_{k=0,\ldots,N_v-1}\sum_{m=0}^{M-1}\left|\tilde{\phi}_m[k]\right|^2$.

Proof. *To prove this theorem, we show that the presented optimal waveform in (14.21) can achieve the upper bound in (14.18). Substituting (14.21) into (14.15), we can obtain (14.20) where the third equality is obtained by inserting the DFT matrix $\mathbf{F}_{N_v}$. By applying the result of* $\mathbf{F}_{N_v}\mathbf{\Phi}_{c,m}\left(\mathbf{1}_Q\otimes\left(\mathbf{F}_{N_v}^{\dagger}\mathbf{e}_{k_{c,\max}}\right)\right) = Q\sqrt{N_v}\tilde{\phi}_m\left[k_{c,\max}\right]\mathbf{e}_{k_{c,\max}}$ *into (14.20), it then yields*

$$E_{c,G} = Q\cdot\sum_{m=0}^{M-1}\left|\tilde{\phi}_m\left[k_{c,\max}\right]\right|^2 + \frac{\sigma_z^2}{P_v}. \tag{14.22}$$

The proof is thus completed. ▲

From Theorem 14.3.1, it can be found that if the waveform length is a multiple of the length of the reference signal, the optimal waveform $\hat{\mathbf{g}}_{c,m}$ is a complex sinusoidal signal, merely composed of a single-frequency component $k_{c,\max}$, no matter what the reference signal is. Moreover, the term $\tilde{\phi}_m^*\left[k_{c,\max}\right]$ represents a power allocation and phase alignment factor for the mth transmit antenna, while the term $\frac{1}{\sqrt{\sum_{m=0}^{M-1}\left|\tilde{\phi}_m\left[k_{c,\max}\right]\right|^2}}$ is a power normalization factor. This simple structure offers an attractive solution for low-complexity implementation of the optimal waveform without the need of executing eigenvalue decomposition, as compared with (14.7) and (14.16). Based on Theorem 14.3.1, two corollaries regarding the achievable energy delivery efficiency gain and the effect of the waveform length on the WPT performance are provided in the following.

Corollary 14.3.1 *When $N_g = Q \cdot N_v$, where Q is a positive integer, the energy delivery efficiency gain achieved by the optimal single-tone waveform $\hat{\mathbf{g}}_{c,m}$ in (14.21) is*

$$E_{c,G} = Q \cdot \sum_{m=0}^{M-1}\left|\tilde{\phi}_m\left[k_{c,\max}\right]\right|^2 + \frac{\sigma_z^2}{P_v}. \tag{14.23}$$

Proof. *This result is directly obtained from (14.22).* ▲

Corollary 14.3.2 *Let J be a positive integer. When $\frac{\sigma_z^2}{P_v}$ approaches to zero, the energy delivery efficiency gain with respect to the optimal single-tone waveform $\hat{\mathbf{g}}_{c,m}$ in (14.21) can be approximately improved by J times, if the waveform length N_g is prolonged by J times.*

Proof. *Consider two waveform lengths $N_{g,1} = Q \cdot N_v$ and $N_{g,2} = J \cdot N_{g,1}$. The energy delivery efficiency gains for the optimal single-tone waveform designs with the length $N_{g,1}$ and $N_{g,2}$ are denoted as E_{c,G_1} and E_{c,G_2}, respectively. From Corollary 14.3.1, we have*

$$\frac{E_{c,G_2}}{E_{c,G_1}} = \frac{JQN_v\sum_{m=0}^{M-1}\left|\tilde{\phi}_m\left[k_{c,\max}\right]\right|^2 + \frac{\sigma_z^2}{P_v}}{QN_v\sum_{m=0}^{M-1}\left|\tilde{\phi}_m\left[k_{c,\max}\right]\right|^2 + \frac{\sigma_z^2}{P_v}}. \tag{14.24}$$

If $\frac{\sigma_z^2}{P_v}$ is sufficiently small, the preceding ratio can be approximated as

$$\frac{E_{c,G_2}}{E_{c,G_1}} \approx J. \tag{14.25}$$

▲

From Corollary 14.3.1 and Corollary 14.3.2, it is found that the energy delivery efficiency gain achieved by the optimal single-tone waveform can be linearly increased by enlarging the waveform length, if $\frac{\sigma_z^2}{P_v}$ is sufficiently small. In general, this condition is true because the transmit power P_v is much larger than the noise power σ_z^2 in the WPT applications. Also the performance gain is determined by $\sum_{m=0}^{M-1}\left|\tilde{\phi}_m\left[k_{c,\max}\right]\right|^2$, which is related to the frequency selectivity of the wireless channels and the reference signal.

In what follows, we investigate the design structure of the optimal reference signal. Let $\tilde{\mathbf{h}}_m = \left[\tilde{h}_m[0], \ldots, \tilde{h}_m[N_v-1]\right]^T = \mathbf{F}_{N_v}\bar{\mathbf{h}}_m$, where $\bar{\mathbf{h}}_m = \left[\mathbf{h}_m^T, \mathbf{0}^T\right]^T$, and define $\bar{\mathbf{R}}$

as a circulant matrix whose first column is $\mathbf{v}$. A theorem regarding the optimal reference signal is provided in the following.

Theorem 14.3.2 *When $N_v \geq L$, the optimal reference signal for the multi-antenna PW system with periodic transmission is given by*

$$\hat{v}[n] = e^{j\frac{2\pi k_{c,\max} n}{N_v}}, \quad n = 0, \ldots, N_v - 1, \tag{14.26}$$

where $k_{c,\max} = \arg \max_{k=0,\ldots,N_v-1} \sum_{m=0}^{M-1} \left|\tilde{h}_m[k]\right|^2$.

Proof. *By the definition of $\mathbf{\Phi}_{c,m} = \mathbf{R}\mathbf{H}_m = \mathbf{1}_Q^T \otimes \bar{\mathbf{\Phi}}_{c,m}$ in (14.17), we have $\phi_m = \bar{\mathbf{R}}\bar{\mathbf{h}}_m$, if $N_v \geq L$. Hence, it implies $\tilde{\phi}_m = \mathbf{F}_{N_v}\phi_m = \mathbf{F}_{N_v}\bar{\mathbf{R}}\bar{\mathbf{h}}_m$. By diagonalizing $\bar{\mathbf{R}}$, it further gives*

$$\tilde{\phi}_m = \mathbf{F}_{N_v}\bar{\mathbf{R}}\mathbf{F}_{N_v}^{\dagger}\mathbf{F}_{N_v}\bar{\mathbf{h}}_m = \sqrt{N_v}Diag\left[\tilde{\mathbf{v}}\right]\tilde{\mathbf{h}}_m, \tag{14.27}$$

where $\tilde{\mathbf{v}} = \mathbf{F}_{N_v}\mathbf{v}$. By using (14.27), we can obtain

$$\begin{aligned}\tilde{\phi}_m \odot \tilde{\phi}_m^* &= Diag\left[\sqrt{N_v}Diag\left[\tilde{\mathbf{v}}\right]\tilde{\mathbf{h}}_m\right] \cdot \sqrt{N_v}Diag\left[\tilde{\mathbf{v}}\right]^*\tilde{\mathbf{h}}_m^* \\ &= N_v Diag\left[\tilde{\mathbf{v}}\right] \cdot Diag\left[\tilde{\mathbf{v}}\right]^* \cdot Diag\left[\tilde{\mathbf{h}}_m\right] \cdot \tilde{\mathbf{h}}_m^* \\ &= N_v\left(\tilde{\mathbf{v}} \odot \tilde{\mathbf{v}}^*\right) \odot \left(\tilde{\mathbf{h}}_m \odot \tilde{\mathbf{h}}_m^*\right).\end{aligned} \tag{14.28}$$

From Corollary 14.3.1, it is known that the optimal reference signal can be found by maximizing $E_{c,G}$ in (14.23), or equivalently, maximizing $\sum_{m=0}^{M-1}\left|\tilde{\phi}_m\left[k_{c,\max}\right]\right|^2$. Because $\sum_{m=0}^{M-1}\tilde{\phi}_m \odot \tilde{\phi}_m^* = \sum_{m=0}^{M-1}\left[\left|\tilde{\phi}_m[0]\right|^2, \ldots, \left|\tilde{\phi}_m[N_v-1]\right|^2\right]^T$, the energy delivery efficiency gain can be maximized by letting $\tilde{\mathbf{v}} = \sqrt{N_v}\mathbf{e}_{k_{c,\max}}$, where $k_{c,\max} =$ $\arg \max_{k=0,\ldots,N_v-1} \sum_{m=0}^{M-1}\left|\tilde{h}_m[k]\right|^2$. The optimal reference signal in the time domain is thus given by $\hat{\mathbf{v}} = \mathbf{F}_{N_v}^{\dagger}\tilde{\mathbf{v}} = \sqrt{N_v}\mathbf{F}_{N_v}^{\dagger}\mathbf{e}_{k_{c,\max}}$. ▲

Theorem 14.3.2 implies that for the scenario of periodic transmission, the optimal reference signal is indeed a complex sinusoidal signal when $N_v \geq L$[4]. Also from (14.27) and (14.28) in the proof of this theorem, we can get $\tilde{\phi}_m[k] = N_v\tilde{h}_m[k]$ and $\left|\tilde{\phi}_m[k]\right|^2 = N_v^2\left|\tilde{h}_m[k]\right|^2$, for $k = 0, \ldots, N_v - 1$, because $\tilde{\mathbf{v}} = \sqrt{N_v}\mathbf{e}_{k_{c,\max}}$. By using Theorem 14.3.1, the optimal waveform, associated with the optimal reference signal, can be explicitly expressed as

$$\hat{\mathbf{g}}_{c,m} = \frac{\tilde{h}_m^*\left[k_{c,\max}\right]}{\sqrt{\sum_{m=0}^{M-1}\left|\tilde{h}_m\left[k_{c,\max}\right]\right|^2}}\left(\frac{1}{\sqrt{Q}}\mathbf{1}_Q \otimes \left(\mathbf{F}_{N_v}^{\dagger}\mathbf{e}_{k_{c,\max}}\right)\right),$$
$$m = 0, \ldots, M-1. \tag{14.29}$$

[4] The complex sinusoidal structure of the optimal reference signal depends on the considered periodic or nonperiodic scenarios, together with the conditions $N_g = QN_v$ and $N_v \geq L$.

Corollary 14.3.3 *When $N_g = Q \cdot N_v$ and $N_v \geq L$, the energy delivery efficiency gain for the optimal reference signal in (14.26) and the corresponding optimal waveform in (14.29) is given by*

$$E_{c,G} = N_g N_v \cdot \sum_{m=0}^{M-1} \left| \tilde{h}_m \left[k_{c,\max} \right] \right|^2 + \frac{\sigma_z^2}{P_v}. \tag{14.30}$$

Proof. *By using $|\tilde{\phi}_m[k]|^2 = N_v^2 |\tilde{h}_m[k]|^2$ and from Corollary 14.3.1, the energy delivery efficiency gain can be derived as $E_{c,G} = QN_v^2 \sum_{m=0}^{M-1} \cdot |\tilde{h}_m[k_{c,\max}]|^2 + \frac{\sigma_z^2}{P_v}$. The proof is thus completed.* ▲

It is addressed by this corollary that under the mentioned conditions and optimal designs, the energy delivery efficiency gain is appropriately proportional to the product of the lengths of the waveform and reference signal, if $\frac{\sigma_z^2}{P_v}$ is small enough.

14.4 Performance Analysis of Multiantenna PW Systems

In this section, we theoretically analyze the WPT performance of the presented multi-antenna PW system. It is tough to analyze the WPT performance under the case of the nonperiodic transmissions of reference signals. Alternatively, the WPT performance is investigated under the case of the periodic transmissions of reference signals. In addition to the average energy delivery efficiency gain, an outage probability of the average harvested energy, which is different from the conventional notation used in wireless information transmission, is taken into consideration to quantify the performance. Specifically, the outage event is defined as follows.

Definition 14.4.1 *A multi-antenna PW system is in outage, if the harvested energy is smaller than or equal to a preset threshold x.*

To make the analysis tractable, it is assumed that $N_g = Q \cdot N_v$ and $N_v = C \cdot L$ throughout this section, where Q and C both take positive integer values. For the WPT applications, this is true because the lengths of the waveforms and the reference signal are in general larger than the channel length in order to achieve a higher energy delivery efficiency gain, as shown in (14.30). Moreover, it is almost impossible to analyze the performance for general channel power delay profiles; instead, we consider a uniform power delay (UPD) channel profile, i.e., $\rho_{m,l} = \frac{1}{L}$ in (14.1), and investigate the impact of the numbers of multipaths and antennas on the WPT performance. From Corollary 14.3.3, one can observe that both the average energy delivery efficiency gain and the outage performance of the average harvested energy are mainly influenced by the channel frequency selective fading effect $\sum_{m=0}^{M-1} \left| \tilde{h}_m \left[k_{c,\max} \right] \right|^2$. To facilitate the analysis, a performance lower bound for the energy delivery efficiency gain in (14.30) under the UPD channel profile is given in the following lemma.

Lemma 14.4.1 *When $N_g = Q \cdot N_v$ and $N_v = C \cdot L$, the energy delivery efficiency gain for the optimal reference signal in (14.26) and the corresponding optimal waveform in (14.29) under the UPD channel profile can be lower bounded by*

$$E_{c,G} \geq N_g L \sum_{m=0}^{M-1} \left|\tilde{u}_m \left[k_{c,\max}\right]\right|^2 + \frac{\sigma_z^2}{P_v}, \tag{14.31}$$

where $\tilde{\mathbf{u}}_m = \left[\tilde{u}_m[0], \ldots, \tilde{u}_m[L-1]\right]^T = \mathbf{F}_L \mathbf{h}_m$, *and* $k_{c,\max} = \arg \max_{k=0,\ldots,L-1} \sum_{m=0}^{M-1} |\tilde{u}_m[k]|^2$. *Moreover, the equality in* (14.31) *holds for* $C = 1$ *or* $L = 1$.

Proof. *From the definition of* $\tilde{\mathbf{h}}_m = \mathbf{F}_{N_v} \bar{\mathbf{h}}_m$ *and* $N_v = C \cdot L$, *it gives*

$$\tilde{h}_m[Ck + l] = \sqrt{\frac{L}{N_v}} \left(\frac{1}{\sqrt{L}} \sum_{n=0}^{L-1} h_m[n] e^{-j\frac{2\pi kn}{L}} e^{-j\frac{2\pi ln}{N_v}} \right),$$
$$k = 0, \ldots, L-1, l = 0, \ldots, C-1. \tag{14.32}$$

By using $\tilde{\mathbf{u}}_m = \mathbf{F}_L \mathbf{h}_m$, *we can obtain the relationship between* $\tilde{h}_m[k]$ *and* $\tilde{u}_m[k]$:

$$\tilde{h}_m[Ck] = \sqrt{\frac{L}{N_v}} \tilde{u}_m[k], \quad k = 0, \ldots, L-1. \tag{14.33}$$

It then implies from (14.33) that

$$\begin{aligned} &\max_{k=0,\ldots,N_v-1} \sum_{m=0}^{M-1} \left|\tilde{h}_m[k]\right|^2 \\ &\geq \max_{k=0,\ldots,L-1} \sum_{m=0}^{M-1} \left|\tilde{h}_m[Ck]\right|^2 \\ &= \max_{k=0,\ldots,L-1} \frac{L}{N_v} \sum_{m=0}^{M-1} |\tilde{u}_m[k]|^2, \end{aligned} \tag{14.34}$$

where the inequality in (14.34) becomes active for $C = 1$. *Note that for* $L = 1$, *the inequality is also active because* $\left|\tilde{h}_m[k]\right|^2 = \frac{1}{N_v} |h_m[0]|^2$ *for all k according to (14.32). From Corollary 14.3.3 and (14.34), the proof is thus completed.* ▲

From Lemma 14.4.1, it is found that the performance lower bound is relevant to the maximum value of $\sum_{m=0}^{M-1} |\tilde{u}_m[k]|^2$, which is the summation of the power of the L-point channel frequency responses over transmit antennas. Furthermore, this lower bound is reached when the length of the reference signal is identical to the number of multipaths or the number of multipaths for each channel link is equal to one. For the convenience of notation, let $\mu_k = \sum_{m=0}^{M-1} |\tilde{u}_m[k]|^2$, for $k = 0, \ldots, L-1$, and $\boldsymbol{\mu} = \left[\mu_0, \ldots, \mu_{L-1}\right]^T$. The characteristic function of the multivariate random vector $\boldsymbol{\mu}$ is provided as follows.

Lemma 14.4.2 *Let* $\boldsymbol{\omega} = \left[\omega_0, \ldots, \omega_{L-1}\right]^T$. *The characteristic function of* $\boldsymbol{\mu}$ *under the UPD channel profile is*

$$\Psi_{\boldsymbol{\mu}}\left(j\boldsymbol{\omega}\right) = \left(\prod_{l=0}^{L-1} \frac{L}{L - j\omega_l}\right)^M . \tag{14.35}$$

Proof. Let $\boldsymbol{\mu}_m = \left[\mu_{m,0}, \ldots, \mu_{m,L-1}\right]^T$, where we define $\mu_{m,k} = \left|\tilde{u}_m\left[k\right]\right|^2$. From the proof of Theorem 5 in [25], the characteristic function of $\boldsymbol{\mu}_m$ under the UPD channel profile can be derived as

$$\Psi_{\boldsymbol{\mu}_m}\left(j\boldsymbol{\omega}\right) = \prod_{l=0}^{L-1} \frac{L}{L - j\omega_l} . \tag{14.36}$$

Because the random vectors $\boldsymbol{\mu}_m$ are independent for different transmit antennas and $\boldsymbol{\mu} = \sum_{m=0}^{M-1} \boldsymbol{\mu}_m$, it results in $\Psi_{\boldsymbol{\mu}}\left(j\boldsymbol{\omega}\right) = \prod_{m=0}^{M-1} \Psi_{\boldsymbol{\mu}_m}\left(j\boldsymbol{\omega}\right)$. Hence, the proof is completed. ▲

By applying (14.15) and Lemma 14.4.1, the average harvested energy during a time period of N_v can be explicitly computed by $E_H \triangleq \frac{1}{N_v}\mathbb{E}\left[\|\mathbf{y}_c\|_2^2\right] = P_v E_{c,G}$, and thus, it is lower bounded by

$$E_H \geq N_g P_v L \cdot \mu_{k_{c,\max}} + \sigma_z^2 . \tag{14.37}$$

Accordingly, a theorem regarding an upper bound for the outage performance of the average harvested energy is provided in the following.

Theorem 14.4.1 *When* $N_g = Q \cdot N_v$ *and* $N_v = C \cdot L$, *the outage performance of the average harvested energy* E_H *for the optimal reference signal in (14.26) and the corresponding optimal waveform in (14.29) under the UPD channel profile is upper bounded by*

$$\Pr\left(E_H \leq x\right) \leq \left(\frac{1}{(M-1)!} \cdot \gamma\left(M, \frac{1}{N_g P_v}\left(x - \sigma_z^2\right)\right)\right)^L , \tag{14.38}$$

where $\gamma\left(s,x\right) = \int_0^x t^{s-1} e^{-t} dt$ *is the lower incomplete Gamma function, and the equality of the upper bound holds for* $C = 1$ *or* $L = 1$.

Proof. By applying the characteristic function in (14.35), the cumulative distribution function (CDF) of $\boldsymbol{\mu}$ can be expressed in terms of $\Psi_{\boldsymbol{\mu}}\left(j\boldsymbol{\omega}\right)$ as follows [33]:

$$\Pr\left(\mu_0 \leq x_0, \mu_1 \leq x_1, \ldots, \mu_{L-1} \leq x_{L-1}\right)$$
$$= \frac{1}{(2\pi)^L} \int_{-\infty}^{\infty} \cdots \int_{-\infty}^{\infty} \Psi_{\boldsymbol{\mu}}\left(j\boldsymbol{\omega}\right) \times \prod_{l=0}^{L-1} \left(\frac{1 - e^{-j\omega_l x_l}}{j\omega_l}\right) d\omega_0 \cdots d\omega_{L-1} , \tag{14.39}$$

where $\mathbf{x} = \left[x_0, \ldots, x_{L-1}\right]^T$. Because $\mu_{k_{c,\max}} = \max_{k=0,\ldots,L-1} \mu_k$, the CDF of the random variable $\mu_{k_{c,\max}}$ is obtained by setting $x_0 = \cdots = x_{L-1} = x$ in (14.39):

$$\Pr\left(\mu_{k_{c,\max}} \leq x\right) = \frac{1}{(2\pi)^L} \int_{-\infty}^{\infty} \cdots \int_{-\infty}^{\infty} \Psi_\mu(j\omega) \times \prod_{l=0}^{L-1} \left(\frac{1 - e^{-j\omega_l x}}{j\omega_l}\right) d\omega_0 \cdots d\omega_{L-1}. \tag{14.40}$$

From (14.35) and (14.40), the CDF of the random variable $\mu_{k_{c,\max}}$ can be explicitly derived as shown in (14.41),

$$\begin{aligned}\Pr\left(\mu_{k_{c,\max}} \leq x\right) &= \frac{1}{(2\pi)^{L-1}} \int_{-\infty}^{\infty} \cdots \int_{-\infty}^{\infty} \prod_{l=1}^{L-1} \left(\left(\frac{L}{L - j\omega_l}\right)^M \cdot \frac{1 - e^{-j\omega_l x}}{j\omega_l}\right) \\ &\quad \cdot \left(\frac{1}{2\pi} \int_{-\infty}^{\infty} \left(\frac{L}{L - j\omega_0}\right)^M \cdot \frac{1 - e^{-j\omega_0 x}}{j\omega_0} d\omega_0\right) d\omega_1 \cdots d\omega_{L-1} \\ &= \left(\frac{1}{(M-1)!} \cdot \gamma(M, Lx)\right)^L,\end{aligned} \tag{14.41}$$

$$\mathbb{E}\left[E_{c,G}\right] \geq N_g L \left(\sum_{l=1}^{L} \binom{L}{l} (-1)^{l+1} \sum_{k=0}^{(M-1)l} b_k(M,L,l) \left(\frac{1}{lL}\right)^{k+1} k!\right) + \frac{\sigma_z^2}{P_v}, \tag{14.42}$$

where the relationship between the CDF function and the characteristic function of the Erlang distribution is applied in the last equality of (14.41) as follows:

$$\frac{1}{2\pi} \int_{-\infty}^{\infty} \left(\frac{\beta}{\beta - j\omega}\right)^M \left(\frac{1 - e^{-j\omega x}}{j\omega}\right) d\omega = \frac{1}{(M-1)!} \cdot \gamma\left(M, \beta x\right), \tag{14.43}$$

where $\gamma(s,x) = \int_0^x t^{s-1} e^{-t} dt$ is the lower incomplete Gamma function. By using (14.37), the outage performance of the average harvested energy E_H is upper bounded by

$$\Pr\left(E_H \leq x\right) \leq \Pr\left(N_g P_v L \cdot \mu_{k_{c,\max}} + \sigma_z^2 \leq x\right). \tag{14.44}$$

From (14.41) and (14.44), it is concluded that the upper bound of the outage performance is given as in (14.38), and by further applying Lemma 14.4.1, we know that the upper bound becomes tight for $C = 1$ or $L = 1$. ▲

Theorem 14.4.2 *When $N_g = Q \cdot N_v$ and $N_v = C \cdot L$, the average energy delivery efficiency gain for the optimal reference signal in (14.26) and the corresponding optimal waveform in (14.29) under the UPD channel profile is lower bounded as shown in (14.42) in which $b_k(M,L,l)$ is the coefficient of x^k, for $k = 0, \ldots, (M-1)l$, in the expansion of*

$$\left(\sum_{k=0}^{M-1}\frac{L^k x^k}{k!}\right)^l, \tag{14.45}$$

and $\binom{n}{k} = \frac{n!}{k!(n-k)!}$.

PROOF. *The lower incomplete Gamma function in (14.41) can be expressed in a form of power series expansion [34]:*

$$\gamma\left(M, Lx\right) = \Gamma\left(M\right) \cdot \left(1 - e^{-Lx}\sum_{k=0}^{M-1}\frac{L^k x^k}{k!}\right), \tag{14.46}$$

where $\Gamma\left(M\right) = \left(M-1\right)!$ *is the Gamma function. Because* $\mu_{k_{c,\max}} \geq 0$*, the mean of the random variable* $\mu_{k_{c,\max}}$ *can be directly computed through its CDF as follows:*

$$\mathbb{E}\left[\mu_{k_{c,\max}}\right] = \int_0^\infty \left(1 - \Pr\left(\mu_{k_{c,\max}} \leq x\right)\right)dx\,. \tag{14.47}$$

By substituting (14.41) and (14.46) into (14.47) and applying the binomial theorem, it leads to

$$\begin{aligned}\mathbb{E}\left[\mu_{k_{c,\max}}\right] &= \int_0^\infty \left(1 - \left(1 - e^{-Lx}\sum_{k=0}^{M-1}\frac{L^k x^k}{k!}\right)^L\right)dx \\ &= \int_0^\infty \left(\sum_{l=1}^{L}\binom{L}{l}(-1)^{l+1}e^{-lLx}\left(\sum_{k=0}^{M-1}\frac{L^k x^k}{k!}\right)^l\right)dx \\ &= \sum_{l=1}^{L}\binom{L}{l}(-1)^{l+1}\sum_{k=0}^{(M-1)l} b_k\left(M, L, l\right)\int_0^\infty e^{-lLx}x^k dx\,.\end{aligned} \tag{14.48}$$

By change of variables, the integral in (14.48) can be further rewritten as

$$\int_0^\infty e^{-lLx}x^k dx = \left(\frac{1}{lL}\right)^{k+1}\int_0^\infty t^k e^{-t}dt = \left(\frac{1}{lL}\right)^{k+1}\Gamma\left(k+1\right). \tag{14.49}$$

Hence, we can get

$$\mathbb{E}\left[\mu_{k_{c,\max}}\right] = \sum_{l=1}^{L}\binom{L}{l}(-1)^{l+1}\sum_{k=0}^{(M-1)l} b_k\left(M, L, l\right)\left(\frac{1}{lL}\right)^{k+1}\Gamma\left(k+1\right). \tag{14.50}$$

From Lemma 14.4.1 and (14.50), the proof is completed. ▲

This theorem gives two immediate remarks for the operation of the presented PW systems in two special cases: (1) multi-antenna PW systems in flat fading channels ($L = 1$) and (2) single-antenna PW systems in frequency selective fading channels

($M = 1$), which provides an important insight into understanding the influence of the numbers of multipaths and transmit antennas on the power transfer performance.

Remark 14.4.1 *For $L = 1$, the average energy delivery efficiency gain is exactly given by $\mathbb{E}\left[E_{c,G}\right] = N_g M + \frac{\sigma_z^2}{P_v}$ because $b_k(M,1,1) = \frac{1}{k!}$, for $k = 0, \ldots, M-1$. It is observed that the efficiency is proportional to the waveform length and the number of antennas, as long as P_v is sufficiently larger than the noise power σ_z^2. Note that the gain N_g mainly comes from the PW, which enables the accumulation of the power of the reference signal. A larger wavelength may require more emitted signal power after waveforming.*

Remark 14.4.2 *For $M = 1$, the average energy delivery efficiency gain is lower bounded by $\mathbb{E}\left[E_{c,G}\right] \geq N_g \sum_{l=1}^{L} \binom{L}{l} (-1)^{l+1} \frac{1}{l} + \frac{\sigma_z^2}{P_v}$ because $b_0(1,L,l) = 1$, for $l = 1, \ldots, L$. Actually, the summation term over the index l is equal to $\sum_{l=1}^{L} \frac{1}{l}$ [25], and we have $\mathbb{E}\left[E_{c,G}\right] \geq N_g \sum_{l=1}^{L} \frac{1}{l} + \frac{\sigma_z^2}{P_v}$, which concludes that the efficiency is increased as the number of multipaths increases. In other words, one can possibly increase the system bandwidth to improve the efficiency by digitally resolving the naturally existing multipaths in wireless environments as many as possible.*

14.5 Simulation Results and Discussions

Computer simulations are conducted to demonstrate the performance of the presented multi-antenna PW systems and to substantiate the analytical findings on the average energy delivery efficiency gain and the outage performance of average harvested energy. We set $P_v = 1$ and normalize the large-scale path loss because our focus is on the WPT performance gain achieved by the multi-antenna PW technology. In addition to a UPD channel profile, i.e., setting $\rho_{m,l} = \frac{1}{L}$ in (14.1), a Saleh-Valenzuela (SV) channel model in IEEE 802.15.4a UWB communication standard is adopted in the simulation [32]. This channel model is typically considered in wideband applications over a central frequency ranging between 2 GHz and 6 GHz. The system bandwidth is set as 125 MHz, i.e., the sampling period $T_S = 8$ ns, and the number of resolvable multipaths is around several dozens with respect to the considered channel bandwidth[5]. Notice that a larger bandwidth is configured for the multi-antenna PW system to digitally resolve the naturally existing multipaths during the channel probing phase, and the estimated CIR is then utilized for the calculation of waveforms and reference signals during the power transfer phase. Otherwise, the estimated CIR is likely to be a single tap despite abundant multipaths in wireless environments. We ignore the noise power by setting $\sigma_z^2 = 0$ because the required signal power for wirelessly charging (at least -10 dBm) is much higher than the common noise power level (-93 dBm at 125 MHz bandwidth). The default values in the stopping criterion of the presented algorithm in Table 14.1 are

[5] The system could be possibly operated on a selective portion of the industrial, scientific and medical radio band (ISM band) band on which the presented waveform concentrates its power, and the remaining bandwidth is reserved for wireless information transmission.

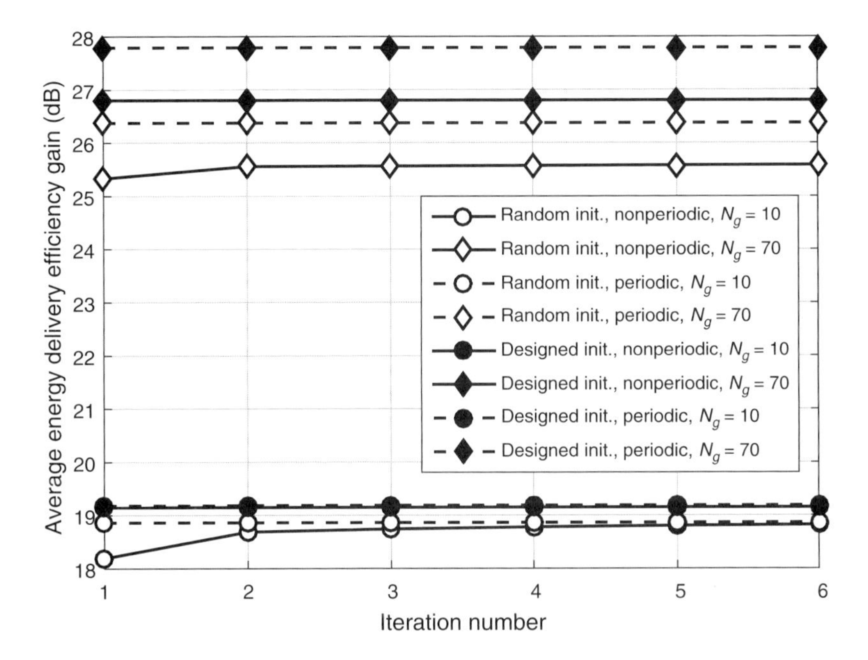

Figure 14.3 Average energy delivery efficiency gain of the presented algorithm with the random and designed initialization for various iteration numbers in the UWB SV channels ($M = 4$ and $N_v = 100$).

set as $\varepsilon = 10^{-3}$ or $I_{\max} = 3$. In addition, the WPT performance of the conventional narrow-band beamforming scheme is included for performance comparison. For this scheme, the total bandwidth B is partitioned into thirty subbands, and the strongest subband is selected to perform the conventional narrow-band beamforming as in [18], which is called frequency-domain approach in this chapter.

Figure 14.3 shows the average energy delivery efficiency gain of the presented algorithm in Table 14.1 with the random and designed initialization for various iteration numbers in the UWB SV channels. The number of transmit antennas and the length of reference signals are given by $M = 4$ and $N_v = 100$, respectively. For the random initialization, the reference signal is initialized with complex binary signals, i.e., $v[n] \in \left\{\pm\frac{1}{\sqrt{2}} \pm j\frac{1}{\sqrt{2}}\right\}$. We can make two observations from this figure. First, for a given initialization scheme and a fixed waveform length, the multi-antenna PW system with the periodic transmission of reference signals can achieve better converged performance than that with the nonperiodic transmission because the former allows for fully concentrating the power on the best selected frequency, whereas the later causes the power leakage to other neighboring frequencies. Second, the presented system with the designed initialization outperforms that with the random initialization, and its performance can quickly get converged within two iterations. Hence, the designed initialization scheme is utilized throughout the following simulation.

Figures 14.4 and 14.5 show the average energy delivery efficiency gain with the nonperiodic and periodic transmissions of reference signals, respectively, for different lengths of waveforms and reference signals in the UWB SV channels. The number of

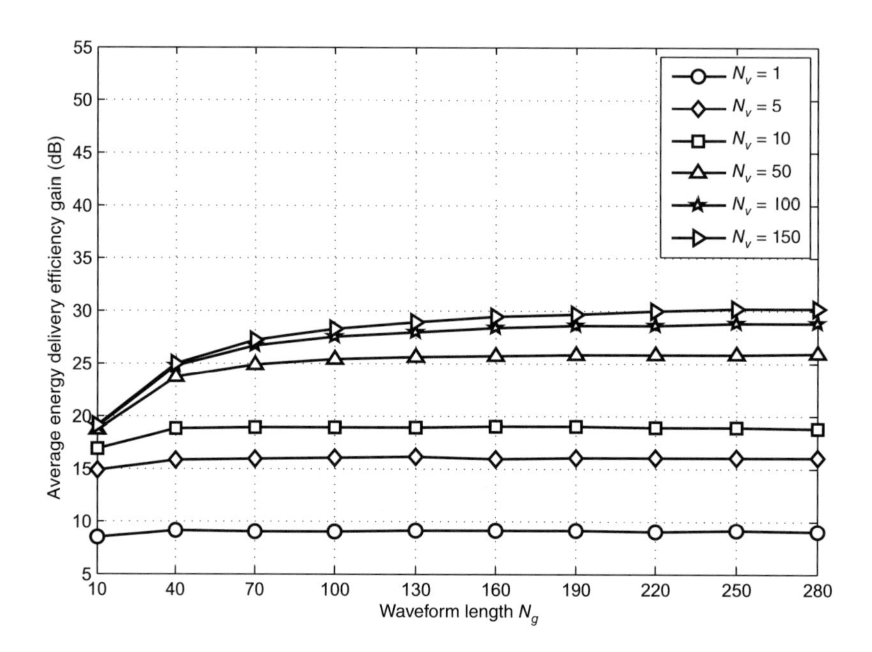

Figure 14.4 Average energy delivery efficiency gain of the presented algorithm with the nonperiodic transmission of reference signals for different lengths of waveforms and reference signals in the UWB SV channels ($M = 4$).

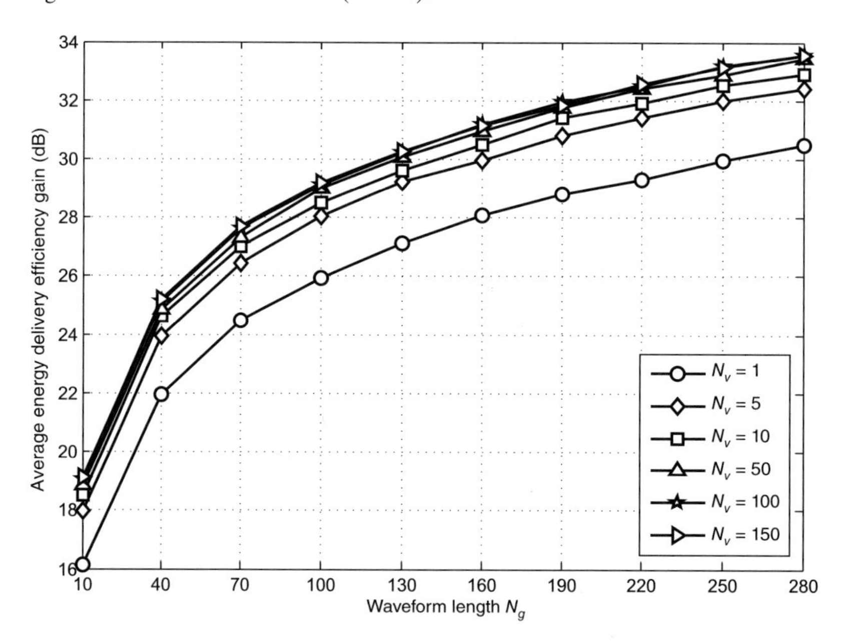

Figure 14.5 Average energy delivery efficiency gain of the presented algorithm with the periodic transmission of reference signals for different lengths of waveforms and reference signals in the UWB SV channels ($M = 4$).

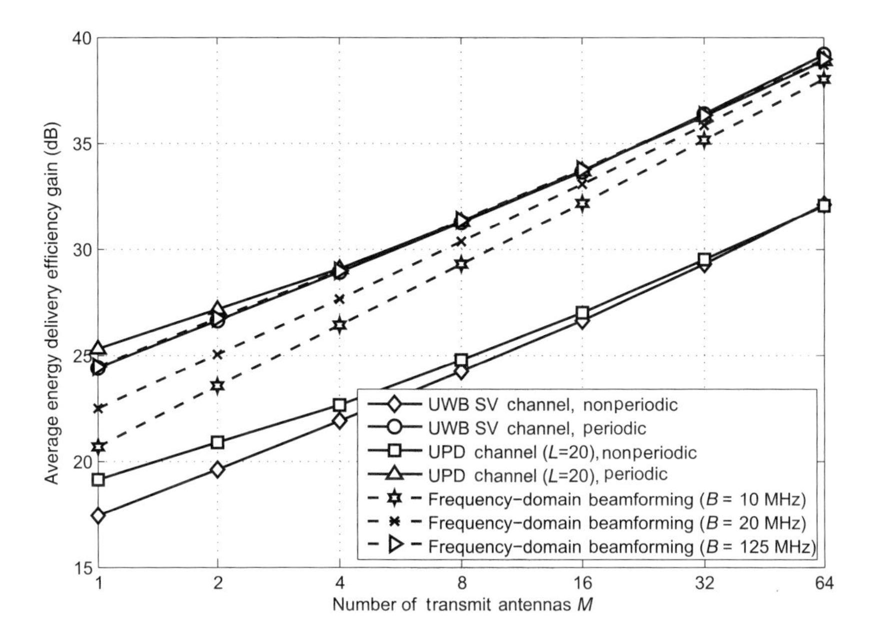

Figure 14.6 Comparisons of average energy delivery efficiency gains between the multi-antenna PW system and the frequency-domain beamforming system for various numbers of transmit antennas in the UWB SV and the UPD channels ($N_g = N_v = 100$).

transmit antennas is given by $M = 4$. For the case of nonperiodic transmission, it is found that the average energy delivery efficiency gain can be dramatically improved by increasing the lengths of waveforms or reference signals. Taking an example of $N_g = 280$, the performance improvement is as large as 22 dB when N_v is increased from 1 to 150. We can also see that the performance improvement becomes moderate as the values of N_v and N_g increase. Similar performance trends can be observed for the case of periodic transmission in Figure 14.5. For $N_v = 150$ and $N_g = 280$, the average energy delivery efficiency gain of the presented multi-antenna PW system is around 34 dB. It is demonstrated from these two figures that for given values of N_v and N_g, the performance of the multi-antenna PW system with the periodic transmission of reference signals is much superior to that with the nonperiodic transmission.

Figure 14.6 compares the performances of the multi-antenna PW system and the frequency-domain beamforming system, in terms of the average energy delivery efficiency gain. The lengths of the waveforms and the reference signals are given by $N_g = N_v = 100$, and the number of multipaths in the UPD channel profile is set as $L = 20$. The frequency-domain beamforming approach is simulated in the UWB SV channel, and its emitted signal power is set the same as the presented scheme for a fair comparison. It exhibits that the energy delivery efficiency gain of the multi-antenna PW system with the periodic transmission of reference signals is 5 dB better than that with the nonperiodic transmission for various numbers of transmit antennas. Furthermore, one can see that when B is increased from 10 MHz to 125 MHz, the performance gap between the frequency-domain approach and the presented time-domain approach with

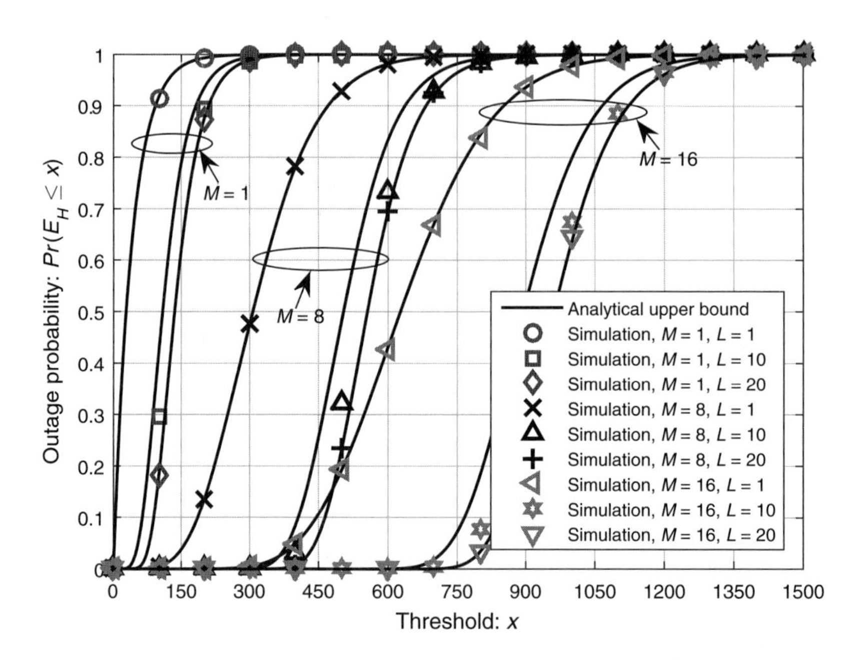

Figure 14.7 Outage performance of the average harvested energy and the analytical upper bound for the multi-antenna PW systems with the periodic transmission of reference signals under the UPD channel profile ($N_v = 20$ and $N_g = 40$).

the periodic transmissions of reference signals becomes narrow in the UWB SV channel because they both utilize the diversity gains of the wireless channels.

Figure 14.7 shows the exact outage probability of the average harvested energy and the derived upper bound given in (14.38) under the UPD channel profile for various numbers of transmit antennas and multipaths. The lengths of periodic reference signals and waveforms are set as $N_v = 20$ and $N_g = 40$, respectively. Obviously, the outage performance gets better when the numbers of transmit antennas and multipaths increase owing to the higher frequency-selective and antenna gains on the combined channel frequency responses. As expected, the analytical results are in close agreement with the simulation results when $N_v = L = 20$ or $L = 1$, thereby validating the correctness of the presented analytical expressions in Theorem 14.4.1. In addition, it is clearly observed that the upper bounds are quite tight for the cases with $L = 10$. In order to verify the tightness of the derived upper bound under different lengths of waveforms and reference signals, the simulation results and analytical results for the outage performance are compared in Figure 14.8, where we set $M = 8$ and $L = 20$ for the UPD channel profile. We can find that the outage probability decreases as the values of N_v and N_g increase. Again the analytical results for the upper bounds are very close to the simulated ones when $N_v = L = 20$. As the value of N_v increases, the difference between the two performance results slightly increases.

Figure 14.9 depicts the average energy delivery efficiency gain as a function of the waveform length and the derived lower bound given in (14.42) under the UPD channel profile for various numbers of transmit antennas and multipaths. The length of reference

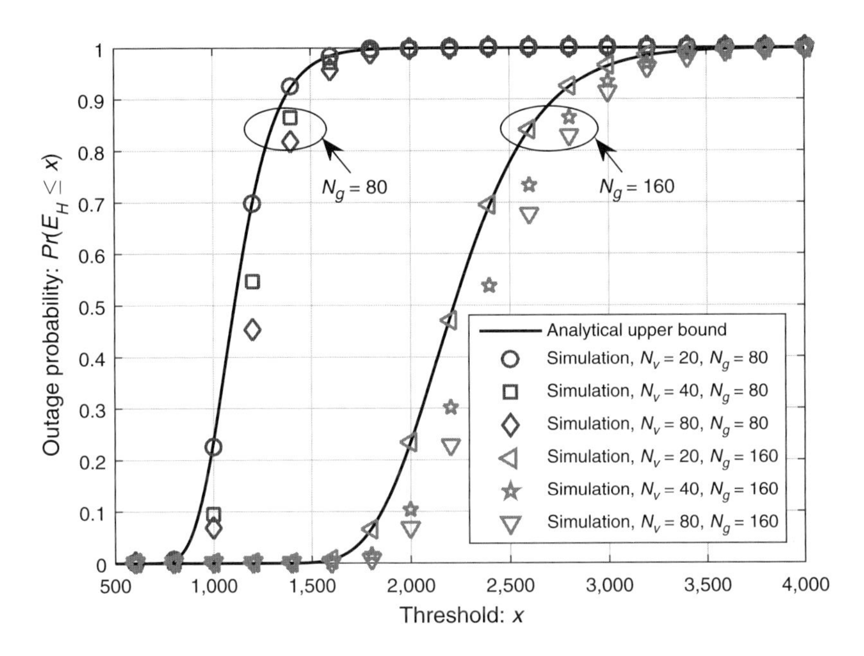

Figure 14.8 Outage performance of the average harvested energy and the analytical upper bound for different lengths of waveforms and reference signals under the UPD channel profile ($M = 8$ and $L = 20$).

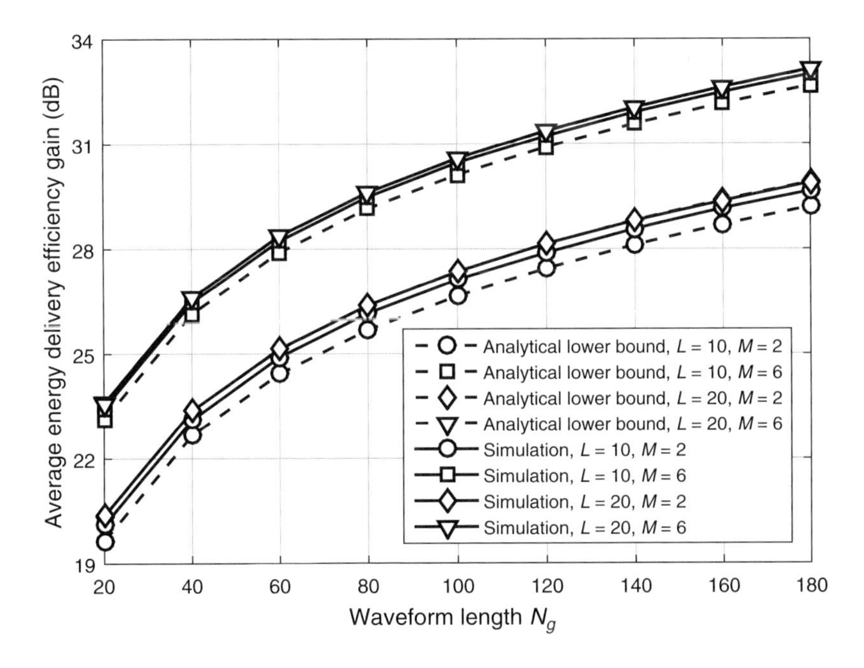

Figure 14.9 Average energy delivery efficiency gain and the analytical lower bound for the multi-antenna PW systems with the periodic transmission of reference signals under the UPD channel profile ($N_v = 20$).

signals is given by $N_\nu = 20$. Our results reveal that a substantial improvement can be achieved by increasing the waveform length, the number of transmit antennas, and the resolvable number of multipaths. It is evident from this figure that the simulation results and the analytical results fit perfectly when $N_\nu = L = 20$, which confirms our theoretical findings in Theorem 14.4.2. The presented lower bound is also very tight in the cases with $L = 10$, and therefore, it can serve as a good lower bound for predicting the performance of the presented multi-antenna PW system in multipath environments.

14.6 Summary

In this chapter, we investigated a joint power waveforming and beamfoming design for WPT, in which the waveforms on multiple transmit antennas driven by a common reference signal are optimized to maximize the gain of energy delivery efficiency. We proved that under the periodic transmission of a reference signal, the optimal waveforms exhibit single-tone structures with appropriate tone selection and power allocation over transmit antennas. It was analytically demonstrated that the corresponding optimal reference signal can be simply determined by picking a frequency component with the largest sum-power of channel frequency responses over transmit antennas. Analytic upper and lower bound expressions for the outage performance of the average harvested energy and the energy delivery efficiency gain were derived in closed forms under the UPD channel profile. The analytic results allowed us to quantify the influence of several system parameters, e.g., the number of transmit antennas, the channel length, the waveform length, on the WPT performance. Simulation results revealed that a 20–30 dB improvement can be achieved by employing the presented scheme compared to the conventional narrow-band beamforming scheme. For related references, readers can refer to [35].

References

[1] J. Jin, J. Gubbi, S. Marusic, and M. Palaniswami, "An information framework for creating a smart city through Internet of Things," *IEEE Internet of Things Journal*, vol. 1, no. 2, pp. 112–121, Apr. 2014.

[2] A. Luigi, I. Antonio, and M. Giacomo, "The Internet of Things: A survey," *Computer Networks*, vol. 54, no. 15, pp. 2787–2805, Oct. 2010.

[3] P. Kamalinejad, C. Mahapatra, Z. Sheng, S. Mirabbasi, V. C. M. Leung, and Y. L. Guan, "Wireless energy harvesting for the Internet of Things," *IEEE Communications Magazine*, vol. 53, no. 6, pp. 102–108, Jun. 2015.

[4] S. Sudevalayam and P. Kulkarni, "Energy harvesting sensor nodes: Survey and implications," *IEEE Communications Surveys and Tutorials*, vol. 13, no. 3, pp. 443–461, Third Quarter 2011.

[5] X. Lu, P. Wang, D. Niyato, D. I. Kim, and Z. Han, "Wireless charging technologies: Fundamentals, standards, and network applications," *IEEE Communications Surveys and Tutorials*, vol. 18, no. 2, pp. 1413–1452, Second Quarter 2016.

[6] N. B. Carvalho, A. Georgiadis, A. Costanzo, H. Rogier, A. Collado, J. A. Garcia, S. Lucyszyn, P. Mezzanotte, J. Kracek, D. Masotti, A. J. S. Boaventura, M. de las Nieves Ruiz Lavin, M. Pinuela, D. C. Yates, P. D. Mitcheson, M. Mazanek, and V. Pankrac, "Wireless power transmission: R&D activities within Europe," *IEEE Transactions on Microwave Theory and Techniques*, vol. 62, no. 4, pp. 1031–1045, Apr. 2014.

[7] M.-L. Ku, W. Li, Y. Chen, and K. J. R. Liu, "Advances in energy harvesting communications: Past, present, and future challenges," *IEEE Communications Surveys and Tutorials*, vol. 18, no. 2, pp. 1384–1412, Second Quarter 2016.

[8] S. Kim, R. Vyas, J. Bito, K. Niotaki, A. Collado, A. Georgiadis, and M. M. Tentzeris, "Ambient RF energy-harvesting technologies for self-sustainable standalone wireless sensor platforms," *Proceedings of the IEEE*, vol. 102, no. 11, pp. 1649–1666, Nov. 2014.

[9] X. Chen, Z. Zhang, H.-H. Chen, and H. Zhang, "Enhancing wireless information and power transfer by exploiting multi-antenna techniques," *IEEE Communications Magazine*, vol. 53, no. 4, pp. 133–141, Apr. 2015.

[10] K. Huang and V. K. N. Lau, "Enabling wireless power transfer in cellular networks: Architecture, modeling and deployment," *IEEE Trans. Wireless Communication*, vol. 13, no. 2, pp. 902–912, Feb. 2014.

[11] G. Yang, C. K. Ho, and Y. L. Guan, "Multi-antenna wireless energy transfer for backscatter communication systems," *IEEE Journal on Selected Areas in Communications*, vol. 33, no. 12, pp. 2974–2987, Dec. 2015.

[12] Z. Wang, L. Duan, and R. Zhang, "Adaptively directional wireless power transfer for large-scale sensor networks," *IEEE Journal on Selected Areas in Communications*, vol. 34, no. 5, pp. 1785–1800, May 2016.

[13] X. Chen, X. Wang, and X. Chen, "Energy-efficient optimization for wireless information and power transfer in large-scale MIMO systems employing energy beamforming," *IEEE Wireless Communications Letters*, vol. 2, no. 6, pp. 667–670, Dec. 2013.

[14] G. Yang, C. K. Ho, R. Zhang, and Y. L. Guan, "Throughput optimization for massive MIMO systems powered by wireless energy transfer," *IEEE Journal on Selected Areas in Communications*, vol. 33, no. 8, pp. 1640–1650, Aug. 2015.

[15] S. Kashyap, E. Bjornson, and E. G. Larsson, "On the feasibility of wireless energy transfer using massive antenna arrays," *IEEE Transactions on Wireless Communications*, vol. 15, no. 5, pp. 3466–3480, May 2016.

[16] L. Liu, R. Zhang, and K.-C. Chua, "Multi-antenna wireless powered communication with energy beamforming," *IEEE Transactions on Communications*, vol. 62, no. 12, pp. 4349–4361, Dec. 2014.

[17] X. Wu, W. Xu, X. Dong, H. Zhang, and X. You, "Asymptotically optimal power allocation for massive MIMO wireless powered communications," *IEEE Wireless Communications Letters*, vol. 5, no. 1, pp. 100–103, Feb. 2016.

[18] Y. Zeng and R. Zhang, "Optimized training for net energy maximization in multi-antenna wireless energy transfer over frequency-selective channel," *IEEE Transactions on Communications*, vol. 63, no. 6, pp. 2360–2373, Jun. 2015.

[19] B. Clerckx and E. Bayguzina, "Waveform design for wireless power transfer," *IEEE Transactions on Signal Processing*, vol. 64, no. 23, pp. 5972–5975, Dec. 2016.

[20] Y. Zeng, B. Clerckx, and R. Zhang, "Communications and signals design for wireless power transmission," *IEEE Transactions on Communications*, vol. 65, no. 5, pp. 2264–2290, May 2017.

[21] K. Huang and E. Larsson, "Simultaneous information and power transfer for broadband wireless systems," *IEEE Transactions on Signal Processing*, vol. 61, no. 23, pp. 5972–5986, Dec. 2013.

[22] S. Lee and R. Zhang, "Distributed wireless power transfer with energy feedback," *IEEE Transactions on Signal Processing*, vol. 65, no. 7, pp. 1685–1699, Apr. 2017.

[23] Y. Zeng and R. Zhang, "Optimized training design for wireless energy transfer," *IEEE Transactions on Communications*, vol. 63, no. 2, pp. 536–550, Feb. 2015.

[24] G. Yang, C. K. Ho, and Y. L. Guan, "Dynamic resource allocation for multiple-antenna wireless power transfer," *IEEE Transactions on Signal Processing*, vol. 62, no. 14, pp. 3565–3577, Jul. 2014.

[25] M.-L. Ku, Y. Han, H.-Q. Lai, Y. Chen, and K. J. R. Liu, "Power waveforming: wireless power transfer beyond time-reversal," *IEEE Transactions on Signal Processing*, vol. 64, no. 22, pp. 5819–5834, Nov. 2016.

[26] J. Xu and R. Zhang, "A general design framework for MIMO wireless energy transfer with limited feedback," *IEEE Transactions on Signal Processing*, vol. 64, no. 10, pp. 2475–2488, May 2016.

[27] J. Xu and R. Zhang, "Energy beamforming with one-bit feedback," *IEEE Transactions on Signal Processing*, vol. 62, no. 20, pp. 5370–5381, Oct. 2014.

[28] X. Zhou, C. K. Ho, and R. Zhang, "Wireless power meets energy harvesting: a joint energy allocation approach in OFDM-based system," *IEEE Transactions on Wireless Communications*, vol. 15, no. 5, pp. 3481–3491, May 2016.

[29] B. Wang, Y. Wu, F. Han, Y.-H. Yang, and K. J. R. Liu, "Green wireless communications: A time-reversal paradigm," *IEEE Journal on Selected Areas in Communications*, vol. 29, no. 8, pp. 1698–1710, Sep. 2011.

[30] S. Budišin "Golay complementary sequences are superior to PN sequences," *IEEE International Conference on Systems Engineering*, pp. 101–104, 1992.

[31] Y. Han, Y. Chen, B. Wang, and K. J. R. Liu, "Time-reversal massive multipath effect: A single-antenna "massive MIMO" solution," *IEEE Transactions on Communications*, vol. 64, no. 8, pp. 3382–3394, Aug. 2016.

[32] A. A. M. Saleh and R. A. Valenzuela, "A statistical model for indoor multipath propagation," *IEEE Journal on Selected Areas in Communications*, vol. 5, no. 2, pp. 128–137, Feb. 1987.

[33] Q. T. Zhang and H. G. Lu "A general analytical approach to multi-branch selection combining over various spatially correlated fading channels," *IEEE Transactions on Communications*, vol. 50, no. 7, pp. 1066–1073, Jul. 2002.

[34] I. S. Gradshteyn and I. M. Ryzhik, *Tables of Integrals, Series and Products*, San Diego, CA: Academic Press, 2007.

[35] M. L. Ku, Y. Han, B. Wang, and K. J. R. Liu, "Joint power waveforming and beamforming for wireless power transfer," *IEEE Transactions on Signal Processing*, vol. 65, no. 24, pp. 6409–6422, Dec. 2017.

Part IV

5G Communications and Beyond

15 Time-Reversal Division Multiple Access

The multipath effect makes high-speed broadband communications a very challenging task due to the severe inter-symbol interference (ISI). By concentrating energy in both the spatial and temporal domains, time-reversal (TR) transmission technique provides a great potential of low-complexity energy-efficient communications. In this chapter, a novel concept of time-reversal division multiple access (TRDMA) is discussed as a wireless channel access method based on its high-resolution spatial focusing effect. It is considered to use TR structure in multiuser downlink systems over multipath channels, where signals of different users are separated solely by TRDMA. Both the single-transmit-antenna scheme and its enhanced version with multiple transmit antennas are developed and evaluated in this chapter. The system performance is investigated in terms of its effective signal-to-interference-plus-noise ratio (SINR), the achievable sum rate and the achievable rates with outage. And some further discussions regarding its advantage over conventional Rake receivers and the impact of spatial correlations between users are given at the end of this chapter. It is shown in both analytical and simulation results that desirable properties and satisfying performances can be achieved in the presented TRDMA multiuser downlink system, which makes TRDMA a promising candidate for future energy-efficient low-complexity broadband wireless communications.

15.1 Introduction

The past decade has witnessed an unprecedented increase of demand for high-speed wireless services, which necessitates the need for future broadband communications. When it comes to broadband, the resolution of perceiving multiple paths increases accordingly. In a rich scattering environment, the adverse multipath effect makes conventional high-speed communications a very challenging task due to the severe intersymbol interference (ISI). To resolve this problem, multicarrier modulation (e.g., OFDM) and/or complicated equalization are needed [1–4] at the receiver to alleviate the ISI. Although the performance is well satisfactory, it often results in a prohibitively high complexity for end-user equipment and wireless terminals in many applications.

On the other hand, time-reversal (TR) signal transmission technique can provide a great potential of low-complexity energy-efficient communications [5], which can make full use of the nature of multipath environments. The history of research on time-reversal transmission technology dates back to the early 1990s [6–10]; however, not much development and interest went beyond the acoustics and ultrasound domains at that time. As found in acoustic physics [6–10] and then further validated in practical underwater propagation environments [11–13], the energy of the TR acoustic waves from transmitters could be refocused only at the intended location with very high spatial resolution (several-wavelength level). Because TR can make full use of multipath propagation and also requires no complicated channel processing and equalization, it was later verified and tested in wireless radio communication systems, especially in ultra-wideband (UWB) systems [14–18].

The single-user TR wireless communications consist of two phases: the *recording phase* and the *transmission phase*. When transceiver A wants to transmit information to transceiver B, transceiver B first sends an impulse that propagates through a scattering multipath environment, and the multi-path signals are received and recorded by transceiver A; then transceiver A simply transmits the time-reversed (and conjugated) waves back through the communication link to transceiver B. By utilizing channel reciprocity [19], the TR waves can retrace the incoming paths, ending up with a "spiky" signal-power spatial distribution focused only at the intended location, as commonly referred to as *spatial focusing effect* [5–17]. Also, from a signal processing point of view, in single-user communications, TR essentially leverages the multipath channel as a facilitating matched filter computing machine for the intended receiver, and concentrates the signal energy in the time domain as well, as commonly referred to as *temporal focusing effect*. It is worth noting that when the channel coherent time is not very small, the transmission phase of a duty cycle can include multiple transmissions of signals without requiring probing the channel before each transmission, which can reasonably maintain the bandwidth efficiency. It is typically the case when TR is used, and was verified by real-life experiments in [5].

In the single-user case, the temporal and spatial focusing effects have been shown to greatly simplify the receiver [14–18, 20, 21] and reduce power consumption and interference while maintaining the quality of service (QoS) [5]. In this chapter, we consider a multiuser downlink system over multipath channels and present a concept of time-reversal division multiple access (TRDMA) as a wireless channel access method by taking advantage of the high-resolution spatial focusing effect of time-reversal structure. In principle, the mechanisms of reflection, diffraction, and scattering in a wireless medium give rise to the uniqueness and independence of the multipath propagation profile of each communication link [19], which are exploited to provide spatial selectivity in spatial division multiple access (SDMA) schemes. Compared with conventional antenna-array-based beamforming SDMA schemes, time-reversal technique makes full use of a large number of multipaths and in essence treats each path as a virtual antenna that naturally exists and is widely distributed in environments.

Thus, with even just one single transmit antenna, time reversal can potentially achieve a very high diversity gain and high-resolution "pinpoint" spatial focusing. The high-resolution spatial focusing effect maps the natural multi-path propagation profile into a unique *location-specific signature* for each link, as an analogy to the artificial "orthogonal random code" in a code-division system. The presented TRDMA scheme exploits the uniqueness and independence of location-specific signatures in the multipath environment, providing a novel low-cost energy-efficient solution for SDMA. Better yet, the TRDMA scheme accomplishes much higher spatial-resolution focusing/selectivity and time-domain signal-energy compression at once, without requiring further equalization at the receiver as the antenna-array based beamforming does.

The potential and feasibility of applying time reversal to multi-user UWB communications were validated by some real-life antenna-and-propagation experiments in [5, 22–24], in which the signal transmit power reduction and interuser interference alleviation as a result of spatial focusing effect were tested and justified for one simplified *one-shot* transmission over deterministic multipath ultra-wideband channels. The idea of TRDMA presented in this chapter was further supported by several important recent works [20, 21, 25]. Reference [20] introduced a TR-based single-user spatial multiplexing scheme for SIMO UWB system, in which multiple data streams are transmitted through one transmit antenna and received by a multi-antenna receiver. Solid simulation results regarding bit-error-ratio (BER) demonstrate the feasibility of applying TR to spatially multiplex data streams. Following [20, 21] took into account the spatial correlation between antennas of the single receiver and numerically investigated through computer simulation its impact to BER performance. Based on [20] and [21, 25] tackled a multiuser UWB scenario with a focus on the impact of channel correlation to the BER performance through simulation. However, there is not much theoretical characterization or proof about system performances found in any of these papers. Furthermore, most of these literatures focus only on BER performances, without looking at the spectral efficiency that is one of the main design purposes for any spatial multiplexing scheme. There is still a lack of system-level theoretical investigation and comprehensive performance analysis of a TR-based multiuser communications system in the literature. Motivated by the high-resolution spatial focusing potential of the time-reversal structure, existing experimental measurements, and supporting literatures, several major developments have been discussed and considered in the TRDMA multiuser communications system. Specifically:

- We discuss the concept of TRDMA as a novel multiuser downlink solution for wireless multipath environments and developed a theoretical analysis framework for the presented scheme.
- We consider a multiuser broadband communication system over multi-path *Rayleigh fading* channels, in which the signals of multiple users are separated solely by TRDMA.
- We define and evaluate a number of system performance metrics, including the effective SINR at each user, achievable sum rate, and achievable rate with $\epsilon-$outage.
- We further investigate the achievable rate region for a simplified two-user case, from which one can see the advantages of TRDMA over its counterpart techniques, due to TR's spatial focusing effect.
- We incorporate and examine quantitatively the impact of spatial correlation of users to system performances for the SISO case to gain more comprehensive understanding of TRDMA.

The rest of this chapter is organized as follows. In Section 15.2, we introduce the channel model and the TRDMA multiuser downlink systems with both a single transmit antenna and multiple antennas. Then, we analyze the effective SINR in Section 15.3. In Section 15.4, both achievable sum rate and $\epsilon-$outage rate are evaluated. Also in Section 15.4, a two-user case achievable rate region is characterized and compared with the Rake-receiver counterparts. In Section 15.5, the impact of spatial correlation between users is investigated and discussed.

15.2 System Model

In this section, we introduce the channel and system model and the TRDMA schemes. We begin with the assumptions and formulations of the channel model. Then, we describe the two phases of the basic TRDMA scheme with a single transmit antenna. Finally, we extend the basic single-input-single-output (SISO) scheme to an enhanced multiple-input-single-output (MISO) TRDMA scheme with multiple transmit antennas at the base station (BS).

15.2.1 Channel Model

In this chapter, we consider a multiuser downlink network over multipath Rayleigh fading channels. We first look at a SISO case where the base station (BS) and all users are equipped with a single antenna. The channel impulse response (CIR) of the communication link between the BS and the ith user is modeled as

$$h_i[k] = \sum_{l=0}^{L-1} h_{i,l}\delta[k-l], \tag{15.1}$$

where $h_i[k]$ is the kth tap of the CIR with length L, and $\delta[\cdot]$ is the Dirac delta function. For each link, we assume that $h_i[k]$'s are independent circular symmetric complex Gaussian (CSCG) random variables with zero mean and variance

$$E[|h_i[k]|^2] = e^{-\frac{kT_S}{\sigma_T}}, \quad 0 \leqslant k \leqslant L-1 \tag{15.2}$$

where T_S is the sampling period of this system such that $1/T_S$ equals the system bandwidth B, and σ_T is the root mean square (rms) delay spread [3] of the channel. Due to the two-phase nature of TR structure, we assume that channels are reciprocal, ergodic, and blockwise-constant with their tap values remaining fixed during at least one duty cycle. Each duty cycle consists of the recording phase and the transmission phase, which occupy the proportions of $(1-\eta)$ and η of the cycle period, with $\eta \in (0,1)$ depending on how fast channels vary over time.

We first assume that the CIRs associated with different users are uncorrelated. While realistic CIRs might not be perfectly uncorrelated, this assumption greatly simplifies the analysis while capturing the essential idea of TRDMA. Moreover, real-life experimental results in [5, 17] show that in a rich-scattering environment the correlation between CIRs associated with different locations decreases to a neglectable level when two locations are even just several wave-lengths apart. A further discussion on the impact of the channel correlation between users to the system performance will be addressed in Section 15.5.

15.2.2 Phase 1: Recording Phase

The block diagram of a SISO TRDMA downlink system is shown in Figure 15.1, in which there are N users receiving statistically independent messages $\{X_1(k), X_2(k), \ldots, X_N(k)\}$ from the BS, respectively. The time-reversal mirror (TRM) shown in the diagram is a device that can record and time-reverse (and conjugate if complex-valued) the received waveform, which will be used to modulate the time-reversed waveform with input signal by convolving them together in the following transmission phase.

During the recoding phase, the N intended users first take turns to transmit an impulse signal to the BS (ideally it can be a Dirac $\delta-$function, but in practice a modified raise-cosine signal can be a good candidate for limited bandwidth for this purpose [5]). Meanwhile, the TRMs at the BS record the channel response of each link and store the time-reversed and conjugated version of each channel response for the transmission phase. For simplicity of analytical derivation, we assume in our analysis that the waveform recorded by TRM reflects the true CIR, ignoring the small corruption caused by thermal noise and quantization noise. Such a simplification was justified and based on the following facts shown in literatures of time reversal:

- The thermal noise (typically modeled as additive white Gaussian noise (AWGN)) can be effectively reduced to a desired level by averaging multiple recorded noisy samples of the same CIR's, provided that channels are slow-varying, as shown in the real-life experiments [5]. This would increase the portion $(1 - \eta)$ of the recording phase in the entire duty cycle, leading to a increased channel probing overhead; but the structure of the analysis for the presented system is not altered.
- The effect of quantization was studied by [26]. It was shown that a nine-bit quantization can be treated as nearly perfect for most applications, and even with one-bit quantization, the TR system can work reasonably well, demonstrating the robustness of the TR-based transmission technique.

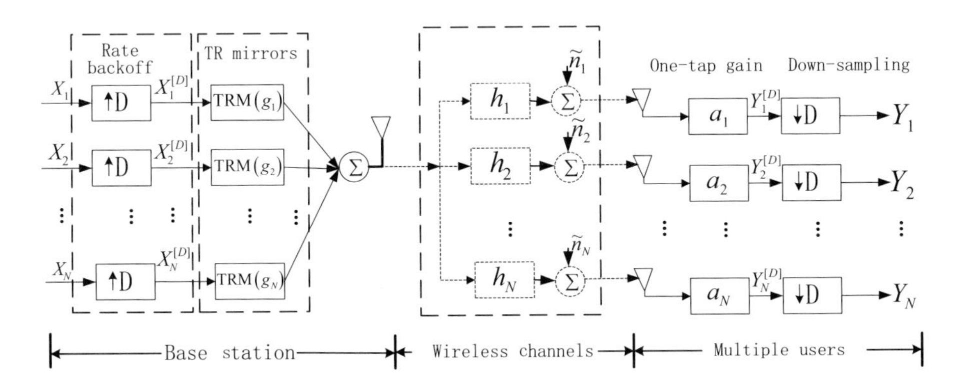

Figure 15.1 The diagram of SISO TRDMA multiuser downlink system with one single antenna.

15.2.3 Phase 2: Transmission Phase

After the channel recording phase, the system starts its transmission phase. At the BS, each of $\{X_1, X_2, \ldots, X_N\}$ represents a sequence of information symbols that are independent complex random variables with zero mean and variance of θ. In other words, we assume that for each i from 1 to N, $X_i[k]$ and $X_i[l]$ are independent when $k \neq l$. As we mentioned earlier, any two sequences of $\{X_1, X_2, \ldots, X_N\}$ are also independent in our model. We introduce the rate backoff factor D as the ratio of the sampling rate to the baud rate, by performing up-sampling and down-sampling with a factor D at the BS and receivers as shown in Figure 15.1. Such a notion of backoff factor facilitates simple rate conversion in the analysis of a TR system.

These sequences are first up-sampled by a factor of D at the BS, and the ith up-sampled sequence can be expressed as

$$X_i^{[D]}[k] = \begin{cases} X_i[k/D], & \text{if } k \bmod D = 0, \\ 0, & \text{if } k \bmod D \neq 0. \end{cases} \tag{15.3}$$

Then the up-sampled sequences are fed into the bank of TRMs $\{g_1, g_2, \ldots, g_N\}$, where the output of the ith TRM g_i is the convolution of the ith up-sampled sequence $\left\{X_i^{[D]}[k]\right\}$ and the TR waveform $\{g_i[k]\}$ as shown in Figure 15.1, with

$$g_i[k] = h_i^*[L-1-k] \Bigg/ \sqrt{E\left[\sum_{l=0}^{L-1} |h_i[l]|^2\right]}, \tag{15.4}$$

which is the normalized (by the average channel gain) complex conjugate of time-reversed $\{h_i[k]\}$. After that, all the outputs of TRM bank are added together, and then the combined signal $\{S[k]\}$ is transmitted into wireless channels with

$$S[k] = \sum_{i=1}^{N} \left(X_i^{[D]} * g_i\right)[k]. \tag{15.5}$$

In essence, by convolving the information symbol sequences with TR waveforms, TRM provides a mechanism of embedding the unique location-specific signature associated with each communication link into the transmitted signal for the intended user.

The signal received at user i is represented as follows

$$Y_i^{[D]}[k] = \sum_{j=1}^{N} \left(X_j^{[D]} * g_j * h_i\right)[k] + \tilde{n}_i[k], \tag{15.6}$$

which is the convolution of the transmitted signal $\{S[k]\}$ and the CIR $\{h_i[k]\}$, plus an additive white Gaussian noise sequence $\{\tilde{n}_i[k]\}$ with zero mean and variance σ^2.

Thanks to the temporal focusing effect, the signal energy is concentrated in a single time sample. The ith receiver (user i) simply performs a one-tap gain adjustment a_i to the received signal to recover the signal and then down-samples it with the same factor D, ending up with $Y_i[k]$ given as follows (for notational simplicity, $L-1$ is assumed to be a multiple of D)

$$Y_i[k] = a_i \sum_{j=1}^{N} \sum_{l=0}^{(2L-2)/D} (h_i * g_j)[Dl] X_j[k-l] + a_i n_i[k], \tag{15.7}$$

where

$$(h_i * g_j)[k] = \sum_{l=0}^{L-1} h_i[l] g_j[k-l] = \frac{\sum_{l=0}^{L-1} h_i[l] h_j^*[L-1-k+l]}{\sqrt{E\left[\sum_{l=0}^{L-1} \left|h_j[l]\right|^2\right]}} \tag{15.8}$$

with $k = 0, 1, \ldots, 2L-2$, and $n_i[k] = \tilde{n}_i[Dk]$, which is AWGN with zero mean and variance σ^2.

15.2.4 TRDMA with Multiple Transmit Antennas

In this section, we generalize the basic TRDMA scheme into an enhanced version with multiple transmit antennas. To maintain low complexity at receivers, we consider a MISO case where the transmitting BS is equipped with M_T antennas together with multiple single-antenna users.

Let $h_i^{(m)}[k]$ denote the kth tap of the CIR for the communication link between user i and the mth antenna of the BS, and we assume it is a circular symmetric complex Gaussian random variable with zero mean and a variance

$$E[|h_i^{(m)}[k]|^2] = e^{-\frac{kT_S}{\sigma_T}}. \tag{15.9}$$

In alignment with the basic SISO case, we also assume that paths associated with different antennas are uncorrelated, i.e., $h_i^{(m)}[k]$ and $h_j^{(w)}[l]$ are uncorrelated for $\forall\, i, j \in \{1, 2, \ldots, N\}$ and $\forall\, k, l \in \{0, 1, \ldots, L-1\}$ when $m \neq w$, where $m, w \in \{1, 2, \ldots, M_T\}$ are the indices of the mth and wth antennas at the BS.

For the MISO TRDMA scheme, each antenna at the BS plays a role similar to the single-antenna BS in the basic scheme. The block diagram for the MISO TRDMA is shown in Figure 15.2. The TR waveform $\{g_i^{(m)}[k]\}$ is the normalized (by the average total energy of MISO channels) complex conjugate of time-reversed $\{h_i^{(m)}[k]\}$, i.e.,

$$g_i^{(m)}[k] = h_i^{(m)*}[L-1-k] \Bigg/ \sqrt{E\left[M_T \sum_{l=0}^{L-1} |h_i^{(m)}[l]|^2\right]}. \tag{15.10}$$

As a result, the average total transmit power at the BS is

$$P = \frac{N \times \theta}{D}, \tag{15.11}$$

which does not depend on the number of the transmit antennas M_T.

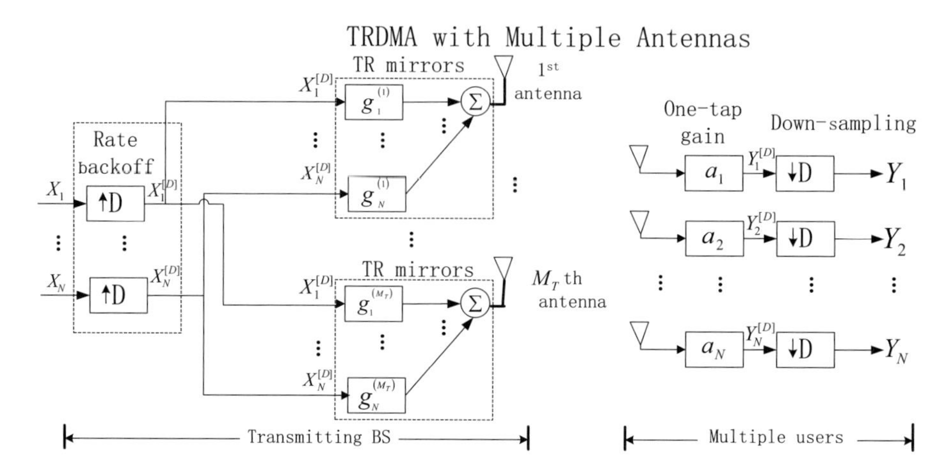

Figure 15.2 The diagram of MISO TRDMA multiuser downlink system with multiple antennas.

The resulting received signal at user i can be similarly represented as

$$Y_i[k] = \sum_{j=1}^{N} \sum_{m=1}^{M_T} \sum_{l=0}^{\frac{2L-2}{D}} \left(h_i^{(m)} * g_j^{(m)} \right) [Dl] X_j[k-l] + n[k], \tag{15.12}$$

where $n[k]$ is additive white Gaussian noise with zero mean and variance σ^2.

Hereafter, we define a modified received signal-to-noise ratio (SNR) ρ for the

$$\rho = \frac{P}{\sigma^2} E\left[\sum_{l=0}^{L-1} |h_i^{(m)}[l]|^2 \right] = \frac{P}{\sigma^2} \frac{1 - e^{-\frac{LT_S}{\sigma_T}}}{1 - e^{-\frac{T_S}{\sigma_T}}}, \tag{15.13}$$

to rule out the potential multipath gain in the system model in the following performance evaluations.

In the following sections, we evaluate the system performance of the presented system in terms of the effective SINR, the achievable sum rate, and the achievable rates with ϵ−outage.

15.3 Effective SINR

In this section, we evaluate the effective SINR of the presented system. Because the basic SISO scheme is just a special case with $M_T = 1$, we analyze the general MISO case with M_T as a parameter in this section.

Note that for $\{(h_i^{(m)} * g_j^{(m)})[k]\}$ in (15.12), when $k = L-1$ and $j = i$, it corresponds to the maximum-power central peak of the autocorrelation function, i.e.,

$$(h_i^{(m)} * g_i^{(m)})[L-1] = \sum_{l=0}^{L-1} |h_i^{(m)}[l]|^2 \Big/ \sqrt{E\left[M_T \sum_{l=0}^{L-1} |h_i^{(m)}[l]|^2 \right]}. \tag{15.14}$$

Subject to the constraint of one-tap receivers, the ith receiver is designed to estimate $X_i[k - \frac{L-1}{D}]$ solely based on the observation of $Y_i[k]$. Then, the remaining components of Y_i can be further categorized into intersymbol interference (ISI), interuser interference (IUI), and noise, as shown here:

$$
\begin{aligned}
Y_i[k] = {} & a_i \sum_{m=1}^{M_T} (h_i^{(m)} * g_i^{(m)})[L-1] X_i\left[k - \frac{L-1}{D}\right] && (Signal) \\
& + a_i \sum_{\substack{l=0 \\ l \neq (L-1)/D}}^{(2L-2)/D} \sum_{m=1}^{M_T} (h_i^{(m)} * g_i^{(m)})[Dl] X_i[k-l] && (ISI) \\
& + a_i \sum_{\substack{j=1 \\ j \neq i}}^{N} \sum_{l=0}^{(2L-2)/D} \sum_{m=1}^{M_T} (h_i^{(m)} * g_j^{(m)})[Dl] X_j[k-l] && (IUI) \\
& + a_i n_i[k]. && (Noise) \qquad (15.15)
\end{aligned}
$$

Note that the one-tap gain a_i does not affect the effective SINR, we consider it as $a_i = 1$ in the subsequent analysis, without loss of generality.

Given a specific realization of the random CIRs, from (15.15), one can calculate the signal power $P_{Sig}(i)$ as

$$
\begin{aligned}
P_{Sig}(i) &= E_X\left[\left|\sum_{m=1}^{M_T} (h_i * g_i)[L-1] X_i\left[k - \frac{L-1}{D}\right]\right|^2\right] \qquad (15.16) \\
&= \theta \left|\sum_{m=1}^{M_T} \left(h_i^{(m)} * g_i^{(m)}\right)[L-1]\right|^2,
\end{aligned}
$$

where $E_X[\cdot]$ represents the expectation over X. Accordingly, the powers associated with ISI and IUI can be derived as

$$
P_{ISI}(i) = \theta \sum_{\substack{l=0 \\ l \neq \frac{L-1}{D}}}^{\frac{2L-2}{D}} \left|\sum_{m=1}^{M_T} \left(h_i^{(m)} * g_i^{(m)}\right)[Dl]\right|^2, \qquad (15.17)
$$

$$
P_{IUI}(i) = \theta \sum_{\substack{j=1 \\ j \neq i}}^{N} \sum_{l=0}^{\frac{2L-2}{D}} \left|\sum_{m=1}^{M_T} \left(h_i^{(m)} * g_j^{(m)}\right)[Dl]\right|^2. \qquad (15.18)
$$

When there exists interference, the SINR is almost always a crucial performance metric used to measure the extent to which a signal is corrupted. It is especially the case for a media-access scheme, where interference management is one of the main design objectives. In this section, we investigate the *effective SINR* at each user in this multiuser network.

We define the average *effective SINR* at user i $SINR_{avg}(i)$ as the ratio of the average signal power to the average interference-and-noise power, i.e.,

$$SINR_{avg}(i) = \frac{E\left[P_{Sig}(i)\right]}{E\left[P_{ISI}(i)\right] + E\left[P_{IUI}(i)\right] + \sigma^2}, \tag{15.19}$$

where each term has been specified in (15.16), (15.17), and (15.18). Note that such defined *effective SINR* in (15.19) bears difference with the quantity $E\left[\frac{P_{Sig}(i)}{P_{ISI}(i)+P_{IUI}(i)+\sigma^2}\right]$ in general. The former can be treated as an approximation of the latter quantity. Such an approximation is especially useful when the calculation of the average SINR using multiple integration is too complex, as is the case in this chapter and literatures such as [18, 27, 28]. The performance of this approximation will be demonstrated in the numerical results shown in Figures 15.3, 15.4, and 15.5.

THEOREM 15.3.1 *For the independent multipath Rayleigh fading channels given in Section 15.2, the expected value of each term for the average effective SINR* (15.19) *at user i can be obtained as shown in* (15.20), (15.21), *and* (15.22).

$$E\left[P_{Sig}(i)\right] = \theta \frac{1+e^{-\frac{LT_S}{\sigma_T}}}{1+e^{-\frac{T_S}{\sigma_T}}} + \theta M_T \frac{1-e^{-\frac{LT_S}{\sigma_T}}}{1-e^{-\frac{T_S}{\sigma_T}}}; \tag{15.20}$$

$$E\left[P_{ISI}(i)\right] = 2\theta \frac{e^{-\frac{T_S}{\sigma_T}}\left(1-e^{-\frac{(L-2+D)T_S}{\sigma_T}}\right)}{\left(1-e^{-\frac{DT_S}{\sigma_T}}\right)\left(1+e^{-\frac{T_S}{\sigma_T}}\right)}; \tag{15.21}$$

$$E\left[P_{IUI}(i)\right] = \theta(N-1)\frac{\left(1+e^{-\frac{DT_S}{\sigma_T}}\right)\left(1+e^{-\frac{2LT_S}{\sigma_T}}\right) - 2e^{-\frac{(L+1)T_S}{\sigma_T}}\left(1+e^{-\frac{(D-2)T_S}{\sigma_T}}\right)}{\left(1-e^{-\frac{DT_S}{\sigma_T}}\right)\left(1+e^{-\frac{T_S}{\sigma_T}}\right)\left(1-e^{-\frac{LT_S}{\sigma_T}}\right)}. \tag{15.22}$$

PROOF. Based on the channel model presented in Section 15.2, the second and fourth moments of $h_i^{(m)}[k]$ are given by[29]

$$E\left[|h_i^{(m)}[k]|^2\right] = e^{-\frac{kT_S}{\sigma_T}}, \tag{15.23}$$

$$E\left[|h_i^{(m)}[k]|^4\right] = 2\left(E\left[|h_i^{(m)}[k]|^2\right]\right)^2 = 2e^{-\frac{2kT_S}{\sigma_T}}. \tag{15.24}$$

Based on (15.23) and (15.24), after some basic mathematical derivations, we obtain the expected values for $\forall i \in \{1,2,\ldots,N\}$ in (15.25), (15.26), and (15.27).

$$E\left[\left|\sum_{m=1}^{M_T}\left(h_i^{(m)} * g_i^{(m)}\right)[L-1]\right|^2\right] = \frac{1+e^{-\frac{LT_S}{\sigma_T}}}{1+e^{-\frac{T_S}{\sigma_T}}} + M_T\frac{1-e^{-\frac{LT_S}{\sigma_T}}}{1-e^{-\frac{T_S}{\sigma_T}}}; \tag{15.25}$$

$$E\left[\sum_{\substack{l=0\\l\neq\frac{L-1}{D}}}^{\frac{2L-2}{D}}\left|\sum_{m=1}^{M_T}\left(h_i^{(m)}*g_i^{(m)}\right)[Dl]\right|^2\right]=2\frac{e^{-\frac{T_S}{\sigma_T}}\left(1-e^{-\frac{(L-2+D)T_S}{\sigma_T}}\right)}{\left(1-e^{-\frac{DT_S}{\sigma_T}}\right)\left(1+e^{-\frac{T_S}{\sigma_T}}\right)},\tag{15.26}$$

$$E\left[\sum_{\substack{j=1\\j\neq i}}^{N}\sum_{l=0}^{\frac{2L-2}{D}}\left|\sum_{m=1}^{M_T}\left(h_j^{(m)}*g_i^{(m)}\right)[Dl]\right|^2\right]$$
$$=(N-1)\frac{\left(1+e^{-\frac{DT_S}{\sigma_T}}\right)\left(1+e^{-\frac{2LT_S}{\sigma_T}}\right)-2e^{-\frac{(L+1)T_S}{\sigma_T}}\left(1+e^{-\frac{(D-2)T_S}{\sigma_T}}\right)}{\left(1-e^{-\frac{DT_S}{\sigma_T}}\right)\left(1+e^{-\frac{T_S}{\sigma_T}}\right)\left(1-e^{-\frac{LT_S}{\sigma_T}}\right)}.\tag{15.27}$$

Therefore, according to (15.16) to (15.18), (15.20) to (15.22) are obtained as shown in Theorem 15.3.1. ▲

From Theorem 15.3.1, one can see that the average interference powers (i.e., ISI and IUI) in (15.26) and (15.27) do not depend on M_T, while the signal power level in (15.25) increases linearly with the number of antennas, which is due to an enhanced focusing capability with multiple transmit antennas leveraging the multipaths in the environment. The enhanced focusing effects monotonically improve the effective SINR. Another interesting observation is that a larger backoff factor D yields higher reception quality of each symbol, which is especially effective in the high SINR regime where interference power dominates the noise power. The asymptotic behavior of the SINR in the high SNR regime with varying D is given by the following theorem.

THEOREM 15.3.2 *In the high SNR regime, when D is small such that $D \ll L$ and $D \ll \sigma_T/T_S$, doubling D leads to approximately a 3dB gain in the average effective SINR.*

PROOF. First note that the signal power does not depend on D and that the noise is negligible in the high SINR regime. Thus, we can focus on the interference powers.

- **For Intersymbol Interference (ISI):**

$$\frac{E\left[P_{ISI}(i,D=d)\right]}{E\left[P_{ISI}(i,D=2d)\right]}=\left(1+e^{-\frac{dT_S}{\sigma_T}}\right)\frac{\left(1-e^{-\frac{(L-2+d)T_S}{\sigma_T}}\right)}{\left(1-e^{-\frac{(L-2+2d)T_S}{\sigma_T}}\right)}\tag{15.28}$$

Because $D \ll L$, then $\frac{\left(1-e^{-\frac{(L-2+d)T_S}{\sigma_T}}\right)}{\left(1-e^{-\frac{(L-2+2d)T_S}{\sigma_T}}\right)} \approx 1$; and because $D \ll \frac{\sigma_T}{T_S}$, then $e^{-\frac{dT_S}{\sigma_T}} \approx 1$. Therefore,

$$\frac{E\left[P_{ISI}(i,D=d)\right]}{E\left[P_{ISI}(i,D=2d)\right]}\approx 2.$$

- **For Interuser Interference (IUI):**

$$\frac{E\left[P_{IUI}(i, D=d)\right]}{E\left[P_{IUI}(i, D=2d)\right]} = \left(1+e^{-\frac{dT_S}{\sigma_T}}\right) \times \frac{\left(1+e^{-\frac{dT_S}{\sigma_T}}\right)\left(1+e^{-\frac{2LT_S}{\sigma_T}}\right) - 2e^{-\frac{(L+1)T_S}{\sigma_T}}\left(1+e^{-\frac{(d-2)T_S}{\sigma_T}}\right)}{\left(1+e^{-\frac{2dT_S}{\sigma_T}}\right)\left(1+e^{-\frac{2LT_S}{\sigma_T}}\right) - 2e^{-\frac{(L+1)T_S}{\sigma_T}}\left(1+e^{-\frac{(2d-2)T_S}{\sigma_T}}\right)}. \tag{15.29}$$

For similar reasons,

$$\frac{E\left[P_{IUI}(i, D=d)\right]}{E\left[P_{IUI}(i, D=2d)\right]} \approx 2.$$

▲

Next, we present some numerical evaluation of the average effective SINR. In this chapter, we mainly consider the broadband systems with frequency bandwidth that typically ranges from hundreds MHz to several GHz, which is much wider that those narrow-band systems specified in 3GPP/3GPP2. In the rich scattering environment, the underlying paths are so many that the number of perceived multiple paths increases quickly with the system bandwidth. For a system with bandwidth B, the minimum resolvable time-difference between two paths is $T_S = 1/B$[4]. Keeping this in mind, we first choose $L = 257$ and $\sigma_T = 128T_S$ from a typical range and evaluate the average effective SINR versus ρ under various system configurations in terms of N (the number of users), M_T (the number of antennas), and D (the rate backoff factor).

In Figures 15.3, 15.4, and 15.5, with $L = 257$ and $\sigma_T = 128T_S$, the solid curves are obtained according to the analytical results given by Theorem 15.3.1,

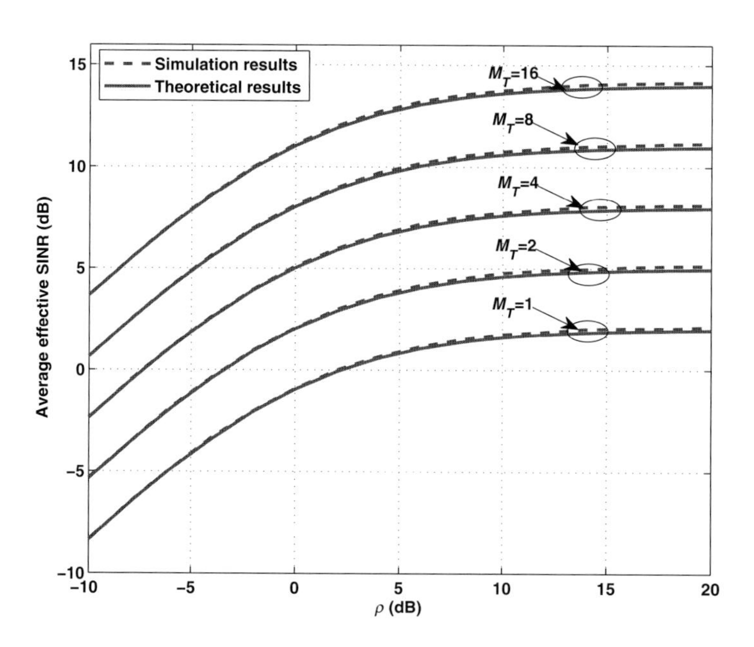

Figure 15.3 The impact of the number of antennas when $D = 8,\ N = 5$.

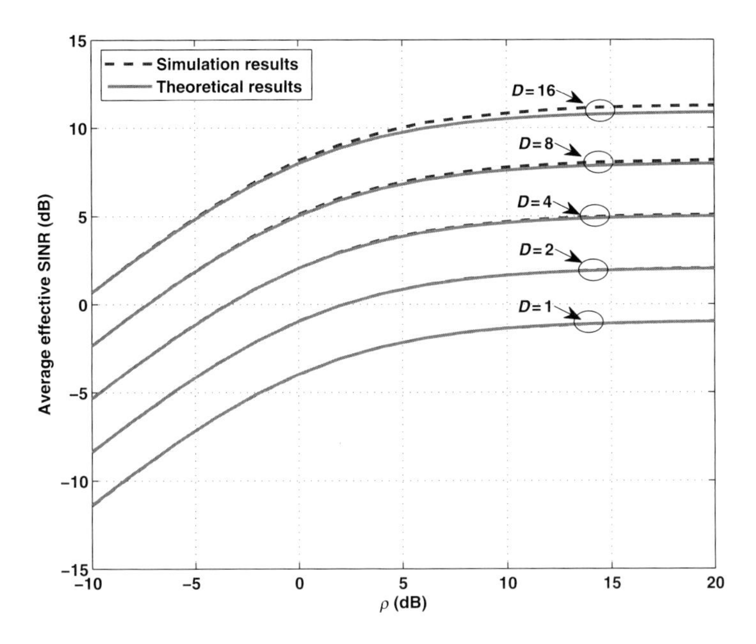

Figure 15.4 The impact of the rate backoff factor when $N = 5$, $M_T = 4$.

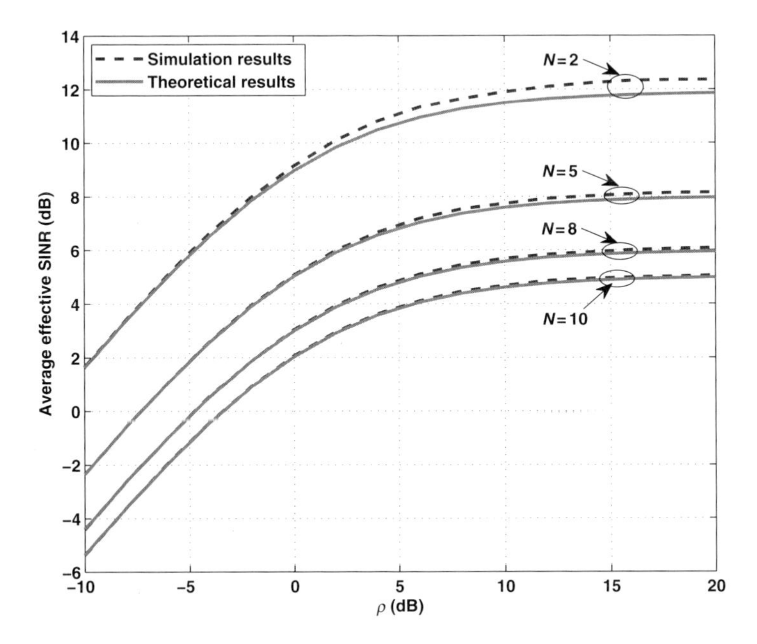

Figure 15.5 The impact of the number of users when $D = 8$, $M_T = 4$.

and the dashed curves are collected from simulation, which numerically computes $E\left[\frac{P_{Sig}(i)}{P_{ISI}(i)+P_{IUI}(i)+\sigma^2}\right]$. One can see that the results shown in Theorem 15.3.1 approximate well the empirical means obtained by simulation, which demonstrates the effectiveness of the definition of effective SINR in the system of interest in this chapter.

Figure 15.3 is plotted with $D = 8$ and $N = 5$, demonstrating the impact of the number of antennas M_T to the effective SINR. From Figure 15.3, one can see that approximately a 3 dB gain is attained as M_T is doubled within a reasonable range. The impact of the rate backoff to the effective SINR is shown with $N = 5$, $M_T = 4$ in Figure 15.4. Both analytical formulas and simulation results show that a lager D can reduce ISI and IUI while maintaining the signal power. In the high SNR regime where interference powers dominates the noise power, approximately a 3 dB gain in effective SINR can be seen when D is doubled in Figure 15.4, as predicted in Theorem 15.3.2. In Figure 15.5, we investigate the impact of the number of users with $D = 8$, $M_T = 4$. Due to the existence of IUI, increasing the number of co-existing users will result in higher interference between users. That implies a trade-off between the network capacity (in terms of number of serviced users) and signal reception quality at each user, as indicated in Figure 15.5.

Furthermore, to demonstrate the usefulness and practical importance of TRDMA, we apply the presented scheme to more practical channel models, the IEEE 802.15.4a outdoor non-line-of-sight (NLOS) channels, operating over bandwidth of $B = 500$ MHz ($T_S = 2$ ns and the typical channel length $L \sim 80$ to 150 taps) and $B = 1$ GHz ($T_S = 1$ ns and the typical channel length $L \sim 200$ to 300 taps), respectively. Figure 15.6 shows the performances of the presented TRDMA scheme over the two aforementioned more practical channel models with $M_T = 4$. Such two practical channel models have comparable system bandwidth and channel lengths with the systems that TRDMA is designed for. From Figure 15.6, one can see that the performances for the practical

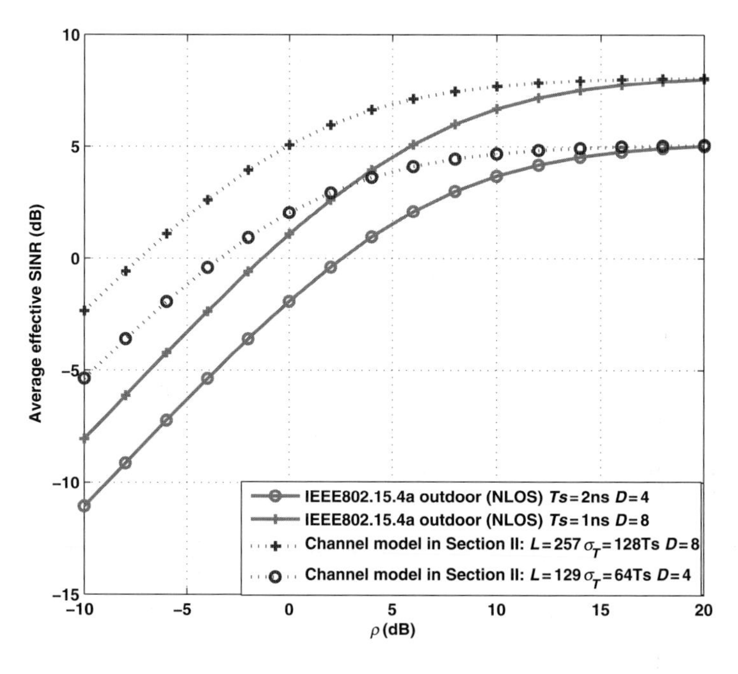

Figure 15.6 Average Effective SINR for IEEE 802.15.4a outdoor NLOS channel models.

channel models well preserve the system performances obtained for our theoretical model, especially in high SNR regime. Note that in Figure 15.6, we set $D = 4$ and 8 for the channels with $T_S = 2$ ns and $T_S = 1$ ns, respectively, to ensure that their baud rates (i.e., B/D) are the same for a fair comparison of the two. As seen from this comparison, a channel's multipath richness (or higher resolution of perceiving multiple paths) due to the broader system bandwidth, gives rise to better user-separation in the presented TRDMA scheme, which in essence increases the degree of freedom of the location-specific signatures.

15.4 Achievable Rates

In this section, we evaluate the presented TRDMA in terms of achievable rates. We first look at its achievable sum rate. Then, two types of achievable rates with $\epsilon-$outage are defined and analyzed. Finally, we derive the two-user achievable rate region of the TR structure and compare it with its Rake-receiver counterparts.

15.4.1 Achievable Sum Rate

The achievable sum rate can be used as an important metric of the efficiency of a wireless downlink scheme, which measures the total amount of information that can be effectively delivered given the total transmit power constraint P.

When the total transmit power is P, the variance of each symbol is limited to $\theta = PD/N$, according to the simple conversion shown in (15.11). For any instantaneous realization of the random channels that we modeled in Section 15.2, one could obtain its corresponding instantaneous effective SINR of user i with symbol variance θ using the following equation

$$SINR(i,\theta) \triangleq \frac{P_{Sig}(i)}{P_{ISI}(i) + P_{IUI}(i) + \sigma^2}, \tag{15.30}$$

where each term is specified in (15.16), (15.17), and (15.18).

Then, under the total power constraint P, the instantaneous achievable rate of user i can be calculated as

$$\begin{aligned} R(i) &= \frac{\eta}{T_S \times B \times D} \log_2\left(1 + SINR(i, PD/N)\right) \\ &= \frac{\eta}{D} \log_2\left(1 + SINR(i, PD/N)\right) \ (bps/Hz), \end{aligned} \tag{15.31}$$

where η serves as a discount factor that describes the proportion of the transmission phase in the entire duty cycle. We normalize the sum rate with bandwidth $B = 1/T_S$, presenting the information rate achieved per unit bandwidth (often referred to as *spectral efficiency*). It is also worth noting that in (15.31), the quantity is divided by D, because of the consequence of rate backoff.

Accordingly, the instantaneous achievable sum rate can be obtained as

$$R = \sum_{i=1}^{N} R(i) = \frac{\eta}{D} \sum_{i=1}^{N} \log_2 \left(1 + SINR(i, PD/N)\right). \tag{15.32}$$

Averaging (15.32) over all realizations of the random ergodic channels, the expected value of the instantaneous achievable sum rate is a good reference of the long-term performance and can be calculated by

$$R_{avg} = E\left[\frac{\eta}{D} \sum_{i=1}^{N} \log_2 \left(1 + SINR(i, PD/N)\right)\right]. \tag{15.33}$$

In the following part of this section, without loss of generality, we use $\eta \approx 1$, ignoring the overhead caused by the recording phase in each duty cycle, which is valid when the fading channels are not varying very fast.

The numerical evaluation of the average achievable sum rate is shown with the CIR length $L = 257$ and delay spread $\sigma_T = 128T_S$ in the system model. We plot this average achievable sum rate (setting $\eta = 1$) in Figure 15.7 with different system configurations. To show how well the scheme performs in more realistic environments, we also include a comparison of the achievable-sum-rate performances for the channel model (with $L = 257$, $\sigma_T = 128T_S$, and $M_T = 4$) introduced in Section 15.2 and the IEEE802.15.4a Outdoor NLOS channel model (with $B = 1$ GHz, $T_S = 1$ ns, $M_T = 4$) in Figure 15.8.

From Figure 15.7, one can see that the sum rate increases monotonically with M_T, as a result of improved SINRs achieved by enhanced spatial focusing. From Figure 15.8, one can see that the IEEE802.15.4a channel model with comparable channel length

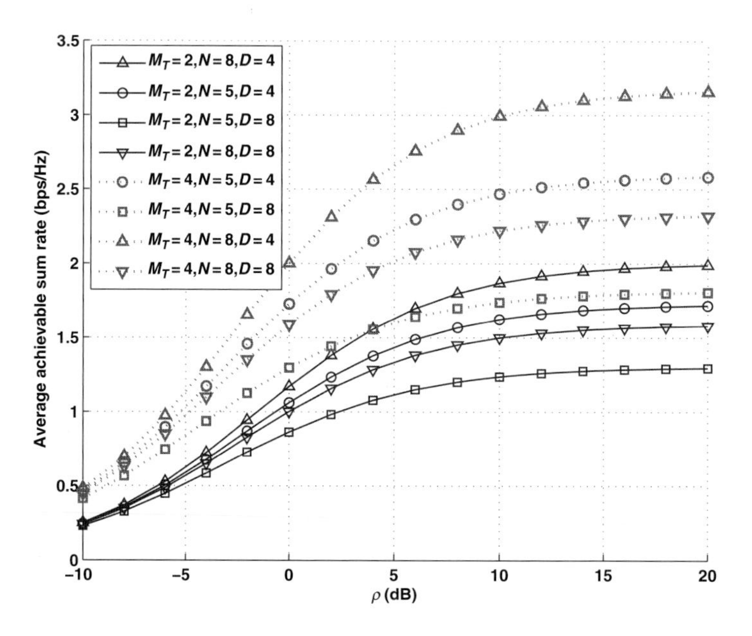

Figure 15.7 The normalized achievable sum rate versus ρ

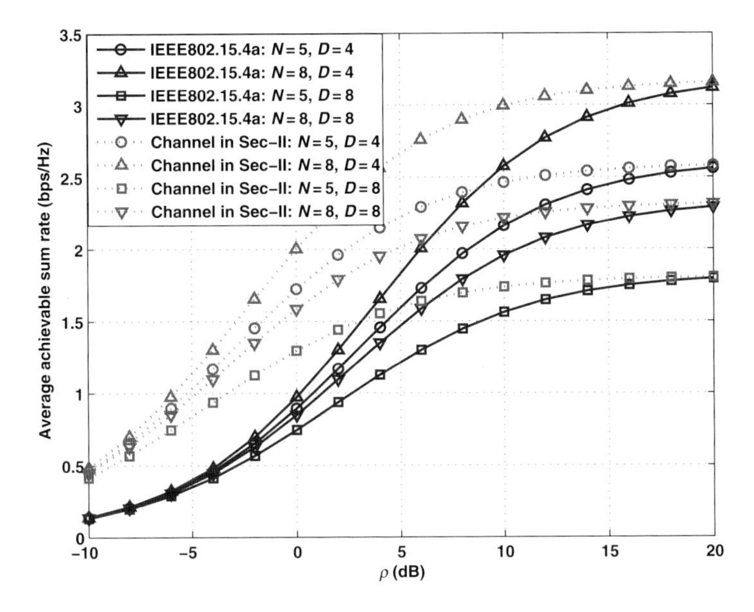

Figure 15.8 The normalized achievable sum rate for IEEE 802.15.4a outdoor NLOS channel models.

($L \sim 200$ to 300 taps) well preserves the achievable sum rates of the theoretical channel model introduced in Section 15.2, especially in high SNR regime. This demonstrates the effectiveness of TRDMA when applied to more practical channels. From both Figures 15.7 and 15.8, one can see that a larger N gives rise to a larger achievable sum rate, and a larger D discounts the achievable sum rate. The mechanisms of how D and N affect the sum rate are summarized as follows:

- A larger N increases the concurrent data streams (or multiplexing order), while it degrades the individual achievable rate of each user due to stronger interference among users. The SINR degradation is inside the logarithm function in (15.32), but the multiplexing order multiplies logarithm function, yielding a higher sum rate when N is larger.
- On the other hand, a larger D improves the reception quality of each symbol as a result of reduced ISI, but it lowers the symbol rate of the transmitter. For similar reasons, the improvement of SINR inside the logarithm function cannot compensate the loss of lowering symbol rate.

Thus, a choice of the pair (D, N) can reveal a fundamental engineering tradeoff between the signal quality at each user and the sum rate of this network.

15.4.2 Achievable Rate with ϵ–Outage

In this section, we look at the achievable rate with $\epsilon-$outage of the TRDMA-based multiuser network. The concept of $\epsilon-$*outage rate* [4, 30] allows bits sent over random channels to be decoded with some probability of errors no larger than ϵ, namely the *outage probability*. Such a concept well applies to slow-varying channels, where the instantaneous achievable rate remains constant over a large number of transmissions, as is typically the case when the TR-structure is applied.

We first define two types of outage events in the TRDMA-based downlink network and then characterize the outage probability of each type.

DEFINITION 15.4.1 (Outage of type I (individual rate outage)) *We say* outage of type I *occurs at user* i *if the achievable rate of user* i*, as a random variable, is less than a given transmission rate* R*, i.e., the outage event of type I can be formulated as* $\left\{\frac{1}{D}\log_2(1+SINR(i,\theta)) < R\right\}$*, and the corresponding outage probability of user* i *for rate* R *is*

$$P_{out\_I}(i) = Pr\left\{\frac{1}{D}\log_2(1+SINR(i,\theta)) < R\right\}, \tag{15.34}$$

where $SINR(i,\theta)$ *is given by* (15.30) *with the variance of each information symbol* $\theta = PD/N$.

DEFINITION 15.4.2 (Outage of type II (average rate outage)) *We say* outage of type II *occurs if the rate achieved per user (averaged over all the users) in the network, as a random variable, is less than a given transmission rate* R*, i.e., the outage event of type II can be formulated as* $\left\{\frac{1}{N}\sum_{i=1}^{N}\frac{1}{D}\log_2\left(1+SINR(i,\theta)\right) < R\right\}$*, and the corresponding outage probability for rate* R *is*

$$P_{out\_II} = Pr\left\{\frac{1}{D\cdot N}\sum_{i=1}^{N}\log_2\left(1+SINR(i,\theta)\right) < R\right\}, \tag{15.35}$$

where $SINR(i,\theta)$ *is given by* (15.30) *with the variance of each information symbol* $\theta = PD/N$.

We present the two types of outage probabilities as functions of the transmission rate R in Figure 15.9. Without loss of generality (due to symmetry), we select user 1's type-I outage probability P_{out_I} as a representative of others. In Figure 15.9, simulation is made with $L = 257$ and $\sigma_T = 128T_S$ under the normalized SNR level $\rho = 10dB$. As one can see, the slopes of curves in Figure 15.9 are all very steep before the outage probabilities approach to 1. This indicates that the TR transmission technology could effectively combat the multipath fading and make the system behave in a more deterministic manner due to the *strong law of large numbers*. Such a property is highly desirable in a broad range of wireless communications, where link stability and reliability are prior concerns. Also, similar discounting effect on the achievable rate of rate backoff D is observed, and a larger N (number of users) would also reduce the individual achievable rate with the same outage probability due to its resulting larger IUI.

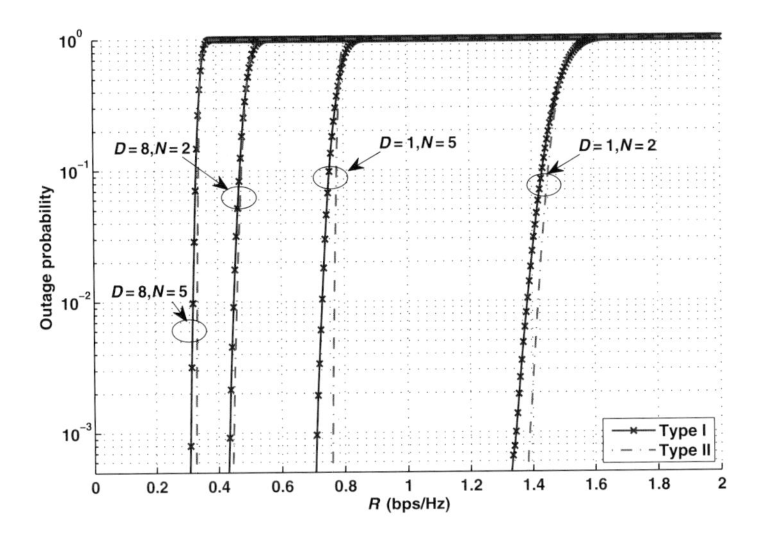

Figure 15.9 The normalized achievable rate with outage.

15.4.3 Achievable Rate Region Improvement over Rake Receivers

In this section, we present TRDMA's improvement of achievable rate region over its counterpart, the Rake receivers. Note that in the single-user case, by shifting the equalization from the receiver to the transmitter, time reversal bears some mathematical similarity to the Rake receivers whose number of fingers is equal or close to the length of channel impulse response. However, as shown in [5], for some broadband communications with typically tens to hundreds of paths, the complexity of Rake receiver with such a large number of fingers is not practical. We demonstrate the advantage of TR structure over Rake receivers in a multiuser scenario where spatial focusing effect of TR structure plays an important role, with the derivation of the two-user achievable rate region (the case of more users can be extended by defining a region in higher dimensional space). Specifically, we look at the TRDMA scheme and Rake-receiver-based schemes in terms of the amount of information delivered (mutual information between input and output) within one single transmission, measured by bits per use of the multipath channel.

Consider a two-user downlink scenario, where the transmitter has two independent information symbols, X_1 and X_2, for two different receivers, respectively. The links between the transmitter and each receiver are modeled as a discrete multipath channel with impulse responses h_1 and h_2 as in Section 15.2. Figure 15.10(a) shows a two-user single-antenna TRDMA scheme as introduced in this chapter; and Figure 15.10(b) shows a two-user Rake-receiver based downlink solution. As we will show later, the presented TRDMA scheme outperforms the Rake-receiver based schemes even when we assume that the number of fingers can be equal to the length of channel impulse response and that the delay, amplitude and phase of each path can be perfectly tracked by the Rake receiver.

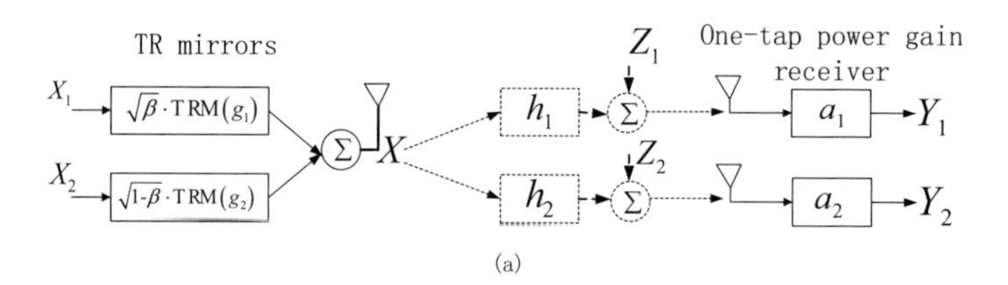

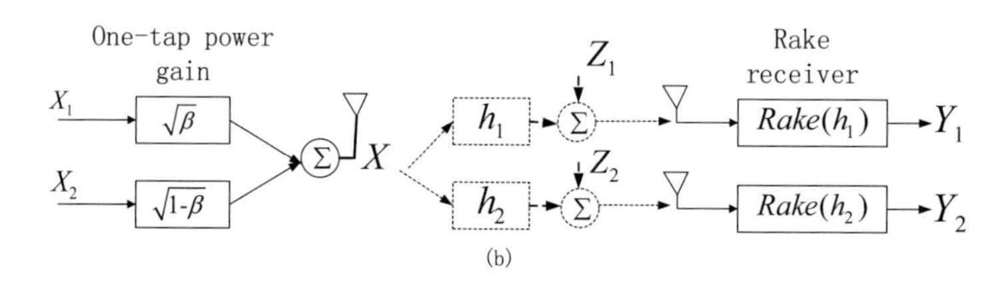

Figure 15.10 Two downlink systems: (a) two-user TRDMA with one single antenna and (b) two-user Rake receivers with one single antenna.

15.4.3.1 Rake Receivers

For the ideal Rake receivers in Figure 15.10(b), the equalized signals can be written as

$$Y_1 = ||h_1||_2 X + Z_1; \quad Y_2 = ||h_2||_2 X + Z_2, \tag{15.36}$$

where $||h_i||_2 = \sqrt{\sum_{l=0}^{L-1} |h_i(l)|^2}$ is the Euclidean norm of the channel impulse response h_i, and Z_i is additive white Gaussian noise with zero-mean and variance σ_i^2. X is the transmitted signal, which is the combination of the two information symbols X_1 and X_2.

One of the most intuitive way of combining X_1 and X_2 is to use orthogonal bases that allocate each user a fraction of the total available degrees of freedom [31]. In the two-user case, suppose that $X(t) = \sqrt{\beta}X_1c_1(t) + \sqrt{1-\beta}X_2c_2(t)$, where $c_1(t)$ and $c_2(t)$ are two orthonormal basis functions that assign a fraction $\alpha \in (0,1)$ of the total available degrees of freedom to user 1 and $(1-\alpha)$ to user 2. We consider the two-user achievable rate region with a total transmit power constraint. Specifically, let us assume that X_1 and X_2 are independent and identically distributed (i.i.d.) random variables with variance Φ, with the power allocation factor β such that the variance of X $var(X) = \left(\sqrt{\beta}\right)^2 \Phi + \left(\sqrt{1-\beta}\right)^2 \Phi = \Phi$.

Then, for the ideal Rake receivers using orthogonal bases, the maximum achievable rate pair (R_1, R_2) in *bits per channel use* is given by [30]

$$\begin{aligned} R_1 &\leq \alpha \log_2\left(1 + \frac{\beta||h_1||_2^2\Phi}{\alpha\sigma_1^2}\right) \\ R_2 &\leq (1-\alpha)\log_2\left(1 + \frac{(1-\beta)||h_2||_2^2\Phi}{(1-\alpha)\sigma_2^2}\right), \end{aligned} \tag{15.37}$$

with all possible values $\alpha \in (0,1)$ and $\beta \in [0,1]$ defining the achievable rate region.

It has been shown that for the input-output correspondence shown in (15.36), the optimal frontier of the concurrently achievable rate pair is characterized by using superposition coding [32–35]. Without loss of generality, we assume that $\frac{\sigma_1^2}{||h_1||_2^2} \leq \frac{\sigma_2^2}{||h_2||_2^2}$, i.e., User 1's channel is advantageous to User 2's. Then the achievable rate region of the superposition coding is given by [30]

$$\begin{aligned} R_1 &\leq \log_2\left(1+\frac{\beta||h_1||_2^2\Phi}{\sigma_1^2}\right) \\ R_2 &\leq \log_2\left(1+\frac{(1-\beta)||h_2||_2^2\Phi}{\beta||h_2||_2^2\Phi+\sigma_2^2}\right) \end{aligned} \tag{15.38}$$

where $\beta \in [0, 1]$ is the power allocation factor that defines the achievable rate region.

15.4.3.2 TRDMA Scheme and Genie-Aided Outer-Bound

For the TRDMA scheme with a single-tap receiver, when just one single transmission is considered, the input-and-output correspondence is reduced to

$$\begin{aligned} Y_1 &= \sqrt{\beta}||h_1||_2 X_1 + \sqrt{1-\beta}\,(h_1 * g_2)\,(L-1)\,X_2 + Z_1; \\ Y_2 &= \sqrt{1-\beta}||h_2||_2 X_2 + \sqrt{\beta}\,(h_2 * g_1)\,(L-1)\,X_1 + Z_2, \end{aligned} \tag{15.39}$$

where $g_i(l) = h_i^*(L-1-l)/||h_i||_2$ implemented by TRMs, and $\left(h_j * g_i\right)$ denotes the convolution of h_j and g_i.

Then, the resulting mutual information is obtained as follows

$$\begin{aligned} R_1 &\leq \log_2\left(1+\frac{||h_1||_2^2\beta\Phi}{|\,(h_1 * g_2)\,(L-1)\,|^2(1-\beta)\Phi+\sigma_1^2}\right) \\ R_2 &\leq \log_2\left(1+\frac{||h_2||_2^2(1-\beta)\Phi}{|\,(h_2 * g_1)\,(L-1)\,|^2\beta\Phi+\sigma_2^2}\right), \end{aligned} \tag{15.40}$$

where $\beta \in [0, 1]$ is the power allocation factor that defines the achievable rate region.

Last, we derive a genie-aided outer-bound for the two-user capacity region, in which case all the interference is assumed to be known and thus can be completely removed. Such a genie-aided outer-bound can be obtained with $\beta \in [0, 1]$ as follows

$$\begin{aligned} R_1 &\leq \log_2\left(1+\frac{||h_1||_2^2\beta\Phi}{\sigma_1^2}\right) \\ R_2 &\leq \log_2\left(1+\frac{||h_2||_2^2(1-\beta)\Phi}{\sigma_2^2}\right) \end{aligned} \tag{15.41}$$

15.4.3.3 Numerical Comparison

We present a numerical comparison of the capacity regions obtained in (15.37), (15.38), (15.40), and (15.41). In particular, we set $\frac{\Phi\ E\left[||h_1||_2^2\right]}{\sigma_1^2} = 10$ dB for User 1

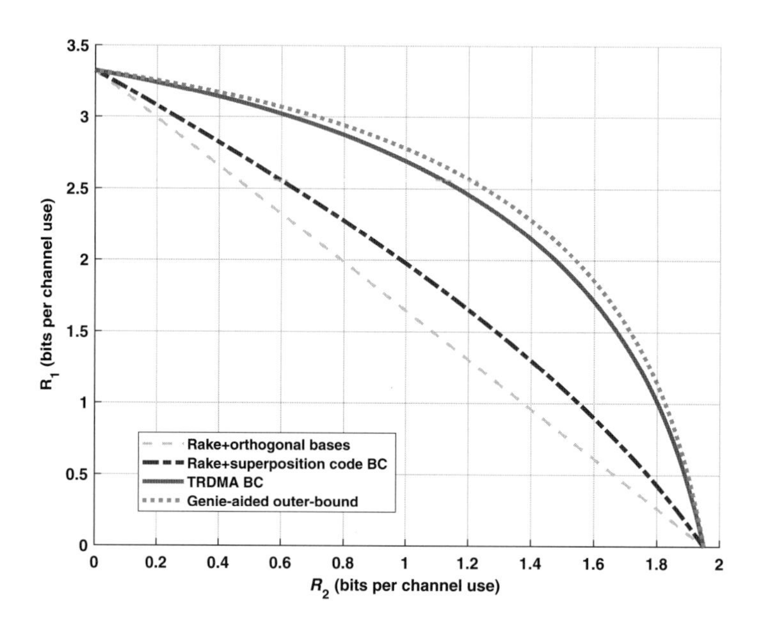

Figure 15.11 Achievable rate region for two-user case.

and $\frac{\Phi\ E[||h_2||_2^2]}{\sigma_2^2} = 5$ dB for User 2. In Figure 15.11, results are obtained by averaging over 1,000 trials of multipath Rayleigh fading channels. Each time, channel impulse responses h_1 and h_2 are randomly generated with parameters $L = 257$ and $\sigma_T = 128T_S$ according to the channel model in Section 15.2.

First, in Figure 15.11, all the schemes achieve the same performances in the degraded single-user case, which corresponds to the two overlapping intersection points on the axes. This is due to the mathematical similarity between TR and Rake receivers in the single-user case and the commutative property of the linear time-invariant (LTI) system. On the other hand, in most cases when both users are active, the presented TRDMA scheme outperforms all the Rake-receiver based schemes as shown in Figure 15.11. Moreover, the frontier achieved by TRDMA is close to the Genie-aided outer-bound. All these demonstrate TRDMA's unique advantage of spatial focusing brought by the preprocessing of embedding location-specific signatures before sending signals into the air. The high-resolution spatial focusing, as the key mechanism of the TRDMA, alleviates interference between users and provides a novel wireless medium access solution for multiuser communications.

15.5 Channel Correlation Effect

In the preceding sections, we assume a model of independent channels, because for rich-scattering multipath profiles associated with reasonably far-apart (typically, several wavelengths) locations, they are often highly uncorrelated [19]. However, channels may become correlated when the environment is less scattering and users are very close to each other. To gain a more comprehensive understanding of TRDMA, it is also interesting and important to develop a quantitative assessment of its performance degradation due to spatial correlation between users.

15.5.1 Spatial Channel Correlation

Although there are many ways to model correlated channel responses, we herein choose to obtain correlated channel responses $\widehat{X}$ and $\widehat{Y}$ by performing element-wise linear combinations of independent channels X and Y as follows [29, 36, 37]

$$\begin{bmatrix} \widehat{X}(i) \\ \widehat{Y}(i) \end{bmatrix} = \begin{bmatrix} \sqrt{\xi} & \sqrt{1-\xi} \\ \sqrt{1-\xi} & \sqrt{\xi} \end{bmatrix} \begin{bmatrix} X(i) \\ Y(i) \end{bmatrix}, \tag{15.42}$$

where the coefficient $\xi \in [0, 1]$.

Before we proceed, we give a definition to *spatial correlation* of two multi-path channel responses.

Definition 15.5.1 *For two multipath channel responses $\widehat{X}$ and $\widehat{Y}$, the spatial correlation of $\widehat{X}$ and $\widehat{Y}$ is defined as*

$$S_{\widehat{X}\widehat{Y}} = \frac{\sum_{i=0}^{L-1} \left| E\left[\widehat{X}(i)\widehat{Y}(i)^*\right] \right|}{\sqrt{\sum_{i=0}^{L-1} E\left[|\widehat{X}(i)|^2\right] \cdot \sum_{j=0}^{L-1} E\left[|\widehat{Y}(j)|^2\right]}}. \tag{15.43}$$

Note that this definition assumes zero-mean channel responses without loss of generality, and $S_{\widehat{X}\widehat{Y}}$ takes values between 0 and 1. Particularly, when $\widehat{X}$ and $\widehat{Y}$ are identical or additive inverse to each other, $S_{\widehat{X}\widehat{Y}} = 1$; when $\widehat{X}$ and $\widehat{Y}$ are uncorrelated, $S_{\widehat{X}\widehat{Y}} = 0$.

15.5.2 Channel Correlation among Users

For simplicity, we look at a two-user SISO case with correlated channel responses. We observe the impact of users' spatial correlation to the system performances.

Let us consider two correlated CIRs, $\widehat{h}_1$ and $\widehat{h}_2$, obtained from the linear combination of two independent CIRs, h_1 and h_2, as shown in (15.42), where $h_i[k]$'s are assumed as in Section 15.2 to be independent circular symmetric complex Gaussian random variables with zero mean and variance $E[|h_i[k]|^2] = e^{-\frac{kT_S}{\sigma_T}}$, for $0 \le k \le L-1$.

Then, the spatial correlation defined in (15.43) for $\widehat{h}_1$ and $\widehat{h}_2$ can be calculated by the simple form

$$S_{\widehat{h}_1\widehat{h}_2} = 2\sqrt{\xi(1-\xi)}. \tag{15.44}$$

Because the spatial correlation only affects the interuser interference power, here we focus on the change of the average power of IUI as a result of channel correlations. Similar to (15.18), the expected value of the new IUI power $\widehat{P}_{IUI}(i)$ at User i in such a two-user SISO case (i.e., $N = 2$ and $M_T = 1$) with the correlated CIRs $\widehat{h}_1$ and $\widehat{h}_2$ can be written as

$$E\left[\widehat{P}_{IUI}(i)\right] = \theta E\left[\sum_{l=0}^{\frac{2L-2}{D}} \left|\left(\widehat{h}_i * \widehat{g}_j\right)[Dl]\right|^2\right], \tag{15.45}$$

where $j \neq i$ $(i, j \in \{1,2\})$, and the TRM

$$\widehat{g}_j[k] = \widehat{h}_j^*[L-1-k] \Big/ \sqrt{E\left[\sum_{l=0}^{L-1} |\widehat{h}_j[l]|^2\right]}$$

corresponds to User j with the CIR $\widehat{h}_j$.

A direct calculation of (15.45) can be tedious. However, by substituting uncorrelated h_1 and h_2 into (15.45) according to the linear transform (15.42), we can utilize the existing results in Section 15.3 and represent the expected value of $\widehat{P}_{IUI}(i)$ in terms of $E\left[P_{Sig}(i)\right]$, $E\left[P_{ISI}(i)\right]$, and $E\left[P_{IUI}(i)\right]$ in (15.20–15.22) calculated with respect to uncorrelated h_1 and h_2, as presented in (15.46).

$$\begin{aligned} E\left[\widehat{P}_{IUI}(i)\right] &= \left[\xi^2 + (1-\xi)^2\right] E\left[P_{IUI}(i)\right] \\ &\quad + 2\xi(1-\xi)\left(E\left[P_{Sig}(i)\right] + E\left[P_{ISI}(i)\right] + \theta E\left[\sum_{l=0}^{L-1} |h_i[l]|^2\right]\right) \\ &= E\left[P_{IUI}(i)\right] + \frac{S^2_{\widehat{h}_1\widehat{h}_2}}{2} \\ &\quad \left(E\left[P_{Sig}(i)\right] + E\left[P_{ISI}(i)\right] - E\left[P_{IUI}(i)\right] + \theta E\left[\sum_{l=0}^{L-1} |h_i[l]|^2\right]\right), \end{aligned} \tag{15.46}$$

Note that in (15.46), the second term $\left(E\left[P_{Sig}(i)\right] + E\left[P_{ISI}(i)\right] - E\left[P_{IUI}(i)\right] + \theta E\left[\sum_{l=0}^{L-1} |h_i[l]|^2\right]\right)$ is always positive, which is a penalty to the system performance due to the two users' spatial correlation. When $S_{\widehat{h}_1\widehat{h}_2} = 0$ (i.e., $\xi = 0$ or $\xi = 1$), $\widehat{h}_1$ and $\widehat{h}_2$ are uncorrelated, and thus $E\left[\widehat{P}_{IUI}(i)\right] = E\left[P_{IUI}(i)\right]$. In the extreme case when $S_{\widehat{h}_1\widehat{h}_2} = 1$ (i.e., $\xi = 0.5$) that maximizes (15.46), $\widehat{h}_1$ and $\widehat{h}_2$ are identical, the IUI achieves its upper-bound

$$E\left[\widehat{P}_{IUI}(i)\right] = \frac{E\left[P_{Sig}(i) + P_{ISI}(i) + P_{IUI}(i)\right] + \theta E\left[\sum_{l=0}^{L-1} |h_i[l]|^2\right]}{2}. \tag{15.47}$$

Because $E\left[P_{Sig}(i) + P_{ISI}(i)\right] = E\left[P_{IUI}(i)\right] + \theta E\left[\sum_{l=0}^{L-1} |h_i[l]|^2\right]$ at $D = 1$, (15.47) can be written as $E\left[\widehat{P}_{IUI}(i)\right] = E\left[P_{Sig}(i)\right] + E\left[P_{ISI}(i)\right]$, when there is no rate backoff.

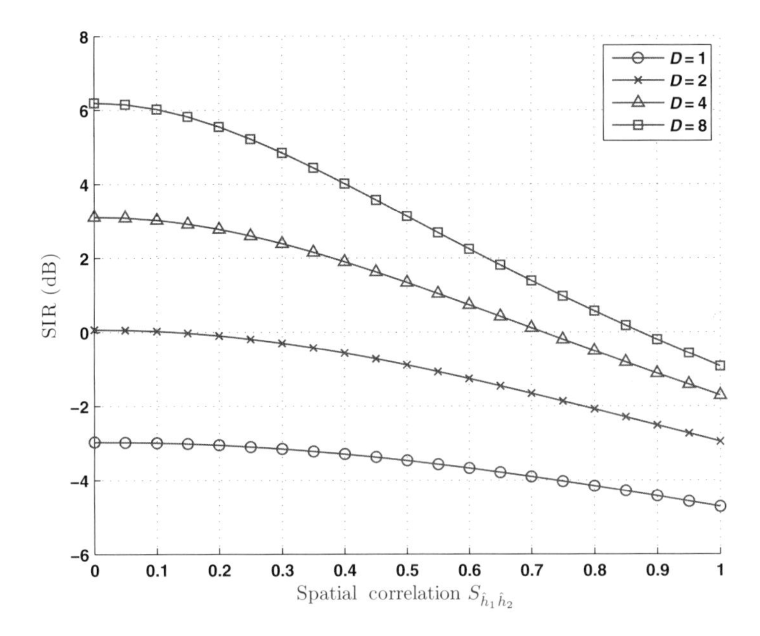

Figure 15.12 SIR vs spatial correlation with $N = 2$ and $M_T = 1$.

The impact of the increased interference would be most prominent in the high SNR regime, where the interference power dominates the noise power. So we evaluate its impact to the system performance in terms of signal-to-interference ratio (SIR), as a close approximation of the effective SINR in high SNR regime. Figure 15.12 shows the influence of the spatial correlation to the SIR with correlated CIRs $\widehat{h}_1$ and $\widehat{h}_2$ of length $L = 257$ and delay spread $\sigma_T = 128T_S$. As one can see in Figure 15.12, the SIR degradation speed varies with different ranges of $S_{\widehat{h}_1\widehat{h}_2}$. In the lower range of $S_{\widehat{h}_1\widehat{h}_2}$ (e.g., from 0 to 0.2) the SIR degrades very slowly. Also, the larger rate backoff D tends to result in a faster performance loss due to spatial correlation as shown in Figure 15.12. However, even for $S_{\widehat{h}_1\widehat{h}_2}$ up to 0.5 which is rare in real-life RF communications over scattering environments, the degraded SIR is preserved within 3 dB away from the performances of uncorrelated channels. This demonstrates the robustness of the presented TRDMA scheme and provides a more comprehensive understanding of its system performances.

15.6 Summary

In this chapter, we present a TRDMA scheme for the multiuser downlink network over multipath channels. Both single-antenna and multi-antenna schemes were developed to utilize the location-specific signatures that naturally exist in the multipath environment. We defined and evaluated both analytically and numerically a variety of performance metrics of including the effective SINR, the achievable sum rate, and achievable rates with outage. We then demonstrated the TRDMA's improvement of achievable rate region over the Rake receivers and investigated the impact of spatial correlations between users to the system performances. Based on the nice properties shown in the analysis and simulation results of this chapter, the presented TRDMA can be a promising technique in the future energy-efficient low-complexity broadband wireless communications. For related references, readers can refer to [38].

References

[1] J. G. Proakis, *Digital Communications*, 4th ed., New York: McGraw-Hill 2001.

[2] G. L. Stuber, *Principles of Mobile Communications*, 2nd ed., Dordrecht: Kluwer, 2001.

[3] A. J. Goldsmith, *Wireless Communication*, New York: Cambridge University Press, 2005.

[4] D. Tse and P. Viswanath, *Fundamental of Wireless Communication*, New York: Cambridge University Press, 2005.

[5] B. Wang, Y. Wu, F. Han, Y. H. Yang, and K. J. R. Liu, "Green wireless communications: A time-reversal paradigm," *IEEE Journal on Selected Areas in Communications*, vol. 29, no. 8, pp.1698–1710, Sep. 2011.

[6] M. Fink, C. Prada, F. Wu, and D. Cassereau, "Self focusing in inhomogeneous media with time reversal acoustic mirrors," *IEEE Ultrasonics Symposium*, vol. 1, pp. 681–686, 1989.

[7] C. Prada, F. Wu, and M. Fink, "The iterative time reversal mirror: A solution to self-focusing in the pulse echo mode," *Journal of the Acoustical Society of America*, vol. 90, pp. 1119–1129, 1991.

[8] M. Fink, "Time reversal of ultrasonic fields. Part I: Basic principles," *IEEE Transactions on Ultrasonic, Ferroelectronic, and Frequency Control*, vol. 39, pp. 555–566, Sep. 1992.

[9] C. Dorme and M. Fink, "Focusing in transmit-receive mode through inhomogeneous media: The time reversal matched filter approach," *Journal of the Acoustical Society of America*, vol. 98, no. 2, part. 1, pp. 1155–1162, Aug. 1995.

[10] A. Derode, P. Roux, and M. Fink, "Robust acoustic time reversal with high-order multiple scattering," *Physical Review Letters*, vol. 75, pp. 4206–4209, 1995.

[11] W. A. Kuperman, W. S. Hodgkiss, and H. C. Song, "Phase conjugation in the ocean: Experimental demonstration of an acoustic time-reversal mirror," *Journal of the Acoustical Society of America*, vol. 103, no. 1, pp. 25–40, Jan. 1998.

[12] H. C. Song, W. A. Kuperman, W. S. Hodgkiss, T. Akal, and C. Ferla, "Iterative time reversal in the ocean," *Journal of the Acoustical Society of America*, vol. 105, no. 6, pp. 3176–3184, Jun. 1999.

[13] D. Rouseff, D. R. Jackson, W. L. Fox, C. D. Jones, J. A. Ritcey, and D. R. Dowling, "Underwater acoustic communication by passive-phase conjugation: Theory and experimental results," *IEEE Journal of Oceanic Engineering*, vol. 26, pp. 821–831, 2001.

[14] A. Derode, A. Tourin, J. de Rosny, M. Tanter, S. Yon, and M. Fink "Taking advantage of multiple scattering to communicate with time-reversal antennas," *Physical Review Letters*, vol. 90, pp. 014301-1–014301-4, 2003.

[15] D. Rouseff, D. R. Jackson, W. L. J. Fox, C. D. Jones, J. A. Ritcey, and D. R. Dowling, "Underwater acoustic communication by passive-phase conjugation: Theory and experimental results," *IEEE Journal of Oceanic Engineering*, vol. 26, pp. 821–831, 2001.

[16] G. F. Edelmann, T. Akal, W. S. Hodgkiss, S. Kim, W. A. Kuperman, and H. C. Song "An initial demonstration of underwater acoustic communication using time reversal," *IEEE Journal of Oceanic Engineering*, vol. 27, pp.602–609, 2002.

[17] S. M. Emami, J. Hansen, A. D. Kim, G. Papanicolaou and A. J. Paulraj, "Predicted time reversal performance in wireless communications," *IEEE Communications Letters*, 2004.

[18] M. Emami, M. Vu, J. Hansen, A. J. Paulraj, and G. Papanicolaou "Matched filtering with rate back-off for low complexity communications in very large delay spread channels," *Proceedings of the Asilomar Conference on Signals, Systems, and Computers*, vol. 1, pp. 218–222, Nov. 2004.

[19] K. F. Sander, G. A. L. Reed, *Transmission and Propagation of Electromagnetic Waves*, 2nd ed., Cambridge, UK: Cambridge University Press, 1986.

[20] H. Nguyen, Z. Zhao, F. Zheng, and T. Kaiser, "Pre-equalizer Design for Spatial Multiplexing SIMO UWB Systems," *IEEE Transactions on Vehicular Technology*, vol. 59, no. 8, pp. 3798–3805, Oct. 2010.

[21] T. K. Nguyen, H. Nguyen, F. Zheng, and T. Kaiser, "Spatial correlation in SM-MIMO-UWB systems using a pre-equalizer and pre-Rake filter," *IEEE International Conference on Ultra-Wideband*, Sep. 2010.

[22] H. T. Nguyen, I. Z. Kovacs, P. C. F. Eggers "A time reversal transmission approach for multiuser UWB communications," *IEEE Transactions on Antennas and Propagation*, vol. 54, no. 11, pp. 3216–3224, Nov. 2006.

[23] I. H. Naqvi, A. Khaleghi, and G. E. Zein, "Performance enhancement of multiuser time reversal UWB communication system," *IEEE International Symposium on Wireless Communication Systems*, Nov. 2007. [http://arXiv:0810.1506v1[cs.NI]].

[24] H. T. Nguyen, "Partial one bit time reversal for UWB impulse radio multi-user communications," *International Conference on Communications and Electronics*, pp. 246-251, Jun. 2008.

[25] T. K. Nguyen, H. Nguyen, F. Zheng, and T. Kaiser, "Spatial correlation in the broadcast MU-MIMO UWB system using a pre-equalizer and time reversal pre-filter," *International Conference on Signal Processing and Communication Systems*, Dec. 2010.

[26] A. Derode, A. Tourin, and M. Fink, "Ultrasonic pulse compression with one-bit time reversal through multiple scattering," *Journal of Applied Physics*, vol. 85, no. 9, May 1999.

[27] P. H. Moose, "A technique for orthogonal frequency division multiplexing frequency offset correction," *IEEE Transactions on Communications*, vol. 42, no. 10, pp. 2908–2914, Oct. 1994.

[28] J. Lee, H.-L. Lou, D. Toumpakaris, and J. M. Cioffi, "SNR analysis of OFDM systems in the presence of carrier frequency offset for fading channels," *IEEE Transactions on Wireless Communications*, vol. 5, no. 12, pp. 3360–3364, Dec. 2006.

[29] A. Papoulis and U. Pillai, *Probability, Random Variables and Stochastic Processes*, 4th ed., New York: McGraw-Hill, 2002.

[30] T. M. Cover, J. A. Thomas, *Elements of Information Theory*, New York: John Wiley & Sons, 2006.

[31] J. L. Massey, "Towards an information theory of spread-spectrum systems," in S. G. Glisic and P. A. Leppanen, editors, *Code Division Multiple Access Communications*, pp. 29–46, Dordrecht: Kluwer Academic Publishers, 1995.

[32] T. M. Cover, "Broadcast channels," *IEEE Transactions on Information Theory*, vol. IT-18, no. 1 pp. 2–14, Jan. 1972.

[33] P. P. Bergmans, "Random coding theorem for broadcast channels with degraded components," *IEEE Transactions on Information Theory*, vol. IT-19, no. 1 pp. 197–207, Mar. 1973.

[34] P. P. Bergmans, "A simple converse for broadcast channels with additive white Gaussian noise," *IEEE Transactions on Information Theory*, vol. IT-20, bo. 2 pp. 279–280, Mar. 1974.

[35] P. P. Bergmans, and T. M. Cover, "Cooperative broadcasting," *IEEE Transactions on Information Theory*, vol. IT-20, no. 3 pp. 317–324, Mar. 1974.

[36] R. B. Ertel and J. H. Reed, "Generation of two equal power correlated Rayleigh fading envelopes," *IEEE Communications Letters*, vol. 2, no. 10. Oct. 1998.

[37] B. Natarajan, C. R. Nassar, and V. Chandrasekhar, "Generation of correlated Rayleigh fading envelopes for spread spectrum applications," *IEEE Communications Letters*, vol. 4, no. 1. Jan. 2000.

[38] F. Han, Y. H. Yang, B. Wang, Y. Wu, and K. J. R. Liu, "Time-reversal division multiple access over multi-path channels," *IEEE Transactions on Communications*, vol. 60, no. 7, pp. 1953–1965, Jul. 2012.

16 Combating Strong–Weak Resonances in TRDMA

In a time-reversal (TR) communication system, the signal-to-noise ratio (SNR) is boosted and the interuser interference (IUI) is suppressed due to the spatial–temporal resonances, commonly known as the focusing effects, of the TR technique when implemented in a rich scattering environment. However, because the spatial–temporal resonances highly depend on the location-specific multipath profile, there exists a strong–weak spatial–temporal resonances effect. In the TR uplink system, different users at different locations enjoy different strengths of spatial–temporal resonances, i.e., the received signal-to-interference-noise ratios (SINRs) for different users vary, and the weak ones can be blocked from correct detection in the presence of strong ones. In this chapter, we formulate the strong–weak spatial–temporal resonances in the multiuser TR uplink system as a max-min weighted SINR balancing problem by joint power control and signature design. Then, a novel two-stage adaptive algorithm that can guarantee the convergence is discussed. In stage I, the original nonconvex problem is relaxed into a Perron Frobenius eigenvalue optimization problem, and an iterative algorithm is discussed to obtain the optimum efficiently. In stage II, the gradient search method is applied to update the relaxed feasible set until the global optimum for the original optimization problem is obtained. Numerical results show that our algorithm converges quickly, achieves a high energy-efficiency, and provides a performance guarantee to all users.

16.1 Introduction

The explosive growth of high-speed wireless services that can support various wireless communication applications with a large number of users calls for future wideband communication solutions. Moreover, the dispersion of a channel in wideband communications will bring in an undesirable phenomenon: intersymbol interference (ISI). Because the resolution of perceiving multiple paths increases significantly with the increase of bandwidth, ISI is more severe in a wideband single-carrier system [1]. To tackle this problem, equalization techniques at the receiver side and/or multicarrier modulations are developed. While the performance of communication is improved, the complexity of terminal devices inevitably increases.

On the other hand, thanks to its inherent nature that fully collects energy of multipath propagation, the time-reversal (TR)-based signal transmission is an ideal paradigm for low-complexity single-carrier broadband communication systems [2]. In essence, by treating each path of the multipath channel in a rich scattering environment as a widely distributed virtual antenna, TR provides a high-resolution spatial–temporal resonance, commonly known as the focusing effect. This focusing effect indeed is the outcome of a resonance of electromagnetic field, in response to the environment, which concentrates energy propagated through the multipath channel onto a particular intended location at a specific time instant. The property of spatial focusing alleviates the interuser interference (IUI) effectively in communications. Meanwhile, the traditional TR signature works as a matched filter at the access point (AP), which brings a temporal focusing and boosts the signal-to-noise ratio (SNR) at the intended location. In [3], a multiuser downlink system over multipath channels using time-reversal division multiple access (TRDMA) method was investigated. TRDMA is capable of achieving a very high diversity gain and of supporting low-cost and low-complexity terminal devices with only one single transmit antenna. As analyzed in [4], the TR-based wideband communication system was further proved to be a desired solution for future wireless communication systems. TR technique is also a promising solution for green Internet of Things (IoTs) in that a typical TR system has a potential of over an order of magnitude of reduction in power consumption and interference alleviation, as well as supporting heterogeneous terminal devices and providing an additional security and privacy guarantee [5]. However, the traditional TR signature is optimal only in the low symbol rate scenario due to ISI. A near optimal waveform design and power allocation solution was proposed in [6], suppressing both ISI and IUI and maximizing the achievable sum rate for the multiuser TRDMA downlink system.

The spatial–temporal resonance of the TR has been proposed as theory and validated through experiments in both acoustic domain and radio frequency (RF) domain. As verified in [7–9], energy of acoustic signal can be refocused with high resolution through a TR procedure after which a divergent wave issued from an acoustic source is converted into a convergent wave focusing on the source. This time-reversal mirror (TRM) is a self-adaptive technique that can be utilized to compensate for propagation distortions. The resonant effects of TR have also been validated through underwater acoustic experiments [10, 11]. In the RF domain, the TR spatial–temporal resonances of electromagnetic (EM) waves were studied in the SISO and MISO schemes, and experiment results showed that the quality of focusing is determined by the bandwidth frequency and spectral correlations of the field [12]. Lerosey et al. conducted experiments on microwaves in an indoor environment to further study the TR spatial–temporal focusing effects [13]. In [14], the TR technique was applied as a prefilter, and its performance was investigated in the channel with large delay spread. Furthermore, the theory of TRM is formally developed for electromagnetic waves in [15]. However, because the quality of spatial–temporal resonances of EM waves highly depends on the propagation environment and transmission bandwidth, there exists a strong–weak spatial–temporal focusing effect in the multiuser TRDMA uplink system. Because of the strong–weak spatial–temporal resonances, the received SNRs of different users can be very distinct, and weak signals can be blocked from correct detection in the presence of strong ones. Note that such a strong–weak spatial–temporal resonance is different from the well-known near–far effect in the CDMA uplink systems. For example, the TR resonances may bring in a signal boosting gain up to 6–10 dB while the near–far effect only causes attenuation in signal strength. Moreover, the reasons for creating the strong–weak spatial–temporal resonances is much more complicated than the near–far effect, which is mainly resulted from the physical distances. The detailed differences between them will be discussed in the following.

The main reason causing the near–far problem in the CDMA uplink systems is that the distances from different transmitters to the same receiver differ a lot, and the signals from the farther transmitters have lower SNR according to the inverse square law [16]. Dynamic power control algorithms have been proposed to alleviate the near–far effect in CDMA uplink systems and other cellular communication systems [17–21]. In [18], a centralized power adjustment scheme for cochannel interference control was proposed and investigated. Alpcan et al. formulated the distributed power control problem in the CDMA system as a noncooperative game where a user-specific utility function based on SIR was designed, and a quantitative criterion for admission control was derived upon possible equilibrium solutions [21]. Furthermore, as the usage of beamforming affected not only the individual link gains but also the power allocation strategy, joint optimization algorithms were proposed to enhance system performance [22–24]. With the uplink and downlink duality [25, 26], dynamic power control has also been applied to ensure multiuser fairness and to improve the total system data capacity in the downlink system [27–33]. The max-min criterion was first proposed in [27] for smart antenna downlink systems, and the Perron Frobenius theorem [34, 35] was involved. Furthermore, a centralized algorithm was proposed in [28] for MISO downlink systems with frequency-flat channels. Recently, a nonlinear Perron Frobenius theorem has been introduced, and a distributed weighted proportional SINR algorithm was proposed to optimize multiuser downlinks [36]. The work has been extended to MIMO downlinks in [33, 37], and a power update problem under multiple power constraints was analyzed in [38]. However, the analysis and algorithms in the previous literature only considered optimizations in downlink systems where constraints are applied to the sum of individual powers. Compared to the total power constraints, to address the nonconvex max-min optimization under multiple individual power constraints is more complicated. Moreover, most existing literature considered the frequency-flat channel model or multicarrier systems, and thus the intersymbol interference (ISI), which is an important issue in the single-carrier system, has been omitted in the analysis.

Unlike the CDMA near–far problem, which resulted solely from distance, the TRDMA uplink systems suffer from the strong–weak spatial–temporal resonances among different users mainly due to the different resonances resulting from location-specific multipath environments. The performance of each link depends on the corresponding signal-to-interference-noise ratio (SINR). To guarantee the performance, we need to combat the strong–weak spatial–temporal resonances and to balance the SINRs among all links. To this end, we formulate the strong–weak spatial–temporal resonances combating problem in the TRDMA uplink system as a max-min weighted

SINR optimization problem by the means of joint power allocation and signature filter design. Note that such a problem is nonconvex and has many individual constraints, which make it even harder to handle. To tackle this challenge, in this chapter, we discuss a two-stage algorithm that solves this problem efficiently in two steps and converges to the global optimum quickly. In stage I, the original optimization problem is relaxed into a Perron Frobenius eigenvalue optimization problem by converting all the individual power constraints into a total power constraint. An iterative algorithm is considered that is targeted to tackle the relaxed problem. In stage II, in order to find the global optimum for the original problem, the gradient search method is applied to shrink the relaxed feasible set. Simulation results demonstrate that the presented algorithm is capable of providing a performance guarantee to all users regardless of their channel gains. Furthermore, the presented algorithm is highly energy efficient compared to the Basic TR schemes [3].

This chapter is organized as follows. In Section 16.2, the system model and problem formulation are described. In Section 16.3, we relax the original TRDMA uplink SINR balancing problem with individual power constraints into an eigenvalue optimization problem and introduce an iterative algorithm, which alternately optimizes the signature matrix and power assignment vector. The two-stage adaptive algorithm for the TRDMA uplink SINR balancing problem with individual constraints is discussed in Section 16.4. Finally, numerical simulation in Section 16.5 demonstrates the performance improvement of the presented scheme compared with traditional methods.

16.2 System Model

As shown in Figure 16.1, in this chapter, we consider a TRDMA uplink system, where K users transmit data through the same media to one single access point (AP) simultaneously. In conventional single-carrier uplink systems, there is a detectability problem where signals with strong SNRs block weaker signals from being detected, e.g., the well-known near–far effect in the CDMA uplink system [1]. In CDMA uplink

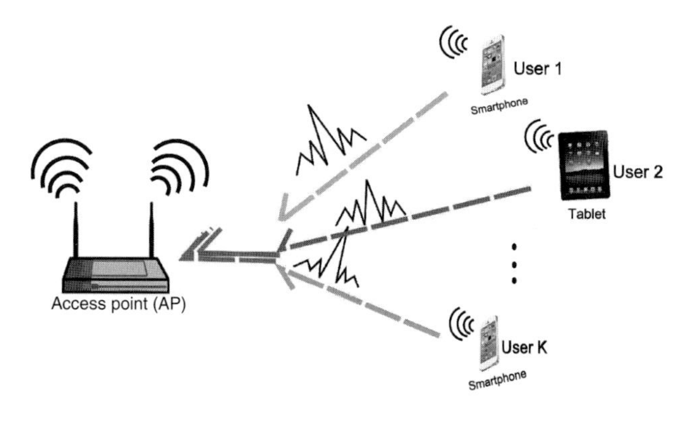

Figure 16.1 Representative case for TRDMA uplink system.

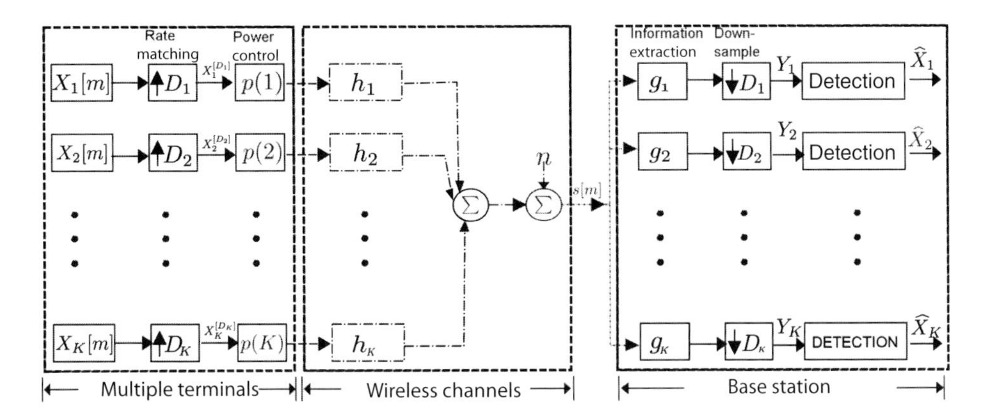

Figure 16.2 Diagram for TRDMA uplink system.

systems, the difference between SNRs of received signal is caused by the variations in transmit power and distance-based propagation attenuation. In the TRDMA uplink system, however, besides the variations in transmit power and distance-based path loss, the multipath channel gain brings in the strong–weak spatial–temporal resonances that cause the SNRs of the received signals to be different. Therefore, the detectability problem also commonly exists in the TRDMA uplink systems and needs to be carefully addressed.

The schematic diagram of a TRDMA uplink system is shown in Figure 16.2 [5]. We can see that the signal of kth user, X_k, is first upsampled by a backoff factor D_k. Then, the upsampled signal $X_k^{[D_k]}$ is boosted by a power control factor $p(k)$ before being transmitted through the location-specific multipath channel h_k. Because all the transmitted signals from different users are combined over the air, the received signal at the AP is a mixture of all transmitted signals and noise. To extract the information of different users, the received signal passes through a user-specific signature filter bank $\mathbf{g}_i$ which is designed according to the channel information obtained in the channel probing phase. In the channel probing phase, devices send an impulse or a pseudorandom noise (PN) sequence to the AP [2], such that the channel estimation can be obtained. We found through real experiments that channel response is rather stationary. In order to combat variations on channels and number of users, TR communication systems must rely on the channel probing phase to update the channel information frequently. This can be done very quickly to be unnoticeable to users under time-varying situation. Then the output is downsampled by a backoff rate D_i, and a series of detectors are performed to detect the information of each user.

However, because the multipath channel gains vary among different users, the SINR of each user in the received signal may be different, and consequently the information of some users may not be detected correctly. From Figure 16.2, we can see that, for a fixed detector, the received SINRs of users are determined jointly by the power control factor and the signature filter. Therefore, in this chapter, our objective is to jointly optimize signature $\mathbf{G} = [\mathbf{g}_1, \ldots, \mathbf{g}_K]$ and power allocation $\mathbf{p} = [p(1), \ldots, p(K)]^T$ to make sure the information of all users can be correctly detected.

16.2.1 Problem Formulation

Now, let us take a close look at this strong–weak spatial–temporal resonances problem and formulate it into an optimization problem. According to the system diagram in Figure 16.2, the received signal at the AP can be written as follows

$$s[m] = \sum_k \sum_l \sqrt{p(k)} h_k[m-l] X_k^{[D_k]}[l] + n[m], \tag{16.1}$$

where $X_k^{[D_k]}$ is the upsampled version of the transmitted signal from the kth user, $p(k)$ is the transmit power, and h_k denotes the channel impulse response of the kth user to AP with channel length L_k, i.e., $h_k[m] = 0$ for $m < 0$ and $m > L_k - 1$.

Then (16.1) can be rewritten in a matrix form as $\mathbf{s} = \sum_k \sqrt{p(k)} \mathbf{H}_k \mathbf{x}_k^{[D_k]} + \mathbf{n}$, where $\mathbf{s}$ is a $(2L-1) \times 1$ vector with $L = \max_k L_k$, $\mathbf{H}_k$ is a $(2L-1) \times L$ Toeplitz matrix with each column being the shifted version of $\mathbf{h}_k$, and $\mathbf{n}$ denotes the additive white Gaussian noise (AWGN) vector, whose elements are identical complex Gaussian with mean zero and variance σ^2.

At the AP side, the received signal $\mathbf{s}$ first passes through a user-specific signature filter bank $\{\mathbf{g}_i, \ \forall\ i\}$ to extract information and to suppress interference. Then it is downsampled to obtain Y_i as follows

$$\begin{aligned} Y_i[m] &= \sum_{j=1}^{K} \sum_l \sqrt{p(j)} X_j[l](h_j * g_i)[mD_i - lD_j] + n[m] \\ &= \sqrt{p(i)} X_i[m](h_i * g_i)[0] \\ &+ \sqrt{p(i)} \sum_{\substack{l=-\lfloor \frac{L-1}{D_i} \rfloor \\ l \neq 0}}^{l=\lfloor \frac{L-1}{D_i} \rfloor} X_i[m-l](h_i * g_i)[D_i l] \\ &+ \sum_{j \neq i} \sqrt{p(j)} \sum_{l=-\lfloor \frac{L-1}{D_j} \rfloor}^{l=\lfloor \frac{L-1}{D_j} \rfloor} X_j[m-l](h_j * g_i)[D_j l] + \tilde{n}_i[m], \end{aligned} \tag{16.2}$$

where $\tilde{n}_i$ is a downsampled version of $n * g_i$.

Based upon (16.2), the uplink SINR for user i is given as,

$$SINR_i^{UL}(\mathbf{G}, \mathbf{p}) = \frac{p(i)\mathbf{g}_i^H \mathbf{R}_i^{(0)} \mathbf{g}_i}{p(i)\mathbf{g}_i^H \hat{\mathbf{R}}_i \mathbf{g}_i + \sum_{j \neq i} p(j) \mathbf{g}_i^H \mathbf{R}_j \mathbf{g}_i + \sigma^2}, \tag{16.3}$$

where $\mathbf{G} = [\mathbf{g}_1, \ldots, \mathbf{g}_K]$ is the signature matrix, $\mathbf{p} = [p(1), \ldots, p(K)]^T$ is the power allocation vector, $\mathbf{R}_i^{(0)} = \mathbf{H}_i^{(L)H} \mathbf{H}_i^{(L)}$ with $\mathbf{H}_i^{(l)}$ being the lth row of $\mathbf{H}_i$ and the superscript H denoting the Hermitian operator, $\mathbf{R}_j = \tilde{\mathbf{H}}_j^H \tilde{\mathbf{H}}_j$, $\tilde{\mathbf{H}}_j$ is the upsampled version of $\mathbf{H}_j$ with factor D_j and sampling center located at $\mathbf{H}_j^{(L)}$, and $\hat{\mathbf{R}}_i = \mathbf{R}_i - \mathbf{R}_i^{(0)}$. The first two terms in the denominator, $p(i)\mathbf{g}_i^H \hat{\mathbf{R}}_i \mathbf{g}_i$ and $\sum_{j \neq i} p(j) \mathbf{g}_i^H \mathbf{R}_j \mathbf{g}_i$, represent ISI and IUI, respectively.

Let us define a crosstalk matrix $\mathbf{\Phi}$ for the TRDMA uplink system whose elements correspond to the ISI and IUI term in $SINR_i^{UL}$ as

$$[\mathbf{\Phi}]_{ij} = \begin{cases} \mathbf{g}_j^H \mathbf{R}_i \mathbf{g}_j & ,i \neq j \\ \mathbf{g}_i^H \hat{\mathbf{R}}_i \mathbf{g}_i & ,i = j \end{cases}.$$

The crosstalk matrix $\mathbf{\Phi}$ is of all positive components, and $\mathbf{\Phi}_i$ is the ith column of matrix $\mathbf{\Phi}$.

Moreover, $\mathbf{D}$ is defined as a diagonal matrix with $[\mathbf{D}]_{ii} = \gamma_i / \mathbf{g}_i^H \mathbf{R}_i^{(0)} \mathbf{g}_i$, and γ_i is the SINR weighted factor for ith user, which supports heterogeneous SINR requirements. Then we can have $SINR_i^{UL}(\mathbf{G}, \mathbf{p})/\gamma_i = p(i)/[\mathbf{D}]_{ii}(\mathbf{\Phi}_i^T \mathbf{p} + \sigma^2)$,

In order to ensure the fairness among all users and to boost the system performance, we jointly design the signature matrix $\mathbf{G} = [\mathbf{g}_1, \cdots, \mathbf{g}_K]$ and power allocation vector $\mathbf{p} = [p(1), \cdots, p(K)]^T$. To balance the $SINR_i^{UL}(\mathbf{G}, \mathbf{p})$ among different users, the max-min fairness is adopted in this chapter as follows

$$\begin{aligned} &\underset{\mathbf{G},\,\mathbf{p}}{\text{maximize}} \;\; \min_j \frac{p(j)}{[\mathbf{D}]_{jj}(\mathbf{\Phi}_j^T \mathbf{p} + \sigma^2)} \\ &\text{subject to } \; \mathbf{p} \succeq \mathbf{0},\; \mathbf{p} \preceq \mathbf{p}_{\max},\; \|\mathbf{g}_i\|_2 = 1,\; i = 1, \ldots, K, \end{aligned} \tag{16.4}$$

where $\mathbf{p}_{\max}$ denotes the vector of individual maximal transmit power, and $\mathbf{0}$ is an all-zero vector with K elements. In an attempt to maximize the entire network throughput, by using the log utility functions, the proportional fairness criterion pulls up the weaker signals thereby giving them QoS protections. However, the QoS protection from proportional fairness is not as strong as that of the max-min criterion. Because the fairness is the top concern in the multiuser uplinks, rather than the entire throughput, in this chapter with the purpose to address the strong–weak spatial–temporal resonances the max-min criterion is selected.

By introducing an auxiliary variable γ and rewriting the K individual SINR expressions into a vector form, we have

$$\begin{aligned} &\underset{\mathbf{G},\,\mathbf{p},\,\gamma}{\text{maximize}} \;\; \gamma \\ &\text{subject to } \; \mathbf{p} \succeq \mathbf{0},\; \mathbf{p} \preceq \mathbf{p}_{\max},\; \mathbf{p} \succeq \gamma \mathbf{D}(\mathbf{\Phi}^T \mathbf{p} + \boldsymbol{\sigma}), \\ &\qquad\qquad \|\mathbf{g}_i\|_2 = 1,\; i = 1, \ldots, K, \end{aligned} \tag{16.5}$$

where $\boldsymbol{\sigma}$ is a $K \times 1$ vector with each element being σ^2.

First of all, the optimization problem in (16.5) is not convex. Moreover, the $4 \times K$ individual constraints make the problem even more challenging. To solve (16.5), we develop a two-stage efficient algorithm as will be shown in the following two sections. Specifically, we first relax the problem into a Perron Frobenius eigenvalue optimization problem through using a total power constraint to enlarge the feasible set and develop an iterative algorithm to find the optimal solution to the relaxed problem. Then, we discuss an adaptive two-stage algorithm to find the global optimal solution to the original optimization problem with individual constraints, based on the relaxed eigenvalue problem.

16.3 Iterative Algorithm with a Total Power Constraint

In this section, we will introduce how to relax the original uplink SINR balancing problem into an equivalent eigenvalue optimization problem and introduce the algorithm where the signature and power assignment are iteratively optimized.

The relaxed version of the original problem in (16.5) is

$$\begin{aligned} &\underset{\mathbf{G},\ \mathbf{p},\ \gamma}{\text{maximize}} \ \gamma \\ &\text{subject to } \mathbf{1}^T\mathbf{p} \le \mathbf{1}^T\mathbf{p}_{\max},\ \mathbf{p} \succeq \gamma\mathbf{D}(\mathbf{\Phi}^T\mathbf{p} + \sigma), \\ &\qquad \mathbf{p} \succeq \mathbf{0},\ \|\mathbf{g}_i\|_2 = 1,\ i = 1, \ldots, K, \end{aligned} \tag{16.6}$$

where **1** is an all-one vector with K elements.

Here we can see, the problem in (16.6) maintains the same objective function but is relaxed by a total power constraint.

16.3.1 Uplink Power Assignment Problem

We first consider the case when the signature matrix is fixed as $\tilde{\mathbf{G}} = [\tilde{\mathbf{g}_1},\ \tilde{\mathbf{g}_2}, \ldots, \tilde{\mathbf{g}_K}]$ with $\|\tilde{\mathbf{g}}_i\|_2 = 1, i = 1, 2, \cdots, K$, and then the problem in (16.6) is reduced to an uplink power allocation problem as

$$\begin{aligned} &\underset{\mathbf{p},\ \gamma}{\text{maximize}} \ \gamma \\ &\text{subject to } \mathbf{p} \succeq \mathbf{0},\ \mathbf{1}^T\mathbf{p} \le \mathbf{1}^T\mathbf{p}_{\max},\ \mathbf{p} \succeq \gamma\mathbf{D}(\mathbf{\Phi}^T\mathbf{p} + \sigma). \end{aligned} \tag{16.7}$$

In Theorem 16.3.1, the necessary condition for the global optimum of problem in (16.7) is introduced.

Theorem 16.3.1 *Given $\tilde{\mathbf{G}}$, if $\boldsymbol{p}^*$ is a global maximizer of the* (16.7)*, then $\boldsymbol{1}^T\boldsymbol{p}^* = \boldsymbol{1}^T\boldsymbol{p}_{\max}$, and $\boldsymbol{p}^* = \gamma^*\boldsymbol{D}(\boldsymbol{\Phi}^T\boldsymbol{p}^* + \sigma)$, where γ^* is the optimum of minimum weighted SINR.*

Proof. The proof of Theorem 16.3.1 is in Appendix A. ▲

Combining the two equations in Theorem 16.3.1, we have an equivalent necessary condition for global optimizer $\mathbf{p}^*$ of (16.7) as $\frac{1}{\gamma^*}\mathbf{1}^T\mathbf{p}_{\max} = \mathbf{1}^T\mathbf{D}(\mathbf{\Phi}^T\mathbf{p}^* + \sigma)$.

Based on the previous analysis, let us define an augmented power vector as $\tilde{\mathbf{p}} = [\mathbf{p}^T,\ 1]^T$ and an augmented matrix, which only depends on **G** and $P_{total} = \mathbf{1}^T\mathbf{p}_{\max}$ as

$$\mathbf{\Lambda}(\mathbf{G}, P_{total}) = \begin{pmatrix} \boldsymbol{D\Phi^T} & \boldsymbol{D\sigma} \\ \frac{1}{P_{total}}\mathbf{1}^T\boldsymbol{D\Phi^T} & \frac{1}{P_{total}}\mathbf{1}^T\boldsymbol{D\sigma} \end{pmatrix}. \tag{16.8}$$

Then, the necessary condition for the global optimum $\mathbf{p}^*$ in (16.7) can be characterized as an eigensystem that $\frac{1}{\gamma^*}\tilde{\mathbf{p}}^* = \Lambda(\mathbf{G}, P_{total})\tilde{\mathbf{p}}^*$. In order to ensure the feasibility of the solution, $\tilde{\mathbf{p}}^*$ should be element-wisely positive as well as γ^*.

In TRDMA uplink systems, due to the existence of ISI and IUI, $\mathbf{\Phi}$ is a nonnegative irreducible cross-talk matrix. As a result, $\mathbf{\Lambda}(\mathbf{G}, P_{total})$ is a matrix with nonnegative entries and is also irreducible. According to the Perron Frobenius Theorem [27, 34, 35], we can have the following properties for $\mathbf{\Lambda}$ as

(i) the maximal eigenvalue $\lambda_{\max}$ is just its spectral radius, and it is simple;
(ii) the eigenvector $\mathbf{v}$ that has all positive entries is the one associated to the largest eigenvalue.

Therefore, there exists a feasible solution to the eigensystem, and the solution is unique. Moreover, based upon the uniqueness and existence, the necessary condition in Theorem 16.3.1 becomes a necessary and sufficient condition for global optimum of (16.7). Thus, the power allocation problem in (16.7) is addressed by finding the Perron Frobenius eigenvector of $\mathbf{\Lambda}(\tilde{\mathbf{G}}, P_{total})$ and the optimal threshold is given by $\gamma^* = 1/\lambda_{\max}(\mathbf{\Lambda}(\tilde{\mathbf{G}}, P_{total}))$.

16.3.2 Joint Signature Design and Power Assignment

From the previous subsection, we know that for any arbitrary matrix $\tilde{\mathbf{G}}$, the optimal power assignment vector $\mathbf{p}^*$ under the total power constraint P_{total} is a $K \times 1$ vector consisting of the first K elements in the scaled $(K+1) \times 1$ dominant eigenvector of $\mathbf{\Lambda}(\tilde{\mathbf{G}}, P_{total})$. Meanwhile the associated optimal threshold γ^* is the reciprocal of the dominant eigenvalue. Therefore, the number of optimization variables in (16.6) is greatly reduced, and the (16.6) is equivalent to an eigenvalue optimization problem as

$$\begin{aligned} &\underset{\mathbf{G}}{\text{minimize}}\ \lambda_{\max}(\mathbf{\Lambda}(\mathbf{G}, P_{total})) \\ &\text{subject to}\ \|\mathbf{g}_i\|_2 = 1, \quad i = 1, \ldots, K, \end{aligned} \tag{16.9}$$

For the Perron Frobenius eigenvalue λ of a matrix $\mathbf{\Lambda}$, it can be represented as [35]

$$\lambda = \min_{\mathbf{y} \succ 0} \max_{\mathbf{x} \succ 0} \frac{\mathbf{x}^T \mathbf{\Lambda} \mathbf{y}}{\mathbf{x}^T \mathbf{y}} = \min_{\mathbf{x} \succ 0} \max_{\mathbf{y} \succ 0} \frac{\mathbf{x}^T \mathbf{\Lambda} \mathbf{y}}{\mathbf{x}^T \mathbf{y}}. \tag{16.10}$$

Recall that $\tilde{\mathbf{p}} = [\mathbf{p}^T, 1]^T$, let us define a cost function as

$$\tilde{\lambda}(\mathbf{G}, \mathbf{p}) = \max_{\mathbf{x} \succ 0} \mathbf{x}^T \mathbf{\Lambda}(\mathbf{G}, P_{total}) \tilde{\mathbf{p}} / \mathbf{x}^T \tilde{\mathbf{p}},$$

and the Perron Frobenius eigenvalue can be represented as $\lambda_{\max}(\mathbf{\Lambda}(\mathbf{G}, P_{total})) = \min_{\mathbf{p} \succ 0} \tilde{\lambda}(\mathbf{G}, \mathbf{p})$.

Then, the optimal threshold for the problem in (16.9) can be obtained as

$$\gamma^* = \frac{1}{\min_{\mathbf{G}} \min_{\mathbf{p} \succ 0} \tilde{\lambda}(\mathbf{G}, \mathbf{p})} = \frac{1}{\min_{\mathbf{G}} \tilde{\lambda}(\mathbf{G}, \mathbf{p}_{opt})}, \tag{16.11}$$

where $\mathbf{p}_{opt} = \arg \min_{\mathbf{p} \succ 0} \tilde{\lambda}(\mathbf{G}, \mathbf{p})$ represents a vector consisting of the first K elements in the dominant eigenvector.

Given the cost function, the problem in (16.9) has an equivalent formulation as

$$\min_{\mathbf{G}} \min_{\mathbf{p} \succ 0} \tilde{\lambda}(\mathbf{G}, \mathbf{p}) \Leftrightarrow \min_{\mathbf{p} \succ 0} \min_{\mathbf{G}} \tilde{\lambda}(\mathbf{G}, \mathbf{p}). \tag{16.12}$$

For the left part in (16.12), when the signature matrix is fixed, the problem is solved as the eigenvalue problem in the previous subsection. Considering the right-hand side of (16.12), when the power allocation vector is fixed, the way to find the corresponding optimal signature matrix $\mathbf{G}^*$ is given in Lemma 16.3.1 as follows.

Lemma 16.3.1 *The optimal signature matrix for a given vector* $\boldsymbol{p}_{ary}$ *is denoted as* $\boldsymbol{G}^*$. *We have* $\boldsymbol{G}^* = arg \min_{\boldsymbol{G}} \tilde{\lambda}(\boldsymbol{G},\boldsymbol{p}_{ary}) = arg \min_{\boldsymbol{G}} \gamma_i / SINR_i^{UL}(\boldsymbol{G},\boldsymbol{p}_{ary})$, $\forall i$. *That is to say, the optimal signature can be obtained by individually maximizing the uplink SINR of each user.*

Proof. The proof of Lemma 16.3.1 is in Appendix B. ▲

Remark 16.3.1 *The SINR maximizing signature can be optimized by*

$$\boldsymbol{g}_i^* = arg \max_{\|\boldsymbol{g}_i\|_2=1} p_{ary}(i)/[\boldsymbol{D}]_{ii}(\boldsymbol{\Phi}_i^T \boldsymbol{p}_{ary} + \sigma^2), \forall i. \tag{16.13}$$

The optimal solution is equivalent to the MMSE beamforming vector, $\boldsymbol{g}_i^* = \alpha_i (\sum_{j=1}^K p_{ary}(j)\boldsymbol{R}_j + \sigma^2 \boldsymbol{I})^{-1} \boldsymbol{H}_i^{(L)H}, \forall i$, *where* α_i *is a normalized factor.*

Furthermore, the necessary and sufficient condition for the global optimum of problem in (16.9) is stated in the following theorem.

Theorem 16.3.2 (*Necessary and Sufficient Condition*) $\boldsymbol{G}^* = [\boldsymbol{g}_1^*, \boldsymbol{g}_2^*, \ldots, \boldsymbol{g}_K^*]$ *is the global optimizer of the problem in* (16.9) *if and only if* $\tilde{\lambda}(\boldsymbol{G}^*,\boldsymbol{p}^*) = \min_{\boldsymbol{G}} \tilde{\lambda}(\boldsymbol{G},\boldsymbol{p}^*)$, *where* $\boldsymbol{p}^* = arg \min_{\boldsymbol{p} \succ \boldsymbol{0}} \tilde{\lambda}(\boldsymbol{G}^*,\boldsymbol{p})$, *i.e.,* $\tilde{\boldsymbol{p}}^*$ *is the Perron Frobenius eigenvector of* $\boldsymbol{\Lambda}(\boldsymbol{G}^*, P_{total})$ *and* $\tilde{\boldsymbol{p}}^{*T} = [\boldsymbol{p}^{*T}, 1]$.

Proof. The proof of Theorem 16.3.2 is in Appendix C. ▲

Theorem 16.3.2 implies that if one of the variable **p** or **G** has reached the optimum, the remaining one can be obtained either by solving the Perron Frobenius eigenpair problem or by independently solving the MMSE problem. Based on this, we consider an iterative algorithm that alternatively optimizes **p** and **G** and eventually converges to the global optimum.

16.3.3 Iterative Algorithm and Convergence

Built upon the previous analysis, we consider an iterative algorithm in Algorithm 4 to jointly optimize the signature matrix and the power assignment vector, given the objective to balance the weighted uplink SINRs for all users.

Theorem 16.3.3 *The sequence* $\{\lambda_{\max}^{(n)}\}_{n=0}^{\infty}$ *generated by the presented algorithm in Algorithm 4 is strictly decreasing and converges to the global optimum of* (16.9), *regardless of the initial value.*

Proof. The proof of Theorem 16.3.3 is in Appendix D. ▲

Algorithm 4 Iterative SINR Balancing Algorithm under Total Power Constraint

---

Require: Given $\{\gamma_i\}_{i=1}^K$, σ^2, $\mathbf{R}_i$ and $\mathbf{R}_i^{(0)}$ $\forall\, i$, $\mathbf{P}_{\max}$. Pick $\epsilon > 0$, $P_{total} = \mathbf{1}^T\mathbf{P}_{\max}$, $\mathbf{p}^{(0)} = \frac{P_{total}}{K}\mathbf{1}$, $\lambda_{\max}^{(0)} = \infty$.

1: **repeat**

2: $n \Leftarrow n + 1$.

3: Calculate the MMSE $\mathbf{g}_i^{(n)}$, $\forall\, i$ under $\mathbf{p}^{(n-1)}$, and normalize it to make $\|\mathbf{g}_i\|_2 = 1$, $\forall\, i$.

4: Build the couple matrix $\mathbf{\Lambda}^{(n)}(\mathbf{G}^{(n)}, P_{total})$ in the way shown in (16.8).

5: Solve the Perron Frobenius eigenpair problem to get $\lambda_{\max}^{(n)}$ and its corresponding eigenvector $\tilde{\mathbf{p}}^{(n)}$ with $\tilde{p}^{(n)}(K+1) = 1$. $\mathbf{p}^{(n)} = \{\tilde{\mathbf{p}}^{(n)}\}_1^K$.

6: **until** $\lambda_{\max}^{(n-1)} - \lambda_{\max}^{(n)} < \epsilon$ or reach the maximal number of iterations.

As we proved, the necessary and sufficient condition for global optimum of (16.9) is equivalent to the condition $\lambda_{max}^{(n+1)} = \lambda_{\max}^{(n)}$, which implies the convergence of $\{\lambda_{\max}^{(n)}\}$ sequence. As a consequence, the algorithm stops as soon as the difference in $\lambda_{\max}^{(n-1)} - \lambda_{\max}^{(n)}$ reaches a predetermined threshold $\epsilon > 0$. On average the algorithm will yield the global optimum of (16.9) in three or four iterations, independent of number of users.

16.4 Two-Stage Adaptive Algorithm with Individual Power Constraints

In the previous section, we discussed how to relax the original problem in (16.5) into a Perron Frobenius eigenvalue optimization problem. For any fixed total power constraint P_{total}, there only exists one pair of optimal signature matrix $\mathbf{G}^*(P_{total})$ and power assignment vector $\mathbf{p}^*(P_{total})$. As we have analyzed, the original problem with individual power constraints has a more tightened feasible set. Hence, we need to gradually shrink the relaxed feasible set until the optimal solution to the original problem is reached.

Lemma 16.4.1 *The optimal power assignment vector generated from iterative SINR balancing algorithm in Algorithm 4 is monotonically increasing as the total power constraint P_{total} increases.*

Proof. The proof is provided in Appendix E. ▲

According to Lemma 16.4.1, we know that as the feasible set is shrunk by reducing the total power constraint in (16.6), the optimal power assignment for each user is monotonically decreasing. Hence, as long as we keep adjusting the feasible set by updating the total power constraint, we will definitely reach a point that is on the boundary of the individual-constrained feasible set. At this point, we achieve the maximum of balanced weighted SINRs. In the SINR balancing scenario under TRDMA uplink systems, the user with the weakest spatial–temporal resonance or a minimal transmit power will restrict the performance of entire system. Here, we define the worst case in a network to be the user who has a minimal difference between its power constraint $p_{\max}(i)$ and its assigned power under a total power constraint P_{total}, i.e., $p_{\max}(i) - p_i^*(P_{total})$. After

Algorithm 5 Two-Stage Adaptive Algorithm for SINR Balancing Problem under Individual Power Constraint

Require: Given $\{\gamma_i\}_{i=1}^K$, σ^2, $\mathbf{R}_i$ and $\mathbf{R}_i^{(0)}$ $\forall\, i$, $\mathbf{P}_{\max}$. Pick $\epsilon > 0$ as the stop criterion or tolerance and $0 < \eta, \mu < 1$ as stepsizes. $P_{total}^{(0)} = \mathbf{1}^T \mathbf{P}_{\max}$, Update $\mathbf{p}^{(0)}$ by Algorithm 4 under $P_{total}^{(0)}$, $\delta \mathbf{p}^{(n)} = \mathbf{P}_{\max} - \mathbf{p}^{(n)}$, $[\text{index}, \delta^{(n)}] = \min(\delta \mathbf{p}^{(n)})$.

1: **repeat**
2: $n \Leftarrow n + 1$
3: $\mathbf{p}$ = Algorithm 4 with $\mu P_{total}^{(n-1)}$, slope $= (p^{(n-1)}(\text{index}) - p(\text{index}))/(1 - \mu)$ P_{total}.
4: $\delta P_{total} = \delta^{(n-1)}/\text{slope}$, $P_{total}^{(n)} = P_{total}^{(n-1)} + \delta P_{total}$.
5: Update $\mathbf{p}^{(n)}$ under $P_{total}^{(n)}$, $\delta \mathbf{p}^{(n)}$, and $[\text{index}, \delta^{(n)}] = \min(\delta \mathbf{p}^{(n)})$
6: **while** $\delta^{(n)} > \epsilon$ **do**
7: $\delta P_{total} = \eta \times \delta P_{total}$, $P_{total}^{(n)} = P_{total}^{(n-1)} + \delta P_{total}$
Update $\mathbf{p}^{(n)}$, $\delta \mathbf{p}^{(n)}$ and $[\text{index}, \delta^{(n)}]$ // Force $\delta^{(n)} \leq \epsilon$
8: **end while**
9: **until** $|\delta^{(n)}| \leq \epsilon$ or reach the maximal number of iterations.

finding the worst case, the relaxed feasible set is updated following the direction of the worst case.

The presented two-stage adaptive algorithm is shown in Algorithm 5, where the SINR balancing problem with individual constraints is solved. In stage I, the original optimization problem in (16.5) is relaxed into an eigenvalue optimization problem, and the corresponding optimum is obtained through the presented iterative algorithm shown in Algorithm 4. In stage II, based on the solution in stage I, the relaxed feasible set is modified by updating the total power constraint using the gradient decent method against the worst case. Before converging to the global optimum, stage I and stage II work alternatively and iteratively.

16.4.1 Analysis on Convergence

Theorem 16.4.1 *The sequence $\{\delta^{(n)}, n = 1, 2, \ldots\}$ generated from Algorithm 5, is a strictly increasing sequence and converges to 0 where the global optimum for problem in* (16.5) *is achieved.*

Proof. The proof is provided in Appendix F. ▲

According to Theorem 16.4.1, we can see that our adaptive algorithm for the SINR balancing problem under individual power constraints is practical, and it always converges to the global optimal solution to the problem in (16.5).

The average number of iterations for Algorithm 1 to converge is $N_1 = 3$, for backward search in Algorithm 2 with stepsize $\eta = 0.8$ is $N_2 = 1$, and for Algorithm 2 with stepsize $\mu = 0.9$ is $N_3 = 3$. Moreover, the complexity for solving MMSE signature is $O(L^2)$, where L is the length of channel. In our real environment measurement, under 125 MHz its typical value is smaller than 30. The computational complexity for solving the Perron Frobenius eigenvalue of the augmented matrix is $O((K+1)^2)$, where K is the number of users in the uplinks. Hence, the total computational complexity for the joint optimization is approximately $N_1 \times N_2 \times (1+N_3) \times (O(L^2)+O((K+1)^2))$. Under the simulation with stepsizes being $\eta = 0.8$ and $\mu = 0.9$, the amount of required computation time is $10 \times (O(L^2)+O((K+1)^2))$.

16.4.2 Properties of SINR Balancing Problem

In this part, we will briefly introduce some properties of the TRDMA SINR balancing problem.

Assumption 16.4.1 *With a small perturbation in total power constraint P_{total}, i.e., $P_{total} \rightarrow (1+\Delta)P_{total}$, $\Delta \ll 1$, the optimal MMSE signature can be viewed as approximately unchanging.*

The detailed explanation for the rationality of Assumption 16.4.1 is shown in Appendix G.

Following Assumption 16.4.1, as $\Delta < 10^{-2}$, because the optimal MMSE signature matrix $\mathbf{G}$ will not change, the augmented matrix $\mathbf{\Lambda}(\mathbf{G}, P_{total})$ in (16.8) will have the same components with $\mathbf{\Lambda}(\mathbf{G}, (1+\Delta)P_{total})$ except the last row. Under this setting, we have

$$\mathbf{\Lambda}(\mathbf{G}, (1+\Delta)P_{total}) = \begin{pmatrix} \mathbf{D\Phi^T} & \mathbf{D}\boldsymbol{\sigma} \\ \frac{\mathbf{1}^T\mathbf{D\Phi^T}}{(1+\Delta)P_{total}} & \frac{\mathbf{1}^T\mathbf{D}\boldsymbol{\sigma}}{(1+\Delta)P_{total}} \end{pmatrix}, \tag{16.14}$$

and $\mathbf{\Lambda}(\mathbf{G}, (1+\Delta)P_{total}) = \mathbf{A} \times \mathbf{\Lambda}(\mathbf{G}, P_{total})$, where $\mathbf{A} = \begin{pmatrix} \mathbf{I} & \mathbf{0} \\ \mathbf{0}^T & \frac{1}{1+\Delta} \end{pmatrix}$.

As mentioned in [39], $\rho(\mathbf{\Lambda}(\mathbf{G}, (1+\Delta)P_{total})) \geq \frac{1}{(1+\Delta)^\alpha}\rho(\mathbf{\Lambda}(\mathbf{G}, P_{total}))$ and $\delta\lambda \geq (\frac{1}{(1+\Delta)^\alpha} - 1)\lambda$, where $\delta\lambda$ is the difference in dominant eigenvalues, $\rho(\cdot)$ denotes the spectral radius of nonnegative matrices, i.e., in our case it is the Perron Frobenius eigenvalue, and $0 < \alpha < 1$ is a coefficient determined by the right and left Perron Frobenius eigenvectors of $\mathbf{\Lambda}(\mathbf{G}, P_{total})$.

Then, we can bound the change of Perron Frobenius eigenvalue under the perturbation of Δ as $0 < -\frac{\delta\lambda}{\lambda} < \frac{\Delta}{\Delta+1}$.

Let us denote $\mathbf{k} = [k_1, k_2, \ldots, k_K]$ as the optimal power assignment ratio vector, where $k_i = p^*(i)/P_{total}$ and $p^*(i)$ is the optimal power assignment to the problem in (16.6) under total power constraint P_{total}. According to the necessary and sufficient condition of optimal power, we have

$$\mathbf{k} = \frac{1}{P_{total}}(\lambda\mathbf{I} - \mathbf{D\Phi}^T)^{-1}\mathbf{D1}, \tag{16.15}$$

where we assume $\boldsymbol{\sigma} = \mathbf{1}$ and thus P_{total} is the SNR. Then we can also have

$$\begin{aligned} \mathbf{k} + \delta\mathbf{k} &= \frac{1}{(1+\Delta)P_{total}}((\lambda+\delta\lambda)\mathbf{I} - \mathbf{D\Phi}^T)^{-1}\mathbf{D1} \\ &= \frac{1}{1+\Delta}\sum_{n\geq 0}\left(-\frac{\delta\lambda}{\lambda}(\mathbf{I} - \frac{1}{\lambda}\mathbf{D\Phi}^T)^{-1}\right)^n \mathbf{k}, \end{aligned} \tag{16.16}$$

where $\delta\mathbf{k}$ denotes the change in power assignment ratio with a total power changes by ΔP_{total}.

Combining the (16.15) and the (16.16), we can represent the change in power assignment ratio vector as $\delta\mathbf{k} = -\frac{\Delta}{1+\Delta}\mathbf{k} + \frac{1}{1+\Delta}\sum_{n\geq 1}(-\frac{\delta\lambda}{\lambda}\mathbf{C}^{-1})^n\mathbf{k}$, where $\mathbf{C}^{-1} = (\mathbf{I} - \frac{1}{\lambda}\mathbf{D}\mathbf{\Phi}^T)^{-1}$, and $\sum_{n\geq 1}(-\frac{\delta\lambda}{\lambda}\mathbf{C}^{-1})^n$ exists. It can be further derived as

$$\begin{aligned}\delta\mathbf{k} &= \frac{\Delta}{1+\Delta}\left(-\mathbf{I} + \frac{1}{\Delta}\sum_{n\geq 1}\left(-\frac{\delta\lambda}{\lambda}\mathbf{C}^{-1}\right)^n\right)\mathbf{k} \\ &= \frac{\Delta}{1+\Delta}\left(\mathbf{I} + \frac{\delta\lambda}{\lambda}\mathbf{C}^{-1}\right)^{-1}\left(-\frac{\Delta+1}{\Delta}\frac{\delta\lambda}{\lambda}\mathbf{C}^{-1} - \mathbf{I}\right)\mathbf{k}.\end{aligned} \tag{16.17}$$

Moreover, owing to the fact that $\mathbf{1}^T\delta\mathbf{k} = 0$, $\mathbf{1}^T\mathbf{k} = 1$, we have $1 = \mathbf{1}^T(-\frac{\Delta+1}{\Delta}\frac{\delta\lambda}{\lambda}\mathbf{C}^{-1})\mathbf{k}$.

Let us define $\mathbf{v} = -\frac{\Delta+1}{\Delta}\frac{\delta\lambda}{\lambda}\mathbf{C}^{-1}\mathbf{k}$, with $0 < v_i, \forall i < 1$ and $\mathbf{1}^T\mathbf{v} = 1$. Equation (16.17) can be rewritten as $\delta\mathbf{k} = \frac{\Delta}{1+\Delta}(\mathbf{I} + \frac{\delta\lambda}{\lambda}\mathbf{C}^{-1})^{-1}(\mathbf{v} - \mathbf{k})$. Dividing both sides by Δ, we have $\frac{\delta\mathbf{k}}{\Delta} = \frac{1}{1+\Delta}(\mathbf{I} + \frac{\delta\lambda}{\lambda}\mathbf{C}^{-1})^{-1}(\mathbf{v} - \mathbf{k})$.

Then taking the limit gives $\operatorname{limit}_{\Delta\to 0}\frac{\delta\mathbf{k}}{\Delta} = (\mathbf{v} - \mathbf{k})$, which implies that $\operatorname{limit}_{\delta P\to 0}\frac{\delta k_i}{\delta P} = \frac{1}{P_{total}}(v_i - k_i), \forall i$, where δP is the perturbation in P_{total}. This demonstrates that the slope $\frac{|\delta k_i|}{\delta P}$ for every user is bounded by $\frac{1}{P_{total}}\max\{k_i, 1-k_i\}$. As a result, within a suitable SNR range where the slope $\frac{|\delta k_i|}{\delta P}$ is quite small, the optimal power assignment ratio will stay stable even when the total power constraint is changing.

16.5 Simulation Results

To evaluate the presented algorithms, we conduct several simulations to demonstrate that the presented weighted SINR balancing algorithm is an ideal solution to ensure fairness and energy-efficiency, as well as to tackle the strong–weak spatial–temporal resonances in the multiuser TRDMA uplink system.

Our simulation settings are described as following:

(i) Channel Model: UWB office non-line-of-sight channel, with bandwidth $B = 500$ MHz and maximal number of channel taps $L = 60$.

(ii) Backoff rate D: The D is set to 4 (16) when there are 3 (20) users.

(iii) Channel Gain $\|\mathbf{H}_i^{(L)}\|_2^2$: uniform distributed among [0, 1] for all users.

(iv) Weighted factors: $\gamma_i = 1, \forall\ i$. Equal maximal power constraint $p_{\max}(i) = p_{\max}, \forall\ i$.

16.5.1 Optimal Power Assignment

In Figure 16.3, we show the optimal power assignment strategy versus the maximal individual power constraint. In this simulation, the channel gains for all three users are predetermined as $\|\mathbf{H}_1^{(L)}\|_2^2 : \|\mathbf{H}_2^{(L)}\|_2^2 : \|\mathbf{H}_3^{(L)}\|_2^2 = 1 : 2 : 3$, i.e., the first user has the

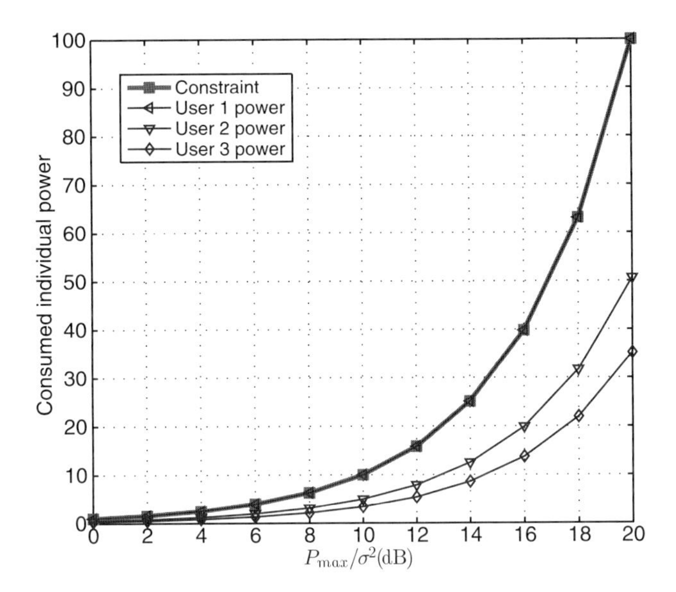

Figure 16.3 Power assignment under three users case.

worst channel response and may suffer the strong–weak spatial–temporal resonances. Under the equal power constraint setting, in order to balance the weighted SINRs, the worst user always uses up all power to boost SINR while others slightly reduce their transmit power to alleviate interference. Meanwhile, the third user takes advantage of its focusing effect such that only smallest power is consumed to maintain the balanced SINR. Moreover, the power assignment ratio between all users is approximately stable when the individual power constraint changes.

16.5.2 Comparison under Different Backoff Rates

In Figure 16.4, 16.5, and 16.6, the performance of the presented algorithm under different backoff rates is studied. For a smaller D, the uplink transmission is conducted more frequently, which causes the ISI to lie close to the signal peak at the receiver side. On the other hand, IUI and ISI can be significantly alleviated by choosing a higher backoff rate D. As BER is a monotonically decreasing function in uplink SINR, users' BER in our scheme is also reducing when the backoff rate increases as shown in Figure 16.4. In Figure 16.5, the balanced SINR becomes higher as the backoff rate increases.

Let us define the achievable sum rate in TR uplinks as $\frac{1}{D}\sum_{i=1}^{K}\log_2(1+SINR_i^{UL})$ bps/Hz. Because the achievable sum rate is normalized by $1/D$, a higher D may result in a larger factor in the denominator outside the logarithm, which leads to a smaller sum rate. From Figure 16.6 we can see that, in low SNR region, a smaller D will always attain a higher sum rate. It is because when the SNR is small, the dominant interference at the receiver side is noise, while ISI is less prominent in the normalized sum rate. On the other hand, as SNR increases, the ISI and IUI become dominant against the noise. Thus, a higher D will reduce that interference efficiently and provide a better sum rate performance.

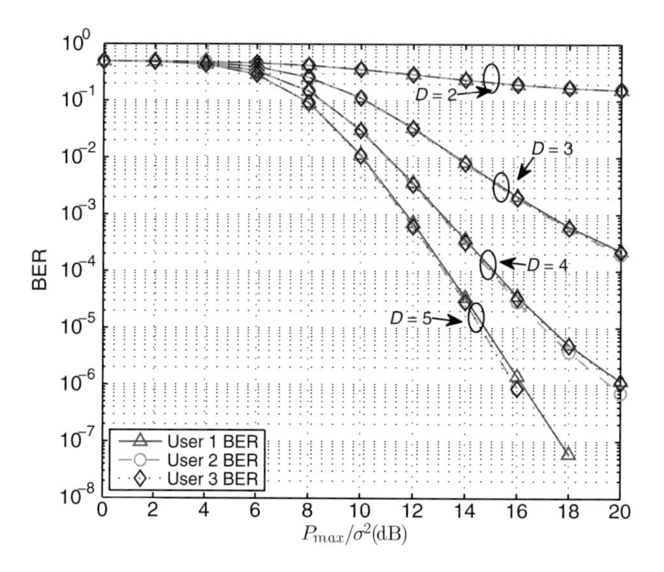

Figure 16.4 BER performance comparison under different backoff rates and three users.

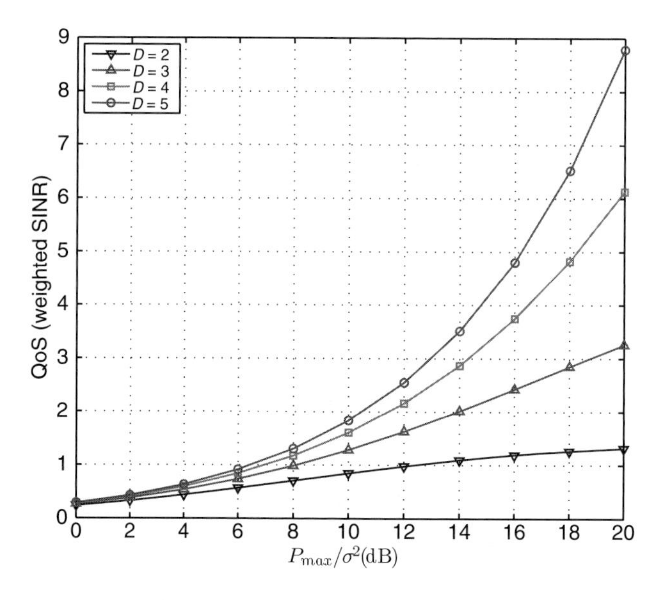

Figure 16.5 Balanced SINR comparison under different backoff rates and three users.

16.5.3 Highly Crowded Network

We then compare the multiuser uplink performance of the presented algorithm and other two schemes: Basic TR and MMSE TR. In the Basic TR scheme, each user transmits with its maximal power, and the signature filter $\mathbf{g}_i$ is the normalized time-reversal conjugate version of its channel response,

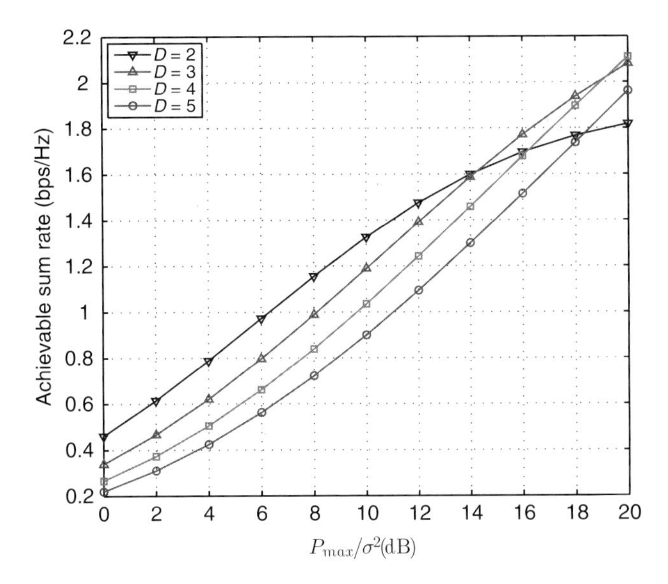

Figure 16.6 Achievable sum rate comparison under different backoff rates and three users.

$$g_i[k] = \frac{h_i^*[L-1-k]}{\sqrt{\sum_{l=0}^{L-1} |h_i[l]|^2}}, \quad k = 0, 1, 2, \ldots, L-1. \tag{16.18}$$

In MMSE TR, each user also transmits in the maximal power but its signature **g** is the MMSE signature that can be calculated by independently maximizing its own SINR.

On the other hand, in our SINR balancing scheme, AP, which has all channel information, controls the transmit power of each user and designs the signature filter based on the optimal power assignment.

In this part, we simulate a highly crowded network where one AP serves 20 users. We assume the 20 users are divided into three groups according to the strength of their spatial–temporal resonances. The channel gain of each user is set to be $\frac{1}{3}$, $\frac{2}{3}$, or 1, and the backoff rate is $D = 16$. Simulations are conducted to compare the BER performance, achievable sum rate, network-level energy-efficiency, and user-level energy-efficiency among Basic TR scheme, MMSE TR scheme, and the presented scheme.

In Figure 16.7, we can see the BER performances of the aforementioned three schemes in a highly crowded network. All users in the Basic TR scheme have such a high BER that the whole system fails to work properly. With MMSE TR, some of users whose channel degradation is severe will have a poor BER performance that is close to the Basic TR BER curves. Thus, even when they are active and transmitting in full power, those users are blocked from getting service due to the low SINRs. Moreover, these blocked users bring high interference to other active users, which degrades the whole network's performance. On the other hand, when applying the presented algorithm to this crowded network, as SINRs are balanced, all users will have almost the same rational BER performance such that all of them can be detected. This indicates that the presented algorithm can support all users, no matter whether

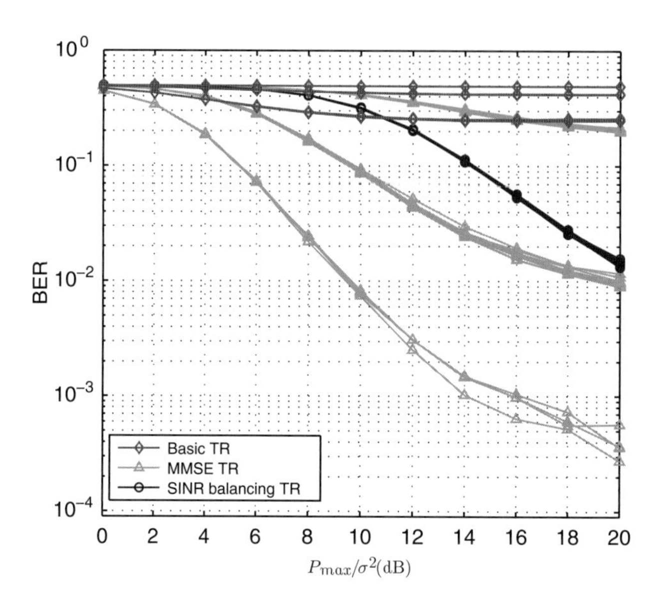

Figure 16.7 BER performance comparison under twenty users case.

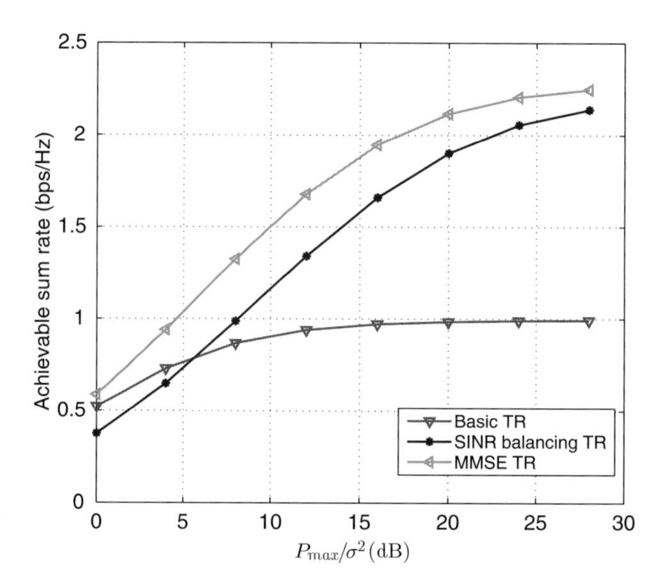

Figure 16.8 Achievable sum rate comparison under twenty users case.

their spatial–temporal resonance is strong or weak and no matter how crowded the network is.

Figure 16.8 shows the curves of achievable sum rate versus the maximal power constraint. Here, the achievable sum rate is normalized by $1/D$, which represents the spectral efficiency. Because Basic TR is aimed to maximize the received signal power regardless of interference, it saturates at a lower rate. In this case, because the interference of other users becomes the dominant factor in individual SINRs and the

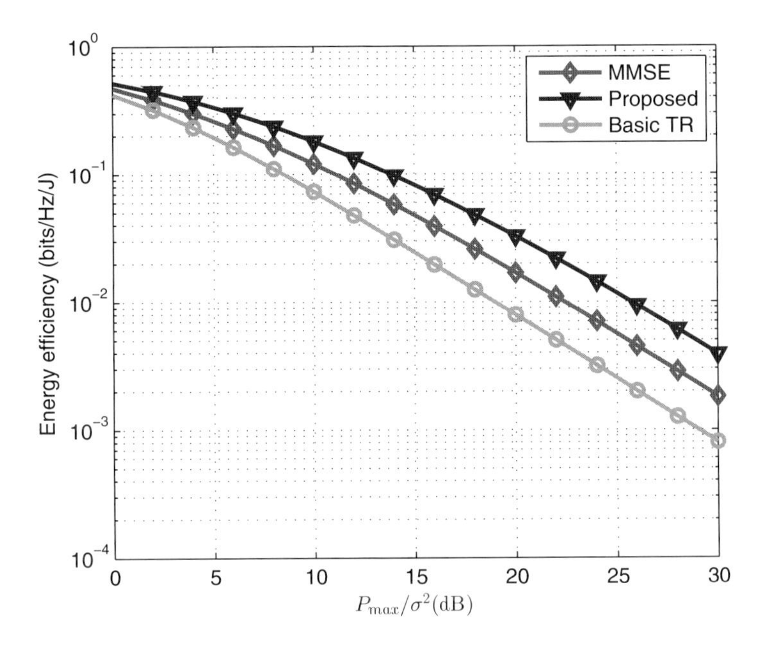

Figure 16.9 Network-level energy-efficiency comparison under 20 users case.

channels become more correlated as the number of users is large, even under maximal power Basic TR fails to support the system with a high quality of service. In the MMSE TR scenario, as everyone transmits in maximal power and MMSE signature is applied at AP to extract information, the SINR is boosted remarkably. As a result, even the network is highly densed, and the interference is large, MMSE signature can suppress the interference and the MMSE TR obtains a higher achievable rate than Basic TR. In a SNR balancing scheme, users with a better spatial–temporal resonance have to sacrifice to achieve the balance, leading to a reduction in the sum rate compared to the MMSE TR. However, the gap between the sum rate of MMSE TR and presented scheme diminishes in high SNR range.

The energy-efficiency feature of the presented scenario is further studied in Figures 16.9 to 16.11. In Figure 16.9, the network-level energy-efficiencies of different schemes are plotted versus the individual power constraints. Here, we define the network-level energy-efficiency as the ratio between achievable sum rate and the total transmit power: Energy-Efficiency (bits/Hz/J) $= \frac{\text{Achievable sum rate (bps/Hz)}}{\text{Total power consumed in the network (W)}}$. Under the same condition, the network-level energy-efficiency of the presented scheme is higher than those of the MMSE TR and the Basic TR. That is because in the MMSE TR and Basic TR, most of the energies are wasted to generate interference, and thus the network performance is contaminated. In the presented algorithm, by the means of joint signature matrix and power allocation optimization, the interference between users is well managed and resources are assigned efficiently.

The trade-off between energy-efficiency and spectral-efficiency is shown in Figure 16.10. As we can see, for a given spectral-efficiency, the energy-efficiency of the

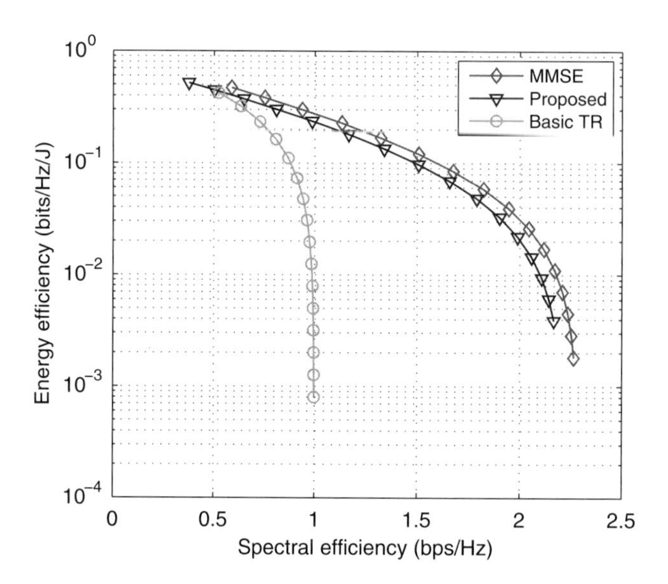

Figure 16.10 Trade-off between energy-efficiency and spectral-efficiency under 20 users case.

presented algorithm is lower than that of the MMSE TR. The reason is that because the presented algorithm is aimed to balance the weighted uplink SINRs, which is restricted by the worst users, the power consumption to achieve the same total network throughput is higher. However, this does not contradict with the result in Figure 16.9 because in order to have a fair comparison we must fix the individual power constraints for all the different schemes to be the same. As we have shown, under the same power constraints, the presented algorithm achieves a higher network-level energy-efficiency than others.

Moreover, the user-level energy-efficiency defined as the ratio of per-user throughput to power consumption of the user is studied. As we can see in Figure 16.11, under the same individual power constraints, the user-level energy-efficiency obtained by the presented algorithm is superior to those of MMSE TR and Basic TR, for all users. In the presented scheme, with joint signature design and power allocation, the interference between users is well managed, and resources are utilized most efficiently with a guarantee on QoS fairness.

Hence, due to the high user-level energy-efficiency, no matter how strong their spatial–temporal resonances are, each user has an incentive and prefers the presented algorithm to the others.

As demonstrated in Figure 16.9 and 16.11, under the same power constraints, both the network-level energy-efficiency and user-level energy-efficiency are improved remarkably through the presented algorithm, compared with MMSE TR and Basic TR. The reason is that by the means of joint signature design and power allocation optimization, the severe interference between users is alleviated, and weak spatial–temporal resonance is also compensated. Moreover, the purpose of the presented algorithm is to balance the weighted uplink SINRs among all users. When equal weighted factors are adopted,

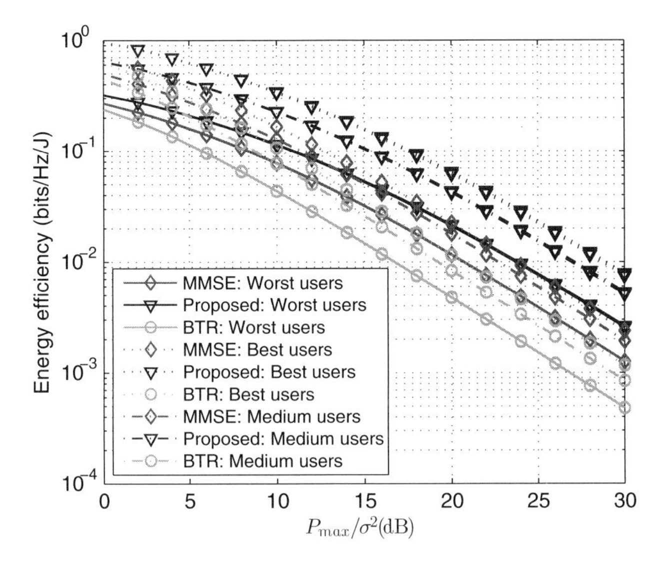

Figure 16.11 User-level energy-efficiency comparison under twenty users case.

the presented algorithm achieves a same uplink SINR for all users, providing a fair QoS guarantee. On the other hand, heterogenous QoS requirements can be achieved by selecting different weighted factors. Hence, the presented algorithm is energy-efficient and QoS-guaranteed in multiuser TRDMA uplinks.

16.6 Summary

In this chapter, we presented a joint power allocation and signature design method to address the strong–weak spatial–temporal resonances in multiuser TRDMA uplink systems. By forming the TRDMA strong–weak spatial–temporal resonances problem into a max-min weighted SINR problem, a two-stage adaptive algorithm that can guarantee to converge to the global optimal solution is discussed. In stage I, the original nonconvex optimization problem is relaxed into a Perron Frobenius eigenvalue optimization problem whose optimum can be efficiently obtained. In stage II, the gradient descent method is applied to adaptively update the relaxed feasible set until the prime global optimum that satisfies all individual constraints is reached. In our simulation, the presented algorithm converges to the global optimum in a few iterations with a relatively low computational complexity. Moreover, the presented algorithm provides a high energy-efficiency and a QoS performance guarantee to all users in network. Simulation results also show that our method is capable of boosting the entire system performance through signature design and power allocation. Therefore, the presented SINR balancing algorithm can be a promising technique for energy-efficient QoS-guaranteed multiuser TRDMA uplink systems. For related references, interested readers can refer to [40].

Appendix A: Proof of Theorem 16.3.1

PROOF. For the problem in (16.7), suppose a global optimizer $\hat{\mathbf{p}}$ is obtained, and $\hat{\gamma}$ is the achievable maximal threshold within the feasible set. For some index n_0, we have $\hat{p}(i)/[\mathbf{D}]_{ii}(\mathbf{\Phi}_i^T\hat{\mathbf{p}}+\sigma^2)=\hat{\gamma}$, $i=n_0$, and $\hat{p}(i)/[\mathbf{D}]_{ii}(\mathbf{\Phi}_i^T\hat{\mathbf{p}}+\sigma^2)\geq\hat{\gamma}$, $\forall i\neq n_0$.

First of all, suppose $\mathbf{1}^T\hat{\mathbf{p}}<\mathbf{1}^T\mathbf{p}_{\max}$, i.e., there is a portion of unused energy at the optimum point as $P_{save}=\mathbf{1}^T\mathbf{p}_{\max}-\mathbf{1}^T\hat{\mathbf{p}}$. If we redistribute P_{save} among all the users such that the updated power assignment vector is $\hat{\mathbf{p}}_{new}=\alpha\hat{\mathbf{p}}$ with $\alpha>1$ and $\mathbf{1}^T\hat{\mathbf{p}}_{new}=\mathbf{1}^T\mathbf{p}_{\max}$. Obviously, $\hat{\mathbf{p}}_{new}$ is in the feasible set defined in (16.7). Thus, we get $\frac{\hat{p}_{new}(i)}{[\mathbf{D}]_{ii}(\mathbf{\Phi}_i^T\hat{\mathbf{p}}_{new}+\sigma^2)}=\frac{\hat{p}(i)}{[\mathbf{D}]_{ii}(\mathbf{\Phi}_i^T\hat{\mathbf{p}}+\frac{\sigma^2}{\alpha})}\geq\hat{\gamma}_{new}>\hat{\gamma}$, which contradicts to the assumption that $\hat{\gamma}$ is the maximum within the feasible set. Hence, when the global optimum γ^* is achieved, we must have $\mathbf{1}^T\mathbf{p}^*=\mathbf{1}^T\mathbf{p}_{\max}$, and the optimal threshold γ^* is a monotonically increasing function in $P_{total}=\mathbf{1}^T\mathbf{p}_{\max}$.

Suppose when γ^* is achieved with the optimal power assignment vector $\mathbf{p}_1$, for some index n_1, we have $\frac{p_1(n_1)}{[\mathbf{D}]_{n_1n_1}(\mathbf{\Phi}_{n_1}^T\mathbf{p}_1+\sigma^2)}>\gamma^*$, and for other users we have $\frac{p_1(i)}{[\mathbf{D}]_{ii}(\mathbf{\Phi}_i^T\mathbf{p}_1+\sigma^2)}=\gamma^*$, $\forall i\neq n_1$. Moreover, $\frac{p(i)}{[\mathbf{D}]_{ii}(\mathbf{\Phi}_i^T\mathbf{p}+\sigma^2)}$ is strictly increasing in $p(i)$ and decreasing in $p(j)$, $j\neq i$, because $\partial\frac{p(i)}{[\mathbf{D}]_{ii}(\mathbf{\Phi}_i^T\mathbf{p}+\sigma^2)}/\partial p(i)=\frac{[\mathbf{D}]_{ii}\sum_{j\neq i}\mathbf{\Phi}_{ji}p(j)+\sigma^2}{([\mathbf{D}]_{ii}(\mathbf{\Phi}_i^T\mathbf{p}+\sigma^2))^2}>0$ and $\partial\frac{p(i)}{[\mathbf{D}]_{ii}(\mathbf{\Phi}_i^T\mathbf{p}+\sigma^2)}/\partial p(j)=-\frac{p(i)[\mathbf{D}]_{ii}\mathbf{\Phi}_{ji}}{([\mathbf{D}]_{ii}(\mathbf{\Phi}_i^T\mathbf{p}+\sigma^2))^2}<0$. If we slightly reduce $p(n_1)$ by δ, then we can obtain a new power assignment vector denoted as $\hat{\mathbf{p}}$, which ensures $\frac{\hat{p}_1(n_1)}{[\mathbf{D}]_{n_1n_1}(\mathbf{\Phi}_{n_1}^T\hat{\mathbf{p}}_1+\sigma^2)}\geq\gamma^*$.

On the other hand, as $p(n_1)$ decreases, we have $\frac{\hat{p}_1(i)}{[\mathbf{D}]_{ii}(\mathbf{\Phi}_i^T\hat{\mathbf{p}}_1+\sigma^2)}>\gamma^*$, $\forall i\neq n_1$. Namely, the minimum of all weighted SINR is not reduced when the total consumed power reduces as $\mathbf{1}^T\hat{\mathbf{p}}_1=\mathbf{1}^T\mathbf{p}_{\max}-\delta$. Hence, as we have proved in the preceding part, by redistributing δ, we can definitely obtain an increase in threshold $\hat{\gamma}^*$ with $\hat{\gamma}^*>\gamma^*$. This contradicts our assumption. ▲

Appendix B: Proof of Lemma 16.3.1

PROOF. Given the definition of cost function $\tilde{\lambda}(\mathbf{G},\mathbf{p})$, for arbitrary $\mathbf{G}$ and $\mathbf{p}$, it can also be formulated as $\tilde{\lambda}(\mathbf{G},\mathbf{p})=\max_{i=1,\cdots,K+1}\frac{\mathbf{e}_i^T\mathbf{\Lambda}(\mathbf{G})\tilde{\mathbf{p}}}{\mathbf{e}_i^T\tilde{\mathbf{p}}}$, where $\mathbf{e}_i$ is a column vector with i^{th} element being one and the rest being zero [28]. From this it follows:

$$\mathbf{e}_i^T\tilde{\mathbf{p}}=\begin{cases}p(i) & ,i\leq K\\ 1 & ,i=K+1\end{cases},\ \text{and}$$

$$\mathbf{e}_i^T\mathbf{\Lambda}(\mathbf{G})\tilde{\mathbf{p}}=\begin{cases}\frac{\gamma_i}{SINR_i^{UL}} & ,i\leq K\\ \frac{1}{P_{total}}\sum_{i=1}^{K}\frac{\gamma_i p(i)}{SINR_i^{UL}} & ,i=K+1.\end{cases}$$

It can be easily verified that $\frac{1}{P_{total}}\sum_{i=1}^{K}\frac{\gamma_i p(i)}{SINR_i^{UL}} \leq \max_i \frac{\gamma_i}{SINR_i^{UL}}$ and $\frac{1}{P_{total}}\sum_{i=1}^{K}\frac{\gamma_i p(i)}{SINR_i^{UL}} \geq \min_i \frac{\gamma_i}{SINR_i^{UL}}$. Thus we have $\tilde{\lambda}(\mathbf{G},\mathbf{p}) = \max_{i=1,\cdots,K+1}\frac{\mathbf{e}_i^T\mathbf{\Lambda}(\mathbf{G})\tilde{\mathbf{p}}}{\mathbf{e}_i^T\tilde{\mathbf{p}}} = \max_{i=1,\cdots,K}\frac{\gamma_i}{SINR_i^{UL}(\mathbf{G},\mathbf{p})}$, and $\min_{\mathbf{x}\succ\mathbf{0}}\frac{\mathbf{x}^T\mathbf{\Lambda}(\mathbf{G},P_{total})\tilde{\mathbf{p}}}{\mathbf{x}^T\tilde{\mathbf{p}}} = \min_{i=1,\cdots,K+1}\frac{\mathbf{e}_i^T\mathbf{\Lambda}(\mathbf{G})\tilde{\mathbf{p}}}{\mathbf{e}_i^T\tilde{\mathbf{p}}} = \min_{i=1,\cdots,K}\frac{\gamma_i}{SINR_i^{UL}(\mathbf{G},\mathbf{p})}$, which will be used in the next section in Appendix C.

For a given $\mathbf{p}_{ary}$, it can be derived that $\tilde{\lambda}(\mathbf{G},\tilde{\mathbf{p}}_{ary}) = \max_{i=1,\cdots,K}\frac{\gamma_i}{SINR_i^{UL}(\mathbf{G},\mathbf{p}_{ary})}$, and the corresponding optimal signature matrix $\mathbf{G}^*$ with fixed power assignment can be obtained by $\mathbf{G}^* = \arg\min_{\mathbf{G}}\max_{i=1,\cdots,K}\frac{\gamma_i}{SINR_i^{UL}(\mathbf{G},\mathbf{p}_{ary})}$. Given a power assignment vector $\mathbf{p}_{ary}$, because the uplink SINR for each user only depends on $\mathbf{g}_i$ and is independent of $\mathbf{g}_j, i \neq j$, $\mathbf{g}_i^* = \arg\min_{\|\mathbf{g}_i\|_2=1}\frac{\gamma_i}{SINR_i^{UL}(\mathbf{G},\mathbf{p}_{ary})}$. ▲

Appendix C: Proof of Theorem 16.3.2

PROOF. Here, we prove the necessary and sufficient condition in Theorem 16.3.2 by contradiction.

(i) To prove the sufficiency: suppose $\mathbf{G}^*$ is not the global optimizer of the problem in (16.9), where $\mathbf{G}^* = \arg\min_{\|\mathbf{g}_i\|_2=1,\forall i}\tilde{\lambda}(\mathbf{G},\mathbf{p}^*)$, and $\tilde{\mathbf{p}}^* = [\mathbf{p}^*,\ 1]$ is the dominant eigenvector of $\mathbf{\Lambda}(\mathbf{G}^*, P_{total})$.

Suppose there exists another $\hat{\mathbf{G}}$ such that $\lambda_{\max}(\mathbf{\Lambda}(\hat{\mathbf{G}}, P_{total})) < \lambda_{\max}(\mathbf{\Lambda}(\mathbf{G}^*, P_{total}))$. Recall (16.10) we can have

$$\begin{aligned}\gamma_{global}^{-1} &= \lambda_{\max}(\mathbf{\Lambda}(\hat{\mathbf{G}}, P_{total}))\\ &= \min_{\mathbf{G}}\lambda_{\max}(\mathbf{\Lambda}(\mathbf{G}, P_{total}))\\ &= \min_{\mathbf{G}}\min_{\mathbf{x}\succ\mathbf{0}}\max_{\mathbf{y}\succ\mathbf{0}}\frac{\mathbf{x}^T\mathbf{\Lambda}(\mathbf{G}, P_{total})\mathbf{y}}{\mathbf{x}^T\mathbf{y}}\\ &\geq \min_{\mathbf{G}}\min_{\mathbf{x}\succ\mathbf{0}}\frac{\mathbf{x}^T\mathbf{\Lambda}(\mathbf{G}, P_{total})\tilde{\mathbf{p}}^*}{\mathbf{x}^T\tilde{\mathbf{p}}^*}\\ &= \min_{\mathbf{G}}\min_{i}\frac{\gamma_i}{SINR_i^{UL}(\mathbf{G},\mathbf{p}^*)}\\ &= \min_{i}\frac{\gamma_i}{SINR_i^{UL}(\mathbf{G}^*,\mathbf{p}^*)}\\ &= \lambda_{\max}(\mathbf{\Lambda}(\mathbf{G}^*, P_{total})).\end{aligned} \tag{16.19}$$

In (16.19), we have $\lambda_{\max}(\mathbf{\Lambda}(\hat{\mathbf{G}}, P_{total})) \geq \lambda_{\max}(\mathbf{\Lambda}(\mathbf{G}^*, P_{total}))$, which contradicts to that $\hat{\mathbf{G}}$ is the global optimum.

(ii) To prove the necessity: suppose $\mathbf{G}^*$ is the global optimizer with the corresponding power vector being $\mathbf{p}^*$ and $\tilde{\lambda}(\mathbf{G}^*,\mathbf{p}^*) > \min_{\mathbf{G}}\tilde{\lambda}(\mathbf{G},\mathbf{p}^*)$. Moreover, suppose there exists an arbitrary feasible $\tilde{\mathbf{G}}$ that satisfies $\tilde{\lambda}(\tilde{\mathbf{G}}, \mathbf{p}^*) = \min_{\mathbf{G}} \tilde{\lambda}(\mathbf{G}, \mathbf{p}^*)$, i.e., $\tilde{\lambda}(\mathbf{G}^*, \mathbf{p}^*) > \tilde{\lambda}(\tilde{\mathbf{G}}, \mathbf{p}^*)$.

Because $\mathbf{G}^*$ is the global optimizer, we have

$$\begin{aligned}
\tilde{\lambda}(\mathbf{G}^*, \mathbf{p}^*) &= \lambda_{\max}(\mathbf{\Lambda}(\mathbf{G}^*, P_{total})) \\
&= \min_{\mathbf{G}} \lambda_{\max}(\mathbf{\Lambda}(\mathbf{G}, P_{total})) \leq \lambda_{\max}(\mathbf{\Lambda}(\tilde{\mathbf{G}}, P_{total})) \\
&= \min_{\mathbf{y}\succ\mathbf{0}} \max_{\mathbf{x}\succ\mathbf{0}} \frac{\mathbf{x}^T \mathbf{\Lambda}(\tilde{\mathbf{G}}, P_{total})\mathbf{y}}{\mathbf{x}^T \mathbf{y}} \\
&\leq \max_{\mathbf{x}\succ\mathbf{0}} \frac{\mathbf{x}^T \mathbf{\Lambda}(\tilde{\mathbf{G}}, P_{total})\tilde{\mathbf{p}}^*}{\mathbf{x}^T \tilde{\mathbf{p}}^*} = \tilde{\lambda}(\tilde{\mathbf{G}}, \mathbf{p}^*).
\end{aligned} \tag{16.20}$$

Then we have $\tilde{\lambda}(\mathbf{G}^*, \mathbf{p}^*) \leq \tilde{\lambda}(\tilde{\mathbf{G}}, \mathbf{p}^*)$, which contradicts to the assumption.

▲

Appendix D: Proof of Theorem 16.3.3

Proof. According to the definition of cost function, we have

$$\begin{aligned}
\tilde{\lambda}(\mathbf{G}^{(n+1)}, \mathbf{p}^{(n)}) &= \min_{\|\mathbf{g}_i\|_2=1, \forall i} \tilde{\lambda}(\mathbf{G}, \mathbf{p}^{(n)}) \\
&\leq \tilde{\lambda}(\mathbf{G}^{(n)}, \mathbf{p}^{(n)}) = \lambda_{\max}^{(n)}.
\end{aligned} \tag{16.21}$$

Moreover, we also have

$$\begin{aligned}
\lambda_{\max}^{(n+1)} &= \min_{\mathbf{p}\succ 0} \tilde{\lambda}(\mathbf{G}^{(n+1)}, \mathbf{p}) \\
&= \min_{\tilde{\mathbf{p}}\succ 0} \max_{\mathbf{x}\succ 0} \frac{\mathbf{x}^T \mathbf{\Lambda}(\mathbf{G}^{(n+1)}, P_{total})\tilde{\mathbf{p}}}{\mathbf{x}^T \tilde{\mathbf{p}}} \\
&\leq \max_{\mathbf{x}\succ 0} \frac{\mathbf{x}^T \mathbf{\Lambda}(\mathbf{G}^{(n+1)}, P_{total})\tilde{\mathbf{p}}^{(n)}}{\mathbf{x}^T \tilde{\mathbf{p}}^{(n)}} \\
&= \tilde{\lambda}(\mathbf{G}^{(n+1)}, \mathbf{p}^{(n)}).
\end{aligned} \tag{16.22}$$

Combining (16.21) and (16.22), it can be induced that $\lambda_{\max}^{(n+1)} \leq \lambda_{\max}^{(n)}, \forall\, n$. The sequence $\{\lambda_{\max}^{(n)}\}$ is nonnegative, lower-bounded by zero, and monotonically decreasing. Hence, the limit $\lambda_{\max}^{\infty} = \lim_{n\to\infty} \lambda_{\max}^{(n)}$ exists.

So far it has been verified that $\lambda_{\max}^{\infty}$ exists, we need to prove that the generated sequence $\{\lambda_{\max}^{(n)}\}$ always converges to the global optimum no matter what the initial point is.

Let us denote the set $X = \{\mathbf{G} \in \mathbb{C}^{L\times K} \mid \mathbf{G} = [\mathbf{g}_1, \cdots, \mathbf{g}_K], \|\mathbf{g}_i\|_2 = 1, \forall i\}$, where K is the number of users, and L is the channel length. Similarly, $Y = \{\mathbf{p} \in \mathbb{R}_+^K \mid \|\mathbf{p}\|_1 = P_{total}\}$. Both sets X and Y are convex compact sets, i.e., bounded and closed. Moreover, set $S = \{\lambda_{\max} : \lambda_{\max} = \rho(\mathbf{\Lambda}(\mathbf{G}, P_{total})), \mathbf{G} \in X\}$ is also compact, due to the compactness of X and the continuity of operator $\rho(\cdot)$ and $\mathbf{\Lambda}(\cdot, P_{total})$. In the algorithm, the tuple $(\mathbf{G}^{(n)}, \mathbf{p}^{(n)})$ is updated in each loop as $(\mathbf{G}^{(n)}, \mathbf{p}^{(n)}) = \mathfrak{T}\{(\mathbf{G}^{(n-1)}, \mathbf{p}^{(n-1)})\}$. The continuous mapping function $\mathfrak{T}$, on the compact space $X \times Y$, is defined as

$$\mathbf{g}_i^{(n)} = \arg \max_{\|\mathbf{g}_i\|_2=1} \frac{SINR_i^{UP}(\mathbf{G}, \mathbf{p}^{(n-1)})}{\gamma_i}, \forall i \tag{16.23}$$

$$\lambda_{\max}^{(n)} \widetilde{\mathbf{p}}^{(n)} = \mathbf{\Lambda}(\mathbf{G}^{(n)}, P_{total})\widetilde{\mathbf{p}}^{(n)}, \tag{16.24}$$

where $\lambda_{\max}^{(n)} = \rho(\mathbf{\Lambda}(\mathbf{G}^{(n)}, P_{total}))$, $\mathbf{G} = [\mathbf{g}_1, \cdots, \mathbf{g}_K]$, and augmented vector $\widetilde{\mathbf{p}}^{(n)} = [\mathbf{p}^{(n)}, 1]$. According to the Schauderï£¡s fixed point theorem, function $\mathfrak{T}$ has at least one fixed point on $X \times Y$, i.e., $\exists\ (\mathbf{G}^{(\infty)}, \mathbf{p}^{(\infty)})$ such that $(\mathbf{G}^{(\infty)}, \mathbf{p}^{(\infty)}) = \mathfrak{T}\{(\mathbf{G}^{(\infty)}, \mathbf{p}^{(\infty)})\}$. Moreover, the generated sequence $\{\lambda_{\max}^{(n)}\}$ is monotonically decreasing, and $\lambda_{\max}^{(\infty)} = \rho(\mathbf{\Lambda}(\mathbf{G}^{(\infty)}, P_{total}))$.

For every potential fixed point $(\mathbf{G}^{(\infty)}, \mathbf{p}^{(\infty)}, \lambda_{\max}^{(\infty)})$, it satisfies both sides of (16.23), (16.24) and

$$\lambda_{\max}^{(\infty)} \widetilde{\mathbf{p}}^{(\infty)} = \rho(\mathbf{\Lambda}(\mathbf{G}^{(\infty)}, P_{total}))\widetilde{\mathbf{p}}^{(\infty)}. \tag{16.25}$$

Moreover, as we have proved in Lemma 1 that $\mathbf{G}^{(\infty)} = \arg \max_{\|\mathbf{g}_i\|_2=1} \frac{SINR_i^{UP}(\mathbf{G}, \mathbf{p}^{(\infty)})}{\gamma_i}\ \forall i$ is equivalent to $\mathbf{G}^{(\infty)} = \arg \min_{\|\mathbf{g}_i\|_2=1} \frac{\gamma_i}{SINR_i^{UP}(\mathbf{G}, \mathbf{p}^{(\infty)})} = \arg \min_{\|\mathbf{g}_i\|_2=1} \tilde{\lambda}(\mathbf{G}^{(\infty)}, \mathbf{p}^{(\infty)})$. Hence, this tuple satisfies the necessary and sufficient condition in Theorem 2 for the global optimality of the joint optimization problem under total power constraint, and thus each locally optimal tuple $(\mathbf{G}^{(\infty)}, \mathbf{p}^{(\infty)}, \lambda_{\max}^{(\infty)})$ is indeed a globally optimal tuple to the relaxed problem that $\lambda_{\max}^{(\infty)} = \lambda_{\max}^{\infty}$. ▲

Appendix E: Proof of Lemma 16.4.1

Proof. Let the optimal power assignment vector in (16.6) with P_{total} be $\mathbf{p}_0$ and the one with $P_{total} + \Delta(\Delta > 0)$ be $\mathbf{p}_1$. Suppose $\exists\ n_0$ such that the optimal power assignment $p_0(n_0) \geq p_1(n_0)$ and for other users $p_0(i) < p_1(i)$.

As shown in the proof of Theorem 16.3.1 in Appendix A, γ^* is a strictly increasing function in P_{total}, i.e., $\gamma_0^* < \gamma_1^*$, where γ_0^* is the optimum under P_{total} and γ_1^* under $P_{total} + \Delta$.

However, as shown in Appendix A, $SINR_i^{UL}$ is strictly increasing in $p(i)$ and decreasing in $p(j), \forall j \neq i$. This gives us that $SINR_{0,n_0}^{UL} \geq SINR_{1,n_0}^{UL}$, where $SINR_{0,n_0}^{UL}$ is the SINR of n_0^{th} user under P_{total} and $SINR_{1,n_0}^{UL}$ is the one under $P_{total} + \Delta$.

According to Theorem 16.3.1, we have $SINR_{0,n_0}^{UL}/\gamma_{n_0} = \gamma_0^*$ and $SINR_{1,n_0}^{UL}/\gamma_{n_0} = \gamma_1^*$. This implies that $\gamma_0^* \geq \gamma_1^*$, which contradicts to the assumption. ▲

Appendix F: Proof of Theorem 16.4.1

PROOF. In the initialization, as $P_{total}^{(0)} = \mathbf{1}^T\mathbf{P}_{\max}$ is an extreme case in our SINR balancing problem, $\delta^{(0)}$ is usually smaller than 0. Except when $\delta^{(0)} = 0$ wherein the optimal power assignment is equal to each individual constraints, our algorithm terminates.

At the nth iteration, if the $\delta^{(n)} > 0$, we use backward search with a coefficient $0 < \eta < 1$ to adjust the stepsize δP_{total} without flip its direction. As a result, we can make sure at the end of nth iteration $\delta^{(n)} < 0$ and the relaxed feasible set is tightened. The stepsize for gradient search is defined as $\delta P_{total} = \frac{\delta^{(n-1)}}{\text{slope}}$, where slope $= \frac{p^{(n-1)}(\text{index}) - p(\text{index})}{(1-\mu)P_{total}^{(n-1)}}$. According to Lemma 16.4.1, the slope is always positive because $\mathbf{p}$ is the optimal power assignment vector with $\mu P_{total}^{(n-1)}$, and $0 < \mu < 1$ is a predefined coefficient. Because $\delta^{(n-1)}$ is always less than or equal to 0 in the presented algorithm, we have stepsize $\delta P_{total} \leq 0$ at the nth iteration, implying $P_{total}^{(n)} = P_{total}^{(n-1)} + \delta P_{total} \leq P_{total}^{(n-1)}$ and $\mathbf{p}^{(n)} \preceq \mathbf{p}^{(n-1)}$. Hence, $\delta^{(n)} = \min_i \{p_{\max}(i) - p^{(n)}(i)\} \geq \delta^{(n-1)}$. Moreover, the equality holds if and only if $\delta^{(n-1)} = 0$. As a consequence, the sequence $\{\delta^{(n)}\}_{n=1}^{\infty}$ is a nonpositive, monotonically increasing sequence and upper-bounded by 0. The limit exists, and $\lim_{n\to\infty} \delta^{(n)} = \delta^{\infty} \leq 0$.

Suppose $\delta^{\infty} = \mu < 0$, and it is achieved for some n_0, i.e., $\delta^{(n)} = \mu$, $\forall\, n \geq n_0$. Then from Algorithm 5, we have $P_{total}^{(n_0+1)} = P_{total}^{(n_0)} + \delta P_{total} \leq P_{total}^{(n_0)}$. As an immediate result $0 \geq \delta^{(n_0+1)} > \delta^{(n_0)} = \mu$, which contradicts to the assumption that the sequence converges to μ. Thus, the limit point of $\{\delta^{(n)}\}_{n=1}^{\infty}$ is $\delta^{\infty} = 0$.

Hence, because the sequence $\{\delta^{(n)}\}_{n=1}^{\infty}$ is strictly increasing before converging, and the limit of $\{\delta^{(n)}, n = 1, 2, \cdots\}$ is $\delta^{\infty} = 0$. Our algorithm will converge to the global optimum. ▲

Appendix G: Explanation for Assumption 16.4.1

To simplify the analysis, let us consider the noise variance $\sigma^2 = 1$. When the constraint is P_{total}, we can have our $\mathbf{g}_i^{MMSE}$ calculated as shown in (16.26).

$$\begin{aligned} \mathbf{g}_i^{MMSE} &= c_i \left(\sum_j p(j)\mathbf{R}_j + \sigma^2\mathbf{I} \right)^{-1} \mathbf{H}_i^{(L)H} \\ &= c_i \mathbf{B}^{-1}(\mathbf{k})\mathbf{H}_i^{(L)H} \frac{1}{P_{total}}. \end{aligned} \tag{16.26}$$

With a slightly perturbation Δ, the updated version of $\mathbf{g}_i^{MMSE}$ is defined as

$$\mathbf{g}_{\Delta i}^{MMSE} = \tilde{c}_i \left(\sum_j p(j)\mathbf{R}_j + \sum_j \delta p(j)\mathbf{R}_j + \sigma^2\mathbf{I} \right)^{-1} \mathbf{H}_i^{(L)H}$$

$$
\begin{aligned}
&= \tilde{c}_i \left(\sum_j k_j \mathbf{R}_j + \frac{1}{P_{total}} \mathbf{I} + \frac{\sum_j \delta p(j) \mathbf{R}_j}{P_{total}} \right)^{-1} \mathbf{H}_i^{(L)H} \frac{1}{P_{total}} \\
&= \frac{\tilde{c}_i}{c_i} \mathbf{g}_i^{MMSE} + \frac{\tilde{c}_i}{c_i} \sum_{n \geq 1} \left(-\frac{\sum_j \delta p(j) \mathbf{R}_j \mathbf{B}^{-1}(\mathbf{k})}{P_{total}} \right)^n \mathbf{g}_i^{MMSE}.
\end{aligned} \tag{16.27}
$$

Then, the difference between $\mathbf{g}_i^{MMSE}$ and $\mathbf{g}_{\Delta i}^{MMSE}$ can be denoted as $\delta \mathbf{g}_i^{MMSE}$ in (16.28).

$$
\begin{aligned}
\delta \mathbf{g}_i^{MMSE} &= \sum_{m \geq 1} \left(\frac{\sum_j \delta p(j) \mathbf{R}_j \mathbf{B}^{-1}(\mathbf{k})}{P_{total}} \right)^{2m-1} \\
&\quad \times \left(\frac{\sum_j \delta p(j) \mathbf{R}_j \mathbf{B}^{-1}(\mathbf{k})}{P_{total}} - \mathbf{I} \right) \mathbf{g}_i^{MMSE}.
\end{aligned} \tag{16.28}
$$

As $\Delta < 10^{-2}$ and $\|\mathbf{g}_i^{MMSE}\|_2 = 1$, we can derive a bound for $\|\delta \mathbf{g}_i^{MMSE}\|_2$ as

$$
\begin{aligned}
\|\delta \mathbf{g}_i^{MMSE}\|_2 &\leq \left\| \sum_{m \geq 1} \left(\frac{\sum_j \delta p(j) \mathbf{R}_j \mathbf{B}^{-1}(\mathbf{k})}{P_{total}} \right)^{2m-1} \right. \\
&\quad \left. \times \left(\frac{\sum_j \delta p(j) \mathbf{R}_j \mathbf{B}^{-1}(\mathbf{k})}{P_{total}} - \mathbf{I} \right) \right\|_2 \\
&\leq \sum_{m \geq 1} \left\| \frac{\sum_j \delta p(j) \mathbf{R}_j \mathbf{B}^{-1}(\mathbf{k})}{P_{total}} \right\|_2^{2m-1} \\
&\quad \times \left\| \frac{\sum_j \delta p(j) \mathbf{R}_j \mathbf{B}^{-1}(\mathbf{k})}{P_{total}} - \mathbf{I} \right\|_2.
\end{aligned} \tag{16.29}
$$

Moreover, as we define $\tilde{k}(i) = \frac{\delta p(i)}{\Delta}$ and $\Delta < 10^{-2}$, we assume $\tilde{k}(i)\Delta \approx k(i)\Delta$. Then we can have

$$
\begin{aligned}
\frac{\sum_j \delta p(j) \mathbf{R}_j \mathbf{B}^{-1}(\mathbf{k})}{P_{total}} &= \Delta \sum_j \tilde{k}(j) \mathbf{R}_j \mathbf{B}^{-1}(\mathbf{k}) \\
&\approx \Delta \sum_j k(j) \mathbf{R}_j \mathbf{B}^{-1}(\mathbf{k}) = \Delta \mathbf{B}^{-1}(\mathbf{k}) \left(\mathbf{B}(\mathbf{k}) - \frac{1}{P_{total}} \mathbf{I} \right) \\
&= \Delta \left(\mathbf{I} - \frac{1}{P_{total}} \mathbf{B}^{-1}(\mathbf{k}) \right).
\end{aligned} \tag{16.30}
$$

Because the eigenvalues of $\frac{1}{P_{total}} \mathbf{B}^{-1}(\mathbf{k})$ lie in $(0,\ 1)$, the eigenvalues of $\frac{\sum_j \delta p(j) \mathbf{R}_j \mathbf{B}^{-1}(\mathbf{k})}{P_{total}}$ will lie in $(0,\ \Delta)$. As a result, $\|\frac{\sum_j \delta p(j) \mathbf{R}_j \mathbf{B}^{-1}(\mathbf{k})}{P_{total}}\|_2 < \Delta$ and $\|\frac{\sum_j \delta p(j) \mathbf{R}_j \mathbf{B}^{-1}(\mathbf{k})}{P_{total}} - \mathbf{I}\|_2 < 1$, which implies $\|\delta \mathbf{g}_i^{MMSE}\|_2 < \sum_{m \geq 1} \Delta^{2m-1} = \frac{\Delta}{1-\Delta^2} \simeq \Delta$. Hence, under a small perturbation in P_{total}, the optimal signature matrix can be approximately viewed as unchanging.

References

[1] A. Goldsmith, *Wireless Communications*. Cambridge, UK: Cambridge University Press, 2005.

[2] B. Wang, Y. Wu, F. Han, Y.-H. Yang, and K. Liu, "Green wireless communications: A time-reversal paradigm," *IEEE Journal on Selected Areas in Communications*, vol. 29, no. 8, pp. 1698–1710, 2011.

[3] F. Han, Y.-H. Yang, B. Wang, Y. Wu, and K. Liu, "Time-reversal division multiple access over multi-path channels," *IEEE Transactions on Communications*, vol. 60, no. 7, pp. 1953–1965, 2012.

[4] Y. Chen, Y.-H. Yang, F. Han, and K. R. Liu, "Time-reversal wideband communications," *IEEE Signal Processing Letters*, vol. 20, no. 12, pp. 1219–1222, 2013.

[5] Y. Chen, F. Han, Y.-H. Yang, H. Ma, Y. Han, C. Jiang, H.-Q. Lai, D. Claffey, Z. Safar, and K. R. Liu, "Time-reversal wireless paradigm for green Internet of Things: An overview," *IEEE Internet of Things Journal*, vol. 1, no. 1, pp. 81–98, 2014.

[6] Y.-H. Yang, B. Wang, W. S. Lin, and K. R. Liu, "Near-optimal waveform design for sum rate optimization in time-reversal multiuser downlink systems," *IEEE Transactions on Wireless Communications*, vol. 12, no. 1, pp. 346–357, 2013.

[7] M. Fink, C. Prada, F. Wu, and D. Cassereau, "Self focusing in inhomogeneous media with time reversal acoustic mirrors," *IEEE Ultrasonics Symposium Proceedings*, pp. 681–686, 1989.

[8] M. Fink, "Time reversal of ultrasonic fields. I. Basic principles," *IEEE Transactions on Ultrasonics, Ferroelectrics, and Frequency Control*, vol. 39, no. 5, pp. 555–566, 1992.

[9] F. Wu, J.-L. Thomas, and M. Fink, "Time reversal of ultrasonic fields. II. Experimental results," *IEEE Transactions on Ultrasonics, Ferroelectrics, and Frequency Control*, vol. 39, no. 5, pp. 567–578, 1992.

[10] W. Kuperman, W. S. Hodgkiss, H. C. Song, T. Akal, C. Ferla, and D. R. Jackson, "Phase conjugation in the ocean: Experimental demonstration of an acoustic time-reversal mirror," *The Journal of the Acoustical Society of America*, vol. 103, no. 1, pp. 25–40, 1998.

[11] D. Rouseff, D. R. Jackson, W. L. Fox, C. D. Jones, J. A. Ritcey, and D. R. Dowling, "Underwater acoustic communication by passive-phase conjugation: Theory and experimental results," *IEEE Journal of Oceanic Engineering*, vol. 26, no. 4, pp. 821–831, 2001.

[12] G. Lerosey, J. De Rosny, A. Tourin, A. Derode, G. Montaldo, and M. Fink, "Time reversal of electromagnetic waves and telecommunication," *Radio Science*, vol. 40, no. 6, pp. 1–10, 2005.

[13] G. Lerosey, J. De Rosny, A. Tourin, A. Derode, and M. Fink, "Time reversal of wideband microwaves," *Applied Physics Letters*, vol. 88, no. 15, p. 154101, 2006.

[14] M. Emami, M. Vu, J. Hansen, A. J. Paulraj, and G. Papanicolaou, "Matched filtering with rate back-off for low complexity communications in very large delay spread channels," *38th Asilomar Conference on Signals, Systems and Computers*, pp. 218–222, 2004.

[15] J. de Rosny, G. Lerosey, and M. Fink, "Theory of electromagnetic time-reversal mirrors," *IEEE Transactions on Antennas and Propagation*, vol. 58, no. 10, pp. 3139–3149, 2010.

[16] T. S. Rappaport, *Wireless Communications: Principles and Practice*. New York: IEEE Press, 1996.

[17] J. Zander, "Performance of optimum transmitter power control in cellular radio systems," *IEEE Transactions on Vehicular Technology*, vol. 41, no. 1, pp. 57–62, 1992.

[18] S. A. Grandhi, R. Vijavan, D. J. Goodman, and J. Zander, "Centralized power control in cellular radio systems," *IEEE Transactions on Vehicular Technology*, vol. 42, no. 4, pp. 466–468, 1993.

[19] R. D. Yates, "A framework for uplink power control in cellular radio systems," *IEEE Journal on Selected Areas in Communications*, vol. 13, no. 7, pp. 1341–1347, 1995.

[20] A. Sampath, S. P. Kumar, and J. M. Holtzman, "Power control and resource management for a multimedia CDMA wireless system," *Personal, Indoor and Mobile Radio Communications*, pp. 21–25, 1995.

[21] T. Alpcan, T. Başar, R. Srikant, and E. Altman, "CDMA uplink power control as a noncooperative game," *Wireless Networks*, vol. 8, no. 6, pp. 659–670, 2002.

[22] D. Gerlach and A. Paulraj, "Base station transmitting antenna arrays for multipath environments," *Signal Processing*, vol. 54, no. 1, pp. 59–73, 1996.

[23] F. Rashid-Farrokhi, K. Liu, and L. Tassiulas, "Transmit beamforming and power control for cellular wireless systems," *IEEE Journal on Selected Areas in Communications*, vol. 16, no. 8, pp. 1437–1450, 1998.

[24] E. Visotsky and U. Madhow, "Optimum beamforming using transmit antenna arrays," *IEEE 49th Vehicular Technology Conference*, pp. 851–856, 1999.

[25] D. N. Tse and P. Viswanath, "Downlink-uplink duality and effective bandwidths," *IEEE International Symposium on Information Theory*, p. 52, 2002.

[26] H. Boche and M. Schubert, "A general duality theory for uplink and downlink beamforming," *IEEE Vehicular Technology Conference (VTC2002-Fall)*, pp. 87–91, 2002.

[27] W. Yang and G. Xu, "Optimal downlink power assignment for smart antenna systems," *IEEE International Conference on Acoustics, Speech and Signal Processing*, pp. 3337–3340, 1998.

[28] M. Schubert and H. Boche, "Solution of the multiuser downlink beamforming problem with individual SINR constraints," *IEEE Transactions on Vehicular Technology*, vol. 53, no. 1, pp. 18–28, 2004.

[29] —, "Iterative multiuser uplink and downlink beamforming under SINR constraints," *IEEE Transactions on Signal Processing*, vol. 53, no. 7, pp. 2324–2334, 2005.

[30] Z. Shen, J. G. Andrews, and B. L. Evans, "Adaptive resource allocation in multiuser OFDM systems with proportional rate constraints," *IEEE Transactions on Wireless Communications*, vol. 4, no. 6, pp. 2726–2737, 2005.

[31] D. W. Cai, T. Q. Quek, and C. W. Tan, "Coordinated max-min SIR optimization in multicell downlink-duality and algorithm," *IEEE International Conference on Communications (ICC)*, pp. 1–6, 2011.

[32] D. W. Cai, C. W. Tan, and S. H. Low, "Optimal max-min fairness rate control in wireless networks: Perron-Frobenius characterization and algorithms," *INFOCOM*, pp. 648–656, 2012.

[33] Y. Huang, C. W. Tan, and B. Rao, "Joint beamforming and power control in coordinated multicell: Max-min duality, effective network and large system transition," *IEEE Transactions on Wireless Communications*, vol. 12, no. 6, pp. 2730–2742, 2013.

[34] E. Seneta, *Non-negative Matrices and Markov Chains*. New York: Springer Science & Business Media, 2006.

[35] A. Arkhangel'Skii, V. Fedorchuk, D. O'Shea, and L. Pontryagin, *General Topology I: Basic Concepts and Constructions Dimension Theory*. New York: Springer Science & Business Media, 2012, vol. 17.

[36] C. W. Tan, M. Chiang, and R. Srikant, "Maximizing sum rate and minimizing MSE on multiuser downlink: Optimality, fast algorithms and equivalence via max-min SINR," *IEEE Transactions on Signal Processing*, vol. 59, no. 12, pp. 6127–6143, 2011.

[37] D. W. Cai, T. Q. Quek, and C. W. Tan, "A unified analysis of max-min weighted SINR for MIMO downlink system," *IEEE Transactions on Signal Processing*, vol. 59, no. 8, pp. 3850–3862, 2011.

[38] D. W. Cai, T. Q. Quek, C. W. Tan, and S. H. Low, "Max-min SINR coordinated multipoint downlink transmission duality and algorithms," *IEEE Transactions on Signal Processing*, vol. 60, no. 10, pp. 5384–5395, 2012.

[39] S. Friedland, S. Karlin *et al.*, "Some inequalities for the spectral radius of non-negative matrices and applications," *Duke Mathematical Journal*, vol. 42, no. 3, pp. 459–490, 1975.

[40] Q. Xu, Y. Chen, and K. J. R. Liu, "Combating strong–weak spatial–temporal resonances in time-reversal uplinks," *IEEE Transactions on Wireless Communications*, vol. 15, no. 1, pp. 568–580, 2016.

17 Time-Reversal Massive Multipath Effect

The explosion of mobile data traffic calls for new efficient 5G technologies. Massive MIMO, which has shown the great potential in improving the achievable rate with a very large number of antennas, has become a popular candidate. However, the requirement of deploying a large number of antennas at the base station may not be feasible in indoor scenarios due to the high implementation complexity. Does there exist a good alternative that can achieve similar system performance to massive MIMO for indoor environment? In this chapter, we show that by using time-reversal signal processing, with a sufficiently large bandwidth, one can harvest the massive multipaths naturally existing in a rich-scattering environment to form a large number of virtual antennas and achieve the desired massive multipath effect with a single antenna. We answer the preceding question by analyzing the time-reversal massive multipath effect (TRMME) and the achievable rate with some waveforms. We also derive the corresponding asymptotic achievable rate under a massive multipath setting. Experiment results based on real indoor channel measurements show the massive multipaths can be revealed with a sufficiently large bandwidth in a practical indoor environment. Moreover, based on our experiments with real indoor measurements, the achievable rate of the time-reversal wideband system is evaluated.

17.1 Introduction

While the past few decades have witnessed the monumental success of mobile and wireless access to the Internet, the proliferation of new mobile communication devices, such as smartphones and tablets, has in turn led to an exponential growth in network traffic. According to the most recent Cisco Visual Networking Index (VNI) annual report [1], the global mobile data traffic grew 69% and the number of mobile devices increased almost half a billion (497 million) in 2014. It is also predicted in the report that the global mobile data traffic will increase nearly tenfold between 2014 and 2019. The demand for supporting the fast-growing consumer data rates urges wireless service providers and researchers to seek a new efficient radio access technology, which is the so-called 5G technology, beyond what current 4G LTE can provide. Besides ultra-densification and mmWave, massive multiple-input multiple-output (MIMO) is one of "big three" 5G technology [2], which can offer multifold benefits such as enormous enhancement in spectral efficiency and power efficiency [3] and simple transmit/receiver structures due to the quasi-orthogonal nature [4]. These benefits make massive MIMO one of the five disruptive technology directions for 5G communication [5].

Even though the benefits of massive MIMO seem very promising, several critical challenges must first be addressed before it can be implemented in practice. First of all, a challenging task is the analog front-end design [6], for example, each tiny antenna needs its own power amplifier and analog-to-digital convertor (ADC). Moreover, the antenna correlation and mutual coupling due to the increasing number of antennas has to be carefully addressed as well [7, 8]. The researchers in Lund University built a 100-antenna MIMO testbed, and the size is $0.8 \times 1.2 \times 1$ m with 300 kg weight and 2.5 kW average power consumption [9]. Considering the requirement of deploying a large amount of antennas, a massive MIMO system has a high implementation cost in indoor scenarios. It is expected that 95% of data traffic will come from indoors in a few years [10], therefore a natural question to ask is: Does there exist a good alternative that can achieve similar system performance to massive MIMO for indoor environments? The answer is yes, and the time-reversal (TR) technology is potentially a counterpart of massive MIMO in indoor scenarios.

It is well known that radio signals will experience many multipaths due to the reflection from various scatters, especially in indoor environments. TR's focusing effect is in essence a spatial–temporal resonance effect that brings all the multipaths to arrive at a particular location at a specific moment. Such a phenomenon will allow us to utilize the naturally existing multipaths as virtual antennas to realize the massive multipath effect, which is a counterpart of massive MIMO effect, even with a single antenna. As shown in Figure 17.1, TR inherently treats the multipaths in the environment as virtual antennas, similar to MIMO that uses multiple antennas for better spatial multiplexing. In essence, if cooperation of users, e.g., cooperative communications, is a distributed way of achieving MIMO effect of high diversity, then TR is similarly a distributed way to achieve the massive MIMO effect through utilizing the multipaths as virtual antennas. The TR waveform is nothing but to control each multipath (virtual antenna).

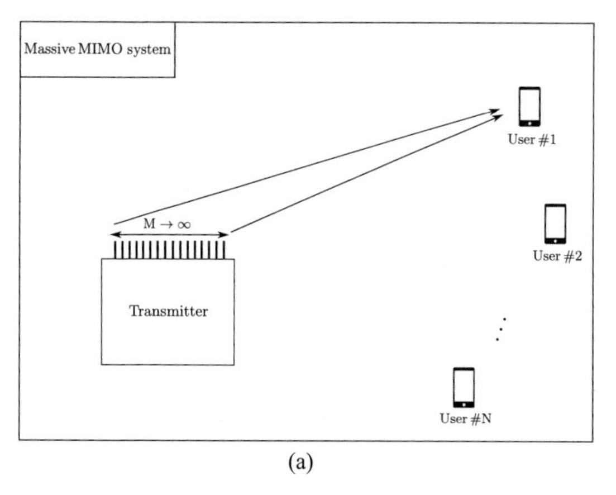

(a)

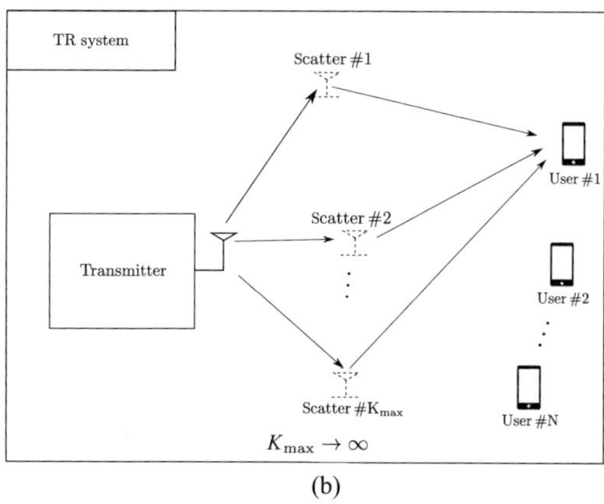

(b)

Figure 17.1 Comparison between (a) massive MIMO and (b) TR system.

In order to harvest the multipaths, the transmit power and bandwidth can be utilized. More specifically, the maximum number of observable multipaths given by an environment increases with the transmit power. Once the power is fixed, the maximum number of observable multipaths is also fixed. In addition, more multipaths can be resolved with the increase of bandwidth because of the better time resolution. Based on the real indoor ultra-wide-band (UWB) channel measurement (both LOS and NLOS) in [11] and [12], around 60–80 independent multipaths can be revealed with a sufficiently large bandwidth. Later in Section 17.6, we will discuss how to realize the massive multipath in a practical indoor environment.

TR technology is a promising candidate for indoor communication, but it requires wide bandwidth to achieve good time resolution. The wideband signal naturally requires the high sampling rate based on the Nyquist sampling theorem, which leads to a heavy computation burden in terms of processing. Fortunately, as indicated by Moore's Law, the more powerful analog-to-digital-converter (ADC) and digital signal processor (DSP) reduce the wideband signal processing cost dramatically [13]. Moreover, researchers and engineers are currently searching for new available wide-band and re-allocating bandwidth for 5G technology [2]. For TR technology, it may use the spectrum of ultra-wide-band (UWB) or mmWave band. Based on the existing study at high frequencies, there still exists a large amount of multipaths, which is essential for TR communication. For example, based on the building penetration and reflection measurements of 28 GHZ in NYC [14], the RF energy is mostly retained within buildings due to low attenuation and high reflection through indoor materials. Moreover, the delay spread for indoor 60 GHz channels ranges between 30 ns and 70 ns [15], which indicates a multipath-rich environment. Even though the spectral efficiency of TR technique is not that high, it becomes more and more important to reduce complexity, operate energy consumption, and offer other benefits given the potential wide bandwidth, especially in indoor scenarios.

By exploiting the massive number of virtual antennas, the TR system can achieve superior focusing effect in the spatial–temporal domain, resulting in the promising performance as an indoor communication candidate for 5G. Moreover, the implementation complexity of a TR system is much lower because it utilizes the environment as the virtual antenna array and computing resource. Specifically, in this chapter, we consider a time-reversal division multiple Access (TRDMA) downlink communication system [16] to demonstrate the TR massive multipath effect (TRMME) under typical waveforms, i.e., the basic TR, zero-forcing (ZF), and minimum mean square error (MMSE) waveforms. We further derive the asymptotic achievable rate performance as the number of observable multipaths grows to infinity. Later, we discuss the approach to realize massive multipaths based on real-world indoor channel measurements. Through the experiments with real indoor measurements, the achievable rate of a TR wideband system with a single antenna is evaluated.

The rest of this chapter is organized as follows. We first discuss the existing related work in Section 17.2. The system model is discussed in Section 17.3. In Section 17.4, the notion of TRMME is introduced, assuming that the TR system has the ability to reveal infinite multipaths in a rich-scattering environment. In Section 17.5, the expected achievable rate of the TR system with typical waveforms is investigated. Moreover, the asymptotic achievable rates with these waveforms is derived in a massive multipath setting. The approach to realize massive multipaths in a practical indoor environment is discussed based on the real-world channel measurements in Section 17.6.

Notations: $|\cdot|$, $(\cdot)^T$, and $(\cdot)^\dagger$ stand for absolute value, transpose, and conjugate transpose, respectively. The boldface lowercase letter **a** and the boldface uppercase letter **A** represent vector in column form and matrix, respectively. $\|\mathbf{a}\|$ denotes the Euclidean norm of a vector. $\mathbf{a}^{[D]}$ denotes the up-sampled vector by inserting $D-1$ zeros between two adjacent elements in the vector **a**. We denote $(\mathbf{a}*\mathbf{b})$ as the linear convolution between two vectors **a** and **b**. $[\mathbf{A}]_{m,n}$ stands for the element in the mth row and the nth column of the matrix **A**. Finally, we denote $\xrightarrow{d}$ as the convergence in distribution.

17.2 Related Work

The TR technology was first introduced to compensate the delay distortion on wired transmission lines by Bogert from Bell Labs in the fifties [17]. Since then, it has been applied in various areas including ultrasonics [18], acoustical imaging [19], electromagnetic imaging [20], and underwater acoustic communication [21]. More recently, TR has drawn more and more attention from researchers in the wireless communications field [22–24]. Under a rich-scattering environment, a TR communication system is shown to have the spatial–temporal focusing effect and thus work as an ideal platform for green wireless communications [25, 26] in terms of lower power consumption and less radio pollution. A time-reversal division multiple access (TRDMA) scheme is proposed in [16], which utilizes the location-specific waveform to separate different users' signal. It is shown in [16, 27] that the TR communication system can be extended to multiple-antenna scenarios easily, and more advanced waveform design can be implemented to further suppress the ISI and interuser interference (IUI) to achieve a higher data rate [28, 29]. The potential application of TR technology in the Internet of Things is discussed in [30].

A closely related technology to TR is the code division multiple access (CDMA). While TR technology utilizes the designed waveforms to distinguish multiple users, CDMA employs spread-spectrum to allocate distinct orthogonal codes to multiple users [31]. Compared with the Rake receiver deployed to counter the effect of multipaths in the CDMA system, TR technology harvests the multipaths in the environment to perform beamforming with the appropriate precoding at the transmitter, which results in much lower complexity at the receiver side.

Time-reversal is not a new terminology in MIMO technology as well. First of all, time-reversal beamforming is well known as conjugate beamforming in MIMO systems when the system bandwidth is small [32]. Then, for wide-band, frequency-selective channel, OFDM can rigorously decompose the channel into parallel independent narrow-bandwidth subcarriers, where TR precoding can be applied [33, 34]. TR can be also employed as the precoding scheme directly for a single carrier wideband system [35, 36].

The focus of this chapter is not on the combination of massive MIMO and TR technologies. Instead, we show that TR technology itself is a promising approach to realize the massive multipath effect, which is similar to massive MIMO effect, for indoor communications. Together with the theoretic analysis of the TRMME, the idea of realizing massive virtual antennas with a single physical one and the approach to resolve the multipaths with the increase of bandwidth in an indoor environment constitute the novelty of this chapter.

17.3 System Model

In this chapter, we consider a time-reversal downlink system where one transmitter simultaneously communicates with N distinct receivers through the TRDMA technique [16]. We assume that both the transmitter and receivers are equipped with one single antenna. However, the results can be easily extended to the multiple-antenna scenario.

17.3.1 Channel Model

Suppose there are totally $K_{\max}$ independent multipaths from the transmitter to the jth receiver, then the channel $h_j(t)$ can be written as

$$h_j(t) = \sum_{k=1}^{K_{\max}} \tilde{h}_{j,k}\delta(t-\tau_k), \tag{17.1}$$

where $\tilde{h}_{j,k}$ and τ_k are the complex channel gain and path delay of the kth path, respectively. Note that the delay spread of the channel is given by $\tau_C = \tau_{K_{\max}}$.

Let W be the bandwidth of the TR system. Through Nyquist sampling, the discrete channel responses is

$$h_j[n] = \int_{n\tau_p-\tau_p}^{n\tau_p} p(n\tau_p-\tau)h_j(\tau)d\tau, \tag{17.2}$$

where $p(t)$ is the pulse with main lobe $\tau_p = 1/W$.

Through (17.2), a L-tap channel $\mathbf{h}_j = [h_j[1], h_j[2], \ldots, h_j[L]]^T$ with $L = round(\tau_C W)$ can be resolved for the link between the transmitter and the jth receiver as follows

$$\mathbf{h}_j = \left[h_{j,1}, h_{j,2}, \ldots, h_{j,L}\right], \tag{17.3}$$

where $h_{j,i}$ is the complex channel gain of the ith tap, and $h'_{j,i}s$ are independent for all $i \in [1, L]$ and $j \in [1, N]$.

Suppose that there are K nonzero elements in the L-tap channel $\mathbf{h}_j$. When the bandwidth W is small, all elements in $\mathbf{h}_j$ are generally nonzero, i.e., $K = L$. On the other hand, when W is sufficiently large, the side lobes of $p(t)$ become negligible and thus there are at most $K = K_{\max} < L$ non-zero elements in $\mathbf{h}_j$. Let $\phi_{K_{\max}}$ be the nonzero multipath set, which reflects the physical patterns of scatters distribution in the environment. Then, $h_j[k] = 0$ for $k \notin \phi_{K_{\max}}$, and for $k \in \phi_{K_{\max}}$, $h_j[k]$ is a complex random variable with zero mean and variance σ_k^2.

Prior to the TR-transmission, a pseudo random sequence is sent to the transmitter from the receiver, based on which the channel state information (CSI) $\mathbf{h}_j$ is estimated. By cross-correlating the received signal with the known pseudo random sequence, the power of CSI is boosted thus maintaining the good CSI quality. Due to the more powerful DSP and the efficient Golay correlator [37], the CSI estimation is obtained quickly in terms of time consumption by converting the multiplication operation into addition/subtraction. In addition, based on the real measurement in [38], the CSI is quite stationary given only slightly changing of the environment, which indicates the channel need not to be reprobed very frequently, and the overhead price of channel probing is very small.

Note that pilot contamination, when a lot of users simultaneously channel probe to the base station, will cause the performance degradation in the TR system due to the nonideal CSI. In the following, we assume the estimated CSI is perfect.

17.3.2 TRDMA Downlink Communication

In the TR system, the transmitter simultaneously communicates with multiple receivers. Specifically, as shown in Figure 17.2, the information to be transmitted to the jth receiver, denoted as X_j, is first up-sampled by a backoff factor D to alleviate the interference, and then precoded by a waveform $\mathbf{g}_j$. Actually, the symbol rate is lower down by D to suppress the ISI caused by the multipath channel. Note that multiple designs of the waveform such as basic TR waveform [25], zero-forcing (ZF) waveform [39], and minimal mean square error (MMSE) waveform [28] can be utilized, and the details will be discussed in the next section. After that, all signals to different receivers are mixed together as follows

$$S[k] = \sum_{i=1}^{N} \left(X_i^{[D]} * \mathbf{g}_i \right) [k], \tag{17.4}$$

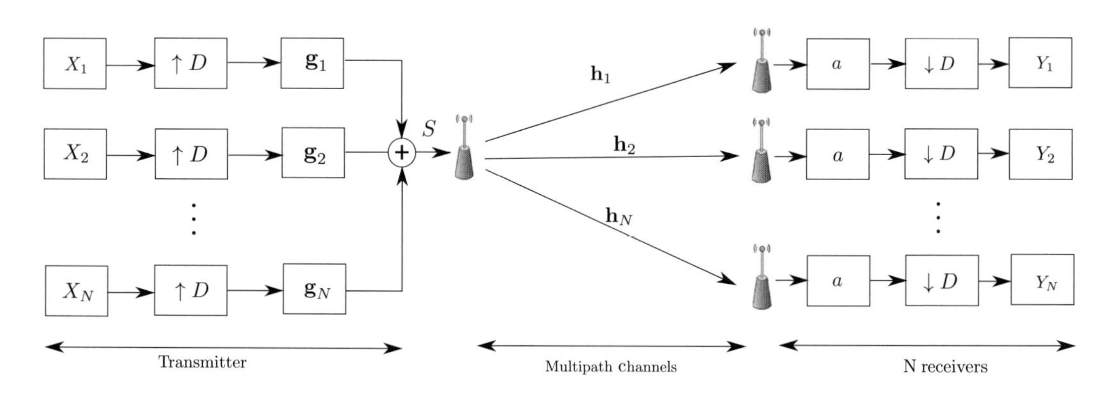

Figure 17.2 TRDMA system.

where

$$X_i^{[D]}[k] = \begin{cases} X_i[k/D], & \text{if } mod(k, D) = 0, \\ 0, & \text{otherwise.} \end{cases} \tag{17.5}$$

The mixed signal is broadcast to all receivers through the rich-scatter environment. At the receiver side, the jth receiver simply scales the received signal and down-samples it to obtain the estimated signal Y_j as follows

$$\begin{aligned} Y_j[k] = (\mathbf{h}_j * \mathbf{g}_j)[L]X_j\left[k - \frac{L}{D}\right] + \sum_{l=1, l \neq L/D}^{\frac{2L-1}{D}} (\mathbf{h}_j * \mathbf{g}_j)[Dl]X_j[k-l] \\ + \sum_{i=1, i \neq j}^{N} \sum_{l=1}^{\frac{2L-1}{D}} (\mathbf{h}_j * \mathbf{g}_i)[Dl]X_i[k-l] + n_j[k]. \end{aligned} \tag{17.6}$$

Without loss of generality, we assume with the typical waveforms $(\mathbf{h}_j * \mathbf{g}_j)$ has the resonating effect at time index L. Then the first term in (17.6) is the desired signal, the second term is the ISI, the third term is the IUI, and the last term is the noise.

The (17.6) can be rewritten by replacing the convolution as inner product as follows

$$\begin{aligned} Y_j[k] = \mathbf{H}_j^{(\frac{L}{D})}\mathbf{g}_j X_j\left[k - \frac{L}{D}\right] + \sum_{l=1, l \neq L/D}^{(2L-1)/D} \mathbf{H}_j^{(l)}\mathbf{g}_j X_j[k-l] \\ + \sum_{l=1}^{(2L-1)/D} \mathbf{H}_j^{(l)} \left(\sum_{i=1, i \neq j}^{N} \mathbf{g}_i X_i[k-l] \right) + n_j[k], \end{aligned} \tag{17.7}$$

where $\mathbf{H}_j^{(m)}$ is the mth row of the $(2L-1)/D \times L$ matrix $\mathbf{H}_j$ decimated by rows of Toeplitz matrix, which can be written in (17.8).

$$\mathbf{H}_j = \begin{pmatrix} h_j[D] & h_j[D-1] & \cdots & h_j[1] & 0 & \cdots & \cdots & 0 \\ h_j[2D] & h_j[2D-1] & \cdots & \cdots & h_j[1] & 0 & \cdots & 0 \\ \vdots & \vdots & \ddots & \ddots & \ddots & \ddots & \ddots & \vdots \\ h_j[L] & h_j[L-1] & \cdots & \cdots & \cdots & \cdots & \cdots & h_j[1] \\ \vdots & \vdots & \ddots & \ddots & \ddots & \ddots & \ddots & \vdots \\ 0 & \cdots & 0 & h_j[L] & \cdots & \cdots & h_j[L-D+1] & h_j[L-2D] \\ 0 & \cdots & \cdots & 0 & h_j[L] & \cdots & h_j[L-D+1] & h_j[L-D] \end{pmatrix}. \tag{17.8}$$

Therefore, $\mathbf{H}_j^{(\frac{L}{D})}$ is the time-reversal channel

$$\mathbf{H}_j^{(\frac{L}{D})} = \left[h_j[L]\, h_j[L-1] \cdots h_j[1]\right]. \tag{17.9}$$

17.3.3 Expected Achievable Rate for Individual User

Let P and P_n be the average transmitting power and noise power, respectively, and $(\cdot)^\dagger$ represent the conjugate transpose operator. According to (17.7) and the uplink-downlink duality [40–42], the achievable rate of the jth receiver can be derived using its dual uplink format, where the uniform power allocation is assumed. Then we take an expectation of the downlink achievable rate as shown in (17.10). In the rest of the chapter, we analyze the expected achievable rate of the TR system.

$$R_j = \frac{W}{D} \mathbb{E}\left[\log_2\left(1+\frac{\frac{P}{N}\mathbf{g}_j^\dagger \mathbf{H}_j^{(\frac{L}{D})\dagger}\mathbf{H}_j^{(\frac{L}{D})}\mathbf{g}_j}{\frac{P}{N}\mathbf{g}_j^\dagger\left(\mathbf{H}_j^\dagger\mathbf{H}_j - \mathbf{H}_j^{(\frac{L}{D})\dagger}\mathbf{H}_j^{(\frac{L}{D})}\right)\mathbf{g}_j + \frac{P}{N}\sum_{i=1,i\neq j}^{N}\mathbf{g}_j^\dagger\mathbf{H}_i^\dagger\mathbf{H}_i\mathbf{g}_j + P_n}\right)\right]. \tag{17.10}$$

17.4 Time-Reversal Massive Multipath Effect

In this section, we derive a time-reversal massive multipath effect (TRMME) for the TR technique in a rich-scattering environment. Similar to the massive MIMO effect in massive MIMO given an excessive amount of antennas [4], the multipath profile of different users in the TR system will also be orthogonalized given massive independent multipaths. Considering the channel delay spread in wideband system, the channel matrix considered in the following is the combination of decimated Toeplitz matrices in (17.8).

Theorem 17.4.1 *(Time-Reversal Massive Multipath Effect): Under the asymptotic setting where* $K = K_{\max} \to \infty$,

$$\begin{cases} [\mathbf{Q}\mathbf{Q}^\dagger]_{m,n} \xrightarrow{d} 0, & \text{if } m \neq n \\ \frac{[\mathbf{Q}\mathbf{Q}^\dagger]_{m,m}}{\lambda_m} \xrightarrow{d} 1, & \text{otherwise} \end{cases}, \tag{17.11}$$

where $\mathbf{Q} = [\mathbf{H}_1^T, \mathbf{H}_2^T, \ldots, \mathbf{H}_N^T]^T$, $\xrightarrow{d}$ *represents the convergence in distribution, and* $\lambda_m = \|\mathbf{h}_j\|^2$ *if* $m = (j-1)(2L-1)/D + L/D$.

The proof for Theorem 17.4.1 is listed in Appendix A. Because $\mathbf{Q}$ is the combination of CSIs from the transmitter to N receivers, the term $\mathbf{Q}\mathbf{Q}^\dagger$ represents the correlation matrix of these N CSIs. Therefore, the derived TRMME implies that the CSIs to N receivers become orthogonal to each other under the rich-multipath setting. Based on the indoor measurements with the TR prototype in [38], TR with 125 MHz bandwidth is capable to formulate a spatial focusing ball as shown in Figure 17.3. With the derived TRMME, the focusing ball of TR naturally shrinks to a pinpoint in a rich-scattering environment with a sufficiently large bandwidth, which is also predicted and observed in the massive MIMO system. Therefore, the derived TRMME is a counterpart of the massive MIMO effect in indoor scenarios.

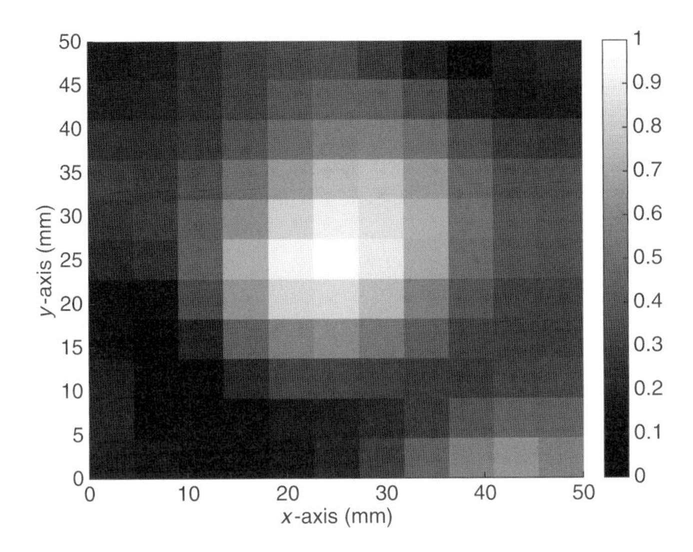

Figure 17.3 Spatial focusing ball with 125 MHz bandwidth.

The assumption that $K_{\max} \rightarrow \infty$ is just for analyzing the asymptotical achievable rate of the TR system as the assumption $M_t \rightarrow \infty$ in early massive MIMO works. In practice, we only need that $K_{\max}$ is large enough to achieve the massive multipath effect. Based on the real indoor measurement in Section 17.6.2, we have demonstrated that the number of the resolved multipaths in a typical indoor environment is large enough given a sufficiently large bandwidth. Even though $K_{\max}$ is a fixed value given the power and environment, there still exist other methods to realize massive multipaths. Because the TR and MIMO technology are not mutually exclusive, the independent multipaths can be easily scaled up by adding a few antennas. How to realize massive multipaths is discussed with real indoor measurement later in Section 17.6. In the following, the asymptotic performance of TR technologies in a rich-scattering environment is derived based on the TRMME.

17.5 Expected Achievable Rate under Different Waveforms

In this section, we analyze the asymptotic rates of TR technology. First, we derive the expected achievable rate under typical waveforms: basic TR waveform [25], ZF waveform [39], and MMSE waveform [28]. Then, the asymptotical achievable rate with the three waveforms is further derived based on the TRMME derived in Section 17.4.

17.5.1 Expected Achievable Rate

The three waveforms are shown in the following,

$$\mathbf{g}_j = \begin{cases} \mathbf{H}_j^{(L/D)\dagger} / \|\mathbf{H}_j^{(L/D)}\|, & \text{Basic TR} \\ c_{ZF}\mathbf{Q}^\dagger(\mathbf{Q}\mathbf{Q}^\dagger)^{-1}\mathbf{e}_{l_j}, & \text{ZF} \\ c_{MMSE}(\mathbf{Q}^\dagger\mathbf{Q} + \frac{1}{p_u}\mathbf{I})^{-1}\mathbf{Q}^\dagger\mathbf{e}_{l_j}, & \text{MMSE} \end{cases} \tag{17.12}$$

where c_{ZF} and c_{MMSE} are normalization constants, $\mathbf{Q} = [\mathbf{H}_1^T, \mathbf{H}_2^T, \cdots, \mathbf{H}_N^T]^T$, $\mathbf{e}_{l_j}$ is an elementary vector with $l_j = (j-1)(2L-1)/D + L/D$, $\mathbf{I}$ is the identity matrix, and p_u is the transmitting signal-to-noise ratio (SNR) of each user defined as

$$p_u = \frac{P}{NP_n}. \tag{17.13}$$

With the definition of $\mathbf{Q}$ and $\mathbf{e}_{l_j}$ earlier, we have

$$\mathbf{Q}^\dagger \mathbf{e}_{l_j} = \mathbf{H}_j^{(L/D)\dagger}. \tag{17.14}$$

Note that under the multipath-rich scenario, ZF waveform can completely cancel out the interference given a large amount of independent multipaths. In addition, MMSE waveform has a simpler closed form solution with the fixed dual uplink power allocation [28].

Considering the more and more power DSP and the small time consumption of channel probing, the overhead of channel probing and waveform design is ignored in the following achievable rate analysis.

THEOREM 17.5.1 ***(Expected Achievable Rate):*** *The expected achievable rate of the TR system with basic TR waveform, ZF waveform, and MMSE waveform can be written as follows*

$$\begin{aligned}
R_j^{Basic} &= \frac{W}{D}\mathbb{E}\left[\log_2\left(1+\frac{p_u\|\mathbf{h}_j\|^4}{p_u([\mathbf{Q}\mathbf{Q}^\dagger\mathbf{Q}\mathbf{Q}^\dagger]_{l_j,l_j} - \|\mathbf{h}_j\|^4) + \|\mathbf{h}_j\|^2}\right)\right],\\
R_j^{ZF} &= \frac{W}{D}\mathbb{E}\left[\log_2\left(1+\frac{p_u}{[(\mathbf{Q}\mathbf{Q}^\dagger)^{-1}]_{l_j,l_j}}\right)\right],\\
R_j^{MMSE} &= \frac{W}{D}\mathbb{E}\left[\log_2\left(\frac{1}{\left[\left(\mathbf{I}+p_u\mathbf{Q}\mathbf{Q}^\dagger\right)^{-1}\right]_{l_j,l_j}}\right)\right].
\end{aligned} \tag{17.15}$$

The proof for Theorem 17.5.1 is listed in Appendix B. Note that the equations in Theorem 17.5.1 are not in closed-form due to the general channel model assumed in this chapter. Theorem 17.5.1 serves as a starting point to derive the asymptotical expected achievable rate, which will be shown in later sections. Even though (17.15) seems similar to those for MIMO MRC/ZF/MMSE receivers, the matrix $\mathbf{Q}$ is different from the channel profile matrix in MIMO system, which results in significantly different derivation of the asymptotical performance in the TR system. More specifically, due to the large channel delay spread in the TR system, there exists ISI. Therefore, backoff factor D is adopted in this chapter, and the channel profile $\mathbf{H}_i$ becomes the decimated Toeplitz matrix, which is much more complicated than that in MIMO system. Furthermore, it is the first work analyzing the asymptotical achievable rate for TR system with various waveform design methods with considering the ISI in practical system.

From Theorem 17.5.1, we can see that the expressions of expected achievable rate under different waveforms are closely related to $\mathbf{QQ}^\dagger$ and $\left[\mathbf{QQ}^\dagger\mathbf{QQ}^\dagger\right]_{l_j,l_j}$. Actually, the asymptotical property of $\mathbf{QQ}^\dagger$ has been studied previously as the TRMME. In the following section, we will further explore the property of $\left[\mathbf{QQ}^\dagger\mathbf{QQ}^\dagger\right]_{l_j,l_j}$ under a massive multipath setting, i.e., when $K_{\max} \to \infty$, and study the corresponding asymptotic expected achievable rates with different waveforms.

17.5.2 Asymptotic Performance

We derive the asymptotic property of $\left[\mathbf{QQ}^\dagger\mathbf{QQ}^\dagger\right]_{l_j,l_j}$ in the following Lemma.

LEMMA 17.5.1 *Under the asymptotic setting where $K = K_{\max} \to \infty$, we have*

$$\limsup_{K_{\max}\to\infty} \frac{\left[\mathbf{QQ}^\dagger\mathbf{QQ}^\dagger\right]_{l_j,l_j} - \|\mathbf{h}_j\|^4}{\sum_{k=1}^{K_{\max}} \sigma_k^2} = \alpha, \tag{17.16}$$

where $\alpha = 2N/D$.

The proof for Lemma 17.5.1 is listed in Appendix C. In a massive MIMO system, when the number of antennas grows large, the random channel vectors between the users and the base station become pairwisely orthogonal [43]. Similarly, in a TR system, when the number of multipaths grows, the random channel vectors between the receivers and the transmitter are also pairwise orthogonal, as shown in the TRMME. Different from the matched filter beamforming in massive MIMO system, basic TR waveform cannot completely remove the interference due to the channel delay spread in the wideband system. Therefore, the analysis of interference in Lemma 17.5.1 is needed for the derivation of the asymptotic expected achievable rate with basic TR waveform.

Based on the TRMME and Lemma 17.5.1, we can analyze the asymptotic expected achievable rate under different waveforms, and the results are summarized in the following Theorem.

THEOREM 17.5.2 *When $K_{\max} \to \infty$, the asymptotic expected achievable rate with the ZF waveform and MMSE waveform satisfy that*

$$\begin{aligned}\lim_{K_{\max}\to\infty} \frac{R_j^{ZF}}{W/D} &= \lim_{K_{\max}\to\infty} \frac{R_j^{MMSE}}{W/D} \\ &= \mathbb{E}\left[\log_2\left(1 + p_u\|\mathbf{h}_j\|^2\right)\right],\end{aligned} \tag{17.17}$$

while the asymptotic expected achievable rate with the basic TR waveform satisfies the following equation,

$$\liminf_{K_{\max}\to\infty} \frac{R_j^{Basic}}{W/D} = \mathbb{E}\left[\log_2\left(1 + \frac{p_u\|\mathbf{h}_j\|^2}{\frac{p_u\alpha\left(\sum_{k=1}^{K_{\max}}\sigma_k^2\right)^2}{\|\mathbf{h}_j\|^2} + 1}\right)\right]. \tag{17.18}$$

The proof for Theorem 17.5.2 is listed in Appendix D. From Theorem 17.5.1, we can see that the ZF and MMSE waveforms generally outperform the basic TR waveform in terms of expected achievable rate. However, when D is sufficiently large so that $\frac{\alpha\left(\sum_{k=1}^{K_{\max}}\sigma_k^2\right)^2}{\|\mathbf{h}_j\|^2}$ goes to zero, this is the case when the ISI and IUI are eliminated, then the basic TR waveform can achieve the same asymptotic expected achievable rate with ZF and MMSE waveforms.

17.6 Simulations and Experiments

In this section, we conduct simulations and experiments to evaluate the expected asymptotical performance of a TR system under various settings. We assume that the N receivers are uniformly, randomly distributed and share the same channel model, which is discussed in Section 17.3. Because more received power will be captured within the multipath-rich environment, we assume the expected channel gain as an increasing function of the number of independent multipaths $K_{\max}$.

17.6.1 Asymptotical Performance

We first validate our theoretical analysis in Theorem 17.5.1. The y-axis is DR_j/W, where R_j is the expected achievable rate of the jth receiver, D is the backoff factor, and W is the system bandwidth. Because the channel gain is assumed to be an increasing function of $K_{\max}$, the asymptotical performance would increase with $K_{\max}$ as well. The case when D is not sufficiently large, e.g., $D = K_{\max}$ is first investigated. The expected asymptotical performance of each receiver is shown in Figure 17.4 with $p_u = 5$ dB and different N. From Figure 17.4, we can observe that the performance using ZF and MMSE waveforms converges to the same limit quickly as $K_{\max}$ increases. Also, there is a gap between the asymptotic limit of ZF/MMSE waveform and the lower bound of the basic TR waveform. This is mainly because when the basic TR waveform is used and the D is not large enough, there exists residual ISI and IUI and $\frac{\alpha\left(\sum_{k=1}^{K_{\max}}\sigma_k^2\right)^2}{\|\mathbf{h}_j\|^2}$ cannot be negligible. By comparing the results with $N = 6$ and $N = 20$, we notice that the gap becomes even larger when N increases, which is due to the increase of $\alpha = 2N/D$.

We also compare the asymptotical performance of basic TR, ZF, and MMSE waveforms with a larger D. It can be seen in Figure 17.8 that the gap between the asymptotic performance of ZF/MMSE waveform and that of the basic TR waveform becomes much smaller when D and $K_{\max}$ are both sufficiently larger. Such a phenomenon is mainly due to less severe ISI and IUI and a much smaller $\frac{\alpha\left(\sum_{k=1}^{K_{\max}}\sigma_k^2\right)^2}{\|\mathbf{h}_j\|^2}$. Therefore, the basic TR waveform can achieve the same optimal asymptotic expected achievable rate with ZF and MMSE waveforms with sufficiently large D.

Note that p_u is fixed as 5 dB in the simulations, which implies that the trend in Figures 17.4 and 17.8 also applies to the energy efficiency. In other words, the energy efficiency of the TR system increases with $K_{\max}$.

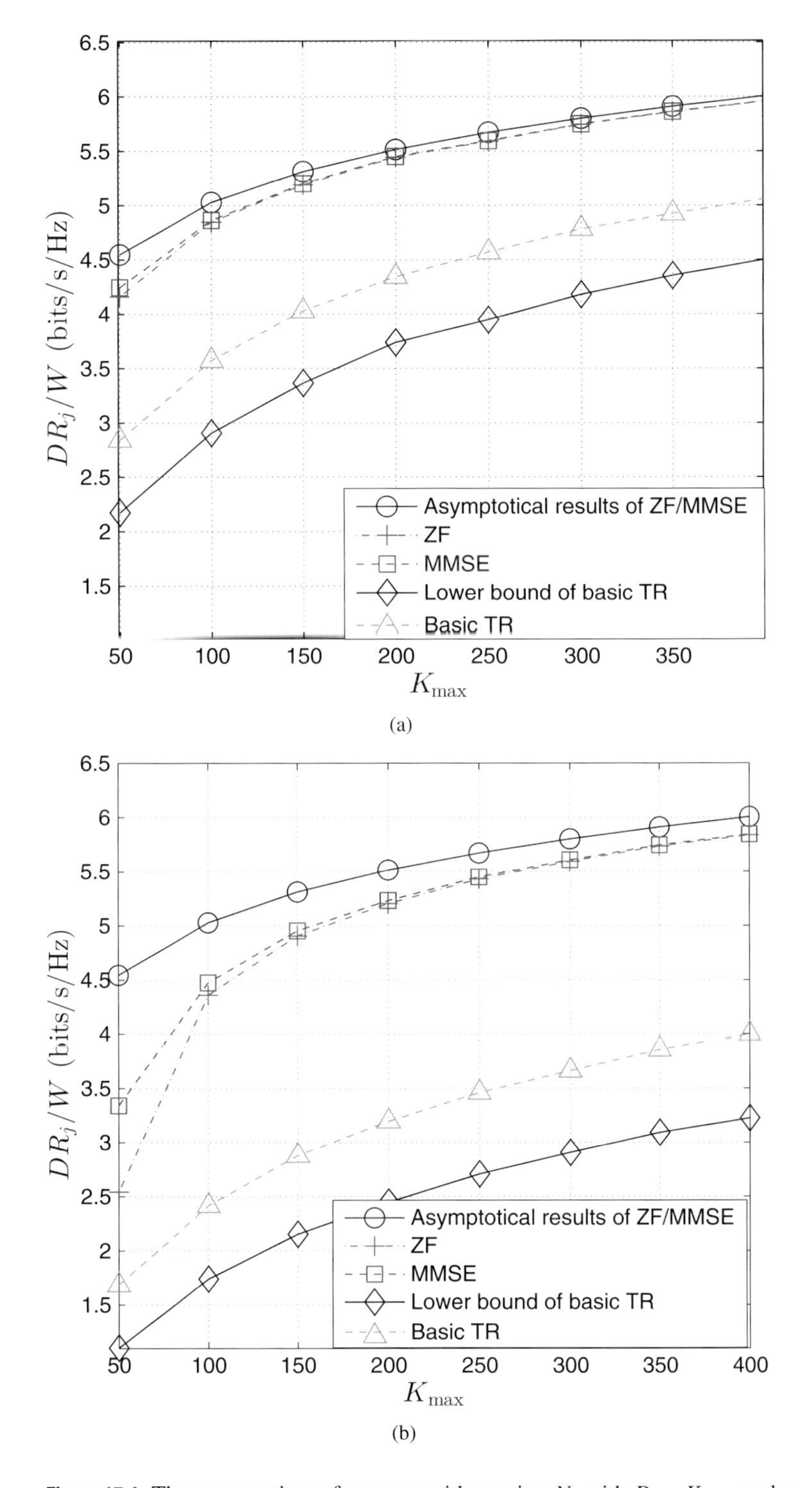

Figure 17.4 The asymptotic performance with varying N, with $D = K_{\max}$ and $p_u = 5$ dB: (a) $N = 6$ and (b) $N = 20$.

17.6.2 The Number of Observable Independent Multipaths K in a Typical Indoor Environment

To achieve the asymptotic performance in Theorem 17.5.1 requires the TR system to operate in a multipath-rich environment. In this section, we investigate the property of K in a typical indoor environment using real-world measurements. First, we demonstrate that, in a typical office, the number of resolvable multipaths is large with a sufficiently large bandwidth. Then, the approach to increase $K_{\max}$ is further discussed and validated through real measurements.

We use two Universal Software Radio Peripherals (USRPs) as channel sounders to probe the channel in a typical office room, whose floor plan is shown in Figure 17.5. As shown in the figure, TX is placed on a grid structure with 5 cm resolution, and RX is placed at the corner. With two USRPs, we scan the spectrum, e.g., from 4.9 GHz to 5.9 GHz, to acquire the channel impulse response with a bandwidth of 10 MHz–1 GHz.

We employ eigenvalue analysis to determine the value of K for any given bandwidth W. First, we estimate the covariance matrix of the measured channels $\mathbf{K}_{h,W}$ using statistical averaging

$$\mathbf{K}_{h,W} = \frac{1}{N}\sum_{i=1}^{N} \mathbf{h}_{i,W}\mathbf{h}_{i,W}^{\dagger}, \tag{17.19}$$

where $\mathbf{h}_{i,W}$ is the channel information obtained at location i with bandwidth W and $N = 100$. Because $\mathbf{K}_{h,W}$ is Hermitian and positive definite, there exists a unitary matrix U such that

$$\mathbf{K}_{h,W} = U\Lambda U^{\dagger} = \sum_{i=1}^{L} \lambda_{i,W}\psi_i\psi_i^{\dagger}, \tag{17.20}$$

where $\lambda_{1,W} \geq \lambda_{2,W} \geq \cdots \geq \lambda_{L,W}$ and $L = \tau_C W$.

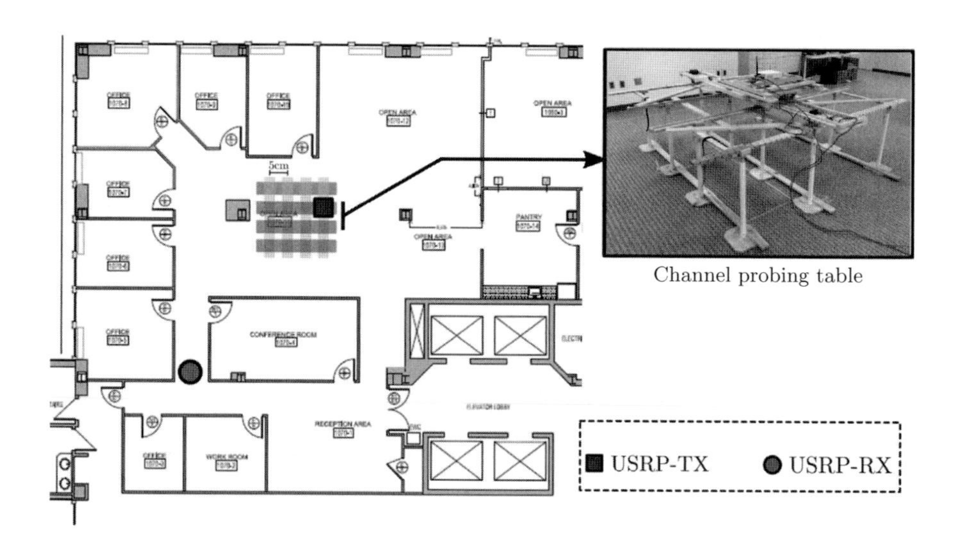

Figure 17.5 Floor plan and experiment setting.

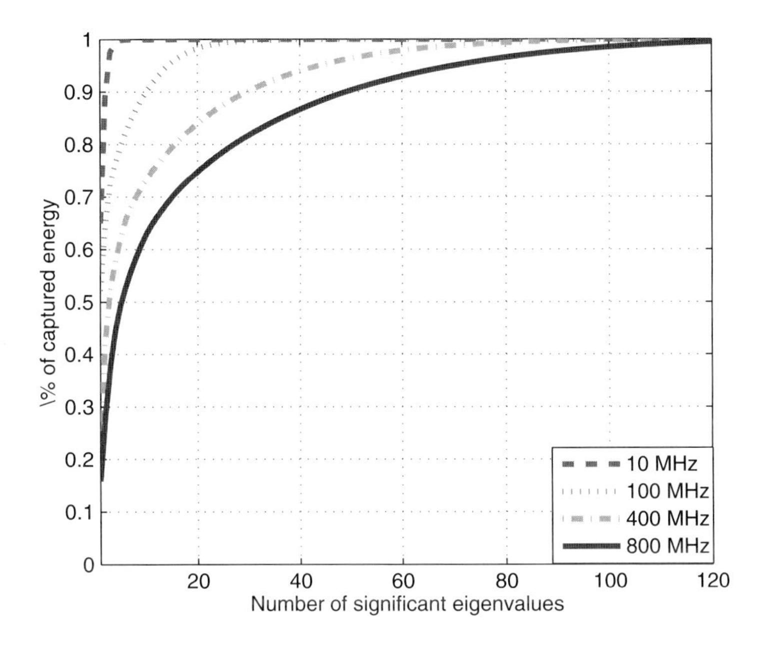

Figure 17.6 Percentage of captured energy versus the number of significant eigenvalues.

In Figure 17.6, we show the percentage of the captured energy E_l versus the number of significant eigenvalues l, with E_l defined as $E_l = \frac{\sum_{i=1}^{l} \lambda_i}{\sum_{i=1}^{L} \lambda_i}$. From Figure 17.6, we can see that the channel energy is concentrated in a small number of eigenvalues when the bandwidth is small, while spread over a large number of eigenvalues as the bandwidth increases. In other words, the degree of freedom K increases as the bandwidth W increases. This is further confirmed in Figure 17.7, where we show the number of significant eigenvalues versus the channel bandwidth by fixing the captured energy at 98%.

From previous measurements, $K_{\max}$ is a large value in a typical indoor environment. Now we discuss an approach to further increase $K_{\max}$ in practical environment. Because the TR and MIMO technology are not mutually exclusive, the degree of freedom can be further scaled up by deploying a couple of antennas to harvest hundreds of virtual antennas as shown in Figure 17.7. As indicated in the figure, it would be easy to realize massive virtual antennas with a few antennas in the TR system instead of installing hundreds of physical antennas.

17.6.3 Achievable Rate Evaluation

The assumption that $K_{\max} \to \infty$ is just for analyzing the asymptotical achievable rate of the TR system as the assumption $M_t \to \infty$ in early massive MIMO works. In practice, we only need that $K_{\max}$ is large enough to achieve massive multipath effect. In the following, we will demonstrate that even with a single antenna, TR wideband system is still capable to reach a promising achievable rate based on our indoor experiment. Our

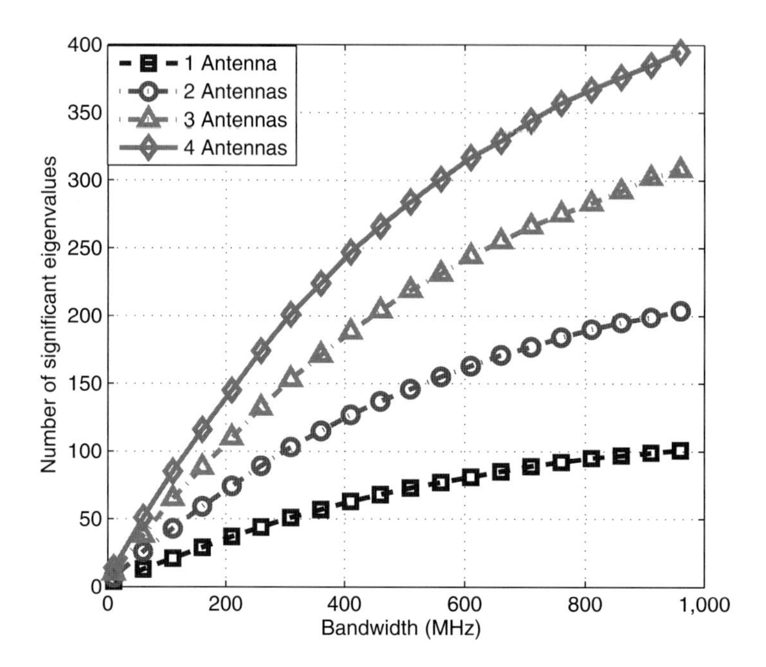

Figure 17.7 Number of significant eigenvalues K at different bandwidth W.

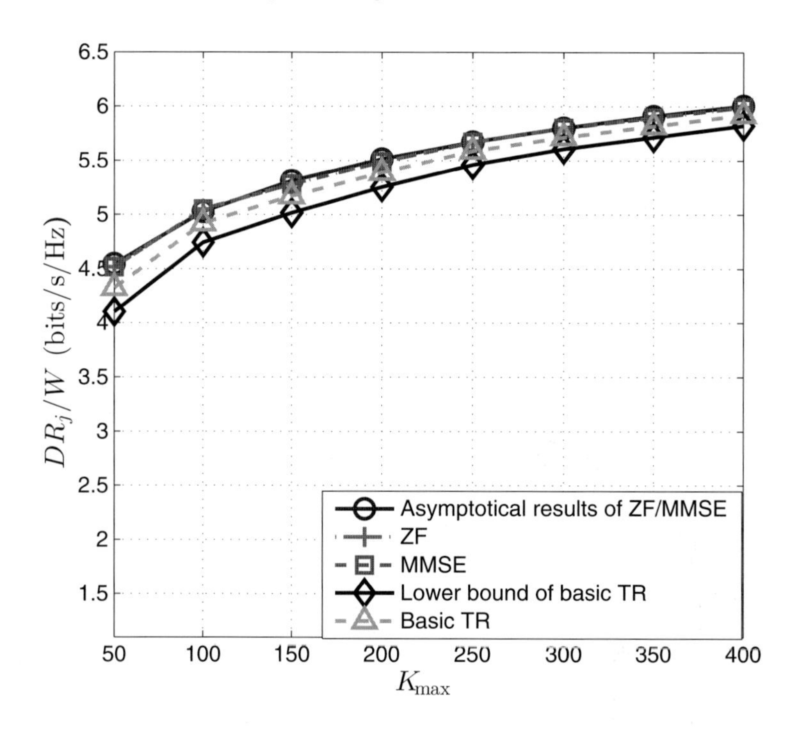

Figure 17.8 The asymptotic performance with $D = K_{\max}^{1.5}$, with $N = 6$ and $p_u = 5$ dB.

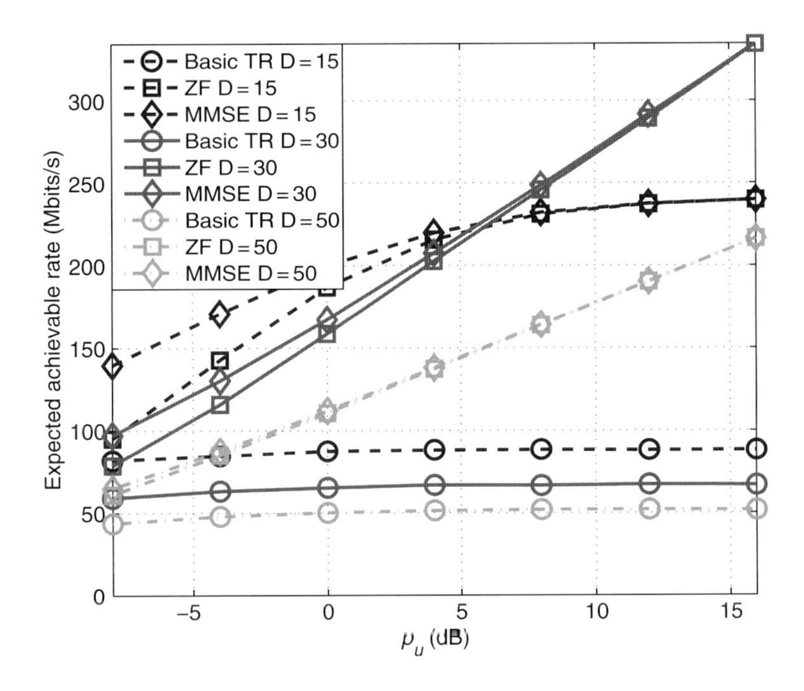

Figure 17.9 Expected achievable rate with $N = 10$ and $W = 1$ GHz.

experiment is conducted with the real indoor channel measurement, and the achievable rate in TR system is calculated based on Theorem 17.5.1.

We first evaluate the expected achievable rate of the TR system in a typical indoor environment using the channel measurements in the previous subsection, with $W = 1$ GHz. Then we compare the performance of the TR system with that of a massive MIMO system. Clearly, there is a tradeoff in selecting a proper D: W/D will decrease as D increases, while both the ISI and IUI are reduced as D increases. In Figure 17.9 we show the expected achievable rate of different waveforms with different D.

We can see that the expected achievable rate of the basic TR waveform saturates quickly as p_u increases because it is interference-limited with $N = 10$ receivers. Increasing D may decrease the expected achievable rate for the basic TR waveform if the decrease in W/D dominates the increase in SINR for a relatively large D as shown in Figure 17.9. The expected achievable rates of ZF and MMSE waveforms also saturate at high p_u with $D = 15$, but can be improved by increasing D, e.g., $D = 30$, to reduce the interference. However, it may hurt the rate performance if we increase D too much, e.g., $D = 50$.

We choose $D = 30$ as the backoff factor used in the TR system to evaluate the achievable rate in a practical indoor environment, as shown in Figure 17.10, which is comparable to a 20 MHz massive MIMO system with around 500 transmit antennas. Note that TR technology pays the price of spectral efficiency loss for the low cost and complexity implementation for indoor communications. For example, as shown in Figure 17.11 [38], the TR prototype is a customized software defined radio (SDR)

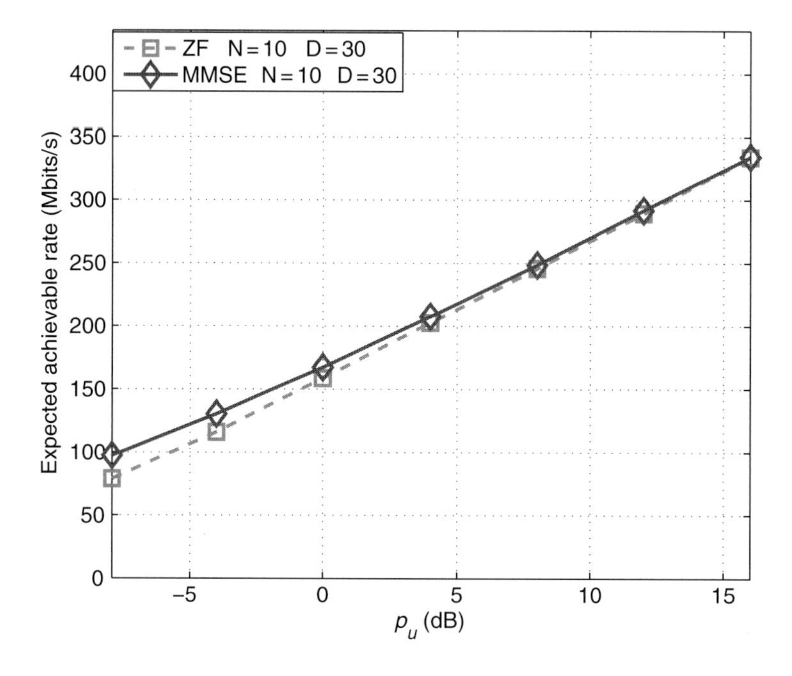

Figure 17.10 Expected achievable rate with $W = 1$ GHz, $N = 10$, and $D = 30$.

Figure 17.11 Time-reversal prototype.

platform for designing and deploying TR-based communication systems. The size of the radio is 5 cm by 17 cm by 23 cm, the weight is about 400 g, and the power consumption is 25 W. Compared with the massive MIMO prototype built in [9], the complexity and operation power consumption are obviously much lower. Considering the potential wide bandwidth available in future (e.g., UWB and mmWave band), the complexity, energy consumption, and other metrics become more and more important compared with the spectral efficiency in indoor scenarios, which makes the TR technology a promising candidate for indoor communication.

17.7 Summary

In this chapter, we demonstrate that the TR technology, through harvesting the naturally existing virtual antennas, can offer a cost-effective solution to realize the massive multipath effect, which is a counterpart of massive MIMO effect in indoor scenarios. With the derived massive multipath effect, we further derive the asymptotic rates of TR technology in a rich-scattering environment. We validate with simulations that the TR system with typical waveforms can asymptotically achieve the limiting achievable rate, where the interference is completely eliminated. Finally, based on the real channel measurements, it is shown that the single-antenna TR wideband system can achieve promising rates in a practical indoor environment. By utilizing the environment as virtual antenna array and computing resource, the low complexity of TR technology is ideal for indoor communications. What a TR system needs is a large enough bandwidth to harvest the multipaths in the environment, which can be made possible with more affordable high-speed ADC and wide spectrum in mmWave band. For related references, readers can refer to [44].

Appendix A: Proof for Theorem 17.4.1

PROOF. In order to reveal all the observable multipaths, e.g., $K = K_{\max}$, the bandwidth of the system W should be large enough so that $L/K_{\max}^{p} \to c$ where c is a constant and $p > 2$. Notice that every element in $\mathbf{QQ}^{\dagger}$ is the sum of multiple independent variables, which converges to a Gaussian random variable in distribution in the asymptotical scenario based on the central limit theorem. Because Gaussian random variable is only determined by the first and second moment and obviously each element in $\mathbf{QQ}^{\dagger}$ has zero mean, we only need to prove the largest variance of off-diagonal element will converge to zero.

Based on the definition of $\mathbf{Q}$, we have

$$\mathbf{QQ}^{\dagger} = \begin{bmatrix} \mathbf{H}_1\mathbf{H}_1^{\dagger} & \mathbf{H}_1\mathbf{H}_2^{\dagger} & \cdots & \mathbf{H}_1\mathbf{H}_N^{\dagger} \\ \mathbf{H}_2\mathbf{H}_1^{\dagger} & \mathbf{H}_2\mathbf{H}_2^{\dagger} & \cdots & \mathbf{H}_2\mathbf{H}_N^{\dagger} \\ \vdots & \vdots & \ddots & \cdots \\ \mathbf{H}_N\mathbf{H}_1^{\dagger} & \mathbf{H}_N\mathbf{H}_2^{\dagger} & \cdots & \mathbf{H}_N\mathbf{H}_N^{\dagger} \end{bmatrix}. \quad (17.21)$$

With (17.21), we can directly obtain

$$\left[\mathbf{Q}\mathbf{Q}^{\dagger}\right]_{l_j,l_j} = \mathbf{H}_j^{(\frac{L}{D})}\mathbf{H}_j^{(\frac{L}{D})\dagger} = \|\mathbf{h}_j\|^2, \tag{17.22}$$

where $[\cdot]_{m,n}$ represents the element in the mth row and the nth column of the matrix.

Then, we prove that $\mathbf{Q}\mathbf{Q}^{\dagger}$ is diagonal by examining the off-diagonal elements. Note that each off-diagonal matrix ($\forall i \neq j$) in (17.21), $\mathbf{H}_i\mathbf{H}_j^{\dagger}$, can be expanded as

$$\mathbf{H}_i\mathbf{H}_j^{\dagger} = \begin{bmatrix} \mathbf{H}_i^{(1)}\mathbf{H}_j^{(1)\dagger} & \mathbf{H}_i^{(1)}\mathbf{H}_j^{(2)\dagger} & \cdots & \mathbf{H}_i^{(1)}\mathbf{H}_j^{(\frac{2L-1}{D})\dagger} \\ \mathbf{H}_i^{(2)}\mathbf{H}_j^{(1)\dagger} & \mathbf{H}_i^{(2)}\mathbf{H}_j^{(2)\dagger} & \cdots & \mathbf{H}_i^{(2)}\mathbf{H}_j^{(\frac{2L-1}{D})\dagger} \\ \vdots & \vdots & \ddots & \cdots \\ \mathbf{H}_i^{(\frac{2L-1}{D})}\mathbf{H}_j^{(1)\dagger} & \mathbf{H}_i^{(\frac{2L-1}{D})}\mathbf{H}_j^{(2)\dagger} & \cdots & \mathbf{H}_i^{(\frac{2L-1}{D})}\mathbf{H}_j^{(\frac{2L-1}{D})\dagger} \end{bmatrix}. \tag{17.23}$$

From (17.23), we can see that each element of $\mathbf{H}_i\mathbf{H}_j^{\dagger}$, $\left[\mathbf{H}_i\mathbf{H}_j^{\dagger}\right]_{m,n} = \mathbf{H}_i^{(m)}\mathbf{H}_j^{(n)\dagger}$, is the sum of multiple independent random variables. Therefore, when K_{max} is sufficiently large, $\left[\mathbf{H}_i\mathbf{H}_j^{\dagger}\right]_{m,n}$ can be regarded as a Gaussian random variable, whose distribution is completely determined by the first and the second moments.

Based on the independence between the channel taps and distinct receivers, it is obvious that

$$\mathbb{E}\left[\mathbf{H}_i^{(m)}\mathbf{H}_j^{(n)\dagger}\right] = 0, \tag{17.24}$$

while the second moment can be upper bounded as follows

$$\begin{aligned} \mathbb{E}\left[|\mathbf{H}_i^{(m)}\mathbf{H}_j^{(n)\dagger}|^2\right] &\overset{(a)}{=} \sum_{l=1}^{L} \mathbb{E}\left[|\mathbf{H}_i^{(m)}(l)|^2|\mathbf{H}_j^{(n)}(l)|^2\right] \\ &\overset{(b)}{\leq} \sum_{l=1}^{L} \mathbb{E}\left[|\mathbf{H}_i^{(\frac{L}{D})}(l)|^2\right]\mathbb{E}\left[|\mathbf{H}_j^{(\frac{L}{D})}(l)|^2\right] \\ &\overset{(c)}{=} \frac{\left(\sum_{k=1}^{K_{\text{max}}}\sigma_k^2\right)^2}{L}, \end{aligned} \tag{17.25}$$

where (a) is obtained directly from the independence, (b) is based on the matrix structure in (17.8) and (c) comes from the fact that the K_{max} multipaths are randomly distributed among the L-tap channel and thus

$$\mathbb{E}\left[|h_j(m)|^2\right] = \mathbb{E}\left[|h_j(n)|^2\right] = \frac{\sum_{k=1}^{K_{\text{max}}}\sigma_k^2}{L}, \quad \forall m, n. \tag{17.26}$$

Note $\sum_{k=1}^{K_{\text{max}}}\sigma_k^2 < K_{\text{max}}$ due to the path loss attenuation. Therefore, (c) will converge to 0 given $L/K_{\text{max}}^p \to c$ where c is a constant and $p > 2$. From (17.24) and (17.25), we can conclude that

$$\left[\mathbf{H}_i\mathbf{H}_j^{\dagger}\right]_{m,n} \overset{d}{\to} 0, \quad \forall m, n, and\ i \neq j. \tag{17.27}$$

Next, let us examine the diagonal submatrix of $\mathbf{QQ}^\dagger$, which can be expanded as

$$\mathbf{H}_j\mathbf{H}_j^\dagger = \begin{bmatrix} \mathbf{H}_j^{(1)}\mathbf{H}_j^{(1)\dagger} & \mathbf{H}_j^{(1)}\mathbf{H}_j^{(2)\dagger} & \cdots & \mathbf{H}_j^{(1)}\mathbf{H}_j^{(\frac{2L-1}{D})\dagger} \\ \mathbf{H}_j^{(2)}\mathbf{H}_j^{(1)\dagger} & \mathbf{H}_j^{(2)}\mathbf{H}_j^{(2)\dagger} & \cdots & \mathbf{H}_j^{(2)}\mathbf{H}_j^{(\frac{2L-1}{D})\dagger} \\ \vdots & \vdots & \ddots & \cdots \\ \mathbf{H}_j^{(\frac{2L-1}{D})}\mathbf{H}_j^{(1)\dagger} & \mathbf{H}_j^{(\frac{2L-1}{D})}\mathbf{H}_j^{(2)\dagger} & \cdots & \mathbf{H}_j^{(\frac{2L-1}{D})}\mathbf{H}_j^{(\frac{2L-1}{D})\dagger} \end{bmatrix}. \tag{17.28}$$

Similarly, each element $\left[\mathbf{H}_j\mathbf{H}_j^\dagger\right]_{m,n} = \mathbf{H}_j^{(m)}\mathbf{H}_j^{(n)\dagger}$ can be regarded as Gaussian variable when $K_{\max}$ is sufficiently large. Because $\mathbf{H}_j^{(m)}$ and $\mathbf{H}_j^{(n)}$ are independent when $m \neq n$, similar to (17.24) and (17.25), we can derive

$$\begin{cases} \mathbb{E}\left[\mathbf{H}_j^{(m)}\mathbf{H}_j^{(n)\dagger}\right] & = 0, \quad m \neq n, \\ \mathbb{E}\left[|\mathbf{H}_j^{(m)}\mathbf{H}_j^{(n)\dagger}|^2\right] & \leq \frac{\left(\sum_{k=1}^{K_{\max}}\sigma_k^2\right)^2}{L}, \quad m \neq n, \end{cases} \tag{17.29}$$

and given $L/K_{\max}^p \to c$ where c is a constant and $p > 2$, we have derived that

$$\left[\mathbf{H}_j\mathbf{H}_j^\dagger\right]_{m,n} \xrightarrow{d} 0, \quad \forall j, and\ m \neq n. \tag{17.30}$$

Therefore, we can conclude that $\mathbf{QQ}^\dagger$ is diagonal. This completes the proof. ▲

Appendix B: Proof of Theorem 17.5.1

Expected Achievable Rate under Basic TR Waveform: From (17.12), the basic TR waveform is $\mathbf{g}_j = \mathbf{H}_j^{(L/D)\dagger}/\|\mathbf{H}_j^{(L/D)}\|$. Therefore, we have

$$\mathbf{g}_j^\dagger\mathbf{H}_j^{(\frac{L}{D})\dagger}\mathbf{H}_j^{(\frac{L}{D})}\mathbf{g}_j = \|\mathbf{H}_j^{(\frac{L}{D})}\|^2, \tag{17.31}$$

and

$$\mathbf{g}_j^\dagger\left(\sum_{i=1}^{N}\mathbf{H}_i^\dagger\mathbf{H}_i\right)\mathbf{g}_j = \frac{\mathbf{H}_j^{(\frac{L}{D})}\mathbf{Q}^\dagger\mathbf{Q}\mathbf{H}_j^{(\frac{L}{D})\dagger}}{\|\mathbf{H}_j^{(\frac{L}{D})}\|^2}. \tag{17.32}$$

According to the definition of $\mathbf{H}_j^{(L/D)}$, we have $\|\mathbf{H}_j^{(\frac{L}{D})}\|^2 = \|\mathbf{h}_j\|^2$, and $\mathbf{H}_j^{(L/D)}$ is the (L/D)th row of $\mathbf{H}_j$ and thus l_j^{th} row of $\mathbf{Q}$. Therefore, (17.31) and (17.32) can be rewritten as

$$\mathbf{g}_j^\dagger\mathbf{H}_j^{(\frac{L}{D})\dagger}\mathbf{H}_j^{(\frac{L}{D})}\mathbf{g}_j = \|\mathbf{h}_j\|^2, \tag{17.33}$$

and

$$\mathbf{g}_j^\dagger\left(\sum_{i=1}^{N}\mathbf{H}_i^\dagger\mathbf{H}_i\right)\mathbf{g}_j = \frac{\left[\mathbf{QQ}^\dagger\mathbf{QQ}^\dagger\right]_{l_j,l_j}}{\|\mathbf{h}_j\|^2}, \tag{17.34}$$

where $[\cdot]_{l_j,l_j}$ is the (l_j,l_j) element of the matrix.

Substituting (17.33) and (17.34) into (17.10), the expected achievable rate of jth receiver with the basic TR waveform can be written as

$$R_j^{Basic} = \frac{W}{D}\mathbb{E}\left[\log_2\left(1 + \frac{p_u\|\mathbf{h}_j\|^4}{p_u([\mathbf{QQ}^\dagger\mathbf{QQ}^\dagger]_{l_j,l_j} - \|\mathbf{h}_j\|^4) + \|\mathbf{h}_j\|^2}\right)\right]. \quad (17.35)$$

Expected Achievable Rate under ZF Waveform: With the ZF waveform $\mathbf{g}_j = c_{ZF}\mathbf{Q}^\dagger(\mathbf{QQ}^\dagger)^{-1}\mathbf{e}_{l_j}$, we have

$$\begin{bmatrix} \mathbf{H}_1\mathbf{g}_j \\ \vdots \\ \mathbf{H}_N\mathbf{g}_j \end{bmatrix} = \mathbf{Qg}_j = c_{ZF}\mathbf{QQ}^\dagger(\mathbf{QQ}^\dagger)^{-1}\mathbf{e}_{l_j} = c_{ZF}\mathbf{e}_{l_j}. \quad (17.36)$$

According to (17.36), we can derive the following

$$\mathbf{g}_j^\dagger\mathbf{H}_i^\dagger\mathbf{H}_i\mathbf{g}_j = \begin{cases} 0, & \forall i \neq j, \\ \mathbf{g}_j^\dagger\mathbf{H}_j^{(\frac{L}{D})\dagger}\mathbf{H}_j^{(\frac{L}{D})}\mathbf{g}_j = c_{ZF}^2, & i = j. \end{cases} \quad (17.37)$$

Substituting (17.37) into (17.10), we can see that both the ISI and IUI are eliminated, and thus the expected achievable rate with ZF waveform can be written as

$$R_j^{ZF} = \frac{W}{D}\mathbb{E}\left[\log_2\left(1 + p_u c_{ZF}^2\right)\right]. \quad (17.38)$$

Because $\mathbf{g}_j^\dagger\mathbf{g}_j = c_{ZF}^2[(\mathbf{QQ}^\dagger)^{-1}]_{l_j,l_j} = 1$, the expected achievable rate with ZF waveform in (17.38) can be rewritten as

$$R_j^{ZF} = \frac{W}{D}\mathbb{E}\left[\log_2\left(1 + \frac{p_u}{[(\mathbf{QQ}^\dagger)^{-1}]_{l_j,l_j}}\right)\right]. \quad (17.39)$$

$$\begin{aligned} R_j^{MMSE} &\overset{(a)}{=} \frac{W}{D}\mathbb{E}\left[\log_2\left(1 + \frac{\mathbf{g}_j^\dagger\mathbf{H}_j^{(\frac{L}{D})\dagger}\mathbf{H}_j^{(\frac{L}{D})}\mathbf{g}_j}{\mathbf{g}_j^\dagger\left(\mathbf{H}_j^\dagger\mathbf{H}_j - \mathbf{H}_j^{(\frac{L}{D})\dagger}\mathbf{H}_j^{(\frac{L}{D})} + \sum_{i\neq j}\mathbf{H}_i^\dagger\mathbf{H}_i + \frac{1}{p_u}\mathbf{I}\right)\mathbf{g}_j}\right)\right] \\ &\overset{(b)}{=} \frac{W}{D}\mathbb{E}\left[\log_2\left(1 + \frac{\mathbf{g}_j^\dagger\mathbf{H}_j^{(\frac{L}{D})\dagger}\mathbf{H}_j^{(\frac{L}{D})}\mathbf{g}_j}{\mathbf{g}_j^\dagger\Lambda_j\mathbf{g}_j}\right)\right] \\ &\overset{(c)}{=} \frac{W}{D}\mathbb{E}\left[\log_2\left(1 + \mathbf{H}_j^{(\frac{L}{D})}\Lambda_j^{-1}\mathbf{H}_j^{(\frac{L}{D})\dagger}\right)\right], \end{aligned} \quad (17.40)$$

Expected Achievable Rate under MMSE Waveform: According to (17.12) and the Woodbury matrix identity [45], the MMSE waveform can be written as

$$\begin{aligned} \mathbf{g}_j &= c_{MMSE}\left(\mathbf{Q}^\dagger\mathbf{Q} + \frac{1}{p_u}\mathbf{I}\right)^{-1}\mathbf{H}_j^{(\frac{L}{D})\dagger} \\ &= \frac{c_{MMSE}\Lambda_j^{-1}\mathbf{H}_j^{(\frac{L}{D})\dagger}}{\mathbf{H}_j^{(\frac{L}{D})}\Lambda_j^{-1}\mathbf{H}_j^{(\frac{L}{D})\dagger} + 1}, \end{aligned} \tag{17.41}$$

where $\Lambda_j \triangleq \mathbf{Q}^\dagger\mathbf{Q} - \mathbf{H}_j^{(\frac{L}{D})\dagger}\mathbf{H}_j^{(\frac{L}{D})} + (1/p_u)\mathbf{I}$.

By multiplying both sides in (17.41) with $\mathbf{H}_j^{(\frac{L}{D})}$, we can derive the following

$$\mathbf{H}_j^{(\frac{L}{D})}\Lambda_j^{-1}\mathbf{H}_j^{(\frac{L}{D})\dagger} = \frac{1}{1 - \mathbf{H}_j^{(\frac{L}{D})}\left(\mathbf{Q}^\dagger\mathbf{Q} + \frac{1}{p_u}\mathbf{I}\right)^{-1}\mathbf{H}_j^{(\frac{L}{D})\dagger}} - 1. \tag{17.42}$$

Moreover, according to (17.41), we have

$$\mathbf{g}_j^\dagger\mathbf{H}_j^{(\frac{L}{D})\dagger}\mathbf{H}_j^{(\frac{L}{D})}\mathbf{g}_j = \frac{c_{MMSE}^2\left(\mathbf{H}_j^{(\frac{L}{D})}\Lambda_j^{-1}\mathbf{H}_j^{(\frac{L}{D})\dagger}\right)^2}{\left(\mathbf{H}_j^{(\frac{L}{D})}\Lambda_j^{-1}\mathbf{H}_j^{(\frac{L}{D})\dagger} + 1\right)^2}, \tag{17.43}$$

and

$$\mathbf{g}_j^\dagger\Lambda_j\mathbf{g}_j = \frac{c_{MMSE}^2\mathbf{H}_j^{(\frac{L}{D})}\Lambda_j^{-1}\mathbf{H}_j^{(\frac{L}{D})\dagger}}{\left(\mathbf{H}_j^{(\frac{L}{D})}\Lambda_j^{-1}\mathbf{H}_j^{(\frac{L}{D})\dagger} + 1\right)^2}. \tag{17.44}$$

Then, the expected achievable rate in (17.10) can be rerewritten in (17.40), where (a) is the direct result from $\mathbf{g}_j^\dagger\mathbf{g}_j = 1$, (b) comes from the definition of Λ_j, and (c) is based on (17.43) and (17.44).

By substituting (17.42) into (17.40), the expected achievable rate with MMSE waveform can be further simplified as

$$\begin{aligned} R_j^{MMSE} &= \frac{W}{D}\mathbb{E}\left[\log_2\left(\frac{1}{1 - \mathbf{H}_j^{(\frac{L}{D})}\left(\mathbf{Q}^\dagger\mathbf{Q} + \frac{1}{p_u}\mathbf{I}\right)^{-1}\mathbf{H}_j^{(\frac{L}{D})\dagger}}\right)\right] \\ &= \frac{W}{D}\mathbb{E}\left[\log_2\left(\frac{1}{1 - \left[\mathbf{Q}\left(\mathbf{Q}^\dagger\mathbf{Q} + \frac{1}{p_u}\mathbf{I}\right)^{-1}\mathbf{Q}^\dagger\right]_{l_j,l_j}}\right)\right] \\ &= \frac{W}{D}\mathbb{E}\left[\log_2\left(\frac{1}{\left[\left(\mathbf{I} + p_u\mathbf{Q}\mathbf{Q}^\dagger\right)^{-1}\right]_{l_j,l_j}}\right)\right], \end{aligned} \tag{17.45}$$

where the second equality comes from the definition of $\mathbf{H}_j^{(\frac{L}{D})}$, and the last equality comes from the following derivation by utilizing the Woodbury matrix identity [45],

$$\left(\mathbf{I}+p_u\mathbf{Q}\mathbf{Q}^\dagger\right)^{-1}=\mathbf{I}-\mathbf{Q}\left(\frac{1}{p_u}\mathbf{I}+\mathbf{Q}^\dagger\mathbf{Q}\right)^{-1}\mathbf{Q}^\dagger. \tag{17.46}$$

Up to now, we have derived the expected achievable rate under different designs of waveform, and the results are summarized in Theorem 17.5.1.

Appendix C: Proof of Lemma 17.5.1

Proof. With the definition of $\mathbf{Q}$ and (17.9), we have

$$\begin{aligned}&\left[\mathbf{Q}\mathbf{Q}^\dagger\mathbf{Q}\mathbf{Q}^\dagger\right]_{l_j,l_j}-\|\mathbf{h}_j\|^4=\\&\sum_{i=1,i\neq j}^{N}\sum_{l=1}^{(2L-1)/D}|\mathbf{H}_i^{(l)}\mathbf{H}_j^{(\frac{L}{D})\dagger}|^2+\sum_{l=1,l\neq(L/D)}^{(2L-1)/D}|\mathbf{H}_j^{(l)}\mathbf{H}_j^{(\frac{L}{D})\dagger}|^2,\end{aligned} \tag{17.47}$$

which is the sum of multiple independent random variables. Therefore, $\left[\mathbf{Q}\mathbf{Q}^\dagger\mathbf{Q}\mathbf{Q}^\dagger\right]_{l_j,l_j}-\|\mathbf{h}_j\|^4$ can be regarded as a Gaussian random variable when $K_{\max}$ is sufficiently large.

Similar to (17.25), we have the following

$$\begin{cases}\mathbb{E}\left[|\mathbf{H}_i^{(l)}\mathbf{H}_j^{(\frac{L}{D})\dagger}|^2\right] &\leq \frac{(\sum_{k=1}^{K_{\max}}\sigma_k^2)^2}{L}, \quad i\neq j,\\ \mathbb{E}\left[|\mathbf{H}_j^{(l)}\mathbf{H}_j^{(\frac{L}{D})\dagger}|^2\right] &\leq \frac{(\sum_{k=1}^{K_{\max}}\sigma_k^2)^2}{L}, \quad l\neq(L/D).\end{cases} \tag{17.48}$$

Therefore, with $K_{\max}\to\infty$, the expectation of $\left[\mathbf{Q}\mathbf{Q}^\dagger\mathbf{Q}\mathbf{Q}^\dagger\right]_{l_j,l_j}-\|\mathbf{h}_j\|^4$ can be bounded by

$$\begin{aligned}\mathbb{E}\left[\left[\mathbf{Q}\mathbf{Q}^\dagger\mathbf{Q}\mathbf{Q}^\dagger\right]_{l_j,l_j}-\|\mathbf{h}_j\|^4\right]&\leq\frac{N(2L-1)}{D}\frac{(\sum_{k=1}^{K_{\max}}\sigma_k^2)^2}{L}\\&\leq\alpha\left(\sum_{k=1}^{K_{\max}}\sigma_k^2\right)^2,\end{aligned} \tag{17.49}$$

with $\alpha=2N/D$.

Similar to the argument in the derivation of (17.26), the fourth moment of $h_j(m)$ can be given as

$$\mathbb{E}\left[|h_j(m)|^4\right]=\mathbb{E}\left[|h_j(n)|^4\right]=2\left(\frac{\sum_{k=1}^{K_{\max}}\sigma_k^2}{L}\right)^2,\quad\forall m,n. \tag{17.50}$$

Then, we have,

$$\begin{cases} \mathbb{E}\left[|\mathbf{H}_i^{(l)}\mathbf{H}_j^{(\frac{L}{D})\dagger}|^4\right] & \leq \frac{4(\sum_{k=1}^{K_{\max}}\sigma_k^2)^4}{L^3}, \quad i \neq j, \\ \mathbb{E}\left[|\mathbf{H}_j^{(l)}\mathbf{H}_j^{(\frac{L}{D})\dagger}|^4\right] & \leq \frac{4(\sum_{k=1}^{K_{\max}}\sigma_k^2)^4}{L^3}, \quad l \neq (L/D). \end{cases} \tag{17.51}$$

As $K_{\max} \to \infty$, $L/K_{\max}^p \to c$ where c is a constant and $p > 2$ holds in the multipath-rich scenario. Therefore, the variance of $\left[\mathbf{Q}\mathbf{Q}^\dagger\mathbf{Q}\mathbf{Q}^\dagger\right]_{l_j,l_j} - \|\mathbf{h}_j\|^4$ goes to zero as $K_{\max} \to \infty$. Combining the first moment in (17.49), we can derive the result in Lemma 17.5.1. This completes the proof. ▲

Appendix D: Proof of Theorem 17.5.2

Proof. According to the TRMME, we have

$$\frac{\left[(\mathbf{Q}\mathbf{Q}^\dagger)^{-1}\right]_{l_j,l_j}}{\|\mathbf{h}_j\|^2} \xrightarrow{d} 1, \tag{17.52}$$

because the inverse of a diagonal matrix should be diagonal.

Then, according to (17.39), the asymptotic expected achievable rate under ZF waveform is

$$\lim_{K_{\max}\to\infty} \frac{R_j^{ZF}}{W/D} = \mathbb{E}\left[\log_2\left(1 + p_u\|\mathbf{h}_j\|^2\right)\right]. \tag{17.53}$$

Similarly, with the TRMME, we can also have

$$\begin{aligned} \left[\left(\mathbf{I} + p_u\mathbf{Q}\mathbf{Q}^\dagger\right)^{-1}\right]_{l_j,l_j} &\xrightarrow{d} \left[(\mathbf{I} + p_u\lambda)^{-1}\right]_{l_j,l_j} \\ &= \frac{1}{1 + p_u\|\mathbf{h}_j\|^2}. \end{aligned} \tag{17.54}$$

By substituting (17.54) into (17.45), the asymptotic expected achievable rate under MMSE waveform is

$$\lim_{K_{\max}\to\infty} \frac{R_j^{MMSE}}{W/D} = \mathbb{E}\left[\log_2\left(1 + p_u\|\mathbf{h}_j\|^2\right)\right]. \tag{17.55}$$

Finally, by substituting (17.16) of Lemma 17.5.1 into (17.35), the asymptotic expected achievable rate under basic TR waveform can be lower bounded as

$$\liminf_{K_{\max}\to\infty} \frac{R_j^{Basic}}{W/D} = \mathbb{E}\left[\log_2\left(1 + \frac{p_u\|\mathbf{h}_j\|^2}{\frac{p_u\alpha\left(\sum_{k=1}^{K_{\max}}\sigma_k^2\right)^2}{\|\mathbf{h}_j\|^2} + 1}\right)\right]. \tag{17.56}$$

This completes the proof. ▲

References

[1] Cisco, "Visual networking index," *Cisco white paper*, 2015.

[2] J. Andrews, S. Buzzi, C. Wan, S. Hanly, A. Lozano, A. Soong, and J. Zhang, "What will 5G be?" *IEEE Journal on Selected Areas in Communications*, vol. 32, no. 6, pp. 1065–1082, 2014.

[3] H. Ngo, E. Larsson, and T. Marzetta, "Energy and spectral efficiency of very large multiuser MIMO systems," *IEEE Transactions on Communications*, vol. 61, no. 4, pp. 1436–1449, 2013.

[4] T. Marzetta, "Noncooperative cellular wireless with unlimited numbers of base station antennas," *IEEE Transactions on Wireless Communications*, vol. 9, no. 11, pp. 3590–3600, 2010.

[5] F. Boccardi, R. Heath, A. Lozano, T. Marzetta, and P. Popovski, "Five disruptive technology directions for 5G," *IEEE Communications Magazine*, vol. 52, no. 2, pp. 74–80, 2014.

[6] J. Liu, H. Minn, and A. Gatherer, "The death of 5G part 2: Will analog be the death of massive MIMO?" www.comsoc.org/ctn/death-5g-part-2-will-analog-be-death-massive-mimo, Jun. 2015.

[7] X. Artiga, B. Devillers, and J. Perruisseau-Carrier, "Mutual coupling effects in multi-user massive MIMO base stations," in *Proceedings of the IEEE APSURSI*, pp. 1–2, 2012.

[8] C. Masouros, M. Sellathurai, and T. Ratnarajah, "Large-scale MIMO transmitters in fixed physical spaces: The effect of transmit correlation and mutual coupling," *IEEE Transactions on Communications*, vol. 61, no. 7, pp. 2794–2804, 2013.

[9] J. Vieira, S. Malkowsky, K. Nieman, Z. Miers, N. Kundargi, L. Liu, I. Wong, V. Owall, O. Edfors, and F. Tufvesson, "A flexible 100-antenna testbed for massive MIMO," in *Proceedings of the IEEE Globecom Workshops*, pp. 287–293, 2014.

[10] M. Panolini, "Beyond data caps: An analysis of the uneven growth in data traffic," www.senzafiliconsulting.com/, Mar. 2011.

[11] R. Saadane, A. Menouni, R. Knopp, and D. Aboutajdine, "Empirical eigenanalysis of indoor UWB propagation channels," in *Proceedings of the IEEE Globecom*, vol. 5, pp. 3215–3219, 2004.

[12] A. M. Hayar, R. Knopp, and R. Saadane, "Subspace analysis of indoor UWB channels," *EURASIP Journal on Applied Signal Processing*, vol. 2005, pp. 287–295, 2005.

[13] A. Inamdar, S. Rylov, A. Talalaevskii, A. Sahu, S. Sarwana, D. Kirichenko, I. Vernik, T. Filippov, and D. Gupta, "Progress in design of improved high dynamic range analog-to-digital converters," *IEEE Transactions on Applied Superconductivity*, vol. 19, no. 3, pp. 670–675, 2009.

[14] T. Rappaport, S. Sun, R. Mayzus, H. Zhao, Y. Azar, K. Wang, G. Wong, J. Schulz, M. Samimi, and F. Gutierrez, "Millimeter wave mobile communications for 5G cellular: It will work!" *IEEE Access*, vol. 1, pp. 335–349, 2013.

[15] P. Smulders and L. Correia, "Characterisation of propagation in 60 GHz radio channels," *Electronics & Communication Engineering Journal*, vol. 9, no. 2, pp. 73–80, Apr. 1997.

[16] F. Han, Y. Yang, B. Wang, Y. Wu, and K. J. R. Liu, "Time-reversal division multiple access over multi-path channels," *IEEE Transactions on Communications*, vol. 60, no. 7, pp. 1953–1965, 2012.

[17] B. P. Bogert, "Demonstration of delay distortion correction by time-reversal techniques," *IRE Transactions on Communications Systems*, vol. 5, no. 3, pp. 2–7, 1957.

[18] M. Fink and C. Prada, "Acoustic time-reversal mirrors," *Inverse Problems*, vol. 17, no. 1, p. R1, 2001.

[19] S. Lehman and A. Devaney, "Transmission mode time-reversal super-resolution imaging," *Journal of the Acoustical Society of America*, vol. 113, no. 5, pp. 2742–2753, 2003.

[20] D. Liu, G. Kang, L. Li, Y. Chen, S. Vasudevan, W. Joines, Q. Liu, J. Krolik, and L. Carin, "Electromagnetic time-reversal imaging of a target in a cluttered environment," *IEEE Transactions on Antennas and Propagation*, vol. 53, no. 9, pp. 3508–3066, 2005.

[21] G. Edelmann, T. Akal, W. Hodgkiss, S. Kim, W. Kuperman, and H. Song, "An initial demonstration of underwater acoustic communication using time reversal," *IEEE Journal of Oceanic Engineering*, vol. 27, no. 3, pp. 602–609, 2002.

[22] A. Derode, P. Roux, and M. Fink, "Robust acoustic time reversal with high order multiple scattering," *Physical Review Letters*, vol. 75, pp. 4206–4209, 1995.

[23] H. Nguyen, J. Anderson, and G. Pedersen, "The potential of time reversal techniques in multiple element antenna systems," *IEEE Communications Letters*, vol. 9, no. 1, pp. 40–42, 2005.

[24] R. de Lacerda Neto, A. Hayar, and M. Debbah, "Channel division multiple access based on high UWB channel temporal resolution," in *Proceedings of the IEEE VTC*, pp. 1–5, 2006.

[25] B. Wang, Y. Wu, F. Han, Y. Yang, and K. J. R. Liu, "Green wireless communications: A time-reversal paradigm," *IEEE Journal on Selected Areas in Communications*, vol. 29, no. 8, pp. 1698–1710, 2011.

[26] M.-A. Bouzigues, I. Siaud, M. Helard, and A.-M. Ulmer-Moll, "Turn back the clock: Time reversal for green radio communications," *IEEE Vehicular Technology Magazine*, vol. 8, no. 1, pp. 49–56, 2013.

[27] Y. Jin, J. Yi, and J. Moura, "Multiple antenna time reversal transmission in ultra-wideband communications," in *Proceedings of the IEEE Globecom*, pp. 26–30, 2007.

[28] Y.-H. Yang, B. Wang, W. S. Lin, and K. J. R. Liu, "Near-optimal waveform design for sum rate optimization in time-reversal multiuser downlink systems," *IEEE Transactions on Wireless Communications*, vol. 12, no. 1, pp. 346–357, 2013.

[29] E. Yoon, S. Kim, and U. Yun, "A time-reversal-based transmission using predistortion for intersymbol interference alignment," *IEEE Transactions on Communications*, vol. 63, no. 2, pp. 455–465, 2014.

[30] Y. Chen, F. Han, Y. Yang, H. Ma, Y. Han, C. Jiang, H. Lai, D. Claffey, Z. Safar, and K. J. R. Liu, "Time-reversal wireless paradigm for green Internet of Things: An overview," *IEEE Internet of Things Journal*, vol. 1, no. 1, pp. 81–98, 2014.

[31] A. J. Viterbi, *CDMA: Principles of Spread Spectrum Communication.* Reading, MA: Addison Wesley Longman Publishing Co., Inc., 1995.

[32] I. Azzam and R. Adve, "Linear precoding for multiuser MIMO systems with multiple base stations," in *Proceedings of the IEEE ICC*, pp. 1–6, 2009.

[33] L. Kewen, M. Zherui, and H. Ting, "A novel TR-STBC-OFDM scheme for mobile WiMAX system," in *Proceedings of the IEEE ISAPE*, pp. 1365–1368, 2008.

[34] M. Maaz, M. Helard, P. Mary, and M. Liu, "Performance analysis of time-reversal based precoding schemes in MISO-OFDM systems," in *Proceedings of the IEEE VTC*, pp. 1–6, 2015.

[35] A. Pitarokoilis, S. K. Mohammed, and E. G. Larsson, "Uplink performance of time-reversal MRC in massive MIMO systems subject to phase noise," *IEEE Transactions on Wireless Communications*, vol. 14, no. 2, pp. 711–723, 2015.

[36] C. Zhou, N. Guo, B. Sadler, and R. Qiu, "Performance study on time reversed impulse MIMO for UWB communications based on measured spatial UWB channels," in *Proceedings of the IEEE MILCOM*, pp. 1–6, 2007.

[37] B. Popovic, "Efficient Golay correlator," *Electronics Letters*, vol. 35, no. 17, pp. 1427–1428, Aug 1999.

[38] Z.-H. Wu, Y. Han, Y. Chen, and K. J. R. Liu, "A time-reversal paradigm for indoor positioning system," *IEEE Transactions on Vehicular Technology*, vol. 64, no. 4, pp. 1331–1339, 2015.

[39] R. Daniels and R. Heath, "Improving on time-reversal with MISO precoding," in *Proceedings of the Eighth International Symposium on Wireless Personal Multimedia Communications Conference*, pp. 18–22, 2005.

[40] D. Tse and P. Viswanath, "Downlink-uplink duality and effective bandwidths," in *Proceedings of the IEEE ISIT*, 2002.

[41] M. Schubert and H. Boche, "Solution of the multiuser downlink beamforming problem with individual SINR constraints," *IEEE Transactions on Vehicular Technology*, vol. 53, no. 1, pp. 18–28, 2004.

[42] R. Hunger and M. Joham, "A general rate duality of the MIMO multiple access channel and the MIMO broadcast channel," in *Proc. IEEE Global Telecommunications Conference*, 2008.

[43] E. Telatar, "Capacity of multiple-antenna gaussian channels," *European Transactions on Telecommunications*, vol. 10, no. 6, pp. 585–595, 1999.

[44] Y. Han, Y. Chen, B. Wang, and K. J. R. Liu, "Time-reversal massive multipath effect: A single-antenna massive MIMO solution," *IEEE Transactions on Communications*, vol. 64, no. 8, pp. 3382–3394, 2016.

[45] G. H. Golub and C. F. Van Loan, "Matrix computations," Baltimore, MD: John Hopkins University Press, 2012.

18 Waveforming

By leveraging the natural multipath propagation of electromagnetic waves, waveforming is proposed as a promising paradigm that treats each multipath component in a wireless channel as a virtual antenna to exploit the spatial diversity. As the most commonly known waveforming technique for wideband systems, the time-reversal (TR) signal transmission produces a TR resonance by coherently combining multipath energy distributed on virtual antennas and thus boosts the received signal strength while reducing interference. The wideband waveforming is, in many ways, similar to the multiple-input multiple-output (MIMO) beamforming, where multiple antennas are deployed to imitate a multipath transmission when the bandwidth is limited. In this chapter, we provide an overview of recent advances on the wideband waveforming, including massive multipath effect, optimal resource allocation, wireless power transfer, and secrecy enhancement for secured communications and compare with the corresponding counterparts of traditional MIMO beamforming.

18.1 Introduction

The nature provides a large number of degrees of freedom by means of radio multipath propagation. As the transmit signal encounters different scatterers in the environment and thus travels through different paths during its propagation, the channel between each transmitter (TX) and receiver (RX) antenna is a multipath channel.

Because an attenuated copy of the original signal is generated and transmitted through a different path when the transmit signal is reflected or scattered by a scatterer, each scatterer in the environment can be viewed as a virtual antenna that transmits directly to the RX, in addition to the TX. Moreover, the channel characteristics between the virtual antenna and the RX antenna are determined by both of the radio paths between the TX and the scatterer and between the scatterer and the RX. An illustration of scatterers as virtual antennas in a multipath propagation environment is depicted in Figure 18.1, where the arrow from the TX directly to the RX represents the line-of-sight (LOS) path, and the other arrows represent paths reflected and scattered by scatterers. The stars in Figure 18.1 represent scatterers in the environment, which can be viewed as virtual antennas that transmit an attenuated signal to the RX. All paths together form a multipath channel between the TX and the RX.

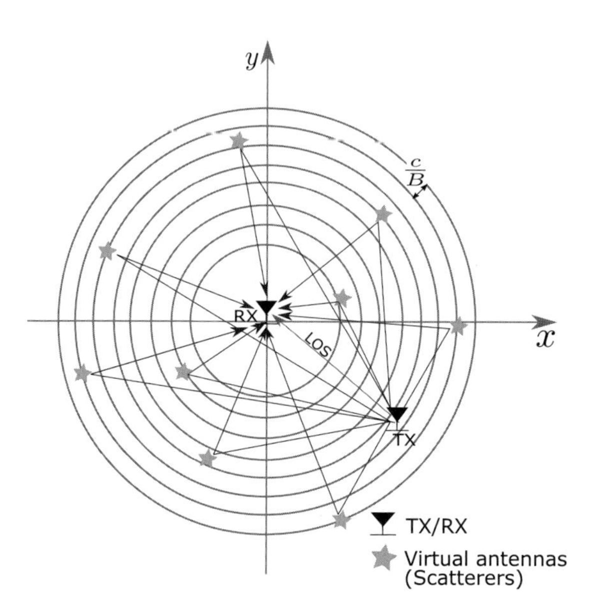

Figure 18.1 Illustration of virtual antennas.

How to control the harvested multipaths as virtual antennas to achieve desired application purposes? *Waveforming* [1–6] has been proposed as a novel technique to control virtual antennas distributed in the environment to take advantages of the spatial diversity and thus achieve a high degree of freedom. When the bandwidth increases, more uncorrelated multipath components are revealed, and thus higher degrees of freedom can be achieved for higher data rates and more reliable communications. The most well-known waveforming technique is the time-reversal (TR) signal transmission that has been considered as a novel paradigm for wideband systems, which treats each multipath component in a multipath channel as a virtual antenna [1]. The TR technique was originally designed for acoustics and ultrasound applications [7–10], and recent studies have been focused on TR wireless communications [11–16]. The TR signal transmission can be easily extended to a multiple-antenna scenario [17–20], and the orthogonal frequency division multiplexing (OFDM) system [21].

By leveraging virtual antennas in a constructive way, TR signal transmission can provide a significant spatial diversity gain and exhibit a TR resonance phenomenon [1, 22, 23]. Inspired by the TR technique, with certain objectives, the waveform design optimally weighs each virtual antenna in a multipath channel and combines the weighted signal from different paths at the receiver. Many works have been focused on optimal waveform design with different optimization criteria, such as interference cancellation, multiuser fairness, and efficient wireless power transfer [2–6, 24]. Waveforming can be designed to address the interference problem, which becomes a bottleneck limiting the system performance with the explosion of wireless traffic and the increase in the number of users in 5G.

In fact, the wideband waveforming is similar to the multiple-input multiple-output (MIMO) beamforming. When the bandwidth is too limited to resolve independent multipaths, MIMO was proposed to intimate a multipath transmission with multiple TX antennas or/and multiple RX antennas, such that high data rate transmission can be provided by exploiting a spatial diversity gain and a spatial multiplexing gain [25–27]. It has been widely deployed in wireless communication standards including Wi-Fi, 3G, and Long-Term Evolution (LTE). In recent years, studies on MIMO systems with a very large number of antennas at the base station (BS), a.k.a. massive MIMO, have shown its potential in reducing interuser interference (IUI) and providing quasi-orthogonal channels and enormous enhancement in spectral efficiency and power efficiency [28]. Hence, as pointed out in [29, 30], massive MIMO technology becomes one of the "big three" technologies for 5G, along with ultradensification and millimeter wave (mmWave). However, massive MIMO also brings up several critical challenges, such as high complexity of hardware, high power consumption, high cost of deploying the massive number of antennas, antenna coupling effect, and difficulty in analog front-end design as well as indoor deployment [31].

In multiuser MIMO systems, beamforming has become a widely used technique that utilizes multiple antennas to form a directional transmission beam pattern, to alleviate IUI. Beamforming can be viewed as a spatial filter, which essentially weighs and sums signals from different TX antennas, such that the desired signal is added coherently at the receiver side [32]. The most common beamforming methods include maximal ratio combining (MRC), zero-forcing (ZF), and minimal mean square error (MMSE). Optimal beamforms can be designed with specific objectives. For example, studies on MIMO beamforming have been conducted thoroughly for optimal resource allocation problems when taking into consideration either the network throughput maximization or the multiuser fairness [33, 34]. Moreover, optimal beamforms have been designed for wireless powered communications to achieve a high-efficiency power transfer, and the optimal simultaneous information and energy delivery [35–38]. Recently, optimal beamforming has been investigated in secured communications where artificial noise is transmitted to degrade eavesdropper's channel with the purpose of not impairing the information delivered to intended receivers [39–42].

This chapter aims to overview the advances in the wideband waveforming and show that it is indeed a dual problem by illustrating its similarities with the well-known narrowband MIMO beamforming, through problems on optimal resource allocation for multiuser wireless communications, wireless power transfer (WPT), and secured communications. In addition, the mathematical models for both systems are studied and compared, considering both the similarities and differences between physical antennas and virtual antennas.

This chapter is organized as follows. In Section 18.2, system models for the narrowband MIMO downlink system and the wideband downlink system are introduced and compared. The similarity and differences between physical antennas in the narrowband MIMO system and virtual antennas exploited from wideband multipath propagation are discussed. TR signal transmission as the first waveforming technique in wideband communications is introduced in Section 18.3. In Section 18.4, we investigate various problems in both the narrowband MIMO beamforming and the wideband waveforming to improve the quality-of-service (QoS), including sum-rate maximization, max-min optimization, and providing robustness. In Section 18.5, how to design optimal beamforming and waveforming in the wireless power transfer is discussed for both systems. The MIMO beamforming and the wideband waveforming, which utilize the artificial noise for secrecy enhancement in wireless communications, are discussed in Section 18.6.

18.2 System Model

In this section, we introduce and compare the system models for narrowband MIMO downlink transmission and wideband downlink transmission. In the MIMO system, beamforming is applied to create a directive transmission pattern by adjusting the weight in each physical antenna. Instead of using multiple transmit antennas at the BS, the wideband waveforming system treats each multipath component in the wireless channel as a virtual antenna. Each multipath component is related to a scatterer or a reflector in the environment, which can be resolved by a large bandwidth. In order to achieve a spatial selectivity in the wideband system, waveforming technique is used to adjust the weight on each multipath component, a.k.a., virtual antenna.

18.2.1 Bandwidth vs Multipath

The multipath can be effectively harvested by adjusting transmission power and bandwidth [31, 43]. On one hand, a higher transmission power can lead to a higher signal-to-noise ratio (SNR). Consequently, the higher the transmission power is, the more observable multipath components are. On the other hand, the spatial resolution in resolving independent multipath components, i.e., the resolution to separate radio paths with different lengths in a multipath propagation, is limited by c/B as marked in Figure 18.1, with c being the speed of light and B being the bandwidth. Therefore, the larger the bandwidth is, the better the spatial resolution is and thus the more multipaths can be revealed. The mathematical explanation of the relationship between bandwidth and multipath resolution is as follows.

In an environment with $K_{\max}$ independent multipath components existing between the TX and the RX, the continuous-time multipath channel $h(t)$ is defined as collections of different radio propagation paths, i.e., $h(t) = \sum_{k=1}^{K_{\max}} \alpha_k \delta(t - \tau_k)$, where $K_{\max}$ is the number of scatterers in the wireless transmission medium. α_k is the multipath coefficients of scatterer k, and τ_k is the time delay associated with α_k. The function $\delta(\cdot)$ is the delta function. Note that the delay spread of the wideband channel is $\tau = \max_k \tau_k$.

However, due to the limited bandwidth W, the estimated discrete-time channel $\mathbf{h}$ at the receiver side is a sampled version of $h(t)$, i.e., $h[l] = \int_{\frac{l-1}{W}}^{\frac{l}{W}} P\left(\frac{l}{W} - t\right) h(t) dt$,

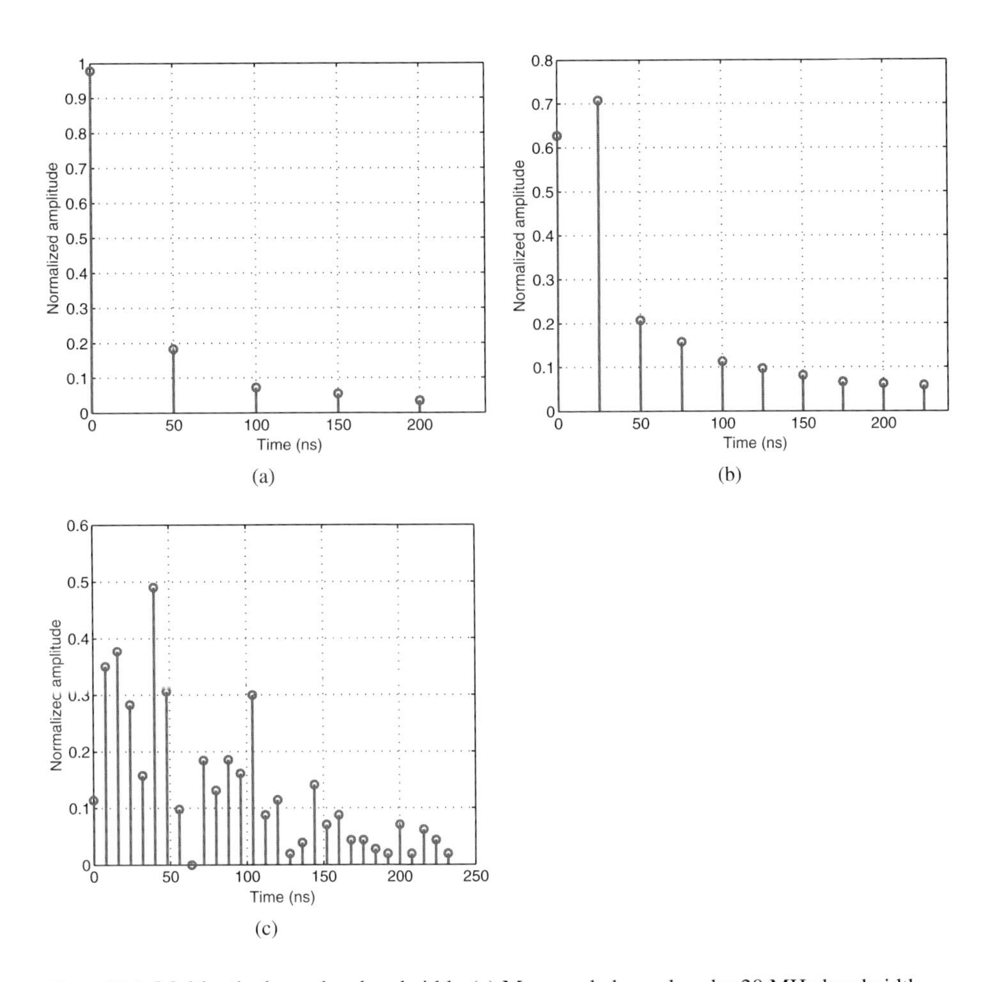

Figure 18.2 Multipath channel vs bandwidth. (a) Measured channel under 20 MHz bandwidth (LTE standard). (b) Measured channel under 40 MHz bandwidth (the IEEE 802.11n standard). (c) Measured channel under 125 MHz bandwidth (entire ISM 5G band).

where $P(\cdot)$ is the window function with length $1/W$. Given a delay spread τ, when $W \leq 1/\tau$, only a single tap is resolved as the integration of all multipaths. When $W > 1/\tau$, multipath signals received within $1/W$ seconds will be integrated into a single tap signal. Consequently, all multipaths can be resolved when $W > 1/\Delta\tau_{\min}$, where $\Delta\tau_{\min}$ represents the smallest difference in time-of-flight (ToF) of consecutively received multipath signals. Because signals with ToF difference equal to or larger than $1/W$ can be separated under a bandwidth W, a larger bandwidth enables a higher sampling rate to sample analog signals received from different paths and resolve more multipath components. Therefore, the number of resolved channel taps, i.e., length of vector $\mathbf{h}$, is determined by $L = \text{round}(\tau W)$, as long as $K_{\max}$ is large enough to provide enough multipaths whose time delay difference $\Delta\tau_{\min}$ is close to 0.

Multipath channels captured under the aforementioned bandwidths at the same location in a rich-scattering environment are plotted and compared in Figure 18.2, demonstrating the relationship between bandwidth and multipath resolution. Comparing

Figures 18.2(a), 18.2(b), and 18.2(c), it is noticed that more multipaths and better resolution are resolved when the bandwidth increases. Therefore, the larger the bandwidth is, the better the spatial resolution is and thus the more multipaths can be revealed. In the LTE standard the bandwidth is 20 MHz, and in Wi-Fi (the IEEE 802.11n standard) systems the bandwidth is 40 MHz. Moreover, the entire industrial, scientific, and medical radio (ISM) 5G band occupies a total of 125 MHz bandwidth. As projected in 5G, high carrier frequencies with larger bandwidths will be adopted in the future wireless communication systems [29], which makes the multipath channel of a good spatial resolution feasible.

18.2.2 Narrowband MIMO System

We consider a single cell multiuser narrowband MIMO downlink system with M antennas at the BS and N users each equipped with a single receive antenna. The system model is illustrated in Figure 18.3(a). Due to the narrowband transmission, each link between a pair of TX and RX antenna is a single-tap channel, i.e., a scalar as the channel coefficient. The channel matrix is defined as the collection of channel vectors from all users:

$$\mathbf{H} = \left[\mathbf{h}_1^T, \mathbf{h}_2^T, \cdots, \mathbf{h}_j^T, \cdots, \mathbf{h}_N^T\right]^T \tag{18.1}$$

where $\mathbf{H} \in \mathbb{C}^{N\times M}$ is an $N \times M$ MIMO channel matrix, and $(\cdot)^T$ denotes the transpose operation. $\mathbf{h}_j$ is the channel vector for user j, $\mathbf{h}_j = [h_{1j}, h_{2j}, \cdots, h_{ij}, \cdots, h_{Mj}]$, where $h_{ij} \in \mathbb{C}$ represents the channel coefficient from antenna i to user j, and each coefficient is independently and identically distributed (i.i.d.).

In the multiuser MIMO downlink system, the received signal $\mathbf{Y}$ at the receiver side can be modeled as

$$\mathbf{Y} = \mathbf{HGs} + \mathbf{n} \tag{18.2}$$

where $\mathbf{s} \in \mathbb{C}^{N\times 1}$ is the transmit signal vector, i.e., $\mathbf{s} = [s_1, s_2, \cdots, s_i, \cdots, s_N]^T$. $\mathbf{n} \in \mathbb{C}^{N\times 1}$ is the additive white Gaussian noise vector with zeros mean and variance σ^2, and

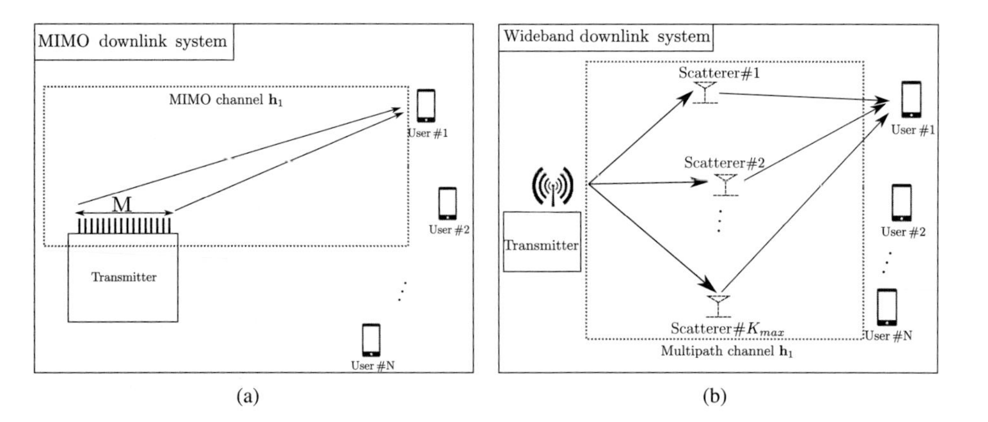

Figure 18.3 Comparison on the system model. (a) MIMO downlink system. (b) Wideband waveforming downlink system [31].

Table 18.1 Summary on system model

	MIMO downlink	Waveforming downlink
Signal	y_j in (18.3)	$y_j[m]$ in (18.4)
Single-user channel	$\mathbf{h}_j$: $1 \times M$ vector	$\mathbf{H}_j$: $\left(2\lfloor\frac{L-1}{D}\rfloor+1\right) \times L$ matrix
All users' channel	$\mathbf{H}$ in (18.1)	$\mathbf{Q}$ in (18.6)
Desired signal	$\sqrt{p_j}\mathbf{h}_j\mathbf{g}_j s_j$	$\sqrt{p_j}\mathbf{H}_j^{(\lfloor\frac{L-1}{D}\rfloor+1)}\mathbf{g}_j s_j\left[m-\lfloor\frac{L-1}{D}\rfloor-1\right]$
IUI	$\sum_{i=1,\, i\neq j}^{N}\mathbf{h}_j\sqrt{p_i}\mathbf{g}_i s_i$	$\sum_{k=1}^{2\lfloor\frac{L-1}{D}\rfloor+1}\mathbf{H}_j^{(k)}\left(\sum_{i=1,\, i\neq j}^{N}\sqrt{p_i}\mathbf{g}_i s_i[m-k]\right)$
ISI		$\sqrt{p_j}\sum_{k=1,\, k\neq\lfloor\frac{L-1}{D}\rfloor+1}^{2\lfloor\frac{L-1}{D}\rfloor+1}\mathbf{H}_j^{(k)}\mathbf{g}_j s_j[m-k]$

$\mathbf{G} \in \mathbb{C}^{M\times N}$ is the MIMO beamforming matrix defined as $\mathbf{G} = [\mathbf{g}_1, \mathbf{g}_2, \cdots, \mathbf{g}_j, \cdots, \mathbf{g}_N]$. Moreover, $\mathbf{g}_j$ in $\mathbf{G}$ is the antenna weight vector for user j, and $\|\mathbf{g}_j\|_2^2 = p_j$, where $\|\ \|_2$ denotes the L2-norm of a vector. In the following, to simplify notation, we assume $\|\mathbf{g}_j\|_2^2 = 1$, and each user has a power allocation factor p_j without loss of generality.

To further analyze the received signal in (18.2), we take user j for an example with its received signal expressed as

$$y_j = \sqrt{p_j}\mathbf{h}_j\mathbf{g}_j s_j + \sum_{\substack{i=1\\ i\neq j}}^{N}\mathbf{h}_j\sqrt{p_i}\mathbf{g}_i s_i + n_j,\ \forall j \tag{18.3}$$

where p_j represents the transmit power assigned to user j, and n_j is the additive white Gaussian noise with $n_j \sim \mathcal{CN}(0, \sigma^2)$ for user j. It is seen from (18.3) that the received signal consists of desired signal component and IUI component as listed in Table 18.1.

18.2.3 Wideband Waveforming System

In this section, we consider a wideband downlink system with N users, and an illustration of the system is plotted in Figure 18.3(b). With a large bandwidth, channels in the wideband waveforming system possess multiple taps, due to the multipath propagation. The channel between the BS and user j is defined as $\mathbf{h}_j = [h_j[1], h_j[2], \cdots, h_j[l], \cdots, h_j[L]]^T$, where $h_j[l] \in \mathbb{C}$ is a scalar representing the wideband channel coefficient on tap l for user j, and each coefficient is independent but not identically distributed.

A diagram shown in Figure 18.4 demonstrates how a wideband waveforming system works in the downlink transmission. During the downlink transmission, the information $\mathbf{s}_j$ to be transmitted to receiver j is first up-sampled by a rate backoff factor D to match the baud rate with the sampling rate [1]. Then the real transmit signal is defined as $s_j^{[D]}[k] = s_j[n]$ if $k = nD$, and $s_j^{[D]}[k] = 0$ otherwise.

Afterwards, the upsampled version of transmit signal is convolved with the waveforming vector $\mathbf{g}_j$ and multiplied with a power coefficient p_j, before combined with

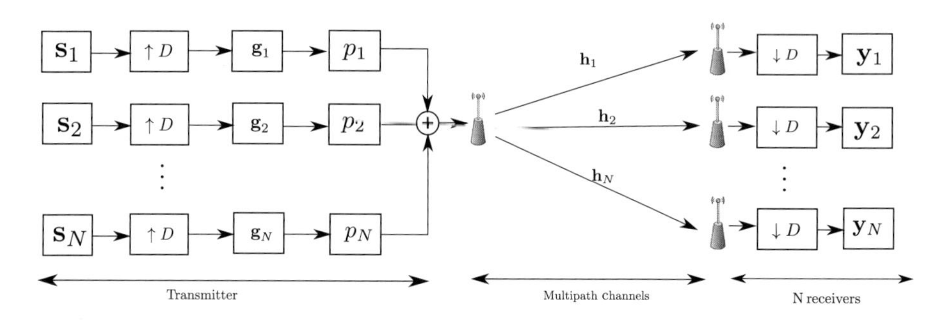

Figure 18.4 Demonstration of wideband downlinks.

other users' downlink transmit signal. $\mathbf{g}_j \in \mathbb{C}^{L\times 1}$ is the waveforming vector for user j, i.e., the time-domain multipath weight vector, with $\|\mathbf{g}_j\|_2 = 1$.

At the receiver side, the received signal of user j is first downsampled by D, and then it can be written as a vector $\mathbf{y}_j$ as

$$\begin{aligned} y_j[m] &= \left(\mathbf{H}_j\Big(\sum_{i=1}^{N}\sqrt{p_i}\mathbf{s}_i^{[D]} * \mathbf{g}_i\Big)\right)[Dm] + n_j[m] \\ &= \sqrt{p_j}\mathbf{H}_j^{(\lfloor\frac{L-1}{D}\rfloor+1)}\mathbf{g}_j s_j\left[m - \left\lfloor\frac{L-1}{D}\right\rfloor - 1\right] \\ &+ \sqrt{p_j}\sum_{\substack{k=1 \\ k\neq\lfloor\frac{L-1}{D}\rfloor+1}}^{2\lfloor\frac{L-1}{D}\rfloor+1}\mathbf{H}_j^{(k)}\mathbf{g}_j s_j[m-k] \\ &+ \sum_{k=1}^{2\lfloor\frac{L-1}{D}\rfloor+1}\mathbf{H}_j^{(k)}\Big(\sum_{\substack{i=1 \\ i\neq j}}^{N}\sqrt{p_i}\mathbf{g}_i s_i[m-k]\Big) + n_j[m] \end{aligned} \tag{18.4}$$

where $\mathbf{H}_j$ is a $\left(2\lfloor\frac{L-1}{D}\rfloor + 1\right) \times L$ Teoplitz convolution matrix generated from $\mathbf{h}_j$ as shown in (18.5). $\mathbf{H}_j^{(k)}$ is the kth row of matrix $\mathbf{H}_j$. $n_j[m]$ is the additive white Gaussian noise with $n_j[m] \sim \mathcal{CN}(0, \sigma^2)$.

$$\mathbf{H}_j = \begin{pmatrix} h_j\left[L - D\lfloor\frac{L-1}{D}\rfloor\right] & \cdots & \cdots & 0 \\ \vdots & \vdots & \ddots & \vdots \\ h_j[L-D] & h_j[L-1-D] & \cdots & 0 \\ h_j[L] & h_j[L-1] & \cdots & h_j[1] \\ 0 & 0 & \cdots & h_j[1+D] \\ \vdots & \vdots & \ddots & \vdots \\ 0 & 0 & \cdots & h_j\left[1 + D\lfloor\frac{L-1}{D}\rfloor\right]. \end{pmatrix} \tag{18.5}$$

As can be seen from (18.4), the received signal contains desired signal parts, interference signals from ISI and IUI, along with noise, which are listed in Table 18.1.

Meanwhile, the channel matrix of all users is defined as

$$\mathbf{Q} = \left[\mathbf{H}_1^T, \mathbf{H}_2^T, \cdots, \mathbf{H}_j^T, \cdots, \mathbf{H}_N^T\right]^T \tag{18.6}$$

where $\mathbf{Q}$ is an $N\left(2\lfloor\frac{L-1}{D}\rfloor + 1\right) \times L$ matrix.

Unlike the MIMO beamforming, which relies on multiple physical antennas for high degrees of freedom, in the wideband waveforming system ample degrees of freedom are provided by the nature where a fine-grained multipath channel is available with a large bandwidth. In practice, each antenna in a wireless communication system requires an independent radio-frequency (RF) chain, which includes a low-noise amplifier, an up/down converter, intermediate-frequency amplifier, an analog-to-digital (A/D) or digital-to-analog (D/A) converter, and some bandpass filters. Because multiple antennas are deployed, the cost and complexity to implement a MIMO beamforming system are much higher than those of a single-antenna wideband system. Hence, waveform design succeeds in making the wideband waveforming system outstanding for future wireless communications by supporting high data rate and reliable services, but with a simple and low-cost hardware implementation.

On the other hand, waveforming can be utilized to detect and track the changes in a multipath propagation environment with a high accuracy and has been proposed for green Internet of Things (IoT) applications, e.g., indoor locationing that achieves a centimeter level accuracy [44–46], indoor speed estimation [47], through-the-wall event detection [48], human recognition [49], and breathing rate estimation [50]. Considering its capability in wireless communications and wireless sensing, waveforming will be a promising technology for 5G and future IoT applications and deserves a thorough study.

We summarize this section and compare the system model between the MIMO downlink system and the waveforming downlink system in Table 18.1.

18.3 Time-Reversal Signal Transmission

In this section, the details of TR signal transmission are discussed, and related works are reviewed.

18.3.1 History of TR Signal Transmission

TR signal processing technique was originally proposed in 1957 to compensate the delay distortion in picture transmission [51]. Then, its applications in acoustic communications were studied, and it has been validated through a series of theoretical and experimental works that the energy of the TR acoustic wave is only focused at the intended locations [7, 8, 10, 52–57].

Later, applications of TR technique in wireless communications have received an increasing attention. In [12, 58–63], the study of TR signal transmission has been extended to applications related to the electromagnetic (EM) field, which demonstrates the TR resonance through experiments and justifies the assumption of channel stationarity and channel reciprocity given that the coherence time is long enough. Meanwhile, the performance of ultra-wideband (UWB) communication system with TR signal transmission has been studied in [11, 13–15, 64]. To further utilize the spatial diversity provided by transmission through multiple antennas, TR signal transmission has been extended to work with MIMO technology [12, 17–20]. Moreover, a TR-based OFDM system where wideband frequency-selective channels are decomposed into independent narrowband subchannels and TR is applied as a precoding method has been studied in [21].

Recently, experiments as well as theoretic analysis have been conducted to illustrate the potential of TR for future green communications [1, 16]. It was shown that a typical TR receiver can achieve over an order of magnitude in power reduction and interference alleviation [1]. Later on, the concept of time-reversal division multiple access (TRDMA) was proposed, and a system-level performance analysis was provided in [22]. In TRDMA, users in a wideband system are separated by the inherent TR resonance, and the ISI and IUI are reduced. In order to further improve the performance of the TRDMA communication system, research has been conducted to suppress interference by means of waveform design [2, 4, 6] and interference precancellation algorithm [3, 5]. A device-to-device communication system based on TR signal transmission was proposed where the optimal power allocation is determined through a Stackelberg game [65]. The performance of a massive multipath TR communication system was studied in [31], where a comprehensive comparison between the massive MIMO system and the massive multipath TR system was conducted both theoretically and through simulations.

18.3.2 How Does TR Work?

Let us consider a wideband downlink system depicted in Figure 18.3(b) with N users. A typical TR signal transmission consists of two phases: (1) the channel probing phase, and (2) the data transmission phase, as shown in Figure 18.5.

18.3.2.1 Channel Probing Phase

Before the BS transmits downlink signals to users, it needs to estimate each user's channel $\mathbf{h}_i$, $\forall\, i$ to generate TR waveforms. In order to estimate the CSI $\mathbf{h}_i$ under bandwidth W, user i first sends out a pulse signal propagating through the continuous-time multipath channel $h_i(t)$. Afterwards, the BS obtained a discrete-time CSI estimation $\mathbf{h}_i$.

18.3.2.2 Data Transmission Phase

As shown in Figure 18.4, during the downlink transmission, the discrete signal $\mathbf{s}_i$ is sent from the BS to user i after upsampling and convolving with the TR waveform $\mathbf{g}_{i,TR}$ that is defined as $\mathbf{g}_{TR,i} = \mathbf{H}_i^{(\lfloor \frac{L-1}{D} \rfloor + 1)\dagger}$, where $(\cdot)^\dagger$ denotes the Hermitian operation, i.e., transpose and conjugate.

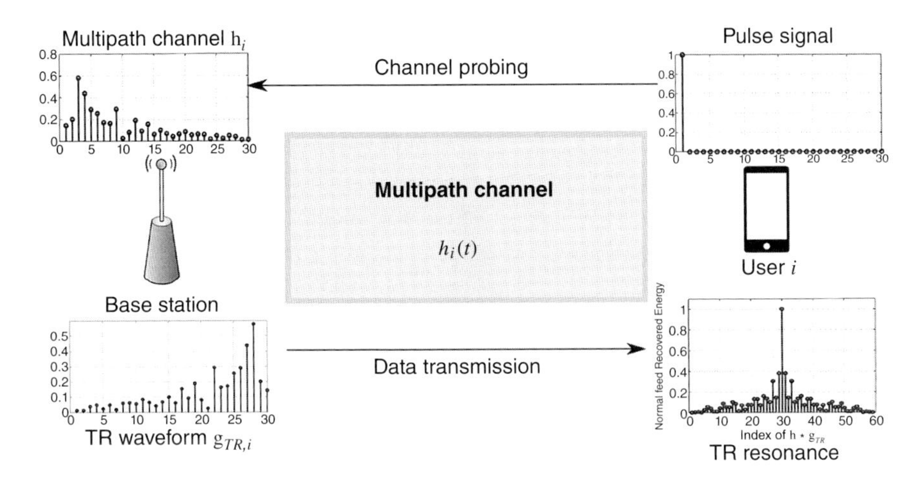

Figure 18.5 Demonstration of TR signal transmission.

Then, according to (18.4), the received signal $y_i[m]$ in a TR-based wideband waveforming system can be written as follows.

$$
\begin{aligned}
y_i[m] &= \sqrt{p_i}\|\mathbf{h}_i\|_2^2 s_i\left[m - \left\lfloor\frac{L-1}{D}\right\rfloor - 1\right] \\
&+ \sqrt{p_i} \sum_{\substack{k=1 \\ k\neq \lfloor\frac{L-1}{D}\rfloor+1}}^{2\lfloor\frac{L-1}{D}\rfloor+1} \mathbf{H}_i^{(k)} \mathbf{H}_i^{(\lfloor\frac{L-1}{D}\rfloor+1)\dagger} s_i[m-k] \\
&+ \sum_{k=1}^{2\lfloor\frac{L-1}{D}\rfloor+1} \mathbf{H}_i^{(k)} \Big(\sum_{\substack{j=1 \\ i\neq j}}^{N} \mathbf{H}_j^{(k)\dagger} \sqrt{p_j} s_j[m-k] \Big) + n_i[m]
\end{aligned}
\tag{18.7}
$$

Because of the TR resonance, energy in the desired signal is boosted by the TR resonating strength $\|\mathbf{h}_i\|_2^2$, while the energy of both ISI and IUI term is typically much smaller. The details of the TR resonance are discussed in the following.

18.3.3 TR Resonance

In a wireless communication system, ISI and IUI deteriorate the QoS of each individual user. On one hand, the spatial focusing effect of the TR resonance suppresses the energy leakage to other receivers and focuses most of the signal energy at the intended user. Consequently, TR signal transmission naturally reduces the required transmit power and lowers the IUI. As depicted in Figure 18.6(a), an example of the spatial focusing effect of the TR resonance is shown with an energy peak at the focused spatial location. On the other hand, due to the temporal focusing effect of the TR resonance, which reduces the intersymbol energy leakage, the ISI can be effectively reduced for high-speed broadband communication with the help of a large backoff factor D. In Figure 18.6(b), an example

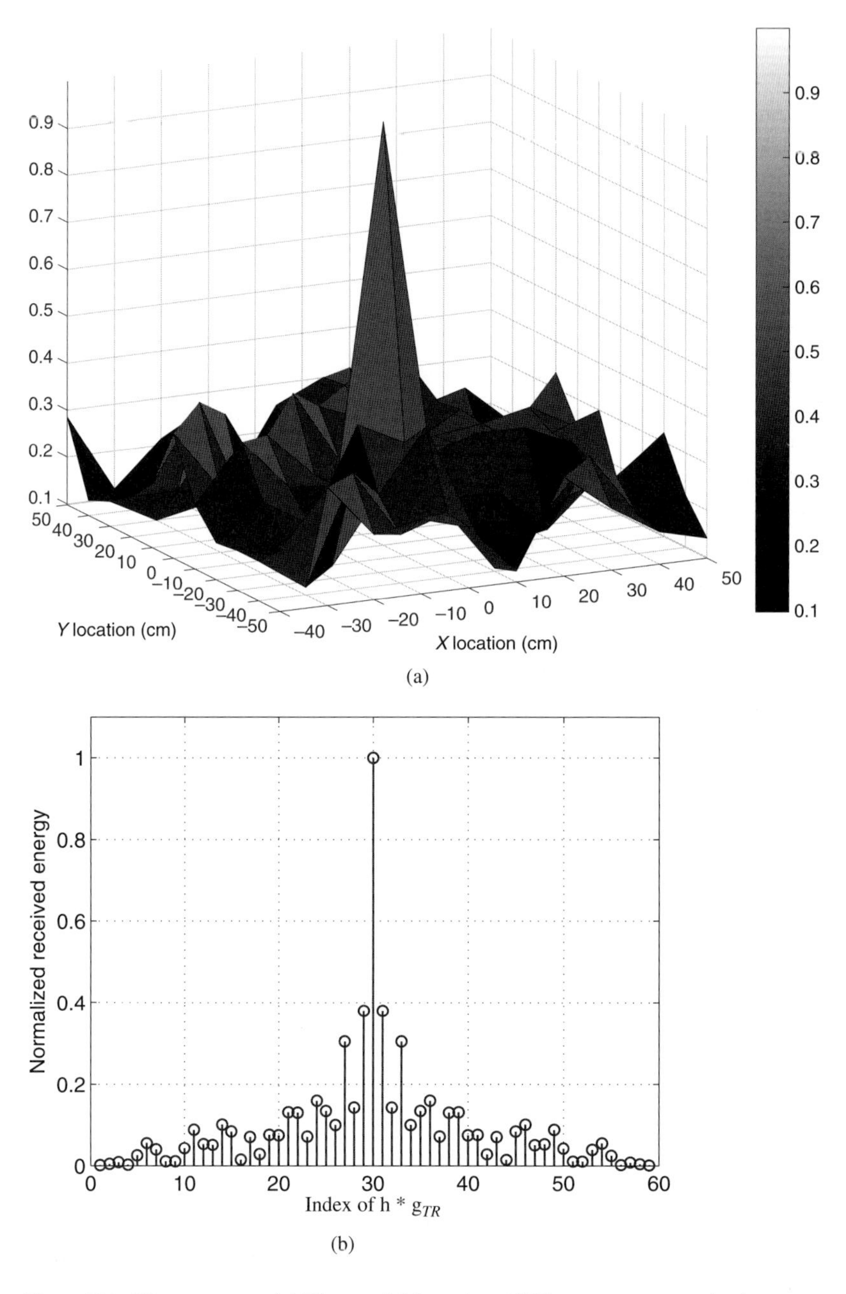

Figure 18.6 TR resonance. (a) The spatial focusing of TR resonance: received energy at different locations. (b) The temporal focusing of TR resonance: received energy at different sampling time ($L = 30$).

of the temporal focusing effect of the TR resonance is shown, where the received energy exhibits a strong peak at the time instance L, corresponding to $\mathbf{h} * \mathbf{g}_{TR}[L]$. In other words, the desired signal component in (18.7) is amplified, outstanding from ISI and IUI components.

Remark 18.3.1 *The TR waveform is the MRC waveform for the wideband communication system [31]. As presented in [66], MRC is a technique to combine received signals from different transmit paths with a weight proportional to each signal strength, which usually is the conjugated version of its corresponding channel coefficient. Consequently, strong signals are further amplified while weak signals are attenuated. Through TR waveforming, signals arrived from different virtual antennas, i.e., multipaths, are added up coherently by convoluting with its TR signature. This is equivalent to multiplying each copy of transmitted signal through different paths with its own conjugated channel coefficient. Hence, TR waveform becomes an MRC waveform.*

Due to the fact that the number of resolved multipath components is proportional to the bandwidth, the resolution of TR resonance highly depends on bandwidth. Examples are depicted in Figures 18.7(a), 18.7(b), and 18.7(c), where the TR resonance under different bandwidths is investigated through simulation, and the studies based on real measurements are in Figures 18.7(d), 18.7(e), and 18.7(f). It is obvious that when the bandwidth increases, the TR resonance has a spikier spatial focusing effect. Moreover, simulation results are consistent with those obtained from the real measured CSI.

Remark 18.3.2 *In order to have a fine resolution for the TR resonance, a large bandwidth W is required to perceive enough independent multipath components in the environment.*

Furthermore, we compare the TR waveforming and MIMO MRC beamforming, in terms of the spacial focusing of the received energy. As plotted in Figures 18.7(a) and 18.7(b), and in Figures 18.7(g) and 18.7(h), TR waveforming achieves a similar spatial focusing effect with MIMO MRC beamforming under a similar consuming bandwidth. Here, the consuming bandwidth is defined as the sum of the transmission bandwidth on each link. As shown by the focusing ball in Figures 18.7(c) and 18.7(i), when using a single TX antenna and 125 MHz bandwidth, TR waveforming outperforms by achieving the similar spatial focusing effect with MIMO MRC beamforming where 100 TX antennas are deployed and each works under 2 MHz, i.e., a consuming bandwidth of 200 MHz.

18.3.4 Massive Multipath Effect

In a wideband waveforming downlink system, channels from the single TX antenna to multiple single-antenna receivers at different locations form a channel matrix $\mathbf{Q}$, which is defined in (18.6). As shown in Figure 18.7(c), the high-resolution spatial focusing effect of TR resonance is supported by the following theorem of massive multipath effect [31].

Theorem 18.3.1 *Massive Multipath Effect [31]:*

Suppose in the channel matrix $\mathbf{Q}$ all channels are normalized as $\|\mathbf{H}_i^{(\lfloor \frac{L-1}{D} \rfloor + 1)}\|_2 = 1$, and the environment is rich with scatterers, i.e., a sufficiently large $K_{\max}$ is guaranteed. Then, when the bandwidth W goes to infinity such that it provides an extremely

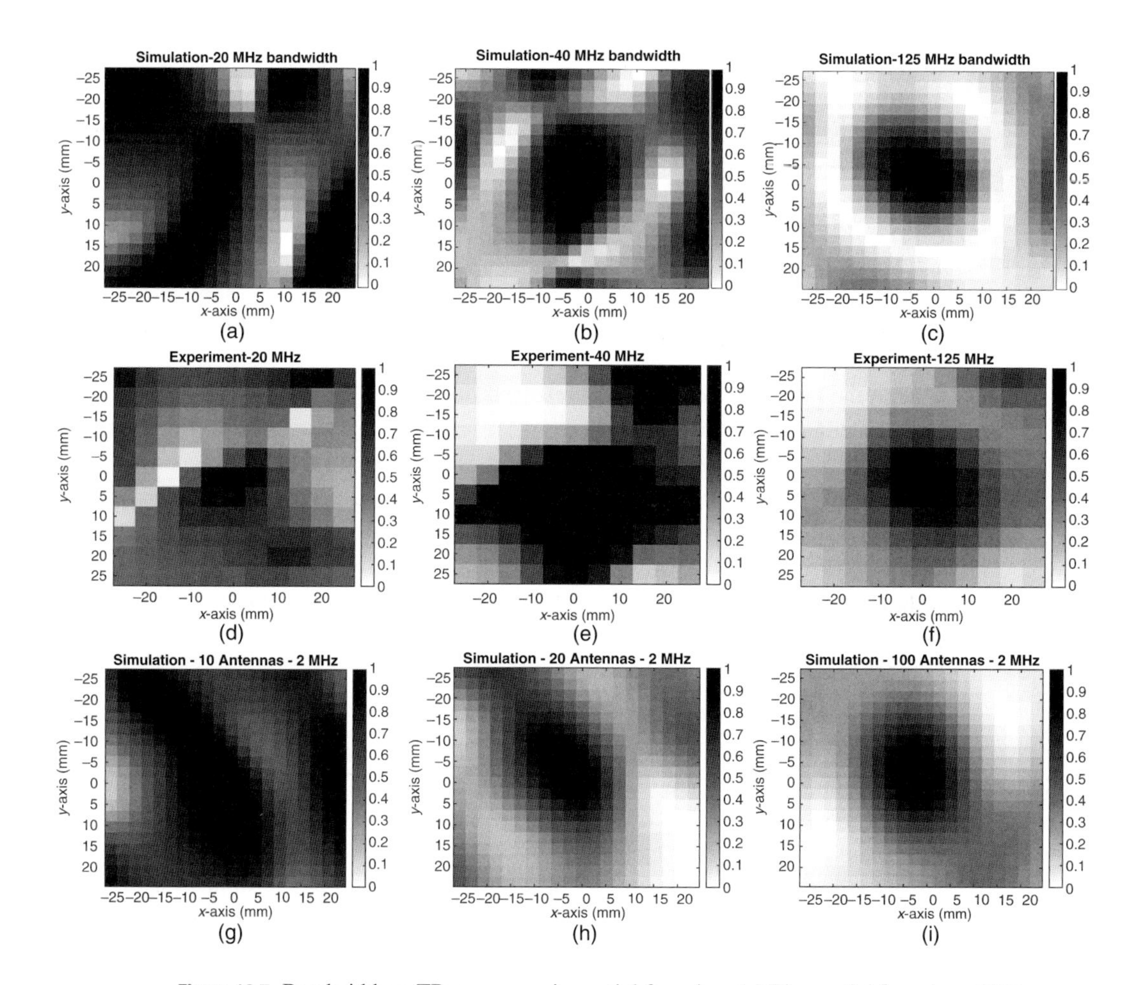

Figure 18.7 Bandwidth vs TR resonance in spatial focusing. (a) The spatial focusing of TR resonance under 20 MHz bandwidth (simulation). (b) The spatial focusing of TR resonance under 40 MHz bandwidth (simulation). (c) The spatial focusing of TR resonance under 125 MHz bandwidth (simulation). (d) The spatial focusing of TR resonance under 20 MHz bandwidth (real measurement). (e) The spatial focusing of TR resonance under 40 MHz bandwidth (real measurement). (f) The spatial focusing of TR resonance under 125 MHz bandwidth (real measurement). (g) The MIMO MRC beamforming of 10 antennas under 2 MHz bandwidth (simulation). (h) The MIMO MRC beamforming of 20 antennas under 2 MHz bandwidth (simulation). (i) The MIMO MRC beamforming of 100 antennas under 2 MHz bandwidth (simulation).

high resolution to resolve all multipath components, we have the following asymptotic behavior for the channel matrix $\mathbf{Q}$ *as*

$$\lim_{W\to\infty} \mathbf{Q}\mathbf{Q}^{\dagger} = \mathbb{I}_{N\times N} \tag{18.8}$$

where $\mathbb{I}_{N\times N}$ *represents an identity matrix with dimension* N.

In other words, the multipath channels become mutually orthogonal if the bandwidth W *to resolve multipath components is large enough under the rich multipath setting.*

Similarly, the spatial pinpoint of the MIMO beamforming as depicted in Figure 18.7(i) follows the massive MIMO effect [67], which shows that the channel matrix **H** of a narrowband MIMO system exhibits an asymptotic behavior when the number of antennas increases. The details are as follows.

Theorem 18.3.2 *Massive MIMO Effect [67]:*

When the number of antennas M grows sufficiently large, the random matrix theory provides a central limit theorem for the distribution of the singular values of **H**, *and thus under the law of large numbers we can have*

$$\lim_{M\to\infty} \frac{1}{M}\mathbf{H}\mathbf{H}^{\dagger} = \mathbb{I}_{N\times N} \tag{18.9}$$

where $(\cdot)^{\dagger}$ *denotes the Hermitian operation, i.e., transpose and conjugate, and* $\mathbb{I}_{N\times N}$ *represents an* $N \times N$ *identity matrix.*

Although it provides the spatial–temporal TR resonating effect, TR waveforming cannot eliminate the ISI and IUI completely. The ISI could be very severe when the symbol duration is smaller than the channel delay spread. As a result, adjacent transmit symbols are contaminated by each other, especially when the transmit power is high. Moreover, because channels between different users locating at different locations are not orthogonal in reality, the IUI under TR waveforming is inevitable in a multiuser wideband waveforming system. Hence, waveform design is indispensable for improving the performance of wideband systems. The optimal waveforming can be designed according to specific performance metrics similar to the MIMO beamforming optimization.

With certain optimization criteria, common waveforming/beamforming methods include the MRC, the ZF, and the MMSE [68]. With the objective to maximize the received signal-to-noise ratio (SNR), the MRC method is proposed where signals from different transmit paths are summed at the receiver after being multiplied by weight factors proportional to the conjugated version of their corresponding propagation channel [66]. Through the MRC waveforming/beamforming, signals are combined linearly and coherently such that strong signals are further amplified while weaker ones are attenuated. Consequently, the receiver SNR is maximized. With the objective to completely eliminate IUI, the ZF beamforming/waveforming has been designed such that the waveform or beamform of one user is orthogonal to other channels such that there is no energy leakage and thus no interference to other users [69]. In an estimation context, in order to minimize the estimate mean square error of the received signal, the MMSE beamforming/waveforming has been proposed [32]. The aforementioned beamform and waveform design are listed in Table 18.2, where **H** is the channel matrix for all users in a MIMO system defined by (18.1) while **Q** denotes that for a wideband waveforming system as defined in (18.6). **G** is the waveforming/beamforming matrix for all users, and $\mathbf{g}_j$ is the waveform for user j. Coefficients c_{MRC}, c_{ZF}, and c_{MMSE} are the normalized factors that guarantee the waveform and beamform to have unit energy. The vector $\mathbf{e}_{l_j}$ is an elementary vector with $l_j = (j-1)\left(2\lfloor\frac{L-1}{D}\rfloor + 1\right) + \lfloor\frac{L-1}{D}\rfloor + 1$. The detailed comparison on waveforming and beamforming are studied in the following sections.

Table 18.2 Typical beamforming vs waveforming

	Beamforming	Waveforming [31]
MRC	$\mathbf{G} = c_{MRC}\mathbf{H}^{\dagger}$	$\mathbf{g}_j = c_{MRC}\mathbf{H}_j^{(\lfloor\frac{L-1}{D}\rfloor+1)\dagger}$
ZF	$\mathbf{G} = c_{ZF}\mathbf{H}^{\dagger}\left(\mathbf{H}\mathbf{H}^{\dagger}\right)^{-1}$	$\mathbf{g}_j = \begin{cases} c_{ZF}\mathbf{Q}^{\dagger}\left(\mathbf{Q}\mathbf{Q}^{\dagger}\right)^{-1}\mathbf{e}_{l_j}, \\ \quad \text{if } \mathbf{Q} \text{ full row rank} \\ c_{ZF}\left(\mathbf{Q}^{\dagger}\mathbf{Q}\right)^{-1}\mathbf{Q}^{\dagger}\mathbf{e}_{l_j}, \\ \quad \text{if } \mathbf{Q} \text{ full column rank} \end{cases}$
MMSE	$\mathbf{G} = c_M\mathbf{H}^{\dagger}\left(\mathbf{H}\mathbf{H}^{\dagger} + \frac{1}{p_u}\mathbb{I}\right)^{-1}$	$\mathbf{g}_j = c_M\mathbf{Q}^{\dagger}\left(\mathbf{Q}\mathbf{Q}^{\dagger} + \frac{1}{p_u}\mathbb{I}\right)^{-1}\mathbf{e}_{l_j}$

18.4 Optimal Resource Allocation

In what follows, the joint problem of power allocation and waveforming/beamforming design is studied with the objective being the sum-rate maximization, max-min optimization, or providing robustness. Moreover, in this section, we use $(\cdot)^*$ to denote an optimal solution.

18.4.1 Sum-Rate Maximization

One important QoS in a multiuser wireless communication system is the total throughput of the network, which is usually defined as the weighted sum-rate of all users. This kind of problem is usually formulated as a joint optimization of power allocation and transmit waveforming/beamforming design, subject to a constraint on total transmit power.

One example of the joint problem can be written as

$$\begin{aligned} \underset{\mathbf{p},\mathbf{G}}{\text{maximize}} \quad & \sum_{i=1}^{N} w_i \log(1 + \text{SINR}_i) \\ \text{subject to} \quad & \|\mathbf{p}\|_1 \leq P,\ p_i \geq 0, \\ & \|\mathbf{g}_i\|_2 = 1,\ \forall i \end{aligned} \tag{18.10}$$

where $\mathbf{p}$ is the power allocation vector with entry p_i as the downlink transmit power assigned to user i, P is the total transmit power constraint, and $\mathbf{g}_i$ is the transmit waveform/beamform of user i. SINR_i represents the received signal-to-interference-plus-noise ratio (SINR) of user i in the downlink transmission, which is a function of the power $\mathbf{p}$ and the matrix $\mathbf{G}$. In order to allocate resources optimally, considering various quality-of-service (QoS) requirements or priorities for different users, weight factors $w_i > 0, \forall i$ are introduced to the optimization problem as predefined hyperparameters to enable heterogeneity. If all users share a homogeneous QoS requirement or priority, $w_i = 1$ for all users. Otherwise, w_i will be different for users with heterogeneous QoS requirements or priorities, and a high priority or QoS requirement corresponds to a large

w_i. Consequently, in the optimal solution users associated with a larger weight w_i will be assigned with more resources to meet their own requirements. Because only the ratio between the weight factor and the sum of all weight factors, i.e., $w_i / \sum_{i=1}^{N} w_i$, affects the optimal power allocation, there is no need to have a constraint on $w_1 + w_2 + \cdots + w_N = 1$. Moreover, because the achievable rate of user i with bandwidth W_i is defined as $W_i \log(1 + \text{SINR}_i)$ and W_i can be incorporated in the weight w_i, we assume $W_i = 1$ in the objective function of (18.10) without loss of generality.

In the MIMO downlink system, according to the signal model in (18.3), the received SINR_i is defined as

$$\text{SINR}_i = \frac{p_i \mathbf{g}_i^\dagger \mathbf{h}_i^\dagger \mathbf{h}_i \mathbf{g}_i}{\sum_{\substack{j=1 \\ j \neq i}}^{N} p_j \mathbf{g}_j^\dagger \mathbf{h}_i^\dagger \mathbf{h}_i \mathbf{g}_j + \sigma^2} \tag{18.11}$$

where the only interference comes from IUI, and $\mathbf{h}_i$ is the MISO channel vector between the multi-antenna BS and the single-antenna user i.

On the other hand, considering the existence of ISI and IUI, the definition of SINR in a multiuser wideband downlink system is different and given in (18.12). In (18.12), inside the denominator, the term $p_i \mathbf{g}_i^\dagger \left(\mathbf{H}_i^\dagger \mathbf{H}_i - \mathbf{H}_i^{(\lfloor \frac{L-1}{D} \rfloor + 1)^\dagger} \mathbf{H}_i^{(\lfloor \frac{L-1}{D} \rfloor + 1)} \right) \mathbf{g}_i$ accounts for the ISI of user i, and $\sum_{j=1, j \neq i}^{N} p_j \mathbf{g}_j^\dagger \mathbf{H}_i^\dagger \mathbf{H}_i \mathbf{g}_j$ is the IUI energy.

$$\text{SINR}_i = \frac{p_i \mathbf{g}_i^\dagger \mathbf{H}_i^{(\lfloor \frac{L-1}{D} \rfloor + 1)^\dagger} \mathbf{H}_i^{(\lfloor \frac{L-1}{D} \rfloor + 1)} \mathbf{g}_i}{p_i \mathbf{g}_i^\dagger \left(\mathbf{H}_i^\dagger \mathbf{H}_i - \mathbf{H}_i^{(\lfloor \frac{L-1}{D} \rfloor + 1)^\dagger} \mathbf{H}_i^{(\lfloor \frac{L-1}{D} \rfloor + 1)} \right) \mathbf{g}_i + \sum_{\substack{j=1 \\ j \neq i}}^{N} p_j \mathbf{g}_j^\dagger \mathbf{H}_i^\dagger \mathbf{H}_i \mathbf{g}_j + \sigma^2} \tag{18.12}$$

Remark 18.4.1 *As mentioned earlier, a backoff factor D is introduced in the wideband system to match rate and combat and eliminate the ISI resulting from the delay spread in a multipath channel. Hence, the actual rate R_j, i.e., the spectral efficiency, for user j in the wideband waveforming system is defined as $R_j = \frac{1}{D} \log(1 + SINR_j)$ [22].*

In (18.10), we combine the constant $\frac{1}{D}$ into the weight factor w_i for the sake of a unified notation.

The solution to the joint optimization problem in (18.10) has been studied for both the MIMO downlink system [70] and the wideband downlink system [2]. In general, despite the different forms of SINR, the optimal power allocation vector $\mathbf{p}^*$ and the optimal waveform or beamform $\mathbf{G}^*$ for the weighted sum-rate maximization problem can be obtained through an iterative optimizing process, described as follows.

(i) The problem is first converted into an uplink optimization through the downlink-uplink duality [71], and it becomes a maximization of the uplink sum-rate with uplink power allocation scheme $\mathbf{q}$ and waveforming/beamforming scheme $\mathbf{S}$.

(ii) By fixing the uplink power allocation scheme $\mathbf{q}$, the optimal uplink waveforming/beamforming scheme $\mathbf{S}$ that maximizes sum-rate can be optimized.

(iii) Given a waveform/beamform matrix $\mathbf{S}$, the best power allocation vector $\mathbf{q}$ is optimized through the *water-filling* algorithm.

(iv) Iterate between step 2 and 3 until the solution converges, then the optimal downlink waveform/beamform is the $\mathbf{G}^*$ obtained directly as the optimal uplink $\mathbf{S}^*$, and then the downlink power allocation is optimized with $\mathbf{G}^*$ and has a closed form.

REMARK 18.4.2 *Given a power allocation* $\mathbf{p}$*, the optimal waveform/beamform* $\mathbf{G}$*, which maximizes the individual rate, has the same form as the MMSE waveform or beamform listed in Table 18.2 [2].*

In the following, we briefly introduce three different optimization problems derived from the weighted sum-rate maximization in (18.10).

18.4.1.1 Transmit Power Minimization

In reality, the performance of a wireless communication system is limited by battery life, which supports the transmission between the BS and users. In order to extend the working time of the system while maintaining a desired QoS, as the dual problem to (18.10), it has been proposed to minimize the total downlink transmit power under the constraint on the minimum of weighted sum-rate. The problem is written as follows.

$$\begin{aligned} \underset{\mathbf{p},\mathbf{G}}{\text{minimize}} \quad & \|\mathbf{p}\|_1 \\ \text{subject to} \quad & \sum_{i=1}^{N} w_i \log(1+\text{SINR}_i) \geq R_{sum}, \\ & p_i \geq 0,\ \|\mathbf{g}_i\|_2 = 1,\ \forall i \end{aligned} \tag{18.13}$$

where $\|\cdot\|_1$ denotes the L1-norm of a vector, i.e., $\|\mathbf{p}\|_1 = \sum_{i=1} |p_i|$. R_{sum} is the minimum requirement of the weighted sum-rate, served as the QoS requirement for this battery-saving system.

The problem in (18.13) has been studied in [70], and it can be converted to a geometric programing (GP) problem involving mean square errors (MSEs), and solved iteratively through downlink-uplink duality [71].

18.4.1.2 Individual Rate Constraint

There is a drawback for the joint optimization problem in (18.10). At the optimal solution, all channel resources, i.e., the transmit power, are allocated to those users with good channel quality so that the corresponding weighted data rates are maximized. Consequently, there will be no or very limited service provided to the rest of users.

To address the inherent unfairness, a variation of the sum-rate maximization problem has been proposed, by adding an additional individual rate constraint for each user. The problem is rewritten as

$$
\begin{aligned}
\underset{\mathbf{p},\mathbf{G}}{\text{maximize}} \quad & \sum_{i=1}^{N} w_i \log(1 + \text{SINR}_i) \\
\text{subject to} \quad & \|\mathbf{p}\|_1 \leq P,\ p_i \geq 0, \\
& \|\mathbf{g}_i\|_2 = 1, \\
& \log(1 + \text{SINR}_i) \geq R_i,\ \forall i
\end{aligned}
\tag{18.14}
$$

where R_i denotes the minimum required individual rate for user i. With a tightened feasible set compared with the one in (18.10), the optimal weighted sum-rate in (18.14) is smaller.

In [72], a distributed iterative algorithm for QoS-constrained weighted sum-rate maximization (QCWSRM) problem was proposed based on the alternating direction method of multipliers (ADMM) for the multiuser MIMO beamforming system, after reformulating the problem into an equivalent weighted minimal mean square error (WMMSE) framework.

18.4.1.3 Dual Problem of the Individual Rate Constraint

As the dual problem to (18.14), a joint optimization that minimizes the transmit power is presented, subject to a constraint on the individual rate. The problem is described as follows.

$$
\begin{aligned}
\underset{\mathbf{p},\mathbf{G}}{\text{minimize}} \quad & \|\mathbf{p}\|_1 \\
\text{subject to} \quad & \log(1 + \text{SINR}_i) \geq R_i, \\
& p_i \geq 0,\ \|\mathbf{g}_i\|_2 = 1,\ \forall i
\end{aligned}
\tag{18.15}
$$

Here, there is no constraint on the weighted sum-rate of the network. An efficient solution to the problem in (18.15) has been proposed in [73], with the individual SINR constraint instead of the individual rate constraint. The problem is solved through the following steps.

(i) The feasibility of (18.15) is first checked to make sure that the SINR requirements are satisfied. If so, the individual SINR is maximized with an arbitrary total transmit power, which yields a corresponding optimal $\mathbf{G}$. This step is similar to the algorithm for solving the max-min SINR problem.

(ii) Then the power allocation $\mathbf{p}$ is obtained through fixing the individual rate to the lower bound R_i and with the waveform/beamform $\mathbf{G}$. The corresponding optimal power allocation will have a closed form, and the total power consumption will be reduced then.

(iii) Repeat steps 1 and 2 with an updated total power constraint $\|\mathbf{p}\|_1$ until the problem converges or the SINR constraints become infeasible.

Remark 18.4.3 *The disadvantage of obtaining the optimal waveform or beamform from a weighted sum-rate maximization is that the system may fail to achieve a fairness* *among all users. With the most aggressive objective as to maximize the system throughput, most of resources will be allocated to only a small portion of the users whose channel qualities and thus SINRs are better compared with the others [6].*

In the following section, an optimization problem is introduced that achieves a balance, a.k.a., fairness, among users in the network.

18.4.2 Max-Min SINR Maximization

In this section, a joint optimization problem that considers the fairness among users in a multiuser wireless communication system is studied. The QoS criterion for the fairness problem is selected as the received SINR.

In general, in order to achieve fairness in a multiuser downlink system, the max-min SINR problem is proposed, whose objective is to maximize the worst received SINR among all users with a total downlink transmit power constraint. This nonconvex optimization problem is given by

$$\begin{aligned} \underset{\mathbf{p},\mathbf{G}}{\text{maximize}} \quad & \min_i \frac{\text{SINR}_i}{\beta_i} \\ \text{subject to} \quad & \|\mathbf{p}\|_1 \leq P,\ p_i \geq 0, \\ & \|\mathbf{g}_i\|_2 = 1,\ \forall i \end{aligned} \tag{18.16}$$

where $\beta_i > 0$ serves as the weight factor to support different priorities or SINR requirements among different users, which is the same as w_i in (18.10). Consequently, a higher power will be allocated to a user with a larger β_i to support a higher SINR in the optimal solution. Meanwhile, the weighted SINR $\frac{\text{SINR}_i}{\beta_i}$ can be viewed as a virtual SINR for user i, taking into account not only the channel quality but also the individual weighted QoS requirement.

To solve the problem in (18.16) efficiently, a SINR lower bound γ is introduced as the slack variable, and the max-min problem can be rewritten into (18.17). The joint optimization problem is now aimed at maximizing the lower bound γ.

$$\begin{aligned} \underset{\mathbf{p},\mathbf{G},\gamma}{\text{maximize}} \quad & \gamma \\ \text{subject to} \quad & \|\mathbf{p}\|_1 \leq P,\ p_i \geq 0, \\ & \|\mathbf{g}_i\|_2 = 1,\ \text{SINR}_i \geq \beta_i \gamma,\ \forall i \end{aligned} \tag{18.17}$$

Remark 18.4.4 *The problem in* (18.16) *is equivalent to the one in* (18.17) *in that the optimum* γ^* *of* (18.17) *is equal to the optimum of the weighted worst SINR in* (18.16), *with the same optimal variables* **G** *and* **p** *[6, 73].*

Subject to the total downlink transmit power constraint, the optimization problem (18.16) and (18.17) can be solved by an iterative optimization process as follows.

(i) First, the downlink problem is converted to an uplink optimization through the downlink-uplink duality [71], with optimal variables **q** and **S**.

(ii) Given an uplink transmit power allocation vector $\mathbf{q}$, the optimal uplink waveforming/beamforming matrix $\mathbf{S}$ is the MMSE waveform/beamform defined in Table 18.2.
(iii) Once the waveforming/beamforming matrix $\mathbf{S}$ is fixed, the optimal uplink power allocation vector $\mathbf{q}$ is obtained through solving a Perron–Frobenius eigen problem.
(iv) Iteratively optimize between step 2 and step 3 until reaching the converging point. Then the optimal waveform/beamform for downlink transmission is $\mathbf{G}^* = \mathbf{S}^*$, and the optimal downlink power allocation vector $\mathbf{p}^*$ has a closed form and can be solved with $\mathbf{G}^*$.

Remark 18.4.5 *At the optimum solution, the virtual SINRs* $\frac{SINR_i}{\beta_i}$ *of all users are balanced, and the fairness is delivered. In other words, the optimal weighted SINRs for all users in the problem* (18.16) *are identical to each other [6, 34].*

Remark 18.4.6 *If the max-min problem has an individual power constraint* $\mathbf{p} \preceq \mathbf{p}_{th}$, *then it is first relaxed to have a total power constraint as* $\|\mathbf{p}\|_1 \leq \|\mathbf{p}_{th}\|_1$. *Afterwards, the relaxed problem has the same formulation as in* (18.16) *and can be solved iteratively. However, in order to find the optimal power allocation with the corresponding optimal* $\mathbf{G}$, *the total power constraint* $\|\mathbf{p}_{th}\|_1$ *should be tightened gradually until the optimal power allocation vector meets the individual power constraint [6].*

The detailed solution to the problem in (18.16) in a MIMO beamforming system is discussed in [34], whereas the details for a wideband waveforming system are studied in [6].

18.4.3 Robustness

All the problems discussed earlier are based on the assumption that the CSI estimation at the BS is perfect and the channel is stationary and reciprocal. However, in the real world, the CSI estimation is hardly accurate because of quantization errors, temporal and frequency offset in reciprocal channels, as well as continuous fading and fluctuation of channels. Hence, it is reasonable and necessary to take the imperfectness of the CSI estimation into consideration.

In this section, robust transmit strategies against imperfect CSI estimation are discussed with optimal power allocation and waveform/beamform design for a single-user downlink system. Aiming at addressing the performance degradation caused by imperfect CSI and improving the robustness of system, the problem is typically formulated as follows.

$$\begin{aligned} \underset{\mathbf{g}}{\text{maximize}} \quad & \min_{\mathbf{H}\in S} \text{SNR} \\ \text{subject to} \quad & \text{Tr}\left(\mathbf{g}^\dagger \mathbf{g}\right) \leq P \end{aligned} \tag{18.18}$$

where SNR denotes the received SNR, $\mathbf{H}$ is the CSI matrix between the transmitter and the receiver, S represents the collection of the possible imperfectness in the CSI estimation, P gives the constraint on the transmit power, and the function $\mathrm{Tr}(\cdot)$ is the operation to take the trace of a matrix.

The definition of SNR is different between the MIMO beamforming system and the wideband waveforming system. Assuming the noise has unit variance, the received SNR in a MIMO beamforming system is defined as $\mathrm{SINR}_{Ea} = \mathbf{g}_I^\dagger \mathbf{H}_{Ea}^\dagger \mathbf{H}_{Ea} \mathbf{g}_I \Big/ \Big(\mathbf{g}_E^\dagger \mathbf{H}_{Ea}^\dagger \mathbf{H}_{Ea} \mathbf{g}_E + \mathbf{g}_{AN}^\dagger \mathbf{H}_{Ea}^\dagger \mathbf{H}_{Ea} \mathbf{g}_{AN} + \sigma_{Ea}^2 \Big)$, and the set $S = \{\mathbf{\Delta}, \|\mathbf{\Delta}\| \leq \epsilon\}$ is the set for possible errors in the CSI estimation, where ϵ denotes the maximal energy of error in the matrix norm [74].

On the other hand, given a zero-mean and unit-variance noise, the received SNR in a wideband waveforming system is given as $\mathrm{SNR} = \Big|(\mathbf{h} * \mathbf{g})[L]\Big|^2 = \mathbf{g}^\dagger \mathbf{H}^{(\lfloor \frac{L-1}{D} \rfloor + 1)^\dagger} \mathbf{H}^{(\lfloor \frac{L-1}{D} \rfloor + 1)} \mathbf{g}$. The set S in (18.18) is the collection of convolution matrices $\mathbf{H}$ defined in (18.5) that are generated from imperfect CSI estimations.

By introducing a slack variable γ, a variation of the problem in (18.18) is written as

$$\begin{aligned} &\underset{\mathbf{g}}{\text{maximize}} \quad \gamma \\ &\text{subject to} \quad \min_{\mathbf{H} \in S} \mathrm{SNR} \geq \gamma, \ \mathrm{Tr}\Big(\mathbf{g}^\dagger \mathbf{g}\Big) \leq P \end{aligned} \tag{18.19}$$

where γ serves as the lower bound on the worst received SNR among all possible imperfect CSI matrices.

After being rewritten into the form of (18.19), the joint optimization problem in (18.18) can be addressed after being relaxed relaxing into a semidefinite programming (SDP) problem, which can be solved efficiently.

Moreover, with the objective of minimizing the total transmit power under a minimum individual SNR requirement, the dual problem to (18.18) is proposed as follows.

$$\begin{aligned} &\underset{\mathbf{g}}{\text{minimize}} \quad \mathrm{Tr}\Big(\mathbf{g}^\dagger \mathbf{g}\Big) \\ &\text{subject to} \quad \min_{\mathbf{H} \in S} \mathrm{SNR} \geq \gamma \end{aligned} \tag{18.20}$$

where γ is no longer an optimal variable.

The optimal wideband waveforming for robustness has been proposed for indoor locationing applications to adjust the size of the TR resonating ball, and thus the locationing resolution, considering the backoff factor D is much larger than the delay spread to eliminate ISI [75]. Due to the large delay spread in the wideband system, the ISI is significant and may degrade the received signal quality measured by the received SINR. Hence, it is worthwhile to evaluate and optimize the robustness in a wideband waveforming system, which is discussed in the following section.

18.4.3.1 Robustness of a Single-User Wideband Waveforming System

To begin with, considering a single-user wideband communication system, the received SINR is defined in (18.12) with only ISI contributes as the interference. Then, the robustness optimization problem for the single-user wideband waveforming system becomes a max-min SINR problem and is given by

$$\begin{aligned} \underset{\mathbf{g}}{\text{maximize}} \quad & \min_{\mathbf{H} \in S} \text{SINR} \\ \text{subject to} \quad & \text{Tr}(\mathbf{g}^{\dagger}\mathbf{g}) \leq P \end{aligned} \tag{18.21}$$

where S is the collection of convolution matrices generated by imperfect CSI estimations.

18.4.3.2 Robustness of a Multiuser Wideband Broadcasting Waveforming System

In a multiuser broadcasting system, because the BS transmits the same signal to all users with the same power, there is no IUI, and users share the same waveform $\mathbf{g}$. Hence, the received SINR for user j is rewritten as follows.

Given the definition of SINR in (18.12), the robustness problem in multi-user broadcasting system is then formulated as

$$\begin{aligned} \underset{\mathbf{g}}{\text{maximize}} \quad & \min_{\mathbf{H}_j \in S_j, j} \text{SINR}_j \\ \text{subject to} \quad & \text{Tr}(\mathbf{g}^{\dagger}\mathbf{g}) \leq P \end{aligned} \tag{18.22}$$

where S_j is the set of channel matrices from all possible imperfect CSI estimations of user j.

18.4.3.3 Robustness of a Multiuser Wideband Multicasting Waveforming System

In this kind of system, both ISI and IUI exist, due to the fact that the transmitted signal is the combination of different messages with different waveforms through the same wireless medium. Consequently, the received SINR for user j is exactly same as in (18.12). Then, the robustness problem becomes a joint optimization problem of the waveform $\mathbf{G}$ and the power $\mathbf{p}$ to maximize the worst SINR among all users against the imperfect CSI estimation. It can be rewritten as

$$\begin{aligned} \underset{\mathbf{G}}{\text{maximize}} \quad & \min_{\mathbf{H}_j \in S_j, j} \text{SINR}_j \\ \text{subject to} \quad & \text{Tr}(\mathbf{G}^{\dagger}\mathbf{G}) \leq P \end{aligned} \tag{18.23}$$

where $\mathbf{G}$ is the waveforming matrix for all users, and the problem in (18.23) has a similar form to the max-min optimization in (18.16).

Remark 18.4.7 *The three aforementioned robustness problems for wideband waveforming systems have a similar problem formulation as the max-min optimization one in* (18.16) *in Section 18.4.2. However, here it adopts a unified weight factor* $\beta_i = 1, \forall i$, *and an additional set of possible SINRs associated with different channels for the same user. Nevertheless, the same algorithms can be applied to efficiently solve the aforementioned robustness problem with SINR as the criterion.*

Table 18.3 summarizes the optimization problems studied in this section.

Table 18.3 Waveform/beamform design for optimal resource allocation

	Problem formulation	Solution
Sum-rate maximization	$\max_{\mathbf{p},\mathbf{G}} \sum_{i=1}^{N} w_i \log(1+\mathrm{SINR}_i)$ s.t. $\|\mathbf{p}\|_1 \le P,\ p_i \ge 0,$ $\|\mathbf{g}_i\|_2 = 1,\ \forall i$	MMSE waveforming Power waterfilling
Max–min SINR	$\max_{\mathbf{p},\mathbf{G}} \min_i \frac{\mathrm{SINR}_i}{\beta_i}$ s.t. $\|\mathbf{p}\|_1 \le P,\ p_i \ge 0,$ $\|\mathbf{g}_i\|_2 = 1,\ \forall i$	MMSE waveforming Perron–Frobenius eigen problem
Robustness	$\max_{\mathbf{g}} \min_{\mathbf{H}\in S} \mathrm{SNR}$ s.t. $\mathrm{Tr}\left(\mathbf{g}^{\dagger}\mathbf{g}\right) \le P$	Semidefinite programming (SDP)

18.5 Wireless Powered Communication

Traditional wireless communications are primarily constrained by the limited accessibility of spectrum resources [76, 77]. Nevertheless, as proposed for the next generation wireless communication system, the network will be moved to a higher frequency band with a larger bandwidth, leading to ample spectrum resources. However, due to the explosive growth of wireless data services and the demand of a significantly high rate in data transmission, wireless devices are subject to limited energy resources, such that convenient and perpetual energy supplies are indispensable. In this section, the comparison between the MIMO beamforming and the wideband waveforming in the wireless powered communication system is discussed.

The wireless powered communication system can operate in three different modes. The first is the WPT system, where the receiver only receives power without decoding the message sent from the transmitter. Moreover, the wireless powered communication can operate in a way such that the power is transferred wirelessly during the downlink, whereas information is sent in the uplink with the energy harvested from the downlink transmission. The third operating mode is called the simultaneous wireless information and power transfer (SWIPT), which simultaneously supports the transmission of both the energy and the information during downlink but under a constraint on the power and spectrum consumption [35–38]. The detailed discussion is as follows.

18.5.1 Wireless Power Transfer System

To begin with, the WPT system is studied in this section with the objective to pursue a maximal reception power under a constraint on the total transmit power. A system

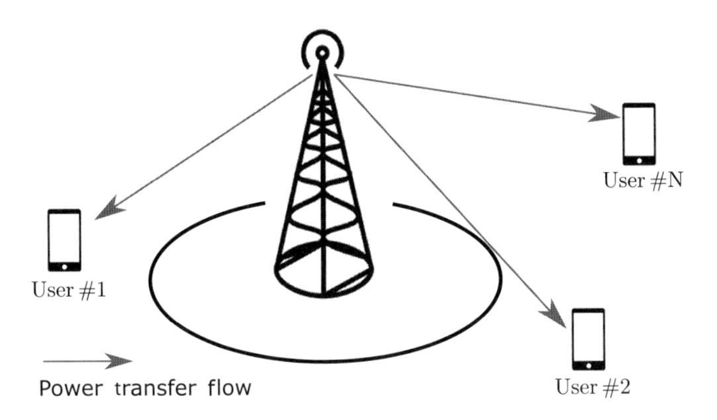

Figure 18.8 A WPT system.

diagram is shown in Figure 18.8. The mathematical formulation of the WPT problem becomes

$$\begin{aligned} \underset{\mathbf{g}}{\text{maximize}} \quad & E\left[\|\mathbf{y}\|_2^2\right] \\ \text{subject to} \quad & \text{Tr}(\mathbf{g}^\dagger \mathbf{g}) \leq P \end{aligned} \tag{18.24}$$

where $\mathbf{y}$ denotes the received signal, $\mathbf{g}$ is the energy waveform or beamform for WPT, and P is the maximum total transmit power. The definition of the received energy $\|\mathbf{y}\|_2^2$ is different in the MIMO beamforming system and the wideband waveforming system. In the MIMO beamforming system, it is defined as $\|\mathbf{y}\|_2^2 = \mathbf{g}^\dagger \mathbf{H}^\dagger \mathbf{H} \mathbf{g}$, where $\mathbf{H}$ is the MIMO channel matrix [35].

On the other hand, in the wideband waveforming system, the definition of $\|\mathbf{y}\|_2^2$ is given in [24] as $\|\mathbf{y}\|_2^2 = \left|(\mathbf{h} * \mathbf{g})[L]\right|^2 = \mathbf{g}^\dagger \mathbf{H}^{(\lfloor \frac{L-1}{D} \rfloor + 1)^\dagger} \mathbf{H}^{(\lfloor \frac{L-1}{D} \rfloor + 1)} \mathbf{g}$.

According to the definition of $\|\mathbf{y}\|_2^2$, it is easy to find out that the optimal energy waveform or beamform for the WPT problem in (18.24) is the principal eigenvector of channel covariance matrix, which is $\mathbf{H}^\dagger \mathbf{H}$ for a MIMO WPT system and $\mathbf{H}^{(\lfloor \frac{L-1}{D} \rfloor + 1)^\dagger} \mathbf{H}^{(\lfloor \frac{L-1}{D} \rfloor + 1)}$ for a wideband WPT system.

Furthermore, the single-tone waveforming/beamforming has been proposed as a simple but suboptimal solution for WPT, where all the power is allocated on the most efficient frequency component of the channel [24]. Considering the fact that the optimal energy waveform or beamform maximizes the average received power in time domain, it is superior to any other schemes. Hence, the harvested energy by means of the single-tone waveform or beamform can serve as a performance lower bound of the optimal energy waveform or beamform.

To achieve the single-tone waveforming in the wideband system, the following steps are followed:

(i) The user channel $\mathbf{h}$ is first converted into the frequency domain channel through the Discrete Fourier transform (DFT) operation DFT($\mathbf{h}$).

(ii) Afterwards, the principal component δ is discovered from $\delta = \text{argmax}_k$ DFT($\mathbf{h}$)[k], where k is the index of frequency component.

(iii) Finally, the single-tone waveform $\mathbf{g}$ is obtained by the inverse-DFT of δ, IDFT(δ).

However, in order to find the single-tone beamform for the MIMO system, a 2-D DFT is required because the channel $\mathbf{H}$ is a matrix.

Remark 18.5.1 *For periodical signals, the single-tone waveform is the same as the optimal energy waveform obtained from* (18.24) *because the DFT matrix becomes a singular matrix of the channel covariance matrix [24].*

18.5.2 Simultaneous Wireless Information and Power Transfer

In this part, the SWIPT system is studied where the WPT and wireless communications are conducted simultaneously. The performance trade-off between the delivered energy and the information capacity for the SWIPT system has been studied in [78, 79]. Huang et al. has investigated a broadband SWIPT system where OFDM and beamform design are deployed to create parallel subchannels to simplify resource allocation [80]. In the following, two major kinds of optimization problems that consider the energy-information trade-off in SWIPT are introduced.

18.5.2.1 Interference as Energy

The first formulation is a joint information and energy waveform/beamform design problem, which treats the interference from information delivery as the transmitted energy. An example of system diagram is shown in Figure 18.9(a). It is designed to maximize the weighted sum-power delivered to all receivers subject to individual SINR constraints [36] as

$$
\begin{aligned}
\underset{\mathbf{g}_E, \mathbf{g}_I}{\text{maximize}} \quad & E\left[\|\mathbf{y}_E\|_2^2\right] + E\left[\|\mathbf{y}_I\|_2^2\right] \\
\text{subject to} \quad & \text{Tr}(\mathbf{g}_E^\dagger \mathbf{g}_E) + \text{Tr}(\mathbf{g}_I^\dagger \mathbf{g}_I) \leq P, \\
& \text{SINR}_I \geq \gamma
\end{aligned}
\tag{18.25}
$$

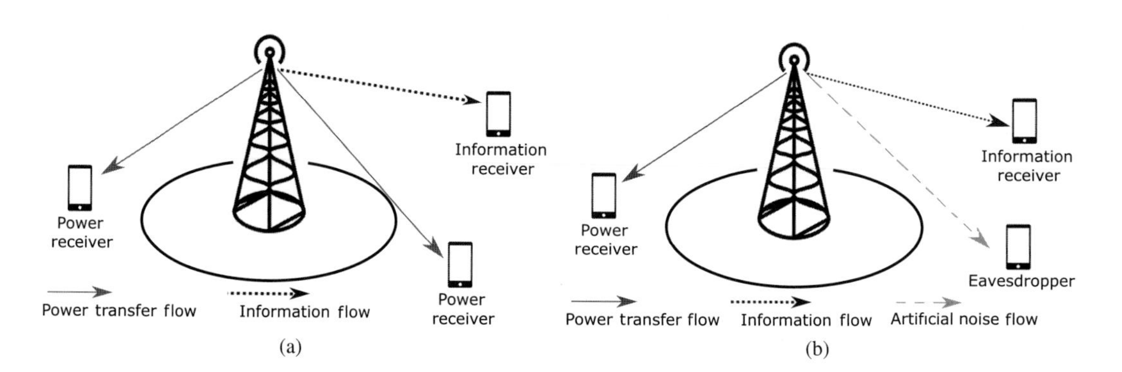

Figure 18.9 Examples of SWIPT. (a) Interference as energy. (b) Interference as security.

where $\mathbf{g}_E$ is the waveform or beamform for power transfer, $\mathbf{g}_I$ represents the waveform or beamform for information delivery, $\mathbf{y}_E$ is the power signal received at the power receiver, $\mathbf{y}_I$ is the information signal received at the power receiver, and SINR_I represents the received SINR of the information signal at the information receiver.

In the MIMO beamforming system, the received energy at the energy receiver contains $\|\mathbf{y}_E\|_2^2 = \mathbf{g}_E^\dagger \mathbf{H}_E^\dagger \mathbf{H}_E \mathbf{g}_E$ and $\|\mathbf{y}_I\|_2^2 = \mathbf{g}_I^\dagger \mathbf{H}_E^\dagger \mathbf{H}_E \mathbf{g}_I$, where $\mathbf{H}_E$ is the channel matrix between the BS and the power receiver and $\mathbf{H}_I$ is the channel matrix between the BS and the information receiver. The received SINR of the information signal at the information receiver, SINR_I, is defined in (18.11).

On the other hand, in a wideband waveforming system, the SINR of the information signal at the information receiver follows the definition in (18.12), with $\mathbf{g}_I$ and $\mathbf{H}_I$ as the intended waveform and channel matrix. Moreover, considering the existence of ISI, the definitions of received energy at the power receiver from the energy waveform is different, i.e., $\|\mathbf{y}_E\|_2^2 = \left|(\mathbf{h}_E * \mathbf{g}_E)[L]\right|^2 = \mathbf{g}_E^\dagger \mathbf{H}_E^{(\lfloor \frac{L-1}{D} \rfloor + 1)^\dagger} \mathbf{H}_E^{(\lfloor \frac{L-1}{D} \rfloor + 1)} \mathbf{g}_E$, compared with the definition in the MIMO beamforming system.

The joint information and power transfer waveform/beamform optimization with the problem formulation in (18.25) is a nonconvex quadratically constrained quadratic program (QCQP). The problem can be addressed in the following steps.

(i) The first is to check the feasibility of (18.25). In particular, the problem is feasible when the SINR constraint for information users are achievable without considering energy users.

(ii) If the problem is feasible, then a semidefinite relaxation (SDR) is applied to convert the nonconvex QCQP problem into a semidefinite program (SDP), which can be solved efficiently.

For a MIMO SWIPT system, it has been proved that for the problem in (18.25), under the condition of independently distributed user channels, there is no need to design dedicated energy beamforming vector. The optimal transmission strategy is to adjust the weights and power allocation of only the information beamforming to maximize the sum-power delivered [36]. Moreover, because the energy waveform/beamform is not for information transmission, an interference precancellation might be adopted at the information receiver to further improve the received SINR.

An alternative formulation of waveform/beamform design that considers the information-energy trade-off is to maximize the information rate under the constraint of the delivered power, as the dual problem to the aforementioned one in (18.25). The mathematical formulation of the dual problem can be written as

$$
\begin{aligned}
\underset{\mathbf{g}_E, \mathbf{g}_I}{\text{maximize}} \quad & \log(1 + \text{SINR}_I) \\
\text{subject to} \quad & \text{Tr}(\mathbf{g}_E^\dagger \mathbf{g}_E) + \text{Tr}(\mathbf{g}_I^\dagger \mathbf{g}_I) \le P, \\
& E\left[\|\mathbf{y}_E\|_2^2\right] + E\left[\|\mathbf{y}_I\|_2^2\right] \ge P_{th}
\end{aligned}
\tag{18.26}
$$

where SINR_I is the received SINR at the information receiver, and $\mathbf{y}_I$ and $\mathbf{y}_E$ are the received signal at the energy receiver from the information transmitter and the energy transmitter, respectively. P_{th} is the the minimum required power delivered to the energy transmitter.

18.5.2.2 Interference as Security

In [41, 81], by using the energy beamform as a secrecy enhancement tool, the resource allocation algorithm is designed for secure communication in MIMO beamforming systems with concurrent wireless information and power transfer. This formulation, named interference-as-security, was proposed to minimize the transmit power, by leveraging both the artificial noise and energy beamforms to facilitate an efficient energy transfer and ensure the communication security. The system demonstration is depicted in Figure 18.9(b), and a generalized and simplified mathematical problem formulation is written as follows.

$$
\begin{aligned}
\underset{\mathbf{g}_E,\mathbf{g}_I,\mathbf{g}_{AN}}{\text{maximize}} \quad & \log(1+\text{SINR}_I) - R \\
\text{subject to} \quad & \text{Tr}(\mathbf{g}_E^\dagger \mathbf{g}_E) + \text{Tr}(\mathbf{g}_I^\dagger \mathbf{g}_I) + \text{Tr}(\mathbf{g}_{AN}^\dagger \mathbf{g}_{AN}) \leq P, \\
& \max\{\log(1+\text{SINR}_E), \log(1+\text{SINR}_{Ea})\} = R, \\
& \text{SINR}_I \geq R_I, \\
& \|\mathbf{y}_I\|_2^2 + \|\mathbf{y}_E\|_2^2 + \|\mathbf{y}_{AN}\|_2^2 \geq P_{th}
\end{aligned}
\tag{18.27}
$$

where SINR_I, SINR_E, and SINR_{Ea} are the received SINRs at the information receiver, the energy receiver, and the eavesdropper, respectively. $\mathbf{g}_E$, $\mathbf{g}_I$, and $\mathbf{g}_{AN}$ are the waveforming/beamforming vectors for the energy transmission, information transmission, and artificial noise, respectively. $\|\mathbf{y}_I\|_2^2 + \|\mathbf{y}_E\|_2^2 + \|\mathbf{y}_{AN}\|_2^2$ denotes the total power delivered at the energy receiver, which can be decomposed into the energy from the information transmission, the energy transfer, and the artificial noise transmission, correspondingly.

In the MIMO beamforming system, given the prior knowledge of energy waveform for interference precancellation at the intended information receiver, the definitions of SINRs in (18.27) are different. For SINR_I and SINR_E, interference only comes from IUI of waveform $\mathbf{g}_{AN}$, i.e., $\text{SINR}_I = \mathbf{g}_I^\dagger \mathbf{H}_I^\dagger \mathbf{H}_I \mathbf{g}_I \Big/ \left(\mathbf{g}_{AN}^\dagger \mathbf{H}_I^\dagger \mathbf{H}_I \mathbf{g}_{AN} + \sigma_I^2\right)$, and $\text{SINR}_E = \mathbf{g}_I^\dagger \mathbf{H}_E^\dagger \mathbf{H}_E \mathbf{g}_I \Big/ \left(\mathbf{g}_{AN}^\dagger \mathbf{H}_E^\dagger \mathbf{H}_E \mathbf{g}_{AN} + \sigma_I^2\right)$. On the other hand, the received SINR at the eavesdropper follows the definition in (18.11), and both $\mathbf{g}_{AN}$ and $\mathbf{g}_E$ contribute to the IUI, i.e., $\text{SINR}_{Ea} = \mathbf{g}_I^\dagger \mathbf{H}_{Ea}^\dagger \mathbf{H}_{Ea} \mathbf{g}_I \Big/ \left(\mathbf{g}_E^\dagger \mathbf{H}_{Ea}^\dagger \mathbf{H}_{Ea} \mathbf{g}_E + \mathbf{g}_{AN}^\dagger \mathbf{H}_{Ea}^\dagger \mathbf{H}_{Ea} \mathbf{g}_{AN} + \sigma_{Ea}^2\right)$. $\mathbf{H}_I$, $\mathbf{H}_E$, and $\mathbf{H}_{Ea}$ represent the channel from the transmitter to the information receiver, the energy receiver, and the eavesdropper, respectively. σ_I^2, σ_E^2, and σ_{AE}^2 are the noise variance at the corresponding receiver side. The total power delivered is given as $\|\mathbf{y}_I\|_2^2 + \|\mathbf{y}_E\|_2^2 + \|\mathbf{y}_{AN}\|_2^2 = \mathbf{g}_I^\dagger \mathbf{H}_E^\dagger \mathbf{H}_E \mathbf{g}_I + \mathbf{g}_E^\dagger \mathbf{H}_E^\dagger \mathbf{H}_E \mathbf{g}_E + \mathbf{g}_{AN}^\dagger \mathbf{H}_E^\dagger \mathbf{H}_E \mathbf{g}_{AN}$.

The solution of this problem has been studied in [41], which was solved through a series of relaxations including SDRs. Then a suboptimal resource allocation scheme with low computational complexity was designed to support both WPT and secured communication.

However, the mathematical details for the problem in (18.27) are more complicated in the wideband waveforming system, and the related study is still open.

Waveform design for problems in WPT and SWIPT studied in this section is summarized in Table 18.4.

Table 18.4 Waveform/beamform design in wireless powered communications

	Problem formulation	Solution
WPT	$\max_{\mathbf{g}} \quad E\left[\|\mathbf{y}\|_2^2\right]$ s.t. $\quad \mathrm{Tr}(\mathbf{g}^\dagger \mathbf{g}) \le P$	Principal eigenvector of channel covariance
SWIPT - Interference as energy	$\max_{\mathbf{g}_E, \mathbf{g}_I} \quad E\left[\|\mathbf{y}_E\|_2^2\right] + E\left[\|\mathbf{y}_I\|_2^2\right]$ s.t. $\quad \mathrm{Tr}(\mathbf{g}_E^\dagger \mathbf{g}_E) + \mathrm{Tr}(\mathbf{g}_I^\dagger \mathbf{g}_I) \le P,$ $\mathrm{SINR}_I \ge \gamma$	SDP after semidefinite relaxations (SDRs)
SWIPT - Interference as security	$\max_{\mathbf{g}_E, \mathbf{g}_I, \mathbf{g}_{AN}} \quad \log(1 + \mathrm{SINR}_I) - R$ s.t. $\quad \mathrm{Tr}(\mathbf{g}_E^\dagger \mathbf{g}_E) + \mathrm{Tr}(\mathbf{g}_I^\dagger \mathbf{g}_I) +$ $\mathrm{Tr}(\mathbf{g}_{AN}^\dagger \mathbf{g}_{AN}) \le P,$ $\max\{\log(1 + \mathrm{SINR}_E),$ $\log(1 + \mathrm{SINR}_{Ea})\} = R,$ $\mathrm{SINR}_I \ge R_I,$ $\|\mathbf{y}_I\|_2^2 + \|\mathbf{y}_E\|_2^2 + \|\mathbf{y}_{AN}\|_2^2 \ge P_{th}$	SDP after SDRs

18.6 Secured Communications

In wireless communications, the broadcasting nature of wireless media grants access to the transmitted information not only for intended receivers but also for eavesdroppers. Hence, security becomes a fundamental and critical problem to the wireless communication system, due to the openness of wireless media that makes it vulnerable to potential eavesdropping.

Recently, plenty of research has been conducted on the information-theoretic physical (PHY) layer security, which exploits the physical characteristics of the wireless fading channel to offer a perfect secrecy [41]. Artificial noise (AN) has been proposed as an effective method to impair the received signals at the eavesdroppers by utilizing the extra degrees of freedom offered by multiple antennas in the MIMO system [39–42].

In a MIMO beamforming system with the presence of the artificial noise, the system model is discussed in [40, 42] and the details are as follows. First, the transmit signal is defined as

$$\mathbf{s} = \mathbf{G}^d \mathbf{x} + \mathbf{G}^{AN} \mathbf{a} \tag{18.28}$$

where $\mathbf{G}^d$ is the data beamform, and $\mathbf{G}^{AN}$ is the AN beamforming vector. $\mathbf{x}$ denotes the information signal, and $\mathbf{a}$ represents the AN signal.

The received signal can be written and decoupled into two groups, one for the intended receiver Bob and the other for the eavesdropper Eve. The received signal for the intended receiver Bob can be written as $\mathbf{y}_b = \mathbf{H}_b \mathbf{s} + \mathbf{n}_b = \mathbf{H}_b \mathbf{G}^d \mathbf{x} + \mathbf{H}_b \mathbf{G}^{AN} \mathbf{a} + \mathbf{n}_b$, and the received signal for eavesdropper Eve is given by $\mathbf{y}_e = \mathbf{H}_e \mathbf{s} + \mathbf{n}_e = \mathbf{H}_e \mathbf{G}^d \mathbf{x} + \mathbf{H}_e \mathbf{G}^{AN} \mathbf{a} + \mathbf{n}_e$.

Here, $\mathbf{H}_b$ is the channel between the transmitter and Bob, $\mathbf{n}_b$ is the noise at the receiver side of Bob, $\mathbf{H}_e$ is the channel between the transmitter and Eve, and $\mathbf{n}_e$ is the noise for Eve. Moreover, $\mathbf{H}_b\mathbf{G}^{AN}\mathbf{a} = \mathbf{0}$, which means the AN precoder $\mathbf{G}^{AN}$ is in the null space of Bob's channel $\mathbf{H}_b$. Given the fact that $\mathbf{H}_b$ is of full row rank, the precoder, i.e., the AN beamform $\mathbf{G}_{AN}$ is defined as follows [40].

$$\mathbf{G}^{AN} = \mathbb{I} - \mathbf{H}_b^\dagger\left(\mathbf{H}_b\mathbf{H}_b^\dagger\right)^{-1}\mathbf{H}_b \tag{18.29}$$

The optimization problem for the MIMO beamforming with AN is studied in [42], where the objective is to maximize the secrecy capacity, subject to a SINR and a power constraint. The performance, measured by achievable ergodic secrecy rates are analyzed with four different data beamforms and three different AN beamforms.

In the wideband waveforming system, each tap in the CSI is viewed as a virtual antenna, and the antenna diversity is achieved by multiple virtual antennas, which enable the AN for the secrecy improvement. According to (18.4), the received signal at the intended information receiver, Bob, can be rewritten as $\mathbf{y}_b = \mathbf{H}_b\mathbf{g}_d * \mathbf{x} + \mathbf{H}_b\mathbf{g}_{AN} * \mathbf{a} + \mathbf{n}_b$, where $\mathbf{H}_b$ is the channel matrix between the BS and Bob, as defined in (18.5). $\mathbf{g}_d$ and $\mathbf{g}_{AN}$ represent the data waveform and the AN waveform, respectively. $\mathbf{x}$ is the transmit information signal, and $\mathbf{a}$ denotes the AN signal. $\mathbf{n}_b$ is a white Gaussian noise at the receiver.

Similarly, the received signal at the eavesdropper is given by $\mathbf{y}_e = \mathbf{H}_e\mathbf{g}_d * \mathbf{x} + \mathbf{H}_e\mathbf{g}_{AN} * \mathbf{a} + \mathbf{n}_e$, where $\mathbf{H}_b$ is the channel matrix between the BS and eavesdropper, and $\mathbf{n}_e$ is the received white Gaussian noise.

Considering the quality of received signal at the intended information receiver, we must have the waveforming vector $\mathbf{g}_{AN}$ lie in the null space of the channel matrix $\mathbf{H}_b$, i.e., $\mathbf{H}_b\mathbf{g}_{AN} = \mathbf{0}$. Moreover, in order to have at least one solution for $\mathbf{g}_{AN}$ feasible, $\mathbf{H}_b$ must have a full row rank. The channel matrix $\mathbf{H}_b$, as defined in (18.5), has a dimension of $N\left(2\lfloor\frac{L-1}{D}\rfloor + 1\right) \times L$. Hence, $\mathbf{H}_b$ is of full row rank if and only if $L \leq \frac{2N-DN}{2N-D}$ when $D < 2N$, or $L \geq \frac{DN-2N}{D-2N}$ when $D > 2N$. If $D = 2N$, then only $N = 1$ yields a full row rank matrix. The related study for wideband waveforming in secured communication systems with AN transmission is still open.

18.7 Summary

By leveraging multipaths in the nature, waveforming is presented as a prominent technique for the wideband system to exploit spatial diversity. Waveforming harvests large degrees of freedom, by treating each multipath component in a multipath channel as a virtual antenna and weighing and combining information coherently. Hence, the wideband waveforming becomes an economic and promising solution to future communication systems, which support low-cost low-complexity implementation and high data rate wireless services. In this chapter, we summarized the novel wideband waveforming system and compared it with the conventional narrowband MIMO beamforming system where high degrees of freedom are achieved by transmitting with multiple antennas. System models and related optimization problems on resource allocation, powered wireless communications, and secured communications were studied in detail. For related references, readers can refer to [82].

References

[1] B. Wang, Y. Wu, F. Han, Y.-H. Yang, and K. J. R. Liu, "Green wireless communications: A time-reversal paradigm," *IEEE Journal on Selected Areas in Communications*, vol. 29, no. 8, pp. 1698–1710, 2011.

[2] Y. H. Yang, B. Wang, W. S. Lin, and K. J. R. Liu, "Near-optimal waveform design for sum rate optimization in time-reversal multiuser downlink systems," *IEEE Transactions on Wireless Communications*, vol. 12, no. 1, pp. 346–357, Jan. 2013.

[3] F. Han and K. J. R. Liu, "A multiuser TRDMA uplink system with 2D parallel interference cancellation," *IEEE Transactions on Communications*, vol. 62, no. 3, pp. 1011–1022, Mar. 2014.

[4] E. Yoon, S. Y. Kim, and U. Yun, "A time-reversal-based transmission using predistortion for intersymbol interference alignment," *IEEE Transactions on Communications*, vol. 63, no. 2, pp. 455–465, Feb. 2015.

[5] Y. H. Yang and K. J. R. Liu, "Waveform design with interference pre-cancellation beyond time-reversal systems," *IEEE Transactions on Wireless Communications*, vol. 15, no. 5, pp. 3643–3654, May 2016.

[6] Q. Xu, Y. Chen, and K. J. R. Liu, "Combating strong–weak spatial–temporal resonances in time-reversal uplinks," *IEEE Transactions on Wireless Communications*, vol. 15, no. 1, pp. 568–580, Jan. 2016.

[7] M. Fink, C. Prada, F. Wu, and D. Cassereau, "Self focusing in inhomogeneous media with time reversal acoustic mirrors," *IEEE Ultrasonics Symposium Proceedings*, pp. 681–686, 1989.

[8] M. Fink, "Time reversal of ultrasonic fields. I. Basic principles," *IEEE Transactions on Ultrasonics, Ferroelectrics, and Frequency Control*, vol. 39, no. 5, pp. 555–566, 1992.

[9] A. Derode, P. Roux, and M. Fink, "Robust acoustic time reversal with high-order multiple scattering," *Physical Review Letters*, vol. 75, no. 23, p. 4206, 1995.

[10] H. C. Song, W. Kuperman, W. Hodgkiss, T. Akal, and C. Ferla, "Iterative time reversal in the ocean," *The Journal of the Acoustical Society of America*, vol. 105, no. 6, pp. 3176–3184, 1999.

[11] H. T. Nguyen, I. Z. Kovcs, and P. C. F. Eggers, "A time reversal transmission approach for multiuser UWB communications," *IEEE Transactions on Antennas and Propagation*, vol. 54, no. 11, pp. 3216–3224, Nov. 2006.

[12] R. C. Qiu, C. Zhou, N. Guo, and J. Q. Zhang, "Time reversal with MISO for ultra-wideband communications: Experimental results," in *2006 IEEE Radio and Wireless Symposium*, Jan. 2006, pp. 499–502.

[13] R. L. D. L. Neto, A. M. Hayar, and M. Debbah, "Channel division multiple access based on high UWB channel temporal resolution," in *IEEE Vehicular Technology Conference*, pp. 1–5, Sep. 2006.

[14] N. Guo, B. M. Sadler, and R. C. Qiu, "Reduced-complexity UWB time-reversal techniques and experimental results," *IEEE Transactions on Wireless Communications*, vol. 6, no. 12, pp. 4221–4226, Dec. 2007.

[15] A. Khaleghi, G. E. Zein, and I. H. Naqvi, "Demonstration of time-reversal in indoor ultra-wideband communication: Time domain measurement," in *2007 4th International Symposium on Wireless Communication Systems*, pp. 465–468, Oct. 2007.

[16] M. A. Bouzigues, I. Siaud, M. Helard, and A. M. Ulmer-Moll, "Turn back the clock: Time reversal for green radio communications," *IEEE Vehicular Technology Magazine*, vol. 8, no. 1, pp. 49–56, Mar. 2013.

[17] H. T. Nguyen, J. B. Andersen, and G. F. Pedersen, "The potential use of time reversal techniques in multiple element antenna systems," *IEEE Communications Letters*, vol. 9, no. 1, pp. 40–42, Jan. 2005.

[18] Y. Jin, Y. Jiang, and J. M. F. Moura, "Multiple antenna time reversal transmission in ultra-wideband communications," in *IEEE Globecom 2007 – IEEE Global Telecommunications Conference*, pp. 3029–3033, Nov. 2007.

[19] C. Zhou, N. Guo, B. M. Sadler, and R. C. Qiu, "Performance study on time reversed impulse MIMO for UWB communications based on measured spatial UWB channels," in *MILCOM 2007 – IEEE Military Communications Conference*, pp. 1–6, Oct. 2007.

[20] A. Pitarokoilis, S. K. Mohammed, and E. G. Larsson, "Uplink performance of time-reversal MRC in massive MIMO systems subject to phase noise," *IEEE Transactions on Wireless Communications*, vol. 14, no. 2, pp. 711–723, Feb. 2015.

[21] M. Maaz, M. Helard, P. Mary, and M. Liu, "Performance analysis of time-reversal based precoding schemes in MISO-OFDM systems," in *2015 IEEE 81st Vehicular Technology Conference (VTC Spring)*, pp. 1–6, May 2015.

[22] F. Han, Y. H. Yang, B. Wang, Y. Wu, and K. J. R. Liu, "Time-reversal division multiple access over multi-path channels," *IEEE Transactions on Communications*, vol. 60, no. 7, pp. 1953–1965, Jul. 2012.

[23] Y. Chen, F. Han, Y. H. Yang, H. Ma, Y. Han, C. Jiang, H. Q. Lai, D. Claffey, Z. Safar, and K. J. R. Liu, "Time-reversal wireless paradigm for green Internet of Things: An overview," *IEEE Internet of Things Journal*, vol. 1, no. 1, pp. 81–98, Feb. 2014.

[24] M. L. Ku, Y. Han, H. Q. Lai, Y. Chen, and K. J. R. Liu, "Power waveforming: Wireless power transfer beyond time reversal," *IEEE Transactions on Signal Processing*, vol. 64, no. 22, pp. 5819–5834, Nov. 2016.

[25] D. Gesbert, M. Shafi, D.-S. Shiu, P. J. Smith, and A. Naguib, "From theory to practice: An overview of MIMO space-time coded wireless systems," *IEEE Journal on Selected Areas in Communications*, vol. 21, no. 3, pp. 281–302, Apr. 2003.

[26] L. Zheng and D. N. C. Tse, "Diversity and multiplexing: A fundamental tradeoff in multiple-antenna channels," *IEEE Transactions on Information Theory*, vol. 49, no. 5, pp. 1073–1096, May 2003.

[27] R. W. Heath and A. J. Paulraj, "Switching between diversity and multiplexing in MIMO systems," *IEEE Transactions on Communications*, vol. 53, no. 6, pp. 962–968, Jun. 2005.

[28] H. Q. Ngo, E. G. Larsson, and T. L. Marzetta, "Energy and spectral efficiency of very large multiuser MIMO systems," *IEEE Transactions on Communications*, vol. 61, no. 4, pp. 1436–1449, Apr. 2013.

[29] J. G. Andrews, S. Buzzi, W. Choi, S. V. Hanly, A. Lozano, A. C. K. Soong, and J. C. Zhang, "What will 5G be?" *IEEE Journal on Selected Areas in Communications*, vol. 32, no. 6, pp. 1065–1082, Jun. 2014.

[30] F. Boccardi, R. W. Heath, A. Lozano, T. L. Marzetta, and P. Popovski, "Five disruptive technology directions for 5G," *IEEE Communications Magazine*, vol. 52, no. 2, pp. 74–80, Feb. 2014.

[31] Y. Han, Y. Chen, B. Wang, and K. J. R. Liu, "Time-reversal massive multipath effect: A single-antenna massive MIMO solution," *IEEE Transactions on Communications*, vol. 64, no. 8, pp. 3382–3394, Aug. 2016.

[32] B. D. V. Veen and K. M. Buckley, "Beamforming: A versatile approach to spatial filtering," *IEEE ASSP Magazine*, vol. 5, no. 2, pp. 4–24, Apr. 1988.

[33] D. P. Palomar, J. M. Cioffi, and M. A. Lagunas, "Joint Tx-Rx beamforming design for multicarrier MIMO channels: A unified framework for convex optimization," *IEEE Transactions on Signal Processing*, vol. 51, no. 9, pp. 2381–2401, Sep. 2003.

[34] C. W. Tan, M. Chiang, and R. Srikant, "Maximizing sum rate and minimizing MSE on multiuser downlink: Optimality, fast algorithms and equivalence via max-min SINR," *IEEE Transactions on Signal Processing*, vol. 59, no. 12, pp. 6127–6143, Dec. 2011.

[35] R. Zhang and C. K. Ho, "MIMO broadcasting for simultaneous wireless information and power transfer," *IEEE Transactions on Wireless Communications*, vol. 12, no. 5, pp. 1989–2001, May 2013.

[36] J. Xu, L. Liu, and R. Zhang, "Multiuser MISO beamforming for simultaneous wireless information and power transfer," *IEEE Transactions on Signal Processing*, vol. 62, no. 18, pp. 4798–4810, Sept. 2014.

[37] Q. Shi, W. Xu, T. H. Chang, Y. Wang, and E. Song, "Joint beamforming and power splitting for MISO interference channel with SWIPT; An SOCP relaxation and decentralized algorithm," *IEEE Transactions on Signal Processing*, vol. 62, no. 23, pp. 6194–6208, Dec. 2014.

[38] H. Lee, S. R. Lee, K. J. Lee, H. B. Kong, and I. Lee, "Optimal beamforming designs for wireless information and power transfer in MISO interference channels," *IEEE Transactions on Wireless Communications*, vol. 14, no. 9, pp. 4810–4821, Sept. 2015.

[39] S. Goel and R. Negi, "Guaranteeing secrecy using artificial noise," *IEEE Transactions on Wireless Communications*, vol. 7, no. 6, pp. 2180–2189, Jun. 2008.

[40] X. Zhou and M. R. McKay, "Secure transmission with artificial noise over fading channels: Achievable rate and optimal power allocation," *IEEE Transactions on Vehicular Technology*, vol. 59, no. 8, pp. 3831–3842, Oct. 2010.

[41] D. W. K. Ng, E. S. Lo, and R. Schober, "Robust beamforming for secure communication in systems with wireless information and power transfer," *IEEE Transactions on Wireless Communications*, vol. 13, no. 8, pp. 4599–4615, Aug. 2014.

[42] J. Zhu, R. Schober, and V. K. Bhargava, "Linear precoding of data and artificial noise in secure massive MIMO systems," *IEEE Transactions on Wireless Communications*, vol. 15, no. 3, pp. 2245–2261, Mar. 2016.

[43] Y. Chen, B. Wang, Y. Han, H. Q. Lai, Z. Safar, and K. J. R. Liu, "Why time reversal for future 5G wireless?" *IEEE Signal Processing Magazine*, vol. 33, no. 2, pp. 17–26, Mar. 2016.

[44] Z.-H. Wu, Y. Han, Y. Chen, and K. J. R. Liu, "A time-reversal paradigm for indoor positioning system," *IEEE Transactions on Vehicular Technology*, vol. 64, no. 4, pp. 1331–1339, Apr. 2015.

[45] C. Chen, Y. Chen, Y. Han, H. Q. Lai, and K. J. R. Liu, "Achieving centimeter-accuracy indoor localization on WiFi platforms: A frequency hopping approach," *IEEE Internet of Things Journal*, vol. 4, no. 1, pp. 111–121, Feb. 2017.

[46] C. Chen, Y. Chen, K. J. R. Liu, Y. Han, and H.-Q. Lai, "High accuracy indoor localization: A WiFi-based approach," in *2016 IEEE International Conference on Acoustics, Speech and Signal Processing (ICASSP)*, pp. 6245–6249, Mar. 2016.

[47] F. Zhang, C. Chen, B. Wang, H. Q. Lai, and K. J. R. Liu, "A time-reversal spatial hardening effect for indoor speed estimation," in *2017 IEEE International Conference on Acoustics, Speech, and Signal Processing*, p. 1, Mar. 2017.

[48] Q. Xu, Y. Chen, B. Wang, and K. J. R. Liu, "TRIEDS: Wireless events detection through the wall," *IEEE Internet of Things Journal*, vol. 4, no. 3, pp. 723–735, Jun. 2017.

[49] —, "Radio biometrics: Human recognition through a wall," *IEEE Transactions on Information Forensics and Security*, vol. 12, no. 5, pp. 1141–1155, May 2017.

[50] C. Chen, Y. Han, Y. Chen, and K. J. R. Liu, "Multi-person breathing rate estimation using time-reversal on WiFi platforms," in *2016 IEEE Global Conference on Signal and Information Processing*, p. 1, Dec. 2016.

[51] B. Bogert, "Demonstration of delay distortion correction by time-reversal techniques," *IRE Transactions on Communications Systems*, vol. 5, no. 3, pp. 2–7, Dec. 1957.

[52] F. Wu, J.-L. Thomas, and M. Fink, "Time reversal of ultrasonic fields. II. Experimental results," *IEEE Transactions on Ultrasonics, Ferroelectrics, and Frequency Control*, vol. 39, no. 5, pp. 567–578, 1992.

[53] C. Dorme, M. Fink, and C. Prada, "Focusing in transmit-receive mode through inhomogeneous media: The matched filter approach," in *IEEE 1992 Ultrasonics Symposium Proceedings*, pp. 629–634 vol. 1, Oct. 1992.

[54] A. Derode, P. Roux, and M. Fink, "Acoustic time-reversal through high-order multiple scattering," in *1995 IEEE Ultrasonics Symposium. Proceedings. An International Symposium*, vol. 2, pp. 1091–1094 vol. 2, Nov. 1995.

[55] W. Kuperman, W. S. Hodgkiss, H. C. Song, T. Akal, C. Ferla, and D. R. Jackson, "Phase conjugation in the ocean: Experimental demonstration of an acoustic time-reversal mirror," *The Journal of the Acoustical Society of America*, vol. 103, no. 1, pp. 25–40, 1998.

[56] D. Rouseff, D. R. Jackson, W. L. Fox, C. D. Jones, J. A. Ritcey, and D. R. Dowling, "Underwater acoustic communication by passive-phase conjugation: Theory and experimental results," *IEEE Journal of Oceanic Engineering*, vol. 26, no. 4, pp. 821–831, 2001.

[57] G. F. Edelmann, T. Akal, W. S. Hodgkiss, S. Kim, W. A. Kuperman, and H. C. Song, "An initial demonstration of underwater acoustic communication using time reversal," *IEEE Journal of Oceanic Engineering*, vol. 27, no. 3, pp. 602–609, Jul. 2002.

[58] B. E. Henty and D. D. Stancil, "Multipath-enabled super-resolution for RF and microwave communication using phase-conjugate arrays," *Physical Review Letters*, vol. 93, no. 24, p. 243904, 2004.

[59] G. Lerosey, J. De Rosny, A. Tourin, A. Derode, G. Montaldo, and M. Fink, "Time reversal of electromagnetic waves," *Physical Review Letters*, vol. 92, no. 19, p. 193904, 2004.

[60] —, "Time reversal of electromagnetic waves and telecommunication," *Radio Science*, vol. 40, no. 6, pp. 1–10, 2005.

[61] G. Lerosey, J. De Rosny, A. Tourin, A. Derode, and M. Fink, "Time reversal of wideband microwaves," *Applied Physics Letters*, vol. 88, no. 15, p. 154101, 2006.

[62] J. de Rosny, G. Lerosey, and M. Fink, "Theory of electromagnetic time-reversal mirrors," *IEEE Transactions on Antennas and Propagation*, vol. 58, no. 10, pp. 3139–3149, 2010.

[63] I. H. Naqvi, G. E. Zein, G. Lerosey, J. D. Rosny, P. Besnier, A. Tourin, and M. Fink, "Experimental validation of time reversal ultra wide-band communication system for high data rates," *IET Microwaves, Antennas Propagation*, vol. 4, no. 5, pp. 643–650, May 2010.

[64] M. Emami, M. Vu, J. Hansen, A. J. Paulraj, and G. Papanicolaou, "Matched filtering with rate back-off for low complexity communications in very large delay spread channels," in *38th Asilomar Conference on Signals, Systems and Computers*. IEEE, pp. 218–222, 2004.

[65] Q. Xu, Y. Chen, and K. J. R. Liu, "Optimal pricing for interference control in time-reversal device-to-device uplinks," in *2015 IEEE Global Conference on Signal and Information Processing (GlobalSIP)*, pp. 1096–1100, Dec. 2015.

[66] T. K. Y. Lo, "Maximum ratio transmission," *IEEE Transactions on Communications*, vol. 47, no. 10, pp. 1458–1461, Oct. 1999.

[67] T. L. Marzetta, "Noncooperative cellular wireless with unlimited numbers of base station antennas," *IEEE Transactions on Wireless Communications*, vol. 9, no. 11, pp. 3590–3600, 2010.

[68] A. Goldsmith, *Wireless Communications*. New York: Cambridge University Press, 2005.

[69] Q. H. Spencer, A. L. Swindlehurst, and M. Haardt, "Zero-forcing methods for downlink spatial multiplexing in multiuser MIMO channels," *IEEE Transactions on Signal Processing*, vol. 52, no. 2, pp. 461–471, Feb. 2004.

[70] S. Shi, M. Schubert, and H. Boche, "Rate optimization for multiuser MIMO systems with linear processing," *IEEE Transactions on Signal Processing*, vol. 56, no. 8, pp. 4020–4030, Aug. 2008.

[71] D. N. C. Tse and P. Viswanath, "Downlink-uplink duality and effective bandwidths," in *Proceedings IEEE International Symposium on Information Theory,*, pp. 52–52, 2002.

[72] T. Ma, Q. Shi, and E. Song, "QoS-constrained weighted sum-rate maximization in multi-cell multi-user MIMO systems: An ADMM approach," in *2016 35th Chinese Control Conference (CCC)*, pp. 6905–6910, Jul. 2016.

[73] M. Schubert and H. Boche, "Solution of the multiuser downlink beamforming problem with individual SINR constraints," *IEEE Transactions on Vehicular Technology*, vol. 53, no. 1, pp. 18–28, Jan. 2004.

[74] J. Wang and D. P. Palomar, "Worst-case robust MIMO transmission with imperfect channel knowledge," *IEEE Transactions on Signal Processing*, vol. 57, no. 8, pp. 3086–3100, Aug. 2009.

[75] F. Zhang, "Report for optimization-based pinpoint beamforming," University of Maryland College Park, Tech. Rep., Jul. 2015.

[76] L. R. Varshney, "Transporting information and energy simultaneously," in *2008 IEEE International Symposium on Information Theory*, Jul. 2008, pp. 1612–1616.

[77] X. Lu, P. Wang, D. Niyato, D. I. Kim, and Z. Han, "Wireless networks with RF energy harvesting: A contemporary survey," *IEEE Communications Surveys Tutorials*, vol. 17, no. 2, pp. 757–789, Second quarter 2015.

[78] P. Grover and A. Sahai, "Shannon meets Tesla: Wireless information and power transfer," in *2010 IEEE International Symposium on Information Theory*, pp. 2363–2367, Jun. 2010.

[79] X. Zhou, R. Zhang, and C. K. Ho, "Wireless information and power transfer: Architecture design and rate-energy tradeoff," *IEEE Transactions on Communications*, vol. 61, no. 11, pp. 4754–4767, Nov. 2013.

[80] K. Huang and E. Larsson, "Simultaneous information and power transfer for broadband wireless systems," *IEEE Transactions on Signal Processing*, vol. 61, no. 23, pp. 5972–5986, Dec. 2013.

[81] M. Tian, X. Huang, Q. Zhang, and J. Qin, "Robust AN-aided secure transmission scheme in MISO channels with simultaneous wireless information and power transfer," *IEEE Signal Processing Letters*, vol. 22, no. 6, pp. 723–727, Jun. 2015.

[82] Q. Xu, C. Jiang, Y. Han, B. Wang, and K. J. R. Liu, "Waveforming: An overview with beamforming," *IEEE Communications Surveys & Tutorials*, vol. 20, no. 1, pp. 132–149, 2018.

19 Spatial Focusing Effect for Networking

Next-generation wireless networks are expected to support exponentially increasing number of users and demands of data, which all rely on the essential media: spectrum. Spectrum sharing among heterogeneous networks is a fundamental issue that determines the network performance. Previous works have focused on a "dynamic spectrum access" mode associated with the cognitive radio technology. Nevertheless, those works are all about the discovery of available spectrum resource either in the time domain or in the frequency domain, i.e., by separating different users' transmission. To initiate a new paradigm of spectrum sharing, the unique characteristics of the technologies should be utilized in the next-generation networks. The 5G networks are featured either by wide bandwidth like mmWave systems, or by large-scale antennas like massive MIMO. Those two trends can lead to a common phenomenon: the spatial focusing effect. Based on this focusing effect, we discuss a general spatial spectrum sharing framework that enables concurrent multiusers spectrum sharing without the requirement of orthogonal resource allocation. Moreover, we design two general network association protocols either in a centralized manner or in a distributed manner. Simulation results show that both time reversal wideband and massive MIMO system can achieve high throughput performance with the spatial spectrum sharing scheme.

19.1 Introduction

Spectrum sharing has been continuously attracting the attention of the research community. It evolved from the original simple division multiple access schemes to some recent progress toward the development of dynamic spectrum access (DSA) in cognitive radio (CR) [1]. Fundamentally, those technologies are all about the discovery of white space in spectrum either in the time domain, e.g., the so-called underlay sharing and overlay sharing [2]; in the frequency domain, e.g., the so-called in-band sharing and out-of-band sharing [3]; or in the spatial domain, i.e., the geo-location databases [4]. When it comes to the emerging 5G era, it is expected to rely on new characteristics of the 5G systems, such as the massive MIMO (multiple-input-multiple-output) effect, which can be utilized for designing a new generation of spectrum sharing solutions.

In this chapter, we focus on a special phenomenon in 5G networks: The spatial focusing effect [5, 6], especially waveforming (such as the time-reversal waveform) is employed when the bandwidth becomes wider and wider as in mmWave or beamforming (such as maximum-ratio combining beamforming) when the scale of antennas becomes larger and larger as in massive MIMO. The spatial focusing effect means that the intended reception signal can be concentrated at the intended location of the receiver with little energy leakage to the others. This focusing effect is fundamentally due to the decreased correlation between the channel states of two different locations. On one hand, the massive number of antennas can create the channel state information with large dimension for each location. By utilizing a matched filter-based precoder or equalizer, the signal energy can concentrate on a corresponding location [7]. On the other hand, the wide bandwidth can help reveal multiple paths in a rich scattering environment, e.g., the scenarios of indoor or dense metropolis. By utilizing the time-reversal (TR) technology [5, 8], the signal energy can also concentrate on an intended location, generating the spatial focusing effect. In TR communications, when transceiver A wants to transmit information to transceiver B, transceiver B first has to send a delta-like pilot pulse that propagates through a scattering and multipath environment, and the signals are received by transceiver A; then, transceiver A simply transmits the time-reversed signals back through the same channel to transceiver B. As a matter of fact, those multipaths can be regarded as virtual antennas, and the TR processing can be regarded as cooperatively controlling those antennas, which can realize the similar effect as massive MIMO [6].

This common spatial focusing phenomenon creates a tunneling effect for each user at his/her own location, such that the interference among different users is rather weak. Thus, multiple users can concurrently conduct data transmission upon the entire spectrum by utilizing their corresponding channel state information as a unique signature. In essence, their locations have ideally separated them in the spatial domain, and each different location is a spatial "white space." Therefore, this emerging spatial focusing effect enables us to develop a general spatial spectrum sharing scheme, which allows the reuse of the entire spectrum, instead of separating users either in time domain or in the spectrum domain like the traditional dynamic spectrum access technology. Note that the traditional so-called spatial domain dynamic spectrum access based on geo-location databases is different from the spatial spectrum sharing presented in this chapter. The functionality of the geo-location databases is to record the spatial white-space locations that allow secondary users to utilize without harmful interference to the primary users. In contrast, our spatial spectrum sharing means every location can be regarded as a white space, and primary/secondary users can concurrently share the entire spectrum all the time wherever they are located. The major challenge of the spatial spectrum sharing lies in the waveforming design in TR or the beamforming design in massive MIMO, according to some specific criterions. In this chapter, instead of focusing on the design of waveformer or beamformer, we are more interested in the spectrum-sharing performance analysis given a specific waveformer or beamformer.

The main points of this chapter can be summarized as follows.

(i) We leverage the spatial focusing to develop the concept of spatial white space. It can be interpreted as a spatial radio resonance phenomenon at a specific location. This phenomenon has been observed in both TR wideband systems and massive MIMO systems.

(ii) We discuss a general spatial spectrum sharing framework for 5G networks including both TR wideband systems and massive MIMO systems, which supports massive users to concurrently share the entire spectrum all the time.
(iii) We theoretically analyze the closed-form signal-to-interference-plus-noise ratio (SINR) expressions for both TR wideband and massive MIMO systems. It is found that the SINR of both systems shares the same expression under the equal power allocation scenario.
(iv) Based on the SINR performance analysis, we discuss two general network association protocols either in a centralized manner or a distributed manner, which are applicable for both TR wideband systems and massive MIMO systems due to exactly the same formulation of the SINR performance.

The remainder of this chapter is organized as follows. Section 19.2 summarizes the related works, and Section 19.3 presents the system models for both TR wideband systems and massive MIMO systems. Based on the system models, we show the spatial focusing characteristic of both systems in Section 19.4. Then, a thorough SINR performance analysis is conducted for both systems in Section 19.5, which leads to two general network association protocols design in Section 19.6. Simulation results are shown in Section 19.7.

19.2 Related Works

In the literature, spectrum sharing for 5G networks has been preliminarily studied in [9–14], where the researchers have attempted to incorporate a range of new emerging technologies to the spectrum sharing issue. The "software defined network" (SDN) technology was applied to the heterogeneous networks (HetNets) spectrum sharing in [9], where the key contribution is the concept of harmonized SDN-enabled framework, relying on distributed input reporting on spectrum availabilities instead of the traditional spectrum-sensing based approach [15, 16]. Meanwhile, the prediction of the spectrum usage was introduced in [10], with the key contribution of exploring the fundamental limits of predictability of spectrum usage patterns in TV bands, ISM bands, cellular bands, and so on. Another prediction of primary users' moving trajectory was proposed in [11], which initiatively transformed the spectrum sensing problem into a primary users' location tracking problem. Moreover, Mitola, contributed a concept of public–private spectrum sharing in [12], as well as an architecture separating the function between control and data. Similarly, Ng et al. proposed to simultaneously utilize licensed and unlicensed bands to improve the energy efficiency and formulated a convex optimization formulation by maximizing the energy efficiency in [13]. Apart from the perspectives of users, the authors in [14] designed a coordination protocol among a set of operators having similar rights for accessing spectrum based on reciprocity modeling from the perspectives of operators.

The medium access control (MAC) layer design for 5G spectrum access has also attracted researchers' attentions. An earlier work in [17] highlighted that due to the small carrier wavelength in mmWave (millimeter wave), narrow beams are essential for overcoming higher path loss, which can alleviate the interuser interference, yet making the carrier sensing infeasible and thus leading to a severe collision problem. The collision probability analysis of mmWave systems was a main contribution in [17]. Similarly, based on the narrow beam characteristics of mmWave networks, the authors in [18] contributed a two-step synchronization and initial access scheme through directional cell search. Meanwhile, in [19], the coverage and rate analysis in mmWave system was derived under a stochastic geometry framework with the knowledge of antenna beamforming pattern and base station density. Recently, a user association scheme for load balancing and fairness in mmWave systems was proposed in [20] based on a convex optimization formulation, and a scheduling algorithm for multihop mmWave system was designed in [21] by modeling three kinds of interference. The scenario of heterogeneous networks equipped with massive MIMO was considered in [22], where the authors contributed two algorithms including a cell association algorithm and an antenna allocation algorithm based on the evolutionary game model, which belongs to the game theoretical cognitive radio networks [23, 24]. Another three user association schemes for massive MIMO networks in [25–27] were all based on the convex optimization model, where the sum rate of the network was optimized under the constraints of fairness and limited resources.

As one can see from the aforementioned works [10–27], the researchers have attempted to re-treat the spectrum sharing problem either by incorporating some new technologies like the SDN and the predictions, or by taking into account some unique characteristics of 5G networks like the narrow beams in mmWave and the large-scale antenna selection in massive MIMO. Different from those existing works, we focus on the spatial focusing effect [5, 6] in this chapter, especially waveforming is employed when the bandwidth becomes wider and wider as in mmWave or beamforming when the scale of antennas becomes larger and larger as in massive MIMO.

19.3 System Models

In this section, we introduce the system and channel models for both TR wideband and massive MIMO systems. Basically, the TR wideband system is featured with a single antenna where each antenna is associated with a wide bandwidth up to hundreds or even thousands of MHz; while the massive MIMO system is featured with a large number of antennas but the bandwidth of each antenna is limited to 20–40 MHz. The different bandwidth settings lead to different channel models for these two systems. Generally, in this chapter, we consider a multiuser downlink network over non-line-of-sight multipath Rayleigh fading channels. Note that in essence, the spatial focusing effect described in this chapter is due to the multipath channel, which provides ample degrees of freedom. Therefore, the channel model has no impact on the results in this chapter, e.g., Winner II channel model.

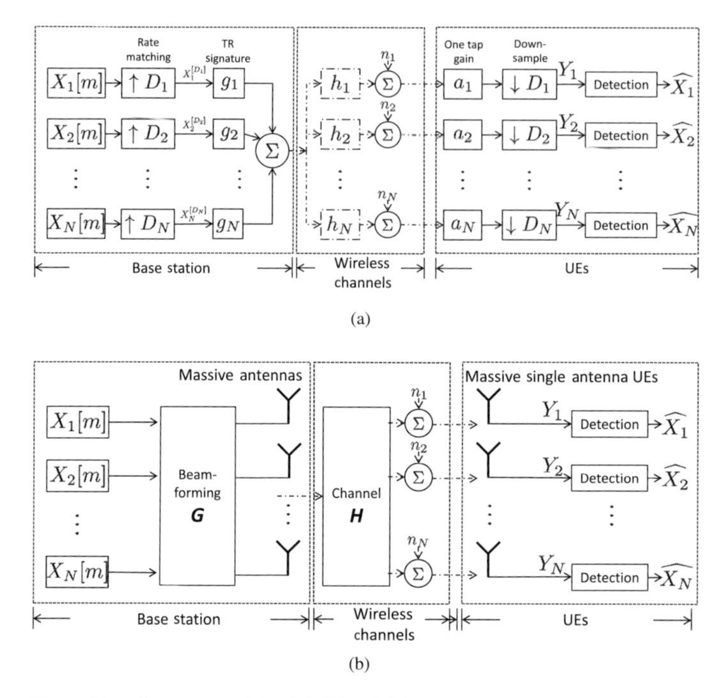

Figure 19.1 System models: (a) TR wideband system, and (b) massive MIMO system.

19.3.1 Time-Reversal Wideband System

19.3.1.1 Downlink Transmission

In the TR wideband system, suppose there are M base stations (BSs) where each BS is equipped with a single antenna, and N user equipment (UEs) where each UE is with a single antenna. The TR wideband downlink system consists of two phases: the channel probing phase and the downlink transmission phase [28]. The channel probing phase is for enabling each BS to obtain the CIRs of each link between it and the associated UEs. This can be done by the following procedure, the N UEs first sequentially transmit an impulse Dirac δ-function signal to the BSs, and the BSs then record and store the CIRs for the transmission phase. In practice, the sequential probing manner can be substituted by designing simultaneous orthogonal probing sequence, and the impulse δ-function signal can be a modified raise-cosine signal. After this channel probing phase, the BSs start the transmission phase. As shown in Figure 19.1(a), for the jth UE, let us denote A_j as the BS that the UE is associating with and denote $\mathcal{N}_{A_j}$ as the current set of UEs that are associating with BS A_j besides the jth UE. The intended message for those UEs belonging to $\mathcal{N}_{A_j}$ can be represented by $\{X_j, X_{j'\in\mathcal{N}_{A_j}}\}$, where each of the sequence of information symbols X_j or $X_{j'}$ are independent complex random variables with zero mean and $\mathbb{E}\left[|X_{j,j'}[k]|^2\right] = 1$. This sequence of message is first up-sampled by the backoff factor D into $\{X_j^{[D]}, X_{j'\in\mathcal{N}_{A_j}}^{[D]}\}$ for the sake of alleviating the intersymbol interference due to the delay spread. The up-sampled sequence is then convoluted with the signature generated by the CIRs. The signature can be the time-reversal waveform, which is a time reversed and conjugated version of each CIR, or can be the enhanced time-reversal waveform, i.e., sum rate optimization [29, 30], or interference cancelation [31, 32]. After that, all the convoluted messages are added up and transmitted into wireless channels. The transmission power of each BS's antenna is regarded as identical P, and the power allocation for the jth TD can be denoted by $P_{A_j,j}$, where it should be satisfied that $P_{A_j,j} + \sum_{j' \in \mathcal{N}_{A_j}} P_{A_j,j'} = P$. In such a case, the transmission signal of BS A_j can be written by

$$S_{A_j}[k] = \sqrt{P_{A_j,j}} \left(X_j^{[D]} * g_{A_j,j} \right) [k] + \sum_{j' \in \mathcal{N}_{A_j}} \sqrt{P_{A_j,j'}} \left(X_{j'}^{[D]} * g_{A_j,j'} \right) [k], \tag{19.1}$$

where $g_{A_j,j}^{\phi}$ is the signature between the ϕth antenna of BS A_j and the jth UE. Note that (19.1) is based on the assumption that the uplink and downlink channels are reciprocal.

19.3.1.2 Channel Model

The TR system is considered to be a single-carrier system with nonorthogonal resource allocation. Because each antenna is considered to operate upon hundreds of MHz bandwidth, multiple taps channel can be considered as follows. The channel between the ith BS and the jth UE can be modeled by the large-scale fading incurred by the distance attenuation combined with the small-scale fading incurred by the multipath environment. The small-scale fading, denoted by $\mathbf{h}_{i,j}$, can be written by

$$\mathbf{h}_{i,j} = \left[h_{i,j}[0], h_{i,j}[1], \ldots, h_{i,j}[L-1] \right], \tag{19.2}$$

where $h_{i,j}[k]$ represents the kth tap of the channel impulse response (CIR) with length L and can be written by

$$h_{i,j}[k] = \sum_{l=0}^{L-1} h_{i,j}^{l} \delta[k-l], \tag{19.3}$$

with $\delta[\cdot]$ as the Dirac delta function. In this chapter, we consider a rich-scattering environment, in which due to the spatial heterogeneity and the rich multipaths, the CIRs of different UEs in different locations are assumed to be uncorrelated.

19.3.1.3 Downlink Reception

For the receiver side of the TR wideband system, the received signal of the jth UE can be represented as follows

$$Y_j^{[D]}[k] = \sum_{i=1}^{M} d_{i,j}^{-\alpha/2} \left(S_i * h_{i,j} \right) [k] + n_j[k], \tag{19.4}$$

where $d_{i,j}$ represents the distance between the ith BS and the jth UE, α is the path loss coefficient, $d_{i,j}^{-\alpha/2}$ represents the aforementioned large-scale fading incurred by the distance attenuation, $h_{i,j}$ represents the small-scale fading between the ith BS and the jth UE, and n_j represents the additive white Gaussian noise with zero mean and variance σ^2.

19.3.2 Massive MIMO System

19.3.2.1 Downlink Transmission

In the massive MIMO system, suppose there are M BSs each associated with a massive number of antennas Φ, and N UEs each with a single antenna with $1 \lll N \ll \Phi$. Similar to the TR wideband system, the massive antenna system also includes two phases: the channel state information (CSI) acquisition phase using an uplink pilot and the data transmission phase [33]. Here, the CSI is assumed to be perfectly known by the BSs so that they can perform beamforming to support multiple UEs' data transmission. As shown in Figure 19.1(b), let us denote the beamforming matrix of the ith BS as $\mathbf{G}_i = \{\mathbf{g}_{i,1}, \mathbf{g}_{i,2}, \ldots, \mathbf{g}_{i,|\mathcal{N}_i|}\}$, where $\mathbf{G}_i \in \mathbb{C}^{\Phi \times |\mathcal{N}_i|}$, $\mathbf{g}_{i,j} \in \mathbb{C}^{\Phi \times 1}$, and $\mathcal{N}_i$ represents the set of UEs that associate with the ith BS, and the number of UEs in $\mathcal{N}_i$ is denoted as $|\mathcal{N}_i|$. Note that the beamforming matrix can be MRC (maximal-ratio combining), ZF (zero-forcing) [34–36], or MMSE (minimum mean square error) [37, 38]. Suppose $\mathbf{s}_i = [s_{i,1}, s_{i,2}, \ldots, s_{i,|\mathcal{N}_i|}]^{\mathrm{T}}$ is the transmit symbols of the ith BS with length determined by the number of UEs that associate with it, where $\mathbf{s}_i \in \mathbb{C}^{|\mathcal{N}_i| \times 1}$ and $\mathbb{E}(\mathbf{s}_i \mathbf{s}_i^*) = \mathbf{I}$.

19.3.2.2 Channel Model

The bandwidth of each antenna is assumed to be regular wideband as 20–40 MHz, where OFDM is employed to provide a group of single-tap subcarriers. For the simplicity of expression, we assume flat fading for each antenna [33] and thus denote the small-scale fading between the ith BS and its associated UEs as follows:

$$\mathbf{H}_i = \{\mathbf{h}_{i,1}^{\mathrm{T}}, \mathbf{h}_{i,2}^{\mathrm{T}}, \ldots, \mathbf{h}_{i,|\mathcal{N}_i|}^{\mathrm{T}}\}, \tag{19.5}$$

where $\cdot^{\mathrm{T}}$ represents the transpose operation. In such a case, we have $\mathbf{H}_i \in \mathbb{C}^{\Phi \times |\mathcal{N}_i|}$ where $\mathbb{C}$ represents the complex domain, i.e., $\mathbf{h}_{i,j} \in \mathbb{C}^{1 \times \Phi}$ is the small-scale fading channel between the ith BS and the jth UE in $\mathcal{N}_i$.

19.3.2.3 Downlink Reception

The received signal of the jth TD can be expressed by

$$Y_j = \sum_{i=1}^{M} \sqrt{P} d_{i,j}^{-\alpha/2} \mathbf{h}_{i,j} \mathbf{G}_i \mathbf{s}_i + n_j, \tag{19.6}$$

where P represents the transmission power of each antenna, $d_{i,j}$ represents the distance between the ith BS and the jth UE, α is the path loss coefficient, $d_{i,j}^{-\alpha/2}$ represents the large-scale attenuation, and n_j represents the additive white Gaussian noise with zero mean and variance σ^2. Note that the intercell interference has been taken into account in (19.6) because the summation means the received signal from all other BSs besides the BS that the jth TD associates with. In order to achieve a similar setting as in the aforementioned TR wideband system, it is also assumed a one-tap receiver in the massive MIMO system.

19.4 Spatial Focusing Effect

In this section, we introduce a special phenomenon in both TR wideband and massive MIMO systems: the spatial focusing effect. This focusing effect means that the signal for an intended location can concentrate only at that specific location with only little energy leakage to other locations, by waveforming (e.g., utilizing an appropriate waveform) based on the channel state information of that specific location. Based on this spatial focusing effect, the multiusers' concurrent data transmission can be readily achieved [28] because the interference among UEs can be alleviated to a large extent due to the energy concentration, which enables the entire spectrum sharing among primary users and secondary users simultaneously. Note that the spatial focusing of TR communications was first mentioned in [8], while later the spatial focusing effect of massive MIMO was mentioned in [7]. Nevertheless, how to leverage the spatial focusing effect to the spectrum sharing has not been investigated in those two works. In this chapter, the main novelty is in presenting a spatial spectrum sharing framework, as well as the performance analysis. There are two major challenges: one is the waveforming design for the TR system and the beamforming design for massive MIMO system, and the other is the CSI acquisition when the number of users is relatively large. In this chapter, we will analyze the spectrum sharing performance of the time-reversal waveforming and MRC beamforming, while the CSI acquisition is not the focus.

19.4.1 Ray-Tracing-Based Simulation

As aforementioned, both TR wideband and massive MIMO systems can exhibit this spatial focusing effect. However, the approaches are different where the TR system relies on a wide bandwidth but the massive MIMO system relies on a large number of antennas. For the TR wideband system in a rich scattering environment with multipath propagation (e.g., dense city or indoor scenarios), the wide bandwidth can help reveal those multiple paths from the transmitter to each specific location. The wider bandwidth can reveal more multipaths for each location, and thus can lead to less correlation between every two different locations' channels. Eventually, by utilizing the time-reversed CIR of a specific location as the waveform of this location, the convolution of the waveform and the channel can generate a unique peak at that specific location, with little energy leakage to the neighboring locations. For the massive MIMO system equipped with a large scale of antennas, those antennas can physically create a large dimension of CSI for each location, which also leads to the small correlation among difference locations. Therefore, by utilizing a simple matched filter precoder based on the CSI of a specific location, the signal energy can be spatially focused at that specific location.

To validate this spatial focusing effect, we construct a ray-tracing-based simulation in a discrete scattering environment. As shown in Figure 19.2, a total number of 400 effective scatters are distributed randomly in a square area with dimension $200\lambda \times 200\lambda$, where λ is the wavelength corresponding to the carrier frequency. In such a case, the wireless channel can be represented as a sum of multipaths using the classical ray-tracing method. Without loss of generality, we adopt a single-bounce ray-tracing model

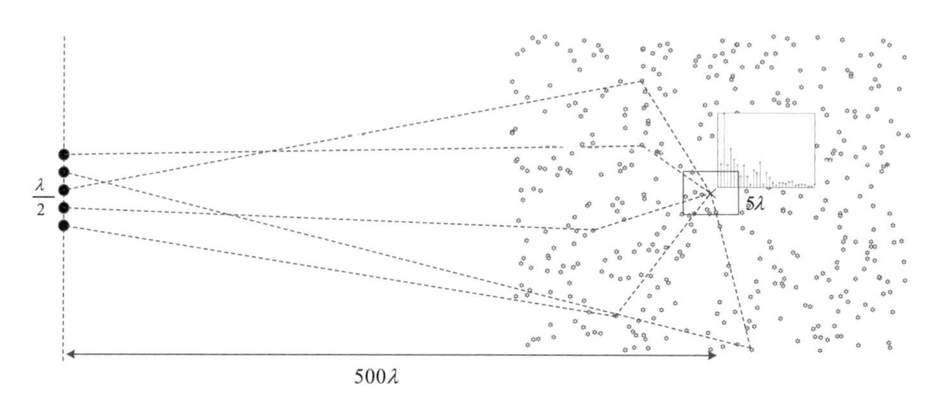

Figure 19.2 Simulation setup for validating spatial focusing effect.

in computing the channel impulse responses in both TR wideband and the large-scale antenna array systems, where both systems are operated on a 5 GHz ISM band. The reflection coefficient of each scatter is chosen to be I.I.D. complex random variables with uniform distribution both in amplitude (from 0 to 1) and phase (from 0 to 2π). For the large antenna array system as show in Figure 19.2, the antennas are placed in a line facing the scattering area, and the interval between each two adjacent antennas is $\lambda/2$. Moreover, the distance between the transmitter and the intended location is chosen to be 500λ for both systems.

In the simulation, for the TR wideband system, it is assumed to be equipped with only a single antenna, and the bandwidth is tuned from 100 MHz to 500 MHz. With a wider bandwidth, more multipaths can be resolved, and thus the CIR has more taps accordingly. By contrast, for the large antenna array system, the number of antennas is adjusted from 20 to 100, while the bandwidth of the system is fixed to be 1 MHz. This narrow-band configuration guarantees that the CIR has a single tap, which is a common assumption in OFDM-based massive MIMO literature [7]. In order to achieve the energy concentration at the intended location, time-reversal mirror precoder and matched filter precoder are applied in TR wideband and large antenna array systems, respectively. As shown in Figure 19.2, we consider the field strengths around the intended location with dimension $5\lambda \times 5\lambda$. Figure 19.3 shows the simulation results of both systems, where the maximal received signal strength is set to be 0 dB for normalization. We can clearly see that for the TR wideband system, the spatial focusing effect can be improved with wider and wider bandwidth. Similarly, for the large antenna array system, the increasing number of antennas can also lead to better spatial focusing effect. Therefore, those simulation results corroborate the existence of spatial focusing phenomenon in future 5G systems when the bandwidth becomes wider as in the mm-Wave system or the number of antennas becomes massive as in the massive MIMO system. Note that in the dynamic scenario, once a user moves out of the spatial focusing spot, the CSI is required to be updated such that the new spatial focusing spot can be established. Therefore, how much users can move in the network is determined by the frequency of CSI updating.

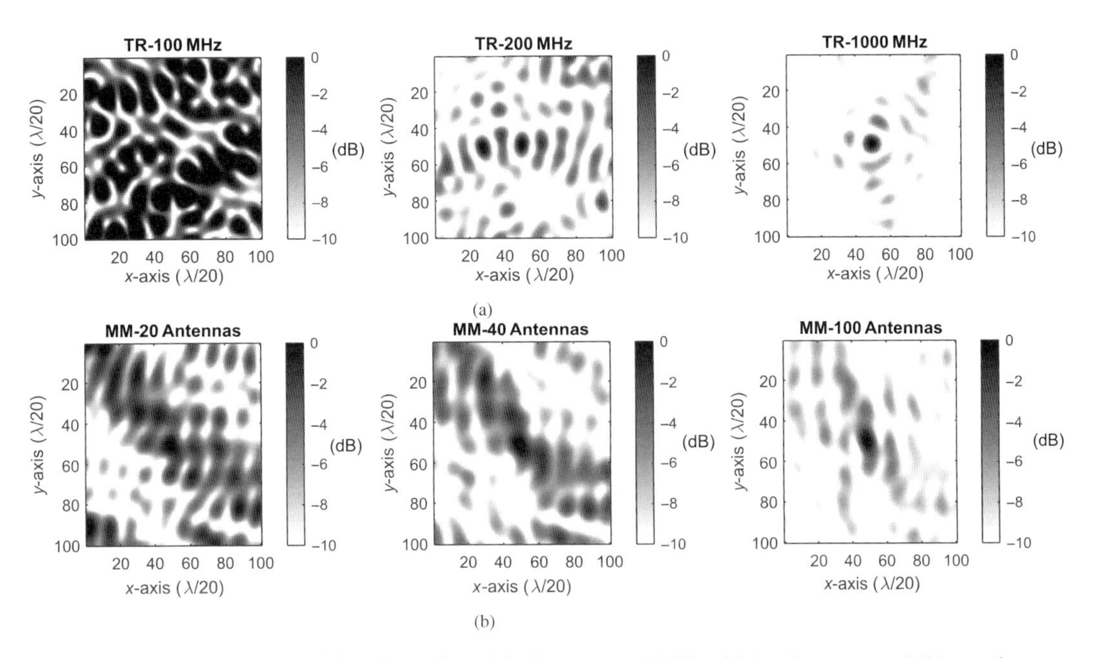

Figure 19.3 Spatial focusing effect of both systems: (a) TR wideband system, and (b) massive MIMO system.

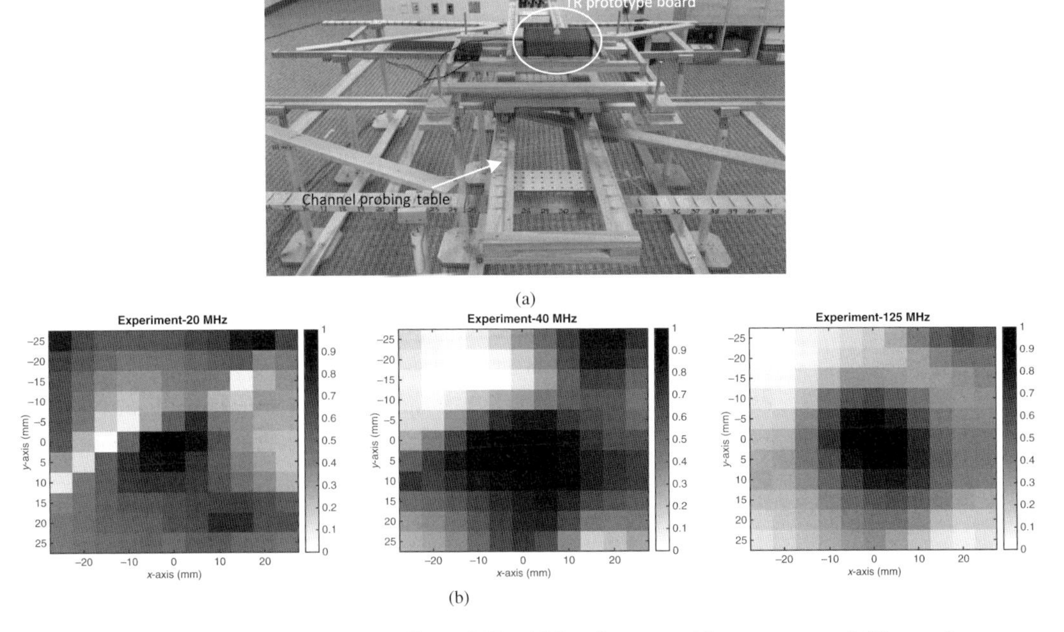

Figure 19.4 Spatial focusing effect of TR wideband system: (a) prototype, and (b) experiment results.

19.4.2 Prototype-Based Experiment

To further verify the spatial focusing effect, we built a prototype of TR wideband system on a customized software-defined radio (SDR) platform, as shown in Figure 19.4(a). The hardware architecture combines a specific designed radio-frequency board covering the ISM band with 125 MHz bandwidth, a high-speed Ethernet port, and an off-the-shelf user-programmable module board. In this experiment, we measure the CIRs of a square region with dimension 5 cm × 5 cm on a channel probing table, which is located in a typical office environment as shown in Figure 19.4. The intended location is chosen to be the center of the measured region, and the corresponding normalized field strength is shown in Figure 19.4(b). We can see that the TR transmission can generate a clear energy focusing around the intended location, even under a not-so-wide bandwidth of 125 MHz. Relying on this spatial focusing effect, the UEs at different locations can be ideally separated, which enables the concurrent data transmission over the entire spectrum.

19.4.3 Spatial Spectrum Sharing

As we can see from both the simulation and experiment results, the spatial focusing effect commonly exists either in wideband systems or in large-scale antenna systems. This common spatial focusing phenomenon enables us to find a general spatial spectrum sharing scheme, where multiple users can concurrently conduct data transmission upon the entire spectrum by utilizing their corresponding channel state information as a unique signature, i.e., their locations have ideally separated them in the spatial domain. The spatial spectrum sharing scheme is rather simple in that, taking the downlink transmission for an example, all the UEs' signals can be added together and transmitted simultaneously by the BS. The system models illustrated in Figure 19.1 have indicated the essence of the spatial spectrum sharing scheme, where for TR wideband system, all UEs' signals are added together after convoluted with their own signatures, and the sum of the signals are transmitted simultaneously; and for massive MIMO systems, all UEs' signals are multiplied by the beamforming matrix and then also transmitted simultaneously. Note that the presented spatial spectrum sharing is independent of the network operators because every pair of transmitter and receiver is barely interfered by other pairs in the physical layer, whether within the same operator or not. In the next section, we will analyze the performance of such a spatial spectrum sharing scheme.

19.5 Spatial Spectrum Sharing Performance

Based on the spatial focusing effect, our further contribution in this chapter is to theoretically analyze the closed-form signal-to-interference-plus-noise ratio (SINR) expressions for both TR wideband and massive MIMO systems. Moreover, two general network association protocols either in a centralized manner or a distributed manner will be discussed, which are also applicable for both systems. In this section, we evaluate the spatial spectrum sharing performance based on the system models described in the previous sections. Although the spatial focusing effect can help separate multiusers' simultaneous transmission, the nonideal uncorrelated channels can still lead to minor inference to each other. Hence, we quantify such interference for both systems in this section as well as the effective SINR performance. SINR is a direct quantified metric reflecting the spatial focusing effect, i.e., a higher SINR represents the signal power is more concentrated at the intended receiver. Moreover, SINR is also a direct indicator to the data rate and the system throughput performance and thus is deduced in detail in this chapter. As we will see later, an interesting phenomenon shows that the SINR expression of both TR wideband and massive MIMO systems share the same formulation, which is fundamentally due to the similar spatial focusing effect existing in both systems.

19.5.1 Time-Reversal Wideband System

Let us consider the multiuser downlink transmission of a TR wideband system. Thanks to the spatial and temporal focusing effect, the signal energy can be concentrated in a single time sample at the location of the intended UE. In such a case, the jth UE just simply performs a one-tap gain adjustment a_j to the received signal to recover the signal and then down-samples it with the same backoff factor D. Thus, the down-sampled received signal of the jth UE, denoted by $Y_j[k]$, can be written as follows:

$$Y_j[k] = a_j \sum_{i=1}^{M} \sum_{l=0}^{(2L-2)/D} d_{i,j}^{-\alpha/2} \left(S_i * h_{i,j}\right)[Dl] + a_j n_j[k], \tag{19.7}$$

where for notational simplicity, we have assumed $L - 1$ to be a multiple of backoff factor D. To better understand the received signal $Y_j[k]$, we further rewrite it into the following separated formulation, from which the intersymbol interference (ISI), interuser interference (IUI), and intercell interference (ICI) can be clearly illustrated.

$$\begin{aligned} Y_j[k] &= a_j \sum_{\phi=1}^{\Phi_t} \sqrt{P_{A_j,j}} X_j[k] \cdot d_{A_j,j}^{-\alpha/2} \left(h_{A_j,j}^{\phi} * g_{A_j,j}^{\phi}\right)[L-1] \\ &+ a_j \sum_{\substack{l=0 \\ l \neq (L-1)/D}}^{(2L-2)/D} \sum_{\phi=1}^{\Phi_t} \sqrt{P_{A_j,j}} X_j[k-l] \cdot d_{A_j,j}^{-\alpha/2} \left(h_{A_j,j}^{\phi} * g_{A_j,j}^{\phi}\right)[Dl] \\ &+ a_j \sum_{j' \in \mathcal{N}_{A_j}} \sum_{l=0}^{(2L-2)/D} \sum_{\phi=1}^{\Phi_t} \sqrt{P_{A_j,j'}} X_{j'}[k-l] \cdot d_{A_j,j}^{-\alpha/2} \left(h_{A_j,j}^{\phi} * g_{A_j,j'}^{\phi}\right)[Dl] \\ &+ a_j \sum_{j' \notin \mathcal{N}_{A_j}} \sum_{l=0}^{(2L-2)/D} \sum_{\phi=1}^{\Phi_t} \sqrt{P_{A_{j'},j'}} X_{j'}[k-l] \cdot d_{A_{j'},j}^{-\alpha/2} \left(h_{A_{j'},j}^{\phi} * g_{A_{j'},j'}^{\phi}\right)[Dl] \\ &+ a_j n_j[k], \end{aligned} \tag{19.8}$$

where the first term is the intended signal for the jth UE, the second term is the ISI caused by channel delay spread, the third term is the IUI caused by the UEs that share the same BS with the jth UE, the fourth term is the ICI from the other nonintended BSs, and the last term is the noise at the receiver. Let us further explain the physical meanings of (19.8). The signal power is the maximum-power central peak of the channel impulse response (CIR) autocorrelation, which is the total power of all multiple paths. For the interference, the ISI is physically due to the channel delay spread and is mathematically due to the cross symbols operation of CIR autocorrelation. The IUI is physically due to the power leakage to an unintended user and is mathematically due to the existence of correlation among different users' CIR as well as the channel delay spread. The ICI is similar to the IUI, which can be interpreted as IUI from other cells.

$$\begin{aligned}
\text{SINR}_j &= \frac{a_j P_{A_j,j} d_{A_j,j}^{-\alpha} C_1}{a_j P_{A_j,j} d_{A_j,j}^{-\alpha} C_2 + a_j (P - P_{A_j,j}) d_{A_j,j}^{-\alpha} C_3 + a_j \sum_{\substack{i=1 \\ i \neq A_j}}^{M} P d_{i,j}^{-\alpha} C_3 + a_j \sigma_j^2} \\
&= \frac{P_{A_j,j} d_{A_j,j}^{-\alpha} C_1}{P_{A_j,j} d_{A_j,j}^{-\alpha} C_2 - P_{A_j,j} d_{A_j,j}^{-\alpha} C_3 + \sum_{i=1}^{M} P d_{i,j}^{-\alpha} C_3 + \sigma_j^2}.
\end{aligned} \quad (19.9)$$

$$\begin{aligned}
\text{SINR}_j &= \frac{\frac{P}{|\mathcal{N}_{A_j}|+1} d_{A_j,j}^{-\alpha} C_1}{\frac{P}{|\mathcal{N}_{A_j}|+1} d_{A_j,j}^{-\alpha} C_2 + \frac{|\mathcal{N}_{A_j}| P}{|\mathcal{N}_{A_j}|+1} d_{A_j,j}^{-\alpha} C_3 + \sum_{\substack{i=1 \\ i \neq A_j}}^{M} P d_{i,j}^{-\alpha} C_3 + \sigma_j^2} \\
&= \frac{C_1}{C_2 - C_3 + \left(|\mathcal{N}_{A_j}| + 1\right) d_{A_j,j}^{\alpha} \left(\sum_{i=1}^{M} d_{i,j}^{-\alpha} C_3 + \sigma_j^2 / P\right)} \\
&= \frac{\xi_1}{\xi_2 + \xi_3 \left(|\mathcal{N}_{A_j}| + 1\right) d_{A_j,j}^{\alpha}},
\end{aligned} \quad (19.10)$$

Here, we employ the definition of effective SINR, which is the ratio of the average signal power to the average interference-and-noise power, as the performance metric. Note that this effective SINR is an approximation of the standard average SINR definition, i.e.,

$$\begin{aligned}
&\mathbb{E}_h \left[\frac{P_{\text{SIG}}}{P_{\text{ISI}} + P_{\text{IUI}} + P_{\text{ICI}} + \sigma^2} \right] \\
&\simeq \frac{\mathbb{E}_h [P_{\text{SIG}}]}{\mathbb{E}_h [P_{\text{ISI}}] + \mathbb{E}_h [P_{\text{IUI}}] + \mathbb{E}_h [P_{\text{ICI}}] + \sigma^2}.
\end{aligned} \quad (19.11)$$

While the standard average SINR calculation in the left-hand side of (19.11) is mathematically intractable due to the complicated convolution and integration operations, in this chapter we utilize the effective SINR, i.e., right-hand side of (19.11), to study the performance, which has been widely utilized in the literature [39–41]. Specifically, it has shown in [28] that this approximation is relatively accurate by comparing the simulation results and approximated theoretical results. Based on the definition of the effective SINR, we should first calculate the average signal power and average interference power, respectively. First, the expected intended signal power for the jth UE from (19.8), denoted by $\mathbb{E}_h [P_{\text{SIG}}]$, can be calculated by

$$\begin{aligned}
\mathbb{E}_h [P_{\text{SIG}}] &= a_j P_{A_j,j} d_{A_j,j}^{-\alpha} \mathbb{E}_h \left[\left| \left(h_{A_j,j} * g_{A_j,j}\right) [L-1] \right|^2 \right] \\
&= a_j P_{A_j,j} d_{A_j,j}^{-\alpha} C_1,
\end{aligned} \quad (19.12)$$

where we use C_1 to represent the expectation of $\left|\left(h_{A_j,j} * g_{A_j,j}\right)[L-1]\right|^2$ over the CIR. Note that the waveform design vector $g_{A_j,j}$ is a function of CIR $h_{A_j,j}$. Similarly, we can have the average ISI $\mathbb{E}_h\left[P_{\text{ISI}}\right]$, and average IUI $\mathbb{E}_h\left[P_{\text{IUI}}\right]$ from (19.8) as follows:

$$\begin{aligned}
&\mathbb{E}_h\left[P_{\text{ISI}}\right] \\
&= a_j P_{A_j,j} d_{A_j,j}^{-\alpha} \mathbb{E}_h\left[\sum_{\substack{l=0\\ l\neq(L-1)/D}}^{(2L-2)/D} \left|\left(h_{A_j,j} * g_{A_j,j}\right)[Dl]\right|^2\right] && (19.13)\\
&= a_j P_{A_j,j} d_{A_j,j}^{-\alpha} C_2, && (19.14)
\end{aligned}$$

$$\begin{aligned}
\mathbb{E}_h\left[P_{\text{IUI}}\right] &= a_j \sum_{j'\in\mathcal{N}_{A_j}} P_{A_j,j'} d_{A_j,j}^{-\alpha} \\
&\quad \cdot \mathbb{E}_h\left[\sum_{l=0}^{(2L-2)/D} \left|\left(h_{A_j,j} * g_{A_j,j'}\right)[Dl]\right|^2\right] \\
&= a_j \sum_{j'\in\mathcal{N}_{A_j}} P_{A_j,j'} d_{A_j,j}^{-\alpha} C_3 \\
&= a_j (P - P_{A_j,j}) d_{A_j,j}^{-\alpha} C_3, && (19.15)
\end{aligned}$$

where we use C_2 and C_3 to represent the expectations of $\sum_{\substack{l=0\\ l\neq(L-1)/D}}^{(2L-2)/D} \left|\left(h_{A_j,j} * g_{A_j,j}\right)[Dl]\right|^2$ and $\sum_{l=0}^{(2L-2)/D} \left|\left(h_{A_j,j} * g_{A_j,j'}\right)[Dl]\right|^2$, respectively. The calculation of the average ICI $\mathbb{E}_h\left[P_{\text{ICI}}\right]$ requires some special transformation. The ICI in the fourth term of (19.8) was written by the summation of interference from all UEs belonging to all other nonintended BSs. Because the interference from the UEs sharing the same BS can be regarded as a single interference source from that BS, we can rewrite the ICI directly by the summation of interference from all other nonintended BSs. Let us denote $\mathcal{N}_i$ as the set of UEs that associate with the ith BS. Then, the ICI, i.e., the fourth item in (19.8), can be written by

$$\text{ICI} = a_j \sum_{\substack{i=1\\ i\neq A_j}}^{M} \sum_{j'\in\mathcal{N}_i} \sum_{l=0}^{(2L-2)/D} X_{j'}[k-l] \cdot d_{i,j}^{-\alpha/2} \left(h_{i,j} * g_{i,j'}\right)[Dl]. \tag{19.16}$$

By taking expectation over the channel, we can have the average ICI as follows

$$\begin{aligned}
\mathbb{E}_h\left[P_{\text{ICI}}\right] &= a_j \sum_{\substack{i=1\\ i\neq A_j}}^{M} \sum_{j'\in\mathcal{N}_i} P_{i,j'} d_{i,j}^{-\alpha} \\
&\quad \cdot \mathbb{E}_h\left[\sum_{l=0}^{(2L-2)/D} \left|\left(h_{i,j} * g_{i,j'}\right)[Dl]\right|^2\right] \\
&= a_j \sum_{\substack{i=1\\ i\neq A_j}}^{M} P d_{i,j}^{-\alpha} C_3, && (19.17)
\end{aligned}$$

where C_3 is same with that in (19.15). By substituting (19.12), (19.13), (19.15), and (19.17) into the right-hand side of (19.11), we can obtain the effective SINR of the jth

UE, SINR_j, as in (19.9). Note that (19.9) is a general expression of the effective SINR evaluation in the TR wideband system. In the following, some special cases will be considered to reveal more characteristics.

First, let us suppose equal power allocation scenario, i.e., the power for each UE belonging to the BS i should be $\frac{P}{|\mathcal{N}_i|}$. In such a case, we can simplify the SINR expression in (19.9) as in (19.10), where for simplicity, we define the parameters ξ_1, ξ_2, and ξ_3 as follows:

$$\xi_1 = C_1, \quad \xi_2 = C_2 - C_3, \quad \xi_3 = \sum_{i=1}^{M} d_{i,j}^{-\alpha} C_3 + \sigma_j^2/P. \tag{19.18}$$

Second, let us consider the time-reversal waveform, i.e.,

$$g_{i,j}[k] = \frac{h_{i,j}^*[L-1-k]}{\sqrt{\mathbb{E}\left[\sum_{l=0}^{L-1} |h_{i,j}[l]|^2\right]}}, \tag{19.19}$$

which is the normalized (by the average channel gain) complex conjugate of time-reversed $\{h_{i,j}^*[k]\}$. This time-reversal waveform can provide the maximum reception power at the intended symbol because the first term in (19.8) corresponds to the maximum-power central peak of the autocorrelation function, i.e.,

$$\left(h_{A_j,j} * g_{A_j,j}\right)[L-1] = \frac{\sum_{l=0}^{L-1} \left|h_{i,j}^*[l]\right|^2}{\sqrt{\mathbb{E}\left[\sum_{l=0}^{L-1} |h_{i,j}[l]|^2\right]}}. \tag{19.20}$$

Furthermore, by considering the exponential-decay multipath channel model, we have

$$\mathbb{E}\left[\left|h_{i,j}[k]\right|^2\right] = e^{-\frac{kT_s}{\sigma_T}}, \quad 0 \le k \le L-1, \tag{19.21}$$

$$\mathbb{E}\left[\left|h_{i,j}[k]\right|^4\right] = 2\left(\mathbb{E}\left[\left|h_{i,j}[k]\right|^2\right]\right)^2 = 2e^{-2\frac{kT_s}{\sigma_T}}, \tag{19.22}$$

where the kth tap between the ith BS and the jth UE, $h_{i,j}[k]$, is assumed to be a circular symmetric complex Gaussian random variable with zero mean, and aforementioned variance, T_s is the sampling period of the system such that $1/T_s$ equals to the system bandwidth B, and δ_T is the root mean square delay spread of the channel. Based on this exponential-decay multi-path channel model, we can calculate the closed-form expressions for C_1, C_2, and C_3 and, in turn, the expressions of ξ_1, ξ_2, and ξ_3 as follows:

$$\xi_1 = \frac{1+e^{-\frac{LT_s}{\sigma_T}}}{1+e^{-\frac{T_s}{\sigma_T}}} + \frac{1-e^{-\frac{LT_s}{\sigma_T}}}{1-e^{-\frac{T_s}{\sigma_T}}}, \tag{19.23}$$

$$\xi_2 = 2\frac{e^{-\frac{T_s}{\sigma_T}}\left(1-e^{-\frac{(L-2+D)T_s}{\sigma_T}}\right)}{\left(1-e^{-\frac{DT_s}{\sigma_T}}\right)\left(1+e^{-\frac{T_s}{\sigma_T}}\right)} - \tag{19.24}$$

$$\frac{\left(1+e^{-\frac{DT_S}{\sigma_T}}\right)\left(1+e^{-\frac{2LT_S}{\sigma_T}}\right)-2e^{-\frac{(L+1)T_S}{\sigma_T}}\left(1+e^{-\frac{(D-2)T_S}{\sigma_T}}\right)}{\left(1-e^{-\frac{DT_S}{\sigma_T}}\right)\left(1+e^{-\frac{T_S}{\sigma_T}}\right)\left(1-e^{-\frac{LT_S}{\sigma_T}}\right)},$$

$$\xi_3 = \sigma_j^2/P + \sum_{i=1}^{M} d_{i,j}^{-\alpha}\cdot \tag{19.25}$$

$$\frac{\left(1+e^{-\frac{DT_S}{\sigma_T}}\right)\left(1+e^{-\frac{2LT_S}{\sigma_T}}\right)-2e^{-\frac{(L+1)T_S}{\sigma_T}}\left(1+e^{-\frac{(D-2)T_S}{\sigma_T}}\right)}{\left(1-e^{-\frac{DT_S}{\sigma_T}}\right)\left(1+e^{-\frac{T_S}{\sigma_T}}\right)\left(1-e^{-\frac{LT_S}{\sigma_T}}\right)}.$$

Combining (19.10) and (19.23–19.25), we have obtained the SINR evaluation of the TR wideband system. In the next subsection, we will analyze the SINR of the massive MIMO system, where it will be seen that the SINR of both systems share exactly the same expression.

19.5.2 Massive MIMO System

Let us consider the multiuser downlink transmission of a massive MIMO system. Because the BSs have performed beamforming based on the CSI, the UE can simply do the one-tap detection to receive the signal. Similar to the TR wideband system, the received signal of the jth UE can be formulated by the summation of the intended signal, IUI and ICI, as follows:

$$\begin{aligned} y_j &= \sum_{i=1}^{M} \sqrt{P} d_{i,j}^{-\alpha/2} \mathbf{h}_{i,j} \mathbf{G}_i \mathbf{s}_i + n_j \\ &= \sqrt{P_{A_j,j}} d_{A_j,j}^{-\alpha/2} \mathbf{h}_{A_j,j} \mathbf{g}_{A_j,j} s_{A_j,j} \\ &\quad + \sum_{j' \in \mathcal{N}_{A_j}} \sqrt{P_{A_j,j'}} d_{A_j,j}^{-\alpha/2} \mathbf{h}_{A_j,j} \mathbf{g}_{A_j,j'} s_{A_j,j'} \\ &\quad + \sum_{\substack{i=1 \\ i \neq A_j}}^{M} \sqrt{P} d_{i,j}^{-\alpha/2} \mathbf{h}_{i,j} \mathbf{G}_i \mathbf{s}_i + n_j, \end{aligned} \tag{19.26}$$

where A_j denotes the BS that the jth TD will associate with, $\mathbf{h}_{A_j,j}$ denotes the CSI between the BS and the jth UE, $s_{A_j,j}$ denotes the intended symbol for the jth UE, $\mathbf{g}_{A_j,j}$ denotes the beamforming vector for the jth UE, $\mathcal{N}_{A_j}$ denotes the current set of TDs that also associate BS A_j besides the jth UE, and n_j represents the noise. In the last equality of (19.26), the first term is the intended signal for the jth UE, while the second term is the IUI and the third term is the ICI for the jth UE. Similarly, the physical meanings of the SINR of massive MIMO systems can be interpreted as follows. The signal power represents the intended symbols from all antennas. Because the OFDM technique is commonly adopted in the MIMO systems, the ISI can be neglected in the narrow-band subcarrier. While the IUI and ICI still exist, they are due to the nonideal uncorrelated assumption among difference users' channels. In such a case, we can write the expectation of the jth UE's signal power and IUI power as follows:

$$\begin{aligned}\mathbb{E}_h\left[P_{\text{SIG}}\right] &= P_{A_j,j} d_{A_j,j}^{-\alpha} \mathbb{E}_h\left[\left|\mathbf{h}_{A_j,j}\mathbf{g}_{A_j,j}\right|^2\right] \\ &= P_{A_j,j} d_{A_j,j}^{-\alpha} C_1',\end{aligned} \tag{19.27}$$

$$\begin{aligned}\mathbb{E}_h\left[P_{\text{IUI}}\right] &= \sum_{j'\in\mathcal{N}_{A_j}} P_{A_j,j'} d_{A_j,j}^{-\alpha} \mathbb{E}_h\left[\left|\mathbf{h}_{A_j,j}\mathbf{g}_{A_j,j'}\right|^2\right] \\ &= \sum_{j'\in\mathcal{N}_{A_j}} P_{A_j,j'} d_{A_j,j}^{-\alpha} C_2' = (P - P_{A_j,j}) d_{A_j,j}^{-\alpha} C_2',\end{aligned} \tag{19.28}$$

where we use constants C_1' and C_2' to represent the expectations of $\left|\mathbf{h}_{A_j,j}\mathbf{g}_{A_j,j}\right|^2$ and $\left|\mathbf{h}_{A_j,j}\mathbf{g}_{A_j,j'}\right|^2$, respectively. For the ICI, it can be rewritten by

$$\begin{aligned}\text{ICI} &= \sum_{\substack{i=1\\ i\neq A_j}}^{M} \sqrt{P} d_{i,j}^{-\alpha/2} \mathbf{h}_{i,j} \mathbf{G}_i \mathbf{s}_i \\ &= \sum_{\substack{i=1\\ i\neq A_j}}^{M} \sum_{j'\in\mathcal{N}_i} \sqrt{P_{i,j'} d_{i,j}^{-\alpha/2}} \mathbf{h}_{i,j} \mathbf{g}_{i,j'} s_{i,j'},\end{aligned} \tag{19.29}$$

and then, we have the power of ICI as follows

$$\begin{aligned}\mathbb{E}_h\left[P_{\text{ICI}}\right] &= \sum_{\substack{i=1\\ i\neq A_j}}^{M} \sum_{j'\in\mathcal{N}_i} P_{i,j'} d_{i,j}^{-\alpha} \mathbb{E}_h\left[\left|\mathbf{h}_{i,j}\mathbf{g}_{i,j'}\right|^2\right] \\ &= \sum_{\substack{i=1\\ i\neq A_j}}^{M} P d_{i,j}^{-\alpha} C_2'.\end{aligned} \tag{19.30}$$

By integrating (19.26), (19.28), and (19.29), we can obtain the effective SINR as follows:

$$\begin{aligned}\text{SINR}_j &= \frac{P_{A_j,j} d_{A_j,j}^{-\alpha} C_1'}{(P - P_{A_j,j}) d_{A_j,j}^{-\alpha} C_2' + \sum_{\substack{i=1\\ i\neq A_j}}^{M} P d_{i,j}^{-\alpha} C_2'} \\ &= \frac{P_{A_j,j} d_{A_j,j}^{-\alpha} C_1'}{-P_{A_j,j} d_{A_j,j}^{-\alpha} C_2' + \sum_{i=1}^{M} P d_{i,j}^{-\alpha} C_2' + \sigma_j^2}.\end{aligned} \tag{19.31}$$

We can see that, the SINR of both TR wideband and massive MIMO systems are quite similar, while the only difference is that there is no ISI in massive MIMO system due to the single-tap channel assumption, i.e., OFDM operation. In the following, we will see that the SINR of both systems share exactly the same expression under some special scenarios.

Similar to the analysis of the TR wideband system, let us also consider the equal power allocation scenario for the massive MIMO system. Then, we can have $P_{i,j} = P/|\mathcal{N}_i|$, and the following derivations

$$\begin{aligned}
\mathrm{SINR}_j &= \frac{\frac{P}{|\mathcal{N}_{A_j}|+1} d_{A_j,j}^{-\alpha} C_1'}{-\frac{P}{|\mathcal{N}_{A_j}|+1} d_{A_j,j}^{-\alpha} C_2' + \sum_{i=1}^{M} P d_{i,j}^{-\alpha} C_2' + \sigma_j^2} \\
&= \frac{C_1'}{-C_2' + \left(|\mathcal{N}_{A_j}| + 1\right) d_{A_j,j}^{\alpha} \left(\sum_{i=1}^{M} d_{i,j}^{-\alpha} C_2' + \sigma_j^2/P\right)} \\
&= \frac{\xi_1'}{\xi_2' + \xi_3' \left(|\mathcal{N}_{A_j}| + 1\right) d_{A_j,j}^{\alpha}},
\end{aligned} \tag{19.32}$$

where for simplicity, we define the parameters ξ_1', ξ_2', and ξ_3' as follows:

$$\xi_1' = C_1', \quad \xi_2' = -C_2', \quad \xi_3' = \sum_{i=1}^{M} d_{i,j}^{-\alpha} C_2' + \sigma_j^2/P. \tag{19.33}$$

By comparing (19.10) and (19.32), it can be seen that both expressions are exactly the same under the equal power allocation scenario, but with different parameters. This indicates that we can design a general network association scheme based on the evaluated SINR, which can be applicable for both TR wideband and massive MIMO systems. This characteristic is fundamentally due to the utilization of multipath effect, where the TR wideband system is through wide band to reveal and harvest existing multiple paths, while the massive MIMO utilizes the physical massive number of antennas to create multiple independent paths. The closed-form expression of (19.32) can be obtained by further considering a specific beamforming scheme. Here, let us consider the beamforming scheme of MRC, then we can have the beamforming vector $\mathbf{g}_{i,j}$ as the normalized conjugate of the CSI, i.e.,

$$\mathbf{g}_{i,j} = \frac{\mathbf{h}_{i,j}^H}{\sqrt{\mathbb{E}\left[|\mathbf{h}_{i,j}|^2\right]}}. \tag{19.34}$$

Moreover, as aforementioned in the system model, the Rayleigh fading channel is considered. In such a case, the CSI $h_{i,j}^{\phi}$ can be regarded as a complex Gaussian random variable with zero mean and variance γ, i.e., $h_{i,j}^{\phi} \sim \mathcal{CN}(0, \gamma)$. Based on this channel model, we can derive the second and fourth moments of $h_{i,j}^{\phi}$ as follows:

$$\mathbb{E}\left[|h_{i,j}^{\phi}|^2\right] = \gamma, \tag{19.35}$$

$$\mathbb{E}\left[|h_{i,j}^{\phi}|^4\right] = 2\left(\mathbb{E}\left[|h_{i,j}|^2\right]\right)^2 = 2\gamma^2, \tag{19.36}$$

$$\mathbb{E}\left[|\mathbf{h}_{i,j}|^2\right] = \sum_{\phi=1}^{\Phi} \mathbb{E}\left[|h_{i,j}^{\phi}|^2\right] = \Phi\gamma, \tag{19.37}$$

$$\mathbb{E}\left[|\mathbf{h}_{i,j}|^4\right] = \mathbb{E}\left[\left|\sum_{\phi=1}^{\Phi} \left|h_{i,j}^{\phi}\right|^2\right|^2\right] = \Phi(\Phi+1)\gamma^2, \tag{19.38}$$

Then, the constants C_1' and C_2' can be calculated by

$$C_1' = \mathbb{E}\left[|\mathbf{h}_{i,j}\mathbf{g}_{i,j}|^2\right] = \frac{\mathbb{E}\left[|\mathbf{h}_{i,j}|^4\right]}{\mathbb{E}\left[|\mathbf{h}_{i,j}|^2\right]} = (\Phi+1)\gamma, \tag{19.39}$$

$$C_2' = \mathbb{E}\left[|\mathbf{h}_{i,j}\mathbf{g}_{i,j'}|^2\right] = \mathbb{E}\left[\left|\sum_{\phi=1}^{\Phi} h_{i,j}^{\phi} g_{i,j'}^{\phi}\right|^2\right] \tag{19.40}$$

$$= \frac{\sum_{\phi=1}^{\Phi} \mathbb{E}\left[|h_{i,j}^{\phi}|^2\right] \mathbb{E}\left[|(h_{i,j'}^{\phi})^*|^2\right]}{\sum_{\phi=1}^{\Phi} \mathbb{E}\left[|h_{i,j'}^{\phi}|^2\right]} = \gamma. \tag{19.41}$$

Thus, we can have the parameters ξ_1', ξ_2', and ξ_3' as follows:

$$\xi_1' = (\Phi+1)\gamma, \quad \xi_2' = -\gamma, \quad \xi_3' = \sum_{i=1}^{M} d_{i,j}^{-\alpha}\gamma + \sigma_j^2/P. \tag{19.42}$$

Finally, by combining (19.32) and (19.42), we have obtained the SINR evaluation of the massive MIMO system. In the next section, we will discuss a network association scheme for both TR wideband and massive MIMO systems, based on the effective SINR evaluation.

19.6 General Network Association Protocols Design

In this section, we design a general UE sharing algorithm that can be applicable for both TR wideband and massive MIMO systems. In the previous section, it was shown that the SINR of TR wideband and massive MIMO systems share the exactly same expression as follows

$$\mathrm{SINR}_j = \frac{\xi_1}{\xi_2 + \xi_3\left(\left|\mathcal{N}_{A_j}\right| + 1\right) d_{A_j,j}^{\alpha}}, \tag{19.43}$$

where only the parameters ξ_1, ξ_2, and ξ_3 are different for the two systems. Based on this general SINR expression, we first discuss a centralized scheme with the target of maximizing all UEs' SINR and then a distributed scheme with the target of maximizing the new arriving UE's SINR. The centralized scheme is more reasonable from the perspective of the overall network performance, but it is also associated with more computation and communication complexities. The distributed scheme is a greedy algorithm with low complexity, but only takes into account the new arrival's performance. Nevertheless, because both systems are inherent with the property of low IUI, the performance of the distributed scheme is quite similar to that of the centralized one, which will be verified in the simulation section.

19.6.1 Centralized Scheme

In practical network systems, UEs register and leave the network sequentially. Therefore, let us consider a dynamic UE arriving and then associating scenario. Suppose there are M BSs and N UEs in the network, where each BS is currently serving some existing UEs. Note that when referring to BS in this section, we will not differentiate between a TR wideband BS and a massive MIMO BS due to the similarity. Considering a new UE arrives at the network, a straightforward problem is which BS the new UE should register and associate with. The new UE should share the entire spectrum resource with all other UEs, but share the power resource only with UEs in the same BS. Apparently, the existing UEs' performance may degrade due to the power sharing and interuser interference from the new UE. Nevertheless, the existing UEs have the priority of maintaining current BS associations as well as the satisfied quality of service (QoS), which is a common assumption in current prevailing systems like 3G and 4G.

When it comes to the centralized scheme, the new UE's association should maximize the overall system throughput, i.e., the social welfare, while given the condition that the existing UEs remain in the associated BSs, and their resultant throughput can still satisfy the corresponding QoS requirement. Let us use j' to represent the existing UEs and j to represent the new UE. Thus, we can model the association problem by the following

$$\begin{aligned} \max \quad & \sum_{j'=1}^{N} R'_{j'} + R'_{j} \\ \text{s.t.} \quad & R'_{j} \geq R^{\text{th}}_{j}, \quad R'_{j'} \geq R^{\text{th}}_{j'}\ \forall j', \end{aligned} \tag{19.44}$$

where $R'_{\cdot}$ represents the UE's expected throughput after the new arrival UE j joins the network, and $R^{\text{th}}_{\cdot}$ represents the minimum QoS requirement of the UE. Because all the existing UEs are supposed to maintain their connections to the current associated BSs, the optimization problem in (19.44) is equivalent to the following problem

$$\max \sum_{j'=1}^{N} R'_{j'} + R'_{j} \Rightarrow \max \sum_{j'=1}^{N} R'_{j'} + R'_{j} - \sum_{j'=1}^{N} R_{j'}, \tag{19.45}$$

where $R_{j'}$ represents the existing UEs' expected throughput before the new UE joins the network.

Moreover, because the overall transmission power of each BS is assumed to be fixed, the expected ICI should not be affected by the new arrival UE, i.e., the new one would not impose more ICI to the UE in a different BS. Instead, the new UE only affects the performance of the UEs that shares the same BS with it due to the power sharing and IUI. Suppose the new UE j accesses BS A_j, we can rewrite (19.45) as follows:

$$\sum_{j'=1}^{N} R'_{j'} + R'_j - \sum_{j'=1}^{N} R_{j'} = R'_j + \sum_{j' \in \mathcal{N}_{A_j}} \left(R'_{j'} - R_{j'} \right)$$
$$= \log\left(1 + \frac{\xi_1}{\xi_2 + \xi_3 \left(\left|\mathcal{N}_{A_j}\right| + 1\right) d^{\alpha}_{A_j,j}}\right)$$
$$+ \sum_{j' \in \mathcal{N}_{A_j}} \left[\log\left(1 + \frac{\xi_1}{\xi_2 + \xi_{3,j'} \left(\left|\mathcal{N}_{A_j}\right| + 1\right) d^{\alpha}_{A_j,j'}}\right) \right.$$
$$\left. - \log\left(1 + \frac{\xi_1}{\xi_2 + \xi_{3,j'} \left|\mathcal{N}_{A_j}\right| d^{\alpha}_{A_j,j'}}\right)\right]$$
$$= \log\left[\frac{\xi_1 + \xi_2 + \xi_3 \left(\left|\mathcal{N}_{A_j}\right| + 1\right) d^{\alpha}_{A_j,j}}{\xi_2 + \xi_3 \left(\left|\mathcal{N}_{A_j}\right| + 1\right) d^{\alpha}_{A_j,j}} \cdot \right.$$
$$\prod_{j' \in \mathcal{N}_{A_j}} \left(\frac{\xi_1 + \xi_2 + \xi_{3,j'} \left(\left|\mathcal{N}_{A_j}\right| + 1\right) d^{\alpha}_{A_j,j'}}{\xi_2 + \xi_{3,j'} \left(\left|\mathcal{N}_{A_j}\right| + 1\right) d^{\alpha}_{A_j,j'}} \right.$$
$$\left.\left. \frac{\xi_2 + \xi_{3,j'} \left|\mathcal{N}_{A_j}\right| d^{\alpha}_{A_j,j'}}{\xi_1 + \xi_2 + \xi_{3,j'} \left|\mathcal{N}_{A_j}\right| d^{\alpha}_{A_j,j'}} \right)\right]$$
$$= \log\left(\frac{\xi_1 + \xi_2 + \xi_3 \left(\left|\mathcal{N}_{A_j}\right| + 1\right) d^{\alpha}_{A_j,j}}{\xi_2 + \xi_3 \left(\left|\mathcal{N}_{A_j}\right| + 1\right) d^{\alpha}_{A_j,j}} \cdot \xi_{4,A_j}\right), \tag{19.46}$$

where ξ_{4,A_j} is

$$\xi_{4,A_j} = \prod_{j' \in \mathcal{N}_{A_j}} \frac{\xi_1 + \xi_2 + \xi_{3,j'} \left(\left|\mathcal{N}_{A_j}\right| + 1\right) d^{\alpha}_{A_j,j'}}{\xi_2 + \xi_{3,j'} \left(\left|\mathcal{N}_{A_j}\right| + 1\right) d^{\alpha}_{A_j,j'}} \cdot \frac{\xi_2 + \xi_{3,j'} \left|\mathcal{N}_{A_j}\right| d^{\alpha}_{A_j,j'}}{\xi_1 + \xi_2 + \xi_{3,j'} \left|\mathcal{N}_{A_j}\right| d^{\alpha}_{A_j,j'}}, \tag{19.47}$$

representing the existing UEs' performance loss due to the admission of the new UE. Thus, the optimization problem in (19.44) can be transformed to the following problem

$$\arg\max_{A_j} \quad \frac{\xi_1 + \xi_2 + \xi_3 \left(\left|\mathcal{N}_{A_j}\right| + 1\right) d^{\alpha}_{A_j,j}}{\xi_2 + \xi_3 \left(\left|\mathcal{N}_{A_j}\right| + 1\right) d^{\alpha}_{A_j,j}} \cdot \xi_{4,A_j} \tag{19.48}$$
$$\text{s.t.} \quad R'_j \geq R^{\text{th}}_j,\ R'_{j'} \geq R^{\text{th}}_{j'}\ \forall j'.$$

Note that there is only one variable A_j in the problem (19.48), and the variable A_j can only be one of the M BSs, solving the problem is as simple as a linear search.

Based on the formulation in (19.48), we can design a centralized network association protocol as follows, as shown in Figure 19.5. When a new UE is intent to join the network, it first broadcasts an impulse to all BSs, such that the BSs can obtain the CIR information as well as estimate the corresponding distance to the new UE according

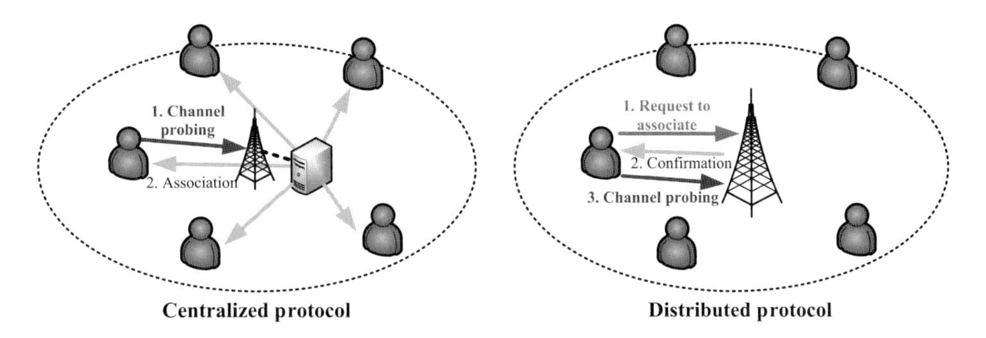

Figure 19.5 Illustration of centralized and distributed protocols.

to the received signal strength. Then, all the BSs report those information and the load status (the number of UEs one BS is currently serving) to a central server through the backhaul, which is in charge of calculating the parameters ξ_1, ξ_2, and ξ_3 in (19.48) as well as the optimal BS that the new UE should access. Notice that this centralized scheme requires a central server to gather all the information from the BSs, which would inevitably incur additional communication costs and delay. In the next section, we will discuss a distributed scheme without requiring such a central server.

19.6.2 Distributed Scheme

The distributed scheme is preferred for the scenario without a central server. Because there is no global information of the entire network in the distributed scheme, the new arrival UE has to make its own decision on which BS it should associate with. Moreover, the performance degradation of the existing UEs is also unknown to the new UE due to the limited information. Nevertheless, the BS can decide whether to admit the new UE for the sake of ensuring that the existing UEs' QoS should be guaranteed. In such a case, a new arrival UE can only make the association decision in a greedy manner, i.e., selecting the BS that can provide the highest expected throughput, which can be easily formulated by the following problem

$$\begin{aligned} &\max \quad R_j = \log\left(1 + \text{SINR}_j\right) \\ \Rightarrow &\max \quad \text{SINR}_j = \frac{\xi_1}{\xi_2 + \xi_3\left(\left|\mathcal{N}_{A_j}\right| + 1\right) d^{\alpha}_{A_j,j}}, \\ &\text{s.t.} \quad R'_j \geq R^{\text{th}}_j,\ R'_{j'} \geq R^{\text{th}}_{j'}\ \forall j', \end{aligned} \tag{19.49}$$

where the definition of the notations are the same with that in (19.44). Because the parameters ξ_1, ξ_2, and ξ_3 are constant for a specific new UE, the problem in (19.49) can be simplified by

$$\begin{aligned} &\arg\min_{A_j} \quad \left(\left|\mathcal{N}_{A_j}\right| + 1\right) d^{\alpha}_{A_j,j} \\ &\text{s.t.} \quad R'_j \geq R^{\text{th}}_j,\ R'_{j'} \geq R^{\text{th}}_{j'}\ \forall j'. \end{aligned} \tag{19.50}$$

Similarly, solving the problem in (19.50) is also a simple linear search. A toy example is for two BSs case: when the load of each BS is given, i.e., the number of existing UEs in each BS N_1 and N_2, the UE can just compare the distance according to the following rule to make the association decision

$$\frac{d_{1,j}}{d_{2,j}} \gtrless \left(\frac{|\mathcal{N}_2|+1}{|\mathcal{N}_1|+1}\right)^{1/\alpha}. \tag{19.51}$$

As a matter of fact, due to the small IUI in TR wideband and massive MIMO systems, the social welfare of the distributed association protocol is quite similar to that of the centralized protocol, which will be verified in the simulation section.

Based on the formulation in (19.50), we can design a distributed network association protocol as follows, as shown in Figure 19.5. The BSs periodically broadcast the beacon signals to the network, including its ID information as well as the current load status, i.e., the number of existing UEs in it. When a new UE is intent to join the network, it can receive beacon signals from the neighboring BSs. The distance between the new UE and the BSs can be roughly estimated by the signal strength and the path loss model. Thus, by calculation (19.50) with the distance and load information, the new UE can determine a priority list of BSs from the high SINR to the low one. Then, the new UE sends a request-to-access packet to the first BS on the list. When a BS receives a request, it would check the constraints in (19.50) to make sure the existing UEs' QoS requirement can be still satisfied. If this can be confirmed, the BS sends a confirmation to the new UE to establish connections. If no confirmation is received within a certain period, the UE would send a request to the second BS on the priority list, so on, and so forth.

19.7 Simulation Results

In this section, we conduct simulation to evaluate the performance of the presented spatial spectrum sharing framework, where both TR wideband systems and massive MIMO systems are considered. For the TR wideband system, we consider the broadband with frequency bandwidth that typically ranges from hundreds of MHz to several GHz, which is much wider than those narrow-band systems specified in 3GPP/4G. In the rich scattering environment, such a broadband system can differentiate rich independent multiple paths, which can help create super-high spatial focusing gain as shown in Figure 19.3(a). On the other hand, for the massive MIMO system, we still consider the traditional narrow-band setting with frequency bandwidth of 20–40 MHz, but with massive number of antennas as to one hundred. Those massive antennas can provide highly independent channel statistics of different locations and also help create super-high spatial focusing gain as shown in Figure 19.3(b). Table 19.1 summarizes the simulation setting of the two systems. For the TR wideband system with 1 GHz bandwidth and the sampling period $T_s = 1$ ns, the channel length is typically $L = 256$, and the root mean square delay spread is typically $\sigma_T = 128T_s$, according to the IEEE 802.15.4a outdoor non-line-of-sight channels. For the massive MIMO system, OFDM technology is adopted for each antenna, and each subcarrier is considered to be with channel length $L = 1$.

Table 19.1 Simulation parameters

Parameters	TR wideband	Massive MIMO
Bandwidth	1 GHz	40 MHz
Number of BSs	$M = 3$	$M = 3$
Number of antennas	1	100
SNR	20 dB	20 dB
Path loss	$\alpha = 4$	$\alpha = 4$
Backoff factor	$D = 16$	$D = 0$
Sampling period	$T_s = 1$ns	$T_s = 1$ns
Channel length	$L = 256$	$L = 1$

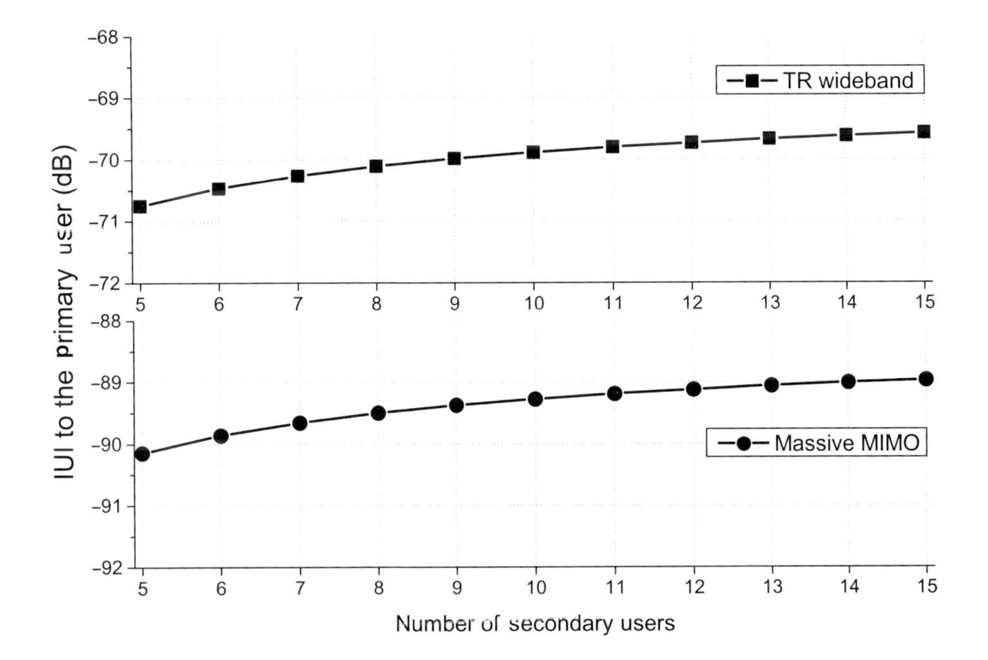

Figure 19.6 Interuser interference in both systems.

Let us first validate the spatial spectrum sharing performance by simulating a BS serving both primary users and secondary users. We consider the scenario that one primary user and 5 secondary users are currently served by one BS, and then 10 more secondary users are sequentially admitted by the same BS sharing the entire spectrum with the existing users. The distance between the primary user and the BS is set as 15 meters, while the locations of the secondary users are randomly generated. Figure 19.6 shows the IUI to the primary user when a different number of secondary users joins the BS, where the y-axis is IUI/Noise in dB. We can see that the IUI to the primary user increases slightly when more and more secondary users are admitted, i.e., approximately only 1 dB loss from 5 to 10 existing secondary users. Therefore, the spatial focusing

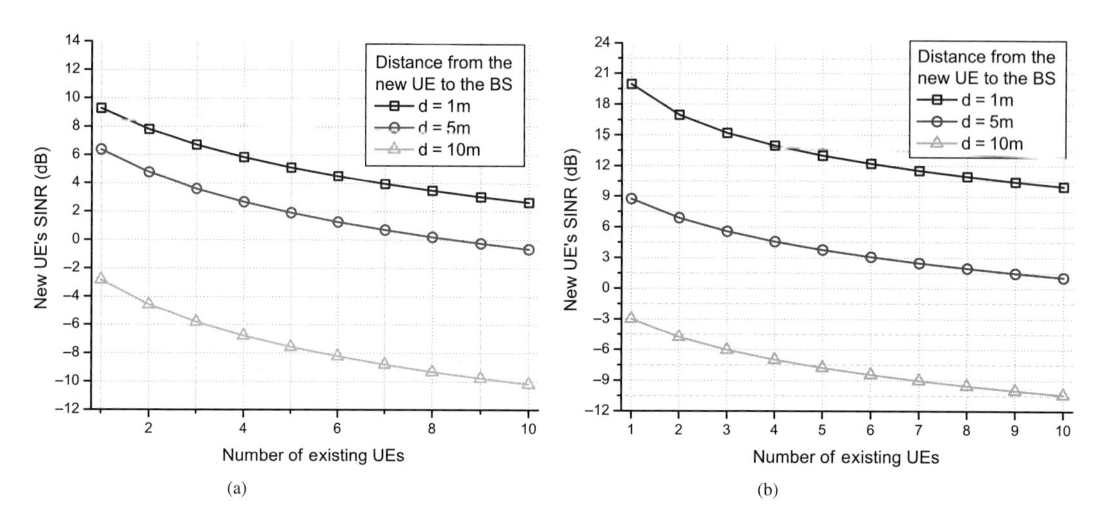

Figure 19.7 SINR performance of the new UE: (a) TR wideband system, and (b) massive MIMO system.

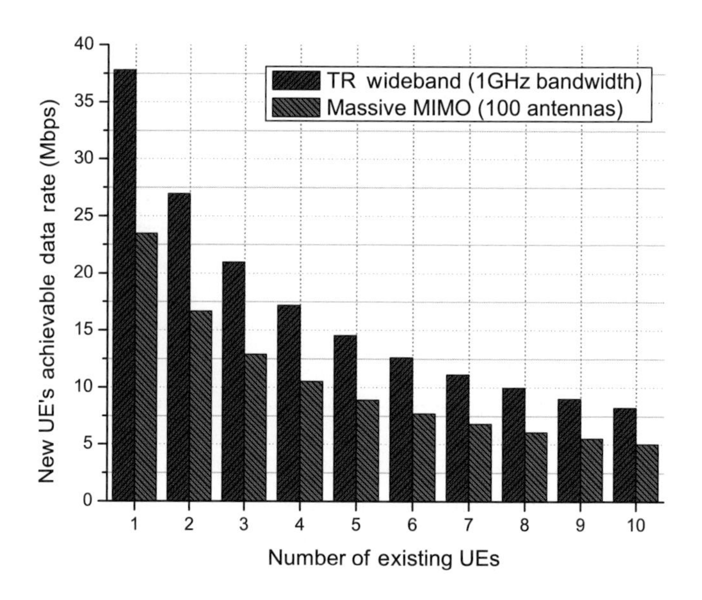

Figure 19.8 Achievable data rate of the new UE in both systems.

effect can greatly decrease the IUI and enable the coexistence of primary users and secondary users on the entire spectrum.

Then, let us investigate the SINR performance of the new arrival UE in both systems, as shown in Figure 19.7 under different numbers of existing UEs in the associated BS, i.e., from 1 UE to 10 UEs. The distance between the new UE and the associated BS is configured as 1 m, 5 m, and 10 m, respectively, which are corresponding to the three curves in each subfigure of Figure 19.7. The distances between the new UE and the other two interference BSs are set as 20 m. Generally, the results in both subfigures show that when the number of existing UEs increases, i.e., the associated BS becomes more and

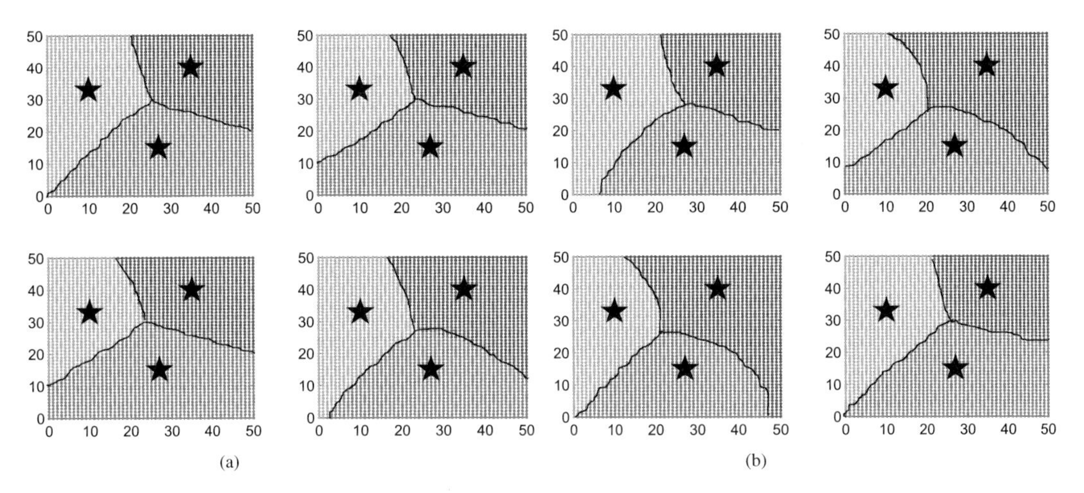

Figure 19.9 Network association illustration: (a) TR wideband system, and (b) massive MIMO system.

more crowded, the SINR of each individual UE decreases due to the power sharing and IUI. Meanwhile, the farther the distance between the new UE and the associated BS can also lead to decreased SINR due to the path loss. Moreover, we also show the numerical achievable data rate of the new UE in Figure 19.8. Note that the TR wideband data rate is calculated by $B/D\log(1+\text{SINR})$, and the massive MIMO data rate is calculated by $B\log(1+\text{SINR})$. It can be seen that both systems can achieve high throughput either by wide bandwidth for TR wideband or by a large number of antennas for massive MIMO.

Based on the SINR performance evaluation, we further simulate the new UE's association status and show the results of both systems in Figure 19.9, where a two-dimensional plane with an area of 50 m $\times$ 50 m is considered. The pentagrams in each area represent three BSs, whose locations are $[10,33]$, $[15,27]$, and $[35,40]$, respectively. For each subfigure, we first generate 20 existing UEs with random locations and simply associate each UE with the corresponding BS. Then, we consider each location in the area as a new arrival UE's location to check which BS should cover this location. The area surrounded by solid-line boundaries in the figure means the region that should be covered by the center BS when a new UE appears. Additionally, for each system, we simulate four different association statuses under different settings of existing UEs. We can see that the existing load of the BSs affects the new UE's association to a large extent. While an interesting phenomenon is that, due to the spatial focusing effect, the boundary between two BSs may vary under different settings of the locations and number of existing UEs in the network. Therefore, in the 5G networks, the traditional fixed boundary between two cells, which is purely determined by the locations and distances, may not hold. On one hand, the spatial focusing gain discussed in Section III can help extend the coverage of one BS. On the other hand, the more UEs share one BS can decrease the power resource of each individual UE and thus lead to the shrinkage of the coverage.

Finally, let us evaluate the presented centralized and distributed network association protocols. In this simulation, the locations of BSs and the 50 m $\times$ 50 m plane are the

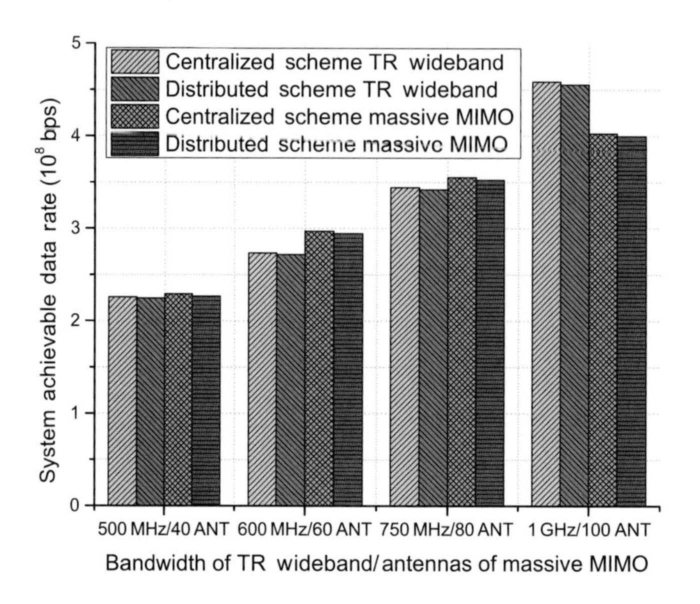

Figure 19.10 Network association performance in both systems.

same as that in Figure 19.9. In each simulation run, 20 UEs are randomly generated within the plane. Once a new UE is generated, the presented association protocols are utilized to determine whether the BS admitting the new UE. We conduct 1,000 independent runs and calculate the average network throughput of both protocols shown in Figure 19.10. For the TR wideband system, the protocols are evaluated under different bandwidths from 500 MHz to 1 GHz; while for the massive MIMO system, the protocols are evaluated under different numbers of antennas from 40 to 100. Surprisingly, the performance of the distributed protocol, which is expected to be worse than the optimal centralized protocol, is indeed similar to that of the centralized one. This phenomenon can be explained by an important parameter ξ_4 defined in (19.47), which represents the existing UEs' performance loss due to the new UE's association. From (19.47), we can see that when the number of existing UEs $|\mathcal{N}_{A_j}|$ is small, the parameters ξ_2 and ξ_3 representing the combination of ISI and IUI are far less than the parameter ξ_1 representing the signal power due to the spatial focusing gain. In such a case, ξ_4 should be approximately equal to 1. However, on the other hand, when the number of existing UEs $|\mathcal{N}_{A_j}|$ is sufficiently large, ξ_4 would approach 1 again. To further validate this, we plot the value of ξ_4 under different numbers of UEs in both systems in Figure 19.11, from which it can be seen that ξ_4 is always approximately 1. Knowing the characteristic of $\xi_4 \approx 1$, let us then review the objective functions of both protocols as follows:

Centralized Scheme:

$$\arg\max_{A_j} \frac{\xi_1 + \xi_2 + \xi_3 \left(|\mathcal{N}_{A_j}| + 1\right) d_{A_j,j}^{\alpha}}{\xi_2 + \xi_3 \left(|\mathcal{N}_{A_j}| + 1\right) d_{A_j,j}^{\alpha}} \cdot \xi_{4,A_j}, \tag{19.52}$$

Distributed Scheme:

$$\arg\min_{A_j} \left(|\mathcal{N}_{A_j}| + 1\right) d_{A_j,j}^{\alpha}. \tag{19.53}$$

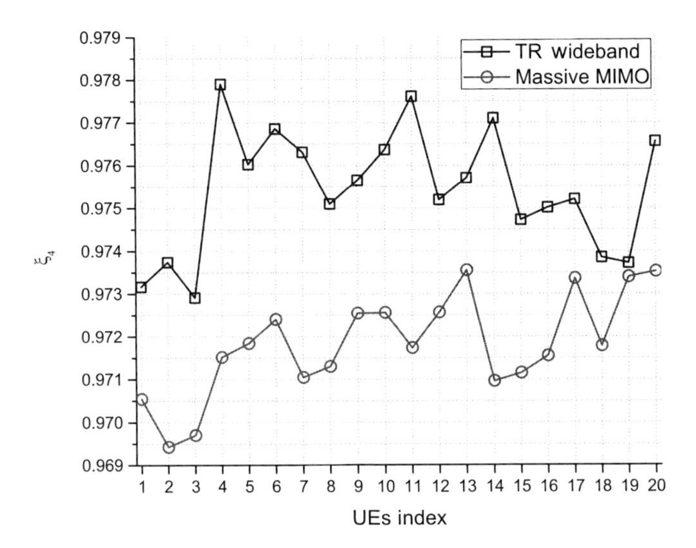

Figure 19.11 The value of ξ_4 in both systems.

We can see that the only parameter that determines the difference of these two schemes is ξ_4. When $\xi_4 \approx 1$, the centralized scheme becomes

$$\begin{aligned}
&\max_{A_j} \frac{\xi_1 + \xi_2 + \xi_3 \left(\left|\mathcal{N}_{A_j}\right| + 1\right) d^{\alpha}_{A_j,j}}{\xi_2 + \xi_3 \left(\left|\mathcal{N}_{A_j}\right| + 1\right) d^{\alpha}_{A_j,j}} \\
&\Rightarrow \max_{A_j} 1 + \frac{\xi_1}{\xi_2 + \xi_3 \left(\left|\mathcal{N}_{A_j}\right| + 1\right) d^{\alpha}_{A_j,j}} \\
&\Rightarrow \min_{A_j} \left(\left|\mathcal{N}_{A_j}\right| + 1\right) d^{\alpha}_{A_j,j},
\end{aligned} \tag{19.54}$$

which leads to the equivalence of these two schemes. Again, due to the spatial focusing effect, which brings the benefit of rather small interference among UEs, the distributed protocol with low complexity can be practically adopted in both TR wideband and massive MIMO systems.

19.8 Summary

In this chapter, we discussed a general spatial spectrum sharing architecture based on the spatial focusing characteristic in both TR wideband and massive MIMO systems. The spatial focusing phenomenon was first shown by simulation results, as well as real-world experiments. Relying on this phenomenon, we analyzed the SINR performance of concurrently spatial spectrum sharing for both TR wideband and massive MIMO systems. It turned out to be that the SINR of both systems shared exactly the same expression under the equal power allocation scenario, which motivated us to design a general network association protocol. The centralized and distributed association schemes were designed, while the performance of the distributed scheme was quite similar to that of the centralized one, due to the benefit of rather small IUI brought by the spatial focusing effect. Because the realization of the spatial focusing effect relies on the accurate CSI of all users, the major challenge of the presented scheme is how to simultaneously collect the users' CSI, especially when the number of users is considerable. To summarize, with the trends of wider bandwidth and larger scale of antennas, spatial focusing effect becomes the uniqueness of the next-generation networks.

In the future, this effect can be a bridge that separates the physical layer and MAC layer design, where the physical layer design should focus on how to strengthen this focusing effect in order to alleviate interference, while the MAC layer design should concentrate on how to utilize this effect to accommodate more users. Therefore, more MAC layer issues should be re-investigated when the spatial focusing effect is taken into account, e.g., the admission control, the handover, as well as the security issue. Meanwhile, the other channel models can be considered, for example Winner II channel model, in which there is a correlation between different paths. For related references, readers can refer to [42].

References

[1] C. Jiang, Y. Chen, K. J. R. Liu, and Y. Ren, "Renewal-theoretical dynamic spectrum access in cognitive radio network with unknown primary behavior," *IEEE Journal on Selected Areas in Communications*, vol. 31, no. 3, pp. 406–416, 2013.

[2] F. R. V. Guimaraes, D. B. da Costa, T. A. Tsiftsis, C. C. Cavalcante, and G. K. Karagiannidis, "Multiuser and multirelay cognitive radio networks under spectrum-sharing constraints," *IEEE Transactions on Vehicular Technology*, vol. 63, no. 1, pp. 433–439, Jan. 2014.

[3] J. Meng, W. Yin, H. Li, E. Hossain, and Z. Han, "Collaborative spectrum sensing from sparse observations in cognitive radio networks," *IEEE Journal on Selected Areas in Communications*, vol. 29, no. 2, pp. 327–337, Feb. 2011.

[4] Q. Wu, G. Ding, J. Wang, and Y.-D. Yao, "Spatial-temporal opportunity detection for spectrum-heterogeneous cognitive radio networks: Two-dimensional sensing," *IEEE Transactions on Wireless Communications*, vol. 12, no. 2, pp. 516–526, Feb. 2013.

[5] B. Wang, Y. Wu, F. Han, Y. H. Yang, and K. J. R. Liu, "Green wireless communications: A time-reversal paradigm," *IEEE Journal on Selected Areas in Communications*, vol. 29, no. 8, pp. 1698–1710, Sep. 2011.

[6] Y. Chen, B. Wang, Y. Han, H.-Q. Lai, Z. Safar, and K. J. R. Liu, "Why time-reversal for future 5G wireless?" *IEEE Signal Processing Magazine*, vol. 33, no. 2, pp. 17–24, Mar. 2016.

[7] F. Rusek, D. Persson, B. K. Lau, E. G. Larsson, T. L. Marzetta, O. Edfors, and F. Tufvesson, "Scaling up MIMO: Opportunities and challenges with very large arrays," *IEEE Signal Processing Magazine*, vol. 30, no. 1, pp. 40–60, Jan. 2013.

[8] C. Oestges, A. D. Kim, G. Papanicolaou, and A. J. Paulraj, "Characterization of space-time focusing in time-reversed random fields," *IEEE Transactions on Antennas and Propagation*, vol. 53, no. 1, pp. 283–293, Jan. 2005.

[9] A. M. Akhtar, X. Wang, and L. Hanzo, "Synergistic spectrum sharing in 5G HetNets: A harmonized SDN-enabled approach," *IEEE Communications Magazine*, vol. 54, no. 1, pp. 40–47, Jan. 2016.

[10] G. Ding, J. Wang, Q. Wu, Y.-D. Yao, R. Li, H. Zhang, and Y. Zou, "On the limits of predictability in real-world radio spectrum state dynamics: From entropy theory to 5G spectrum sharing," *IEEE Communications Magazine*, vol. 54, no. 7, pp. 178–183, Jul. 2015.

[11] B. Li, S. Li, A. Nallanathan, and C. Zhao, "Deep sensing for future spectrum and location awareness 5G communications," *IEEE Journal on Selected Areas in Communications*, vol. 33, no. 7, pp. 1331–1344, Jul. 2015.

[12] J. Mitola, J. Guerci, J. Reed, Y.-D. Yao, Y. Chen, T. C. Clancy, J. Dwyer, H. Li, H. Man, R. McGwier, and Y. Guo, "Accelerating 5G QoE via public-private spectrum sharing," *IEEE Communications Magazine*, vol. 52, no. 5, pp. 77–85, May. 2014.

[13] D. W. K. Ng, M. Breiling, C. Rohde, F. Burkhardt, and R. Schober, "Energy-efficient 5G outdoor-to-indoor communication: SUDAS over licensed and unlicensed spectrum," *IEEE Transactions Wireless Communications*, vol. 15, no. 5, pp. 3170–3186, May 2016.

[14] B. Singh, S. Hailu, K. Koufos, A. A. Dowhuszko, O. Tirkkonen, R. Jäntti, and R. Berry, "Coordination protocol for inter-operator spectrum sharing in co-primary 5G small cell networks," *IEEE Communications Magazine*, vol. 53, no. 7, pp. 34–40, Jul. 2015.

[15] C. Jiang, Y. Chen, Y. Gao, and K. J. R. Liu, "Joint spectrum sensing and access evolutionary game in cognitive radio networks," *IEEE Transactions on Wireless Communications*, vol. 12, no. 5, pp. 2470–2483, 2013.

[16] C. Jiang, Y. Chen, and K. J. R. Liu, "Multi-channel sensing and access game: Bayesian social learning with negative network externality," *IEEE Transactions on Wireless Communications*, vol. 13, no. 4, pp. 2176–2188, 2014.

[17] S. Singh, R. Mudumbai, and U. Madhow, "Interference analysis for highly directional 60-GHz mesh networks: The case for rethinking medium access control," *IEEE/ACM Transactions on Networking*, vol. 19, no. 5, pp. 1513–1527, May 2011.

[18] H. Shokri-Ghadikolaei, C. Fischione, G. Fodor, P. Popovski, and M. Zorzi, "Millimeter wave cellular networks: A MAC layer perspective," *IEEE Transactions on Wireless Communications*, vol. 63, no. 10, pp. 3437–3458, Oct. 2015.

[19] T. Bai and R. Heath, "Coverage and rate analysis for millimeter-wave cellular networks," *IEEE Transactions on Wireless Communications*, vol. 14, no. 2, pp. 1100–1114, Feb. 2015.

[20] G. Athanasiou, P. C. Weeraddana, C. Fischione, and L. Tassiulas, "Optimizing client association for load balancing and fairness in millimeter-wave wireless networks," *IEEE/ACM Transactions on Networking*, vol. 23, no. 3, pp. 836–850, Jun. 2015.

[21] J. Garcia-Rois, F. Gomez-Cuba, M. Riza Akdeniz, F. Gonzalez-Castano, J. Burguillo-Rial, S. Rangan, and B. Lorenzo, "On the analysis of scheduling in dynamic duplex multihop mmWave cellular systems," *IEEE Transactions on Wireless Communications*, vol. 14, no. 11, pp. 6028–6042, Nov. 2015.

[22] P. Wang, W. Song, D. Niyato, and Y. Xiao, "QoS-aware cell association in 5G heterogeneous networks with massive MIMO," *IEEE Network*, vol. 29, no. 6, pp. 76–82, Dec. 2015.

[23] C. Jiang, Y. Chen, K. J. R. Liu, and Y. Ren, "Network economics in cognitive networks," *IEEE Communications Magazine*, vol. 53, no. 5, pp. 75–81, 2015.

[24] C. Jiang, Y. Chen, Y. Yang, C. Wang, and K. J. R. Liu, "Dynamic Chinese restaurant game: Theory and application to cognitive radio networks," *IEEE Transactions on Wireless Communications*, vol. 13, no. 4, pp. 1960–1973, 2014.

[25] D. Bethanabhotla, O. Y. Bursalioglu, H. C. Papadopoulos, and G. Caire, "Optimal user-cell association for massive MIMO wireless networks," *IEEE Transactions on Wireless Communications*, vol. 15, no. 3, pp. 1835–1850, Mar. 2016.

[26] D. Liu, L. Wang, Y. Chen, T. Zhang, K. K. Chai, and M. Elkashlan, "Distributed energy efficient fair user association in massive MIMO enabled HetNets," *IEEE Communication Letters*, vol. 19, no. 10, pp. 1770–1773, Oct. 2015.

[27] N. Wang, E. Hossain, and V. K. Bhargava, "Joint downlink cell association and bandwidth allocation for wireless backhauling in two tier HetNets with large-scale antenna arrays," *IEEE Transactions on Wireless Communications*, vol. 15, no. 5, pp. 3251–3268, May 2016.
[28] F. Han, Y. H. Yang, B. Wang, Y. Wu, and K. J. R. Liu, "Time-reversal division multiple access over multi-path channels," *IEEE Transactions on Communications*, vol. 60, no. 7, pp. 1953–1965, Jul. 2012.
[29] Y. H. Yang, B. Wang, W. S. Lin, and K. J. R. Liu, "Near-optimal waveform design for sum rate optimization in time-reversal multiuser downlink systems," *IEEE Transactions on Wireless Communications*, vol. 12, no. 1, pp. 346–357, Jan. 2013.
[30] Q. Y. Xu, Y. Chen, and K. J. R. Liu, "Combating strong–weak spatial–temporal resonances in time-reversal uplinks," *IEEE Transactions on Wireless Communications*, vol. 15, no. 1, pp. 1953–1965, Jan. 2016.
[31] F. Han and K. J. R. Liu, "A multiuser TRDMA uplink system with 2D parallel interference cancellation," *IEEE Transactions on Communications*, vol. 62, no. 3, pp. 1011–1022, Mar. 2014.
[32] Y. H. Yang and K. J. R. Liu, "Waveform design with interference pre-cancellation beyond time-reversal system," *IEEE Transactions on Wireless Communications*, vol. 15, no. 5, pp. 3643–3654, May 2016.
[33] H. Q. Ngo, E. G. Larsson, and T. L. Marzetta, "Energy and spectral efficiency of very large multiuser MIMO systems," *IEEE Transactions on Communications*, vol. 61, no. 4, pp. 1436–1449, Apr. 2013.
[34] S. Jin, X. Wang, Z. Li, K.-K. Wong, Y. Huang, and X. Tang, "On massive MIMO zero-forcing transceiver using time-shifted pilots," *IEEE Transactions on Wireless Communications*, vol. 65, no. 1, pp. 59–74, Jan. 2016.
[35] T. Cui, F. Gao, T. Ho, and A. Nallanathan, "Distributed space-time coding for two-way wireless relay networks," *IEEE Transactions on Signal Processing*, vol. 57, no. 2, pp. 658–41 671, 2009.
[36] F. Gao, R. Zhang, and Y.-C. Liang, "Optimal channel estimation and training design for two-way relay networks," *IEEE Transactions on Communications*, vol. 57, no. 10, pp. 3024–3033, 2009.
[37] M. R. McKay, I. B. Collings, and A. M. Tulino, "Achievable sum rate of MIMO MMSE receivers: A general analytic framework," *IEEE Transactions on Information Theory*, vol. 56, no. 1, pp. 396–410, Jan. 2010.
[38] F. Gao, T. Cui, and A. Nallanathan, "On channel estimation and optimal training design for amplify and forward relay network," *IEEE Transactions on Wireless Communications*, vol. 7, no. 5, pp. 1907–1916, 2008.
[39] M. Emami, M. Vu, J. Hansen, A. J. Paulraj, and G. Papanicolaou, "Matched filtering with rate back-off for low complexity communications in very large delay spread channels," in *Proceedings of the Asilomar Conference on Signals, Systems and Computers (ACSSC)*, vol. 1, pp. 218–222, Nov. 2004.
[40] P. H. Moose, "A technique for orthogonal frequency division multiplexing frequency offset correction," *IEEE Transactions on Communications*, vol. 42, no. 10, pp. 2908–2914, Oct. 1994.
[41] J. Lee, H.-L. Lou, D. Toumpakaris, and J. M. Cioff, "SNR analysis of OFDM systems in the presence of carrier frequency offset for fading channels," *IEEE Transactions on Wireless Communications*, vol. 5, no. 12, pp. 3360–3364, Dec. 2006.
[42] C. Jiang, B. Wang, Y. Han, Z.-H. Wu, and K. J. R. Liu, "Exploring spatial focusing effect for spectrum sharing and network association," *IEEE Transactions on Wireless Communications*, vol. 16, no. 7, pp. 4216–4231, 2017.

20 Tunneling Effect for Cloud Radio Access Network

The explosion of today's wireless traffic requires operators to deploy more access points (APs) and design an efficient collaboration mechanism to alleviate the interference among them. However, the collaborative techniques cannot work efficiently due to the high latency and low bandwidth interface between the APs in traditional networks. To address this challenge, cloud radio access network (C-RAN) is proposed, where a pool of baseband units (BBUs) are connected to the distributed remote radio heads (RRHs) via high bandwidth and low latency links (i.e., the front-haul) and are responsible for all the baseband processing. But the limited front-haul link capacity may prevent the C-RAN from fully utilizing the benefits made possible by the centralized baseband processing. As a result, the front-haul link capacity becomes a bottleneck. To address this challenge, in this chapter, we propose to use the time-reversal (TR)-based communication as the air interface in C-RAN. Due to the unique spatial and temporal focusing effects of TR-based communications, multiple terminal devices (TDs) are naturally separated by their location-specific signatures. Such a property allows signals to be combined to deliver without demanding more bandwidth. Therefore, the TR-based communication in essence creates a "tunneling" effect such that the baseband signals for all the TDs can be efficiently combined and transmitted in the front-haul. We study the performance of the presented C-RAN architecture in terms of spectral efficiency and front-haul rate, based on extensive measurements of the wireless channel in a real-world environment. It is shown that with nearly the same amount of traffic load in the front-haul, more information can be transmitted when there are more TDs. The presented TR tunneling effect can help deliver more information in the C-RAN and alleviate the burden of the front-haul caused by network densification.

20.1 Introduction

With the proliferation of new mobile devices and applications, the demand for ubiquitous wireless services has increased dramatically in recent years. It has been projected that by the year 2020, the volume of the wireless traffic will rise to about 1,000 times that of the year 2010 [1]. This explosion of wireless traffic will be a new challenge to wireless networks. On one hand, the huge number of wireless devices and the ever-growing data rate driven by a broad range of mobile applications lead to an unprecedented demand for network throughput, which may cause a huge deficit of spectrum. On the other hand, with a large number of coexisting wireless devices competing for network service, the scheduling delay will significantly deteriorate the user experience in many delay-sensitive applications, due to the much smaller chance of being scheduled in either coordinated networks or random access networks. In fact, people have started to feel the impact in some places, such as at the airport, conferences, and stadiums, where it is difficult to access the wireless network with hundreds of other devices around.

To accommodate the massive devices, heterogeneous and small cell networks (HetSNets) have been considered a promising solution. HetSNets are expected to boost data capacity through coverage expansion and load balancing. Nevertheless, with an increasing number of small cells or access points (APs) deployed, interference problems become more severe, and it is necessary for multiple APs to collaborate closely in order to mitigate the interference. However, efficient collaborative radio techniques cannot work efficiently due to the high latency and low bandwidth interface between the APs in traditional wireless networks.

To address the aforementioned challenge, cloud-based radio access network (C-RAN) has been proposed as a viable solution recently [2–5]. It is a novel type of RAN architecture, where a pool of baseband units (BBUs) are connected to the distributed remote radio heads (RRHs) via high bandwidth and low latency links. The BBUs are responsible for all the baseband processing through high performance computing. In this centralized structure, many coordinated communication schemes become possible or more efficient. For example, the coordinated multiple-point process (CoMP) in the LTE-A standard [6] can be implemented in the C-RAN to improve network capacity and energy efficiency [7]. In addition, by moving the baseband processing to the cloud, the RRHs need only support the basic transmission/reception functionalities, which further reduces their energy consumption and deployment cost.

Nevertheless, the limited front-haul link capacity [8] between the BBU and the RRH may prevent the C-RAN from fully utilizing the benefits made possible by concentrating the processing intelligence. In most of the current C-RAN structures, the data transmitted in the front-haul is proportional to the aggregate traffic of all the terminal devices (TDs) [9, 10]. As a result, the front-haul link capacity becomes a bottleneck when there are massive TDs in the network. To tackle this challenge, several solutions have been proposed. One of them is to use compression where the baseband signal is compressed before the front-haul transmission and then decompressed after the front-haul transmission [11–13]. Although signal compression can alleviate the traffic in the front-haul under certain cases, it introduces extra computation complexity at the RRH side, which makes this approach less cost effective. Moreover, although the compression reduces the data rate consumption of each individual TD, because the total data rate consumption is the aggregate of all the TDs, there is still a deficit of the front-haul link capacity in a dense network. An alternative solution is the sparse beamforming [9, 10, 14] where each TD is associated with a cluster of APs. However, the data rate in the front-haul link is related to the cluster size, and a larger cluster requires a higher front-haul link capacity [9]. As a result, the limited front-haul link capacity makes it impossible to fully take advantage of the available spatial diversity, which is one of the main benefits of the C-RAN structure.

Time-reversal (TR) wireless communication has been known for some time [15–17]; however, its applications have been mainly considered as a specialty use for extreme multipath environment where other wireless communication techniques do not work well. Recently, more features of TR-based wireless communication had been discovered. In [18], it is illustrated that TR-based wireless communication is a "green" broadband wireless communication technique that can provide energy-efficient transmissions in the multipath-rich environments such as indoor and urban areas. Due to the unique spatial and temporal focusing effects of TR-based communications, all the TDs are naturally separated by the location-specific signatures in both downlink [19] and uplink [20]. As a result, the TR-based communication becomes an efficient scheme in terms of good spatial multiplexing and much lower implementation complexity to utilize larger bandwidth, by which the achievable data rate of TR communications can outperform other wireless communication technologies, for example LTE [21]. These facts make TR a promising candidate in the future broadband wireless communication solutions, which has been illustrated in various applications, for example cognitive radio networks [22] and the Internet of Things (IoT) [23].

Because all the TDs are naturally separated by their signatures, the baseband signals for all the TDs can be efficiently combined and transmitted if TR communication is used as the air interface. We aim to leverage this unique feature of the TR-based communications to create in essence a tunneling effect between the BBU and RRH to alleviate the traffic load in the front-haul link of C-RAN. Specifically, in this chapter, we propose a C-RAN framework using TR communication as the air interface. The architectures for both downlink and uplink data transmissions are designed. We analyze the performance of both the downlink and uplink transmissions in terms of the spectral efficiency and the data rate consumption in the front-haul link. To illustrate the effectiveness of the presented system, we conduct experiments to measure the multipath channel information in the real-world environment, based on which we show that the TR-based C-RAN creates a unique "tunneling effect" such that more information can be transmitted in the front-haul link with the same amount of bits/energy when there are more TDs in the system. This feature is highly desirable in the C-RAN system because it significantly alleviates the traffic load in the front-haul link caused by the network densification, which makes the presented TR-based C-RAN a perfect candidate to work in the multipath-rich environments such as indoor and urban areas to provide wireless connections to massive TDs using relatively broad bandwidth.

The rest of this chapter is organized as follows: in Section 20.2, the system model and the working phases are introduced. We analyze the performance of the downlink and uplink schemes in Sections 20.3 and 20.4, respectively. In Section 20.5, the real-world channel measurement setting is introduced, followed by numerical results demonstrating the effectiveness of the downlink and uplink schemes.

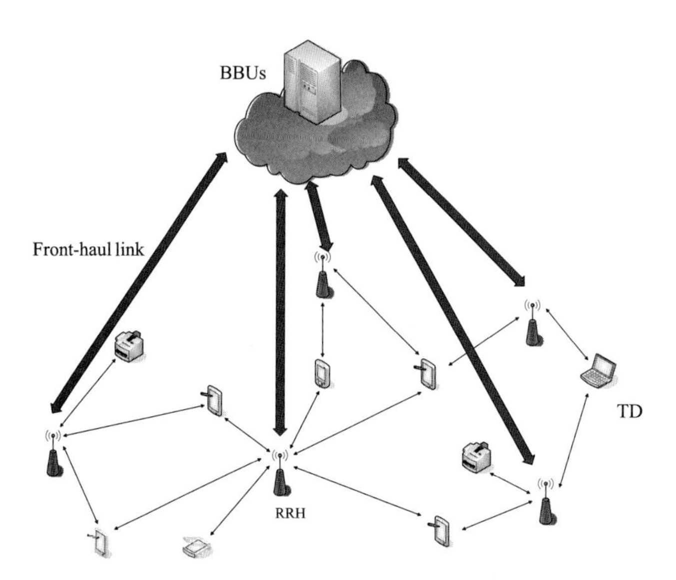

Figure 20.1 The system model.

20.2 System Model

In this chapter, we consider a C-RAN in the indoor environments to accommodate the massive terminal devices (TDs) in both uplink and downlink data transmissions. The presented system consists of multiple RRHs that connect to the cloud via fronthaul links. As shown in Figure 20.1, multiple RRHs are distributed in an area and transmit/receive data to/from various TDs in this area. Each of the RRHs uses TR-based communication to communicate with the TDs. All the RRHs work in the same spectrum.

In the following, we first briefly introduce the basic TR-based wireless communication and then the channel model adopted in this chapter. After that, we will describe the three working phases of the system in detail: the channel probing phase, the downlink data transmission phase and the uplink data transmission phase.

20.2.1 Channel Model

In the indoor broadband wireless communication, the signal suffers from the multipath effect caused by the reflections of the indoor environment. Instead of trying to avoid the multipath effect, TR-based communication utilizes all the multipaths to act like a matched filter to achieve spatial and temporal focusing effects. More details about the basic TR-based transmission can be referred to Chapter 1.

Without loss of generality, we assume that each RRH is equipped with M_T antennas, and each TD is equipped with one antenna. We assume a multi-path Rayleigh fading channel and the channel impulse response (CIR) of the communication link between the mth antenna of the ith RRH and the jth TD is modeled as

$$h_{i,j}^{(m)}[k] = \sum_{l=0}^{L-1} h_{i,j}^{(m),(l)} \cdot \delta[k-l], \tag{20.1}$$

where $h_{i,j}^{(m),(l)}$ is the complex amplitude of the lth tap of the CIR with length L and $h_{i,j}^{(m)}[k]$ is the kth tap of the CIR. In practice, the $h_{i,j}^{(m)}$ is an equivalent channel, which is a combination of the multipath environment, the raised-cosine filter, and the antenna. Because the raised-cosine filter and the antenna remain the same for the same radio, there is no need to counter the effects of them for TR to work. In the rest of this chapter, we treat $h_{i,j}^{(m)}$ as the channel for the presented system.

20.2.2 The TR-based C-RAN Channel Probing Phase

In the C-RAN, all the RRHs work together to serve the TDs in downlink and uplink. To achieve this, the BBUs first need to gather all the necessary information of all the TDs. In the TR-based communication, the TDs are separated naturally by their CIRs as the location-specific signature. Therefore, the BBUs need to collect the CIR information of all the TDs before all the TDs can be served. We propose the channel probing phase in the C-RAN where the BBUs get the channel information $h_{i,j}^{(m)}$'s of from all the RRHs to the TDs. In the presented system, the system periodically switches between the channel probing, downlink transmission, and uplink transmission phases. The downlink and uplink transmissions work by time division duplexing (TDD) such that the channel information can be shared by the downlink and uplink. In the following, we will first introduce the channel probing phase, which is common and necessary for both downlink and uplink transmissions, after which we describe in detail the downlink and uplink transmission phases, respectively.

Let $\mathbf{R}$ denote the set of indices of all the RRHs, $\mathbf{T}$ the set of indices of all the TDs, $\mathbf{T}_i$ the set of indices of all the TDs subscribed to the RRH i, and $\mathbf{R}_j$ the set of the indices of all the RRHs that the jth TD is subscribed to. Note that we have $\mathbf{T}_i \subseteq \mathbf{T}$, $\mathbf{R}_j \subseteq \mathbf{R}$.

In the channel probing phase, the N TDs first take turns to transmit a channel probing signal to all the RRHs, and the RRHs transmit the received channel probing signal through the front-haul links to the BBUs, where the channel information is extracted. The channel probing signal can be an impulse signal or a predefined pseudo random code known by the BBUs beforehand. Because all the RRHs work in the same band, the channel probing signal transmitted by user j can be received by all the corresponding RRHs simultaneously, and the BBUs can extract the channel information between each TD and all its corresponding RRHs using various methods. For instance, the predefined pseudo random sequence can be the Golay sequence [24], and the channel information $h_{i,k}^{(m)}$ can be obtained by calculating the cross correlation of the transmitted Golay sequence and the sequence received by the mth antenna of RRH i. At the end of the channel probing phase, the BBUs have the channel knowledge between all the TDs and their corresponding RRHs. Because the downlink and uplink transmissions use the TDD, the channel information works for both uplink and downlink. Moreover, because all the baseband processing is conducted in the BBUs, the TDs do not need to have the channel information in either downlink or uplink. Therefore, no feedback is needed to deliver the channel information back to the TDs.

To understand the overhead caused by the channel probing phase in the presented system, we analyze it quantitatively. In [25], it was shown through experiments that the channel information in the indoor environment does not change in hours for a TD that does not move. On the other hand, in our experiment, we discover that the channel information changes much as one moves more than 3 cm. Therefore, frequent channel information update is needed for those TDs that are moving. For example, we consider a typical handheld device. As the typical walking speed is 1.4 m per s, the TD needs to do channel probing every 18 ms. In the experiment described in Section 20.5, we use Golay sequence of total length 2048 as the channel probing sequence, and the sampling rate is 125 MHz. The time needed for a single channel probing is 16 μs. Compared with the channel updating period, which is 18 ms, the channel probing overhead is less than 0.1% of the total time for a typical moving TD, which is comparable to the channel estimation overhead in the LTE systems [26]. While the C-RAN system has to handle the channel probing for every single TD, the system-wide overhead will be the aggregation of the individual overheads.

An implicit feature of the channel probing phase is that each TD is only subscribed to the RRHs close enough to it. The RRHs far away from the TD j cannot get the channel probing signal and will not add it to the subscription list. The searching range of the TD j can be adjusted by tuning the power of the channel probing signal. Increasing the power will extend the searching range so that the TD can possibly subscribe to more RRHs. On the RRH side, this feature enables automatic power management. If an RRH is far away from all the active TDs, using this RRH to serve the TDs might not be energy efficient. In the presented method, this RRH does not get any channel probing signal and does not use any power to transmit data to the TDs, while the TDs are better served by other RRHs closer to them.

20.2.3 The Downlink Transmission Architecture

After the channel probing phase, the BBUs start to utilize the collected channel information to serve the downlink and uplink data transmissions. The system uses TDD to support downlink and uplink transmissions. In this subsection, we describe the downlink transmission phase.

In the downlink transmission, as shown in Figure 20.2, there are two steps: In step (1), the transmitted signals are calculated at the BBUs and quantized before they are transmitted through the front-haul to the RRH; in step (2), the RRH converts the baseband signals to the RF signals and transmits them to the TDs through the multipath channel.

As shown in Figure 20.2(a), the intended symbol sequence $X_j[k]$ for the jth TD transmitted from the mth antenna of the ith RRH is first up-sampled by the backoff factor D in order to alleviate the intersymbol interference (ISI) and then convolved with the signature $g_{i,j}^{(m)}[k]$ of the channel $h_{i,j}^{(m)}[k]$, which is

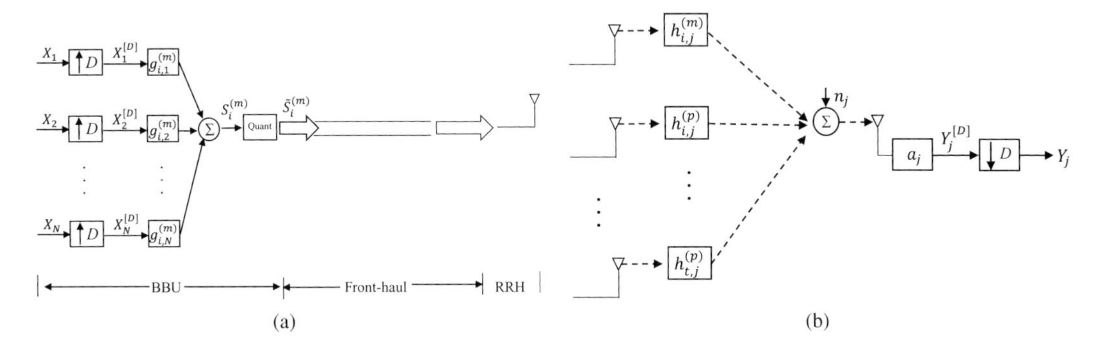

Figure 20.2 The two steps in the downlink data transmission phase: (a) BBU to RRH through front-haul, and (b) RRHs to subscribed TD j.

$$g_{i,j}^{(m)}[k] = \frac{h_{i,j}^{(m)*}[L-1-k]}{\sqrt{\sum_{t\in\mathbf{T}_i}\sum_{m=1}^{M_T}\sum_{l=0}^{L-1}|h_{i,j}^{(m)}[l]|^2}}, \quad (20.2)$$
$$k = 0, 1, \cdots, L-1$$

where $h_{i,j}^{(m)*}[L-1-k]$ denotes the conjugate of $h_{i,j}^{(m)}[L-1-k]$.

After that, the intended signals for all the subscribed TDs at the mth antenna of RRH i are combined as

$$S_i^{(m)}[k] = \sum_{j\in\mathbf{T}_i}\left(X_j^{[D]} * g_{i,j}^{(m)}\right)[k]. \quad (20.3)$$

where $\left(X_j^{[D]} * g_{i,j}^{(m)}\right)[k]$ denotes the convolution of $X_j^{[D]}[k]$ and $g_{i,j}^{(m)}[k]$.

The average power of the baseband signal $S_i^{(m)}[k]$ can be calculated as

$$E[\|S_i^{(m)}[k]\|^2] = \frac{\theta}{D} \quad (20.4)$$

where $\theta = E[\|X_j[k]\|^2]$.

Then the $S_i^{(m)}$ is quantized, and the BBUs transmit the quantized $\tilde{S}_i^{(m)}[k]$ through the front-haul with a limited capacity. The quantization of $S_i^{(m)}$ can be modeled as

$$\tilde{S}_i^{(m)}[k] = S_i^{(m)}[k] + q_i^{(m)}[k], \quad (20.5)$$

where $q_i^{(m)}[k]$ is the quantization noise at the mth antenna of RRH i. By (20.3), $S_i^{(m)}[k]$ is a summation of multiple independent variables and can be approximated as a complex Gaussian random variable by the law of large numbers, and $q_i^{(m)}[k]$ can be approximated as a complex random variable whose real and imaginary parts are uniformly distributed in the range $(-\frac{Q_i^{(m)}}{2}, \frac{Q_i^{(m)}}{2})$, where $Q_i^{(m)} = \frac{2K_i^{(m)}}{2^{B_i^{(m)}}}$ is the quantization level [27] of the baseband signal at the mth antenna of ith AP, $B_i^{(m)}$ is the number of bits used to represent the real/imaginary part of $S_i^{(m)}[k]$, and $[-K_i^{(m)}, K_i^{(m)}]$ is the dynamic range of the real/imaginary part of $S_i^{(m)}[k]$.

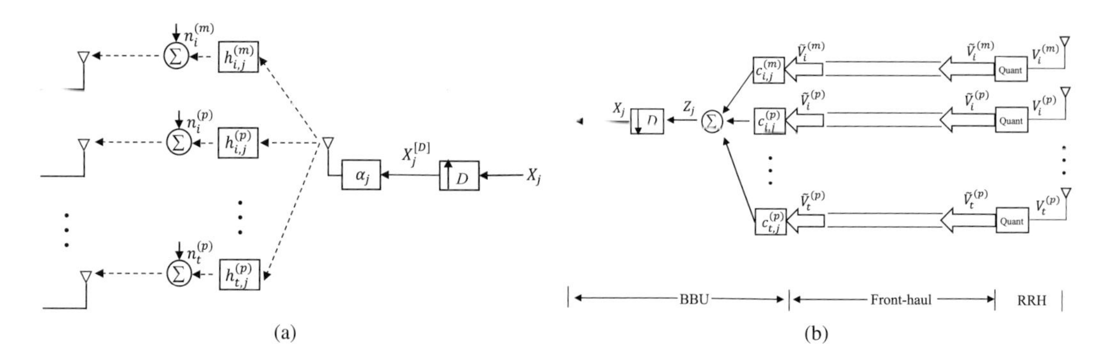

Figure 20.3 The two steps in the uplink data transmission phase: (a) TD j to all corresponding RRHs, and (b) RRHs to BBUs through front-haul link.

In step (2), each RRH i simultaneously transmits the baseband signal $\tilde{S}_i^{(m)}[k]$ via the mth antenna over the air to all the subscribed TDs for $m = 1, 2, \cdots, M_T$, and each subscribed TD will receive the signal from all the corresponding RRHs simultaneously. The received signal is a combination of the intended signal and the interference contaminated by noise. The TD j then first amplifies the received signal with a_j and then down-samples it with the factor D, obtaining the received sequence Y_j. The noise is assumed to be zero-mean additive white Gaussian noise with variance $E[|n_j[k]|^2] = \sigma^2$, $\forall j, k$. In the next section, we will investigate the received signal Y_j to analyze the performance of the presented system in the downlink phase.

20.2.4 The Uplink Transmission Architecture

In the uplink transmission, as shown in Figure 20.3, there are two similar steps in the opposite direction: In step (1), the TDs simultaneously transmit the data through the multipath channels to the corresponding RRHs; in step (2), the RRHs convert the RF signal into baseband signal and quantize them before they are transmitted via the front-haul to the BBUs. The BBUs jointly process the received baseband signal to extract the uplink data.

In step 1, all the TDs simultaneously transmit the symbol sequences over the air to the corresponding RRHs. The intended symbol sequence $X_j[k]$ from the jth TD is first up-sampled by the backoff factor D in order to alleviate the ISI and then scaled by the factor α_j before being transmitted through the multipath channel to all the corresponding RRHs. The purpose of the scaling factor α_j is to realize the power control. It is assumed that the values of α_j's are calculated by the BBUs and signaled to the TDs through the feedback/control channel. The signal received at the mth antenna of RRH i is a combination of the signals from all the TDs that can reach RRH i and contaminated by the white Gaussian noise upon receiving,

$$V_i^{(m)}[k] = \sum_{j \in \mathbf{T}_i} \alpha_j \left(X_j^{[D]} * h_{i,j}^{(m)} \right) [k] + n_i^{(m)}[k] \tag{20.6}$$

where $n_i^{(m)}[k]$ is the AWGN with variance $E[|n_i^{(m)}[k]|^2] = \sigma^2$, $\forall i, k$.

The average power of the baseband signal $V_i^{(m)}[k]$ can be calculated as

$$E[\| V_i^{(m)}[k]\|^2] = \frac{\theta \cdot \sum_{j\in \mathbf{T}_i} \alpha_j^2 \sum_{l=0}^{L-1} \|h_{i,j}^{(m)}[l]\|^2}{D} + \sigma^2. \tag{20.7}$$

The signal received at the RRH is then quantized and transmitted to the BBU pool through the front-haul, which can be represented as

$$\tilde{V}_i^{(m)}[k] = V_i^{(m)}[k] + q_i^{(m)}[k], \tag{20.8}$$

where $q_i^{(m)}[k]$ is the quantization noise. Similar to the downlink case, $q_i^{(m)}[k]$ can be approximated as a complex random variable whose real and imaginary parts are uniformly distributed in the range $(-\frac{Q_i^{(m)}}{2}, \frac{Q_i^{(m)}}{2})$. Note that although we use the same notation for the downlink and uplink quantization noise, they can be different due to distinction in the signal dynamic range and number of bits used.

Upon receiving the transmitted baseband signals from all the RRHs, the BBUs work together to extract the data of each TD. As shown in Figure 20.3(b), the data of TD j is extracted by combining the baseband signals from all the antennas of all the corresponding RRHs. The signal from mth antenna of the ith RRH is first convolved with $c_{i,j}^{(m)}$, where

$$c_{i,j}^{(m)}[k] = \frac{h_{i,j}^{(m)*}[L-1-k]}{\sqrt{\sum_{l=0}^{L-1} \|h_{i,j}^{(m)*}[l]\|^2}}, \qquad k = 0, 1, \cdots, L-1. \tag{20.9}$$

After that, the processed signal from all the antennas of all the corresponding RRHs are combined as

$$Z_j[k] = \sum_{i\in R_j} \sum_{m=1}^{M_T} c_{i,j}^{(m)} * \tilde{V}_i^{(m)}[k]. \tag{20.10}$$

In Section 20.4, we will look into the signal $Z_j[k]$ to investigate the uplink performance of the presented system.

20.3 Downlink Performance Analysis

In this section, we analyze the performance of the presented system from two perspectives, the spectral efficiency and the data rate in the front-haul. The spectral efficiency indicates how efficiently the presented system uses the available spectrum, and the data rate in the front-haul evaluates how much capacity is necessary to deploy the presented system.

20.3.1 Spectral Efficiency

Because all the RRHs and TDs work in the same spectrum, the signal received by each TD is a mixture of the intended signal, interference and noise. The signal received by

TD j can be represented as

$$\begin{aligned}
Y_j[k] &= a_j \sum_{i\in\mathbf{R}_j}\sum_{m=1}^{M_T} \tilde{S}_i^{(m)} * h_{i,j}^{(m)}[k] + a_j n_j[k] \\
&= a_j \sum_{i\in\mathbf{R}_j}\sum_{m=1}^{M_T}\sum_{t\in\mathbf{T}_i} X_t^{[D]} * g_{i,t}^{(m)} * h_{i,j}^{(m)}[k] \\
&\quad + a_j \sum_{i\in\mathbf{R}_j}\sum_{m=1}^{M_T} q_i^{(m)} * h_{i,j}^{(m)}[k] + a_j n_j[k], \qquad (20.11)
\end{aligned}$$

where, in the last equality, the first term is the intended signal combined with the interference, the second term is the received quantization noise, and the third term is the white Gaussian noise. In the following, we will analyze the first and second terms subsequently.

The first term can be further written as

$$\begin{aligned}
& a_j \sum_{i\in\mathbf{R}_j}\sum_{m=1}^{M_T}\sum_{t\in\mathbf{T}_i} X_t^{[D]} * g_{i,t}^{(m)} * h_{i,j}^{(m)}[k] \\
&= a_j \sum_{i\in\mathbf{R}_j}\sum_{m=1}^{M_T} X_j^{[D]} * g_{i,j}^{(m)} * h_{i,j}^{(m)}[k] \\
&\quad + \sum_{i\in\mathbf{R}_j}\sum_{m=1}^{M_T}\sum_{\substack{l=0\\ l\neq\frac{L-1}{D}}}^{\frac{2L-2}{D}} X_j\left[k+\frac{L-1}{D}-l\right]\cdot g_{i,j}^{(m)} * h_{i,j}^{(m)}[Dl] \\
&\quad + \sum_{i\in\mathbf{R}_j}\sum_{m=1}^{M_T}\sum_{\substack{t\in\mathbf{T}_i\\ t\neq j}}\sum_{l=0}^{\frac{2L-2}{D}} X_t\left[k+\frac{L-1}{D}-l\right]\cdot g_{i,t}^{(m)} * h_{i,j}^{(m)}[Dl]. \qquad (20.12)
\end{aligned}$$

The first term is the intended signal for TD j, the second term is the ISI, and the third term is the interuser interference (IUI). Note that by the channel reciprocity in the channel probing phase, for any RRH u with $u \notin \mathbf{R}_j$, TD j and RRH u cannot reach each other, and therefore TD j does not suffer from the interference from RRH u.

Because the one-tap gain a_j does not affect the SINR, we assume it as $a_j = 1$ in the subsequent analysis, without loss of generality.

The signal power in the downlink can be written as

$$\begin{aligned}
P_{sig}^{(dl)} &= E_X\left[\| \sum_{i\in\mathbf{R}_j}\sum_{m=1}^{M_T} X_j[k]\cdot g_{i,j}^{(m)} * h_{i,j}^{(m)}[L-1]\|^2\right] \\
&= \theta\left\|\sum_{i\in\mathbf{R}_j}\sum_{m=1}^{M_T} g_{i,j}^{(m)} * h_{i,j}^{(m)}[L-1]\right\|^2. \qquad (20.13)
\end{aligned}$$

Accordingly, the ISI and IUI power can be written as

$$P_{isi}^{(dl)} = E_X \left[\left\| \sum_{i \in \mathbf{R}_j} \sum_{m=1}^{M_T} \sum_{\substack{l=0 \\ l \neq \frac{L-1}{D}}}^{\frac{2L-2}{D}} X_j \left[k + \frac{L-1}{D} - l \right] g_{i,j}^{(m)} * h_{i,j}^{(m)}[Dl] \right\|^2 \right]$$

$$= \theta \sum_{\substack{l=0 \\ l \neq \frac{L-1}{D}}}^{\frac{2L-2}{D}} \left\| \sum_{i \in \mathbf{R}_j} \sum_{m=1}^{M_T} g_{i,j}^{(m)} * h_{i,j}^{(m)}[Dl] \right\|^2, \tag{20.14}$$

and

$$P_{iui}^{(dl)} = E_X \left[\left\| \sum_{i \in \mathbf{R}_j} \sum_{m=1}^{M_T} \sum_{\substack{t \in \mathbf{T}_i \\ t \neq j}} \sum_{l=0}^{\frac{2L-2}{D}} X_t \left[k + \frac{L-1}{D} - l \right] g_{i,t}^{(m)} * h_{i,j}^{(m)}[Dl] \right\|^2 \right]$$

$$= \theta \sum_{l=0}^{\frac{2L-2}{D}} \left\| \sum_{i \in \mathbf{R}_j} \sum_{m=1}^{M_T} \sum_{\substack{t \in \mathbf{T}_i \\ t \neq j}} g_{i,t}^{(m)} * h_{i,j}^{(m)}[Dl] \right\|^2. \tag{20.15}$$

Next, we analyze the quantization noise in the received signal. From (20.11), we can have the quantization noise power as

$$\sigma_{q,(dl)}^2 = E \left[\left\| \sum_{i \in \mathbf{R}_j} \sum_{m=1}^{M_T} q_i^{(m)} * h_{i,j}^{(m)}[k] \right\|^2 \right]$$

$$= E \left[\sum_{i \in \mathbf{R}_j} \sum_{m=1}^{M_T} \left\| \sum_{l=0}^{L-1} h_{i,j}^{(m)}[l] \cdot q_i^{(m)}[k-l] \right\|^2 \right]$$

$$= \sum_{i \in \mathbf{R}_j} \sum_{m=1}^{M_T} \sum_{l=0}^{L-1} \left\| h_{i,j}^{(m)}[l] \right\|^2 \cdot E \left[\left\| q_i^{(m)}[k] \right\|^2 \right], \tag{20.16}$$

because we assume that $q_i^{(m)}$ is independent of $h_{i,j}^{(m)}$. By [27], we have

$$E \left[\left\| q_i^{(m)}[k] \right\|^2 \right] = \frac{(Q_i^{(m)})^2}{12} + \frac{(Q_i^{(m)})^2}{12} = \frac{(Q_i^{(m)})^2}{6}, \tag{20.17}$$

which is the summation of the quantization noise power in the real and imaginary parts.

The spectral efficiency of the TD j can be defined as

$$r_j^{(dl)} = \log_2\left(1 + \frac{P_{sig}^{(dl)}}{P_{isi}^{(dl)} + P_{iui}^{(dl)} + \sigma_{q,(dl)}^2 + \sigma^2}\right)/D. \tag{20.18}$$

20.3.2 Front-Haul Rate

In this section, we analyze the front-haul rate in the presented system in the downlink mode. As shown in Figure 20.2(a), in the downlink mode, the quantized signal $\tilde{S}_i^{(m)}[k]$ is transmitted from the BBUs to the RRH through the front-haul. The data rate in the front-haul connecting the BBUs and the ith RRH can be expressed as

$$R_{fh,i} = 2 \cdot W \cdot \sum_{m=1}^{M_T} B_i^{(m)} \tag{20.19}$$

where W is the bandwidth of the system. It can be seen that $R_{fh,i}$ is solely dependent on the number of bits used for each symbol given the bandwidth of the system and the number of transmitting antennas. If $B_i^{(m)}$ is large, the power of the quantization noise goes down while the data rate in the front-haul increases, and vice versa if $B_i^{(m)}$ is small.

By (20.17), if the dynamic range $K_i^{(m)}$ of the signal grows, it needs to increase $B_i^{(m)}$ to keep the same quantization noise level. In Section 20.5, we will show through numerical results that $K_i^{(m)}$ does not change much as the number of TDs in the system grows, and therefore the presented system has a "tunneling" effect such that the front-haul rate keeps almost constant while serving more TDs.

20.4 Uplink Performance Analysis

In this section, we analyze the uplink performance of the presented system in two perspectives: the spectral efficiency and the data rate in the front-haul.

20.4.1 Spectral Efficiency

The combined signal can be written as

$$\begin{aligned} Z_j[k] &= \sum_{i\in\mathbf{R}_j}\sum_{m=1}^{M_T} \tilde{V}_i^{(m)} * c_{i,j}^{(m)}[k] \\ &= \sum_{i\in\mathbf{R}_j}\sum_{m=1}^{M_T}\sum_{t\in\mathbf{T}_i} \alpha_t X_t^{[D]} * h_{i,t}^{(m)} * c_{i,j}^{(m)}[k] \\ &\quad + \sum_{i\in\mathbf{R}_j}\sum_{m=1}^{M_T} q_i^{(m)} * c_{i,j}^{(m)}[k] + \sum_{i\in\mathbf{R}_j}\sum_{m=1}^{M_T} n_i^{(m)} * c_{i,j}^{(m)}[k] \end{aligned} \tag{20.20}$$

where the first term is the mixture of the intended signal and interference for TD j, the second term is caused by the quantization noise, and the third term is caused by the white Gaussian noise. In the following, we will analyze them subsequently.

The first term can be further written as

$$\begin{aligned}
&\sum_{i\in\mathbf{R}_j}\sum_{m=1}^{M_T}\sum_{t\in\mathbf{T}_i}\alpha_t X_t^{[D]} * h_{i,t}^{(m)} * c_{i,j}^{(m)}[k] \\
&= \sum_{i\in\mathbf{R}_j}\sum_{m=1}^{M_T}\alpha_j X_j^{[D]} * h_{i,j}^{(m)} * c_{i,j}^{(m)}[k] \\
&\quad + \sum_{i\in\mathbf{R}_j}\sum_{m=1}^{M_T}\sum_{\substack{l=0\\ l\neq\frac{L-1}{D}}}^{\frac{2L-2}{D}}\alpha_j X_j\left[k+\frac{L-1}{D}-l\right]\cdot h_{i,j}^{(m)} * c_{i,j}^{(m)}[Dl] \\
&\quad + \sum_{i\in\mathbf{R}_j}\sum_{m=1}^{M_T}\sum_{\substack{t\in\mathbf{T}_i\\ t\neq j}}\sum_{l=0}^{\frac{2L-2}{D}}\alpha_t X_t\left[k+\frac{L-1}{D}-l\right]\cdot h_{i,t}^{(m)} * c_{i,j}^{(m)}[Dl].
\end{aligned} \tag{20.21}$$

The first term in (20.21) is the intended signal from TD j, the second term is the ISI, and the third term is the IUI.

The signal power in the uplink can be written as

$$\begin{aligned}
P_{sig}^{(ul)} &= E_X\left[\left\|\sum_{i\in\mathbf{R}_j}\sum_{m=1}^{M_T}\alpha_j X_j[k]\cdot h_{i,j}^{(m)} * c_{i,j}^{(m)}[L-1]\right\|^2\right] \\
&= |\alpha_j|^2\theta\left\|\sum_{i\in\mathbf{R}_j}\sum_{m=1}^{M_T}h_{i,j}^{(m)} * c_{i,j}^{(m)}[L-1]\right\|^2,
\end{aligned} \tag{20.22}$$

where $\theta = E[\|X_j[k]\|^2]$. Accordingly, the ISI and IUI power can be written as

$$\begin{aligned}
P_{isi}^{(ul)} &= E_X\left[\left\|\sum_{i\in\mathbf{R}_j}\sum_{m=1}^{M_T}\sum_{\substack{l=0\\ l\neq\frac{L-1}{D}}}^{\frac{2L-2}{D}}\alpha_j X_j\left[k+\frac{L-1}{D}-l\right]h_{i,j}^{(m)} * c_{i,j}^{(m)}[Dl]\right\|^2\right] \\
&= |\alpha_j|^2\theta\sum_{\substack{l=0\\ l\neq\frac{L-1}{D}}}^{\frac{2L-2}{D}}\left\|\sum_{i\in\mathbf{R}_j}\sum_{m=1}^{M_T}h_{i,j}^{(m)} * c_{i,j}^{(m)}[Dl]\right\|^2,
\end{aligned} \tag{20.23}$$

and

$$P_{iui}^{(ul)} = E_X \left[\left\| \sum_{i \in \mathbf{R}_j} \sum_{m=1}^{M_T} \sum_{\substack{t \in \mathbf{T}_i \\ t \neq j}} \sum_{l=0}^{\frac{2L-2}{D}} \alpha_t X_t \left[k + \frac{L-1}{D} - l \right] h_{i,t}^{(m)} * c_{i,j}^{(m)}[Dl] \right\|^2 \right]$$

$$= \theta \sum_{l=0}^{\frac{2L-2}{D}} \left\| \sum_{i \in \mathbf{R}_j} \sum_{m=1}^{M_T} \sum_{\substack{t \in \mathbf{T}_i \\ t \neq j}} \alpha_t \cdot h_{i,t}^{(m)} * c_{i,j}^{(m)}[Dl] \right\|^2 . \tag{20.24}$$

In this chapter, we assume the α_j is chosen as

$$\alpha_j = \frac{\eta}{\sum_{i \in \mathbf{R}_j} \sum_{m=1}^{M_T} h_{i,j}^{(m)} * c_{i,j}^{(m)}[L-1]} \tag{20.25}$$

where η is a scalar common for all the TDs. In this way, $\alpha_j \cdot \sum_{i \in \mathbf{R}_j} \sum_{m=1}^{M_T} h_{i,j}^{(m)} * c_{i,j}^{(m)} [L-1]$ is common for all the TDs, which ensures that the signal power for all the TDs are the same according to (20.22). The parameter η can be adjusted according to the maximum transmitting power allowed at each TD.

Next, we analyze the quantization noise in the received signal. From (20.20), we can have the quantization noise power as

$$\sigma_{q,(ul)}^2 = E \left[\left\| \sum_{i \in \mathbf{R}_j} \sum_{m=1}^{M_T} q_i^{(m)} * c_{i,j}^{(m)}[k] \right\|^2 \right]$$

$$= E \left[\sum_{i \in \mathbf{R}_j} \sum_{m=1}^{M_T} \left\| \sum_{l=0}^{L-1} c_{i,j}^{(m)}[l] \cdot q_i^{(m)}[k-l] \right\|^2 \right]$$

$$= \sum_{i \in \mathbf{R}_j} \sum_{m=1}^{M_T} \sum_{l=0}^{L-1} \left\| c_{i,j}^{(m)}[l] \right\|^2 \cdot E \left[\left\| q_i^{(m)}[k] \right\|^2 \right], \tag{20.26}$$

because we can assume that $q_i^{(m)}$ is independent of $c_{i,j}^{(m)}$. Similar to the downlink scenario, we have

$$E[\|q_i^{(m)}[k]\|^2] = \frac{(Q_i^{(m)})^2}{12} + \frac{(Q_i^{(m)})^2}{12} = \frac{(Q_i^{(m)})^2}{6}. \tag{20.27}$$

The last term in (20.20) is the AWGN collected from all the corresponding antennas of TD j. Its power can be calculated by

$$\sigma_{n,(ul)}^2 = E \left[\left\| \sum_{i \in \mathbf{R}_j} \sum_{m=1}^{M_T} n_i^{(m)} * c_{i,j}^{(m)}[k] \right\|^2 \right] = |\mathbf{R}_j| * M_T * \sigma^2 \tag{20.28}$$

where $|\mathbf{R}_j|$ stands for the cardinality of the set $\mathbf{R}_j$. We can see that the AWGN functions differently in the downlink and uplink. In the downlink, because the AWGN affects the

TD when the TD receives the signal, it does not depend on the number of corresponding antennas. On the other hand, in the uplink, because the AWGN is gathered when each of the corresponding antennas receives the signal, the noise power scales up with the number of corresponding antennas.

The spectral efficiency of the TD j in the uplink can be defined as

$$r_j^{(ul)} = \log_2\left(1 + \frac{P_{sig}^{(ul)}}{P_{isi}^{(ul)} + P_{iui}^{(ul)} + \sigma_{q,(ul)}^2 + \sigma_{n,(ul)}^2}\right)/D. \tag{20.29}$$

20.4.2 Front-Haul Rate

In this section, we analyze the front-haul rate in the presented system in the uplink mode. As shown in Figure 20.3(a), in the uplink mode, the quantized signal $\tilde{V}_i[k]$ is transmitted from the RRH to the BBU through the front-haul. Similar to the downlink case, the data rate in the front-haul can be expressed as

$$R_{fh,i} = 2 \cdot W \cdot \sum_{m=1}^{M_T} B_i^{(m)}. \tag{20.30}$$

Similar to the downlink mode, $R_{fh,i}$ is solely dependent on the number of bits used for each symbol, given the bandwidth of the system and the number of transmitting antennas. If $B_i^{(m)}$ is large, the power of the quantization noise goes down while the data rate in the front-haul increases, and vice versa if $B_i^{(m)}$ is small.

By (20.27), if the dynamic range $K_i^{(m)}$ of the signal grows, it needs to increase $B_i^{(m)}$ to keep the same quantization noise level. In the uplink, the total baseband signal power is dependent on the number of TDs, which is different from the downlink scenario. Therefore, when the number of TDs increases in the system, the dynamic range $K_i^{(m)}$ grows, where more bits are needed to keep the same quantization noise level.

In the next section, we will show through numerical results how $K_i^{(m)}$ and $B_i^{(m)}$ change with the number of TDs in the system.

20.5 Performance Evaluation

In this section, we evaluate the performance of the presented system using measured channels. We first introduce the experimental setting where we measure the multipath channels. Then we show some numerical results obtained by using the measured channels.

20.5.1 Channel Measurement

We build a TR radio prototype to measure the multipath channels. A snapshot of the radio stations of our prototype is illustrated in Figure 20.4, where a single antenna is attached to a small cart with RF board and computer installed on the cart. The tested

Figure 20.4 The TR radio prototype.

signal bandwidth spans from 5.3375 GHz to 5.4625 GHz, centered at 5.4 GHz. An office room in the J. H. Kim Engineering Building at the University of Maryland is used as the measurement site. As shown in Figure 20.5(a), the RRHs are placed at six locations across the room, while the TDs are placed in multiple locations in the small room marked with "A." The layout of room "A" and an example of the placement of the TDs are shown in Figure 20.5(b). We measure the multipath channels from each RRH to the TDs at all the possible locations. In this experiment, we have 800 possible TD locations and 6 possible AP locations, from which 4,800 independent multipath channel measurements are obtained. In the following subsections, the performance of the presented system is evaluated using the measured channels.

20.5.2 Downlink Front-Haul Rate and Spectral Efficiency

In this section, we show the unique features of the presented system in the downlink through numerical results. We first show that the front-haul rate keeps almost constant independent of the number of TDs in the system, after which we show the achievable spectral efficiency of the system under various settings in deployment and load. We also compare the result with the C-RAN based on LTE to show the advantage of the presented system in being able to utilize the wireless channel more efficiently.

In Figures 20.6 and 20.7, we generate the baseband signal $S_i^{(m)}[k]$ according to (20.3) using the measured channels and show the complementary cumulative distribution function (CCDF) [28] of the peak to average power ratio (PAPR) of the signal $S_i^{(m)}[k]$ in-band (I) and quadrature (Q) parts under various conditions. The $X_j[k]$'s are QPSK modulated. Let $N_i = \|\mathbf{T}_i\|$ denote the number of TDs subscribed to RRH i, and it is shown that the PAPR of the $S_i^{(m)}[k]$ does not change much with N_i. For example, if we look at the dotted horizontal line at $CCDF = 0.05$, it always crosses the curves

Table 20.1 The 5% PAPR for different N_i's in downlink (db)

	$N_i = 1$	$N_i = 4$	$N_i = 8$	$N_i = 16$
D = 1, (I)	5.74	5.86	6.06	6.26
D = 1, (Q)	5.74	5.88	6.06	6.26
D = 2, (I)	5.86	6.02	6.22	6.36
D = 2, (Q)	5.86	6.04	6.22	6.36
D = 4, (I)	5.88	6.04	6.24	6.36
D = 4, (Q)	5.86	6.04	6.22	6.36
D = 8, (I)	5.88	6.04	6.24	6.36
D = 8, (Q)	5.88	6.04	6.22	6.36

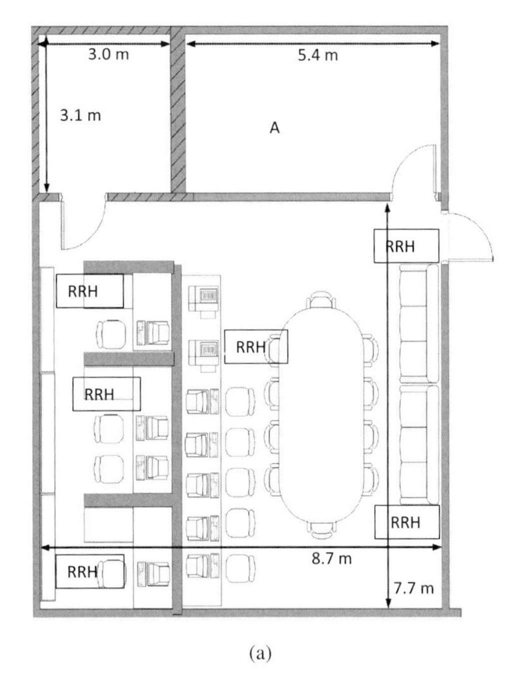

(a)

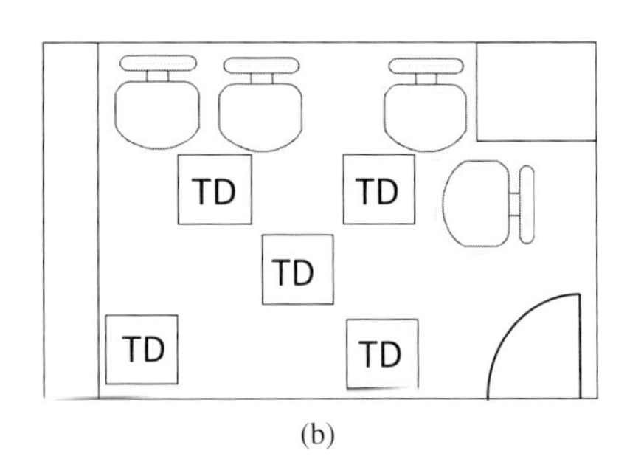

(b)

Figure 20.5 The floor plans of the testing sites: (a) the testing room, and (b) Room A.

at around 6 db. The detailed statistics are shown in Table 20.1. It means 95% of the baseband symbols $S_i^{(m)}[k]$ have no more than four times the average power. By (20.4), the average power of $S_i^{(m)}[k]$ is only dependent on θ and D. Therefore, the dynamic range $[-K_i^{(m)}, K_i^{(m)}]$ of $S_i^{(m)}[k]$ changes very little for different N_i's, where some $B_i^{(m)}$ can be used to always maintain the same level of quantization noise power. As a result, the data rate in each front-haul link is constant.

Next, we evaluate the spectral efficiency of the system with different number of RRHs and TDs. For each single-channel realization, the $P_{sig}^{(dl)}$, $P_{isi}^{(dl)}$, $P_{iui}^{(dl)}$, and $\sigma_{q,(dl)}^2$ can be calculated by (20.13), (20.14), (20.15), and (20.16). By plugging them into (20.18), the spectral efficiency of each individual TD can be calculated. By averaging over all the

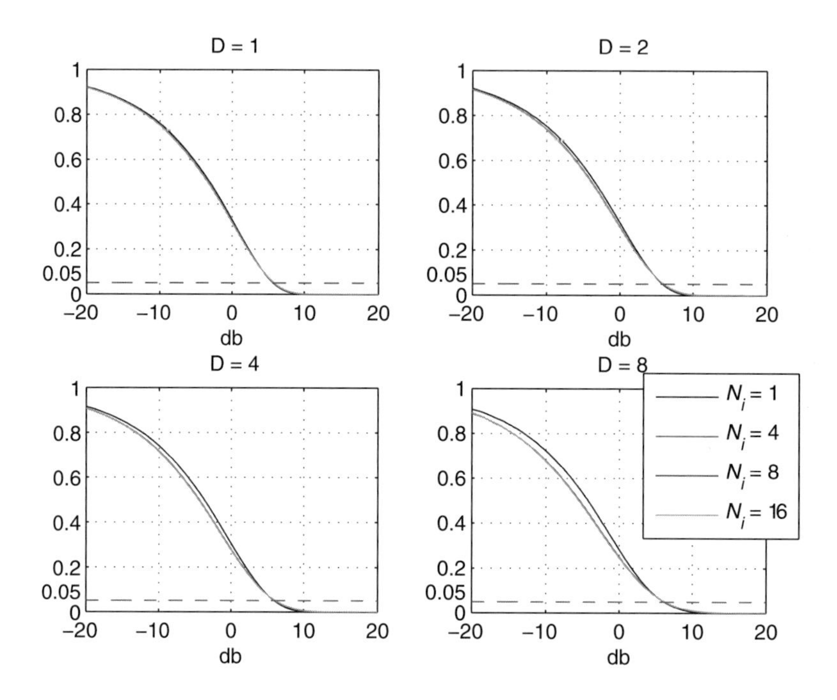

Figure 20.6 The CCDF of downlink QPSK baseband signal PAPR (I).

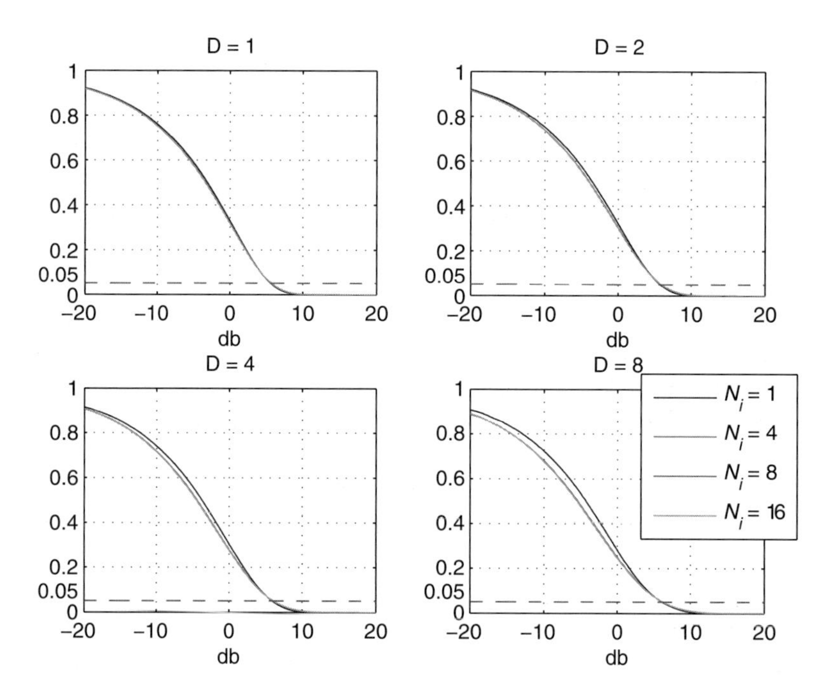

Figure 20.7 The CCDF of downlink QPSK baseband signal PAPR (Q).

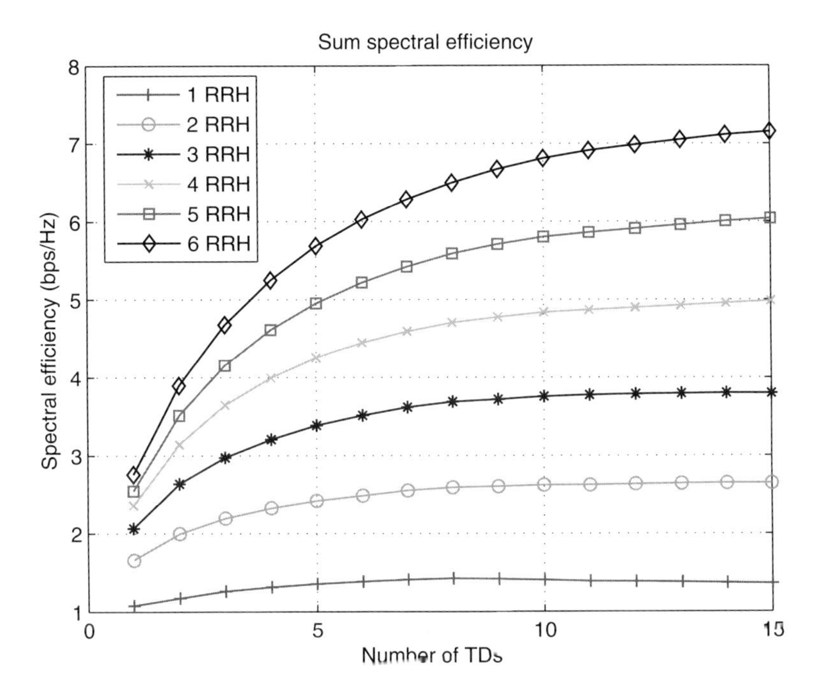

Figure 20.8 The sum spectral efficiency (D = 1).

channel realizations, we show the average and sum spectral efficiency in Figures 20.8 through 20.11.

As shown in the figures, for each given number of RRHs in this system, the individual spectral efficiency decreases with more TDs in the system, while the sum spectral efficiency increases with more TDs in the system. Note that by previous analysis, the data rate in the front-haul keeps constant. It means that more information can be transmitted in the front-haul link with the same amount of bits and energy consumed. The reason is that by using TR-based air interface, multiple TDs are naturally separated by the location-specific signatures. Therefore, even though the baseband signals for multiple TDs are mixed together in the front-haul link, they can still be separated when transmitted through the air interface. In other words, with the TR-based air interface, we are able to create a "tunnel" in the front-haul link such that the baseband signals can be efficiently combined to alleviate the traffic in front-haul.

Moreover, it can be observed that both the individual and sum spectral efficiency are improved if more RRHs are added into this system. The new RRHs contribute both extra power and degree of freedom to the system. The extra power alleviates the influence of the quantization and environmental noise, while the extra degree of freedom helps by enhancing the focusing effect [18] and thus mitigating the interference. Usually, in a dense wireless network, the interference is the dominating factor that limits the performance of the system. To illustrate this phenomenon, we compare the effect of adding in more RRHs with that of increasing the power of a single RRH. As it can be seen from Figure 20.12, the curve with circle markers shows that by just increasing the

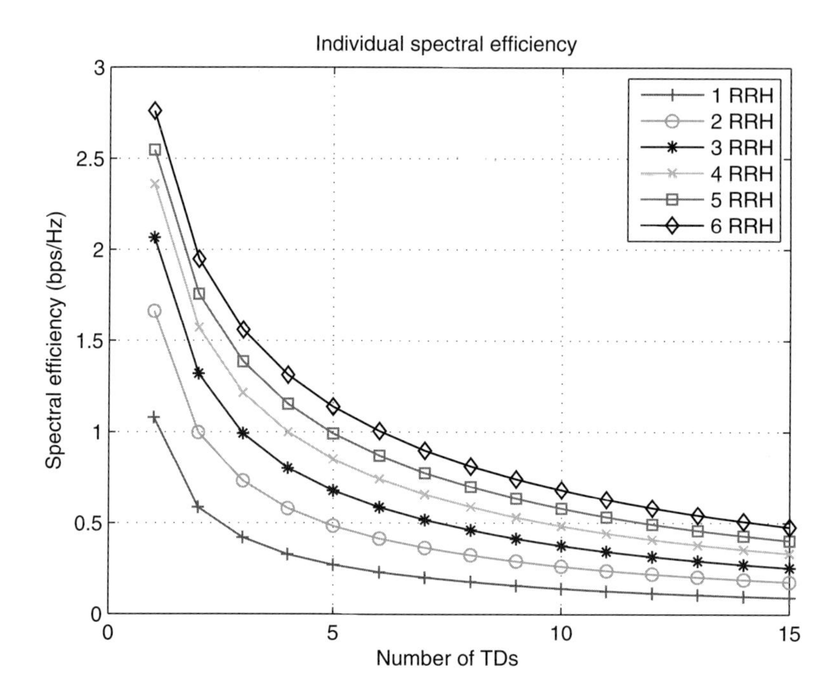

Figure 20.9 The individual spectral efficiency (D = 1).

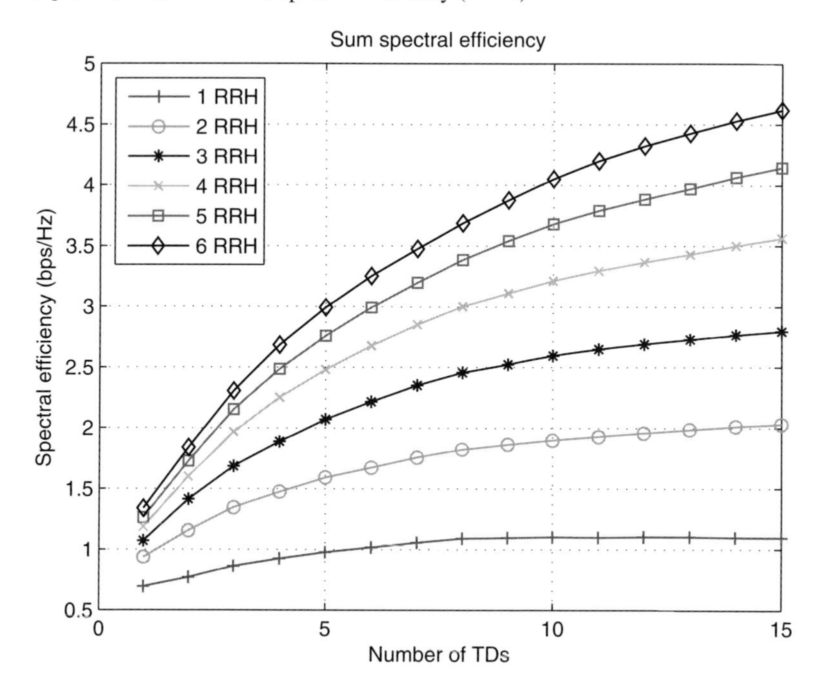

Figure 20.10 The sum spectral efficiency (D = 4).

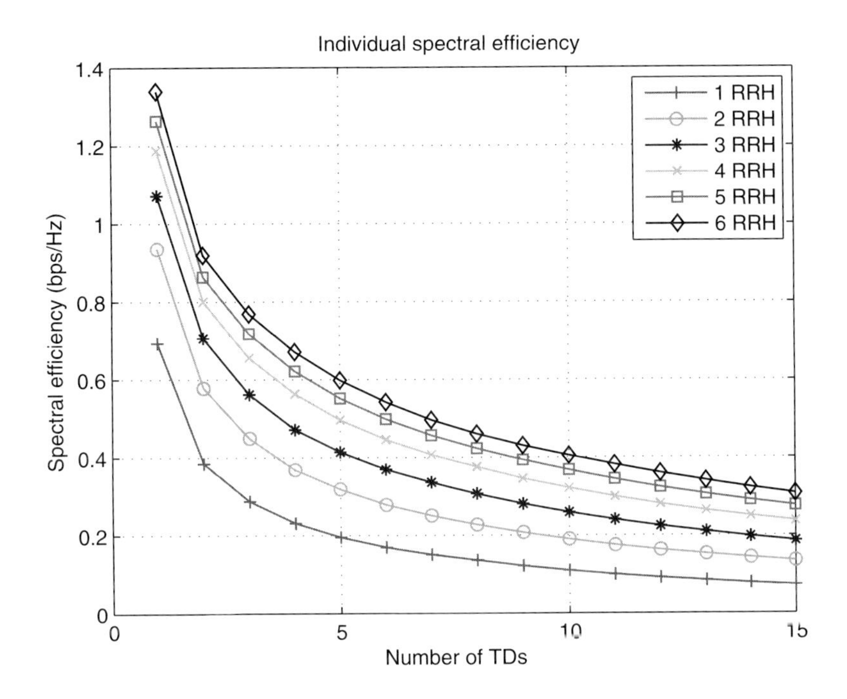

Figure 20.11 The Individual Spectral Efficiency (D = 4)

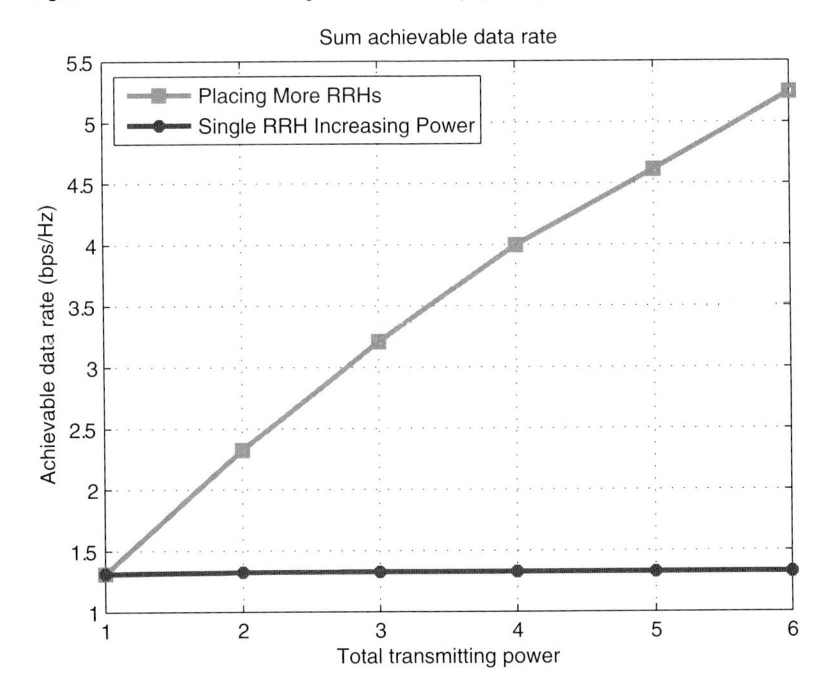

Figure 20.12 The comparison between adding more RRHs and single RRH increasing power.

power of one single RRH, the spectral efficiency keeps almost constant, while the curve shows that by adding in more RRHs, the spectral efficiency is improved significantly.

20.5.3 Uplink Front-Haul Rate and Spectral Efficiency

In this section, we use numerical results to illustrate the effectiveness of the presented system in the uplink.

In Figures 20.13 and 20.14, we generate the baseband signal $V_i^{(m)}[k]$ according to (20.6) using the measured channels and show the CCDF of the PAPR of both I and Q parts of the signal $V_i^{(m)}[k]$ under various conditions. The $X_j[k]$'s are QPSK modulated. It is shown that the PAPR of $V_i^{(m)}[k]$ does not change much with the number of TDs subscribed to the RRH i. Similarly, the $CCDF = 0.05$ line always crosses the CCDF curves at around 6 db. The detailed statistics are shown in Table 20.2. It means 95% of the baseband symbols $V_i^{(m)}[k]$ have no more than four times the average power. By plugging (20.25) into (20.7), the average power of $V_i^{(m)}[k]$ can be calculated as

$$E\left[\|V_i^{(m)}[k]\|^2\right] = \frac{\eta^2\theta}{D} \cdot \sum_{j\in\mathbf{T}_i} \cdot \frac{\sum_{l=0}^{L-1} \|h_{i,j}^{(m)}\|^2}{\left(\sum_{t\in\mathbf{R}_j} \sum_{m=1}^{M_T} h_{t,j}^{(m)} * c_{t,j}^{(m)}[L-1]\right)^2} + \sigma^2, \quad (20.31)$$

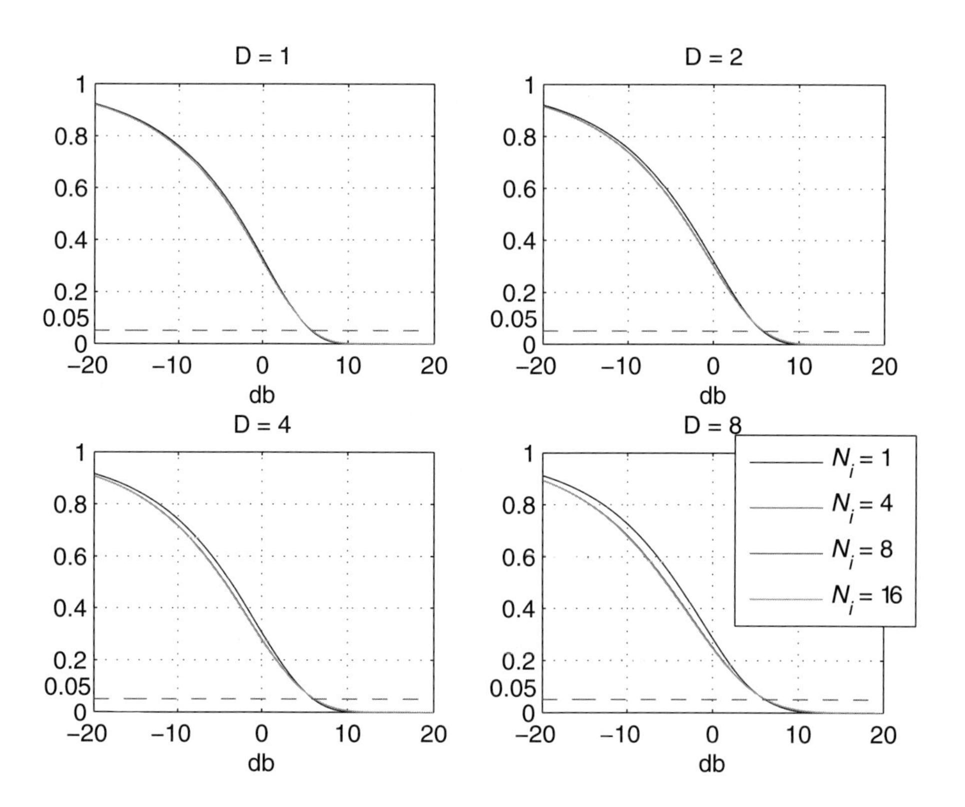

Figure 20.13 The CCDF of uplink QPSK baseband signal PAPR (I).

Table 20.2 The 5% PAPR for different N_i's in uplink (db)

	$N_i = 1$	$N_i = 4$	$N_i = 8$	$N_i = 16$
D = 1, (I)	5.74	5.86	6.06	6.22
D = 1, (Q)	5.74	5.88	6.06	6.22
D = 4, (I)	5.86	6.02	6.22	6.34
D = 4, (Q)	5.86	6.02	6.22	6.36
D = 8, (I)	5.88	6.04	6.22	6.36
D = 8, (Q)	5.86	6.04	6.22	6.36
D = 16, (I)	5.88	6.04	6.24	6.36
D = 16, (Q)	5.88	6.04	6.22	6.36

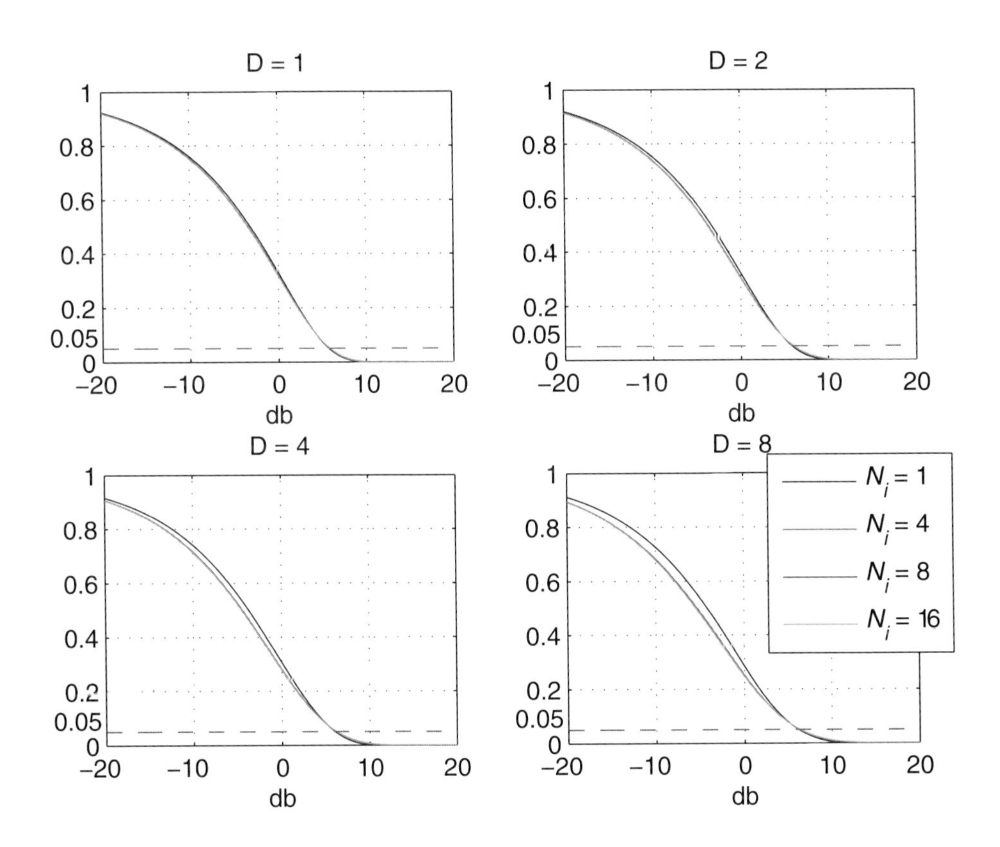

Figure 20.14 The CCDF of uplink QPSK baseband signal PAPR (Q).

which grows approximately linearly with N_i. Therefore, the dynamic range $[-K_i^{(m)}, K_i^{(m)}]$ of $V_i^{(m)}[k]$ increases linearly with $\sqrt{N_i}$. In order to maintain the same level of quantization noise power, more bits are needed to represent $V_i^{(m)}[k]$. However, because $Q_i^{(m)} = \frac{2K_i^{(m)}}{2^{B_i^{(m)}}}$, in order to maintain the same level of quantization noise power, $B_i^{(m)}$ grows to the order of $\log_2(K_i^{(m)})$ and thus $\log_2(\sqrt{N_i})$. For example, $E[\|q_i^{(m)}[k]\|^2]$ is

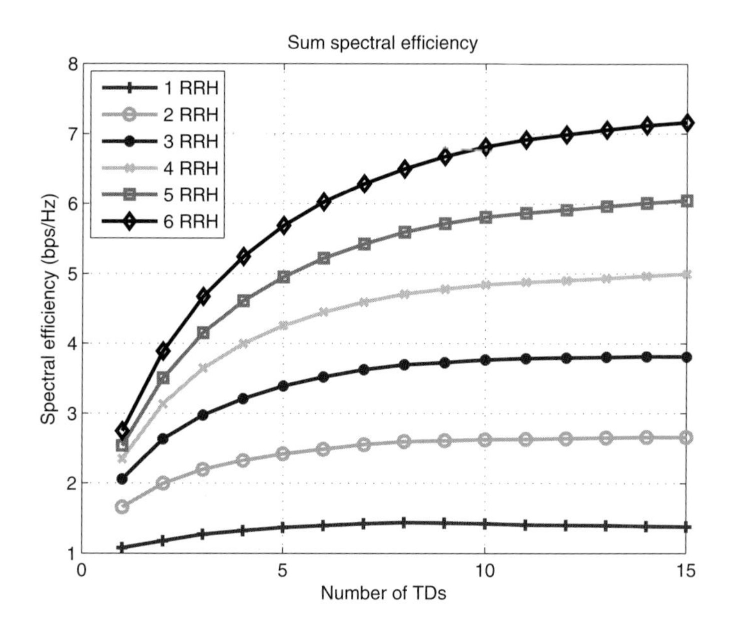

Figure 20.15 The sum spectral efficiency (D = 1).

the same for $B_i^{(m)} = 12, N_i = 1$, and $B_i^{(m)} = 14, N_i = 16$. In other words, when there are more TDs in the system, it is necessary to use slightly more bits in the front-haul link, which is much less significant compared to the increase of the number of TDs.

Similar to the downlink scenario, we evaluate the spectral efficiency of uplink by averaging over all the channel realizations. We slightly increase $B_i^{(m)}$ when it is necessary to keep the same level of quantization noise power. We show the average and sum spectral efficiencies under various conditions in Figures 20.15 through 20.18 and observe similar trends as in downlink.

Similar to the downlink scenario, the "tunneling" effect is also observed in the uplink case. When there are more TDs in the system, with using almost the same amount of bits in the front-haul link, more information can be extracted at the BBU side.

20.5.4 Comparison with LTE-Based C-RAN

To illustrate the advantage of the TR "tunneling" effect in the C-RAN, in this section, we compare the presented system with LTE-based C-RAN in multiple scenarios.

Suppose we have a certain number of RRHs distributed in an area where there are N TDs, and each of them has Ω bits of data to be transmitted. We first consider the downwlink. In the LTE-based C-RAN, all the RRHs work in separate bands, and each of the RRHs is responsible for serving part of the TDs. Because each RRH serves multiple TDs by dividing the time and/or frequency resource, the baseband signals for multiple TDs cannot be mixed together. As a result, the total amount of baseband signal is proportional to $N \times \Omega$, which can be approximated by $\phi_N^{(dl)} = N \times \Omega \times \lambda^{(dl)}$, where $\lambda^{(dl)}$ is some constant accounting for the modulation and channel coding. On the

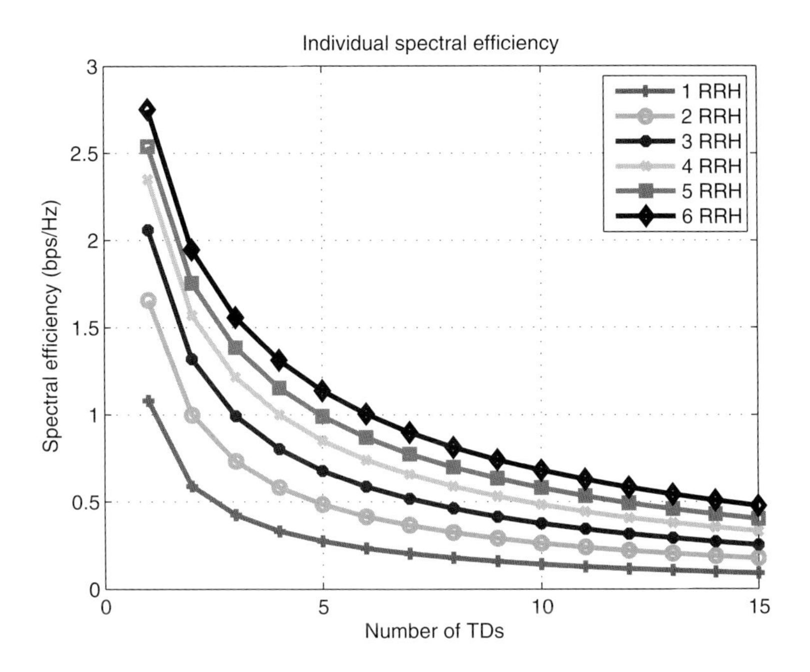

Figure 20.16 The individual spectral efficiency (D = 1).

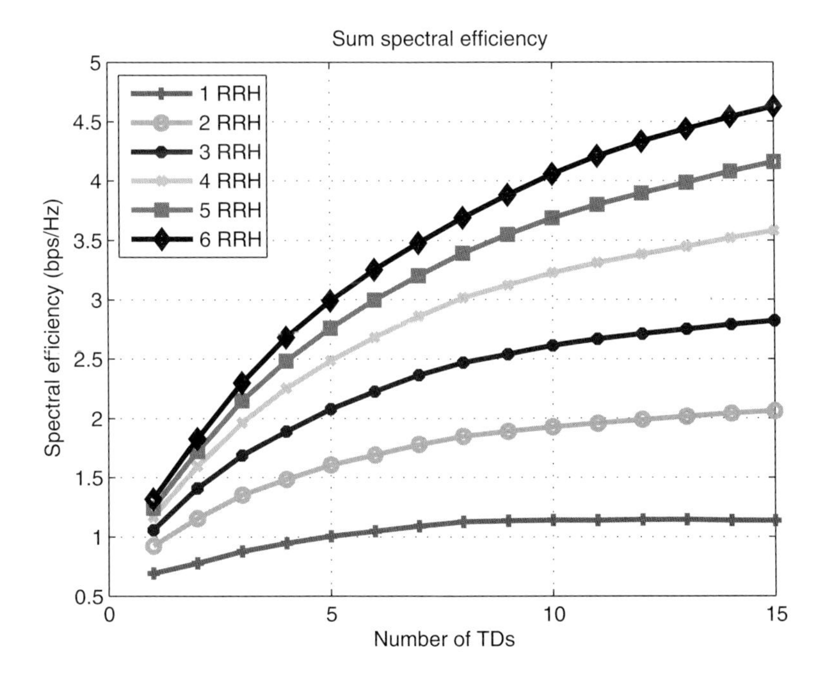

Figure 20.17 The sum spectral efficiency (D = 4).

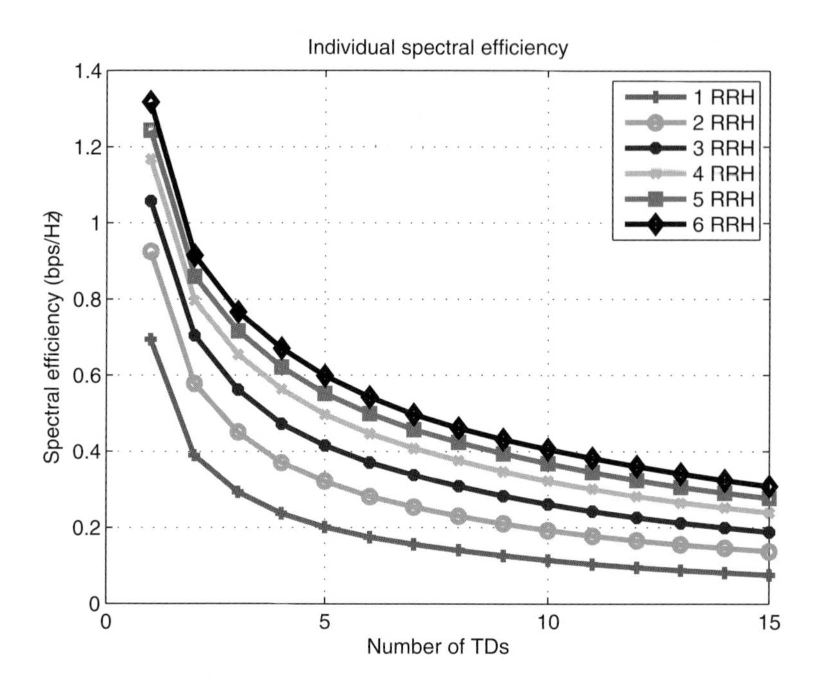

Figure 20.18 The individual spectral efficiency (D = 4).

other hand, in the TR-based C-RAN, as analyzed in Section 20.5.2, the data for multiple TDs can be efficiently combined without increasing the traffic in the front-haul. As a result, the total amount of baseband signal transmitted in the front-haul link is constant independent of N, which can be approximated by $\varphi_N^{(dl)} = \Omega \times \mu^{(dl)}$, where $\mu^{(dl)}$ is some constant accounting for the modulation and channel coding. We define $\tau_N^{(dl)} = \frac{\phi_N^{(dl)}}{\phi_1^{(dl)}}$ and $\upsilon_N^{(dl)} = \frac{\varphi_N^{(dl)}}{\varphi_1^{(dl)}}$ to characterize the growth of data transmitted in the front-haul caused by increasing the number of TDs.

In the uplink, the LTE works by single-carrier frequency division multiple access (SC-FDMA) [29], where multiple TDs are separated by the division of the frequency resource. As a result, the aggregate data transmitted in the front-haul is also proportional to $N \times \Omega$, and $\tau_N^{(ul)}$ is the same as the $\tau_N^{(dl)}$. On the other hand, in the TR-based C-RAN, as analyzed in Section 20.5.3, the data transmitted in the front-haul only increases slightly. We define $\varphi_N^{(ul)} = \Omega \times B_N^{(ul)} \times \mu^{(ul)}$ where $B_N^{(ul)}$ is the average number of bits to represent a baseband signal symbol when there are N TDs, and similarly $\upsilon_N^{(ul)} = \frac{\varphi_N^{(ul)}}{\varphi_1^{(ul)}}$. In this example, we have $B_1^{(ul)} = 12$, and $B_N^{(ul)}$ is increased when necessary according to the analysis in Section 20.4.2.

In Figure 20.19, we show the $\tau_N^{(dl)}$, $\tau_N^{(ul)}$, $\upsilon_N^{(dl)}$, and $\upsilon_N^{(ul)}$ with different N's. It is illustrated that in both downlink and uplink, the total amount of data transmitted in the front-haul of LTE-based C-RAN increases linearly with the number of TDs. In contrast, the total amount of data transmitted in the front-haul of TR-based C-RAN keeps

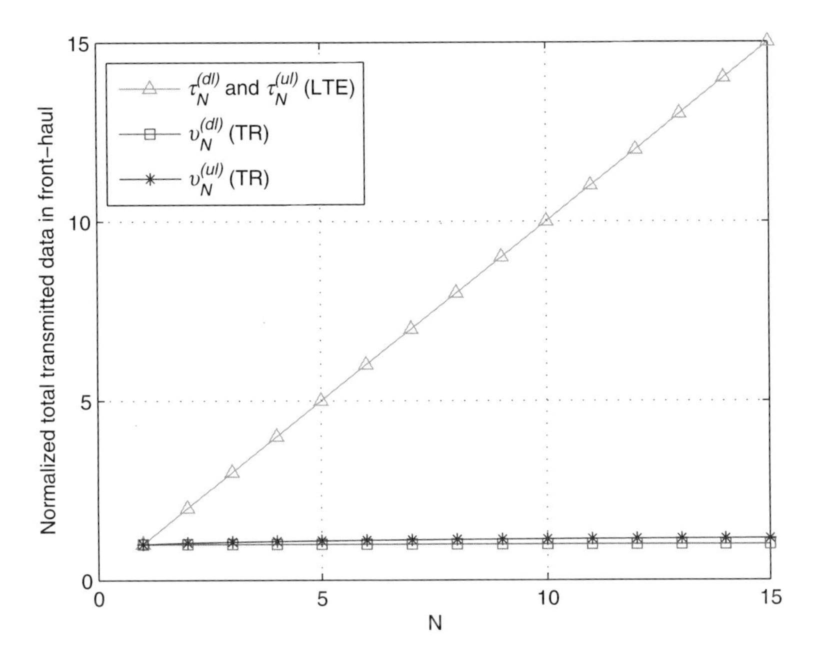

Figure 20.19 The comparison of normalized total transmitted data in front-haul between TR-based C-RAN and LTE-based C-RAN.

constant in the downlink regardless of the number of TDs, while only slightly increases in the uplink. It is due to the unique "tunneling" effect such that more information can be transmitted with nearly the same amount of bits consumed in the front-haul.

We also compare the spectral efficiency of the presented system with that of LTE-based C-RAN. Suppose there are N TDs distributed in an area to be served. We gradually add in extra RRHs to the C-RAN. In the TR-based C-RAN, the extra RRHs mainly help enhance the focusing effect and thus improve the spectral efficiency. More specifically, we define $r_{M,N}^{(avg)}$ as the average spectral efficiency of an individual TD when there are M RRHs and N TDs in the system. To evaluate the effect of adding in extra RRHs and TDs, we define

$$\xi_{M,N} = \frac{r_{M,N}^{(avg)}}{r_{1,1}^{(avg)}}, \tag{20.32}$$

which normalizes the average spectral efficiency to that when the wireless channel is exclusively used by a single pair of RRH and TD.

On the other hand, in the LTE-based C-RAN, we assume that each TD is associated with only one RRH, and multiple RRHs work in separate bands. The extra RRHs will offload part of the TDs from the existing RRHs, and thus each TD has an improved chance of being scheduled. Similar to the TR-based C-RAN, we define the effective average spectral efficiency $\delta_{M,N}^{(avg)} = C_{M,N}^{(avg)} \cdot \beta_{M,N}^{(avg)}$ where $\beta_{M,N}^{(avg)}$ is the average portion of time and frequency resource that a single TD shares when there are M RRHs and N TDs, and $C_{M,N}^{(avg)}$ is the average spectral efficiency of a TD conditioning if it is scheduled

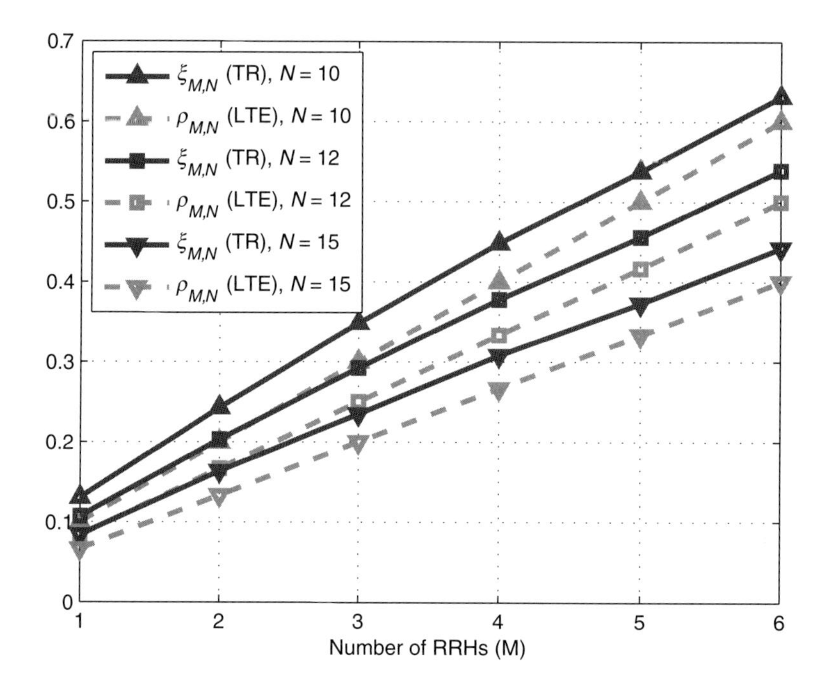

Figure 20.20 The comparison of normalized effective individual spectral efficiency between TR-based C-RAN and LTE-based C-RAN.

and allocated the entire time and frequency resource of an RRH. Because the RRHs work in separate bands, $C_{M,N}^{(avg)}$ does not change with M or N, and $\beta_{M,N}^{(avg)} = \min(\frac{M}{N}, 1)$. Similarly, we define

$$\rho_{M,N} = \frac{\delta_{M,N}^{(avg)}}{\delta_{1,1}^{(avg)}} \tag{20.33}$$

in order to evaluate the effect of adding in extra RRHs and TDs.

We plot $\xi_{M,N}$ and $\rho_{M,N}$ for multiple combinations of M and N in Figure 20.20. It is observed that $\xi_{M,N}$ is above $\rho_{M,N}$ for each combination. Note that the achievable data rate of a TD is the product of the spectral efficiency and the bandwidth. We define the achievable downlink data rate of a single TD in the TR-based C-RAN as $R_{M,N}^{TR} = r_{M,N}^{(avg)} \times W^{TR}$ where W^{TR} is the bandwidth that a TR-based TD utilizes, and the achievable downlink data rate of a single TD in the LTE-based C-RAN as $R_{M,N}^{LTE} = \delta_{M,N}^{(avg)} \times W^{LTE}$ where W^{LTE} is the bandwidth that a LTE-based TD utilizes. In the TR-based system, because larger bandwidth can be easily utilized with much reduced cost [21], the $R_{1,1}^{TR}$ can be greater than $R_{1,1}^{LTE}$ by utilizing larger bandwidth. For example, in this experiment, $W^{TR} = 125MHz$ and $W^{LTE} = 20MHz$. By $\xi_{M,N}$ and $\rho_{M,N}$ in Figure 20.20 where $r_{M,N}^{(avg)}$ and $\delta_{M,N}^{(avg)}$ are normalized to $r_{1,1}^{(avg)}$ and $\delta_{1,1}^{(avg)}$, $R_{M,N}^{TR}$ will be greater than $R_{M,N}^{LTE}$ for any M and N. It means that the TR-based C-RAN is able to more efficiently utilize the wireless channel in the multiple-RRH and multiple-TD setting.

20.6 Summary

In this chapter, we discussed a time-reversal (TR)-based cloud radio access network (C-RAN) architecture. Both the downlink and uplink working schemes were designed and analyzed. Through analysis, we discovered the TR "tunneling" effect in the presented C-RAN architecture, i.e., the baseband signals for multiple terminal devices (TDs) can be efficiently combined and transmitted in the front-haul link to alleviate the traffic load. We built a TR radio prototype to measure the wireless channel in the real-world environment, with which we illustrated the "tunneling" effect in both downlink and uplink of the presented C-RAN architecture. It is observed that for both downlink and uplink, the sum spectral efficiency of multiple TDs increases with the number of TDs in the system while the front-haul data rate keeps almost constant. Based on the nice properties shown in this chapter, the presented TR-based C-RAN architecture serves as a promising solution to tackle the challenge to the C-RAN front-haul link capacity caused by network densification. For related references, interested readers can refer to [30].

References

[1] Qualcomm, "1000x mobile data challenge," Technical Report, Nov. 2013.

[2] ChinaMobile, "C-RAN: The road towards green RAN," *White Paper*, Oct. 2011.

[3] M. Webb, Z. Li, P. Bucknell, T. Moulsley, and S. Vadgama, "Future evolution in wireless network architectures: Towards a 'cloud of antennas'," in *2012 IEEE Vehicular Technology Conference (VTC Fall)*, pp. 1–5, Sep. 2012.

[4] C.-L. I, J. Huang, R. Duan, C. Cui, J. Jiang, and L. Li, "Recent progress on C-RAN centralization and cloudification," *IEEE Access*, vol. 2, pp. 1030–1039, 2014.

[5] Y. Beyene, R. Jantti, and K. Ruttik, "Cloud-RAN architecture for indoor DAS," *IEEE Access*, vol. 2, pp. 1205–1212, 2014.

[6] A. Ghosh, R. Ratasuk, B. Mondal, N. Mangalvedhe, and T. Thomas, "LTE-advanced: Next-generation wireless broadband technology [invited paper]," *IEEE Wireless Communications*, vol. 17, no. 3, pp. 10–22, Jun. 2010.

[7] J. Lorca and L. Cucala, "Lossless compression technique for the fronthaul of LTE/LTE-advanced cloud-RAN architectures," in *2013 IEEE 14th International Symposium and Workshops on a World of Wireless, Mobile and Multimedia Networks (WoWMoM)*, pp. 1–9, Jun. 2013.

[8] R. Wang, H. Hu, and X. Yang, "Potentials and challenges of C-RAN supporting multi-RATS toward 5G mobile networks," *IEEE Access*, vol. 2, pp. 1187–1195, 2014.

[9] B. Dai and W. Yu, "Sparse beamforming and user-centric clustering for downlink cloud radio access network," *IEEE Access*, vol. 2, pp. 1326–1339, 2014.

[10] R. Zakhour and D. Gesbert, "Optimized data sharing in multicell MIMO with finite backhaul capacity," *IEEE Transactions on Signal Processing*, vol. 59, no. 12, pp. 6102–6111, Dec. 2011.

[11] Y. Zhou and W. Yu, "Optimized backhaul compression for uplink cloud radio access network," *IEEE Journal on Selected Areas in Communications*, vol. 32, no. 6, pp. 1295–1307, Jun. 2014.

[12] X. Rao and V. Lau, "Distributed fronthaul compression and joint signal recovery in cloud-RAN," *IEEE Transactions on Signal Processing*, vol. 63, no. 4, pp. 1056–1065, Feb. 2015.

[13] S.-H. Park, O. Simeone, O. Sahin, and S. Shamai, "Inter-cluster design of precoding and fronthaul compression for cloud radio access networks," *IEEE Wireless Communications Letters*, vol. 3, no. 4, pp. 369–372, Aug. 2014.

[14] O. Simeone, O. Somekh, H. V. Poor, and S. Shamai, "Downlink multicell processing with limited-backhaul capacity," *EURASIP Journal on Advances in Signal Processing*, vol. 2009, pp. 3:1–3:10, Feb. 2009.

[15] T. Strohmer, M. Emami, J. Hansen, G. Papanicolaou, and A. J. Paulraj, "Application of time-reversal with MMSE equalizer to UWB communications," in *IEEE Global Telecommunications Conference, 2004 (GLOBECOM'04)*, vol. 5. IEEE, pp. 3123–3127, 2004.

[16] C. Oestges, J. Hansen, S. M. Emami, A. D. Kim, G. Papanicolaou, and A. J. Paulraj, "Time reversal techniques for broadband wireless communication systems," in *European Microwave Conference (Workshop)*, pp. 49–66, 2004.

[17] R. C. Qiu, C. Zhou, N. Guo, and J. Q. Zhang, "Time reversal with MISO for ultrawideband communications: Experimental results," *IEEE Antennas and Wireless Propagation Letters*, vol. 5, no. 1, pp. 269–273, 2006.

[18] B. Wang, Y. Wu, F. Han, Y.-H. Yang, and K. J. R. Liu, "Green wireless communications: A time-reversal paradigm," *IEEE Journal on Selected Areas in Communications*, vol. 29, no. 8, pp. 1698–1710, Sept. 2011.

[19] F. Han, Y.-H. Yang, B. Wang, Y. Wu, and K. J. R. Liu, "Time-reversal division multiple access over multi-path channels," *IEEE Transactions on Communications*, vol. 60, no. 7, pp. 1953–1965, Jul. 2012.

[20] F. Han and K. J. R. Liu, "A multiuser TRDMA uplink system with 2D parallel interference cancellation," *IEEE Transactions on Communications*, vol. 62, no. 3, pp. 1011–1022, Mar. 2014.

[21] Y. Chen, Y.-H. Yang, F. Han, and K. J. R. Liu, "Time-reversal wideband communications," *IEEE Signal Processing Letters*, vol. 20, no. 12, pp. 1219–1222, Dec. 2013.

[22] H. Ma, F. Han, and K. J. R. Liu, "Interference-mitigating broadband secondary user downlink system: A time-reversal solution," in *2013 IEEE Global Communications Conference (GLOBECOM)*, pp. 884–889, Dec. 2013.

[23] Y. Chen, F. Han, Y.-H. Yang, H. Ma, Y. Han, C. Jiang, H.-Q. Lai, D. Claffey, Z. Safar, and K. J. R. Liu, "Time-reversal wireless paradigm for green Internet of Things: An overview," *IEEE Internet of Things Journal*, vol. 1, no. 1, pp. 81–98, Feb. 2014.

[24] M. J. Golay, "Complementary series," *IRE Transactions on Information Theory*, vol. 7, no. 2, pp. 82–87, Apr. 1961.

[25] Z.-H. Wu, Y. Han, Y. Chen, and K. J. Liu, "A time-reversal paradigm for indoor positioning system," *IEEE Transactions on Vehicular Technology*, vol. 64, no. 4, pp. 1331–1339, 2015.

[26] F. Weng, C. Yin, and T. Luo, "Channel estimation for the downlink of 3GPP-LTE systems," in *2010 2nd IEEE International Conference on Network Infrastructure and Digital Content*, pp. 1042–1046, Sep. 2010.

[27] B. Widrow and I. Kollár, *Quantization Noise: Roundoff Error in Digital Computation, Signal Processing, Control, and Communications*. New York: Cambridge University Press, 2008.

[28] J. Gentle, *Computational Statistics*, ser. Statistics and Computing. New York: Springer, 2009.

[29] J. Zyren and W. McCoy, "Overview of the 3GPP long term evolution physical layer," *Freescale Semiconductor Inc., white paper*, 2007.

[30] H. Ma, B. Wang, Y. Chen, and K. J. R. Liu, "Time-reversal tunneling effects for cloud radio access network," *IEEE Transactions on Wireless Communications*, vol. 15, no. 4, pp. 3030–3043, 2016.

Part V

IoT Connections

21 Time Reversal for IoT

In this chapter, we present an overview of the time-reversal wireless paradigm for green Internet of things (IoT). It is shown that the time-reversal technique is a promising technique that focuses signal waves in both time and space domains. The unique asymmetric architecture significantly reduces the cost of the terminal devices, the total number of which is expected to be very large for IoT. The focusing effect of the time-reversal technique can harvest the energy of all the multipaths at the receiver, which improves the energy efficiency of the wireless transmission and thus the battery life of terminal devices in IoT. Facilitated by the high-resolution spatial focusing, the time-reversal division multiple access scheme leverages the uniqueness of the multipath profiles in the rich-scattering environment and maps them into location-specific signatures, so that spatial multiplexing can be achieved for multiple users operating on the same spectrum. In addition, the time-reversal system can easily support heterogeneous terminal devices by providing various quality-of-service (QoS) options through adjusting the waveform and rate backoff factor. Finally, the unique location-specific signature in time-reversal system can provide additional physical-layer security and thus can enhance the privacy and security of customers in IoT. All the advantages show that the time-reversal technique is a promising paradigm for IoT.

21.1 Introduction

During the past decade, the "Internet of Things (IoT)" has drawn great attention from both academia and industry because it offer the challenging notion of creating a world where all the *things*, known as smart objects [1] around us are connected, typically in a wireless manner, to the Internet and communicate with each other with minimum human intervention [2–4]. An example of IoT system is shown in Figure 21.1. The ultimate goal of IoT is to create a better world where things around us know what we like, what we want, and what we need and act accordingly without explicit instructions [5], and thus improve the quality of our lives and consistently reduce the ecological impact of mankind on the planet [6].

The term "Internet of Things (IoT)" was first proposed by Kevin Ashton in his presentation at Procter&Gamble (P&G) in 1999 [7]. During the presentation, Ashton envisioned the potential of IoT by stating "The Internet of Things has the potential to

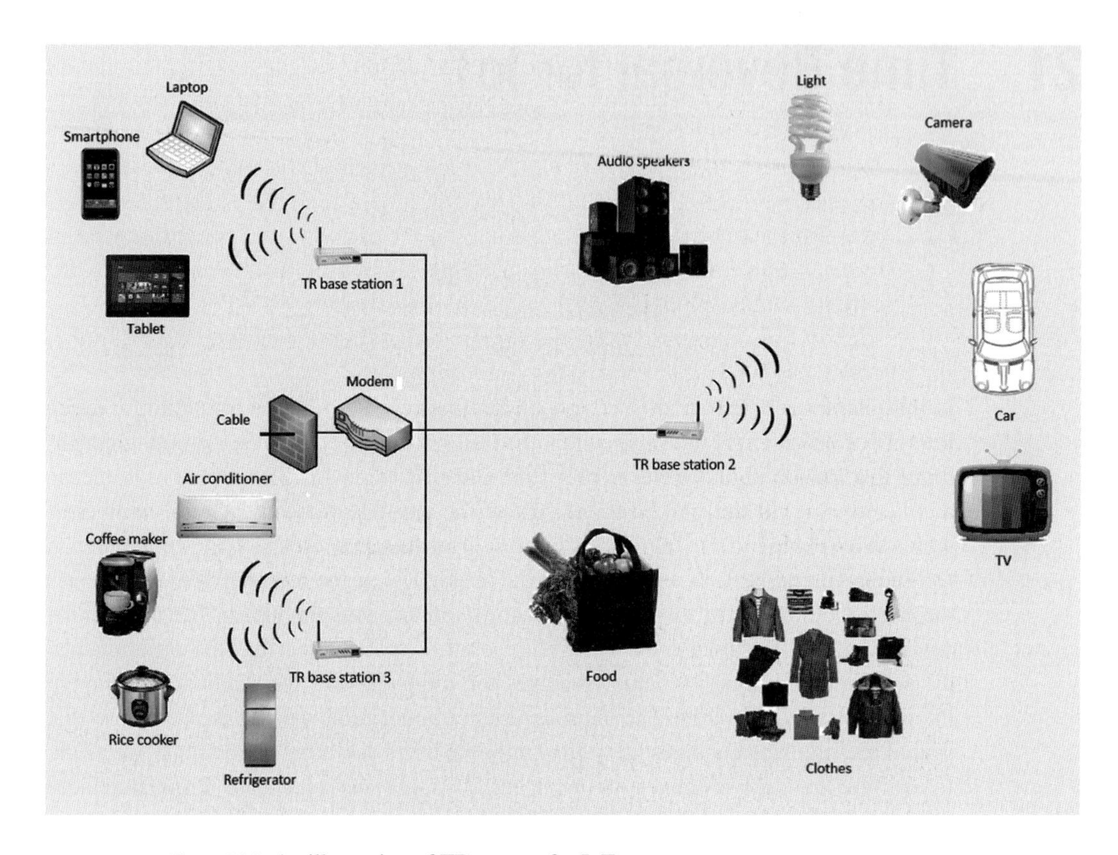

Figure 21.1 An illustration of TR system for IoT.

change the world, just as the Internet did. Maybe even more so." Such a concept then became popular when the MIT Auto-ID center presented their vision of IoT in 2001 [8]. In 2005, IoT was formally introduced by the International Telecommunication Union (ITU) through the ITU Internet report [9].

With a very broad vision, IoT has shown its great potential to improve the quality of our lives. However, research into the IoT is still in its infancy, and there are still a lot of challenges that need to be addressed before the realization of IoT. In the following, we summarize some key technical challenges from the perspective of wireless communication, which is the essential technology of IoT to allow people and things to connect to Internet anywhere at anytime [10].

- *Better battery life*: Typically the things in IoT are powered with small batteries, due to which the power consumption is low and thus requires low computational complexity of the wireless communication techniques.
- *Multiple active things*: The IoT is expected to have many concurrent active things transmitting data, which leads to severe interference among things. Thus, low-interference wireless technologies are desired.

- *Low-cost terminal devices*: For widespread adoption of the IoT technology, the cost at the terminal devices, i.e., things, needs to be low. Therefore, simple processing at the terminal devices' side is preferred.
- *Heterogeneous terminal devices*: Different from current wireless systems that have a collection of rather uniform devices, it is to be expected that the IoT will exhibit a much higher level of heterogeneity, as things that are totally different in terms of functionality, technology and application fields will belong to the same communication environment. Thus, the wireless solution to IoT should be able to support heterogeneous terminal devices with different quality-of-service (QoS) options such as from very low bit rate to very high.
- *Scalability*: The density of the things in IoT may be very high or low, which requires the wireless technology to be highly scalable to provide satisfactory QoS for low- to high-density areas.
- *Privacy and security*: Because everything in IoT has a unique identification, there is a need to have a technically sound solution to guarantee privacy and the security of the customers in order to have a widespread adoption of IoT.

To qualify as a good wireless communication solution to IoT, a technology should be able to handle the challenges raised in the preceding list. Currently existing wireless technologies for IoT can be classify into two groups: (1) wireless technologies for low-data-rate and low-power applications such as remote control [11, 12] and (2) wireless technologies for high-data-rate applications such as video streaming [13–17]. Note that the technologies suitable for low-data-rate applications may not be able to meet the requirements of the high-data-rate applications.

A typical wireless communication technology suitable for low-power, low-data-rate applications is ZigBee [11]. Mainly based on IEEE 802.15.4, ZigBee can operate in the 868 MHz, 915 MHz, and 2.4 GHz bands with respective data rates of 20 kb/s, 40 kb/s, and 250 kb/s. A similar technology is Z-Wave [12], whose main purpose is to enable short message transmission from a control node to multiple nodes. The maximum speed of Z-Wave is 200 kb/s working at 2.4 GHz band. The most significant advantage of ZigBee and Z-Wave is the low price [18, 19]. For instance, there exist $3–$5 chips including RF module, the digital baseband module, and a programmable microcontroller. Both of these technologies were designed for low-power applications in battery-operated devices. Moreover, ZigBee even includes a sleep mode mechanism to reduce power consumption. The complexity of hardware is quite low: 32–128 kbytes of memory is enough to implement the system including the higher layers. On the other hand, the most obvious disadvantage of ZigBee and Z-Wave is their low data rate. Moreover, the 2.4 GHz frequency band is already crowded with interfering devices, e.g., microwave ovens, Wi-Fi equipment, and cordless phones. The sub-GHz electromagnetic (EM) waves propagate very far, so very high node density may not be achievable due to the high interference levels created by other similar devices.

The most popular technologies for high-data-rate applications are Bluetooth [13] and Wi-Fi [14]. Bluetooth, based on IEEE 802.15.1, is a wireless technology for exchanging data over short distance. Compared with ZigBee and Z-Wave, the data rate could be increased to Megabit per s (Mbps). Wi-Fi, based on IEEE 802.11, is a popular technology that allows an electronic device to exchange data or connect to the Internet wirelessly. The speed of Wi-Fi can achieve up to several Gigabits per second (Gbps) according to IEEE 802.11ac with the help of MIMO and very high order modulation. The most important advantage of these two technologies is the high data rate. However, they require higher power consumption, higher complexity of hardware (MIMO in Wi-Fi), and thus higher price [20]. Because both transmitter and receiver use the same architecture, i.e., symmetric architecture is used, the power consumption of terminal devices is high. In addition, a large number of Wi-Fi access points (APs) deployed close to each other operating in the same or adjacent channels will severely interfere with each other. Thus, these technologies do not seem to offer robust performance in interference-limited scenarios even with costly terminal devices. Another possible technology is the 3G/4G mobile communications [21–23]. However, the poor indoor coverage of 3G/4G signals greatly limits its application to IoT, where communications mostly happen in indoor environments.

From the preceding discussions, we can see that existing technologies can only address partial challenges while leaving the rest unaddressed, e.g., both the heterogeneity and scalability challenges cannot be handled by existing technologies. A natural question to ask is: Is there a wireless communication technique that can address most, if not all, challenges? As pointed out in [24], time-reversal (TR) signal transmission is an ideal paradigm for low-complexity, low-energy-consumption green wireless communication because of its inherent nature to fully harvest energy from the surrounding environment by exploiting the multipath propagation to recollect all the signal energy that could be collected as the ideal Rake receiver. The theoretic analysis in [24] shows that a typical TR system has a potential of over an order of magnitude of reduction in power consumption and interference alleviation, which means that TR system can provide better battery life and support multiple concurrent active users. Moreover, with the asymmetric TR architecture, only one-tap detection is needed at the receiver side [25], thus the computational complexity at the terminal devices is low, which means the cost of the terminal devices is also low. Note that the achievable rate can still be very high when the bandwidth is wide enough, as shown in [26]. In addition, the TR system can easily support heterogeneous terminal devices by providing various QoS options through adjusting the waveform and backoff factor [25, 27]. Finally, the unique location-specific signature in TR system can provide additional physical-layer security and thus can enhance the privacy and security of customers in IoT [24]. Overall, we will provide an overview to show that time-reversal technique is an ideal paradigm for IoT.

The rest of the chapter is organized as follows. In Section 21.2, we introduce some basic concepts of time-reversal technique. Then, we discuss in Section 21.3 an asymmetric time-reversal division multiple access (TRDMA) architecture and discuss in details why TR is an ideal paradigm for IoT. In Section 21.4, we discuss other challenging issues and future directions, including advanced waveform design, MAC layer issues, and low-cost high-speed ADC and DAC.

21.2 Some Basics of Time Reversal

21.2.1 The Basic Principles of Time Reversal

TR signal processing is a technology to focus the power of signal waves in both time and space domains. The research of time reversal can date back to early 1970s, when phase conjugation was first observed and studied by Zel'dovich et al. [28]. Unlike the phase conjugation that uses an holographic or parametric pumping [29], the time reversal uses transducers to record the signal waves and enables signal processing on the recorded waveforms.

The time-reversal signal processing was applied by Fink et al. in 1989 [30], followed by a series of theoretical and experimental works in acoustic communications [31–38]. As found in acoustic physics [30–34] and then further validated in practical underwater propagation environments[35–37], the energy of the TR acoustic waves from transmitters could be refocused only at the intended location with very high spatial resolution. Because TR can make full use of multipath propagation and also requires no complicated channel processing and equalization, it was later verified and tested in wireless radio communication systems. Experimental validations of TR technique with EM waves have been conducted in [24, 39–45], including the demonstration of spatial and temporal focusing properties [24, 39–45] and channel reciprocity [24, 41]. The feasibility of applying TR technique into ultra-wideband (UWB) communications have been studied in [46–48] with the focus on the bit error rate (BER) performance through simulation. A system-level theoretical investigation and comprehensive performance analysis of a TR-based multiuser communication system was conducted in [25], where the concept of time-reversal division multiple access (TRDMA) was proposed. To improve the performance of the TRDMA systems, interference suppression through waveform design [27, 49] and interference cancellation [50] are proposed. The implementation complexity issue is studied in [26, 51, 52]. Moreover, as shown in [53, 54], with random scatterers, TR can achieve focusing that is far beyond the diffraction limit, i.e., half wavelength.

The principle of time reversal transmission is very simple, which can be referred to Chapter 1. There are two basic assumptions for the time-reversal communication system to work

- Channel reciprocity: The impulse responses of the forward link channel and the backward link channel are assumed to be identical.
- Channel stationarity: The channel impulse responses are assumed to be stationary for at least one probing-and-transmitting cycle.

These two assumptions generally hold in reality, especially for indoor environment, as validated through real experiments in [24, 55]. In [55], Qiu et al. conducted experiments in a campus lab area and showed that the correlation between the impulse response of the forward link channel and that of backward link channel is as high as about 0.98, which means that the channel is highly reciprocal. In [24], Wang et al. showed with experimental results that the multipath channel of an office environment is actually

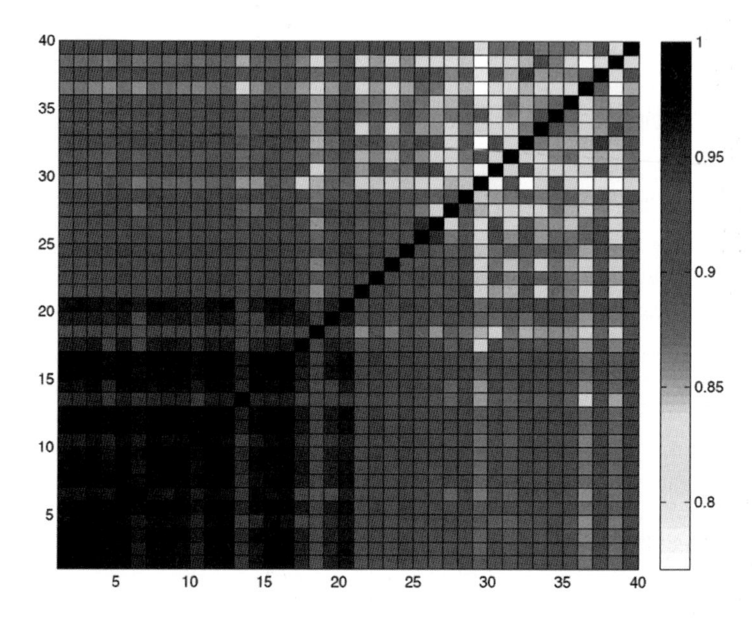

Figure 21.2 Correlation of channel responses at different time epochs [24].

not changing a lot. In their experiment, Wang et al. measured the channel information every 1 min, and a total of forty channel snapshots were taken and stored, where the first 20 snapshots correspond to a static environment, snapshots 21-30 correspond to a moderately varying environment, and snapshots 31-40 correspond to a varying environment. The experimental results are shown in Figure 21.2, where we can see that most of the correlation coefficients between different snapshots are higher than 0.8, and those between static snapshots are above 0.95, which means that the channel is highly stationary.

By utilizing channel reciprocity, the reemitted TR waves can retrace the incoming paths, ending up with a constructive sum of signals of all the paths at the intended location and a "spiky" signal-power distribution over the space, as commonly referred to as *spatial focusing effect*. Also from the signal processing point of view, in the point-to-point communications, TR essentially leverages the multipath channel as a matched filter and focuses the wave in the time domain as well, as commonly referred to as *temporal focusing effect*. By treating the environment as a facilitating matched filter computing machine, the complexity of TR systems is significantly reduced, which is ideal for IoT applications, as we will discuss later.

21.2.2 Temporal Focusing and Spatial Focusing of Time Reversal

In principle, the mechanisms of reflection, diffraction, and scattering in wireless medium give rise to the uniqueness and independence of the channel impulse response of each multipath communication link [56]. Obtained from real indoor experiments [24], Figures 21.3 and 21.4 show that, when the reemitted TR waves from transceiver

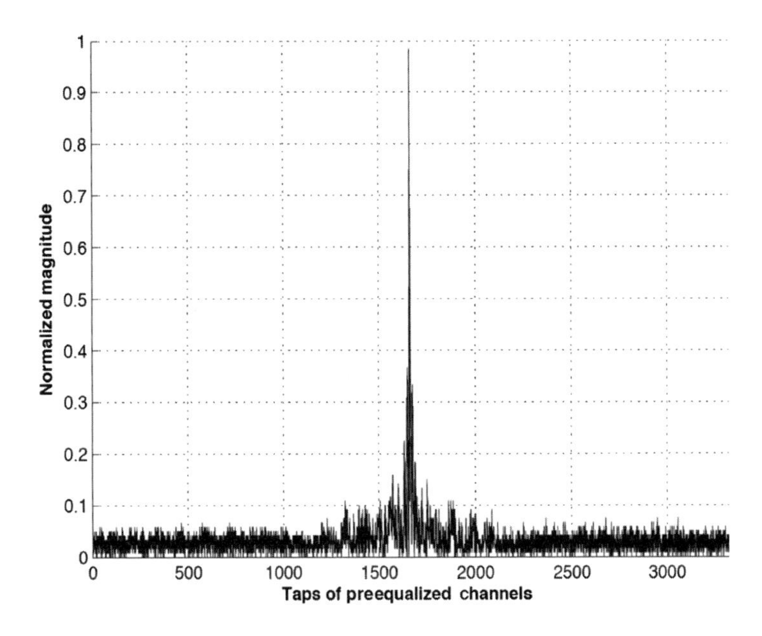

Figure 21.3 Temporal focusing effect obtained from experiments [24].

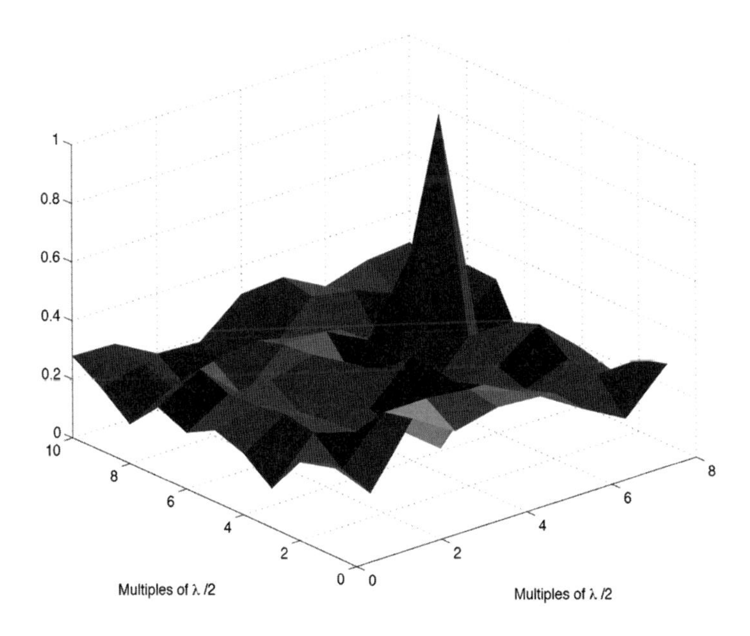

Figure 21.4 Spatial focusing effect obtained from experiments [24].

A propagate in the wireless medium, the location of transceiver B is the only location that is associated with the reciprocal channel impulse response. That is to say that given the reemitted TR waveform from transceiver A that is specific to the channel impulse response between transceiver A and B, the environment will serve as a natural matched-filter only for the intended transceiver B. As a result, the temporal focusing effect of the specific reemitted TR waveform can be observed only at the location of transceiver B. It means that at the time instance of time focusing, the signal power not only exhibits a strong peak in the time domain at transceiver B as shown in Figure 21.3, but also concentrates spatially only at the location of transceiver B in the rich multipath environments as shown in Figure 21.4. A physical meaning is that at this moment, it creates a resonating effect due to the time-reversal transmission.

Experimental results in both acoustic/ultrasound domain and radio frequency (RF) domain further verified the temporal focusing and spatial focusing effects of the time-reversal transmission, as predicted by theory. Authors of [30–34] found that acoustic energy can be refocused on the source with very high resolution (wavelength level). In [35–37], acoustics experiments in the ocean were conducted to validate the focusing effects of time reversal in real underwater propagation environments. In the RF domain, experiments in [39, 40, 46] demonstrated the spatial and temporal focusing properties of electromagnetic signal transmission with time reversal by taking measurements in RF communications. Furthermore, a TR-based interference canceler to mitigate the effect of clutter was presented in [57], and target detection in a highly cluttered environment using TR was investigated in [58, 59]. In [24], real-life RF experiment results were obtained in typical indoor environments, which shows the great potential of TR as a new paradigm of the green wireless communications.

In the context of communication systems, the temporal focusing effect concentrates a large portion of the useful signal energy of each symbol within a short time interval, which effectively suppresses the intersymbol interference (ISI) for high-speed broadband communications. The spatial focusing effect allows the signal energy to be harvested at the intended location and reduces leakage to other locations, leading to a reduced required transmit power consumption and lower co-channel interference to other locations. The benefits and unique advantages of TR-based communication systems due to the temporal and spatial focusing effects promise a great potential for the applications of IoT, as will be discussed in the remaining parts of this chapter.

21.2.3 Time-Reversal Communication System

A very simple TR-based communication system is shown in Figure 21.5. The channel impulse response (CIR) between the two transceivers is modeled as

$$h(t) = \sum_{v=1}^{V} h_v \delta(t - \tau_v), \tag{21.1}$$

where h_v is the complex channel gain of the vth path of the CIR, and τ_v is the corresponding path delay, and the V is the total number of the underlying multipaths (assuming infinite system bandwidth and time resolution). Without loss of generality, we assume that $\tau_1 = 0$ in the following discussion, i.e., the first path arrives at time $t = 0$, and as a result, the delay spread of the multipath channel T is given by $T = \tau_V - \tau_1 = \tau_V$.

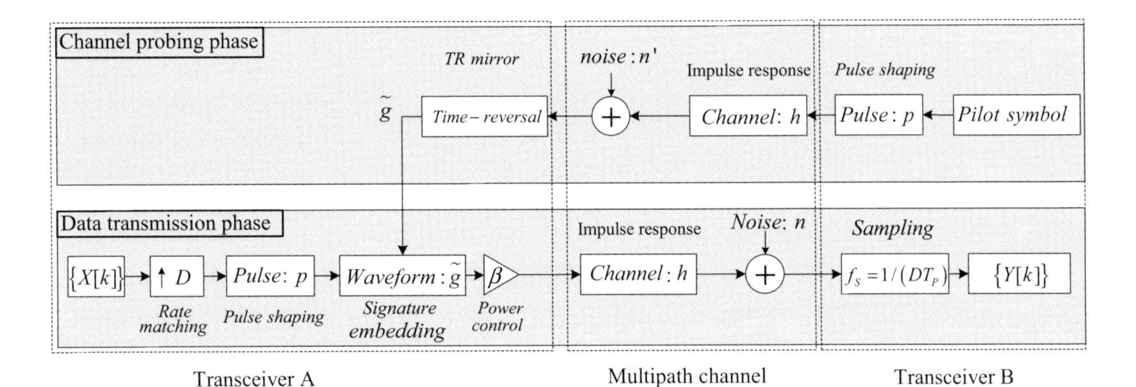

Figure 21.5 The basic time reversal communications.

Constrained by the limited bandwidth of practical communication system, pulse shaping filters are typically used to limit the effective bandwidth of the transmission. Generally, the duration of the pulse T_P is limited by the available bandwidth B through the simple relation $T_P = 1/B$.

21.2.3.1 Channel Probing Phase

Prior to transceiver A's TR-transmission, transceiver B first sends out a pulse $p(t)$ of duration T_P (other than an ideal impulse, which demands infinite bandwidth), which propagates to transceiver A through the multipath channel $h(t)$, where transceiver A keeps a record of the received waveform $\tilde{h}(t)$, which is the convolution of $h(t)$ and $p(t)$, represented as follows:

$$\tilde{h}(t) = \int_{t-T_P}^{t} p(t-\tau)h(\tau)d\tau, \quad 0 \leqslant t \leqslant T + T_p \;\&\; T_P \ll T, \tag{21.2}$$

where $\tilde{h}(t)$ can be treated as an equivalent channel response for the system with a limited bandwidth B. From (21.2), one can see that for those paths whose time differences are less than the pulse duration T_P, they are mixed together due to the limited system bandwidth B. Also for $|t_1 - t_2| > T_P$, the received value $\tilde{h}(t_1)$ and $\tilde{h}(t_2)$ are determined by completely different sets of paths. Therefore, given a limited bandwidth B, the corresponding pulse duration T_P determines the time-domain resolution to resolve two adjacent paths. In other words, from the system's perspective, those paths whose time differences are within the duration T_P are treated like one path in the equivalent channel response $\tilde{h}(t)$.

21.2.3.2 Data Transmission Phase

Upon receiving the waveform, transceiver A time-reverses (and conjugate, when complex-valued) the received waveform $\tilde{h}(t)$, and uses the normalized TR waveform as a basic signature waveform $\{\tilde{g}(t)\}$, i.e.,

$$\tilde{g}(t) = \frac{\tilde{h}^*(-t)}{\sqrt{\int_0^{T+T_P} \left|\tilde{h}(\tau)\right|^2 d\tau}} = \frac{\int_t^{t+T_P} p^*(-t+\tau)h^*(-\tau)d\tau}{\sqrt{\int_0^{T+T_P} \left|\tilde{h}(\tau)\right|^2 d\tau}}. \quad (21.3)$$

Defining $g(t) \triangleq h^*(-t)$ and $q(t) \triangleq p^*(-t)$, $\tilde{g}(t)$ in (21.3) can be represented as

$$\tilde{g}(t) = (g * q)(t). \quad (21.4)$$

At transceiver A, there is a sequence of information symbols $\{X[k]\}$ to be transmitted to transceiver B. Typically, the symbol rate can be much lower than the system chip rate.[1] Therefore, a rate backoff factor D is introduced to match the symbol rate with the chip rate by inserting $(D-1)$ zeros between two symbols [24, 25, 46, 60]. Applying the pulse shaping filter $p(t)$,

$$W(t) = \sum_{k \in \mathbb{Z}^+} X[k] \cdot p(t - kDT_P), \quad (21.5)$$

and the transmitted signal[2] can be expressed as

$$S(t) = \beta\,(W * \tilde{g})(t) = \beta \sum_{k \in \mathbb{Z}^+} X[k]\,(p * q * g)(t - kDT_P). \quad (21.6)$$

The signal received at transceiver B is the convolution of $S(t)$ and $h(t)$, plus additive white Gaussian noise (AWGN) $\tilde{n}(t)$ with zero-mean and variance σ_N^2, i.e.,

$$\begin{aligned} Y(t) &= (S * h)(t) + \tilde{n}(t) \\ &= \tilde{n}(t) + \beta \sum_{k \in \mathbb{Z}^+} X[k]\,(p * q * g * h)(t - kDT_P) \\ &= \tilde{n}(t) + \beta \sum_{k \in \mathbb{Z}^+} X[k] \left(\tilde{h} * \tilde{g}\right)(t - kDT_P), \end{aligned} \quad (21.7)$$

where $\tilde{h}(t) = (p * h)(t)$, and $\tilde{g}(t) = (q * g)(t)$.

Thanks to the temporal focusing, when $t = kDT_P$, the power of $\left(\tilde{h} * \tilde{g}\right)(t - kDT_P)$ achieves its maximum for $X[k]$, i.e.,

$$\left(\tilde{h} * \tilde{g}\right)(0) = \int_0^{T+T_P} \tilde{h}(\tau)\tilde{g}(-\tau)d\tau = \sqrt{\int_0^{T+T_P} \left|\tilde{h}(\tau)\right|^2 d\tau} \quad (21.8)$$

As the receiver, transceiver B simply samples the received signal every DT_P s at $t = kDT_P$,[3] for $k = 1, 2, \cdots$, in order to detect the symbol $X[k]$,

[1] The duration of each chip is T_P.

[2] Note that in this chapter, the baseband system model is considered. As a result, no RF components are included in the system diagrams.

[3] It is assumed here that the synchronization has been achieved at a reference time t=0, without loss of generality.

$$Y[k] = Y(t = kDT_P) = \beta \sum_{l=-\lfloor \frac{T+T_P}{DT_P} \rfloor}^{\lfloor \frac{T+T_P}{DT_P} \rfloor} X[k+l] \left(\tilde{g} * \tilde{h}\right)(lDT_P) + \tilde{n}(kDT_P)$$

$$= \underbrace{\beta \left(\tilde{h} * \tilde{g}\right)(0) X[k]}_{\text{Signal}} + \underbrace{\beta \sum_{\substack{l=-\lfloor \frac{T+T_P}{DT_P} \rfloor \\ l \neq 0}}^{\lfloor \frac{T+T_P}{DT_P} \rfloor} X[k+l] \left(\tilde{g} * \tilde{h}\right)(lDT_P)}_{\text{ISI}} + \underbrace{n[k]}_{\text{Noise}}, \tag{21.9}$$

where $n[k] \triangleq \tilde{n}(kDT_P)$.

Consequently, the resulting signal-to-interference-plus-noise ratio (SINR) is obtained as

$$SINR = \frac{\beta^2 \int_0^{T+T_P} \left|\tilde{h}(\tau)\right|^2 d\tau}{\beta^2 \sum_{\substack{l=-\lfloor \frac{T+T_P}{DT_P} \rfloor \\ l \neq 0}}^{\lfloor \frac{T+T_P}{DT_P} \rfloor} \left|\left(\tilde{g} * \tilde{h}\right)(lDT_P)\right|^2 + \sigma_N^2} \tag{21.10}$$

assuming that each information symbol $X[k]$ has unit power.

21.2.3.3 An Equivalent System Model with Limited Bandwidth

Based on (21.2)–(21.10), one can come up with an equivalent system model shown in Figure 21.6 for the system with limited system bandwidth as shown in Figure 21.5. In the equivalent system model, $\tilde{h}(t) = (h * p)(t)$ is treated as the effective channel response for such a finite-bandwidth system, taking into account the use of the band-limiting pulse shaping filter $p(t)$. Accordingly, the time-reversed (and conjugated) version of

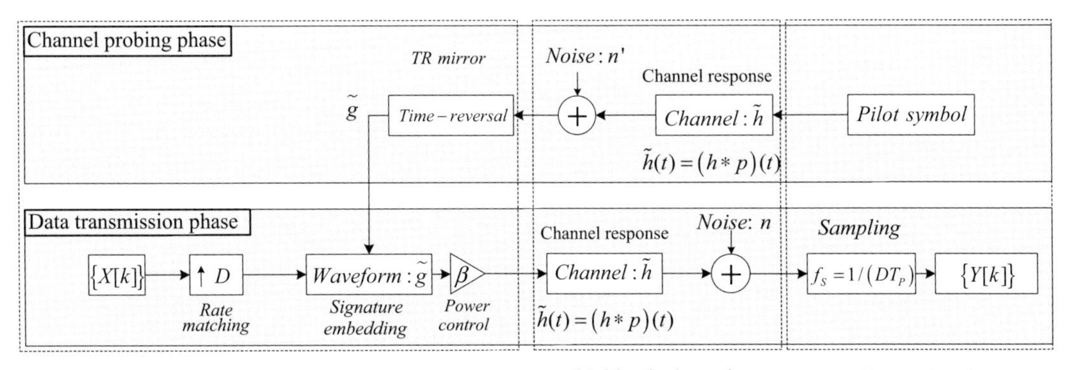

Figure 21.6 The basic time reversal communications with equivalent channel response $\tilde{h}(t)$.

the equivalent channel response $\tilde{g}(t) = \tilde{h}^*(-t)$, is the corresponding TR signature waveform for the equivalent model.

In the following discussion of the TRDMA scheme in this chapter, we use the simpler equivalent model by looking at the effective channel response $\tilde{h}(t) = (h * p)(t)$, which can be verified by comparing Figures 21.5 and 21.6.

21.3 Asymmetric TRDMA Architecture for IoT

Based on TR technique, we recently introduced in [25] a novel multiuser media access scheme, time-reversal division multiple access (TRDMA), for wideband communication. Leveraging the unique temporal and spatial focusing effects of the time reversal (TR) technique [24, 61], the TRDMA exploits the spatial degrees of freedom of the environment and uses the multipath channel profile associated with each user's location as a location-specific signature for the user. Moreover, such channel profiles may be further improved by mixing spatial degrees of freedom and temporal degrees of freedom as shown in [62, 63].

With the concept of TRDMA, in this section, we present an asymmetric TRDMA architecture for IoT, where most of the computational complexity is concentrated at the more powerful base station (BS), resulting in a minimal complexity and cost at the terminal devices in both uplink and downlink. As shown in Figure 21.1, the presented IoT system consists of multiple TR BSs, and each BS serves multiple heterogeneous terminal devices, which ranges from laptop and TV to light and clothes. In the following, we will first focus on the single BS scenario and then discuss the multiple-BS scenario in Section 21.3.5.

21.3.1 Channel Probing Phase

Consider a wireless broadband multiuser network that consists of one BS and N terminal users.[4] Each user communicates with BS simultaneously over the same spectrum. Assuming a rich-scattering environment, each user's location is associated with a unique (effective) channel response $\tilde{h}_i(t)$, $i = 1, 2, \ldots, N$.

The channel probing occurs when a terminal user joins the network and periodically afterwards.[5] The channel probing process is performed for one user at a time. For the ith user's channel probing, the terminal user first sends a pulse pilot signal $p(t)$ to the BS, so that the TR mirror at the BS can record and time reverse (and conjugate, if complex-valued) the received waveform $\tilde{h}_i(t)$, and use the TR waveform $\tilde{g}_i(t)$ as the basic signature waveform, given by the following[6]

$$\tilde{g}_i(t) = \frac{\tilde{h}_i^*(-t)}{\sqrt{\int_0^{T+T_P} \left|\tilde{h}_i(\tau)\right|^2 d\tau}}. \tag{21.11}$$

[4] In this chapter, users and devices are interchangeable.
[5] In general, the probing period depends on how fast the channel may vary.
[6] As we mentioned in Section 21.2, we use the effective channel response for finite-bandwidth system.

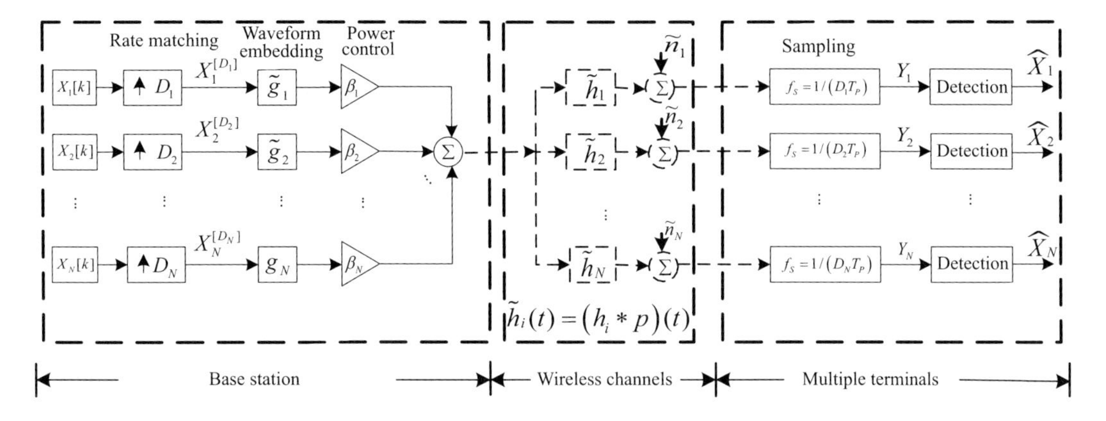

Figure 21.7 The basic diagram of the TRDMA downlink.

21.3.2 Data Transmission Phase – Downlink

After the channel recording phase, the system starts its data transmission phase. We first introduce the downlink scheme in this part. In the downlink scheme, at the BS, each of $\{X_1[k], X_2[k], \ldots, X_N[k]\}$ represents a sequence of information symbols that are independent complex random variables with zero mean. As shown in Figure 21.7, we allow different users to adopt different rate backoff factors to accommodate the heterogeneous QoS requirement of the applications of Internet of Things.

To implement the rate backoff, the ith sequence is first up-sampled by a factor of D_i at the BS, and the ith up-sampled sequence can be expressed as

$$X_i^{[D_i]}[k] = \begin{cases} X_i[k/D_i], & \text{if } k \; mod \; D_i = 0, \\ 0, & \text{if } k \; mod \; D_i \neq 0. \end{cases} \quad (21.12)$$

Then the up-sampled sequences are used to modulate the signature waveforms $\{\tilde{g}_1, \tilde{g}_2, \ldots, \tilde{g}_N\}$, by calculating the convolution of the ith up-sampled sequence $\{X_i^{[D_i]}[k]\}$ and the TR waveform $g_i(t)$ as shown in Figure 21.7.

After that, all the signals are combined together, and the combined signal $S(t)$ to be transmitted is given by

$$\begin{aligned} S(t) &= \sum_{k \in \mathbb{Z}^+} \sum_{j=1}^{N} \beta_j X_j^{[D_j]}[k] \tilde{g}_j(t - kT_P) \\ &= \sum_{k \in \mathbb{Z}^+} \sum_{j=1}^{N} \beta_j X_j[k] \tilde{g}_j(t - kD_j T_P). \end{aligned} \quad (21.13)$$

In essence, by convolving the information symbol sequences with TR waveforms, the TR structure provides a mechanism of embedding the unique location-specific signature associated with each communication link into the transmitted signal for the intended user.

The signal received at user i is represented as follows

$$\begin{aligned} Y_i(t) &= \left(S * \tilde{h}_i\right)(t) + \tilde{n}_i(t) \\ &= \sum_{k\in\mathbb{Z}^+}\sum_{j=1}^{N} \beta_j X_j[k]\left(\tilde{h}_i * \tilde{g}_j\right)(t - kD_jT_P) + \tilde{n}_i(t), \end{aligned} \tag{21.14}$$

which is the convolution of the transmitted signal $S(t)$ and the channel response $\tilde{h}_i(t)$, plus an additive white Gaussian noise sequence $\tilde{n}_i(t)$ with zero mean and variance σ_N^2.

Thanks to the temporal focusing effect, the ith receiver (user i) simply samples the received signal every D_iT_P s at $t = kD_iT_P$, ending up with $Y_i[k]$ given as follows

$$\begin{aligned} Y_i[k] = {} & \beta_i X_i[k]\left(\tilde{h}_i * \tilde{g}_i\right)(0) && \text{Signal} \\ & + \beta_i \sum_{\substack{l=-\lfloor\frac{T+T_P}{D_iT_P}\rfloor \\ l\neq 0}}^{\lfloor\frac{T+T_P}{D_iT_P}\rfloor} X_i[k+l]\left(\tilde{h}_i * \tilde{g}_i\right)(lD_iT_P) && \text{ISI} \\ & + \sum_{\substack{j=1 \\ j\neq i}}^{N} \beta_j \sum_{l=-\lfloor\frac{T+T_P}{D_jT_P}\rfloor}^{\lfloor\frac{T+T_P}{D_jT_P}\rfloor} X_j[k+l]\left(\tilde{h}_i * \tilde{g}_j\right)(lD_jT_P) && \text{IUI} \\ & + n_i[k], \end{aligned} \tag{21.15}$$

where $n_i[k] = \tilde{n}_i(kD_iT_P)$, and

$$(\tilde{h}_i * \tilde{g}_j)(lD_jT_P) = \begin{cases} \dfrac{\int_{lD_jT_P}^{T+T_P} \tilde{h}_i(\tau)\tilde{h}_j(\tau - lD_jT_P)d\tau}{\sqrt{\int_0^{T+T_P} \left|\tilde{h}_j(\tau)\right|^2 d\tau}}, & \text{if } 0 \leqslant l \leqslant \lfloor\frac{T+T_P}{D_jT_P}\rfloor, \\ \dfrac{\int_0^{T+T_P+lD_jT_P} \tilde{h}_i(\tau)\tilde{h}_j(\tau - lD_jT_P)d\tau}{\sqrt{\int_0^{T+T_P} \left|\tilde{h}_j(\tau)\right|^2 d\tau}}, & \text{if } \lfloor\frac{T+T_P}{D_jT_P}\rfloor \leqslant l < 0. \end{cases} \tag{21.16}$$

Thanks to the spatial focusing effect, in (21.16) when $i \neq j$, the power of $(\tilde{h}_i * \tilde{g}_j)(lD_jT_P)$ is typically very small compared to the power of the $(\tilde{h}_i * \tilde{g}_i)(0)$, which suppresses the interuser interference (IUI) for the TRDMA downlink.

Consequently, based on (21.15), the resulting SINR for user i in the TRDMA downlink is given by

$$SINR_{DL}^{(i)} = \frac{P_{sig}^{DL}(i)}{P_{ISI}^{DL}(i) + P_{IUI}^{DL}(i) + \sigma_N^2}, \tag{21.17}$$

where

$$P_{Sig}^{DL}(i) = \beta_i^2 \int_{0}^{T+T_P} \left|\tilde{h}_i(\tau)\right|^2 d\tau, \tag{21.18}$$

$$P_{ISI}^{DL}(i) = \beta_i^2 \sum_{\substack{l=-\lfloor \frac{T+T_P}{D_i T_P} \rfloor \\ l \neq 0}}^{\lfloor \frac{T+T_P}{D_i T_P} \rfloor} \left|\left(\tilde{h}_i * \tilde{g}_i\right)(l D_i T_P)\right|^2, \tag{21.19}$$

and

$$P_{IUI}^{DL}(i) = \sum_{\substack{j=1 \\ j \neq i}}^{N} \beta_j^2 \sum_{l=-\lfloor \frac{T+T_P}{D_j T_P} \rfloor}^{\lfloor \frac{T+T_P}{D_j T_P} \rfloor} \left|\left(\tilde{h}_i * \tilde{g}_j\right)(l D_j T_P)\right|^2. \tag{21.20}$$

21.3.3 Data Transmission Phase – Uplink

In this part, we describe the TRDMA uplink scheme, which facilitated, together with the downlink scheme, the asymmetric TRDMA architecture for IoT. Given the asymmetric complexity distribution between BS and terminal users in the downlink, the design philosophy of such an uplink is to keep the complexity of terminal users at minimal level.

In the TRDMA uplink, N users simultaneously transmit independent messages $\{X_1[k], X_2[k], \ldots, X_N[k]\}$ to the BS through the multipath channels. Similar to the downlink scheme, the rate backoff factor D is introduced to match the symbol rate with the system's chip rate. For any user U_i, $i \in \{1, 2, \ldots, N\}$, the rate matching process is performed by up-sampling the symbol sequence $\{X_i[k]\}$ by a factor D_i, as shown in Figure 21.8. The up-sampled sequence of modulated symbols for user i can be expressed as

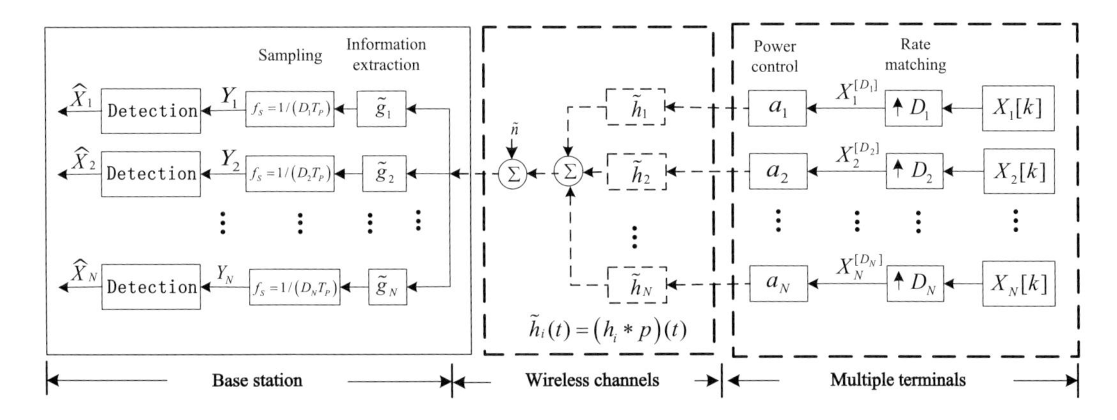

Figure 21.8 The basic diagram of the TRDMA uplink.

$$X_i^{[D_i]}[k] = \begin{cases} X_i[k/D_i], & \text{if } k \bmod D_i = 0, \\ 0, & \text{if } k \bmod D_i \neq 0. \end{cases} \tag{21.21}$$

The scaling factors a_i, for $i \in \{1, 2, \ldots, N\}$ in Figure 21.8, are used to implement the transmit power control, whose values are assumed to be instructed by the BS through the feedback/control channel. After multiplying with scaling factor, the sequence of $a_i X_i^{[D_i]}[k]$ for all $i \in \{1, 2, \ldots, N\}$ is transmitted through the corresponding multipath channel $\tilde{h}_i(t)$.

When the sequence $\{a_i X_i^{[D_i]}[k]\}$ propagates through its wireless channel $\tilde{h}_i(t)$, the convolution between $\{a_i X_i^{[D_i]}[k]\}$ and the effective channel response $\{h_i[k]\}$ is automatically taken as the channel output for user i. Then, all of the channel outputs for the N users are mixed together in the air plus the additive white Gaussian noise (AWGN) $\tilde{n}[k]$ at the BS with zero mean and variance σ_N^2, as illustrated in Figure 21.8. Consequently, the mixed signal received at the BS can be written as

$$S(t) = \sum_{k \in \mathbb{Z}^+} \sum_{i=1}^{N} a_i X_i[k] \tilde{h}_i(t - kD_i T_P) + \tilde{n}(t). \tag{21.22}$$

Upon receiving the mixed signal as shown in (21.22), the BS passes this mixed signal through a bank of N filters, each of which performs the convolution between its input signal $S(t)$ and the user's signature waveform $\tilde{g}_i(t)$ that has been calculated for the downlink. Such a convolution using the signature waveform extracts the useful signal component and suppresses the signals of other users. As the output of the ith filter, i.e., the convolution of $S(t)$ and the signature of user i ,$\tilde{g}_i(t)$, can be represented as

$$Y_i(t) = \sum_{k \in \mathbb{Z}^+} \sum_{j=1}^{N} a_j X_j[k] \left(\tilde{g}_i * \tilde{h}_j\right)(t - kD_j T_P) + (\tilde{g}_i * \tilde{n})(t), \tag{21.23}$$

in which the highest gain for user i's symbol $X_i[k]$ is achieved at the temporal focusing time $t = kD_i T_P$.

Sampling $Y_i(t)$ every $D_i T_P$ s at $t = kD_i T_P$, we have

$$\begin{aligned} Y_i[k] = {} & a_i X_i[k] \left(\tilde{g}_i * \tilde{h}_i\right)(0) && \text{Signal} \\ & + a_i \sum_{\substack{l=-\lfloor \frac{T+T_P}{D_i T_P} \rfloor \\ l \neq 0}}^{\lfloor \frac{T+T_P}{D_i T_P} \rfloor} X_i[k+l] \left(\tilde{h}_i * \tilde{g}_i\right)(l D_i T_P) && \text{ISI} \\ & + \sum_{\substack{j=1 \\ j \neq i}}^{N} a_j \sum_{l=-\lfloor \frac{T+T_P}{D_j T_P} \rfloor}^{\lfloor \frac{T+T_P}{D_j T_P} \rfloor} X_j[k+l] \left(\tilde{h}_j * \tilde{g}_i\right)(l D_j T_P) && \text{IUI} \\ & + n_i[k], \end{aligned} \tag{21.24}$$

where $n_i[k] = (\tilde{g}_i * \tilde{n})(kD_iT_P)$ is a sample of the colored noise after the $\tilde{g}_i(t)$ filtering, which is still a Gaussian random variable with zero mean and the same variance σ_N^2, because $\tilde{g}_i$ is a normalized waveform as shown in (21.11).

Examining (21.15) and (21.24), the same mathematical structure can be found by switching the roles of the signature waveforms $\tilde{g}_i$'s and the channel responses $\tilde{h}_i$'s in the convolution (and ignoring the scaling factor a_i and noise term.) Therefore, mathematically,[7] a virtual spatial focusing effect as observed in the downlink can be seen in the user's signature domain of the presented uplink scheme. Such a virtual spatial focusing effect enables the BS to use the user's signature waveform to extract the useful component out of the combined received signals, allowing multiple users to access the BS simultaneously.

Consequently, based on (21.24), the resulting SINR for user i in the TRDMA uplink is given by

$$SINR_{UL}^{(i)} = \frac{P_{sig}^{UL}(i)}{P_{ISI}^{UL}(i) + P_{IUI}^{UL}(i) + \sigma_N^2}, \tag{21.25}$$

where

$$P_{Sig}^{UL}(i) = a_i^2 \int_0^{T+T_P} \left|\tilde{h}_i(\tau)\right|^2 d\tau, \tag{21.26}$$

$$P_{ISI}^{UL}(i) = a_i^2 \sum_{\substack{l=-\lfloor \frac{T+T_P}{D_iT_P} \rfloor \\ l \neq 0}}^{\lfloor \frac{T+T_P}{D_iT_P} \rfloor} \left|\left(\tilde{h}_i * \tilde{g}_i\right)(lD_iT_P)\right|^2, \tag{21.27}$$

and

$$P_{IUI}^{UL}(i) = \sum_{\substack{j=1 \\ j \neq i}}^{N} a_j^2 \sum_{l=-\lfloor \frac{T+T_P}{D_jT_P} \rfloor}^{\lfloor \frac{T+T_P}{D_jT_P} \rfloor} \left|\left(\tilde{h}_j * \tilde{g}_i\right)(lD_jT_P)\right|^2. \tag{21.28}$$

21.3.4 Performance of TRDMA

In this section, we compare the performance of the presented TRDMA system with that of the UWB impulse radio system in terms of different metrics, where we assume that the UWB impulse radio system uses the ideal Rake receiver that collects all the taps of channel information. We first compare the average achievable data rate of each user when the power consumption is the same for two systems. As shown in Figure 21.9, the TRDMA system is able to provide higher achievable data rate for each user than the UWB impulse radio system.

[7] Unlike the *physical* spatial focusing effect observed in the downlink in which the useful signal power is concentrated at different physical locations, in the uplink, the signal power concentration in the users' signature waveform space is achieved mathematically at the BS.

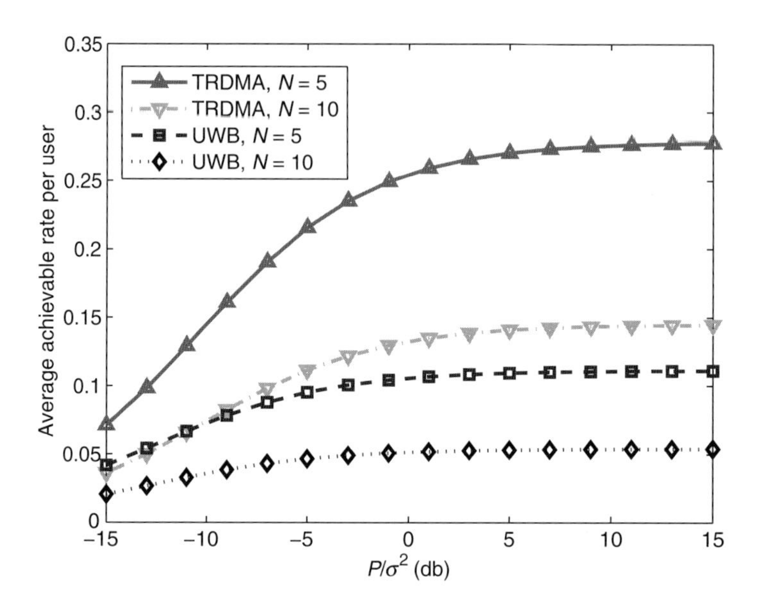

Figure 21.9 The performance comparison between TRDMA and UWB in terms of average achievable data rate per user.

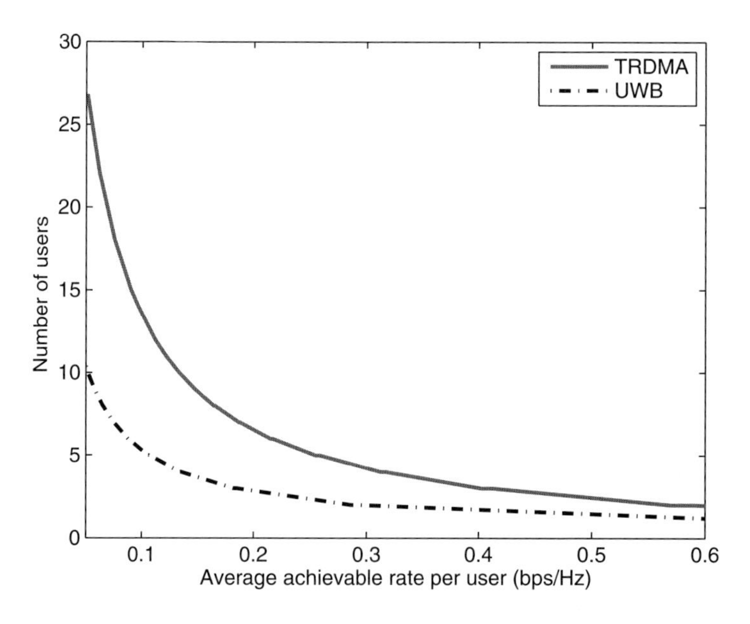

Figure 21.10 The performance comparison between TRDMA and UWB in terms of number of supported users.

We then evaluate the number of users each system can support. Because TRDMA mitigates the interference among users, it is expected to be able to support more users. In Figure 21.10, we show the number of supported users versus the average achievable rate of each user. We can see that, as we have anticipated, the TRDMA system is able

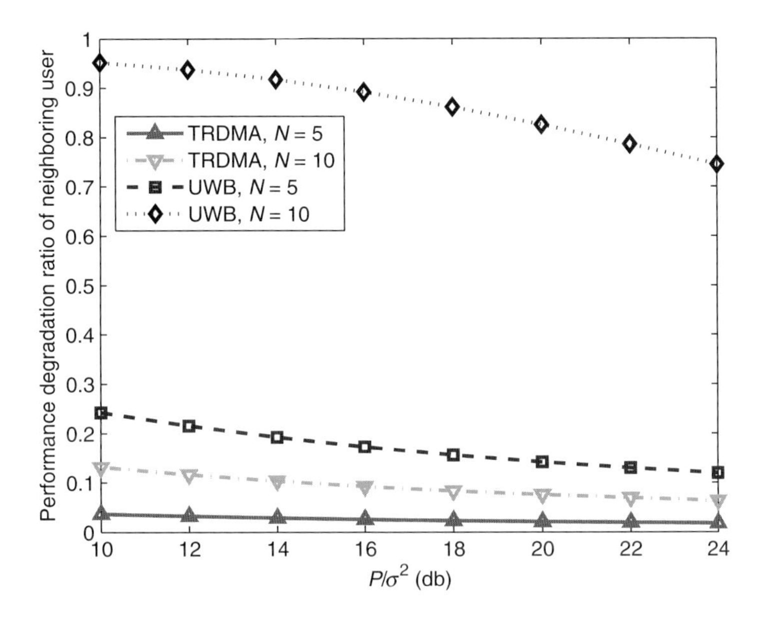

Figure 21.11 Impact to other users outside the system.

to support more users than the UWB impulse radio system. For example, if the required data rate of each user is 0.1 bps/Hz, which is equivalent to 10 Mbps if the bandwidth is 100 MHz, then the TRDMA system can support about 20 users while the UWB impulse radio system can support only 5 users.

On the other hand, if the achievable data rate of each user is fixed, the TRDMA system has less impact on the neighboring users, i.e., causing less interference to users outside the system. As shown in Figure 21.11, when we fix the achievable rate of each user as 0.1 bps/Hz, the performance degradation due to the TRDMA system is much less than that of UWB impulse radio system. Therefore, the TRDMA system has the potential to admit more users and thus is a much better solution to the IoT.

Finally, we show the achievable rate region of two-user case in Figure 21.12, where we further compare the presented TRDMA system with ideal Rake-receiver schemes with orthogonal bases and superposition codes [25]. We can see that the TRDMA scheme outperforms all the Rake-receiver based schemes, and the frontier achieved by TRDMA scheme is close to the Genie-aided outer-bound where all the interference is assumed to be known and thus can be completely removed. These results demonstrate TRDMA's unique advantage of spatial focusing brought by the preprocessing of embedding location-specific signatures before sending signals into the air. The high-resolution spatial focusing, as the key mechanism of the TRDMA, alleviates interference among users and provides a promising multiuser wireless communication solution for IoT.

21.3.5 Scalability

In the previous sections, we have shown that a single TRDMA BS has the potential to serve a lot of users while maintaining little interference to other wireless users. However,

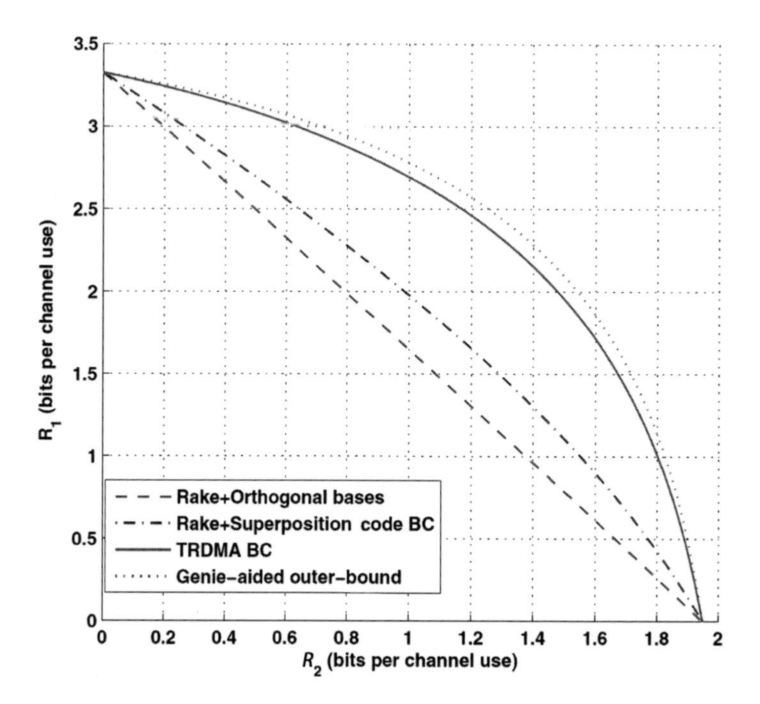

Figure 21.12 Achievable capacity region for two-user case [25].

in IoT applications, the density of users may be so high that one single BS is insufficient to support all of them. One possible solution is to add more BSs, and we will show that the TRDMA system is highly scalable, and extra BSs can be easily installed whenever necessary.

Different from other wireless communication systems where extra mechanism is needed to prevent or alleviate the interference introduced by adding more BSs, the TRDMA system does not need extra effort on suppressing the interference introduced by more BSs due to the spatial focusing effect. As an example shown in Figure 21.13, if six more BSs are added surrounding the original one, all of them could use the full spectrum as the original one in the TRDMA system, while in other systems the spectrum needs to be re-allocated so that no adjacent BSs share the same band. This ease of scalability also increases the spectrum efficiency by fully reusing spectrum among BSs.

Figure 21.14 shows the aggregate achievable data rate versus the number of users at different number of BSs. We can see that given a specific number of BSs, the aggregate achievable rate increases as the number of users increases, but saturates when the number of users is large. Nevertheless, such saturation can be resolved by increasing the number of BSs, which means that adding more BSs can bring significant gain. This is partially because although different BSs share the same spectrum, they are nearly orthogonal with each other. Such orthogonality is not in the traditional fashion such as time, code, or frequency divisions that are achieved by extra effort, but in a natural spatial division that is only utilized by TRDMA system.

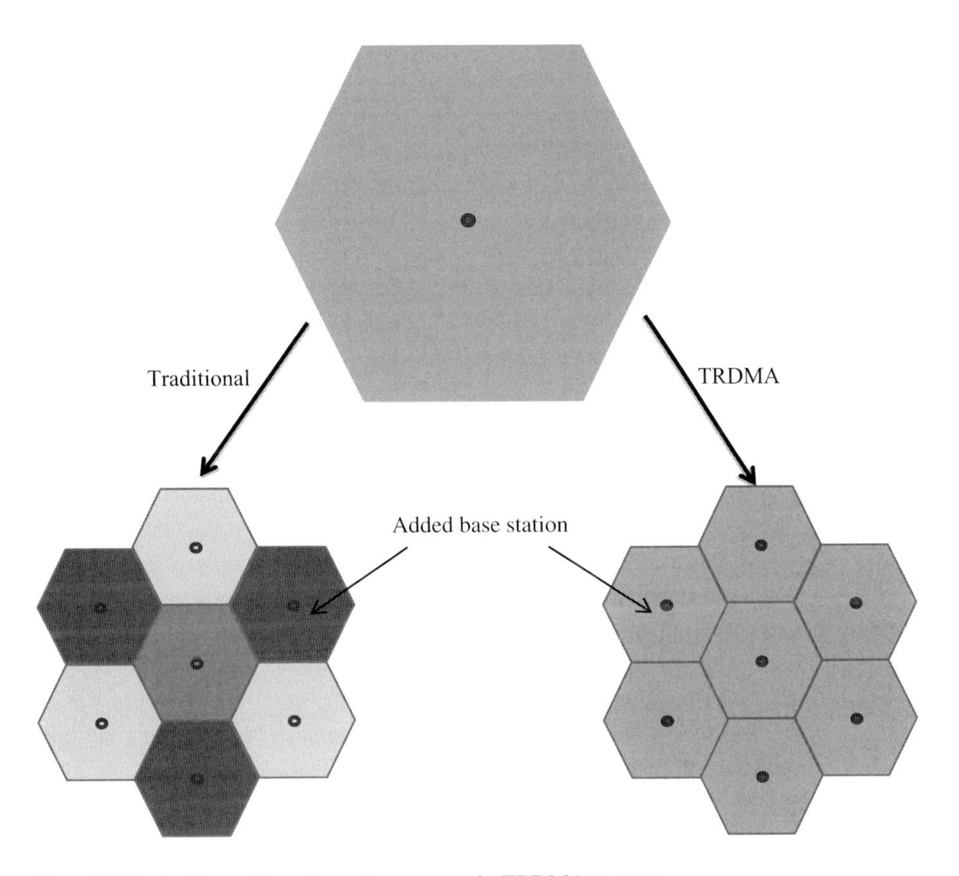

Figure 21.13 An illustration of spectrum re-use in TRDMA system.

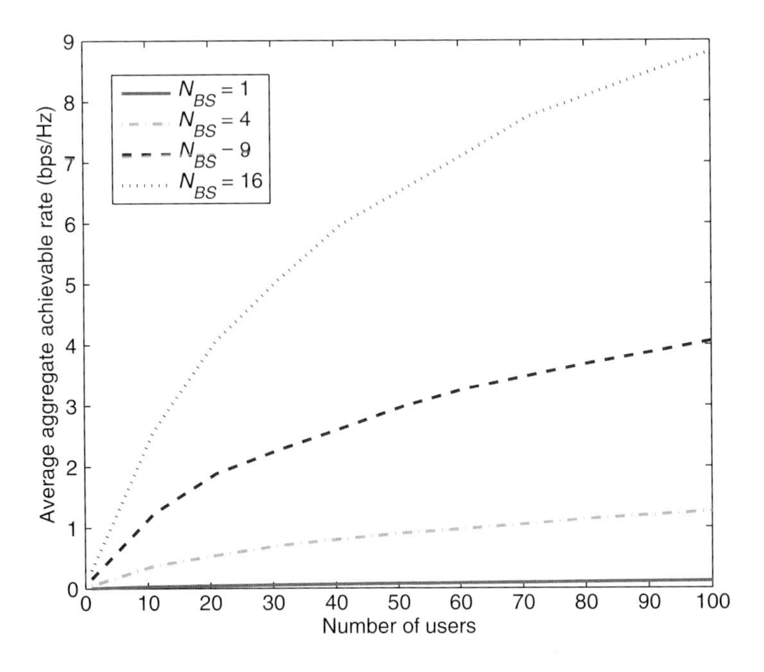

Figure 21.14 Scalability performance of the TRDMA system.

21.3.6 Physical-Layer Security

Based on the unique location-specific multipath profile, the TRDMA system can be exploited to enhance system security. In a rich scattering wireless environment, multiple paths are formed by numerous surrounding reflectors. For terminal devices at different locations, the received waveforms undergo different reflecting paths and delays, and hence the multipath profile can be viewed as a unique location-specific signature. As this information is only available to the BS and the intended terminal device, it is very difficult for other unauthorized users to infer or forge such a signature. It has been shown in [64] that even when the eavesdroppers are close to the target terminal device, the received signal strength is much lower at the eavesdroppers than at the target terminal device in an indoor application because the received signals are added incoherently at the eavesdroppers.

Our TRDMA system is somehow like the direct sequence spread spectrum (DSSS)-based secret communications. In DSSS communications, the energy of an original data stream is spread to a much wide spectrum band by using a pseudorandom sequence, and the signal is hidden below the noise floor. It is only those who know the pseudo-random sequence that could recover the original sequence from the noise-like signals. However, if the pseudo-random sequence has been leaked to a malicious user, that user is also capable of decoding the secret message. Nevertheless, for the presented TRDMA system, this would no longer be a problem because the underlying spreading sequence is not a fixed choice but instead a location-specific signature. For the intended terminal device, the multipath channel automatically serves as a decipher that recovers the original data sent by the BS; and for all other ineligible users at different locations, the signal that propagates to them would be noise-like and probably is hidden below the noise floor. Therefore, malicious users are unable to recover the secret message because the security is inherent in the physical layer.

21.3.7 Discussions and Remarks

From the analysis and discussions in the previous sections, we can see that the asymmetric TRDMA system is an ideal wireless solution to the IoT because it can handle the challenges of IoT including providing better battery life, supporting multiple active things, dealing with low-cost terminal devices, accommodating heterogeneous terminal devices, being highly scalable, and providing extra physical-layer security as summarizing below

- In both downlink and uplink, the BS consumes most of the complexity, while keeping the complexity of terminal users at a minimal level. This is a very desirable feature for the solution to IoT because it can provide much better battery life and reduce the cost of the terminal devices and thus the entire system as a whole.
- Both downlink and uplink can support simultaneous transmissions of multiple users because the TRDMA system in essence forms a virtual massive MISO technology that leverages the large number of multipaths in the rich-scattering

environment. The downlink has a physical spatial focusing effect, whereas the uplink has a virtual spatial focusing effect due to the mathematical duality between the TRDMA uplink and downlink.

- Different users can adopt different rate backoff factors to achieve heterogeneous QoS requirements, i.e., the TRDMA system can accommodate heterogeneous terminal devices for IoT.
- More BSs can be easily added in the TRDMA system without extra mechanism for preventing or alleviating the interference introduced, i.e., the TRDMA system is highly scalable.
- Based on the unique location-specific multipath profile, the TRDMA system can provide extra system security in the physical layer.

21.4 Other Challenging Issues and Future Directions

21.4.1 Advanced Waveform Design

In our discussion of TRDMA in the previous section, the TR CIR serves as the transmit signature waveform to modulate symbols. The received signal is the transmitted waveform convolving with the multipath channel with additive noise. Such a time-reversed waveform is essentially the matched-filter [65], which guarantees the optimal BER performance by virtue of its maximum signal-to-noise ratio (SNR). However, in a high-data-rate scenario such as video streaming, when the symbol duration is smaller than the channel delay spread, the transmit waveforms are overlapped and thus interfere with each other. When the symbol rate is very high, such ISI can be notably severe and causes crucial performance degradation, i.e., the BER performance can be very poor with a basic time-reversed waveform. Further, in a multiuser downlink scenario, the TR BS uses each user's particular CIR as its specific waveform to modulate the symbols intended for that user. Despite the inherent randomness of the channel impulse responses, as long as they are not orthogonal to each other, which is almost always the case, these waveforms will inevitably interfere with each other when transmitted concurrently. Hence, the performance of TRDMA can be impaired and even limited by the interuser interference (IUI).

Based on given design criteria such as system performance, QoS constraints, or fairness among users, the waveform design can be formulated as an optimization problem with the transmitted waveforms as the optimization valuables. The basic idea of waveform design is to carefully adjust the amplitude and phase of each tap of the waveform based on the channel information, such that after convolving with the channel, the received signal at the receiver retains most of the intended signal strength and rejects or suppresses the interference as much as possible.

To rewrite (21.15) in a vector from, we define the following notations. The multipath channel between the base-station and the jth user is denoted by a vector $\mathbf{h}_j$, a column vector of L elements where $L = \left\lfloor \frac{T+T_P}{T_P} \right\rfloor$ and $[\mathbf{h}_j]_k = \tilde{h}_j(t)|_{t=kT_P}$. Let X_j denote an information symbol for user j, and $\mathbf{g}_j$ be the transmit waveform for user j, where

$[\mathbf{g}_j]_k = \tilde{g}_j(t)|_{t=kT_P}$ in (21.15). The length of $\mathbf{g}_k$ is also L. The received signal vector $\mathbf{y}_i$ at user i, where $[\mathbf{y}_i]_k = Y_i[k]$ in (21.15), is given by

$$\mathbf{y}_i = \mathbf{H}_i \sum_{j=1}^{N} \mathbf{g}_j X_j + \mathbf{n}_i, \tag{21.29}$$

where $\mathbf{H}_i$ is the Toeplitz matrix of size $(2L-1) \times L$ with the first column being $[\mathbf{h}_i^T \; \mathbf{0}_{1\times(L-1)}]^T$, and $\mathbf{n}_i$ denotes the additive white Gaussian noise (AWGN) with $[\mathbf{n}_i]_k = n_i[k]$. User i estimates the symbol X_i by the sample $[y_k]_L$. Note that (21.29) represents the received signal when the rate backoff factor $D > L$. When $D < L$, the received waveforms of different symbols overlap with each other and give rise to the ISI. To characterize the effect of ISI, the decimated channel matrix of size $(2L_D - 1) \times L$, where $L_D = \lfloor \frac{L-1}{D} \rfloor + 1$, is defined as

$$\tilde{\mathbf{H}}_i = \sum_{l=-L_D+1}^{L_D-1} \mathbf{e}_{L_D+l} \mathbf{e}_{L+lD}^T \mathbf{H}_i, \tag{21.30}$$

where $\mathbf{e}_l$ is the lth column of a $(2L-1) \times (2L-1)$ identity matrix. In other words, $\tilde{\mathbf{H}}_i$ is obtained by decimating the rows of $\mathbf{H}_i$ by D, i.e., centering at the Lth row, every Dth row of $\mathbf{H}_i$ is kept in $\tilde{\mathbf{H}}_i$, while the other rows are discarded. The center row index of $\tilde{\mathbf{H}}_i$ is L_D. Then the sample for symbol estimation can be written as

$$\begin{aligned}[\mathbf{y}_i]_L &= \mathbf{h}_{iL}^H \mathbf{g}_i X_i[L_D] + \mathbf{h}_{iL}^H \sum_{j\neq i} \mathbf{g}_j X_j[L_D] \\ &+ \sum_{l=1, l\neq L_D}^{2L_D-1} \mathbf{h}_{il}^H \sum_{j=1}^{N} \mathbf{g}_j X_j[l] + n_i[L],\end{aligned} \tag{21.31}$$

where the $\mathbf{h}_{il}^H = \mathbf{e}_l^T \tilde{\mathbf{H}}_i$ denotes the lth row of $\tilde{\mathbf{H}}_i$, and $X_j[l]$ denotes user j's lth symbol. It can be seen from (21.31) that the symbol $X_i[L_D]$, the L_Dth symbol of user i, is interfered by the previous $L_D - 1$ symbols and the later $L_D - 1$ symbols as well as other users' $K(2L_D - 1)$ symbols, and also corrupted by the noise. The design of waveforms $\{\mathbf{g}_i\}$ has critical influence to the symbol estimation and thus the system performance.

It can be observed that the mathematical structure of waveform design is similar to the beamforming problem, which is also known as the multi-antenna precoder design [66–70]. Therefore, beamforming approaches such as singular value decomposition (SVD), zero forcing (ZF), and minimal mean square error (MMSE) can be analogously employed in waveform design. In the literature, there have been many studies investigating the problems of designing advanced waveforms to suppress the interference [27, 46, 71–76]. If the basic TR waveforms are adopted, i.e., $\mathbf{g}_i = \mathbf{h}_{iL}$, then the intended signal power for each user is maximized but without considering the interference caused by other symbols. As such, the performance is limited by the interference when the transmit power is high. Another possible waveform design is ZF [77], which minimizes all the interference signal power but without taking into account the intended signal power. Thus, the resulting SNR can be very low and causes severe performance degradation,

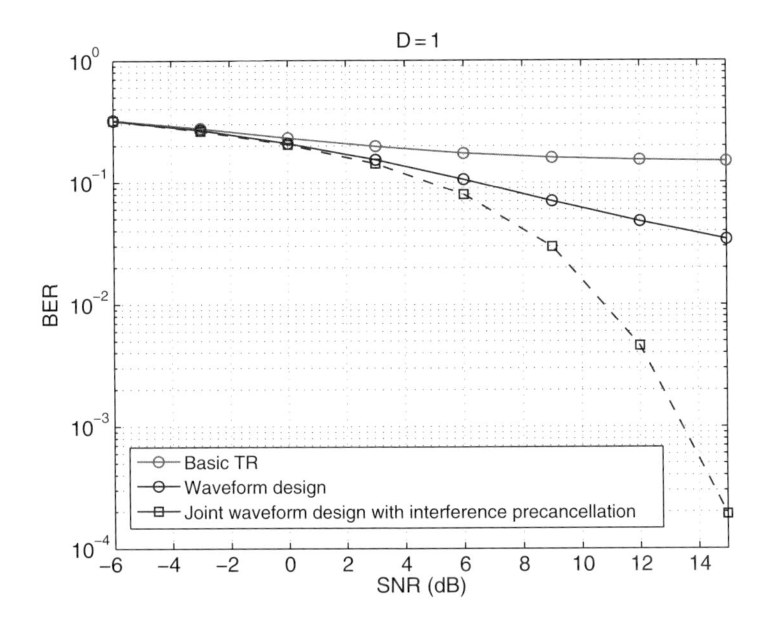

Figure 21.15 BER performance comparison using basic TR waveform, waveform design, and joint waveform design and interference precancellation.

especially when the transmit power is relatively low. In [27], it has been shown that well-designed waveforms can strike a balance between enhancing the intended signal power and suppressing the interference power.

Besides the channel information, another important side information the transmitter can exploit in waveform design is the transmitted symbol information. The waveform of one symbol, when arriving at the receiver, induces ISI to the previous symbols as well as the following symbols. Given what has been transmitted, the causal part of ISI can be cancelled in advance in designing the waveform of the current symbol. Such a design philosophy is analogous to the transmitter-based interference presubtraction [78–80] in the nonlinear precoding literature. A notable distinction for TR systems is that only the causal part of ISI can be cancelled while the anticausal part of ISI cannot be cancelled and needs to be suppressed by the waveform design based on channel information [49].

Figure 21.15 shows the BER performance for a single-user TR system when $D = 1$ using different waveforms, including basic TR waveform, the waveform design in [27], and the joint waveform design and interference pre-cancellation in [49]. It can be seen that when $D = 1$, the ISI is so severe that the BER curve of the basic TR waveform starts to saturate at even middle SNR, which is unacceptable. The waveform design in [27] is able to suppress the interference and make the BER keep decreasing when SNR increases. The joint waveform design and interference precancellation in [49] can further improve the performance significantly because it makes use of more information, i.e., the transmitted symbols, to cancel the ISI in advance. It is evident that the dramatic performance improvement brought from the waveform design demonstrates its inevitable necessity in TR systems.

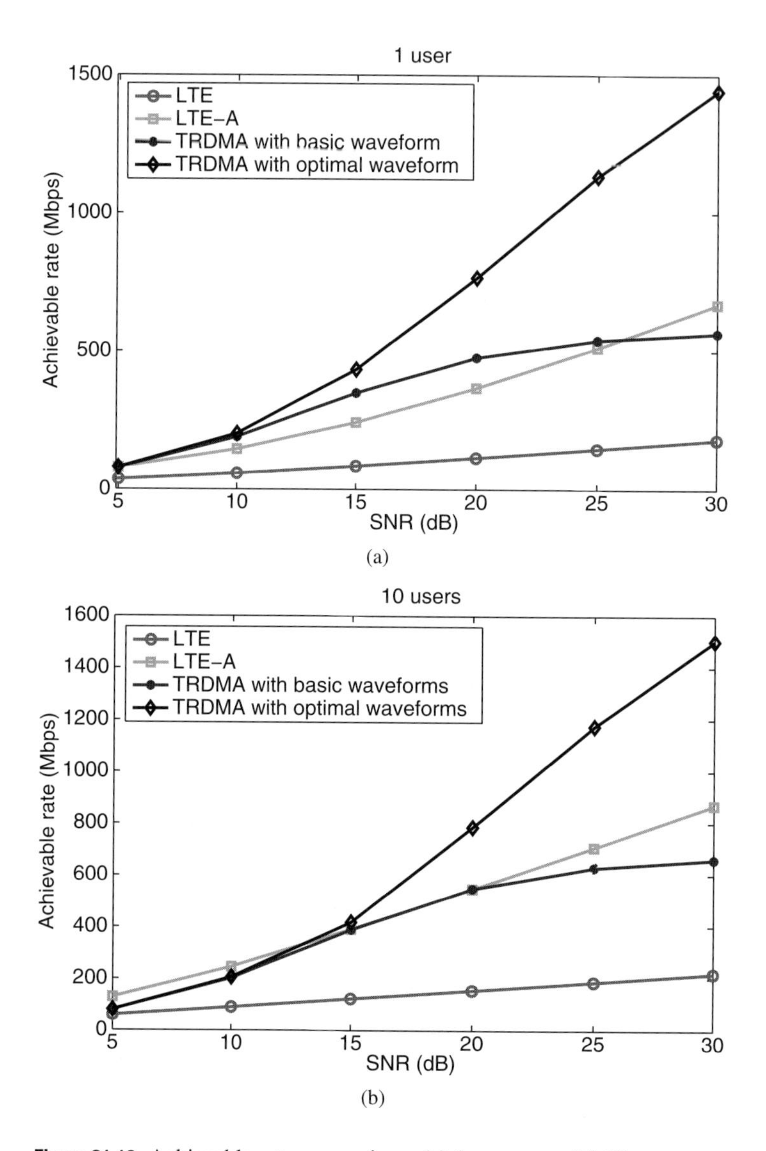

Figure 21.16 Achievable rate comparison: (a) 1 user case; (b) 10 users case.

Figure 21.16 shows the performance comparison in terms of achievable rate of the TRDMA system with 500 MHz bandwidth with two OFDM systems: one is LTE system with 20 MHz bandwidth and the other is LTE-A system with 100 MHz bandwidth. We can see that for one user case, even with basic TR waveform, the TRDMA scheme can achieve much better performance than LTE in all SNR region and better performance than LTE-A in most SNR region. With optimal waveform, the performance of TRDMA can be further improved. When there are ten users, due to the selectivity among different users, the achievable rate of LTE and LTE-A can be enhanced, due to which LTE-A can achieve comparable and even slightly better performance than TRDMA with basic TR waveform. Nevertheless, with optimal waveform, TRDMA can still outperform LTE and LTE-A in most SNR regions, which demonstrates that TRDMA can achieve higher throughput than OFDM systems when the bandwidth is wide enough, e.g., five times as in the simulations.

21.4.2 MAC Layer Issue

The medium access control (MAC) layer provides addressing and channel access control mechanisms that make it possible for several terminals or network nodes to communicate within a multiple access network that incorporates a shared medium [81]. In the MAC layer design, coordination is the most basic and important function, which manages multiple users to access the network with the objective of both efficiency and fairness. Most existing prevailing systems, such as IEEE 802.11 Wi-Fi and IEEE 802.15.4 ZigBee, are based on the contention scheme. For example, in Wi-Fi systems, distributed coordination function (DCF) is adopted with carrier sensing multiple access (CSMA) and collision avoidance (CA) [82]. When a Wi-Fi user has packet to transmit, it first senses the channel, i.e., ŞListen-Before-TalkŤ. After detecting the channel as idle, the Wi-Fi user has to keep sensing the channel for an additional random time [83], i.e., random backoff and only when the channel remains idle for this additional random time period, the station is allowed to initiate its transmission. If there is a collision, the user needs to backoff and repeat this procedure again. Under such a scheduling, there is only one Wi-Fi user talking with the access point (AP) at one time. However, when the number of users is large, no one can access the network due to the contention failure and extremely long backoff. We have all seen and experienced such a phenomenon in highly dense-population area including airports and conference halls. A typical example is that Steve Jobs failed to demo the Wi-Fi function of the new released iPhone due to the overwhelming connections in the conference room [84]. Therefore, such a contention-based coordination function of the MAC layer is a bottleneck for accommodating a large number of users, which generally exists in IoT.

The most prominent characteristic of the TR system is that it does not require such coordination function, where users are naturally separated by their locations. There are two phases in the TR systems: channel probing phase and data transmission phase. In the channel probing phase, all the users can transmit their unique pilots (e.g., pseudo noise sequences) to the BS for channel estimation. In the data transmission phase, BS can communicate with all the users simultaneously through location-specific signatures. Therefore, there is no need for the BS in TR systems to perform the coordination function, which simplifies the MAC layer design to a large extent. In addition to coordination, some additional functionalities required by the MAC are also needed in TR systems, including accepting MAC service data units (MSDUs) from higher layers and adding headers and trailers to create MAC protocol data unit (MPDU) for physical layer, fragmenting one frame into several frames to increase the delivery probability, and encrypting MAC layer packet to ensure security and privacy [83]. Nevertheless, we would like to emphasize again that the location-specific signature in TR systems can provide additional physical-layer security.

21.4.3 Low-Cost High-Speed ADC and DAC

The inherent nature of time-reversal communications is to fully harvest energy from the surrounding environment by exploiting the multipath propagation to recollect all the signal energy that could be collected. To have superior advantage, the time-reversal communications need to operate in a rich multipath environment, which generally requires wide bandwidth. As a consequence, the sampling rate is typically high. Moreover, to avoid missing the peak during the sampling and simplify the synchronization process, a two to four times oversampling is generally required, which makes the sampling even more challenging. Therefore, one key implementation issue in time-reversal communications is the high sampling rate of the analog-to-digital converter (ADC). Fortunately, due to the advance of semiconductor technologies and the continuous drive from the emerging wideband communication applications, the performance of ADC has been improved a lot during the past decade in terms of both sampling rate and resolution. For example, there are 17 different commercial off-the-shelf ADCs from Texas Instruments with a sampling rate at least 1 GHz and resolution at least eight bits [85]. However, the price of such ADCs is typically high, e.g., an ADC with two-channel 2 GHz sampling rate and eight bits resolution costs US$329, which may be a barrier to the application of time-reversal technique on IoT. In such a case, there is a need to find cheaper solutions to high sampling rate ADCs.

There are two possible ways to reduce the cost of high sampling rate ADCs. The first way is to implement the ADC on chip. In such a case, the cost of ADC comes from the silicon cost, which depends on silicon wafer cost and the size of the ADC on silicon. In general, with silicon implementation, the cost of ADC can be reduced from several hundred dollars to several cents without considering the capital cost. Nevertheless, because the capital cost is typically very high, such a kind of implementation is only suitable for large volume productions. The other way to reduce the cost is using a set of low sampling rate cheap ADCs to achieve the high sampling rate. Because the price of commercial ADCs grows exponentially as the increase of sampling rate, by replacing the high sampling rate ADCs with a set of low sampling rate ADCs, the cost can be reduced dramatically. One straightforward way is to use time interleaving [86, 87]. In such an approach, the input signal is passed through a series of parallel interleaved low sampling rate ADCs where the interleaving is achieved through the time shifts. After the sampling, the samples are passed through the de-interleaver to generate the high sampling rate signal. However, the front-end of a commercial ADC has an inherent analog bandwidth limitation [88, 89], due to which the time interleaving approach is not practical. The second approach is to use parallel bandpass sampling approach [90, 91] where the input signal is passed through a series of filter banks before the ADCs and the reconstruction method depends on the corresponding filters in the filter bank. This kind of approach can resolve the analog bandwidth limitation but necessitates sophisticated digital algorithms for accurate frequency synchronization. Another approach is to use random demodulation [92, 93] where the input signal is passed through parallel channels. In each channel, the input signal is first multiplied by a periodic random waveform in the analog domain, then lowpass filtered, and finally sampled using low sampling rate ADC. The random demodulation approach overcomes the disadvantages of the time interleaving approach and the parallel bandpass sampling approach, but is limited by the technology for generating the periodic random waveforms. Note that in all these approaches, the costly high sampling rate ADC is traded with the computation complexity for reconstruction in digital domain, which is relatively cheap.

21.5 Summary

In this chapter, we provided an overview to show that the TR technique is an ideal paradigm for IoT. Because of the inherent nature to fully harvest energy from the surrounding environment by exploiting the multipath propagation to recollect all the signal energy, the TR system has a potential of over an order of magnitude of reduction in power consumption and interference alleviation, which means that TR system can provide better battery life and support multiple concurrent active users. The unique asymmetric architecture discussed in this chapter can significantly reduce the computational complexity and thus the cost of the terminal devices, the total number of which is typically very large for IoT. Moreover, through adjusting the waveform and rate backoff factor, various QoS options can be easily supported in TR systems. Finally, the unique location-specific signature in TR systems can provide additional physical-layer security and thus can enhance the privacy and security of customers in IoT. All these advantages, including providing better battery life, supporting multiple active things, dealing with low-cost terminal devices, accommodating heterogeneous terminal devices, being highly scalable, and providing extra physical-layer security, show that the TR technique is an ideal paradigm for IoT.

Recently, researchers have started to envision the next major phase of mobile telecommunications standards beyond current 4G standards, known as 5G. According to [94], key concepts of 5G include new modulation techniques such as nonorthogonal multiple access schemes, massive distributed MIMO, advanced interference management, and efficient support of machine-type devices to enable the IoT with potentially higher numbers of connected devices. Based on the discussion in this chapter, it turns out that TR can easily resolve these issues, which means that TR is potentially a promising 5G technology. For related references, interested readers can refer to [95].

References

[1] G. Kortuem, F. Kawsar, D. Fitton, and V. Sundramoorthy, "Smart objects as building blocks for the Internet of Things," in *IEEE Internet Computing*, vol. 14, no. 1, pp. 44–51, Feb. 2010.

[2] C. Institutes, "Smart networked objects and Internet of Things," in *Information Communication Technologies and Micro Nano Technologies Alliance, White Paper*, January 2011.

[3] L. Atzori, A. Iera, and G. Morabito, "The Internet of Things: A survey," in *Computer Networks*, vol. 54, no. 15, pp. 2787–2805, Oct. 2010.

[4] D. Le-Phuoc, A. Polleres, M. Hauswirth, G. Tummarello, and C. Morbidoni, "Rapid prototyping of semantic mash-ups through semantic web pipes," in *Proceedings of the 18th International Conference on World Wide Web*, pp. 581–590, 2009.

[5] A. Dohr, R. Modre-Opsrian, M. Drobics, D. Hayn, and G. Schreier, "The Internet of Things for ambient assisted living," in *International Conference on Information Technology: New Generations (ITNG)*, pp. 804–809, 2010.

[6] E. Commission, "Internet of Things in 2020 road map for the future," in *Working Group RFID of the ETP EPOSS, Technical Report*, May 2008.

[7] K. Ashton, "That 'Internet of Things' thing in the real world, things matter more than ideas," in *RFID Journal*, Jun. 2009.

[8] D. L. Brock, "The electronic product code (EPC) a naming scheme for physical objects," in *Auto-ID Center, White Paper*, Jan. 2001.

[9] I. T. Union, "ITU internet reports 2005: The Internet of Things," in *International Telecommunication Union, Workshop Report*, Nov. 2005.

[10] P. Guillemin and P. Friess, "Internet of Things strategic research roadmap," in *The Cluster of European Research Projects, Technical Report*, Sep. 2009.

[11] C. Gomez and J. Paradells, "Wireless home automation networks: A survey of architectures and technologies," in *IEEE Communications Magazine*, pp. 92–101, Jun. 2010.

[12] J. Berman, "Z-wave chip aims to cut implementation cost," in *EDN Network*, pp. 92–101, Apr. 2005.

[13] "IEEE 802.15.1-2002- IEEE standard for telecommunications and information exchange between systems – LAN/MAN – specific requirements – Part 15: Wireless medium access control (MAC) and physical layer (PHY) specifications for wireless personal area networks (WPANs)," in URL: http://ieeexplore.ieee.org/xpl/mostRecentIssue.jsp?punumber=7932.

[14] "IEEE 802.11: Wireless LAN medium access control (MAC) and physical layer (PHY) specification," in URL: http://standards.ieee.org/getieee802/download/802.11-2012.pdf.

[15] L. Li, X. Hu, K. Chen, and K. He, "The applications of WiFi-based wireless sensor network in Internet of Things and smart grid," in *IEEE Conference on Industrial Electronics and Applications (ICIEA)*, 2011.

[16] M. Ha, S. H. Kim, H. Kim, K. Kwon, N. Giang, and D. Kim, "Snail gateway: Dual-mode wireless access points for WiFi and IP-based wireless sensor networks in the Internet of Things," in *IEEE Consumer Communications and Networking Conference (CCNC)*, 2012.

[17] X. Xie, D. Deng, and X. Deng, "Design of embedded gateway software framework for heterogeneous networks interconnection," in *International Conference on Electronics and Optoelectronics (ICEOE)*, 2011.

[18] "Digi-key corporations website," in URL: www.digikey.com/product-detail/en/CC2531F256RHAT/296-25186-2-ND/2171344?WT.mc_id=PLA_2171344.

[19] "Insteon compared," in *White Paper*, 2013, URL: www.insteon.com/pdf/insteoncompared.pdf.

[20] C. Corporation, "Wi-Fi radio characteristics and the cost of WLAN implementation," in URL: www.connect802.com/download/techpubs/2005/commercial_radios_E0523-15.pdf, 2005.

[21] H.-C. Hsieh and C.-H. Lai, "Internet of Things architecture based on integrated PLC and 3G communication networks," in *IEEE International Conference on Parallel and Distributed Systems*, pp. 853–856, 2011.

[22] Z. Shi, K. Liao, S. Yin, and Q. Ou, "Design and implementation of the mobile Internet of Things based on TD-SCDMA network," in *IEEE International Conference on Information Theory and Information Security*, pp. 954–957, 2010.

[23] J.-M. Liang, J.-J. Chen, H.-H. Cheng, and Y.-C. Tseng, "An energy-efficient sleep scheduling with QoS consideration in 3GPP LTE-advanced networks for Internet of Things," *IEEE Journal on Emerging and Selected Topics in Circuits and Systems*, vol. 3, pp. 13–22, Mar. 2013.

[24] B. Wang, Y. Wu, F. Han, Y.-H. Yang, and K. J. R. Liu, "Green wireless communications: A time-reversal paradigm," *IEEE Journal of Selected Areas in Communications, Special Issue on Energy-Efficient Wireless Communications*, vol. 29, no. 8, pp. 1698–1710, Sep. 2011.

[25] F. Han, Y.-H. Yang, B. Wang, Y. Wu, and K. J. R. Liu, "Time-reversal division multiple access over multi-path channels," *Communications, IEEE Transactions on*, vol. 60, no. 7, pp. 1953–1965, 2012.

[26] Y. Chen, Y. Yang, F. Han, and K. J. R. Liu, "Time-reversal wideband communications," *IEEE Signal Processing Letters*, vol. 20, no. 12, pp. 1219–1222, Dec. 2013.

[27] Y.-H. Yang, B. Wang, W. S. Lin, and K. J. R. Liu, "Near-optimal waveform design for sum rate optimization in time-reversal multiuser downlink systems," *IEEE Transactions on Wireless Communications*, vol. 12, no. 1, pp. 346–357, Jan. 2013.

[28] B. Y. Zeldovich, N. F. Pilipetsky, and V. V. Shkunov, *Principles of Phase Conjugation*. Berlin: Springer-Verlag, 1985.

[29] A. P. Brysev, L. M. Krutyanskii, and V. L. Preobrazhenskii, "Wave phase conjugation of ultrasonic beams," *Physics-Uspekhi*, vol. 41, no. 8, pp. 793–805, 1998.

[30] M. Fink, C. Prada, F. Wu, and D. Cassereau, "Self focusing in inhomogeneous media with time reversal acoustic mirrors," in *IEEE Ultrasonics Symposium*, pp. 681–686 vol. 2, 1989.

[31] C. Prada, F. Wu, and M. Fink, "The iterative time reversal mirror: A solution to self-focusing in the pulse echo mode," *The Journal of the Acoustical Society of America*, vol. 90, no. 2, pp. 1119–1129, 1991.

[32] M. Fink, "Time reversal of ultrasonic fields. I. Basic principles," *IEEE Transactions on Ultrasonics, Ferroelectrics and Frequency Control*, vol. 39, no. 5, pp. 555–566, 1992.

[33] C. Dorme and M. Fink, "Focusing in transmit–receive mode through inhomogeneous media: The time reversal matched filter approach," *The Journal of the Acoustical Society of America*, vol. 98, no. 2, pp. 1155–1162, 1995.

[34] A. Derode, P. Roux, and M. Fink, "Robust acoustic time reversal with high-order multiple scattering," *Physical Review Letters*, vol. 75, pp. 4206–4209, Dec. 1995.

[35] W. A. Kuperman, W. S. Hodgkiss, H. C. Song, T. Akal, C. Ferla, and D. R. Jackson, "Phase conjugation in the ocean: Experimental demonstration of an acoustic time-reversal mirror," *The Journal of the Acoustical Society of America*, vol. 103, no. 1, pp. 25–40, 1998.

[36] H. C. Song, W. A. Kuperman, W. S. Hodgkiss, T. Akal, and C. Ferla, "Iterative time reversal in the ocean," *The Journal of the Acoustical Society of America*, vol. 105, no. 6, pp. 3176–3184, 1999.

[37] D. Rouseff, D. Jackson, W. L. J. Fox, C. Jones, J. Ritcey, and D. Dowling, "Underwater acoustic communication by passive-phase conjugation: Theory and experimental results," *IEEE Journal of Oceanic Engineering*, vol. 26, no. 4, pp. 821–831, 2001.

[38] G. Edelmann, T. Akal, W. Hodgkiss, S. Kim, W. Kuperman, and H. C. Song, "An initial demonstration of underwater acoustic communication using time reversal," *IEEE Journal of Oceanic Engineering*, vol. 27, no. 3, pp. 602–609, 2002.

[39] B. E. Henty and D. D. Stancil, "Multipath-enabled super-resolution for RF and microwave communication using phase-conjugate arrays," *Physical Review Letters*, vol. 93, p. 243904, Dec. 2004.

[40] G. Lerosey, J. de Rosny, A. Tourin, A. Derode, G. Montaldo, and M. Fink, "Time reversal of electromagnetic waves," *Physical Review Letters*, vol. 92, p. 193904, May 2004.

[41] R. C. Qiu, C. Zhou, N. Guo, and J. Q. Zhang, "Time reversal with MISO for ultrawideband communications: Experimental results," *IEEE Antennas and Wireless Propagation Letters*, vol. 5, pp. 269–273, 2006.

[42] G. Lerosey, J. de Rosny, A. Tourin, A. Derode, G. Montaldo, and M. Fink, "Time reversal of electromagnetic waves and telecommunications," *Radio Science*, vol. 40, pp. 1–10, 2005.

[43] G. Lerosey, J. de Rosny, A. Tourin, A. Derode, and M. Fink, "Time reversal of wideband microwaves," *Applied Physics Letters*, vol. 88, p. 154101, Apr. 2006.

[44] I. H. Naqvi, G. E. Zein, G. Lerosey, J. de Rosny, P. Besnier, A. Tourin, and M. Fink, "Experimental validation of time reversal ultrawide-band communication system for high data rates," *IET Microwaves, Antennas & Propagation*, vol. 4, pp. 643–650, 2010.

[45] J. de Rosny, G. Lerosey, and M. Fink, "Theory of electromagnetic time-reversal mirrors," *IEEE Transactions on Antennas and Propagation*, vol. 58, pp. 3139–3149, Oct. 2010.

[46] M. Emami, M. Vu, J. Hansen, A. J. Paulraj, and G. Papanicolaou, "Matched filtering with rate back-off for low complexity communications in very large delay spread channels," in *Conference Record of the Thirty-Eighth Asilomar Conference on Signals, Systems and Computers*, vol. 1, pp. 218–222, 2004.

[47] H. T. Nguyen, I. Z. Kovacs, and P. C. F. Eggers, "A time reversal transmission approach for multiuser UWB communications," *IEEE Transactions on Antennas and Propagation*, vol. 54, pp. 3216–3224, Nov. 2006.

[48] N. Guo, B. M. Sadler, and R. C. Qiu, "Reduced-complexity UWB time-reversal techniques and experimental results," *IEEE Transactions on Wireless Communications*, vol. 6, pp. 4221–4226, Dec. 2007.

[49] Y. H. Yang and K. J. R. Liu, "Waveform design with interference pre-cancellation beyond time-reversal system," *IEEE Transactions on Wireless Communications*, vol. 15, no. 15, pp. 3643–3654, May 2016.

[50] F. Han and K. J. R. Liu, "A multiuser TRDMA uplink system with 2D parallel interference cancellation," *IEEE Transactions on Communications*, vol. 62, no. 3, pp. 1011–1022, Mar. 2014.

[51] P. Kyritsi and G. Papanicolaou, "One-bit time reversal for WLAN applications," in *IEEE International Symposium on Personal, Indoor and Mobile Radio Communications*, pp. 532–536, 2005.

[52] D.-T. Phan-Huy, S. B. Halima, and M. Helard, "Frequency division duplex time reversal," in *IEEE Globecom*, 2011.

[53] G. Lerosey, J. de Rosny, A. Tourin, and M. Fink, "Focusing beyond the diffraction limit with far-field time reversal," *Science*, vol. 315, pp. 1120–1122, Feb. 2007.

[54] F. Lemoult, G. Lerosey, J. de Rosny, and M. Fink, "Resonant metalenses for breaking the diffraction barrier," *Physical Review Letters*, vol. 104, p. 203901, May 2010.

[55] R. C. Qiu, C. Zhou, N. Guo, and J. Q. Zhang, "Time reversal with MISO for ultra-wideband communications: Experimental results," *IEEE Antenna and Wireless Propagation Letters*, vol. 5, pp. 269–273, 2006.

[56] K. F. Sander and G. A. L. Reed, *Transmission and Propagation of Electromagnetic Waves*, 2nd ed. New York: Cambridge University Press, 1986.

[57] J. M. F. Moura and Y. Jin, "Time reversal imaging by adaptive interference canceling," *IEEE Transactions on Signal Processing*, vol. 56, no. 1, pp. 233–247, 2008.

[58] —, "Detection by time reversal: Single antenna," *IEEE Transactions on Signal Processing*, vol. 55, no. 1, pp. 187–201, 2007.

[59] Y. Jin and J. M. F. Moura, "Time-reversal detection using antenna arrays," *IEEE Transactions on Signal Processing*, vol. 57, no. 4, pp. 1396–1414, 2009.

[60] F. Han, Y.-H. Yang, B. Wang, Y. Wu, and K. J. R. Liu, "Time-reversal division multiple access in multi-path channels," in *Global Telecommunications Conference*, pp. 1–5, 2011.

[61] M. Lienard, P. Degauque, V. Degardin, and I. Vin, "Focusing gain model of time-reversed signals in dense multipath channels," *IEEE Antennas and Wireless Propagation Letters*, vol. 11, pp. 1064–1067, 2012.

[62] G. Montaldo, G. Lerosey, A. Derode, A. Tourin, J. de Rosny, and M. Fink, "Telecommunication in a disordered environment with iterative time reversal," *Waves Random Media*, vol. 14, pp. 287–302, May 2004.

[63] F. Lemoult, G. Lerosey, J. de Rosny, and M. Fink, "Manipulating spatiotemporal degrees of freedom of waves in random media," *Physical Review Letters*, vol. 103, p. 173902, Oct. 2009.

[64] X. Zhou, P. Eggers, P. Kyritsi, J. Andersen, G. Pedersen, and J. Nilsen, "Spatial focusing and interference reduction using MISO time reversal in an indoor application," in *IEEE Workshop on Statistical Signal Processing (SSP)*, 2007.

[65] J. Proakis and M. Salehi, *Digital Communications*, 5th ed. New York: McGraw-Hill, 2008.

[66] F. Rashid-Farrokhi, K. J. R. Liu, and L. Tassiulas, "Transmit beamforming and power control for cellular wireless systems," *IEEE Journal on Selected Areas in Communications*, vol. 16, no. 8, pp. 1437–1450, Oct. 1998.

[67] F. Rashid-Farrokhi, L. Tassiulas, and K. J. R. Liu, "Joint optimal power control and beamforming in wireless networks using antenna arrays," *IEEE Transactions on Communications*, vol. 46, no. 10, pp. 1313–1324, Oct. 1998.

[68] Y.-H. Yang, S.-C. Lin, and H.-J. Su, "Multiuser MIMO downlink beamforming based on group maximum SINR filtering," *IEEE Transactions on Signal Processing*, vol. 59, no. 4, pp. 1746–1758, Apr. 2011.

[69] H. Sampath, P. Stoica, and A. Paulraj, "Generalized linear precoder and decoder design for MIMO channels using the weighted MMSE criterion," *IEEE Transactions on Communications*, vol. 49, no. 12, pp. 2198–2206, 2001.

[70] F. Dietrich, R. Hunger, M. Joham, and W. Utschick, "Linear precoding over time-varying channels in TDD systems," in *Proceedings of the ICASSP'03*, vol. 5, pp. 117–120, 2003.

[71] Z. Ahmadian, M. Shenouda, and L. Lampe, "Design of pre-Rake DS-UWB downlink with pre-equalization," *IEEE Transactions on Communications*, vol. 60, no. 2, pp. 400–410, Feb. 2012.

[72] Y. Jin, J. M. Moura, and N. O'Donoughue, "Adaptive time reversal beamforming in dense multipath communication networks," in *Proceedings of the 42nd Asilomar Conference on Signals, Systems and Computers*, pp. 2027–2031, Oct. 2008.

[73] R. C. Daniels and R. W. Heath, "Improving on time-reversal with MISO precoding," in *Proceedings of the Eighth International Symposium on Wireless Personal Communications Conference*, 2005.

[74] L.-U. Choi and R. D. Murch, "A transmit preprocessing technique for multiuser MIMO systems using a decomposition approach," *IEEE Transactions on Wireless Communications*, vol. 3, no. 1, pp. 2–24, Jan. 2004.

[75] P. Kyritsi, G. Papanicolaou, P. Eggers, and A. Oprea, "Time reversal techniques for wireless communications," in *IEEE Vehicular Technology Conference*, no. 4, pp. 47–51, 2004.

[76] M. Brandt-Pearce, "Transmitter-based multiuser interference rejection for the downlink of a wireless CDMA system in a multipath environment," *IEEE Journal on Selected Areas in Communications*, vol. 18, no. 3, pp. 407–417, Mar. 2000.
[77] P. Kyritsi, P. Stoica, G. Papanicolaou, P. Eggers, and A. Oprea, "Time reversal and zero-forcing equalization for fixed wireless access channels," in *the 39th Asilomar Conference on Signals, Systems and Computers*, pp. 1297–1301, 2005.
[78] C. Windpassinger, R. F. H. Fischer, T. Vencel, and J. Huber, "Precoding in multiantenna and multiuser communications," *IEEE Transactions on Wireless Communications*, vol. 3, no. 4, pp. 1305–1316, 2004.
[79] W. Yu, D. Varodayan, and J. Cioffi, "Trellis and convolutional precoding for transmitter-based interference presubtraction," *IEEE Transactions on Communications*, vol. 53, no. 7, pp. 1220–1230, 2005.
[80] M. H. M. Costa, "Writing on dirty paper," *IEEE Transactions on Information Theory*, vol. 29, no. 3, pp. 439–441, 1983.
[81] "Media access control," in http://en.wikipedia.org/wiki/Media_access_control.
[82] G. Bianchi, "Performance analysis of the IEEE 802.11 distributed coordination function," *IEEE Journal of Selected Areas in Communications*, vol. 18, no. 3, Mar. 2000.
[83] M. Ergen, "IEEE 802.11 tutorial," in http://wow.eecs.berkeley.edu/ergen/docs/ieee.pdf.
[84] "Even Steve Jobs has demo hiccups," in http://news.cnet.com/8301-31021_3-20007009-260.html.
[85] "A/D converters, Texas Instruments 2013," www.ti.com/lsds/ti/data-converters/analog-to-digital-converter-products.page/.
[86] A. Kohlenberg, "Exact interpolation of band-limited functions," *Journal of Applied Physics*, vol. 24, no. 12, pp. 1432–1436, 1953.
[87] Y.-P. Lin and P. Vaidyanathan, "Periodically nonuniform sampling of bandpass signals," *IEEE Transactions on Circuits and Systems II: Analog and Digital Signal Processing*, vol. 45, no. 3, pp. 340–351, 1998.
[88] M. El-Chammas and B. Murmann, "General analysis on the impact of phase-skew in time-interleaved ADCs," *IEEE Transactions on Circuits and Systems I: Regular Papers*, vol. 56, no. 5, pp. 902–910, 2009.
[89] P. Nikaeen and B. Murmann, "Digital compensation of dynamic acquisition errors at the front-end of high-performance A/D converters," *IEEE Journal of Selected Topics in Signal Processing*, vol. 3, no. 3, pp. 499–508, 2009.
[90] Y. Eldar and A. Oppenheim, "Filterbank reconstruction of bandlimited signals from nonuniform and generalized samples," *IEEE Transactions on Signal Processing*, vol. 48, no. 10, pp. 2864–2875, 2000.
[91] Y. Tian, D. Zeng, and T. Zeng, "Design and implementation of multifrequency front end using bandpass over sampling," in *IET International Radar Conference 2009*, pp. 1–4, 2009.
[92] J. Tropp, J. Laska, M. Duarte, J. Romberg, and R. Baraniuk, "Beyond Nyquist: Efficient sampling of sparse bandlimited signals," *Information Theory, IEEE Transactions on*, vol. 56, no. 1, pp. 520–544, 2010.
[93] M. Mishali and Y. Eldar, "From theory to practice: Sub-Nyquist sampling of sparse wideband analog signals," *IEEE Journal of Selected Topics in Signal Processing*, vol. 4, no. 2, pp. 375–391, 2010.
[94] "5G wiki," in http://en.wikipedia.org/wiki/5G.
[95] Y. Chen, F. Han, Y.-H. Yang, H. Ma, Y. Han, C. Jiang, H.-Q. Lai, D. Claffey, Z. Safar, and K. J. R. Liu, "Time-reversal wireless paradigm for green Internet of Things: An overview," *IEEE Internet of Things Journal*, vol. 1, no. 1, pp. 81–98, 2014.

22 Heterogeneous Connections for IoT

With the pervasive presence of massive smart devices, Internet of Things (IoT) is enabled by wireless communication technology. The devices in IoT usually have very diverse bandwidth capabilities and thus are in need of many communication standards. To facilitate communications between these heterogeneous bandwidths of devices, middlewares have often been developed. However, they are often not suitable for resource-constrained scenarios due to their complexity. It leads us to ask, is there a unified approach that can support the communication between the devices with heterogeneous bandwidths? In this chapter, we discuss the time-reversal (TR) approach to answer such a question. A novel TR-based heterogeneous system is presented, which can address the bandwidth heterogeneity and maintain the benefit of TR at the same time. Although there is an increase in complexity, it concentrates mostly on the digital processing of the access point (AP), which can be easily handled with more powerful digital signal processor (DSP). Because there is no middleware in the presented system and the additional physical layer complexity concentrates on the AP side, the presented TR approach better satisfies the requirement of low-complexity and energy-efficiency for terminal devices (TDs). We further conduct the theoretical analysis of the interference in the presented system. Simulations show the bit-error-rate (BER) performance can be significantly improved with appropriate spectrum allocation. Finally, Smart Homes is chosen as an example of IoT applications to evaluate the performance of the presented system.

22.1 Introduction

Ubiquitous RFID tags, sensors, actuators, mobile phones, and etc. cut across many areas of modern-day living, which offers the ability to measure, infer, and understand the environmental indicators. The proliferation of these devices creates the term the Internet of Things (IoT), wherein these devices blend seamlessly with the environment around us, and the information is shared across the whole platform [1].

The notion of IoT dates back to 1999, when it was first proposed by Ashton [2]. Even though logistic is the originally considered application, in the past decade, the coverage of IoT has been extended to a wide range of applications including healthcare, utilities, transportation, etc. [3]. Thanks to the significant maturity and market size of wireless communication technologies such as ZigBee, Bluetooth, Wi-Fi, and near-field communication (NFC), IoT is on the path of transforming the current static Internet into a fully integrated future Internet [4]. Due to its high impact on several aspects of everyday life and behavior of the potential users [5], IoT is listed as one of six "Disruptive Civil Technologies" by the US National Intelligence Council with potential impacts on US national power [6].

Considering the massive amount of devices and various application scenarios in the IoT, the devices within the IoT are highly heterogeneous. From the perspective of communication, one of the significant heterogeneities is the bandwidth heterogeneity and thus the corresponding radio-frequency (RF) front-end. To address the bandwidth heterogeneity, various communication standards such as ZigBee, Bluetooth, and Wi-Fi are adopted simultaneously in the current IoT platform, which leads to a wild growth of co-located wireless communication standards [7]. When multiple wireless communication standards are operated in the same geographical environment, the devices often suffer from harmful interference. Furthermore, the communication between devices with different communication standards is only possible through the use of gateway nodes, resulting in the fragmentation of the whole network, hampering the objects, interoperability, and slowing down the development of a unified reference model for IoT [8].

To enable the connectivity between devices with various bandwidths, some existing works build middlewares to hide the technical details of different communication standards from the application layer. In [9], service oriented device architecture (SODA) is proposed as a promising approach to integrate service oriented architecture (SOA) principles into the IoT. An effective SOA-based integration of IoT is illustrated in enterprise service [10]. Business Process Execution Language (BPEL) has been widely used as the process language in the middleware [11]. However, these technologies used to realize middleware architectures are often not suitable for resource-constrained scenarios due to their complexity.

Instead of building middlewares, is there any other more effective approach to enable the connectivity between the devices with different bandwidths? We try to answer this question by proposing the time-reversal (TR) approach. It is well known that radio signals will experience many multipaths due to the reflection from various scatters, especially in indoor environments. Through time-reversing (and conjugate, when complex-valued) the multipath profile as the beamforming signature, TR technique can constructively add up the signals of all the paths at the intended location, ending up with a spatio-temporal resonance effect [12]. As pointed out in [12], the TR technique is an ideal candidate for low-complexity, low-energy consumption green wireless communication because of its inherent nature to fully harvest energy from all the paths. A TR-based multiuser media access scheme is proposed in [13], where only the simple detection based on a single received symbol is needed at the device side resulting in low computational complexity and low cost of the terminal devices. With the signature determined by the physical location, TR technique can provide additional physical-layer security and thus can enhance the privacy and security of customers in IoT. An overview of the TR wireless paradigm for green IoT has been presented in [14] summarizing all the promising features of TR technique. However, they cannot be directly applied to

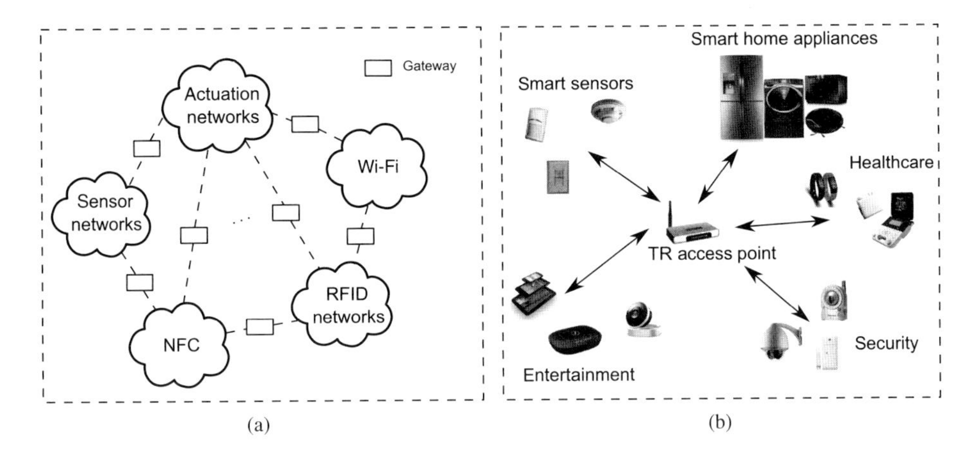

Figure 22.1 Comparison between (a) existing IoT approach and (b) heterogeneous TR-based IoT approach.

address the bandwidth heterogeneity in IoT because of the implicit assumption that all terminal devices share the same bandwidth and thus the RF front-end.

In order to support devices with various bandwidths in IoT, a novel TR-based heterogeneous system is discussed in this chapter, where a bank of various pulse-shaping filters are implemented to support data streams of different bandwidths. By integrating the multirate signal processing into TR technique, the presented system is capable of supporting these heterogeneous devices with a single set of RF front-end, therefore it is a unified framework for connecting devices of heterogeneous bandwidths. As shown in Figure 22.1, instead of connecting devices with different wireless communication standards through gateways and middlewares, the TR-based heterogeneous system in this chapter directly links the devices together. The increase of complexity in the presented system lies in the digital processing at the access point (AP), instead of at the devices' ends, which can be easily handled with a more powerful digital signal processor (DSP). Meanwhile, the complexity of the terminal devices stays low and therefore satisfies the low-complexity and scalability requirement of IoT. Because there is no middleware in the presented scheme and the additional physical layer complexity concentrates on the AP side, the presented heterogeneous TR system better satisfies the low-complexity and energy-efficiency requirement for the terminal devices (TDs) compared with the middleware approach. Theoretical analysis of the interference is further conducted to predict the system performance. Simulation results show that the presented system can support the devices of heterogeneous bandwidths with a reasonable bit-error-rate (BER) performance. In addition, the BER performance can be significantly improved with the appropriate spectrum allocation.

The rest of this chapter is organized as follows. We first discuss the system architecture and working scheme of the existing homogeneous TR system in Section 22.2. Based on the existing TR system, a TR-based heterogeneous system is developed in Section 22.3. In Section 22.4, theoretical analysis regarding the interference in the presented system is derived. Simulation results about the BER performance of the system are discussed in Section 22.5.

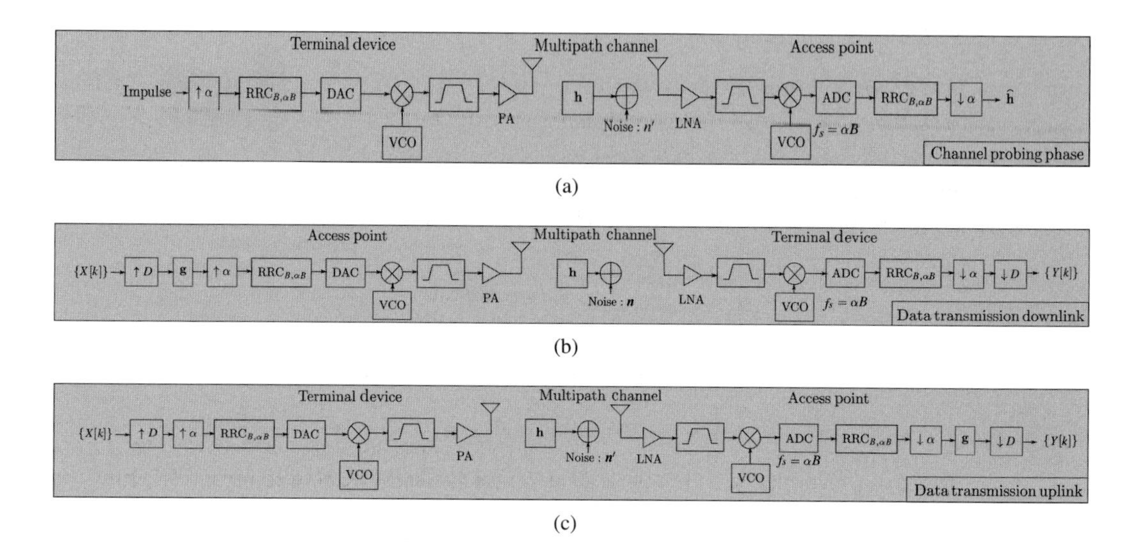

Figure 22.2 Typical homogeneous TR system: (a) channel probing phase, (b) data transmission downlink, and (c) data transmission uplink.

22.2 Typical Homogeneous Time-Reversal System

In this section, we will first introduce the system architecture and working mechanism of the TR-based homogenous system, where the AP and all terminal devices (TDs) share the same spectrum, and thus the bandwidth and ADC sample rate.

A typical TR-based homogenous system is shown in Figure 22.2 [14]. The channel impulse response (CIR) between the two transceivers is modeled as

$$h(t) = \sum_{v=1}^{V} h_v \delta(t - \tau_v) \tag{22.1}$$

where h_v is the complex channel gain of vth path of the CIR, τ_v is the corresponding path delay, and V is the total number of the independent multipaths in the environment (assuming infinite system bandwidth and time resolution). Without loss of generality, we assume that $\tau_1 = 0$ in the rest of the chapter, i.e., the first path arrives at time $t = 0$, and as a result, the delay spread of the multipath channel τ_C is given by $\tau_C = \tau_V - \tau_1 = \tau_V$.

Considering the practical communication system with limited bandwidth, pulse shaping filters are typically deployed to limit the effective bandwidth of transmission. In practice, raised-cosine filter is typically utilized as a pulse-shaping filter, which minimizes the intersymbol interference (ISI) [15]. Generally, the raised-cosine filter is split into two root-raised-cosine filters $\mathbf{RRC}_{B,f_s}[n]$ and deployed at each side of the transceivers, where B is the available bandwidth and f_s is the sample rate of the system. Based on the Nyquist rate [16], an α-times oversampling (i.e., $f_s = \alpha B$) is practically implemented to counter the sampling frequency offset (SFO).

22.2.1 Channel Probing Phase

As shown in Figure 22.2(a), prior to AP's TR-transmission, an impulse is upsampled by α, filtered by $\mathbf{RRC}_{B,\,f_s}[n]$, and transmitted out after going through the RF components at the TD side. The transmitted signal propagates to AP through the multipath channel $h(t)$, where AP samples the received signal. Then the sampled signal goes through RF components, which later is filtered by another $\mathbf{RRC}_{B,\,f_s}[n]$, downsampled by α, and finally recorded as the estimated CIR $\widehat{\mathbf{h}}$.

With sample rate $f_s = \alpha B$, the discrete CIR can be written as

$$\overline{h}[n] = \sum_{v=1}^{V} h_v \delta[nT_s - \tau_v], \tag{22.2}$$

where $T_s = 1/(\alpha B)$. Assuming perfect channel estimation (noise and interference are ignored in the channel probing phase), the equivalent CIR between two RRC filters in Figure 22.2(a) is written as

$$\widetilde{\mathbf{h}} = \left(\mathbf{RRC}_{B,\,f_s} * \overline{\mathbf{h}} * \mathbf{RRC}_{B,\,f_s}\right). \tag{22.3}$$

Based on the polyphase identity [17], the equivalent CIR (between the expander and decimator) for the system with bandwidth B can be represented as

$$\widehat{\mathbf{h}} = \left(\mathbf{RRC}_{B,\,f_s} * \overline{\mathbf{h}} * \mathbf{RRC}_{B,\,f_s}\right)_{[\alpha]}, \tag{22.4}$$

where $(\cdot)_{[\alpha]}$ represents α-times decimation. From (22.4), one can see that those paths in (22.2), whose time differences are within the main lobe of raised-cosine filter, are mixed together for the system with a limited bandwidth B.

22.2.2 Data Transmission Phase

Upon acquiring the equivalent CIR $\widehat{\mathbf{h}}$, different designs of signature waveforms (e.g., basic TR signature [12], ZF signature [18], and MMSE signature [19]) can be implemented at the AP side. With no loss of generality, the basic TR signature is considered in the rest of chapter. In other words, the AP time-reverses (and conjugate, when complex-valued) the equivalent CIR $\widehat{\mathbf{h}}$, and uses the normalized TR waveform as the basic TR signature $\mathbf{g}$, i.e.,

$$g[n] = \frac{\widehat{h}^*[L-1-n]}{\|\widehat{\mathbf{h}}\|} \tag{22.5}$$

where L is the number of taps in $\widehat{\mathbf{h}}$.

According to Figure 22.2(b), there is a sequence of information symbols $\{X[k]\}$ to be transmitted to the TD. Typically, the symbol rate can be much lower than the system chip rate $(1/B)$. Therefore, a rate backoff factor D is introduced to match the symbol rate with chip rate by inserting $(D-1)$ zeros between two symbols [12, 13, 20], i.e.,

$$X^{[D]}[k] = \begin{cases} X[k/D], & \text{if } (k \bmod D) = 0 \\ 0, & \text{if } (k \bmod D) \neq 0. \end{cases} \tag{22.6}$$

where $(\cdot)^{[D]}$ denotes the D-times interpolation. Consequently, the signature embedded symbols before the α-times expander can be written as

$$S[k] = \left(\mathbf{X}^{[D]} * \mathbf{g}\right)[k]. \tag{22.7}$$

Based on the previous derivation in the channel probing phase, the system components between the expander and decimator in Figure 22.2(b) can be replaced by $\widehat{\mathbf{h}}$. Therefore, the signal received at the TD side before the decimator with rate D is the convolution of $S[k]$ and $\widehat{\mathbf{h}}$, plus additive white Gaussian noise (AWGN) $\widetilde{n}[k]$ with zero-mean and variance σ_N^2, i.e.,

$$Y^{[D]}[k] = \left(\mathbf{S} * \widehat{\mathbf{h}}\right)[k] + \widetilde{n}[k]. \tag{22.8}$$

Then, TD decimates the symbols with backoff factor D in order to detect the information symbols $\{X[k]\}$, i.e.,

$$\begin{aligned} Y[k] = & \sqrt{p_u}(\widehat{\mathbf{h}} * \mathbf{g})[L-1]X[k - \frac{L-1}{D}] \\ & + \sqrt{p_u} \sum_{l=0, l \neq (L-1)/D}^{(2L-2)/D} (\widehat{\mathbf{h}} * \mathbf{g})[Dl]X[k-l] + n[k], \end{aligned} \tag{22.9}$$

where $n[k] \triangleq \widetilde{n}[Dk]$ and p_u stands for the power amplifier.

Benefiting from temporal focusing, the power of $(\widehat{\mathbf{h}} * \mathbf{g})$ achieves its maximum at $(L-1)$ for $X[k - \frac{L-1}{D}]$, i.e.,

$$(\widehat{\mathbf{h}} * \mathbf{g})[L-1] = \frac{\sum_{l=0}^{L-1} \widehat{h}[l]\widehat{h}^*[l]}{\|\widehat{\mathbf{h}}\|} = \|\widehat{\mathbf{h}}\|. \tag{22.10}$$

Consequently, the resulting signal-to-interference-plus-noise ratio (SINR) is obtained as

$$\text{SINR} = \frac{p_u \|\widehat{\mathbf{h}}\|^2}{p_u \sum_{l=0, l \neq (L-1)/D}^{(2L-2)/D} |(\widehat{\mathbf{h}} * \mathbf{g})[Dl]|^2 + \sigma_N^2}, \tag{22.11}$$

assuming that each information symbol $X[k]$ has unit power.

Regarding the uplink, the previously designed signature waveform $\mathbf{g}$ serves as the equalizer at the AP side as shown in Figure 22.2(c). Similar to the signal flow in the downlink scheme, the AP can detect the information symbol based on the temporal focusing of $(\widehat{\mathbf{h}} * \mathbf{g})$ in the uplink. Such a scheme of both downlink and uplink is defined as the asymmetric architecture, which provides the asymmetric complexity distribution between the AP and TD. In other words, the design philosophy of uplink is to keep the complexity of terminal users at minimal level.

Note that the homogeneous TR system can be easily extended to multi-user scenario according to the previous work [13], which exploits the spatial degrees of freedom in the environment and uses the multipath profile associated with each user's location as a location-specific signature for the user. In addition, different users are allowed to adopt different rate backoff factors to accommodate the heterogeneous QoS requirements for various applications in the IoT.

Remark: Even though the homogeneous TR system can support different QoS through varying D, all devices in the system must share the same bandwidth and thus the sample rate, which increases not only the hardware cost but also the computation burden for those low-end TDs. Besides the heterogenous QoS required by very diverse applications, the definition of heterogeneity in IoT should also cover the heterogeneous hardware capabilities (such as bandwidth, sample rate, computational and storage power, etc.), which apparently is not supported by the homogeneous TR system. The more general heterogeneity requirement in the IoT motivates the heterogenous TR paradigm in this chapter.

22.3 Heterogeneous Time-Reversal System

Even though the homogeneous TR system cannot handle bandwidth heterogeneity, the majority of challenges in the IoT can be tackled simultaneously through the TR technique [14]. Does there exist an efficient way to modify the existing homogenous TR system to handle the bandwidth heterogeneity while maintaining the most benefit of the TR technique? The answer is yes, and the heterogeneous TR system is potentially the best candidate to address the issue.

In contrast with the same spectrum occupation of all devices in the homogenous setting, N types of TDs with distinct spectrum allocation and bandwidths are supported simultaneously by a single AP in the heterogenous TR system. In other words, different types of TDs have the distinct carrier frequency (f_{c_i}) and bandwidth (B_i) as shown in Figure 22.3.

22.3.1 Modifications on Homogenous TR System

In order to support the heterogenous TDs, several modifications need to be conducted at both AP and TD sides of the existing homogeneous TR system.

22.3.1.1 TD Side

As stated before, heterogeneous TDs of different types have distinct f_{c_i}'s and B_i's. First of all, the radio-frequency (RF) components of different types have to be distinct. Specifically, the oscillation frequency of the voltage-controlled oscillator (VCO) at type i TD is set to f_{c_i}, and the bandwidth of analog bandpass filter is B_i. Then, the ADC deployed for type i TDs has the sample rate of $f_{s_i} = \alpha B_i$ based on the

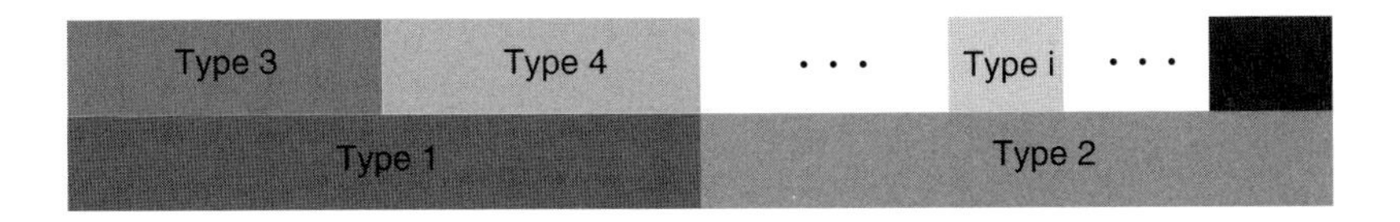

Figure 22.3 Spectrum occupation of heterogeneous TDs.

previous discussion. Furthermore, various root-raised-cosine filters for different types are required, i.e., $\mathbf{RRC}_{B_i, f_{s_i}}$.

22.3.1.2 AP Side

In order to support heterogeneous TDs simultaneously, the bandwidth of AP, denoted as B_{AP}, is the aggregated version of the bandwidth of all heterogeneous TDs. Even though more complicated digital signal processing is enforced to handle different data streams for various types, only one set of RF components is needed at the AP side. The digital signal processing includes frequency shift, rate convertor and root-raised-cosine filter. More specifically, a frequency shift component $\exp^{j\omega_i n}$ is implemented for each type to support multiple carrier frequencies. A distinct sample rate convertor (expander or decimator) with rate $\alpha B_{AP}/B_i$ is deployed for each type i to enable the multirate processing. The root-raised-cosine filter $\mathbf{RRC}_{B_i,\alpha B_{AP}}$ for type i is utilized to limit the effective bandwidth of signals for the heterogeneous TDs.

In the following, the detailed system mechanism together with the modified system architecture is developed for the presented heterogeneous TR system.

22.3.2 Channel Probing Phase

The channel probing phase of a type i TD is shown in Figure 22.4. Compared with the one in Figure 22.2(a), there exists some differences mentioned in the last subsection. Prior to the data transmission phase, an impulse is upsampled by α, filtered by $\mathbf{RRC}_{B_i,\alpha B_i}[n]$, and transmitted out after going through the RF components at the TD side. The transmitted signal propagates to AP through the multipath channel $h_i(t)$, where AP samples the received signal with a higher sample rate $f_s = \alpha B_{AP}$, shifts the signal to baseband (based on the difference between f_{c_i} and $f_{c_{AP}}$), filters it through the other matched $\mathbf{RRC}_{B_i,\alpha B_{AP}}[n]$, downsamples the waveform by $\alpha B_{AP}/B_i$, and finally records the downsampled waveform as $\widehat{\mathbf{h}}_i$.

With sample rate $f_s = \alpha B_{AP}$, the discrete CIR can be written as

$$\overline{h}_i[n] = h_i(nT_s), \tag{22.12}$$

where $T_s = 1/(\alpha B_{AP})$.

Because the digital-to-analog convertor (DAC) serves the interpolator, the transmitted signal of the TD shown in Figure 22.4 is mathematically equivalent to that generated through the following process, i.e., upsampled by $\alpha B_{AP}/B_i$, filtered by

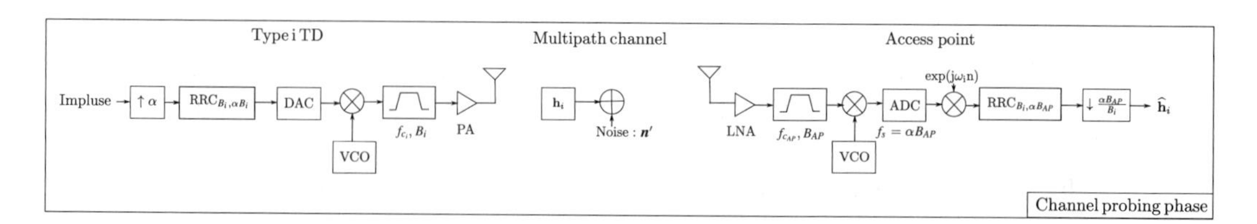

Figure 22.4 Channel probing of type i TD in heterogeneous TR system.

$\mathbf{RRC}_{B_i,\alpha B_{AP}}[n]$, and converted to analog signal by the DAC. Therefore, similarly according to the polyphase identity, the equivalent CIR for the type i TD with bandwidth B_i can be expressed as

$$\widehat{\mathbf{h}}_i = \sqrt{\beta_i}\left(\mathbf{RRC}_{B_i,\alpha B_{AP}} * \overline{\mathbf{h}}_i * \mathbf{RRC}_{B_i,\alpha B_{AP}}\right)_{[\alpha\beta_i]}, \tag{22.13}$$

where $\beta_i = B_{AP}/B_i$, and $\sqrt{\beta_i}$ is used to compensate the power difference between $\mathbf{RRC}_{B_i,\alpha B_i}[n]$ and $\mathbf{RRC}_{B_i,\alpha B_{AP}}[n]$.

Even though the channel probing of a single type is evaluated here, it can be extended straightforward to multitype TDs by deploying different digital processing for multitype in parallel, e.g., frequency shift, RRC filtering, and downsampling with type-specific factor. In other words, the AP can support heterogeneous TDs with one single set of RF components but more complicated digital processing.

22.3.3 Data Transmission Phase

Suppose N types of TDs are communicating with the AP simultaneously, where the number of TDs in type i is denoted as M_i. Upon acquiring the equivalent CIRs, the signature waveform $\mathbf{g}_{i,j}$ is designed for j^{th} TD in the type i with various existing design methods. Take the basic TR signature design for example, i.e.,

$$g_{i,j}[n] = \frac{\widehat{h}^*_{i,j}[L-1-n]}{\|\widehat{\mathbf{h}}_{i,j}\|}, \tag{22.14}$$

where $\widehat{\mathbf{h}}_{i,j}$ is defined in (22.13).

First, the downlink data transmission is considered. As shown in Figure 22.5(a), let $\{X_{i,j}[k]\}$ be the the sequence of information symbols transmitted to the jth TD in type i. Similar to the case in the homogeneous TR system, a rate backoff factor $D_{i,j}$ is introduced to adjust the symbol rate, i.e., the symbol rate for j^{th} TD in the type i is $(B_i/D_{i,j})$. Then, the signature $\mathbf{g}_{i,j}$ is embedded into the TD-specific data stream $\mathbf{X}^{[D_{i,j}]}_{i,j}$, and the signature embedded symbols of the same type i are merged together as $\mathbf{S}_i$, e.g.,

$$\mathbf{S}_i = \sum_{j=1}^{M_i}\left(\mathbf{X}^{[D_{i,j}]}_{i,j} * \mathbf{g}_{i,j}\right). \tag{22.15}$$

Later, the merged symbols $\mathbf{S}_i$ go through the type-specific digital signal processing, i.e., upsampled with factor $\alpha B_{AP}/B_i$, filtered by $\mathbf{RRC}_{B_i,\alpha B_{AP}}$, and carried to the type-specific digital frequency with frequency shift $\exp(-j\omega_i n)$. In the end, the processed data streams of N types are mixed together and broadcasted to all the heterogeneous TDs through one set of RF components at the AP.

Regarding the receiver side, the jth TD in type i is taken as an example. The broadcast signal propagates to the TD through the multipath profile $h_{i,j}(t)$. Later, the signal passes through the analog bandpass filter centering at f_{c_i} with bandwidth B_i. Note that the filtered signal includes not only the intended signal but also the interference, e.g., the interuser inference (IUI) from the TDs within the same type and the intertype interference (ITI) from the other types (whose spectrum overlaps with type i). Thanks to

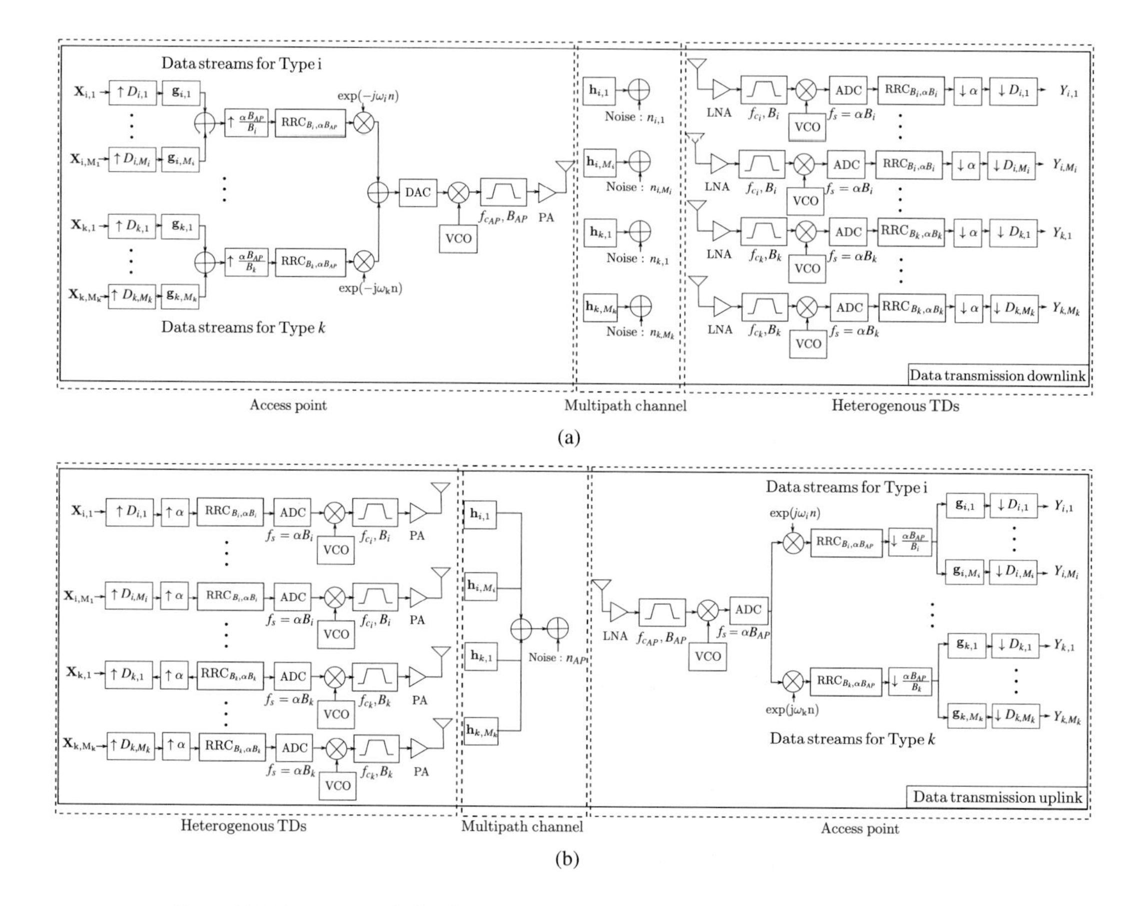

Figure 22.5 Data transmission in heterogeneous TR system: (a) downlink, and (b) uplink.

the spatio-temporal focusing effect, the interference is suppressed due to the unique multipath profile. Afterwards, the signal is carried to baseband and sampled with a sample rate $f_{s_i} = \alpha B_i$, which is much smaller than that at the AP for the low-end TDs. In the end, the sampled signal goes through $\mathbf{RRC}_{B_i,\alpha B_i}$, and the rate matching decimator to generate symbols $\{Y_{i,j}[k]\}$, based on which $\{X_{i,j}[k]\}$ are detected. The theoretical analysis regarding the signal-to-interference-plus-noise ratio (SINR) will be derived in the next section.

The system architecture of uplink is shown in Figure 22.5(b). From the figure, the property of asymmetric architecture is preserved in the heterogeneous TR system. Same as the homogeneous TR system, the precoding signatures $\mathbf{g}_{i,j}$'s in the downlink serve as the equalizers in the uplink. After converting the signal into digital domain through a single set of RF components at the AP, multiple parallel digital processing (e.g., frequency shift, RRC filtering, and rate conversion) is required to support N types of TDs simultaneously.

Remark: Compared with the existing homogeneous TR system, the heterogenous TR system maintains the capability to support different QoS through not only varying the backoff factor $D_{i,j}$ but also by providing the flexibility for TDs to select various B_i's. More importantly, the heterogeneous TR system architecture further promotes the benefit of the asymmetric complexity. In other words, the new modifications enhance the concentration of the complexity at the AP side. Regarding the AP, a single set of RF components is required. Even though more complicated parallel digital signal processing is needed, it can be easily satisfied with a more powerful DSP unit at an affordable cost and complexity. Regarding the heterogeneous TDs, the ADC sample rate is reduced significantly for those devices with smaller bandwidth, which lowers the cost of hardware dramatically for the low-end TDs. In addition, the lower sample rate naturally decreases the computational burden as well.

Compared with the middleware approach, the presented TR approach has two main advantages. First, the presented TR approach serves a unified system model for IoT, while middleware leads to the fragmentation of the whole network due to the coexistence of different communication standards. Moreover, by concentrating the complexity at the AP, the presented TR approach better satisfies the requirement of low-complexity and energy-efficiency at the TDs because no middleware needs to be implemented on the TD side.

22.4 Performance Analysis of Heterogeneous TR System

In this section, we conduct some theoretic analysis on the presented heterogeneous TR system and evaluate the SINR for the individual TD. Without loss of generality, the downlink scenario is investigated here. Due to the asymmetric architecture and channel reciprocity, the uplink scenario can be analyzed similarly. In the following, two special cases in the heterogeneous TR system are first studied. Then, the analysis of a specific TD in the general setting is derived through extending the results of the special cases.

22.4.1 Overlapping Case

First, a special case of heterogenous TR system is considered. Suppose there are only two types of TDs in the system, e.g., type i and type k. As shown in Figure 22.6, both types share the same carrier frequency with AP, whose spectrum is overlapped. Without loss of generality, only a single TD is assumed to exist within each type.

In this special case, the downlink system architecture in Figure 22.5(a) can be significantly simplified. In the first place, the frequency shift can be removed due to the same carrier frequency. Moreover, the analog bandpass filter could also be ignored in the analysis because the effective bandwidth has already been limited by the RRC filters.

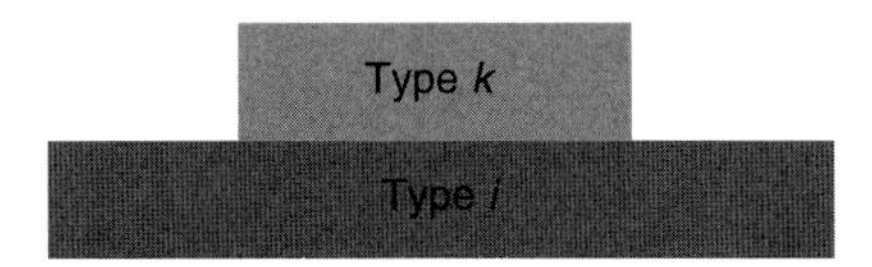

Figure 22.6 Spectrum occupation of case I.

Denote $\widehat{\mathbf{h}}_{a,b}$ as the equivalent CIR for the type a symbols sent from the AP to the type b TD. Based on (22.4), we have

$$\widehat{\mathbf{h}}_{a,a} = \sqrt{\beta_a}\left(\mathbf{RRC}_{B_a,\alpha B_{AP}} * \overline{\mathbf{h}}_a * \mathbf{RRC}_{B_a,\alpha B_{AP}}\right)_{[\alpha\beta_a]}. \tag{22.16}$$

In addition, the equivalent CIR for interference can be derived as follows through utilizing the noble identities [17],

$$\widehat{\mathbf{h}}_{a,b} = \left(\mathbf{RRC}_{B_a,\alpha B_{AP}} * \overline{\mathbf{h}}_b * \mathbf{RRC}^{[\beta_b]}_{B_b,\alpha B_b}\right)_{[\alpha]}, \tag{22.17}$$

where $a, b \in \{i, k\}$, $\beta_a = B_{AP}/B_a$, and $\overline{\mathbf{h}}_a$ is the discrete CIR from the AP to the type a TD with sample rate $f_s = \alpha B_{AP}$.

Upon acquiring the equivalent CIRs, the signature for each type is designed, e.g.,

$$g_a[n] = \frac{\widehat{h}^*_{a,a}[L-1-n]}{\|\widehat{\mathbf{h}}_{a,a}\|}, \tag{22.18}$$

where $a \in \{i, k\}$. Note that there exists focusing effect of the term $(\mathbf{g}_a * \widehat{\mathbf{h}}_{a,a})$ based on (22.18). Therefore, the simplified system model is shown in Figure 22.7 based on the preceding equations.

From the figure, the received symbols at type i TD can be expressed as

$$\begin{aligned} Y_i[n] = {} & \frac{\sqrt{p_u}}{\beta_i}\left(\mathbf{g}_i * \widehat{\mathbf{h}}_{i,i}\right)[L_i - 1]X_i\left[n - \frac{L_i - 1}{D_i}\right] \\ & + \frac{\sqrt{p_u}}{\beta_i}\sum_{l=0, l\neq (L_i-1)/D_i}^{(2L_i-2)/D_i}\left(\mathbf{g}_i * \widehat{\mathbf{h}}_{i,i}\right)[D_i l]X_i[n-l] \\ & + \sqrt{p_u}\sum_{l=0}^{(L_{k,i}-1)/(\beta_k D_k)}\left(\mathbf{g}_k^{[\beta_k]} * \widehat{\mathbf{h}}_{k,i}\right)[\beta_k D_k l]\,X_k[n-l] \\ & + n_i[n], \end{aligned} \tag{22.19}$$

where p_u is the power amplifier, $L_i = \text{length}\left(\widehat{\mathbf{h}}_{i,i}\right)$, and $L_{k,i} = \text{length}(\mathbf{g}_k^{[\beta_k]} * \widehat{\mathbf{h}}_{k,i})$.

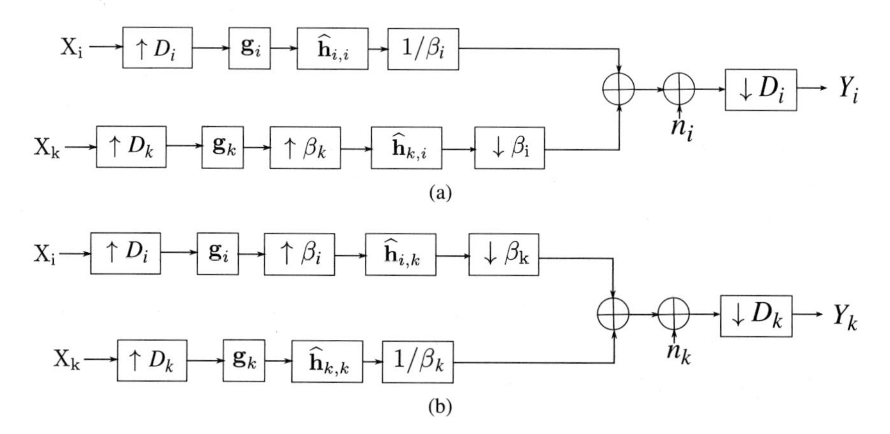

Figure 22.7 Equivalent architecture in case I: (a) equivalent data stream for type i, and (b) equivalent data stream for type k.

Figure 22.8 Spectrum occupation of case II.

In (22.19), the first and second terms are the typical expected signal term and ISI term, respectively. In addition, the third term is the ITI from the type k TD. Moreover, based on the spatio-temporal focusing effect in the TR system, the ITI is suppressed naturally with the location-specific signature. A similar equation of the received symbols can be derived for the type k TD.

Because the frequency shift $\exp^{-j\omega_i n}$ has unitary power, the analysis in (22.20) is also applied for the case where the carrier frequency of type i and k are different as long as their spectrum is overlapped.

22.4.2 Nonoverlapping Case

Another special case, where the spectrum of different types is nonoverlapped, is considered in this subsection. Suppose two types of TDs exist in the system, e.g., type i and type k, each of which contains a single TD. As shown in Figure 22.8, there are no ITIs between nonoverlapped types due to the corresponding analog bandpass filters and RRC filters. Therefore, the analysis becomes straightforward, and the received symbols at type i TD are derived as,

$$\begin{aligned} Y_i[n] = & \frac{\sqrt{p_u}}{\beta_i}\left(\mathbf{g}_i * \widehat{\mathbf{h}}_{i,i}\right)[L_i - 1]X_i\left[n - \frac{L_i - 1}{D_i}\right] \\ & + \frac{\sqrt{p_u}}{\beta_i}\sum_{l=0, l\neq(L_i-1)/D_i}^{(2L_i-2)/D_i}\left(\mathbf{g}_i * \widehat{\mathbf{h}}_{i,i}\right)[D_i l]X_i[n-l] \\ & + n_i[n], \end{aligned} \qquad (22.20)$$

which is well studied in the homogeneous TR system [12].

22.4.3 Mixed Case

Based on the previous analysis of two special cases, the heterogeneous TR system is analyzed under the general scenario, where N types of TDs are supported in the system, and the number of type i TD is M_i. The spectrum of different types is shown in Figure 22.3.

As discussed in Section 22.3, $\{X_{i,j}[k]\}$ denotes the information symbols for the jth TD in type i, and $D_{i,j}$ and $\mathbf{g}_{i,j}$ are the backoff factor and the embedded signature for the symbols $\{X_{i,j}[k]\}$, respectively. Based on Section 22.4.1, the TDs in type i suffer ITI from type k TDs, where $k \in T_i$ and T_i denote the set of types whose spectrum is overlapped with type i. In other words, the data streams of type k, where $k \notin T_i$, causes no interference to type i TDs according to Section 22.4.2.

Regarding the CIR, denote $\overline{\mathbf{h}}_{i,j}$ as the discrete CIR from the AP to the jth TD in type i with sample rate $f_s = \alpha B_{AP}$. Moreover, let $\widehat{\mathbf{h}}_{i_m,k_n}$ be the equivalent CIR for the data stream of mth TD in type i between the AP to n^{th} TD in type k. Similar to (22.16) and (22.17), the equivalent CIR for data streams can be derived as

$$\widehat{\mathbf{h}}_{i_m,k_n} = \begin{cases} \sqrt{\beta_i}\left(\mathbf{RRC}_{B_i,\alpha B_{AP}} * \overline{\mathbf{h}}_{i,n} * \mathbf{RRC}_{B_i,\alpha B_{AP}}\right)_{[\alpha\beta_i]} & i = k \\ \left(\mathbf{RRC}_{B_i,\alpha B_{AP}} * \overline{\mathbf{h}}_{k,n} * \mathbf{RRC}^{[\beta_k]}_{B_k,\alpha B_k}\right)_{[\alpha]} & i \neq k, \end{cases} \tag{22.21}$$

where $\beta_i = B_{AP}/B_i$. From (22.21), the length of the equivalent CIR solely depends on the types of data stream and the receiving TD. Once the CIRs are estimated, various signature design methods can be deployed. Take the basic TR signature of the jth TD of type i for example, i.e.,

$$g_{i,j}[n] = \frac{\widehat{h}^{*}_{i_j,i_j}[L-1-n]}{\|\widehat{\mathbf{h}}_{i_j,i_j}\|}. \tag{22.22}$$

Thus the received symbols at the jth of type i TD $\mathbf{Y}_{i,j}$ can be expressed as

$$\begin{aligned} Y_{i,j}[n] &= \frac{\sqrt{p_u}}{\beta_i}\left(\mathbf{g}_{i,j} * \widehat{\mathbf{h}}_{i_j,i_j}\right)[L_i - 1]X_{i,j}\left[n - \frac{L_i - 1}{D_{i,j}}\right] \\ &+ \frac{\sqrt{p_u}}{\beta_i} \sum_{l=0, l\neq (L_i-1)/D_{i,j}}^{(2L_i-2)/D_{i,j}} \left(\mathbf{g}_{i,j} * \widehat{\mathbf{h}}_{i_j,i_j}\right)[D_{i,j}l]X_{i,j}[n-l] \\ &+ \frac{\sqrt{p_u}}{\beta_i} \sum_{\substack{m=1 \\ m\neq j}}^{M_i} \sum_{\substack{l=0 \\ l\neq (L_i-1)/D_{i,m}}}^{(2L_i-2)/D_{i,m}} \left(\mathbf{g}_{i,m} * \widehat{\mathbf{h}}_{i_m,i_j}\right)[D_{i,m}l]X_{i,m}[n-l] \\ &+ \sqrt{p_u} \sum_{k\in T_i} \sum_{m=1}^{M_k} \sum_{l=0}^{\frac{L_{k,i}-1}{\beta_k D_{k,m}}} \left(\mathbf{g}^{[\beta_k]}_{k,m} * \widehat{\mathbf{h}}_{k_m,i_j}\right)\left[\beta_k D_{k,m} l\right] X_{k,m}[n-l] \\ &+ n_{i,j}[n], \end{aligned} \tag{22.23}$$

where $L_i = \text{length}\left(\widehat{\mathbf{h}}_{i_*,i_*}\right)$, $\beta_i = B_{AP}/B_i$, and $L_{k,i} = \text{length}\left(\mathbf{g}^{[\beta_k]}_{k,*} * \widehat{\mathbf{h}}_{k_*,i_*}\right)$.

In (22.23), the first term is the intended signal, the second and third terms represent the ISI and the IUI within the same type, and the ITI from overlapped types ($k \in T_i$) is expressed as the fourth term. Based on (22.23), the SINR for the jth TD in type i within the general heterogeneous TR system can be calculated correspondingly like (22.11).

22.5 Simulation Results

In this section, we conduct simulation to demonstrate the ability of presented TR approach to support heterogeneous bandwidth devices with a reasonable BER performance. We assume that N types of TDs coexist in the system with single or multiple

Table 22.1 Features of one HD video and two HD audio

Device name	Bandwidth	Backoff factor	Mod.	Cod. rate	Waveform
HD Video 1	150 MHz	8	QPSK	1/2	Basic TR
HD Audio 1	50 MHz	12	QPSK	1/2	Basic TR
HD Audio 2	50 MHz	12	QPSK	1/2	Basic TR

devices within each type. Different types of devices have heterogenous bandwidths, spectrum occupation, hardware capabilities, and QoS requirements. The CIR used in the simulation is based on the ultra-wide band (UWB) channel model of IEEE P802.15 [21], which makes the simulation in the following a good predication of the system performance.

22.5.1 TDMA and Spectrum Allocation

We first consider three devices in the heterogeneous TR system, whose features are listed in Table 22.1. According to the table, the bit rates of the HD video and the HD audio are around 18 Mbits/s and 4 Mbits/s. Based on previous discussion, the bandwidth of AP is assumed to be 150 MHz to support simultaneous data transmission to these three devices.

We first consider the case that three devices are categorized into two types, where Type 1 includes the HD video device and Type 2 consists of two HD audio devices. The BER performance of three devices under such a scenario is shown in Figure 22.9. Inferred from the figure, the BER performance of the two HD audio devices is much worse compared with the BER of HD video. The reason behind this is that the suppression of IUI in the TR system heavily depends on the number of resolved independent multipaths, which increases with the bandwidth. Because the bandwidth of two HD audio is much narrower, the IUI from the other devices becomes severer with the basic TR signature. In order to tackle the IUI for the devices with a narrower bandwidth, along with the TR technology, other techniques have to be considered in the heterogenous TR system as well.

We first consider applying time division multiple access (TDMA) to the heterogeneous TR system. In other words, the AP supports one HD audio at a time. To maintain the same QoS requirement in terms of bit rate, either adjusting the coding rate or decreasing the backoff factor is adopted in the system. The improved BER performance of three devices with simple waveform design is shown in Figure 22.10, where (a) removes the channel coding and (b) decreases the backoff rate to maintain the same bit rate for the HD audio. Compared with the BER in Figure 22.9, the BER performance is improved significantly with the simple waveform design. Moreover, decreasing the backoff factor to maintain the bit rate seems to be a better strategy for the devices with narrow bandwidth through comparing (a) and (b). Note that there are some waveform design techniques [19] that potentially can be implemented in the heterogeneous TR system with even better performance.

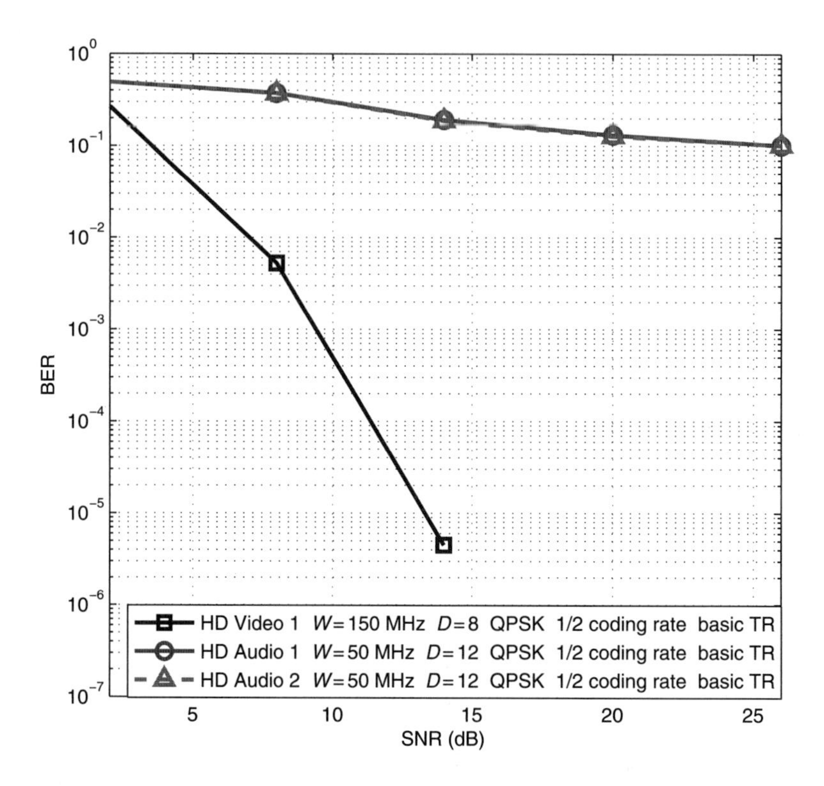

Figure 22.9 BER performance of three devices where two HD audio devices are included in the same type with basic TR signature.

Even though a narrow bandwidth decreases the number of resolved independent multipaths, thus resulting in severer IUI, a narrow bandwidth on the other hand provides more flexibility for spectrum allocation. Therefore, another way to improve the BER performance in Figure 22.9 is to arrange the spectrum occupation smartly, thus removing the unnecessary interference. For example, three devices in Table 22.1 can be categorized into three distinct types, where two HD audio devices are allocated into two spectrally nonoverlapped types. Then the improved BER performance with the spectrum allocation is shown in Figure 22.11.

22.5.2 Heterogeneous TR System versus Homogeneous TR System

As discussed in the previous section, appropriate spectrum allocation can significantly improve the BER performance even with a narrow bandwidth in the heterogeneous TR system. In other words, the narrow bandwidth under heterogeneous setting does not necessarily lead to worse BER performance compared with the wide bandwidth under homogeneous setting. Inspired by that, we investigate the BER performance of the devices in both homogeneous and heterogeneous TR system under the same bit rate.

Assume there exists three devices with bit rate requirement of 12.5 Mbits/s supported by a TR AP with 150 MHz bandwidth. Suppose the devices have flexible hardware capabilities, i.e., the carrier frequency and the bandwidth. To support these devices, two

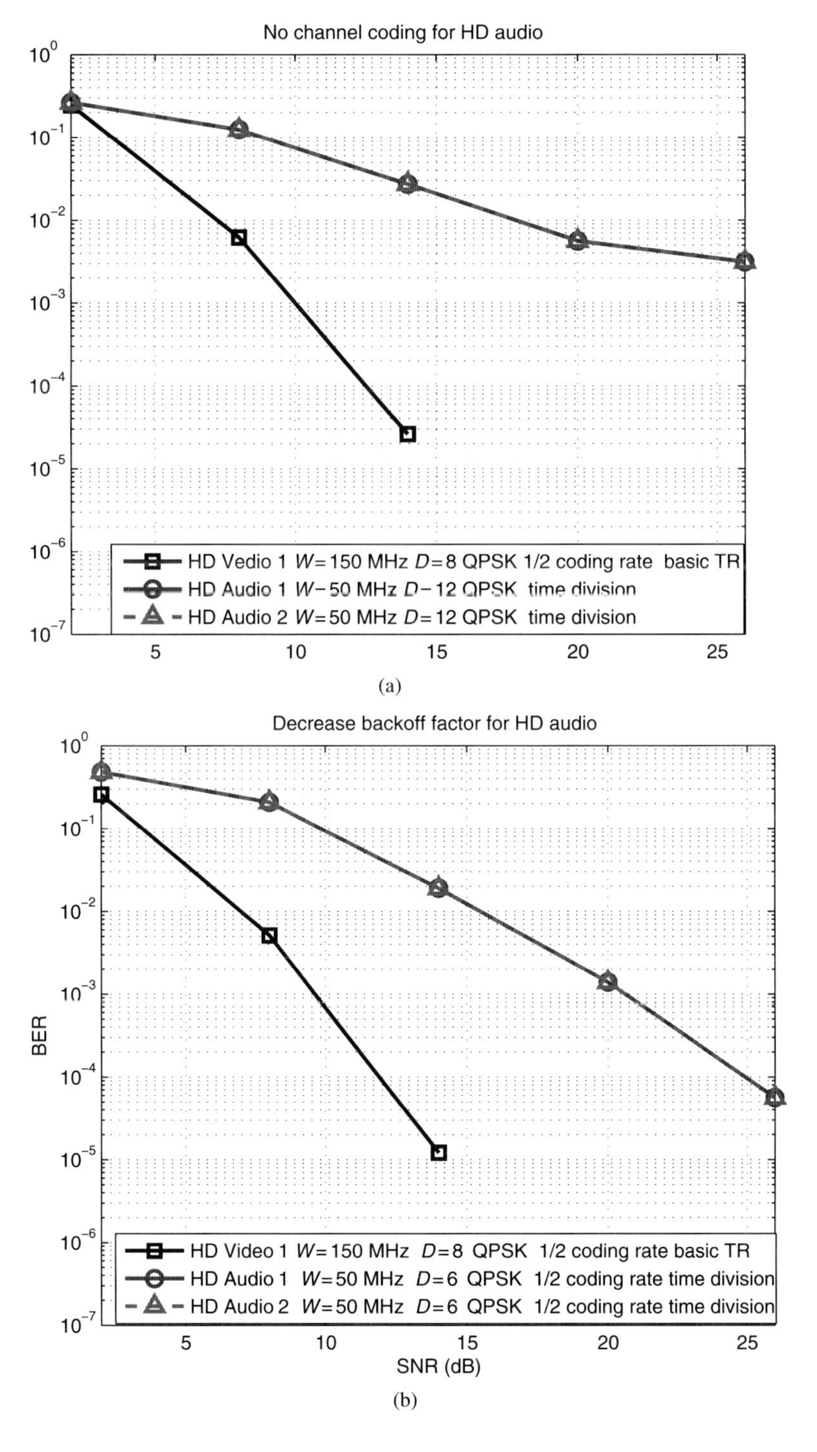

Figure 22.10 Improved BER performance with TDMA for the HD audio. (a) No channel coding to maintain the same bit rate. (b) Decrease backoff factor to maintain the same bit rate.

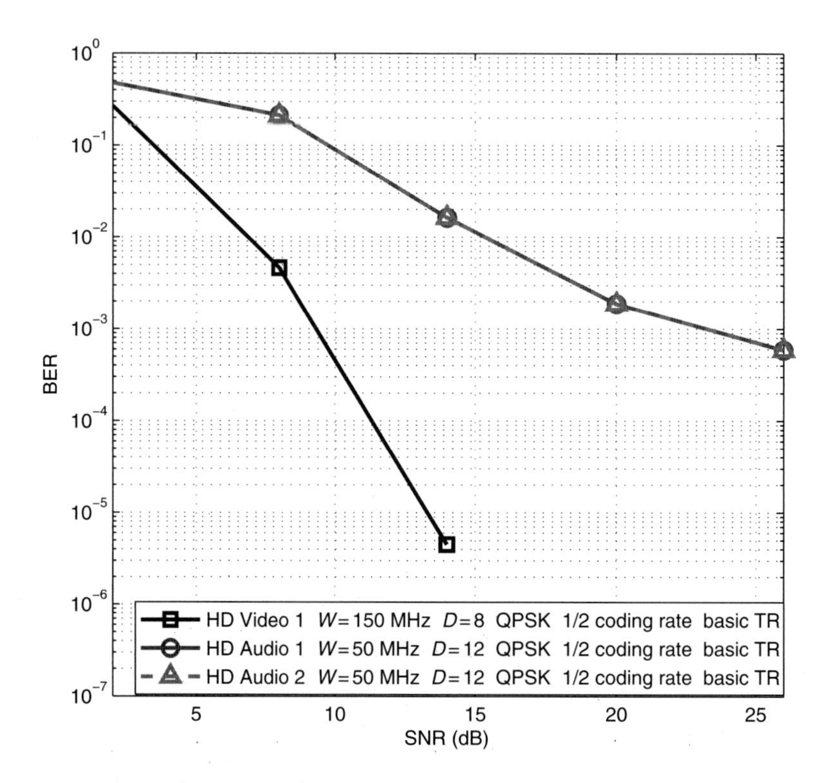

Figure 22.11 Improved BER performance with spectrum allocation for the HD audio.

potential paradigms, homogeneous paradigm and heterogeneous paradigm, are available. For the sake of fairness, the basic TR signature is adopted in both paradigms.

In the homogeneous setting, all three devices occupy the 150 MHz spectrum with QPSK modulation and backoff factor $D = 12$. A channel coding with 1/2 coding rate is employed. In the heterogeneous setting, the devices are categorized into three nonoverlapped types. More specifically, three devices with 50 MHz bandwidth are allocated into three nonoverlapped spectrums. To maintain the same bit rate, a backoff factor $D = 4$ is implemented. Their BER performance is shown in Figure 22.13. From the figure, the BER performance of the homogeneous paradigm saturates fast, which is due to the well-known fact that ISI and IUI would dominate the noise with the basic TR signature at high SNR region [19]. However, the IUI is better tackled in the heterogenous paradigm with smart spectrum allocation even though the number of independent multipaths resolved by the narrower bandwidth becomes fewer. Therefore, the performance of the heterogeneous paradigm can be even better than that of the homogeneous paradigm with additional techniques like spectrum allocation.

22.5.3 Heterogeneous TR System Case Study: Smart Homes

In this section, we choose smart homes as an example of the IoT application to test the BER performance with the heterogeneous TR paradigm. Instrumenting buildings with the IoT technologies will help in not only reducing resource (electricity, water) consumption, but also in improving the satisfaction level of humans. Typically, the

Table 22.2 Features of devices in the smart homes

Device name	Bandwidth	Backoff factor	Mod.	Cod. rate	Waveform
HD video 1	150 MHz	8	QPSK	1/2	Basic TR
HD audio 1	50 MHz	12	QPSK	1/2	Basic TR
Smart sensor 1	10 MHz	10	QPSK	1/2	Basic TR
Smart sensor 2	10 MHz	10	QPSK	1/2	Basic TR
Smart sensor 3	10 MHz	10	QPSK	1/2	Basic TR
Smart sensor 4	10 MHz	10	QPSK	1/2	Basic TR
Smart sensor 5	10 MHz	10	QPSK	1/2	Basic TR

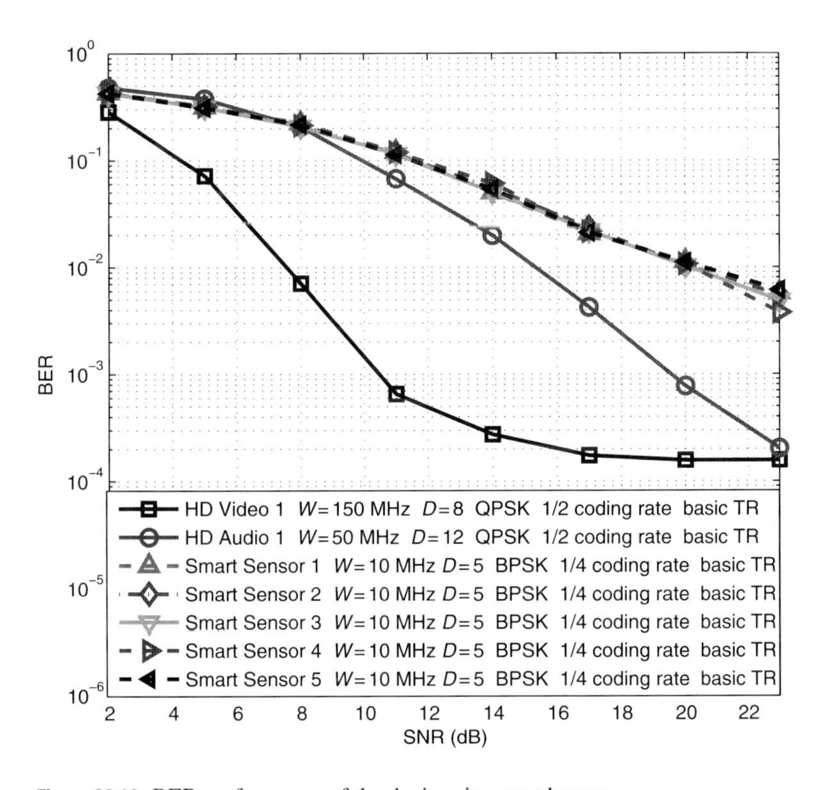

Figure 22.12 BER performance of the devices in smart homes

HD video and HD audio are employed in smart homes for both security monitoring and entertainment. Moreover, smart sensors are used in smart homes to both monitor resource consumptions as well as to proactively detect the users' need. Therefore, in the following simulation, we assume one HD video, one HD audio, and five smart sensors in the smart home are supported by the heterogeneous TR paradigm. The specific features of these devices are listed in Table 22.2, and the corresponding BER performance is shown in Fig 22.12. Note the saturation of the BER for the HD video is due to the dominant IUI with the basic TR signature. In addition, the slight difference in the BER for the smart sensors comes from the frequency-selectivity of the channel.

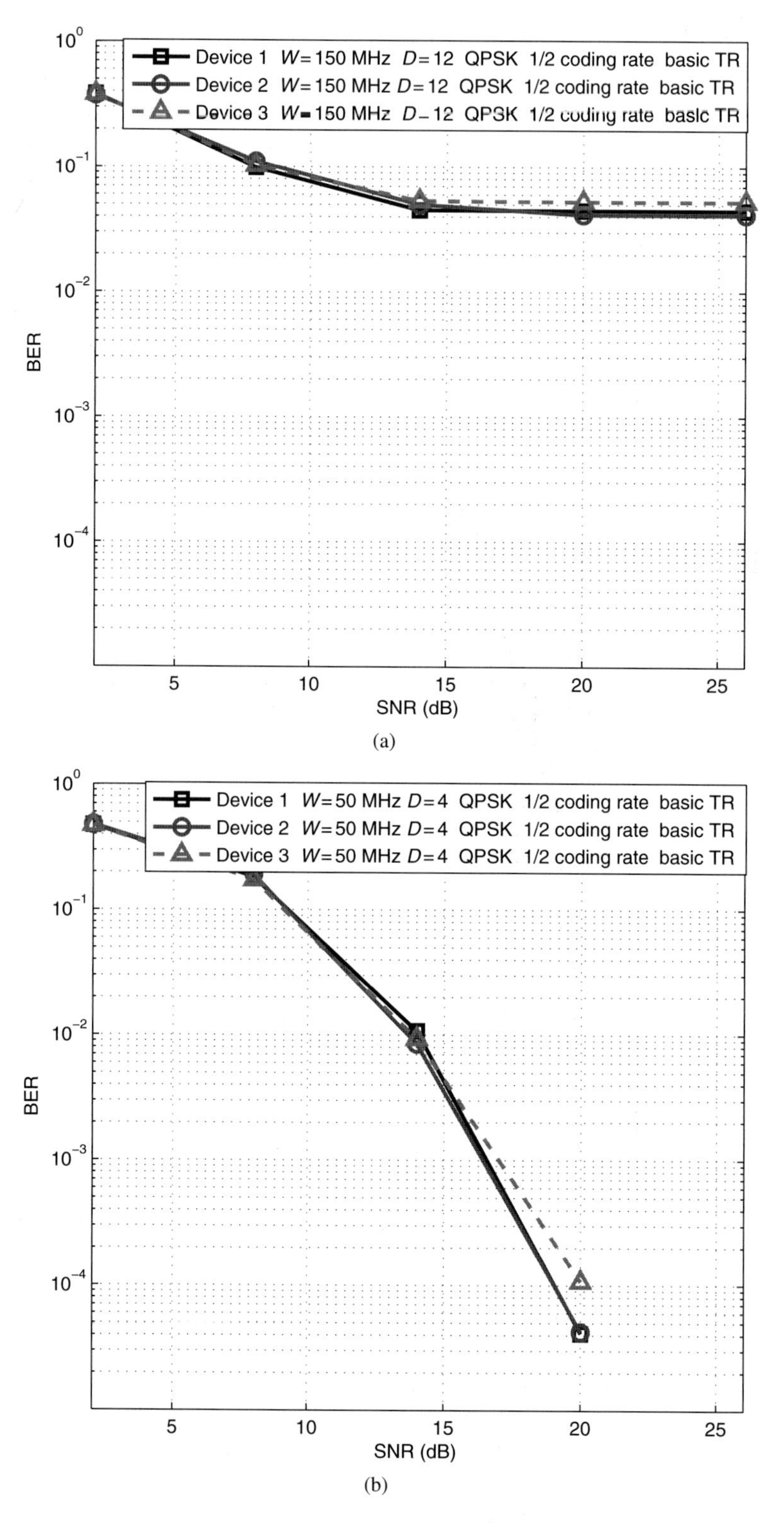

Figure 22.13 BER comparison of (a) homogeneous paradigm and (b) heterogeneous paradigm with basic TR signature.

22.6 Summary

A novel TR-based heterogeneous communication system is discussed that can support devices with various bandwidths in IoT. Different from building middlewares, the presented approach enables connectivity between devices with heterogeneous bandwidth requirement by means of multirate signal processing. In this way, the complexity of the presented system mostly lies in the parallel digital processing at the AP side, which can be easily handled with more powerful DSP, while maintaining low complexity of the TDs. Therefore, compared with middleware approach, the presented TR approach better satisfies the requirement of low-complexity and energy-efficiency for terminal devices in IoT. System performance is evaluated through both theoretical analysis and simulations, which show that the presented system can serve the devices of heterogeneous bandwidths with a reasonable BER performance, and the BER performance can be improved significantly with appropriate spectrum allocation. For related references, interested readers can refer to [22].

References

[1] J. Gubbi, R. Buyya, S. Marusic, and M. Palaniswami, "Internet of Things (IoT): A vision, architectural elements, and future directions," *Future Generation Computer Systems*, vol. 29, no. 7, pp. 1645–1660, 2013.

[2] K. Ashton, "That 'Internet of Things' thing," *RFiD Journal*, vol. 22, no. 7, pp. 97–114, 2009.

[3] H. Sundmaeker, P. Guillemin, P. Friess, and S. Woelfflé, *Vision and Challenges for Realising the Internet of Things*, Brussels: European Commission, 2010.

[4] L. Yan, Y. Zhang, L. Yang, and H. Ning, *The Internet of Things: From RFID to the Next-Generation Pervasive Networked Systems.* Boca Raton, FL: Auerbach Publications, 2006.

[5] L. Atzori, A. Iera, and G. Morabito, "The Internet of Things: A survey," *Computer Networks*, vol. 54, no. 15, pp. 2787–2805, 2010.

[6] National Intelligence Council, "Disruptive civil technologies – six technologies with potential impacts on US interests out to 2025," Technical Report, Apr. 2008.

[7] E. De Poorter, I. Moerman, and P. Demeester, "Enabling direct connectivity between heterogeneous objects in the Internet of Things through a network-service-oriented architecture," *EURASIP Journal on Wireless Communications and Networking*, vol. 2011, no. 1, pp. 1–14, 2011.

[8] M. Zorzi, A. Gluhak, S. Lange, and A. Bassi, "From today's intranet of things to a future Internet of Things: A wireless-and mobility-related view," *IEEE Wireless Communications,* vol. 17, no. 6, pp. 44–51, 2010.

[9] S. de Deugd, R. Carroll, K. Kelly, B. Millett, and J. Ricker, "SODA: Service oriented device architecture," *IEEE Pervasive Computing,* vol. 5, no. 3, pp. 94–96, Jul. 2006.

[10] P. Spiess, S. Karnouskos, D. Guinard, D. Savio, O. Baecker, L. Souza, and V. Trifa, "SOA-based integration of the Internet of Things in enterprise services," in *Proceedings of the 2009 IEEE International Conference on Web Services*, pp. 968–975, Jul. 2009.

[11] J. Pasley, "How BPEL and SOA are changing web services development," *IEEE Internet Computing,* vol. 9, no. 3, pp. 60–67, May 2005.

[12] B. Wang, Y. Wu, F. Han, Y.-H. Yang, and K. J. R. Liu, "Green wireless communications: A time-reversal paradigm," *IEEE Journal on Selected Areas in Communications*, vol. 29, no. 8, pp. 1698–1710, 2011.

[13] F. Han, Y.-H. Yang, B. Wang, Y. Wu, and K. J. R. Liu, "Time-reversal division multiple access over multi-path channels," *IEEE Transactions on Communications*, vol. 60, no. 7, pp. 1953–1965, 2012.

[14] Y. Chen, F. Han, Y.-H. Yang, H. Ma, Y. Han, C. Jiang, H.-Q. Lai, D. Claffey, Z. Safar, and K. J. R. Liu, "Time-reversal wireless paradigm for green Internet of Things: An overview," *IEEE Internet of Things Journal*, vol. 1, no. 1, pp. 81–98, Feb. 2014.

[15] I. Glover and P. M. Grant, *Digital Communications*. Harlow, UK: Pearson Education, 2010.

[16] A. V. Oppenheim, R. W. Schafer, J. R. Buck et al., *Discrete-Time Signal Processing*. Englewood Cliffs, NJ: Prentice-Hall, 1989.

[17] P. P. Vaidyanathan, *Multirate Systems and Filter Banks*. Englewood Cliffs, NJ: Pearson Education, 1993.

[18] R. Daniels and R. Heath, "Improving on time reversal with MISO precoding," *Proceedings of the Eighth International Symposium on Wireless Personal Communications Conference*, pp. 18–22, 2005.

[19] Y.-H. Yang, B. Wang, W. S. Lin, and K. J. R. Liu, "Near-optimal waveform design for sum rate optimization in time-reversal multiuser downlink systems," *IEEE Transactions on Wireless Communications*, vol. 12, no. 1, pp. 346–357, 2013.

[20] M. Emami, M. Vu, J. Hansen, A. Paulraj, and G. Papanicolaou, "Matched filtering with rate back-off for low complexity communications in very large delay spread channels," *Conference Record of the Thirty-Eighth Asilomar Conference on Signals, Systems and Computers*, vol. 1, pp. 218–222, Nov. 2004.

[21] J. Foerster, "Channel modeling sub-committee report final," *IEEE P802. 15-02/368r5-SG3a*, 2002.

[22] Y. Han, Y. Chen, B. Wang, and K. J. R. Liu, "Enabling heterogeneous connectivity in Internet of Things: A time-reversal approach," *IEEE Internet of Things Journal*, vol. 3, no. 6, pp. 1036–1047, 2016.

Index